DEFORMATION OF CONCRETE STRUCTURES

McGRAW-HILL
INTERNATIONAL
BOOK COMPANY

New York
St Louis
San Francisco
Auckland
Bogotá
Düsseldorf
Johannesburg
Madrid
London
Mexico
Montreal
New Delhi
Panama
Paris
São Paulo
Singapore
Sydney
Tokyo
Toronto

DAN E. BRANSON
Professor of Civil Engineering
University of Iowa

Deformation of Concrete Structures

Library of Congress Cataloging in Publication Data

Branson, Dan E

Deformation of concrete structures.

Includes bibliographical references and index.

1. Reinforced concrete construction.
2. Deformations (Mechanics) I. Title.

TA683.B816 624'.1834 77-23173

ISBN 0-07-007240-X

DEFORMATION OF CONCRETE STRUCTURES

1 2 3 4 MPMP 7 9 8 7

CONTENTS

PREFACE

The objective of this book is to provide a comprehensive yet concise and practical treatment of the subject of initial and time-dependent deformation of reinforced and prestressed concrete structures that is not otherwise available to students, research engineers, and practicing engineers alike. The emphasis is on structural serviceability and esthetic criteria as opposed to safety or strength aspects of design.

The development of this material parallels the work of the author over the past decade and a half as it relates to his research and class lectures, and to his service on various committees of the American Concrete Institute and the Prestressed Concrete Institute. Procedures by different authors and agencies are studied, including reference to the ACI Code, ACI Manual of Concrete Practice, ACI SP-27, ACI SP-43, AASHTO, AISC, BPR, CEB, FIP, PCA, PCI Handbook, British Code, and codes of other countries.

Systematic procedures for computing the material properties of different weight concrete, and the initial and time-dependent deformation of noncomposite and composite reinforced and prestressed concrete structures are presented. Equations are provided for all needed material parameters and deformation calculations so that the procedures readily lend themselves to computer solution. Tables and charts are also provided in many cases for hand calculations. Detailed numerical examples and comparisons between different methods and with experimental results are included. A discussion of experimental results is also presented in each chapter. The principal topics describe the prediction or estimation of elastic, cracking, creep, and shrinkage effects in concrete structures as related to reinforced beams and one-way slabs, two-way construction, prestressed members, and composite members. The control of such structural deformation is also studied.

Chapter 1 refers primarily to the elastic, cracking, creep, shrinkage, temperature, and relaxation effects of the materials. A study of differential shrinkage, creep, and temperature effects in composite beams is presented in Chap. 2. Chapter 3 includes the subjects of short-time, creep, shrinkage, and total time-dependent deflection of reinforced beams and one-way slabs, and both steel and reinforced concrete composite beams. The deflection of two-way slab systems with and without beams is presented in Chap. 4. The topics in Chap. 5 include the loss of prestress, camber, and deflection of noncomposite and composite prestressed members, including the special effects in partially prestressed members. Chapter 6 refers to the control of deflections by allowable values and span–depth ratios. Although information from various chapters, such as Chaps. 1, 2, and 6, is used in other parts of the book, each chapter stands alone and may be used independently.

This material is primarily concerned with deformation that occurs at service load levels under static conditions and may not apply for loads with strong dynamic characteristics such as those due to earthquakes, transient winds, and machinery. In the case of deflections, the deformation is measured below the level of supports for horizontal members, with consideration also being given to camber of both nonprestressed and prestressed members.

Most of the rational, semiempirical, and empirical deterministic formulations herein represent a nominal approach for design purposes, and not definitive or statistical results by any means. Probabilistic methods are needed for the accurate estimation of the variability of deformational behavior, including confidence intervals for computed deformation. Because of the variability of concrete structural deformation, it appears to be not only feasible but essential that relatively simple procedures be used so that engineers will guard against placing undue reliance on the computed results.

With regard to the approximate procedures described, it is suggested that more complex deterministic methods may improve results for test specimens under controlled laboratory conditions and with prescribed loading in some cases, but such may not generally be the case for actual structures under imprecise or unknown loading histories (including construction loads), variable creep and shrinkage behavior, etc. For these and other reasons, it may normally be preferable to consider computed deformation as "characteristic or representative values" and not "predicted values" in the case of actual structures, especially when the computed values are to be compared to standard allowable values which may themselves be poorly defined.

It is expected that future refinements and improvements will be made in the various parameters and procedures described, but it is hoped that the contents of the book can provide in one document a basis for the practical consideration of the subject matter presented.

The principal units used are English units, with extensive conversion to both mks and SI units included. Appropriate conversion factors are also tabulated separately.

The author would like to thank his university and technical committee colleagues, and the many authors too numerous to mention who contributed to this work. He especially appreciates the assistance of Messrs. C. I. Chen, M. L. Christiason, K. M. Kripanarayanan, S. Kundapur, B. L. Meyers, M. V. Pregnoff, F. Roll, C. G. Salmon, and A. F. Shaikh for their review and suggestions pertaining to various parts of the manuscript.

DAN E. BRANSON

Iowa City, Iowa

NOTATION

A = cross-sectional area
A = air content in percent; also subscript (in roman type) denoting air content
A_g = area of gross section, neglecting the steel
A_{ns} = area of nonprestressed tension steel in prestressed member
A'_{ns} = area of compression steel (in compression zone) in prestressed member
A_{ps} = area of prestressing steel (in tension zone) in prestressed member
A_s = area of tension steel in reinforced member
A'_s = area of compression steel in reinforced member
A_{sh} = shrinkage coefficient in Eq. (3-56)
A_t = area of transformed section
a = depth of equivalent rectangular stress block
a = distance from end of beam to the nearest of two symmetrical diaphragms; also used as the distance from end to harped point in two-point harping
a, a_1, a_2, a_3 = constants
B = number of bags or sacks (94 lb = 42.5 kg = 418 N) of cement per cubic yard of concrete; also subscript (in roman type) denoting cement content

b = width of compression face of member; also used as a constant
b_{11}, b_{12}, b_{22} = parameters defined in prestress loss and nonprestressed steel stress equations
b_0 = periphery of critical section for slabs
b_w = web width
C = creep coefficient defined as the ratio of creep strain to initial strain; also used as a constant
C = cross-sectional torsion constant
C.F. = correction factor
CR_c = ultimate prestress loss due to concrete creep
CR_s = ultimate prestress loss due to steel relaxation for pretensioning
CR_{sp} = ultimate prestress loss due to steel relaxation for post-tensioning
CS = slab column strip
C_L = slab creep coefficient based on the slab age when the permanent live load, etc., is applied
C_s = creep coefficient at the time of slab casting
C_{s1} = slab creep coefficient based on the slab age when the shores are struck
C_{s2} = creep coefficient of the precast beam concrete at the time of slab casting
C_t = creep coefficient at any time t
C_{t1} = creep coefficient for the slab loading, where the age of the precast beam concrete at the time of slab casting is considered
C_{t2} = creep coefficient of the precast beam concrete at any time after slab casting, due to the precast beam dead load
C_{ts} = creep coefficient, where the age of the beam concrete at the time of load application is taken into account
C_u = ultimate (in time) creep coefficient
c = creep strain in millionths
c = specific creep or unit creep strain in millionths per psi
c_1, c_2 = size of column in two directions
D = slab flexural rigidity; also used to denote dead load
D = differential shrinkage and creep strain or coefficient
d = constant; also used (in roman type) to denote days
d = effective depth of section, or distance from extreme compression fiber to centroid of tension steel
d' = distance from extreme compression fiber to centroid of compression steel
E_1 = moment of inertia of slab section of a composite beam
E_2 = moment of inertia of steel or precast beam section of a composite beam
ES = ultimate prestress loss due to elastic shortening
E_c = modulus of elasticity of concrete
E_{cb}, E_{cc}, E_{cs} = modulus of elasticity of concrete beams, columns, slabs, respectively
E_{ci} = modulus of elasticity of concrete at the time of initial loading
E_{ct} = $E_c/(1 + C_t)$

E'_{ct} $\doteq E_c/(1 + T)$
$(E_c)_t$ = modulus of elasticity of concrete at any time t
E_{ns} = modulus of elasticity of nonprestressed steel
E_{ps} = modulus of elasticity of prestressing steel
E_s = modulus of elasticity of steel
E'_s = modulus of elasticity of nonprestressed tension steel
e = napierian base = 2.7183
e = eccentricity of steel; also used as a general constant
e_c = eccentricity of steel at center of beam; also used to denote the eccentricity of steel in composite section
e_e = eccentricity of steel at end of beam
e_{ns} = eccentricity of nonprestressed steel
e_{ps} = eccentricity of prestressing steel
F = differential shrinkage and creep force
F = percent of fine aggregate to total aggregate by weight; also subscript denoting percent fines
F_{ns} = force in nonprestressed steel
f = constant
f_c = concrete stress
f_{cds} = concrete stress at the steel centroid due to differential shrinkage and creep
f_{ci} = initial concrete stress (concrete stress due to initial loading)
f_{cs} = concrete stress at the steel centroid due to superimposed sustained load
f_{csl} = concrete stress at the steel centroid due to slab dead load (plus diaphragm, etc., dead load when applicable)
f'_c = compressive strength of concrete
f'_{ci} = compressive strength of concrete at the time of initial loading
$\sqrt{f'_c}$ = square root of the compressive strength of concrete—same units as f'_c (psi) in empirical expressions
$(f'_c)_t$ = compressive strength of concrete at any time t
$(f'_c)_{7d}, (f'_c)_{28d}$ = compressive strength of concrete at age 7, 28 days, etc.
$(f'_c)_u$ = ultimate (in time) compressive strength of concrete
f_D = concrete stress due to dead load
f_{ns} = stress in nonprestressed steel
f_0 = stress in prestressing steel at transfer (after elastic loss)
f_{pe} = compressive stress in concrete due to prestress only after all losses, at the extreme fiber of a section at which tensile stresses are caused by applied loads
f_{ps} = stress in prestressing steel
f_r = modulus of rupture of concrete
f_s = steel stress
Δf_s = loss of prestress
f_{sds} = steel stress due to differential shrinkage and creep
f_{si} = initial or tensioning stress in steel

f_{sr}	= steel stress relaxation
f_t'	= direct tensile strength of concrete
f_{to}	= total computed tensile stress
f_y	= yield point or yield strength of steel
f_y'	= 0.1% offset yield strength of steel
H	= ambient relative humidity in percent; also subscript denoting relative humidity
h	= overall thickness of member; also used for story height
h_f	= flange thickness of T-beams
I	= moment of inertia (second moment of the area) of a section
I	= impact factor
I_1	= moment of inertia of slab
I_2	= moment of inertia of steel beam or precast beam of a composite section
I_{2s}	= moment of inertia of the precast beam, such as I_e in Eq. (3-19) in which M_a refers to the precast beam plus slab dead load and M_{cr} to the precast beam, as in Eqs. (3-87) and (3-89)
I_a, $(I_e)_a$	= average effective moment of inertia
I_b, I_c, I_s	= moment of inertia of beams, columns, slabs, respectively
I_c	= moment of inertia of composite section with transformed slab; the slab width is divided by E_2/E_1
I_{cr}	= moment of inertia of cracked transformed section
I_{ct}, I'_{ct}	= moment of inertia of cracked transformed section using E_{ct}, E'_{ct}, respectively
I_{e1}, I_{e2}, I_m	= moments of inertia at two ends of span and at midspan
I_{eff}, I_e	= effective moment of inertia
I_g	= moment of inertia of gross concrete section, neglecting the steel
I_{rep}	= effective moment of inertia for repeated loading
I_{ucr}	= moment of inertia of uncracked transformed section, including the steel
jd	= internal lever arm by the elastic theory
K	= concrete age when loaded
K	= prestress wobble coefficient
K	= deflection coefficient. For prismatic beams uniformly loaded, ideal end conditions, and the equation $\Delta = KML^2/EI$ (used in Eq. (3-1) and Chap. 5, for example):

cantilever beam	$K = 1/4$	$(-)\ M = wL^2/2$
simple beam	$K = 5/48$	$(+)\ M = wL^2/8$
hinged–fixed beam	$K = 8/185$	$(-)\ M = wL^2/8$
fixed–fixed beam	$K = 1/32$	$(-)\ M = wL^2/12$

For other cases and for nonprismatic beams, see various handbooks and papers, such as Ref. 36 by the Portland Cement Association. Alternatively, see Eqs. (3-39) to (3-42) for K values of cantilever, simple, and continuous beams with uniform and concentrated loading.

K_b, K_c, K_{ec}, K_s = flexural stiffness (moment per unit rotation) of beams, columns, equivalent columns, slabs, respectively
K_s = slab deflection coefficient
K_{sh} = shrinkage (also temperature change) deflection coefficient
K_t = torsional stiffness (moment per unit rotation) of torsional member
k = creep and shrinkage correction factors
kd = depth of neutral axis by the elastic theory
k_r = reduction factor
$(k_r)_a$ = average value of k_r for the span
$(k_r)_s$ = reduction factor for the effect of slab compression steel in a composite section, such as Term (10) of Eq. (3-87)
k_t = prestress loss ratio
L, l = span length; also used as the long side of a rectangular slab panel
L_1 = span length in the direction moments are being determined for slabs, measured center-to-center of supports
L_2 = span length transverse to L_1, measured center-to-center of supports
L_n = clear span
LDF = lateral (moment) distribution factor for slabs
M = bending moment
M_1 = maximum bending moment under slab dead load; also used as a beam end moment
M_{1D} = bending moment between symmetrically placed diaphragms
M_2 = maximum bending moment under steel or precast beam dead load; also used as a beam end moment
M_a = maximum service load moment (unfactored moment) at the stage for which deflections are being considered
M_{cr} = cracking moment
M_{e1}, M_{e2} = beam end moments
M_m = net moment at midspan
M_0 = beam statical moment; also used for statical moment of slab panel
M_P = moment due to permanent load supported by a composite beam
$M_{S,Di}$ = bending moment due to slab or slab plus diaphram, etc., dead load
M_u = ultimate moment
MS = slab middle strip
m = modular ratio at the time of additional sustained load application, including the time of slab casting of a composite beam
m = $f_y/0.85\,f_c'$
n = modular ratio at the time of loading, such as the time of prestressing or at 28 days—$n = E_s/E_{ci}$ or E_s/E_c
n_{ct} = increased modular ratio due to creep; $n_{ct} = E_s/E_{ct} = n(1 + C_t)$
n'_{ct} = increased modular ratio due to creep and shrinkage; $n'_{ct} = n(1 + T)$
P = concentrated load
P, P_{ps} = prestress force after losses
ΔP = loss of prestress force
P_i = initial tensioning force

P_0	= prestress force at transfer (after elastic loss)
P_s	= prestress force at jacking end
ΔP_s	= total loss of prestress at the time of slab casting minus the initial elastic loss
ΔP_t	= total loss of prestress at any time minus the initial elastic loss
ΔP_u	= total ultimate loss of prestress minus the initial elastic loss
P_x	= prestress force at point x
PL	= prestress loss due to steel relaxation in percent
p	= tension steel percentage = $100A_s/bd$
p'	= compression steel percentage = $100A'_s/bd$
Q	= differential shrinkage and creep force; also transverse load
q	= uniformly distributed loading, such as force per unit area
r	= radius of gyration = $\sqrt{I/A}$
S	= section modulus; also used as the short side of a rectangular slab panel
S	= slump in inches; also subscript (in roman type) denoting a slump or superimposed load
S	= summation coefficient defined in Eqs. (5-36) to (5-39)
SH	= ultimate prestress loss due to shrinkage
T	= multiplier for additional long-time deflections due to time-dependent effects, such as creep and shrinkage
T	= temperature or thickness, including its use as a subscript (in roman type for thickness) (T = thickness in creep and shrinkage correction equations)
T	= tension force, such as that due to shrinkage in Eq. (3-50)—T_{sh}
t	= time in general, time in hours in the steel relaxation equations, time in days in other equations, and subscript denoting time-dependence
t	= flange thickness of T-beams
t_{LA}	= age of concrete when loaded
V	= shear force
v	= shear stress
v/s	= volume–surface ratio
W	= total load on a span
w	= uniformly distributed load
w	= unit weight of concrete, for example in pounds per cubic foot (pcf)
x	= variable distance along a beam; also slab coordinate direction
y	= slab coordinate direction
y	= vertical distance measured from the centroidal axis of a beam; see Fig. 2-1 for composite beam values, for example
y_{cs}	= distance from centroid of composite section to centroid of slab
y_t	= distance from centroidal axis of gross section, neglecting the steel, to the extreme fiber in tension
Z	= section modulus
α	= thermal coefficient; also used as slope or angle change, such as the angle change of steel profile in radians from jacking end to any point x.

α = ratio of flexural stiffness of beam section to the flexural stiffness of a width of slab bounded laterally by the centerline of the adjacent panel, if any, on each side of the beam = $E_{cb}I_b/E_{cs}I_s$

α, α_1 = ratio of creep coefficient at any time to ultimate creep coefficient = C_t/C_u; representative values are given in Table 3-20

α_1, α_2 = α in the direction of L_1, L_2, respectively

α_{ec} = ratio of flexural stiffness of the equivalent column to the combined flexural stiffness of the slabs and beams at a joint taken in the direction moments are being determined = $K_{ec}/\Sigma(K_s + K_b)$

α_m = average value of α for all beams on the edges of a panel

α_s = ratio of creep coefficient at the time of sustained load application, such as slab dead load, to $C_u = C_{ts}/C_u$; representative values are given in Table 5-3

$(\alpha_s)_{cp}$ = ratio of precast beam dead load creep that takes place before the slab is cast to the ultimate creep = C_{s2}/C_u; representative values are given in Table 3-25

$(\alpha_s)_{sh}$ = ratio of precast beam shrinkage that takes place before the slab is cast to $(\varepsilon_{sh})_u = (\varepsilon_{sh})_s/(\varepsilon_{sh})_u$; representative values are given in Table 3-25

β = creep correction factor for the precast beam concrete age when loaded

β = ratio of clear spans in long to short direction of two-way slab systems

β_s = creep correction factor for the precast beam concrete age when a sustained load (such as the slab dead load for a composite beam) is applied

β_s = ratio of length of continuous edges to total perimeter of a slab panel

β_t = ratio of torsional stiffness of edge beam section to the flexural stiffness of a width of slab equal to the span length of the beam, center-to-center of supports = $E_{cb}C/2E_{cs}I_s$

Δ = maximum deflection or camber

Δ_{DS} = differential shrinkage and creep deflection

Δ_i = initial deflection, camber

$(\Delta_i)_1$ = initial deflection under slab dead load

$(\Delta_i)_{1D}$ = initial deflection due to diaphragm dead load

$(\Delta_i)_2$ = initial deflection under steel beam or precast beam dead load

$(\Delta_i)_D$ = initial dead-load deflection

$(\Delta_i)_{P_0}$ = initial camber due to the initial prestress force, P_0

Δ_L = live-load deflection

Δ_t = time-dependent deflection, camber

Δ_u = ultimate deflection, camber

δ_t = specific creep or unit creep strain defined as creep strain per unit stress

δ_u = ultimate specific creep

ε = unit strain

ε_{ci} = initial concrete strain such as at the level of prestressing steel centroid

ε'_{ci} = initial concrete strain at the level of nonprestressed steel centroid

$(\varepsilon_{sh})_t$ = shrinkage strain at any time t

$(\varepsilon_{sh})_u$ = ultimate (in time) shrinkage strain
η = relaxation coefficient
θ = slope of the deflection curve
μ, ν = Poisson's ratio
μ = coefficient of friction
ρ = tension steel ratio = A_s/bd
ρ' = compression steel ratio = A'_s/bd
ρ_b = steel ratio producing a balanced condition
ρ_{ns} = nonprestressed steel ratio = A_{ns}/A_g
ρ_p = prestressing steel ratio = A_{ps}/bd
ρ_{ps} = prestressing steel ratio = A_{ps}/A_g
ρ_{pt} = prestressing steel ratio = A_{ps}/A_t
σ = unit stress
ϕ = curvature
ϕ = capacity reduction factor; also a power in Eq. (3-45)
ψ = coefficient in Eq. (3-5); also a coefficient in Eqs. (3-33) and (5-74)
ψ_1 = coefficient in Eq. (3-9)

Subscripts

1 = cast-in-place slab of a composite beam or the effect of the slab such as due to slab dead load
2 = precast beam
A = air content
a = average value
B = bags of cement, cement content; also used to denote bottom fiber
b = beam; also bottom fiber
c = composite section; also concrete
c = column; also slab column strip
ci = initial concrete value
cp = concrete creep
cr = cracking
D = dead load
DS, ds = differential shrinkage and creep
e1, e2 = two ends of a beam span
F = percent fines
H = relative humidity
I = slab–beam interface of a composite beam
i = initial value; also slab and beam interface in composite beams
L = live load; also long slab dimension
LA = loading age
m = midspan; also slab middle strip
ns = nonprestressed steel
ps = prestressing steel
S = superimposed sustained load; also short slab dimension

S	= slump; also composite slab centroid
s	= slab; also used for steel
sh	= shrinkage
T	= temperature
T	= thickness as average thickness for creep and shrinkage correction factors; also used to denote top fiber
t	= time-dependent
t	= top fiber
u	= ultimate value
x, *y*	= coordinate directions

METRIC CONVERSION

Length

1 inch (in. or ″) = 2.540 centimeters (cm) = 0.02540 meters (m)
1 foot (ft or ′) = 30.48 centimeters (cm) = 0.3048 meters (m)
1 yard (yd) = 91.44 centimeters (cm) = 0.9144 meters (m)

Area

1 square inch (in^2) = 6.452 square centimeters (cm^2)
1 square foot (ft^2) = 929.0 square centimeters (cm^2)
= 0.09290 square meters (m^2)

Volume

1 cubic inch (in^3) = 16.39 cubic centimeters (cm^3)
1 cubic foot (ft^3) = 2.832×10^4 cubic centimeters (cm^3)
= 0.02832 cubic meters (m^3)
1 cubic yard (yd^3) = 7.646×10^5 cubic centimeters (cm^3)
= 0.7646 cubic meters (m^3)

Moment of Inertia

$1\ in^4 = 41.62\ cm^4$

Temperature

$C = (5/9)(F - 32)$, as $C = (5/9)(212 - 32) = 100°C$
$F = (9/5)(C) + 32$, as $F = (9/5)(100) + 32 = 212°F$

Load—Force

1 pound (lb or #) = 453.6 gram-force (g) = 0.4536 kilogram-force (kg)
= 4.448 newtons (N) = 0.004448 kilonewtons (kN)
1 kilopound = 1 kip = 1000 lb = 453.6 kg = 4448 N = 4.448 kN
1 kilogram-force (kg) = 9.8067 newtons (N)

Distributed Load—Distributed Force

1 pound per foot (lb/ft) = 1.488 kg/m = 14.59 N/m
1 pound per square foot (psf) = 4.882 kg/m^2 = 47.88 N/m^2 = 0.04788 kN/m^2
1 pound per cubic foot (pcf) = 16.02 kg/m^3 = 157.1 N/m^3 = 0.1571 kN/m^3

(Example: w = 145 pcf = 2320 kg/m^3 = 22.8 kN/m^3)

Moment

1 foot-pound (ft-lb) = 0.1383 kg-m = 1.356 N-m
1 foot-kip (ft-k) = 138.3 kg-m = 1356 N-m = 1.356 kN-m

(Example: M = 100 ft-k = 13 830 kg-m = 135 600 N-m = 135.6 kN-m)

Stiffness

1 lb-in^2 = 2.927 kg-cm^2 = 28.700 N-cm^2

Curvature

1 1/in. = 0.3937 1/cm

Stress or Pressure

1 pound per square inch (psi) = 0.07031 kg/cm^2 = 0.006895 N/mm^2
= 6895 N/m^2 = 6895 Pa

(Examples: f'_c = 3000 psi = 211 kg/cm^2 = 20.7 N/mm^2 = 20.7 × 10^6 Pa
f_y = 40 000 psi = 2810 kg/cm^2 = 276 N/mm^2 = 276 × 10^6 Pa)

1 kip per square inch (ksi) = 70.31 kg/cm^2 = 6.895 N/mm^2
1 pascal (Pa) = 1 N/m^2
1 bar = 14.5 psi = 10^5 N/m^2 (atmospheric pressure)
1 N/mm^2 = 145.04 psi

Modulus of Elasticity

1 psi = 0.07031 kg/cm^2 = 6.895 × 10^{-6} kN/mm^2

(Examples: E_c = 3.31 × 10^6 psi = 233 000 kg/cm^2 = 22.8 kN/mm^2
E_s = 29 × 10^6 psi = 2.04 × 10^6 kg/cm^2 = 200 kN/mm^2)

CHAPTER
ONE

MATERIAL PROPERTIES

1-1 INTRODUCTION

The principal aspects of material behavior related to the deformation of concrete structures are normally referred to as elastic, cracking, creep, shrinkage, temperature, and relaxation effects. The corresponding parameters most commonly used to define these effects or properties are the modulus of elasticity of concrete and steel, Poisson's ratio, modulus of rupture and direct tensile strength of concrete, concrete creep coefficient or specific creep strain, shrinkage strain, thermal coefficient, and steel relaxation stress or percentage. The general quality of concrete and the influence of the time-dependent hydration process (as determined for practical purposes in most cases by the compressive strength) are important functions of concrete deformation primarily as they relate to the above properties. Additional factors such as environmental conditions, member size and shape, stress history, concrete mix, etc., affect the deformational behavior of concrete structures as well.

The material parameters are defined in this chapter, with numerical values given for practical application. Continuous time functions are presented so that the equations readily lend themselves to computer solution. Representative values of these equations are tabulated in each case for visual reference. Ranges of variation between results computed by these procedures and experimental results are shown in the form of computed versus experimental graphs, and also by graphs and tables showing the probability of agreement. Numerical examples of the prediction of creep and shrinkage are included, as well as a design aid to facilitate these calculations.

The procedure followed in this book for determining material properties to be used in computing concrete deformation is to approximate *average* values for the material parameters. This is generally consistent with current design practice in most cases, as it relates to the magnitude of design factors in common use. It also follows a philosophy of "unit factor of safety" in estimating material properties for the prediction of serviceability conditions, as advocated by ACI Committee 435 in their June 1968 *ACI Journal* paper on Allowable Deflections. The variability of the problem can then be taken into account in the limitations placed on the predicted results.

The subject of prestress loss calculations might be considered a safety condition and thus an exception to this average-value philosophy, although, even in this case, normal design practice tends to compute average prestress losses, and not upper limit values, for example. From an ultimate strength point of view, the question of safety is essentially independent of the loss of prestress (at least for bonded members).

In addition to the procedures described in the text for predicting initial and time-dependent concrete properties, reviews of different equations and procedures for determining creep and shrinkage and for analyzing concrete creep under constant and variable stress are included in the Appendices.

1-2 STRENGTH AND ELASTIC PROPERTIES

Concrete Compressive Strength

A study of concrete strength versus time[1,2,3*] for the data of Refs. 2 to 8 indicates an appropriate general equation in the form of Eq. (1-1) for predicting compressive strength at any time.

$$(f_c')_t = \frac{t}{a + bt}(f_c')_{28\text{d}} \tag{1-1}$$

where a and b are constants, $(f_c')_{28\text{d}}$ is 28-day strength, and t is time. The ranges of a and b in Eq. (1-1) for the normal-weight, sand-lightweight and all-lightweight concretes (using both moist and steam curing, and Types I and III cement) of Refs. 2 to 8 (88 specimens) are $a = 0.50$ to 8.00, $b = 0.67$ to 1.10. Note that a value of b greater than unity represents a retrogression in strength after 28 days. The following equations[1,2,3,9] approximate average values for these data.

Moist-cured concrete, Type I (normal) cement

$$(f_c')_t = \frac{t}{4.00 + 0.85t}(f_c')_{28\text{d}} \tag{1-2}$$

For example, $(f_c')_{7\text{d}} = 0.70(f_c')_{28\text{d}}$ and $(f_c')_u = 1.18(f_c')_{28\text{d}}$.

* Superscripts refer to reference numbers at the end of each chapter.

Moist-cured concrete, Type III (high-early strength) cement

$$(f_c')_t = \frac{t}{2.30 + 0.92t}(f_c')_{28\text{d}} \tag{1-3}$$

For example, $(f_c')_{7\text{d}} = 0.80(f_c')_{28\text{d}}$ and $(f_c')_\text{u} = 1.09(f_c')_{28\text{d}}$.

Steam-cured concrete, Type I (normal) cement

$$(f_c')_t = \frac{t}{1.00 + 0.95t}(f_c')_{28\text{d}} \tag{1-4}$$

For example, $(f_c')_{2.0\text{d}} = 0.69(f_c')_{28\text{d}}$ and $(f_c')_\text{u} = 1.05(f_c')_{28\text{d}}$.

Steam-cured concrete, Type III (high-early strength) cement

$$(f_c')_t = \frac{t}{0.70 + 0.98t}(f_c')_{28\text{d}} \tag{1-5}$$

For example, $(f_c')_{2.0\text{d}} = 0.75(f_c')_{28\text{d}}$ and $(f_c')_\text{u} = 1.02(f_c')_{28\text{d}}$.

In Eqs. (1-2) to (1-5), t is the age of concrete in days and $(f_c')_\text{u}$ refers to an ultimate (in time) value. These equations were developed[1] by plotting observed

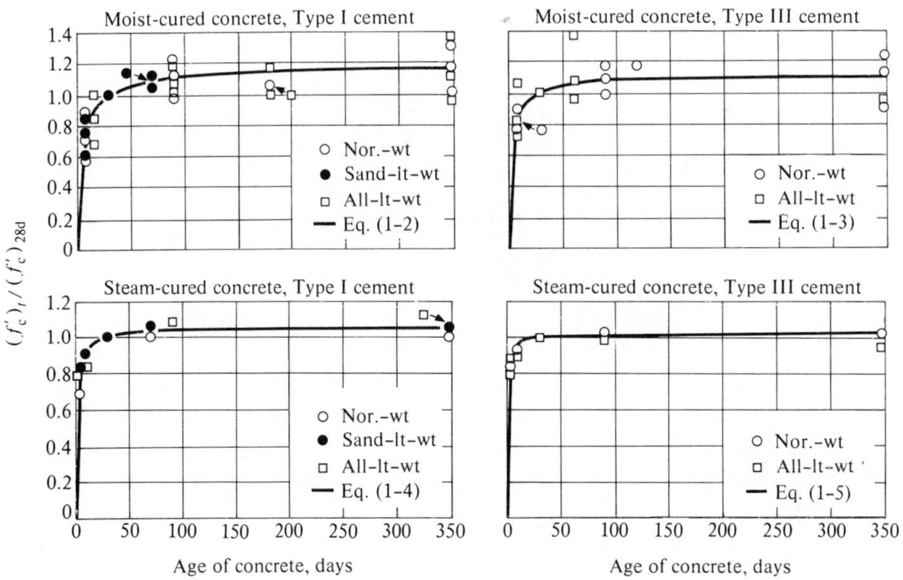

FIGURE 1-1
Concrete strength versus time curves using Eqs. (1-2) to (1-5), and comparison with experimental data from Refs. 2 to 8 for different weight concretes (using both moist and steam curing, and Types I and III cement). Where three data points are shown for a given age, they refer to the upper and lower limits and the average value for a given set of data. Where only one data point is shown, the range is too small to indicate. Data for 88 specimens are included.[1]

values of $(f_c')_t/(f_c')_{28d}$ versus time. Curves which fit the average data were then determined by trial and error. The basis for the equations is seen to be the 28-day strength. However, in the case of Eqs. (1-4) and (1-5), the type of steam curing may affect substantially the strength–time ratio in the early days following curing.

Equations (1-2) to (1-5) are compared in Fig. 1-1 with the experimental data from Refs. 2 to 8 (88 specimens). Note that these equations refer to average values only with the ranges of variation in the data indicated. The effect of steam curing and Type III cement can also be seen in Fig. 1-1 relative to the "degree of flatness" of the strength–time curve. These data fall within about ± 20 percent of the average-value curves, although most of the data are in closer agreement than this.

For some 253 specimens made of different weight concrete, it is shown[10] in Figs. 1-2, 1-3, and Table 1-1 that about 62 percent of the data agrees within ± 10 percent of the average-curve Eqs. (1-2) to (1-5), 93 percent within ± 20 percent, 97 percent within ± 30 percent, and 100 percent within ± 40 percent. In general, the use of normal-weight, sand-lightweight or all-lightweight aggregate does not appreciably influence these results. The plus-minus "discrepancy bands" referred to here, in Figs. 1-2 (see $X = \pm 20$ percent in Fig. 1-2, for example), 1-3, and elsewhere herein are obtained in each case as the difference in computed and

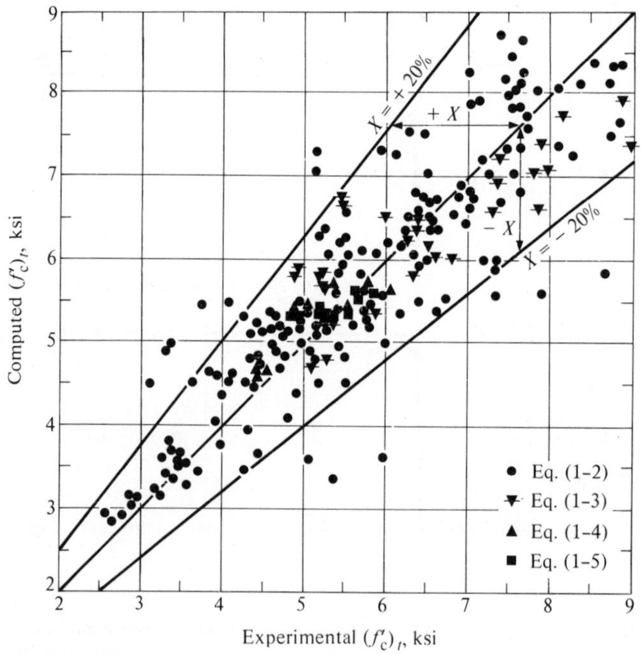

FIGURE 1-2
Computed versus experimental values of compressive strength, $(f_c')_t$, for different weight concrete, cement, and curing conditions, using Eqs. (1-2) to (1-5) and 253 specimens.[10]

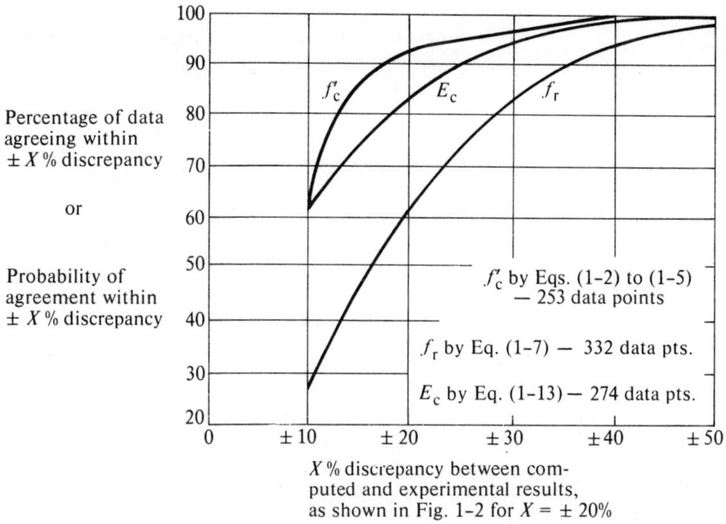

FIGURE 1-3
Curves showing the probability of agreement with experimental data involving different weight concrete, for calculated results of f'_c, f_r, and E_c.

measured value divided by the *computed* value for plus X, and divided by the *measured* value for minus X. This is the same procedure as that used by ACI Committee 435[11] and Ref. 3.

Modulus of Rupture, Direct Tensile Strength and Modulus of Elasticity of Concrete

The modulus of rupture (bending tensile strength), f_r, direct tensile strength, f'_t, and modulus of elasticity, E_c, of concrete are not accurate functions of compressive strength and unit weight. Considerable variation is frequently found between computed results by various simplified methods and actual results, as is indicated

Table 1-1 Tabulated values from Fig. 1-3 showing the probability of agreement with experimental data involving different weight concrete, for calculated results of f'_c, f_r, and E_c

	Percentage of data agreeing within $\pm X\%$ discrepancy or Probability of agreement within $\pm X\%$ discrepancy		
$\pm X\%$ discrepancy	f'_c by Eqs. (1-2) to (1-5)—253 data pts.	f_r by Eq. (1-7) —332 data pts.	E_c by Eq. (1-13) —274 data pts.
±10%	62%	27%	62%
±20%	93%	61%	83%
±30%	97%	83%	95%
±40%	100%	94%	99%
±50%	100%	98%	100%

in this section. Nevertheless, for most practical purposes the methods given by Eqs. (1-6) to (1-13) are considered satisfactory. For metric units by these equations, see the footnotes to Tables 1-2 and 1-3.

A 1966 ACI Committee 435 paper[11] noted the following variation for the modulus of rupture: for normal-weight concrete and continuously moist lightweight concrete,[4,12] $f_r = 7.5\sqrt{f'_c}$ to $12\sqrt{f'_c}$; and for lightweight concrete in a drying condition,[12,13] $f_r = 5.0\sqrt{f'_c}$ to $11\sqrt{f'_c}$. The smaller factors tend to apply to higher-strength concretes and the larger factors to lower-strength concretes. A 1971 ACI Committee 209 paper[9] suggested Eq. (1-6) for different weight concretes.

$$f_r = a_1\sqrt{wf'_c}, \text{ psi} \qquad w \text{ in pcf and } f'_c \text{ in psi} \qquad (1\text{-}6)$$

where $a_1 = 0.60$ to 1.00. Thus, $f_r = 6.0\sqrt{f'_c}$ to $12\sqrt{f'_c}$, when $w = 100$ pcf to 145 pcf. The report stated that a conservative value of $a_1 = 0.60$ may be used although a value of $a_1 = 0.65$ to 0.70 is more realistic in most cases.

Equation (1-7) is recommended in Refs. 3 and 14, and herein for general use when actual test results are not available.

Recommended for normal-weight, sand-lightweight, and all-lightweight concrete

$$f_r = 0.65\sqrt{wf'_c}, \text{ psi} \qquad w \text{ in pcf and } f'_c \text{ in psi} \qquad (1\text{-}7)$$

The provisions of the ACI Code (318-71)[15] are given by Eqs. (1-8).

$$\left.\begin{array}{ll}\text{Normal-weight concrete} & f_r = 7.5\sqrt{f'_c} \\ \text{Sand-lightweight concrete} & f_r = 6.4\sqrt{f'_c} \\ \text{All-lightweight concrete} & f_r = 5.6\sqrt{f'_c}\end{array}\right\} \text{psi} \qquad f'_c \text{ in psi} \qquad (1\text{-}8)$$

As can be seen in Table 1-2, results by Eq. (1-7) are only slightly less conservative (higher) than the corresponding values by Eqs. (1-8). For some 332 specimens made of different weight concrete, it is shown[10] in Figs. 1-3, 1-4, and Table 1-1 that about 27 percent of the data agrees within ± 10 percent of the computed results using Eq. (1-7), 61 percent within ± 20 percent, 83 percent within ± 30 percent, 94 percent within ± 40 percent, and 98 percent within ± 50 percent.

Equation (1-9) has been suggested by ACI Committee 209[9] for computing the direct tensile strength of different weight concrete.

$$f'_t = \tfrac{1}{3}\sqrt{wf'_c}, \text{ psi} \qquad w \text{ in pcf and } f'_c \text{ in psi} \qquad (1\text{-}9)$$

The well-known Eq. (1-10), which was developed by Pauw[16] in 1960, has been used in the ACI Codes (318-63)[17] and (318-71)[15] for computing the modulus of elasticity of concrete.

$$E_c = 33\sqrt{w^3 f'_c}, \text{ psi} \qquad w \text{ in pcf and } f'_c \text{ in psi} \qquad (1\text{-}10)$$

Equation (1-11) was later developed by the Portland Cement Association[18] on the basis of a broader range of concrete weights, strengths, etc.

$$E_c = a_2\sqrt{w^3 f'_c}, \text{ psi} \qquad w \text{ in pcf and } f'_c \text{ in psi} \qquad (1\text{-}11)$$

Table 1-2 Various values of the modulus of rupture and direct tensile strength of concrete, psi*

Parameter	Compressive strength, f'_c, psi						
	2500	3000	4000	5000	6000	7000	8000
Modulus of rupture, f_r							
ACI Comm. 435 ranges:[11]							
$f_r = 5.0\sqrt{f_c'}$ to	250	274	316	354	387	418	447
$12\sqrt{f_c'}$ †	600	657	759	849	930	1004	1073
Values recommended, Eq. (1-7) $f_r = 0.65\sqrt{wf'_c}$:[3]							
N.wt. ($w = 145$), $f_r = 7.84\sqrt{f'_c}$	392	429	496	555	606	655	700
S.lt.wt. ($w = 120$), $f_r = 7.12\sqrt{f'_c}$	356	390	450	504	551	596	636
A.lt.wt. ($w = 100$), $f_r = 6.50\sqrt{f'_c}$	325	356	411	460	503	544	581
ACI Code (318-71) values, Eqs. (1-8):[15]							
Normal-weight, $f_r = 7.5\sqrt{f'_c}$	375	411	475	531	581	628	670
Sand-lightweight, $f_r = 6.4\sqrt{f'_c}$	320	350	405	453	495	535	572
All-lightweight, $f_r = 5.6\sqrt{f'_c}$	280	307	354	396	434	468	501
Direct tensile strength, f'_t							
Eq. (1-8) $f'_t = \frac{1}{3}\sqrt{wf'_c}$:[9]							
N.wt. ($w = 145$), $f'_t = 4.02\sqrt{f'_c}$	201	220	255	284	312	337	359
S.lt.wt. ($w \doteq 120$), $f'_t = 3.65\sqrt{f'_c}$	183	200	231	258	283	306	326
A.lt.wt. ($w = 100$), $f'_t = 3.33\sqrt{f'_c}$	167	182	211	235	258	278	298

* For example: $f'_c = 4000$ psi $= 281$ kg/cm^2 $= 27.6$ N/mm^2
$f_r = 496$ psi $= 34.9$ kg/cm^2 $= 3.42$ N/mm^2
(for other values, see metric conversion factors).

† $12\sqrt{f'_c}$ is also the maximum computed tensile stress allowed for partially prestressed beams in the ACI Code (318-71).

where $a_2 = 36, 35, 33, 31, 29, 27, 25$ when $f'_c = 2500, 3000, 4000, 5000, 6000, 7000, 8000$ psi, respectively. Thus it is seen that Eq. (1-11) predicts results which are higher, the same, lower than those of Eq. (1-10) when f'_c is lower, equal to, higher than 4000 psi, respectively. It might be noted in Fig. 1-5 and Table 1-3 that the results by Eq. (1-11) are considerably lower than those of Eq. (1-10), and even decreasing, for increasingly high concrete strengths. Such low values (but not the decreasing aspect) appear to be valid for lightweight concrete, and possibly a little *too* low for normal-weight concrete, especially of higher strength.

Equation (1-12) was recently obtained[1] from a least squares analysis in which measured values of E_c were plotted against computed values of $\sqrt{w^3 f'_c}$. The same trend for high strength concrete as mentioned above can be seen in Fig. 1-5 for Eq. (1-12).

$$E_c = 164\,000 + 30\sqrt{w^3 f'_c}\text{, psi} \qquad w \text{ in pcf and } f'_c \text{ in psi} \tag{1-12}$$

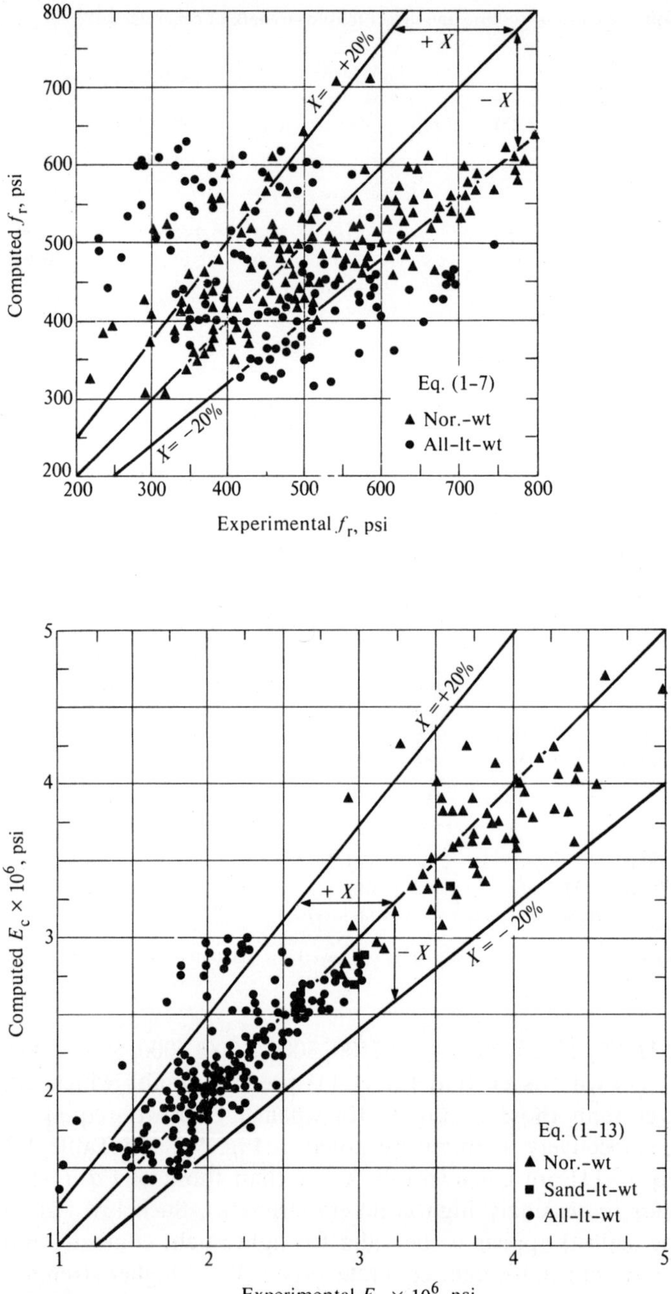

FIGURE 1-4
Computed versus experimental values of modulus of rupture, f_r (using Eq. (1-7) and 332 data points) and modulus of elasticity, E_c (using Eq. (1-13) and 274 data points) for different weight concrete, w. Both equations are expressed in terms of w and f'_c.[10]

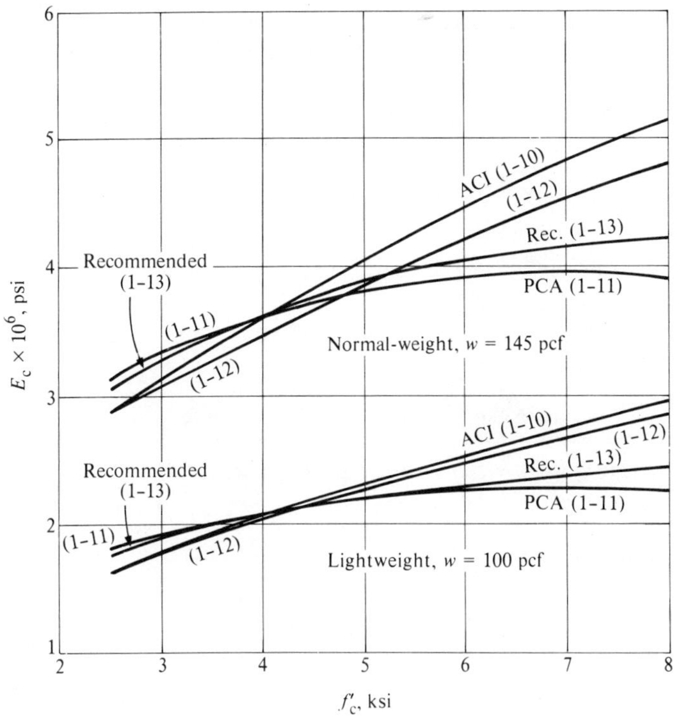

1 pcf = 16.02 kg/m^3 = 0.1571 kN/m^3
1 ksi = 70.31 kg/cm^2 = 6.895 N/mm^2
1 psi = 0.07031 kg/cm^2 = 6.895 × 10^{-6} kN/mm^2

FIGURE 1-5
Modulus of elasticity, E_c, versus compressive strength, f'_c, for normal-weight and lightweight concrete, by Eqs. (1-10) to (1-13).

As can be seen in Fig. 1-5 and Table 1-3, results by Eq. (1-13)[10] are between those of Eqs. (1-10) and (1-11), and very close to Eq. (1-11) for most of the strength range shown. The curves in Fig. 1-6 indicate[10] that the probability of agreement with some 274 data points is essentially the same for Eqs. (1-11) and (1-13), and somewhat lower than these for Eqs. (1-10) and (1-12). Only a limited number of strengths in the range of 7000 psi (492 kg/cm^2, 48.3 N/mm^2) and greater were included. Equation (1-13) is thought to be preferable over the entire strength range up to about 10 000 psi (703 kg/cm^2, 69.0 N/mm^2), and is recommended herein.

Recommended for normal-weight, sand-lightweight, and all-lightweight concrete

$$E_c = a_3\sqrt{w^3 f'_c}, \text{ psi} \qquad w \text{ in pcf and } f'_c \text{ in psi} \tag{1-13}$$

where $a_3 = 39.0 - 0.0015f'_c$.

It should be noted in Fig. 1-5 and Table 1-3, however, that computed values of E_c by any of these four equations do not differ greatly, especially in the usual strength range of f'_c = 2500 psi to 6000 psi (176 to 422 kg/cm^2, 17.2 to 41.4 N/mm^2).

Table 1-3 Various values of the modulus of elasticity of concrete, psi*

Parameter	Compressive strength, f_c', psi*						
	2500	3000	4000	5000	6000	7000	8000
Modulus of elasticity, $E_c \times 10^6$							
ACI Code (318-71) values, Eq. (1-10)							
$E_c = 33\sqrt{w^3 f_c'}$:[15]							
N.wt. ($w = 145$), $E_c = 57\,600\sqrt{f_c'} =$	2.88	3.16	3.64	4.08	4.46	4.82	5.15
S.lt.wt. ($w = 115$ to 120), $E_c = 41\,550\sqrt{f_c'} =$	2.08	2.28	2.63	2.94	3.22	3.48	3.72
A.lt.wt. ($w = 100$), $E_c = 33\,000\sqrt{f_c'} =$	1.65	1.81	2.09	2.34	2.55	2.76	2.95
PCA, Eq. (1-11) $E_c = a_2\sqrt{w^3 f_c'}$:[18]							
$a_2 =$	36	35	33	31	29	27	25
Normal-weight ($w = 145$), $E_c =$	3.14	3.35	3.64	3.83	3.92	3.95	3.90
Sand-lightweight ($w = 115$ to 120), $E_c =$	2.27	2.42	2.63	2.76	2.83	2.85	2.82
All-lightweight ($w = 100$), $E_c =$	1.80	1.92	2.09	2.20	2.24	2.26	2.24
Eq. (1-12) $E_c = 164\,000 + 30\sqrt{w^3 f_c'}$:[1]							
$E_c = 164\,000 +$							
N.wt. ($w = 145$), $52\,400\sqrt{f_c'} =$	2.87	3.04	3.47	3.86	4.21	4.54	4.84
S.lt.wt. ($w = 115$ to 120), $37\,800\sqrt{f_c'} =$	2.05	2.23	2.55	2.83	3.09	3.32	3.54
A.lt.wt. ($w = 100$), $30\,000\sqrt{f_c'} =$	1.66	1.81	2.06	2.29	2.48	2.67	2.85
Values recommended, Eq. (1-13)							
$E_c = a_3\sqrt{w^3 f_c'}$, $a_3 = 39 - 0.0015 f_c'$:[10]							
$a_3 =$	35.25	34.5	33.0	31.5	30.0	28.5	27.0
Normal-weight ($w = 145$), $E_c =$	3.08	3.31	3.64	3.89	4.05	4.16	4.21
Sand-lightweight ($w = 115$ to 120), $E_c =$	2.22	2.39	2.63	2.81	2.93	3.01	3.04
All-lightweight ($w = 100$), $E_c =$	1.76	1.89	2.09	2.23	2.32	2.38	2.41

* For example: $f_c' = 4000$ psi $= 281$ kg/cm^2 $= 27.6$ N/mm^2
$E_c = 3.64 \times 10^6$ psi $= 256\,000$ kg/cm^2 $= 25.1$ kN/mm^2
(for other values, see metric conversion factors).

Also, it must be kept in mind that the modulus of elasticity of concrete is a rather elusive property in general because of its significant variation with different mixes, ages, stress levels, rates of load application, ways in which it is defined, etc. Results by these equations are intended to approximate both initial tangent moduli, and secant moduli up to working stress levels and perhaps overload levels as well.

For some 274 data points for different weight concrete, it is shown[10] in Figs. 1-3, 1-4, 1-6, and Table 1-1 that about 62 percent of the data agrees within ± 10 percent of the computed results using Eq. (1-13), 83 percent within ± 20 percent, 95 percent within ± 30 percent, 99 percent within ± 40 percent, and 100 percent within ± 50 percent.

Since the probability information in Figs. 1-3, 1-6, and Table 1-1 is presented on the basis of plus-minus "discrepancy bands" (see Figs. 1-2, 1-4), one might

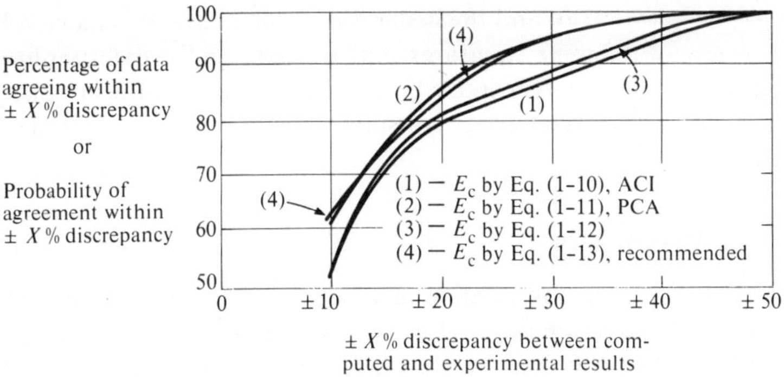

FIGURE 1-6
Curves showing the probability of agreement with experimental data involving different weight concrete, for calculated results of E_c, using Eqs. (1-10) to (1-13) and 274 data points.

assume that approximately half of the remaining data for a given plus-minus percentage would fall above, and half below, the plus-minus band. For the case of the modulus of elasticity, E_c, in Table 1-1 for example, 83 percent of the data agrees within ± 20 percent of the predicted results, which means that $\frac{17}{2}$ = about 9 percent of the time the actual results would fall above, and 9 percent below, the ± 20 percent range of agreement.

Finally, as one would imagine the probability of agreement between computed and experimental results is shown in Fig. 1-3 and Table 1-1 to be greatest for the compressive strength, f'_c, using Eqs. (1-2) to (1-5), next for the modulus of elasticity, E_c, using Eq. (1-13), and least for the modulus of rupture, f_r, using Eq. (1-7). These equations for f_r and E_c, which are functions of concrete weight, w, appear in Fig. 1-4 to be equally applicable to different weight concrete.

Modulus of Elasticity of Steel

The ACI Code (318-71)[15] recommends: $E_s = 29 \times 10^6$ psi (2.04×10^6 kg/cm^2, 200 kN/mm^2) for nonprestressed steel, and the modulus of elasticity of prestressing steel to be determined by tests or supplied by the manufacturer. Values commonly used[19] for prestressing steel are: $E_s = 29 \times 10^6$ psi for wires and bars, $E_s = 28 \times 10^6$ psi (1.97×10^6 kg/cm^2, 193 kN/mm^2) for 270 K, 7-wire strand, $E_s = 27 \times 10^6$ psi (1.90×10^6 kg/cm^2, 186 kN/mm^2) for 250 K, 7-wire strand, and $E_s = 25 \times 10^6$ psi (1.76×10^6 kg/cm^2, 172 kN/mm^2) for large strand.

1-3 CREEP AND SHRINKAGE OF CONCRETE

The principal variables that affect concrete creep and shrinkage are outlined and evaluated in Appendix A, along with a general procedure for determining the various effects, including the use of a simple design aid. Comparisons with experimental data, and numerical examples of the simplified procedure outlined here in the text (example in text also) and the general procedure, are also included in Appendix A. A review of different equations and procedures for determining shrinkage and creep is presented in Appendix B, along with numerical examples. Methods for analyzing concrete creep under constant and variable stress are also described in Appendix B.

Creep and Shrinkage Parameters

The design approach to be used for predicting creep and shrinkage refers to "standard conditions" and correction factors for "nonstandard conditions." Based on information from Refs. 1–3, 6–8, and 20–28, the general Eqs. (1-14) and (1-15) and the standard Eqs. (1-16) to (1-18), which were developed by the author and colleagues,[1,2,3] and used by ACI Committee 209,[9] U.S. Naval Civil Eng. Lab.,[29] Taylor Woodrow Constr. Ltd. Matls. Lab., England,[30] and others, are recommended for computing a creep coefficient (defined as the ratio of creep strain to initial strain) and unrestrained shrinkage of concrete at any time. A similar approach with standard conditions and correction factors has been used by Wagner,[31] Jones *et al.*,[6] and the CEB,[32] as described in Appendix B, which also includes comparative examples.

General equations

$$C_t = \frac{t^c}{d + t^c} C_u \tag{1-14}$$

$$(\varepsilon_{sh})_t = \frac{t^e}{f + t^e} (\varepsilon_{sh})_u \tag{1-15}$$

where $c, d, e,$ and f are constants, C_u = ultimate creep coefficient, $(\varepsilon_{sh})_u$ = ultimate shrinkage strain, and t = time.

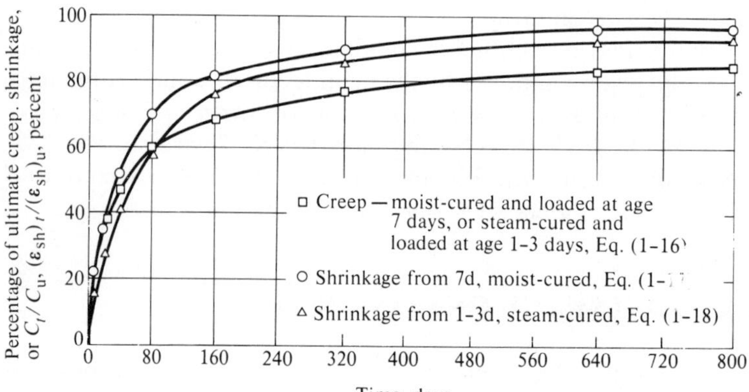

FIGURE 1-7
Standard creep and shrinkage versus time curves using Eqs. (1-16), (1-17) and (1-18).

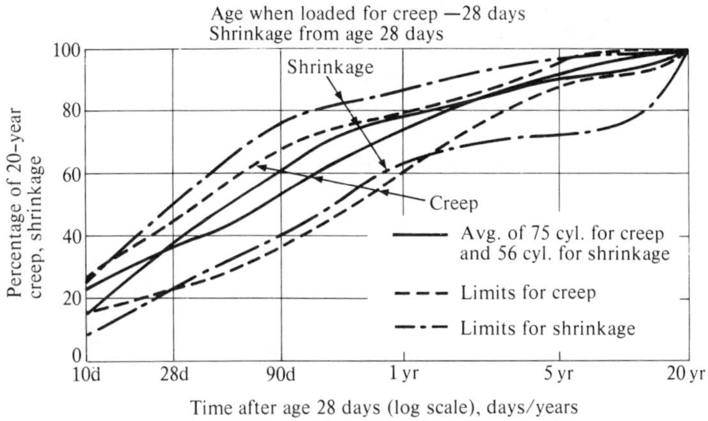

FIGURE 1-8
Creep and shrinkage versus time ratio curves for 20-year normal-weight concrete data, by Troxell, Raphael, and Davis. (*From ASTM Proc.*,[22] *with permission.*)

When the power of t is unity, Eqs. (1-14) and (1-15) reduce to the familiar hyperbolic Eqs. (B-5) of Ross[20] and (B-7) of Lorman,[21] in slightly different form. The form of these equations is thought to be convenient for design purposes, in which the concept of the ultimate (in time) value is modified by the "time-ratio" term to yield the desired result. Primarily because the increase in creep after say 100 to 200 days is usually more pronounced than shrinkage (or in percent of the ultimate value, shrinkage usually increases more rapidly than creep during the first few months), appropriate powers of t in Eqs. (1-14) and (1-15) were found[1,2,3] to be unity for shrinkage (flatter hyperbolic form) and 0.60 for creep (steeper curve for larger values of t). This can be seen in Figs. 1-7, 1-8, and Table 1-4. The delaying effect of steam curing on shrinkage can also be seen in Fig. 1-7. The ranges of the constants in the time-ratio terms were: $c = 0.40$ to 0.80, $d = 6$ to 30, $e = 0.90$ to 1.10, and $f = 20$ to 130.

The creep and shrinkage curves in Fig. 1-8 are based on 20-year data of Troxell, Raphael, and Davis[22] for normal-weight concrete with an initial time of 28 days. The results in Figs. 1-7 and 1-8 are roughly comparable.

Standard equations and standard conditions—ambient relative humidity $\leq$40 percent, average thickness of member (part under consideration) 6 in. (15.2 cm)

Creep coefficient for loading age 7 days for moist-cured concrete and 1–3 days for steam-cured concrete

$$C_t = \frac{t^{0.60}}{10 + t^{0.60}} C_u \qquad (1\text{-}16)$$

From Fig. 1-9 for the standard conditions, the average $C_u = 2.35$. From Eq. (1-22) and Table 1-7 for $H = 70$ percent, $C_u = 0.80(2.35) = 1.88$, for example.

Shrinkage after age 7 days for moist-cured concrete

$$(\varepsilon_{sh})_t = \frac{t}{35 + t}(\varepsilon_{sh})_u \tag{1-17}$$

Shrinkage after age 1–3 days for steam-cured concrete

$$(\varepsilon_{sh})_t = \frac{t}{55 + t}(\varepsilon_{sh})_u \tag{1-18}$$

From Figs. 1-10 and 1-11 for the standard conditions, the average $(\varepsilon_{sh})_u = 780 \times 10^{-6}$ in/in (cm/cm) for both Eqs. (1-17) and (1-18). From Eq. (1-23) and Table 1-7 for $H = 70$ percent, $(\varepsilon_{sh})_u = 0.70(780 \times 10^{-6}) = 546 \times 10^{-6}$ in/in (cm/cm), for example.

In Eqs. (1-16) to (1-18), t is time in days after loading for creep and, for shrinkage, time after initial shrinkage is considered. The standard conditions in both cases are 7 days for moist-cured and 1–3 days for steam-cured concrete. However, the time-ratio terms can also be used for other than the standard starting times, with appropriate correction factors for nonstandard conditions being applied to the ultimate values, C_u and $(\varepsilon_{sh})_u$. In this procedure, the time-ratio terms define the curvature, and the ultimate values the magnitude, of the creep–time and shrinkage–time curves.

These equations and average ultimate values were developed using normal-weight, sand-lightweight, and all-lightweight concrete (with both moist and steam curing, and Types I and III cement); and are considered by the author and ACI Committee 209[9] to apply equally well to these different cases. That is, for

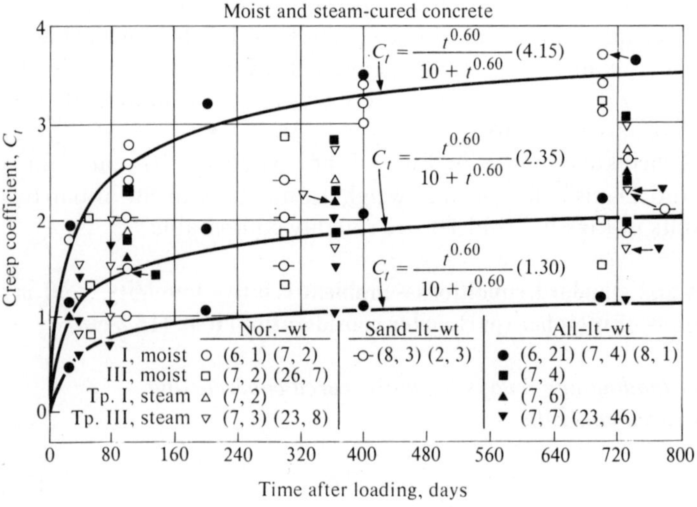

FIGURE 1-9
Creep coefficient versus time, comparing Eq. (1-16) with published data. Upper, lower, and average values are plotted for the curves and each set of data. All data were reduced to "standard" conditions using the correction factors herein. (6, 21) indicates Ref. 6 and 21 data points.[3]

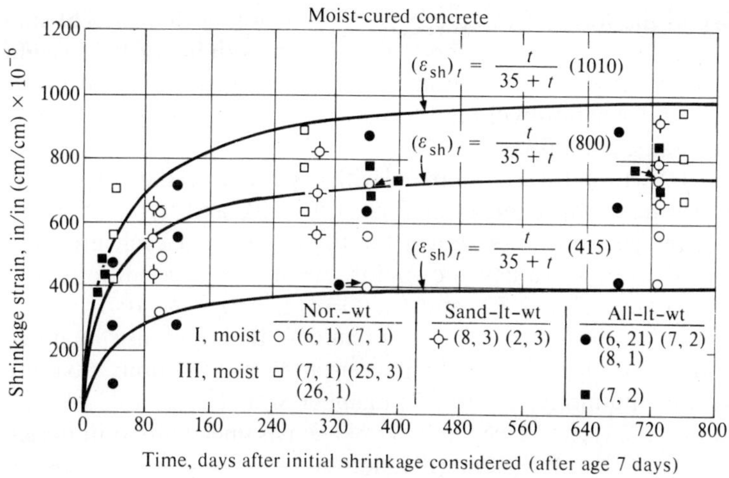

FIGURE 1-10
Shrinkage strain versus time, comparing Eq. (1-17) with published data. Upper, lower, and average values are plotted for the curves and each set of data. All data were reduced to "standard" conditions using the correction factors herein. (6, 21) indicates Ref. 6 and 21 data points.[3]

design purposes in most cases no consistent distinction seems to be found for creep and shrinkage of different weight concrete, since the cement gel apparently does most of the shrinking and creeping, with only nominal effects from the aggregate. However, especially with regard to shrinkage, the possibility of exceptional situations should be guarded against. In connection with creep, as found also by Neville,[33] there does not seem to be any fundamental difference

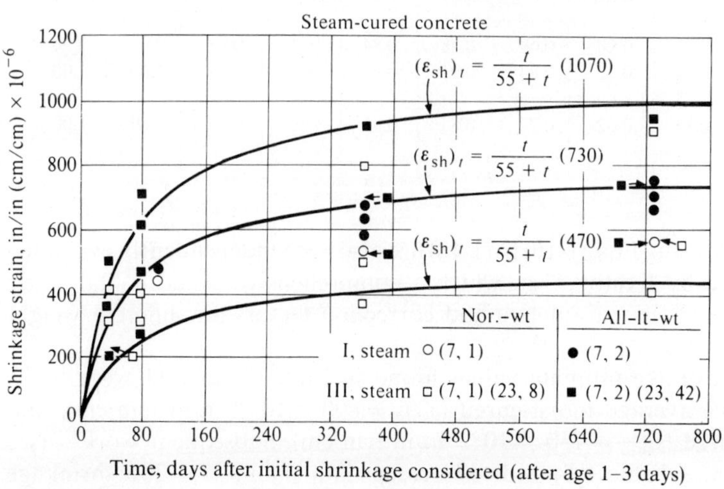

FIGURE 1-11
Shrinkage strain versus time, comparing Eq. (1-18) with published data. Upper, lower, and average values are plotted for the curves and each set of data. All data were reduced to "standard" conditions using the correction factors herein. (23, 8) indicates Ref. 23 and 8 data points.[3]

between creep (say in the form of the creep coefficient) for normal-weight and lightweight concrete. However, due to the lower elastic modulus of lightweight concrete and correspondingly higher initial strains for a given load or stress level, the total creep deformation will tend to be greater in lightweight concrete structures under the same conditions.

Results by Eqs. (1-16) to (1-18) are compared with experimental data (originally 120 creep and 95 shrinkage specimens) in Figs. 1-9, 1-10, and 1-11, in which upper, lower, and average values are shown for the curves and each group of data. The ranges and average values shown for the standard conditions are: $C_u = 1.30$ to 4.15, average, $C_u = 2.35$; and $(\varepsilon_{sh})_u = 415 \times 10^{-6}$ to 1070×10^{-6} in/in (cm/cm), average $(\varepsilon_{sh})_u = 780 \times 10^{-6}$ in/in (cm/cm). It should also be noted that the correction factors for the principal nonstandard conditions (loading age, Table 1-6; relative humidity, Table 1-7; member size, Tables 1-8, 1-9) are less than unity, so that the typical creep and shrinkage parameters used in design are usually less than the values for standard conditions.

As can be seen in Figs. 1-9 to 1-11, the level of a given curve is defined by the ultimate value, with the same time-ratio term used in general for each case. The validity of this is demonstrated in Ref. 28 where 28-day data are extrapolated to complete time curves quite well for creep, and reasonably well for shrinkage, using these time-ratio terms. See Table 1-4, or Fig. 1-7 for values of the time-ratio terms.

Table 1-4 Time-ratio values for creep and shrinkage

Creep and shrinkage ratios	Time								
	28 d	3 mth	6 mth	1 yr	2 yr	5 yr	10 yr	20 yr	30 yr
C_t/C_u Eq. (1-16)	0.42	0.60	0.69	0.78	0.84	0.90	0.93	0.95	0.96
$(\varepsilon_{sh})_t/(\varepsilon_{sh})_u$ Eq. (1-17)	0.44	0.72	0.84	0.91	0.95	0.98	0.99	1.00	1.00
$(\varepsilon_{sh})_t/(\varepsilon_{sh})_u$ Eq. (1-18)	0.34	0.62	0.77	0.87	0.93	0.97	0.99	0.99	1.00

The procedure using Eqs. (1-16) to (1-18) has also been independently evaluated and recommended by Keeton,[29] in which a comprehensive experimental study was made of the various parameters and correction factors for different weight concrete.

The original average ultimate values found in Figs. 1-9 to 1-11 were:[1,2,3,9] average $C_u = 2.35$, average moist-cured $(\varepsilon_{sh})_u = 800 \times 10^{-6}$ in/in (cm/cm), and average steam-cured $(\varepsilon_{sh})_u = 730 \times 10^{-6}$ in/in (cm/cm). Subsequent work[10] (see Figs. 1-12, 1-13, and Table 1-5) with 479 creep data points and 356 shrinkage data points resulted in the same average $C_u = 2.35$, but a new average $(\varepsilon_{sh})_u = 780 \times 10^{-6}$ in/in (cm/cm) for both moist and steam-cured concrete. It was found that no consistent distinction in the ultimate shrinkage strain was apparent

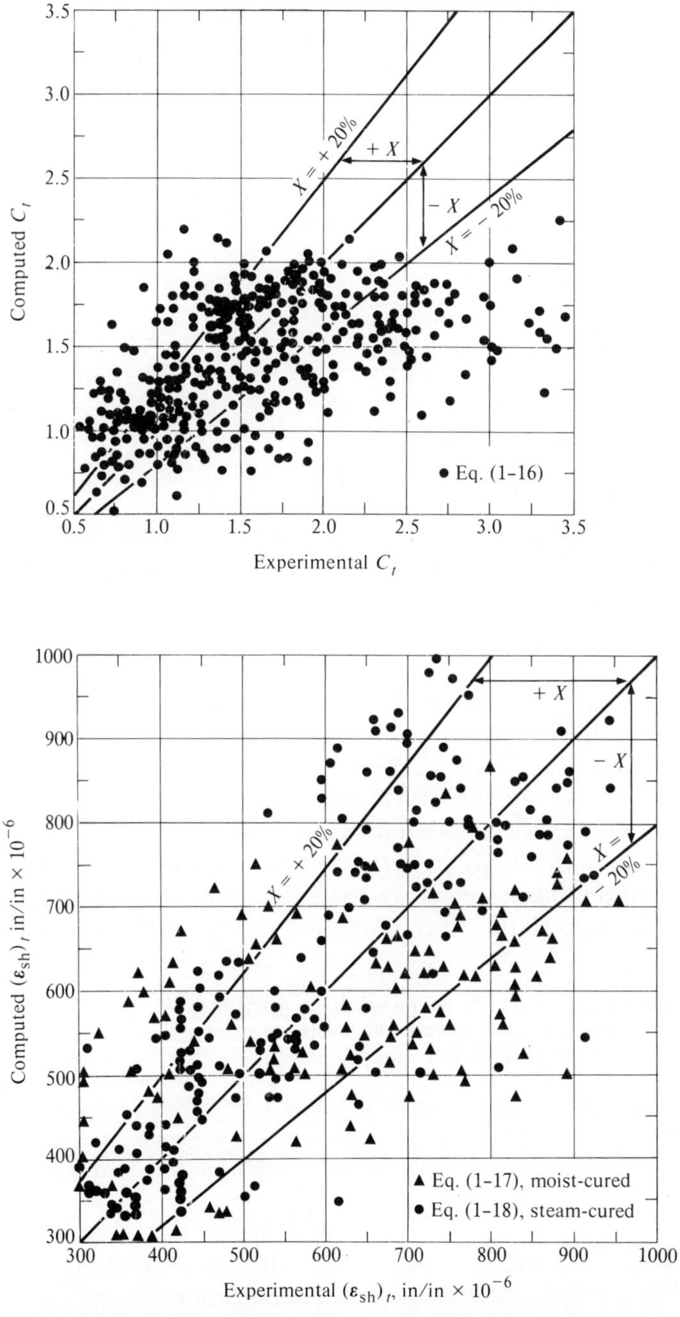

FIGURE 1-12
Computed versus experimental values of creep coefficient, C_t (using Eq. (1-16) with avg. $C_u = 2.35$ and 479 data points) and shrinkage strain, $(\varepsilon_{sh})_t$ (using Eqs. (1-17) and (1-18) with avg. $(\varepsilon_{sh})_u = 780 \times 10^{-6}$ in/in (cm/cm) and 356 data points) for different weight concrete.[10]

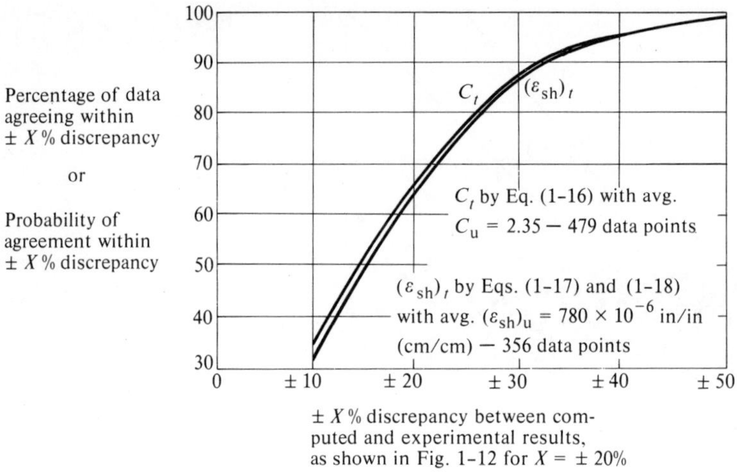

FIGURE 1-13
Curves showing the probability of agreement with experimental data involving different weight concrete, for calculated results of C_t and $(\varepsilon_{sh})_t$ using Eqs. (1-16) to (1-18) with avg. C_u and $(\varepsilon_{sh})_u$.

for moist and steam-cured concrete, even though different time-ratio terms (Eqs. (1-17) and (1-18)) are required, and different starting times used, for these two cases. Of course, for any given set of data and conditions, the ultimate values in Eqs. (1-16) to (1-18) will vary somewhat. However, the same time-ratio terms are applicable quite generally, as the extrapolation results of Ref. 28 have shown.

For the rather large number of data points used, the curves in Fig. 1-13 indicate essentially the same probability of agreement for creep and shrinkage prediction using the average ultimate values of $C_u = 2.35$ and $(\varepsilon_{sh})_u = 780 \times 10^{-6}$ in/in (cm/cm) and Eqs. (1-16) to (1-18). Thus it is seen[10] in Figs. 1-12, 1-13, and

Table 1-5 Tabulated values from Fig. 1-13 showing the probability of agreement with experimental data involving different weight concrete, for calculated results of C_t and $(\varepsilon_{sh})_t$ using Eqs. (1-16) to (1-18) with average C_u and $(\varepsilon_{sh})_u$

	Percentage of data agreeing within $\pm X\%$ discrepancy or Probability of agreement within $\pm X\%$ discrepancy	
$\pm X\%$ discrepancy	C_t by Eq. (1-16) with avg. C_u—479 data pts.	$(\varepsilon_{sh})_t$ by Eqs. (1-17) and (1-18) with avg. $(\varepsilon_{sh})_u$—356 data pts.
±10%	34%	32%
±20%	66%	64%
±30%	88%	87%
±40%	96%	96%
±50%	99%	99%

Table 1-5 for different weight concrete, that about 33 percent of the *creep* data (479 data points) and the *shrinkage* data (356 data points) agrees within ± 10 percent of the computed results using Eqs. (1-16) to (1-18) with the average C_u and average $(\varepsilon_{sh})_u$, 65 percent within ± 20 percent, 87.5 percent within ± 30 percent, 96 percent within ± 40 percent, and 99 percent within ± 50 percent.

Since the probability information in Figs. 1-13 and Table 1-5 is presented on the basis of plus-minus "discrepancy bands" (see Fig. 1-12), one might assume that approximately half of the remaining data for a given plus-minus percentage would fall above, and half below, the plus-minus band. For the case of the creep coefficient, C_t, and shrinkage strain, $(\varepsilon_{sh})_t$, in Table 1-5 for example, about 87.5 percent of the data agrees within ± 30 percent of the predicted results, which means that $12.5/2 =$ about 6 percent of the time the actual results would fall above, and 6 percent below, the ± 30 percent range of agreement, etc.

Creep coefficient versus specific creep or unit creep strain In computing initial and creep deformation, the use of the specific creep or unit creep strain, δ_t (creep per unit stress), or creep coefficient, C_t (ratio of creep strain to initial strain), can yield the same results, since they are related by Eq. (1-19). This is seen from the relations: creep strain $= (\sigma_{\text{constant}})\delta_t = (\varepsilon_{\text{initial}})C_t$, and $E_{ci} = (\sigma_{\text{constant}})/(\varepsilon_{\text{initial}})$.

$$C_t = \delta_t E_{ci} \tag{1-19}$$

In such cases the choice of C_t or δ_t is a matter of convenience depending on whether it is desired to apply the creep factor to stress (such as for columns) or deformation (such as camber and deflection). However, when using an experimental value of C_t or δ_t to calculate creep, C_t may be a more fundamental parameter, since the initial strain, ε_i, through the use of the stress, σ, and stiffness or elastic modulus, E_{ci}, must also be computed. When calculating creep using δ_t, only the stress, σ, is involved. This distinction is apparent when considering normal and lightweight concrete, with their significantly different values of E_{ci}, for example. It is noted that C_t defined by Eq. (1-19) is based on the assumption of a constant E_{ci}. This essentially conforms to the experimental determination of C_t.

Simplified Procedure for Determining Correction Factors[1,2,3,9]

A general method for determining various effects on creep and shrinkage is described in Appendix A, along with a study of those effects that are usually relatively small and tend to offset each other. Neglecting these effects, the resulting simplified procedure for determining appropriate correction factors is presented below. These two procedures are illustrated in Appendix A by means of a numerical example for typical conditions. Comparisons with experimental data are also shown in Appendix A for the correction-factor equations used in both general and simplified procedures.

All correction factors are applied to the ultimate values in the standard Eqs. (1-16) to (1-18). However, since creep and shrinkage for any period in these equations are linear functions of the ultimate values, the correction factors in this procedure may be applied to shorter-term creep and shrinkage as well. Also, within reasonable limits, the time-ratio term in Eqs. (1-16) to (1-18) can be applied to other than the standard starting times. As previously explained, the time-ratio term defines the curvature, and the ultimate values—C_u and $(\varepsilon_{sh})_u$—the magnitude, of the creep–time and shrinkage–time curves.

Loading ages later than 7 days for moist-cured concrete and later than 1–3 days for steam-cured concrete Use Eqs. (1-20) and (1-21) for the creep correction factors.

$$\text{Creep (C.F.)}_{LA} = 1.25t_{LA}^{-0.118} \quad \text{for moist-cured concrete} \qquad (1\text{-}20)$$

$$\text{Creep (C.F.)}_{LA} = 1.13t_{LA}^{-0.095} \quad \text{for steam-cured concrete} \qquad (1\text{-}21)$$

where t_{LA} is the loading age in days. Representative values are shown in Table 1-6.

As can be seen in Fig. A-2*a*, Ref. 33, and elsewhere, one finds a rather wide scatter in concrete creep data as it relates to the effect of loading age. According to Neville,[33] creep is a function of age of loading primarily because it is a function of the stress–strength ratio, and therefore the influence of loading age on creep does not change greatly after about 28 days. This is reflected in the results by Eqs. (1-20) and (1-21), which are shown in Table 1-6. These results tend to approximate average values, as can be seen in Fig. A-2*a*, and are somewhat higher than those used by the CEB[32] and those found for the 7-day base in Ref. 33. For example, at the rather late loading age of 90 days, as compared to the 7-day standard condition base, $(\text{C.F.})_{LA} = 0.74$ in Table 1-6 versus 0.63 for the data in Ref. 33 and 0.53 by CEB. On the other hand, for another set of data in Ref. 33, for a loading age of 90 days, as compared to a 3-day base (for which the correction factor would be expected to be lower than the 7-day base, not higher), $(\text{C.F.})_{LA} =$ about 0.83.

Two important points on the above discussion are the following: (1) the author believes it appropriate to attempt to use average values consistently in estimating creep and shrinkage deformation, as discussed at the beginning of this chapter; and (2) lower limit, or lower than average, values tend to underestimate (on the unsafe side) predicted time-dependent (in this case creep) deformation. These same comments also apply to the effects of relative humidity (Table 1-7) and member size (Tables 1-8, 1-9).

Shrinkage considered from other than 7 days for moist-cured concrete and other than 1–3 days for steam-cured concrete Determine the differential in Eqs. (1-17) and (1-18) for any period starting after this time. That is, shrinkage say between 28 days and 1 year would be equal to 7-day to 1-year shrinkage minus 7-day to 28-day shrinkage (see example in Appendix A). For shrinkage of moist-cured concrete from day 1 (can be used to determine differential shrinkage in composite beams, for example), use shrinkage C.F. = 1.20. An estimated interpolation can be used between 1.20 at 1 day and 1.00 at 7 days. In this procedure for moist-cured concrete, the concrete is assumed to have been cured for approximately 5 days to 7 days.

Table 1-6 Creep correction factors for various non-standard loading ages, computed by Eqs. (1-20) and (1-21)

Loading age, t_{LA} days	Creep $(C.F.)_{LA}$, moist-cured (1-20)	Creep $(C.F.)_{LA}$, steam-cured (1-21)
7	1.00	0.97
10	0.95	0.90
20	0.87	0.85
28	0.84	0.83
60	0.77	0.76
90	0.74	0.74

Table 1-7 Creep and shrinkage correction factors for non-standard relative humidity, computed by Eqs. (1-22) to (1-24)

Relative humidity, *H*, percent	Creep $(C.F.)_H$ (1-22)	Shrinkage $(C.F.)_H$ (1-23), (1-24)
40% or less	1.00	1.00
50%	0.94	0.90
60%	0.87	0.80
70%	0.80	0.70
80%	0.73	0.60
90%	0.67	0.30
100%	0.60	0.00

Ambient relative humidity greater than 40 percent Use Eqs. (1-22) to (1-24) for the creep and shrinkage correction factors.[1,25,34]

$$\text{Creep } (C.F.)_H = 1.27 - 0.0067H \quad \text{when } H \geq 40\% \tag{1-22}$$

$$\text{Shrinkage } (C.F.)_H = 1.40 - 0.010H \quad \text{when } 40\% \leq H \leq 80\% \tag{1-23}$$

$$\text{Shrinkage } (C.F.)_H = 3.00 - 0.030H \quad \text{when } 80\% \leq H \leq 100\% \tag{1-24}$$

where *H* is ambient relative humidity in percent. Representative values are shown in Table 1-7.

The values in Table 1-7 parallel those used by CEB[32] for shrinkage and also for creep, with the exception of the creep correction factor when *H* is very high. For example, when $H = 60$, 70, 80, 90, and 100 percent, $(C.F.)_H = 0.85$, 0.75, 0.62, 0.49, and 0.33 by CEB (when converted to the 40 percent base herein), respectively, which are less than the values in Table 1-7 by 0.02 at 60 percent, 0.05 at 70 percent, 0.11 at 80 percent, 0.18 at 90 percent, and 0.27 at 100 percent. The noticeable differences begin at about 80 percent relative humidity.

It should be noted that the correction factors herein were developed from data that did not include completely saturated (submerged) specimens. In addition the method is designed to apply to creep of field structures subjected to changes in humidity, which is always greater than creep of equivalent structures subjected to a constant humidity. As in the case of the effect of loading age, the lower values may tend to underestimate (or be on the unsafe side of) creep deformation. The data (Fig. A-2*b*) used to obtain Eq. (1-22) and the values in Table 1-7 were based on 3 in. (7.6 cm) by 9 in. (22.9 cm) and 6 in. (15.2 cm) by 18 in. (45.7 cm) specimens.

Average thickness of member (part under consideration) greater than 6 in. (15.2 cm) Use Eqs. (1-25) to (1-28) or Eqs. (1-29) and (1-30). These methods offer two approaches for estimating the effect of member size on creep and shrinkage. The average-thickness method tends to compute values that are higher, or on the safe side, as compared to the volume–surface ratio method. The average-thickness method is a little simpler, and is recommended for members up to about 12 to 15 in. (30 to 38 cm) average thickness. The volume–surface ratio method is recommended for larger members.

The method of treating the effect of member size in terms of the average thickness is based on information from Refs. 1, 2, 6, and 26. The volume–surface ratio equations were adapted from a paper by Hansen and Mattock,[26] and parallel values of the PCI Committee on Prestress Losses.[35]

Average-thickness method (recommended for average thicknesses up to about 12 to 15 in. or 30 to 38 cm)

$$\text{Creep (C.F.)}_T = 1.14 - 0.023T \quad \text{for } \leq 1 \text{ year loading} \tag{1-25}$$

$$\text{Creep (C.F.)}_T = 1.10 - 0.017T \quad \text{for ultimate values} \tag{1-26}$$

$$\text{Shrinkage (C.F.)}_T = 1.23 - 0.038T \quad \text{for } \leq 1 \text{ year drying} \tag{1-27}$$

$$\text{Shrinkage (C.F.)}_T = 1.17 - 0.029T \quad \text{for ultimate values} \tag{1-28}$$

where T is the average thickness of the part of the member under consideration, in inches. Representative values are shown in Table 1-8. It may be noted in

Table 1-8 Creep and shrinkage correction factors for average thickness of members ≥2 in. (5.1 cm), computed by Eqs. (1-25) to (1.28) above 6 in. (15.2 cm) thick*†

Average thickness of member, T, inches (part being considered)		Creep $(\text{C.F.})_T$		Shrinkage $(\text{C.F.})_T$	
		≤1 yr.	ult. value	≤1 yr.	ult. value
in.	cm	(1-25)	(1-26)	(1-27)	(1-28)
2†	5.1†	1.30	1.30	1.35	1.35
3†	7.6†	1.17	1.17	1.25	1.25
4†	10.2†	1.11	1.11	1.17	1.17
5†	12.7†	1.04	1.04	1.08	1.08
6	15.2	1.00	1.00	1.00	1.00
8	20.3	0.96	0.96	0.93	0.94
10	25.4	0.91	0.93	0.85	0.88
12	30.5	0.86	0.90	0.77	0.82
15	38.1	0.80	0.85	0.66	0.74
20	50.8	0.68	0.76	0.47	0.59

* This method is recommended for average thicknesses up to about 12 to 15 in. (30 to 38 cm).

† The creep and shrinkage correction factors for members less than 6 in. (15.2 cm) thick were obtained from CEB.[32]

Table 1-9 Creep and shrinkage correction factors for volume–surface ratios ≥1.5 in. (3.8 cm), computed by Eqs. (1-29) and (1.30)*†

Volume–surface ratio, v/s†		Creep $(\text{C.F.})_T$	Shrinkage $(\text{C.F.})_T$
in.	cm	(1-29)	(1-30)
1.5	3.8	1.00	1.00
2	5.1	0.96	0.96
3	7.6	0.88	0.87
4	10.2	0.80	0.78
5	12.7	0.72	0.69
6	15.2	0.64	0.60
8	20.3	0.48	0.42

* This method is recommended for average thicknesses greater than about 12 to 15 in. (30 to 38 cm).

† Examples: for rectangular sections 6 by 12 in. (15.2 by 30.5 cm) and 18 by 36 in. (45.7 by 91.4 cm), $v/s = 2.0$ in. (5.1 cm) and 6.0 in. (15.2 cm), respectively. For typical prestressed concrete sections, see Fig. C-1 for values of v/s.

Table 1-8 that the correction factors are closer to unity (smaller size effect) for ultimate values than for 1-year values because of the longer time for internal adjustment in the case of the ultimate values.

Volume–surface ratio method (recommended for average thicknesses greater than about 12 to 15 in. or 30 to 38 cm)

$$\text{Creep } (\text{C.F.})_{\text{T}} = 1.12 - 0.08v/s \qquad v/s \geq 1.5 \text{ in. } (3.8 \text{ cm}) \tag{1-29}$$

$$\text{Shrinkage } (\text{C.F.})_{\text{T}} = 1.14 - 0.09v/s \qquad v/s \geq 1.5 \text{ in. } (3.8 \text{ cm}) \tag{1-30}$$

where v/s is the volume–surface ratio of the member in inches. Representative values are shown in Table 1-9.

Average thickness of member (part under consideration) less than 6 in. (15.2 cm) Creep and shrinkage may be noticeably higher for very thin members, say 2 in. (5.1 cm) and 3 in. (7.6 cm) thick, as compared to 6 in. (15.2 cm). For members less than 6 in. (15.2 cm) thick the correction factors in Table 1-8 should be used. These correspond to the CEB[32] values for small members.

Bazant[36] has pointed out that for 100 percent ambient humidity, creep does not depend on thickness or size. Thus it may be desirable to interpolate linearly between the effect of member size at $H = 50$ percent and no size effect at $H = 100$ percent.

However, it is felt by the author that this refinement can usually be considered in the category of those effects that are relatively small and tend to cancel other small effects (when using the correction factors for humidity and size as given herein) in most cases. In the case of humidities at or near 100 percent, it would be reasonable to eliminate the member-size correction factor in design calculations. There are clearly other such coupled or interdependent effects on creep and shrinkage that are beyond the scope of this treatment.

These comments have referred to creep and not shrinkage at high humidity, since shrinkage is zero when $H = 100$ percent (shrinkage $\text{C.F.}_{\text{H}} = 0$ in Table 1-7).

A summary of ultimate (in time) creep coefficients and shrinkage strains including the effects of concrete age, relative humidity, moist and steam curing, but not member size, is tabulated in Table 1-10.

Slumps greater than 4 in. (10.2 cm) See Appendix A for the creep and shrinkage correction factors. Note that this applies only for high slumps, and normally will not be applicable to many concrete structures.

Air contents greater than 7 percent See Appendix A for the creep correction factor. No shrinkage correction factor is required for the effect of air content.

Other correction factors for creep and shrinkage, which are usually not excessive and tend to offset each other, are described in Appendix A. For design purposes in most cases, these may normally be neglected.

Table 1-10 Summary of ultimate (in time) creep coefficients and shrinkage strains, including the effects of relative humidity, moist and steam curing, and concrete age effects*

	Average relative humidity, moist or steam curing, ultimate creep coefficient or shrinkage strain																							
	≥90%				80%				70%				60%				50%				≤40%			
	M. cured		S. cured		M. cured		S. cured		M. cured		S. cured		M. cured		S. cured		M. cured		S. cured		M. cured		S. cured	
Age*	C_u	†	C_u	†	C_u	†	C_u	†	C_u	†	C_u	†	C_u	†	C_u	†	C_u	†	C_u	†	C_u	†	C_u	†
1d	—	281	1.57	234	—	562	1.72	468	—	655	1.88	546	—	749	2.04	624	—	842	2.21	702	—	936	2.35	780
7d	1.57	234	1.53	209	1.72	468	1.66	418	1.88	546	1.82	487	2.04	624	1.98	557	2.21	702	2.14	626	2.35	780	2.28	696
10d	1.50	182	1.42	198	1.63	364	1.54	396	1.79	425	1.69	462	1.94	485	1.84	528	2.10	546	1.99	594	2.23	607	2.12	660
20d	1.37	149	1.34	172	1.49	298	1.46	343	1.64	347	1.60	400	1.78	397	1.74	458	1.92	447	1.88	515	2.05	496	2.00	572
28d	1.32	130	1.31	155	1.44	260	1.42	310	1.58	303	1.56	362	1.72	347	1.70	414	1.86	390	1.83	465	1.97	433	1.95	517
60d	1.21	86	1.20	112	1.32	172	1.30	224	1.45	201	1.43	261	1.57	230	1.55	298	1.70	259	1.68	336	1.81	287	1.79	373
90d	1.17	66	1.17	89	1.27	131	1.27	178	1.39	153	1.39	207	1.51	175	1.51	237	1.63	197	1.63	266	1.74	218	1.74	296

See Tables 1-8 and 1-9 for correction factors for the effect of member size. All of the effects in the simplified procedure herein are included in this table except the size effect. These are average values that apply nominally to different weight and strength concretes, and Types I and III cement. See Fig. 1-13 and Table 1-5 for an indication of the variability of creep and shrinkage, plus-minus.

See Eqs. (1-16) to (1-18) for multipliers to obtain values at times other than ultimate (such as at 28 days, 3 months, 6 months, 1 year, and 5 years, as given in Table 1-4).

* Concrete age referred to in the table—loading *age* for creep, and period from the *age* indicated to ultimate for shrinkage.

† Ultimate shrinkage micro-strain, $(\varepsilon_{sh})_u$, as 281×10^{-6} in/in (cm/cm).

Comments on the creep and shrinkage procedure and correction factors The author would like to emphasize two points with regard to the determination of creep and shrinkage in the design of structures:

1. As described above and in Appendix B, the effect of any differences in correction factors between the various methods is normally not large. But what can be important is the base or, in the case of the procedure herein, the average value to be used in Eqs. (1-16) to (1-18), as shown in Figs. 1-9 to 1-11. Such base or standard values should be established by tests for local concrete and conditions, or selected on the basis of local experience, with the correction factors then being applied for varying or nonstandard conditions. Thus for better control of time-dependent deformation of concrete structures, a catalog of such creep and shrinkage information should be compiled for concrete using local aggregates and under local conditions. This could be accomplished for representative jobs at minimal cost.

2. For important or special projects, long-term creep and shrinkage should be extrapolated from short-term tests, as described in the next section.

The correction factors for the principal nonstandard conditions, based on the Simplified Procedure, are summarized as follows: for loading age, Eqs. (1-20), (1-21), and Table 1-6; for relative humidity, Eqs. (1-22) to (1-24), and Table 1-7; and for member size, Eqs. (1-25) to (1-30), and Tables 1-8, 1-9. It can be seen in these tables that all correction factors, except for members less than 6 in. (15.2 cm) thick, are less than unity, so that the resulting creep and shrinkage design values will normally be less than the values for standard conditions. The correction factors in the above Simplified Procedure and the CEB method were compared and found to be in good agreement in most cases. Two exceptions are the creep correction factors for age at loading and high humidity, which are both higher (on the safe side) by the procedure herein, as previously noted. The procedures of the author and colleagues, Wagner, Jones *et al.*, and CEB are presented and compared in Appendix B.

Numerical Example of Simplified Procedure for Predicting Creep and Shrinkage

Assume the following typical conditions for a prestressed bridge girder: steam-cured concrete, 70 percent ambient relative humidity, age of loading and age from which shrinkage is considered 1 day, average thickness of member (part under consideration—say lower half of girder for loss and camber calculations) 10 inches (25.4 cm), slump 2.5 inches (6.4 cm), and 6 percent air content.

Required: compute the one-year and ultimate creep coefficient and shrinkage strain, using the average ultimate values given herein.

Average values for standard conditions (see Eqs. 1-16, 1-18, and Table 1-4)

$$C_u = C_{1\text{d to ult}} = \underline{2.35} \qquad (\varepsilon_{sh})_u = (\varepsilon_{sh})_{1\text{d to ult}} = \underline{780 \times 10^{-6}\ \text{in/in (cm/cm)}}$$

$$C_{1\,\text{d to 1 yr}} = \frac{365^{0.60}}{10 + 365^{0.60}} 2.35 = 0.78 \times 2.35$$
$$= \underline{1.83}$$
$$(\varepsilon_{sh})_{1\,\text{d to 1 yr}} = \frac{365}{55 + 365}(780 \times 10^{-6}) = 0.87(780 \times 10^{-6})$$
$$= \underline{679 \times 10^{-6}\ \text{in/in (cm/cm)}}$$

Solution for the conditions of the example The following correction factors apply. For creep:

From Eq. (1-22), Table 1-7, $(C.F.)_H = 0.80$
From Eq. (1-25), Table 1-8, $(C.F.)_T = 0.91$ for 1 yr. value
From Eq. (1-26), Table 1-8, $(C.F.)_T = 0.93$ for ult. value

For shrinkage:

From Eq. (1-23), Table 1-7, $(C.F.)_H = 0.70$
From Eq. (1-27), Table 1-8, $(C.F.)_T = 0.85$ for 1 yr. value
From Eq. (1-28), Table 1-8, $(C.F.)_T = 0.88$ for ult. value

The resulting 1-year and ultimate creep and shrinkage parameters are:

$$C_{1\,\text{d to ult}} = (0.80 \times 0.93)(2.35) = 0.744 \times 2.35 = \underline{1.75} \qquad \text{ANS}$$
$$C_{1\,\text{d to 1 yr}} = (0.80 \times 0.91)(1.83) = 0.728 \times 1.83 = \underline{1.33} \qquad \text{ANS}$$
$$(\varepsilon_{sh})_{1\,\text{d to ult}} = (0.70 \times 0.88)(780 \times 10^{-6}) = \underline{480 \times 10^{-6}\ \text{in/in (cm/cm)}} \qquad \text{ANS}$$
$$(\varepsilon_{sh})_{1\,\text{d to 1 yr}} = (0.70 \times 0.85)(679 \times 10^{-6}) = \underline{404 \times 10^{-6}\ \text{in/in (cm/cm)}} \qquad \text{ANS}$$

Both general and simplified procedures for predicting creep and shrinkage are illustrated in more detail in Appendix A, including the use of a simple design aid for determining the correction factors (Fig. A-4).

Prediction of Creep and Shrinkage From Short-Term Tests

The prediction of long-term strains from short-term tests is more difficult for shrinkage than for creep. For example, using Eqs. (1-16) and (1-17), Meyers, Branson, and Schumann[28] obtained the following results for a wide variety of data in which 28-day data was extrapolated to one-year and two-year results: for *creep,* 90 percent of all calculated values were within 10 percent, and 97 percent within 15 percent of the observed results; and for *shrinkage,* 45 percent of the calculated values were within 10 percent, and 82 percent within 20 percent of the one-year observed results, while 27 percent of the calculated values were within 10 percent and 82 percent within 20 percent of the two-year observed results, with all calculated values falling within 30 percent of the observed shrinkage results. Thus for design purposes in most cases, the determination of long-term creep and shrinkage using Eqs. (1-16) to (1-18) and a 28-day experimental program can be recommended.

On the other hand, ACI Committee 209[75] has suggested that for reasonable accuracy it is desirable to conduct shrinkage tests for as long as possible, and two months is considered the minimum acceptable testing period. Neville and Meyers[24] found a coefficient of variation of 15 percent for a two-months test period and 10 percent for a six-months test period for both creep and shrinkage using a hyperbolic equation.

1-4 STEEL RELAXATION

Steel Relaxation Equations

On the basis of the results of an extensive number of relaxation tests at constant length on cold-drawn stress-relieved wire by Magura, Sozen, and Siess,[37] as well as on other data, the PCI Committee on Prestress Losses[35] has recommended Eqs. (1-31) and (1-32) for predicting loss of prestress due to steel relaxation. It is concluded that the relaxâtion characteristics of strand can also be predicted by these equations.

Stress-relieved steel

$$f_s = f_{si}\left[1 - \frac{\log_{10} t}{10}\left(\frac{f_{si}}{f_y} - 0.55\right)\right] \qquad f_{si} \geq 0.60 f_y \tag{1-31}$$

Stabilized (low-relaxation) steel

$$f_s = f_{si}\left[1 - \frac{\log_{10} t}{45}\left(\frac{f_{si}}{f_y} - 0.55\right)\right] \qquad f_{si} \geq 0.60 f_y \tag{1-32}$$

where f_s = the remaining stress at any time t after prestressing, f_{si} = the initial stress, f_y = stress at 1 percent elongation, and t = time after initial stressing in hours. The original equation[37] used f_y' = 0.1 percent offset yield strength instead of f_y in Eq. (1-31). The effect of using f_y (as above) rather than f_y' is to reduce the constant-length relaxation values, which is consistent with relaxation behavior in prestressed beams due to elastic, creep and shrinkage shortening. These equations can also be written in the form of Eqs. (1-33) and (1-34).

Stress-relieved steel

$$PL_r = \frac{\log_{10} t}{10}\left(\frac{f_{si}}{f_y} - 0.55\right)100 \qquad f_{si} \geq 0.60 f_y \tag{1-33}$$

Stabilized (low-relaxation) steel

$$PL_r = \frac{\log_{10} t}{45}\left(\frac{f_{si}}{f_y} - 0.55\right)100 \qquad f_{si} \geq 0.60 f_y \tag{1-34}$$

where PL_r is the steel relaxation loss in percent. Representative values of Eqs. (1-33) and (1-34) are shown in Table 1-11.

Table 1-11 Ultimate loss of prestress due to steel relaxation by Eqs. (1-33) and (1-34)*

f_{si}/f_y	Stress-relieved steel Eq. (1-33) $(PL_r)_u$, %	Stabilized steel Eq. (1-34) $(PL_r)_u$, %
0.60	2.5%	0.6%
0.65	5.0%	1.1%
0.70	7.5%	1.7%
0.75	10.0%	2.2%
0.80	12.5%	2.8%
0.85	15.0%	3.3%
0.90	17.5%	3.9%

* Using $t = 10^5$ hours (11.4 years) for the ultimate values.

Table 1-12 Ultimate loss of prestress due to steel relaxation by Eqs. (1-37) and (1-38)*†

Ultimate values of $SH + ES + CR_c$, %	Strand pretensioned Eq. (1-37) $(PL_r)_u$, %	Strand post-tensioned Eq. (1-38) $(PL_r)_u$, %
10%	9.2%	9.5%
15%	8.6%	9.1%
20%	8.0%	8.6%
25%	7.5%	8.1%
30%	6.7%	7.6%

* Used with 270 K strand by AASHTO–PCI.[39]
† Approximate values for the case of $SH = ES = CR_c$.

Other types of steel The PCI Committee suggests that relaxation losses for other types of prestressing reinforcement shall be based on manufacturer's recommendations supported by adequate test data.

The method used in the PCI Handbook[38] for predicting relaxation losses is based on the PCI Committee Method, with the magnitude of prestretching, and elastic, creep and shrinkage losses included in the procedure by means of a table of coefficients.

The AASHTO–PCI Joint Committee on Prestressed Concrete[39] has recommended Eqs. (1-35) and (1-36) for computing relaxation losses for 270 K seven-wire stress-relieved strand and normal-weight concrete.

Pretensioned strands

$$CR_s = 20\,000 - 0.125(SH + ES + CR_c) \tag{1-35}$$

Post-tensioned strands

$$CR_{sp} = 20\,000 - 0.125(0.8SH + 0.5ES + CR_c) \tag{1-36}$$

where CR_s, CR_{sp}, SH, ES, and CR_c are the ultimate losses in psi due to relaxation for pretensioning, relaxation for post-tensioning, shrinkage, elastic shortening, and creep, respectively. Equation (1-35) is arrived at by taking approximately 10 percent of $0.70f'_s$, where $f'_s = 270$ ksi (19 000 kg/cm^2, 1860 N/mm^2), and deducting one-eighth of the shrinkage, elastic shortening and concrete creep losses to compensate for their effect on the prestressing steel. In the post-tensioned case, it is felt that only 0.8 of the shrinkage loss affects post-tensioned concrete due to the increased curing time thus decreasing the amount of concrete shrinkage. The same reasoning applies for the elastic shortening loss, reducing its contribution to one-half.

These equations can also be written in the form of Eqs. (1-37) and (1-38), using $f_{si} = 190$ ksi (13 400 kg/cm^2, 1310 N/mm^2) for 270 K strand.

Pretensioned strands

$$(PL_r)_u = 10.5 - 0.125(SH + ES + CR_c) \tag{1-37}$$

Post-tensioned strands

$$(PL_r)_u = 10.5 - 0.125(0.8SH + 0.5ES + CR_c) \tag{1-38}$$

where $(PL_r)_u$ = the ultimate relaxation loss in percent and the other terms refer to the respective ultimate losses in percent. The author believes Eqs. (1-37) and (1-38) are equally applicable for 250 K strand and lightweight concrete, since there is little difference between 250 K and 270 K constant-length relaxation in percent of initial stress, and the appropriate elastic, shrinkage and creep losses are substituted in the equations. Representative values of Eqs. (1-37) and (1-38) are shown in Table 1-12.

Based on studies by the author and his colleagues,[2,3] Eqs. (1-39) and (1-40) are suggested for the usual range of $f_{si}/f_y = 0.65$ to 0.80.

Stress-relieved steel

$$PL_r = 1.5 \log_{10} t \quad \text{and} \quad (PL_r)_u = 7.5\% \quad \text{at } t = 10^5 \text{ hr} = 11.4 \text{ yr} \tag{1-39}$$

Stabilized (low-relaxation) steel

$$PL_r = 0.70 \log_{10} t \quad \text{and} \quad (PL_r)_u = 3.5\% \quad \text{at } t = 10^5 \text{ hr} = 11.4 \text{ yr} \tag{1-40}$$

in which $t = 10^5$ hours = 11.4 years is assumed to be a reasonable time for considering the ultimate value. This follows the suggestion of Antill[40] that steel relaxation is probably insignificant beyond 10^5 hours. Equations (1-39) and (1-40) employ a linear prestress-loss versus log-time relation as shown for the constant-length relaxation data for seven-wire strand in Fig. 1-14. This is the same time-function as that used with Eqs. (1-31) to (1-34). It is noted that Eqs. (1-39) and (1-40) could be used for any desired ultimate value by simply modifying the constants 1.5 and 0.70.

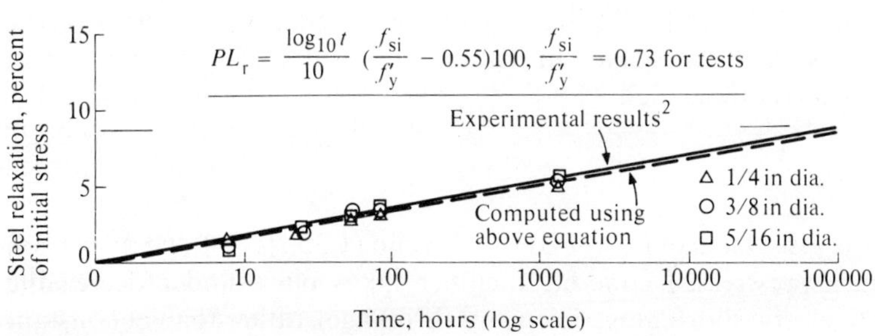

FIGURE 1-14
Measured and computed steel relaxation data for constant-length tests with seven-wire strand at the University of Iowa,[2] using 250 K steel with $f_{si} = 172$ ksi (12 100 kg/cm^2, 1190 N/mm^2) and $f'_y = 235$ ksi (16 500 kg/cm^2, 1620 N/mm^2).

Table 1-13 Loss of prestress due to steel relaxation at different times by Eqs. (1-33), (1-34), (1-39), and (1-40)*

	t in hours				
Values	10 hr	100 hr = 4.2 days	1000 hr = 1.4 months	10 000 hr = 1.14 yr	100 000 hr = 11.4 yr
Eq. (1-33), (1-34), (1-39), (1-40): $(PL_r)_t/(PL_r)_u =$	0.20	0.40	0.60	0.80	1.00
Eq. (1-39)—stress rel.: $PL_r = 1.5 \log_{10} t =$	1.5%	3.0%	4.5%	6.0%	7.5%
Eq. (1-40)—stabilized: $PL_r = 0.70 \log_{10} t =$	0.7%	1.4%	2.1%	2.8%	3.5%

* $t = 10^5$ hours = 100 000 hours = 11.4 years was assumed to be a reasonable time for considering an ultimate value.

Equations (1-39) and (1-40) are somewhat simpler than the other relaxation-loss equations presented herein and may represent a degree of refinement that is more consistent with the approximate nature of one's ability to predict creep and shrinkage losses in most cases, for example. Other effects, such as the magnitude and time of prestretching, variation in the actual value of f_{si}/f_y, etc., may also be uncertain at the time of preparing design calculations. Representative values of Eqs. (1-39) and (1-40) are shown in Table 1-13.

Recommended Procedure for Predicting Loss of Prestress Due to Steel Relaxation

For both pretensioning and post-tensioning using different weight concrete, when $f_{si}/f_y = 0.65$ to 0.80 (a typical range), the author prefers to use Eqs. (1-39) and (1-40) for most design calculations (results in Table 1-13). This is thought to be the simplest method and one that is consistent with the nature of the problem. For other ranges of f_{si}/f_y, and for additional checks on pre and post-tensioning relaxation loss calculations for different weight concrete members, refer to Eqs. (1-31) to (1-34), and (1-35) to (1-38). Results by these equations in terms of the relaxation loss in percent are shown in Tables 1-11 and 1-12. The practice of prestretching is also recommended with these procedures.

Discussion

As was noted in the discussion of Eqs. (1-31), (1-32) and (1-35), (1-36), the relaxation of steel stress in a prestressed concrete member takes place under decreasing steel strain (due to elastic shortening, creep and shrinkage), rather than at constant length as in a relaxation test. Thus the constant-length tests tend to overestimate steel relaxation when applied to prestressed beams. In fact, the decrease in strain or stress from causes other than steel relaxation is of the order of 10 percent to 30 percent, which may reduce the usual applied stress ratios to a value approaching the 0.55 factor used in Eqs. (1-31) to (1-34), for example.

The recommendations by CEB–FIP[32] suggest an equation similar to Eq. (1-41) for determining the relation between steel relaxation in a prestressed concrete beam and a constant-length test.

$$\frac{PL_r \text{ for prestressed member}}{PL_r \text{ in constant-length test}} = \left(1 - 3\,\frac{SH + CR_c}{f_{si}}\right) \tag{1-41}$$

When $(SH + CR_c)/f_{si}$ = say 0.10 to 0.20 (10 to 20 percent) in Eq. (1-41), the beam relaxation is 70 to 40 percent of the constant-length relaxation, respectively.

It should also be noted that relaxation increases rapidly with temperature.[32] Thus the higher temperatures (as 150–160°F for approx. 18 hours[35]) involved in steam-curing cause higher relaxation losses, as compared to moist-curing. The PCI Committee on Prestress Loss[35] suggests the following for steam-cured members: the computed relaxation loss from the time of initial stress until stress transfer shall be multiplied by 1.3 when f_{si}/f_y is equal to 0.80, by 1.9 when f_{si}/f_y is equal to 0.70, and by 2.5 when f_{si}/f_y is equal to 0.60. A linear interpolation may be used when f_{si}/f_y is between these values.

For all of the reasons of uncertainty, and the fact that the prediction of relaxation losses should tend toward average values, in order to be consistent with other procedures normally used in computing prestress losses, the simple Eqs. (1-39) and (1-40) and the ultimate relaxation losses of 7.5 percent for stress-relieved steel and 3.5 percent for low-relaxation steel are considered reasonable for most design calculations.

This ultimate value of $(PL_r)_u = 7.5$ percent is compared with results by the other equations presented for computing prestress loss due to steel relaxation, with the constant-length University of Illinois[37] data for those cases using pre-stretching, and with common values used in the past, as follows.

1. From Table 1-11, Eq. (1-33), when $f_{si}/f_y = 0.65$ to 0.80:

$$\underline{(PL_r)_u = 5.0\% \text{ to } 12.5\%}$$

2. From Table 1-12, Eqs. (1-37), (1-38), when $SH + ES + CR_c = 10\%$ to 30%:

$$\underline{(PL_r)_u = 6.7\% \text{ to } 9.2\%} \qquad \text{for pretensioning}$$

and

$$\underline{(PL_r)_u = 7.6\% \text{ to } 9.5\%} \qquad \text{for post-tensioning}$$

3. From the constant-length test results in Table 1-14, when $f_{si}/f_y = 0.65$ to 0.80:

$$\underline{(PL_r)_u = 3.4\% \text{ to } 8.3\% \qquad \text{with an average of } 6.2\%}$$

4. Common steel relaxation losses used in the past:

$$\underline{(PL_r)_u = 3\% \text{ to } 5\%}$$

Table 1-14 Results of constant-length relaxation tests on cold-drawn stress-relieved wire at the University of Illinois for those specimens that were prestretched[37]

Specimen mark*	Initial stress f_{si}, ksi*	Initial offset f_{si}/f'_y, %	Initial 1% stress f_{si}/f_y, %	Prestretch: Stress, ksi*	Prestretch: Time, min	Final measurement: Time t, hr	Final measurement: Relaxation loss $(PL_r)_t$, %	Loss extrapolated to 10^5 hr = 11.4 yr† $(PL_r)_u$, %
OR202-P	142.7	65.4%	60.3%	153.7	15	11 948	3.7%	4.5%
OR204-P	152.8	70.1%	64.5%	165.2	15	11 903	3.6%	4.4%
OR205-P	161.0	73.9%	67.9%	176.0	15	1 011	2.5%	4.1%
OR208-P	168.5	77.3%	71.1%	186.9	15	4 442	5.6%	7.7%
OR209-P	179.0	82.1%	75.5%	205.0	15	1 229	4.9%	7.9%
OR211-P	186.8	85.6%	78.8%	204.5	15	944	3.5%	5.9%
OR212-P	194.0	89.0%	81.9%	220.7	15	4 370	7.7%	10.6%
OR213-P	209.0	95.8%	88.2%	229.7	15	1 205	6.1%	9.9%
OR301-P	142.0	67.6%	66.0%	153.5	10	5 040	2.5%	3.4%
OR302-P	145.0	69.0%	67.4%	158.5	10	5 040	4.1%	5.5%
OR303-P	151.0	71.9%	70.2%	167.5	10	28 201	5.8%	6.5%
OR306-P	170.0	81.0%	79.0%	187.0	10	28 321	7.4%	8.3%
OR307-P	202.7	96.6%	94.4%	224.4	10	68 930	11.6%	11.9%
OR309-P	190.0	90.5%	88.4%	207.5	10	68 270	12.6%	13.1%
OR401-P	212.6	95.4%	91.0%	234.4	10	21 746	11.9%	13.7%
OR403-P	188.4	84.5%	80.5%	205.2	10	68 240	13.0%	13.5%

By Magura, Sozen, and Siess (from *PCI Journal*,[37] with permission).

* For these 16 specimens, the ultimate strength varied from 255 ksi to 266 ksi, the 0.1% offset stress, f'_y, varied from 210 ksi to 223 ksi, and the stress at $\varepsilon_s = 1\%$ elongation, f_y, varied from 215 ksi to 237 ksi. The notation, f'_y and f_y, has been reversed from that used in Ref. 36 in order to be consistent with the PCI Committee. (1 ksi = 70.31 kg/cm^2 = 6.895 N/mm^2.)

† Extrapolated losses were computed using a log-time relation as in Eqs. (1-31) to (1-34) and (1-39), (1-40), with $t = 10^5$ hours = 11.4 years assumed to be a reasonable time for considering an ultimate value.

When it is desired to compute the relaxation increment for any time interval (such as from the time of initial stress until stress transfer, for example), the term $\log_{10} t$ in Eqs. (1-31) to (1-34), and (1-39), (1-40) should be replaced with the term $(\log_{10} t - \log_{10} t_1)$, which is similar to that used by VanHorn,[41] where t and t_1 are the limits of the time interval in hours. For mathematical consistency, the value for t_1 at the time of anchorage of the prestressing steel should be taken as 1 hour, so that $\log_{10} t_1 = 0$.

1-5 THERMAL EFFECTS AND AVERAGE RELATIVE HUMIDITY

Coefficient of Thermal Expansion of Concrete

The following values have been suggested for the linear thermal expansion coefficient of concrete, where α is in inches per inch per degree Fahrenheit (F) or millionths per degree F. Multiply by $\frac{9}{5} = 1.8$ for α in millionths per degree Celsius (C).

Normal-weight concrete

Martin[42]—average $\alpha = 5.5$ millionths per °F (9.9 millionths per °C).
AASHTO[43]—design $\alpha = 6.0$ millionths per °F (10.8 millionths per °C).

Lightweight concrete

ACI Committee 213[44]—$\alpha = 4$ to 6 millionths per °F (7.2 to 10.8 millionths per °C).

CEB[32]—$\alpha = 2.7$ to 6.1 millionths per °F (5 to 11 millionths per °C), and recommends 4.4 millionths per °F (8 millionths per °C).

Martin[45]—uses $\alpha = 4.4$ millionths per °F, but points out that this value may change considerably according to whether the fine aggregate is lightweight or normal-weight siliceous or calcareous sand.

The following average values are suggested for most design purposes:

Normal-weight concrete

$$\alpha = 5.5 \times 10^{-6} \text{ in/in (cm/cm) per °F (10 per °C)} \tag{1-42}$$

Sand-lightweight concrete

$$\alpha = 5.0 \times 10^{-6} \text{ in/in (cm/cm) per °F (9 per °C)} \tag{1-43}$$

All-lightweight concrete

$$\alpha = 4.4 \times 10^{-6} \text{ in/in (cm/cm) per °F (8 per °C)} \tag{1-44}$$

Effect of Temperature Change

Normal daily maximum and minimum temperatures and design temperature differentials proposed by Martin[42,45] are given in Table 1-14. Since there is a time lag between concrete temperature and ambient temperature (greater lag in less dense or lightweight concrete than normal-weight concrete) the design temperature differentials in Table 1-15 for normal-weight concrete are based on two-thirds of the normal daily variations, with lightweight concrete differentials taken as 95 percent of the normal-weight values.

The AASHTO[43] design temperature differentials for concrete structures are:

Moderate climate	Temp. rise 30°F, temp. fall 40°F.
Cold climate	Temp. rise 35°F, temp. fall 45°F.

Average Relative Humidity

Typical average relative humidities from two different sources are also given in Table 1-15.

Table 1-15 Environmental and design conditions in the United States

		Values from Martin[42,45]				Values from ACI Com 435[46]			
						Average relative humidity			
		Normal temp.		Design temp.					
State	City or locality	Daily max. °F	Daily min. °F	dif-ferential* Nor.-wt con. °F	Avg. rel. hum. %	Jan. out-side %	Jan. in-doors† %	July out-side %	Mean annual outside %
	Birmingham	92.5	36.3	37	67	73	38	77	71
Ala	Mobile	92.0	43.7	32	73				
	Montgomery	92.4	37.6	37	72				
	Anchorage	66.0	4.3	41	70				
Alask	Fairbanks	71.7	−21.4	62	66				
	Juneau	62.7	20.0	28	81				
	Flagstaff	81.4	14.1	45	54				
Ariz	Phoenix	104.6	35.3	46	41				
	Tucson	98.5	37.0	41	38	48	44	45	44
Ark	Little Rock	92.7	30.5	41	70				
	Bakersfield	100.3	37.4	42	51				
	Long Beach	82.6	40.8	28	66				
	L.A. Airport	75.9	45.0	21	70	65	53	75	70
	L.A. Civ. Ctr.	83.3	46.6	24	63				
Calif	Oakland	74.6	40.6	23	75				
	Sacramento	93.4	37.2	37	67				
	San Diego	78.0	45.4	22	69	71	52	81	75
	S.F. Airport	73.8	41.7	21	73	79	44	84	78
	S.F. Fed. Bldg.	68.9	45.5	16	73				
	Colo. Springs	84.2	15.2	46	51				
Colo	Denver	88.4	14.8	49	53	56	21	50	52
	Pueblo	92.1	14.7	52	48				
Conn	Bridgeport	82.6	22.7	40	70				
	Hartford	85.0	17.3	45	68				
Dela	Wilmington	86.2	25.2	41	70	76	23	82	78
DC	Washington	87.0	29.4	38	63	71	27	75	72
	Jacksonville	92.0	45.0	31	75	81	52	80	79
	Key West	88.0	65.4	15	75				
Flori	Miami	89.7	57.9	21	74	79	77	75	76
	Tallahassee	90.5	42.7	32	76				
	Tampa	90.3	51.0	26	75				
	Atlanta	87.0	37.1	33	70				
Geor	Macon	92.8	37.9	37	70				
	Savannah	91.2	40.2	34	74				

Table 1-15 Continued

State	City or locality	Values from Martin[42,45] Normal temp. Daily max. °F	Daily min. °F	Design temp. differential* Nor.-wt con. °F	Avg. rel. hum. %	Values from ACI Com 435[46] Average relative humidity Jan. outside %	Jan. indoors† %	July outside %	Mean annual outside %
Hawaii	Honolulu	85.2	65.8	13	68				
Idaho	Boise	91.4	22.1	46	58				
Ill	Chicago	84.1	19.0	43	66	79	17	69	73
	Peoria	87.0	17.6	46	72				
	Springfield	89.6	20.5	46	70				
Ind	Fort Wayne	85.1	19.6	44	71				
	Indianapolis	86.0	21.0	43	72				
	South Bend	84.5	18.1	44	71				
Iowa	Des Moines	87.3	11.3	51	71	79	15	66	71
	Sioux City	89.5	9.5	53	71				
Kans	Topeka	91.6	19.0	48	69				
	Wichita	92.8	22.3	47	65				
Kenty	Louisville	88.5	26.0	42	70				
Louis	New Orleans	90.7	44.8	31	77	79	52	78	78
	Shreveport	94.8	38.3	38	70				
Maine	Portland	79.5	11.7	45	75				
Mary	Baltimore	87.2	25.3	41	67				
Mass	Boston	81.9	23.0	39	67				
Mich	Det. City Airp.	83.9	20.4	42	69	83	17	68	75
	Det. Met. Airp.	83.9	18.9	43	71				
	Flint	83.5	16.7	45	70				
	Grand Rapids	84.5	16.4	45	73				
	Lansing	83.4	16.5	45	73				
Minn	Duluth	77.1	−0.6	52	70	83	9	73	77
	Minn.–St. Paul	83.9	2.3	54	70	79	11	67	71
Miss	Jackson	93.1	37.4	37	75				
Misso	Kansas City	91.9	23.4	46	65	74	21	66	69
	St. Louis	89.2	23.5	44	68				
	Springfield	90.3	23.8	44	70				

Table 1-15 Continued

		Values from Martin[42,45]				Values from ACI Com 435[46]			
		Normal temp.		Design temp. dif-ferential* Nor.-wt con. °F	Avg. rel. hum. %	Average relative humidity			
State	City or locality	Daily max. °F	Daily min. °F			Jan. out-side %	Jan. in-doors† %	July out-side %	Mean annual outside %
Monta	Billings	88.5	13.1	50	55				
	Great Falls	84.7	12.5	48	55				
	Havre	85.3	3.6	54	60				
	Helena	84.4	8.4	51	59	69	14	47	59
Nebr	Lincoln	92.3	16.2	51	—				
	Omaha	89.7	12.9	51	68	76	15	65	69
	Elko	90.8	10.0	54	54				
Nev	Las Vegas	103.5	32.0	48	40				
	Reno	89.4	16.2	49	49				
N.H.	Concord	82.8	10.6	48	71				
N.J.	Atlantic City	83.8	26.1	38	72				
	Newark	86.1	24.7	41	64				
	Trenton	85.2	25.9	40	—				
N.M.	Albuquerque	91.2	23.5	45	42				
	Roswell	95.5	20.6	50	49				
N.Y.	Albany	83.7	14.4	46	70				
	Buffalo	80.1	17.2	42	73	80	17	72	75
	N.Y. Cent. Pk.	85.3	26.4	39	65	72	23	72	71
	N.Y. Kenn. Air.	84.7	24.7	40	68				
	N.Y. LaGua. Fl.	84.8	26.8	39	61				
	Rochester	82.5	16.9	44	70				
	Syracuse	82.4	16.5	44	71				
N.C.	Charlotte	88.8	34.0	37	68				
	Raleigh	88.1	31.4	38	70				
N.D.	Bismark	85.7	0.1	57	64	76	9	66	71
	Fargo	83.8	−2.8	58	71				
Ohio	Akron	83.2	20.7	42	71				
	Cincinnati	87.4	22.7	43	69				
	Cleveland	82.4	20.8	41	71				
	Columbus	86.6	22.0	43	71	79	22	68	72
	Dayton	85.3	22.2	42	69				
	Toledo	85.1	18.4	44	72				
	Youngstown	81.7	19.7	41	74				
Okla	Oklahoma City	93.5	28.1	44	66	73	28	70	70
	Tulsa	93.1	26.5	44	66				

Table 1-15 Continued

State	City or locality	Values from Martin[42,45] Normal temp. Daily max. °F	Daily min. °F	Design temp. differential* Nor.-wt con. °F	Avg. rel. hum. %	Values from ACI Com 435[46] Average relative humidity Jan. outside %	Jan. indoors† %	July outside %	Mean annual outside %
Ore	Portland	78.6	33.0	30	75				
	Salem	81.7	31.5	33	76				
Pac I	Guam	87.3	72.2	10	83				
Penn	Allentown	85.2	20.4	43	71				
	Erie	80.1	18.8	41	71				
	Harrisburg	87.0	23.6	42	66				
	Philadelphia	85.9	24.3	41	67	73	24	69	70
	Pittsburgh	83.3	20.7	42	69				
	Wilkes-Barre	81.6	21.4	40	69				
R.I.	Providence	81.4	21.0	40	67				
S.C.	Charleston	89.2	38.3	34	76				
	Columbia	92.5	34.5	39	72				
	Greensville	87.0	34.7	35	68				
S.D.	Rapid City	87.7	9.8	52	59				
	Sioux Falls	86.0	5.2	54	69				
Tenn	Memphis	91.1	32.4	39	69				
	Nashville	90.7	30.9	40	69	75	30	71	71
Texas	Amarillo	94.2	23.5	47	55				
	Austin	95.6	40.5	37	66				
	Corpus Chris.	93.8	47.4	31	77				
	Dallas	95.0	36.0	39	64	71	37	61	67
	Houston	92.8	43.6	33	74				
	San Antonio	94.2	41.6	35	67	67	48	67	67
	Wichita Falls	98.3	31.6	44	63				
Utah	Salt Lake City	94.1	17.5	51	55	72	19	36	52
Verm	Burlington	81.9	6.9	50	71	78	13	68	71
Virgi	Norfolk	87.9	32.2	37	72				
	Richmond	89.4	29.0	40	72				
	Roanoke	88.0	29.1	39	63				
Wash	Seattle	75.6	33.0	28	75	85	31	70	77
	Spokane	85.6	19.2	44	64				
W. In.	San Juan, Pu. R.	85.5	70.0	10	79				

Table 1-15 Continued

State	City or locality	Values from Martin[42,45] Normal temp. Daily max. °F	Normal temp. Daily min. °F	Design temp. differential* Nor.-wt con. °F	Avg. rel. hum. %	Values from ACI Com 435[46] Average relative humidity Jan. outside %	Jan. indoors† %	July outside %	Mean annual outside %
W.V.	Charleston	85.6	27.9	38	70				
Wisc	Madison	82.2	9.1	49	73				
	Milwaukee	78.9	12.8	44	73	80	15	71	75
Wyom	Casper	87.1	13.9	49	54				
	Cheyenne	85.2	13.6	48	52				

(From *ACI Journal*,[42,45] with permission.)

$°C = (\frac{5}{9})(°F - 32)$.

* Values computed as two-thirds of the normal daily maximum and minimum temperatures in the table, as reported by the Environmental Data Service. 95% of these design temperature differentials were recommended for lightweight concrete.

† The lowest relative humidity which will produce condensation on surfaces which are the same temperature as the mean outdoor temperature. These can be considered the lowest values to be expected.

1-6 POISSON'S RATIO AND EXTENSIBILITY OF CONCRETE

Poisson's Ratio of Concrete

The Poisson-ratio effect, defined simply as the ratio of lateral strain to longitudinal strain, is normally neglected in practical concrete deformation calculations. This is due to its relatively small elastic value to begin with, the fact that creep Poisson's ratio tends to be even smaller, and such structures as slabs and tanks are relatively insensitive to the effect, which is given by $(1 - \mu^2)$.

Elastic values of Poisson's ratio normally range[47] from 0.15 to 0.25, with a typical value for design taken as

$$\mu = 0.20 \tag{1-45}$$

Values of creep Poisson's ratio have been reported as essentially the same as the elastic value to essentially zero.[33]

Extensibility of Concrete

The extensibility of concrete, or its ability to slowly deform in tension (extend) without cracking, is normally not taken into account in deformation calculations, but should be recognized. For example, for specimens loaded in tension so that it took two to three months to produce failure, the extensibility or strains have been reported[47] to range from 1.2 to 2.5 times the strains for similar specimens under rapid loading.

Differential shrinkage represents an example of such a slowly applied increasing tensile stress that may produce cracking. However, the effect is primarily due to creep and can be approximated or included in creep deformation calculations.

1-7 THE EFFECTIVE OR SUSTAINED MODULUS METHOD

A reduced concrete modulus of elasticity usually referred to as the effective or sustained modulus for computing initial plus creep deformation is defined by Eq. (1-46) for concrete under constant sustained stress.

$$E_{ct} = \frac{\sigma_{\text{constant}}}{\varepsilon_{\text{initial}} + \varepsilon_{\text{creep}}} = \frac{\sigma_{\text{constant}}}{(1 + C_t)\varepsilon_{\text{initial}}} = \frac{E_c}{(1 + C_t)} \tag{1-46}$$

By substituting Eq. (1-46) in the following modular ratio form, Eq. (1-47) defining an initial-plus-creep modular ratio is obtained.

$$n_{ct} = \frac{E_s}{E_{ct}} = \frac{E_s}{E_c}(1 + C_t) = n(1 + C_t) \tag{1-47}$$

where n is the initial or short-time modular ratio, E_s/E_c.

It is noted that the calculation of initial plus creep deformation as $(1 + C_t)$ times the initial deformation (method used in this book for primary calculations such as deflections) yields the same results as that obtained using the effective or sustained modulus of elasticity, E_{ct}, provided n (at the time of loading) and not n_{ct} is used if the transformed section properties are considered. This can be seen from the fact that E_c for computing initial deformation (as $\Delta = KML^2/EI$) is replaced by $E_{ct} = E_c/(1 + C_t)$ for computing initial plus creep deformation, in which the factor 1 corresponds to the initial deformation.

The author believes that the effective or sustained modulus of elasticity is useful in some secondary calculations (such as for initial plus creep effects in composite slabs including differential shrinkage, where the effects of creep under variable stress are approximated), but he does not endorse the method for use in primary calculations (such as deflections) involving the modular ratio, n_{ct}, and transformed section properties.

Among his reasons are the following: the method does not seem to predict experimental results satisfactorily. For example, for the 30 beams of Washa and Fluck,[48] and the six beams of Yu and Winter[49] ($I_g/I_{cr} = 1.9$ to 3.5 for singly and doubly reinforced rectangular beams and T-beams),[11] the measured time-dependent deflections were of the order of 1 to 2 times the initial deflections. However, the effective or sustained modulus method requires a creep coefficient of the order of 6 to 8 to predict these deflections, with shrinkage warping being computed separately by any of the suitable methods available. Also, a practical inconvenience is the fact that each time a new creep factor is to be considered, a new neutral axis location and transformed moment of inertia must be computed.

Effects of creep under variable stress, such as the loss of prestress due to creep under the decreasing prestress force, are approximated in the procedures used herein. When it is desired to consider the effects of variable stress or load history in more detail, the use of the rate of creep method[50] or the superposition method[51] is suggested when creep-time and load-history information are available. Various methods for analyzing concrete creep under constant and variable stress are described in Appendix B.

REFERENCES

1. D. E. Branson and M. L. Christiason, Time-Dependent Concrete Properties Related to Design—Strength and Elastic Properties, Creep and Shrinkage (based on M.S. thesis, University of Iowa, Jan. 1970), in "Designing for Effects of Creep, Shrinkage and Temperature in Concrete Structures," 257–277, ACI Publication SP 27-13 (symposium volume), 1971.
2. D. E. Branson, B. L. Meyers, and K. M. Kripanarayanan, Time-Dependent Deformation of Noncomposite and Composite Prestressed Concrete Structures (paper presented by D.E.B. at the 40th Annual Meeting, Highway Research Board, Washington, D.C., Jan. 1970; based on Iowa State Highway Commission and University of Iowa Research Reports Nos. 69-1, Feb. 1969, and 70-6, Aug. 1970), *National Academy of Sciences–National Academy of Engineering, Highway Research Record,* No. 324, *Symposium on concrete deformation,* 15–43, 1970.
3. D. E. Branson and K. M. Kripanarayanan, Loss of Prestress, Camber and Deflection of Non-Composite and Composite Prestressed Concrete Structures (four papers presented by D.E.B. on various aspects at the 6th Congress, Fédération Internationale de la Précontrainte, Prague, June 1970; and at Design Seminars sponsored by the Prestressed Concrete Institute, Chicago, Jan. 1971 and June 1971, and by the California Division of Highways and Ceramic Lightweight Aggregate Association, Sacramento, Mar. 1971), *PCI Journal,* **16,** 22–52, Sept.–Oct. 1971.
4. J. J. Shideler, Lightweight Aggregate Concrete for Structural Use, *ACI Journal,* Proceedings **54,** 4, 299–328, Oct. 1957.
5. Paul Klieger, Long-Time Study of Cement Performance in Concrete: Chapter 10—Progress Report on Strength and Elastic Properties in Concrete, *ACI Journal,* Proceedings **54,** 6, 481–504, Dec. 1957.
6. T. R. Jones, T. J. Hirsch, and H. K. Stephenson, "The Physical Properties of Structural Quality Lightweight Aggregate Concrete," 1–46, Texas Transportation Institute, Texas A & M University, College Station, Texas, Aug. 1959.
7. J. A. Hanson, Prestress Loss as Affected by Type of Curing, *PCI Journal,* **9,** 2, 69–93, Apr. 1964.
8. D. W. Pfeifer, Sand Replacement in Structural Lightweight Concrete—Creep and Shrinkage Studies, *ACI Journal,* Proceedings **65,** 2, 131–142, Feb. 1968.
9. Subcommittee 2, ACI Committee 209, Subcommittee Chairman, D. E. Branson, Committee 209 Chairman, J. R. Keeton, Prediction of Creep, Shrinkage and Temperature Effects in Concrete Structures (paper presented by D.E.B. at the Annual ACI Convention, New York, Apr. 1970), in "Designing for Effects of Creep, Shrinkage and Temperature in Concrete Structures," 51–93, ACI Publication SP 27-3 (symposium volume), 1971.
10. C. I. Chen and D. E. Branson, Design Procedures for Predicting and Evaluating the Time-Dependent Deformation of Reinforced, Partially Prestressed and Fully Prestressed Structures of Different Weight Concrete, *Research Report,* Civil Engineering Department, University of Iowa, Aug. 1972.

11. ACI Committee 435, Chairman, D. E. Branson, Deflections of Reinforced Concrete Flexural Members, *ACI Journal,* Proceedings **63,** 6, 637–674, June 1966. Also published in "ACI Manual of Concrete Practice," Part 2: Structural Design, Structural Specifications, Structural Analysis, 1967, 1972.
12. G. Winter, L. C. Urquhart, C. E. O'Rourke, and A. H. Nilson, "Design of Concrete Structures," 7th edn, 1–660, McGraw-Hill Book Co., New York, 1964.
13. J. A. Hanson, Tensile Strength and Diagonal Tension Resistance of Structural Lightweight Concrete, *ACI Journal,* Proceedings **58,** 1, 1–39, July 1961.
14. D. E. Branson, Compression Steel Effect on Long-Time Deflections, *ACI Journal,* Proceedings **68,** 8, 555–559, Aug. 1971.
15. ACI Committee 318, Building Code Requirements for Reinforced Concrete (ACI 318-71), 1–78, American Concrete Institute, Detroit, 1971.
16. Adrian Pauw, Static Modulus of Elasticity of Concrete as Affected by Density, *ACI Journal,* Proceedings **57,** 6, 679–687, Dec. 1960.
17. ACI Committee 318, Building Code Requirements for Reinforced Concrete (ACI 318-63), 1–144, American Concrete Institute, Detroit, 1963.
18. D. W. Pfeifer, Revised Equation for Computing the Modulus of Elasticity of Concrete, Portland Cement Association, private communication, May 1970.
19. T. Y. Lin, "Design of Prestressed Concrete Structures," 2nd edn, 1–614, John Wiley & Sons, New York, 1963.
20. A. D. Ross, Concrete Creep Data, *The Structural Engineer,* **15,** 8, 314–326, Aug. 1937.
21. W. R. Lorman, The Theory of Concrete Creep, *ASTM Proceedings,* **40,** 1082–1102, 1940.
22. G. E. Troxell, J. M. Raphael, and R. W. Davis, Long Time Creep and Shrinkage Tests of Plain and Reinforced Concrete, *ASTM Proceedings,* **58,** 1–20, 1958.
23. T. W. Reichart, Creep and Drying Shrinkage of Lightweight and Normal-Weight Concretes, National Bureau of Standards, NBS Monograph 74, 1–30, Mar. 1964.
24. A. M. Neville and B. L. Meyers, Creep of Concrete: Influencing Factors and Prediction, *ACI Special Publication* SP-9, 1–33, 1964.
25. J. R. Keeton, Study of Creep in Concrete, Phases 1–5, *Naval Civil Engineering Laboratory, Technical Reports* R 33—I, II, III, Port Hueneme, California, 1965.
26. T. C. Hansen and A. H. Mattock, Influence of Size and Shape of Member on Shrinkage and Creep of Concrete, *ACI Journal,* Proceedings **63,** 2, 267–289, Feb. 1966.
27. A. Pauw and J. W. Chai, Creep and Creep Recovery of Plain Concrete, *Missouri Cooperative Highway Research Programme, Report* 67-8, 1967.
28. B. L. Meyers, D. E. Branson, and C. G. Schumann, Prediction of Creep and Shrinkage Behavior for Design from Short-Term Tests, *PCI Journal,* **17,** 3, 29–45, May–June 1972.
29. John R. Keeton, Creep and Shrinkage of Reinforced Thin-Shell Concrete, *Naval Civil Engineering Laboratory, Technical Report* R 704, 1–58, Port Hueneme, California, Nov. 1970.
30. B. L. Meyers, Simplified Method of Predicting Creep and Shrinkage Properties of Concrete, *Taylor Woodrow Materials Laboratory, England, Technical Note* 692, Aug. 1971.
31. O. Wagner, Das Kriechen Unbewehrten Betons, *Deutscher Ausschuss für Stahlbeton, Bulletin* 131, Berlin, 1958.
32. Comité Européen du Béton—Fédération Internationale de la Pré-Contrainte, "International Recommendations for the Design and Construction of Concrete Structures," 1–80, Cement and Concrete Association, London, June 1970.
33. A. M. Neville, "Creep of Concrete: Plain, Reinforced, and Prestressed," 1–622, North-Holland Publishing Company, Amsterdam, 1970.
34. The California Producers Committee on Volume Change and Affiliated Technical Organizations, Drying Shrinkage of Concrete, *Concrete Research, Oakland,* 1–40, Mar. 1966.
35. PCI Committee on Prestress Losses, Chairman, D. D. Magura, Recommendations for Estimating Prestress Losses, *PCI Report,* 1–13, 1970.
36. Z. P. Bazant, discussion of Ref. 9, in "Designing for Effects of Creep, Shrinkage and Temperature in Concrete Structures," 86, ACI Publication SP 27-3 (symposium volume), Fall 1971.
37. D. D. Magura, M. A. Sozen, and C. P. Siess, A Study of Relaxation in Prestressing Reinforcement, *PCI Journal,* **9,** 2, 13–58, Apr. 1964.

38. Prestressed Concrete Institute, "PCI Design Handbook," 1-1 to 12-8, Prestressed Concrete Institute, Chicago, 1971.
39. AASHTO–PCI Joint Committee, Chairmen, W. Henneberger and W. B. Bennett, Jr., "Interim Specification, Prestressed Concrete, Section 6," 1–38, Prestressed Concrete Institute, Chicago, Nov. 1970.
40. J. M. Antill, Relaxation Characteristics of Prestressing Tendons, *Civil Engineering Transactions,* Institute of Engineers, Australia, **CE 7,** 2, 1965.
41. D. A. VanHorn, Prestressed Concrete Materials Related to Loss in Prestress Force, *PCI Design Seminar, Chicago,* 1–38, Jan. 1971 and June 1971.
42. Ignacio Martin, Effect of Environmental Conditions on Thermal Variations and Shrinkage of Concrete Structures in the United States, in "Designing for Effects of Creep, Shrinkage and Temperature in Concrete Structures," 279–300, ACI Publication SP 27-14, symposium volume, Fall 1971.
43. AASHTO, "Standard Specifications for Highway Bridges," 11th edn, 1–467, *AASHTO,* Washington D.C., 1973.
44. ACI Committee 213, Chairman, J. A. Hanson, Guide for Structural Lightweight Aggregate Concrete, *ACI Journal,* Proceedings **64,** 8, 433–469, Aug. 1967.
45. Ignacio Martin, Environment Effect on Thermal Variations and Shrinkage of Lightweight Concrete Structures, *ACI Journal,* Proceedings **69,** 3, 179–184, Mar. 1972.
46. ACI Committee 335, Chairmen, M. V. Pregnoff, D. E. Branson, Deflection of Concrete Beams, *ACI Committee 335 Report,* 2nd draft, 1–93, Feb. 1962.
47. G. E. Troxell and H. E. Davis, "Composition and Properties of Concrete," 1–434, McGraw-Hill Book Co., Inc., New York, 1956.
48. G. W. Washa and P. G. Fluck, The Effect of Compressive Reinforcement on the Plastic Flow of Reinforced Concrete Beams, *ACI Journal,* Proceedings **49,** 8, 89–108, Oct. 1952.
49. Wei-Wen Yu and George Winter, Instantaneous and Long-Time Deflections of Reinforced Concrete Beams Under Working Loads, *ACI Journal,* Proceedings **57,** 1, 29–50, July 1960.
50. A. D. Ross, Creep of Concrete Under Variable Stress, *ACI Journal,* Proceedings **54,** 9, 739–758, Mar. 1958.
51. Douglas McHenry, A New Aspect of Creep in Concrete and Its Application to Design, *Proceedings ASTM,* **43,** 1969–1984, 1943.
52. D. E. Branson, B. L. Meyers, and K. M. Kripanarayanan, Loss of Prestress, Camber and Deflection of Noncomposite and Composite Structures Using Different Weight Concretes, *Final Report of Project HR-137,* 1–229, Iowa State Highway Commission, Report 70-6, Aug. 1970.
53. B. L. Meyers and D. E. Branson, Design Aid for Predicting Creep and Shrinkage Properties of Concrete, *ACI Journal,* Proceedings **69,** 9, 551–555, Sept. 1972.
54. D. E. Branson and M. G. Houg, "Early Creep–Time Equations," research studies, University of Iowa, 1972.
55. H. Schorer, Prestressed Concrete Design Principles and Reinforcing Units, *ACI Journal,* Proceedings **39,** 4, 493–528, July 1943.
56. I. Lyse, The Shrinkage and Creep of Concrete, *Magazine of Concrete Research,* **11,** 33, 143–150, Nov. 1960.
57. E. M. Wallo and C. E. Kesler, Prediction of Creep in Structural Concrete, *University of Illinois Eng. Exp. St., Bulletin* 498, Nov. 1968.
58. J. R. Shank, The Plastic Flow of Concrete, *Ohio State University, Engineering Experiment Station, Bulletin* 91, Sept. 1935.
59. J. R. Shank, The Mechanics of Plastic Flow of Concrete, *Proceedings, American Concrete Institute,* **32,** 149–180, 1936.
60. G. E. Large, "Basic Reinforced Concrete Design: Elastic and Creep," 2nd edn, 1–527, Ronald Press Company, New York, 1957.
61. Bureau of Reclamation, A 10-year Study of Creep Properties of Concrete, *U.S. Bureau of Reclamation Report* SP-38, Denver, Colo., 1953.
62. U.S. Army Engineer Waterways Experiment Station, Corps of Engineers, Investigation of Creep

in Concrete; Review of Literature on Creep in Concrete, *Vicksburg, Miss., Miscellaneous Paper* 6-132, Report 1, June 1955.

63. I. I. Ulitskii, A Method of Computing Creep and Shrinkage Deformation of Concrete for Practical Purposes, *Beton i Zhelezobeton,* 4, 174–180, 1962. Translation No. 6030, Commonwealth Scientific and Industrial Research Organization, Melbourne, Australia.
64. F. G. Thomas, A Conception of the Creep of Unreinforced Concrete and an Estimation of the Limiting Values, *The Structural Engineer,* **11,** 2, 69–73, 1933.
65. A. M. Neville, M. A. Ward, and G. G-S. Kwei, Basic and Drying Creep of Concrete, *Materials and Structures,* **2,** 8, 117–124, 1969.
66. V. J. Rao and W. Dilger, "Time-Dependent Behavior of Composite Prestressed Concrete Beams," University of Calgary, Calgary, Alberta, Canada, 1973.
67. A. Pauw, Time-Dependent Deformations of Concrete, *Missouri State Highway Department, Study* 71-1, 1–275, University of Missouri, Sept. 1971.
68. H. Trost, Implications of the Superposition Principle in Creep and Relaxation Problems for Concrete and Prestressed Concrete, *Beton- und Stahlbetonbau,* Berlin-Wilmersdorf, 10, 230–238, 261–269, 1967.
69. Z. Bazant, Prediction of Concrete Creep Effects Using Age-Adjusted Effective Modulus Method, *ACI Journal,* Proceedings **69,** 4, 212–217, Apr. 1972.
70. W. H. Glanville, Studies in Reinforced Concrete III—Creep or Flow of Concrete Under Load, *Dept. of Scientific and Industrial Research, London, Building Research Technical Paper* 12, 1930. See also *Structural Engineer,* **11,** 54, 1933; and *1st International Congress of Plain and Reinforced Concrete,* Liége, 1930.
71. F. Dischinger, Untersuchungen über die Knicksicherheit, die elastische Verformung und das Kriechen des Betons bei Bogenbrucken, *Der Bauingenieur,* **18,** 487–520, 539–552, 595–621, 1937.
72. F. Dischinger, Elastische und Plastische Verformungen bei Eisenbetontragwerke, *Der Bauingenieur,* **20,** 53–63, 286–294, 426–437, 1939.
73. G. L. England and J. M. Illston, Methods of Computing Stress in Concrete from a History of Measured Strain, *Civil Engineering and Public Works Review,* April, May, June, 1965.
74. V. Mossiossian and W. L. Gamble, "Time-Dependent Behavior of Noncomposite and Composite Prestressed Concrete Structures Under Field and Laboratory Conditions, *Civil Engineering Studies, Structural Research Series* No. 385, University of Illinois, May 1972.
75. Subcommittee 1, ACI Committee 209, Subcommittee Chairman, B. L. Meyers, Committee 209 Chairman, J. R. Keeton, Effects of Concrete Constituents, Environment, and Stress on Creep and Shrinkage of Concrete (paper presented by BLM at the Annual ACI Convention, New York, Apr. 1970), in "Designing for Effects of Creep, Shrinkage and Temperature in Concrete Structures," 1–42, ACI Publication SP 27-1 (symposium volume), 1971.
76. T. W. Duffy, "Concrete Shrinkage in the Early Stages After Casting," 1–58, M.S. thesis, University of Iowa, July 1975.

APPENDIX

A

VARIABLES AFFECTING CREEP AND SHRINKAGE, AND COMPUTATION PROCEDURES

The principal variables that affect creep and shrinkage in most cases have been discussed by Jones *et al.*,[6] Ross,[20] Neville and Meyers,[24] Pauw and Chai,[27] the CEB–FIP,[32] Neville,[33] and others.[52] These effects are outlined and evaluated in Fig. A-1 and below. The corresponding correction factors, based on the standard conditions specified in the text, were obtained by the author and Christiason.[1] These nominal correction factors are given by Eqs. (1-20) to (1-30) and (A-1) to (A-10), and are compared with experimental data in Figs. A-2 and A-3. The variables included are:

- Time of initial loading, and time initial shrinkage is considered
- Environmental humidity
- Size of member
- Water–cement content in the form of slump and cement content
- Mix proportions in the form of percent fines and air content

General Procedure for Determining Creep and Shrinkage Correction Factors

Loading age and age initial shrinkage considered See text, Eqs. (1-20) and (1-21), and Table 1-6.

Humidity See text, Eqs. (1-22), (1-23), (1-24), and Table 1-7.

Size of member (for average thickness of member—part under consideration—other than 6 in. or 15.2 cm See text, Eqs. (1-25) to (1-30), and Tables 1-8, 1-9.

Parameters affecting creep and shrinkage concrete strains

1. Member size	5. Length of curing	9. Environment temp.	12. No. of load cycles
2. Water-cement ratio	6. Curing temp.	10. Time of init. load and time init. shrinkage considered	13. Unloading period
3. Mix proportions	7. Curing humidity	11. Duration of load period	14. Stress distr.
4. Type of aggregate	8. Environment hum.		15. Stress magnitude
			16. Stress rate

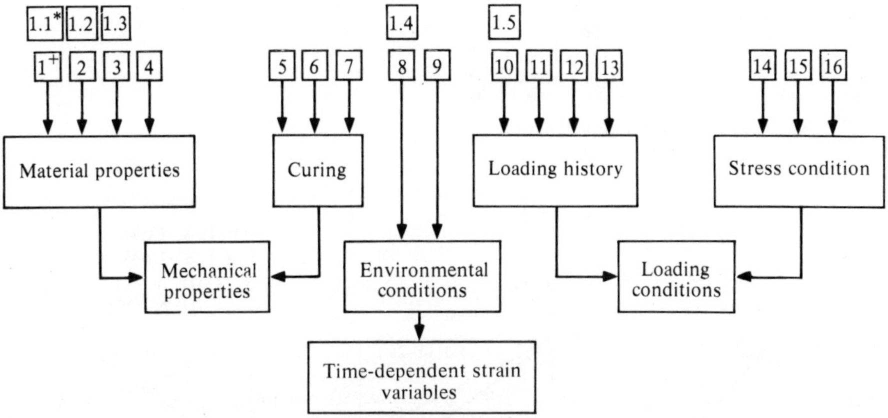

*Parameters used in the general procedure as correction factors.
+These numbers refer to the parameters listed above.

FIGURE A-1
Time-dependent strain variables. (*From Pauw and Chai,*[27] *and Pauw,*[66] *with permission.*)

Slump

$$\text{Creep } (\text{C.F.})_S = 0.82 + 0.067S \tag{A-1}$$

$$\text{Shrinkage } (\text{C.F.})_S = 0.89 + 0.041S \tag{A-2}$$

where S is the slump in inches.

Cement content

$$\text{Creep } (\text{C.F.})_B = 1.00 \tag{A-3}$$

$$\text{Shrinkage } (\text{C.F.})_B = 0.75 + 0.034B \tag{A-4}$$

where B is the number of bags of cement per cubic yard of concrete.

Percentage fines

$$\text{Creep } (\text{C.F.})_F = 0.88 + 0.0024F \tag{A-5}$$

$$\text{Shrinkage } (\text{C.F.})_F = 0.30 + 0.0140F \qquad \text{for } F \leq 50\% \tag{A-6}$$

$$= 0.90 + 0.0020F \qquad \text{for } F \geq 50\% \tag{A-7}$$

where F is the percentage of fine aggregate by weight.

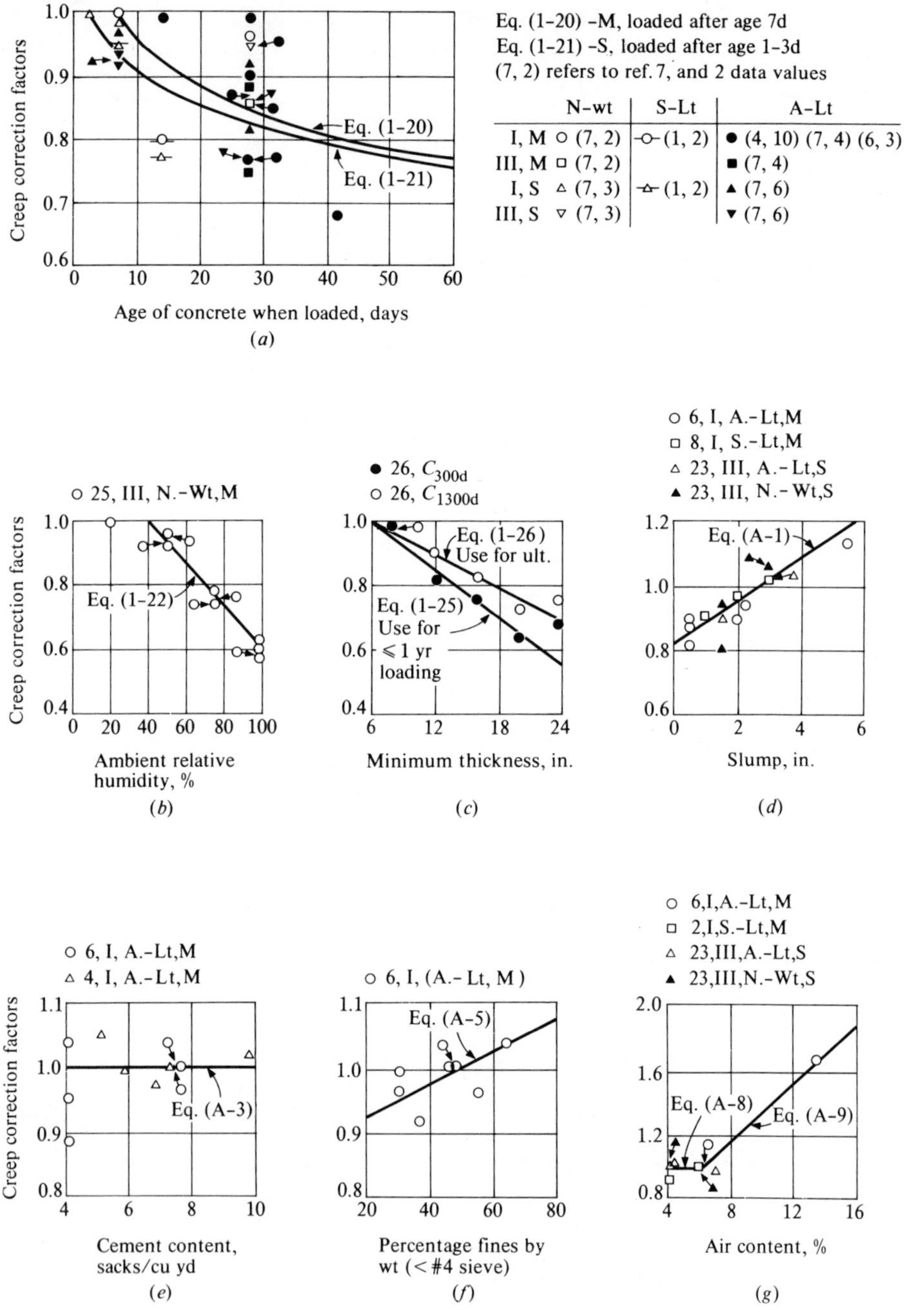

FIGURE A-2
Nominal creep correction factors for the parameters shown.[1] Notation: 3—reference no.; I, III—type cement; N.-Wt, S.-Lt, A.-Lt—normal, sand, and all-lightweight concrete; M, S—moist, steam curing.

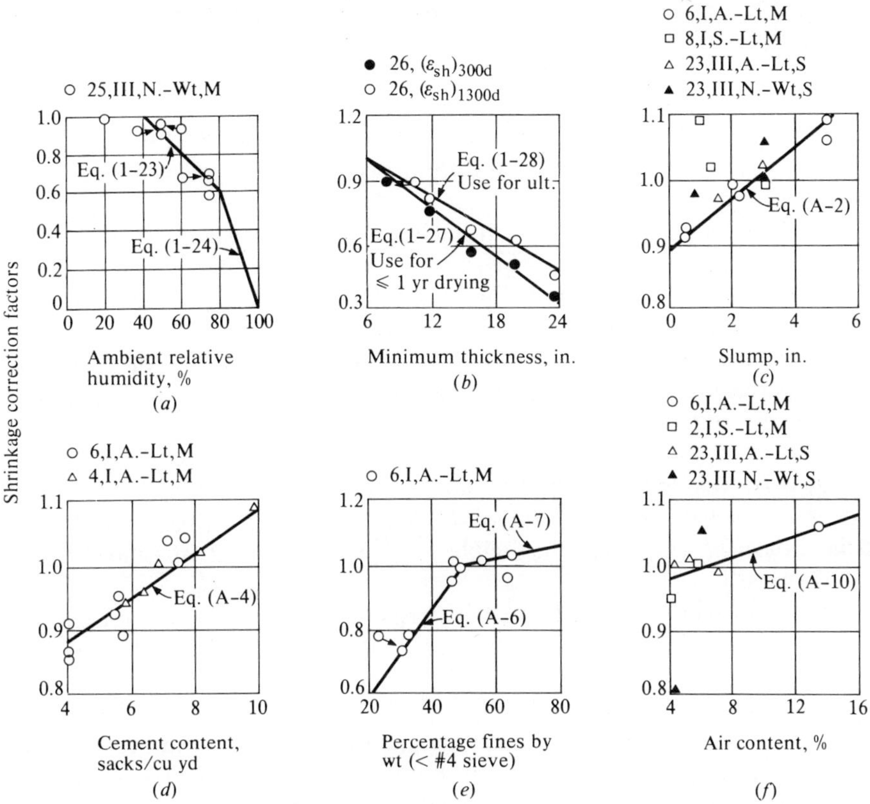

FIGURE A-3
Nominal shrinkage correction factors for the parameters shown.[1] Notation: 3—reference no.; I, III—type cement; N.-Wt, S.-Lt, A.-Lt—normal, sand, and all-lightweight concrete; M, S—moist, steam curing.

Air content

$$\text{Creep (C.F.)}_A = 1.00 \qquad \text{for } A \leq 6\% \tag{A-8}$$

$$= 0.46 + 0.090A \qquad \text{for } A > 6\% \tag{A-9}$$

$$\text{Shrinkage (C.F.)}_A = 0.95 + 0.0080A \tag{A-10}$$

where A is the air content in percent.

Evaluation of Correction Factors for the General Procedure, Which Results in the Simplified Procedure Used in the Text

The following comments refer to the nominal correction factors for creep and shrinkage (Eqs. (A-1) to (A-10)), which are usually not excessive and tend to offset each other. For design purposes in most cases, these may normally be neglected. The resulting simplified procedure used in the text is defined by Eqs. (1-20) to (1-30).

Creep correction factors (equations and curves shown in Fig. A-2)
Slump: $(C.F.)_S = 0.95$ for 2 in., 1.00 for 2.7 in., 1.02 for 3 in., 1.09 for 4 in., 1.16 for 5 in. *Comment*—May be marginal but normally can be neglected for slumps up to 4 in. (10.2 cm).

Cement content (sacks/cubic yard): $(C.F.)_B = 1.00$. *Comment*—No correction factor is required.

Percentage fines (by weight): $(C.F.)_F = 0.95$ for 30 percent, 1.00 for 50 percent, 1.05 for 70 percent. *Comment*—Normally negligible.

Air content (in percent): $(C.F.)_A = 1.00$ up to 6 percent, 1.09 for 7 percent, 1.17 for 8 percent. *Comment*—May be marginal but normally can be neglected for air contents up to 7 percent.

Shrinkage correction factors (equations and curves shown in Fig. A-3)
Slump: $(C.F.)_S = 0.97$ for 2 in., 1.00 for 2.7 in., 1.01 for 3 in., 1.05 for 4 in., 1.09 for 5 in. *Comment*—Normally negligible for slumps up to 4 in., and perhaps marginal above 4 in. (10.2 cm).

Cement content (sacks/cubic yard): $(C.F.)_B = 0.87$ for 4 sacks, 0.95 for 6 sacks, 1.00 for 7.5 sacks, 1.09 for 10 sacks. *Comment*—Normally negligible for say 5 to 9 sacks/cubic yard.

Percentage fines (by weight): $(C.F.)_F = 0.86$ for 40 percent, 1.00 for 50 percent, 1.04 for 70 percent. *Comment*—May be marginal but normally can be neglected.

Air content (in percent): $(C.F.)_A = 0.98$ for 4 percent, 1.00 for 6 percent, 1.03 for 10 percent. *Comment*—Normally negligible.

The curves and equations summarized in Appendix A tend to represent average values for the data cited. The correction factors in the simplified procedure are further discussed in the text with Eqs. (1-20) to (1-30), as they relate to other recommendations in the literature. Although other sets of data may differ somewhat from these in individual cases, it is doubtful if substantial refinement is feasible for this type of analysis, especially when one applies this procedure to the prediction of creep and shrinkage behavior in the design of structures.

Numerical Example of General Procedure for Predicting Creep and Shrinkage

Assume the following typical conditions: moist-cured concrete, 70 percent ambient relative humidity, age of loading 28 days, shrinkage considered from both 7 days and 28 days, average thickness of member 12 in. (30.5 cm), slump 3 in. (7.6 cm), 60 percent fines, 8 bags of cement per cubic yard of concrete, and 6 percent air content.

Required: compute the one-year and ultimate creep coefficient and shrinkage strain, using the average values given in the text.

***Table A-1 Tabulation of creep and shrinkage correction factors using Eqs. (1-20) to (1-28) and (A-1) to (A-10)**

Condition	Creep corr. factors				Shrinkage corr. factors			
	Eq.	Table	Fig.	Factor	Eq.	Table	Fig.	Factor
28d loading age to ult.	1-20	1-6	A-2*a*	0.84	—	—	—	—
70% humidity	1-22	1-7	A-2*b*	0.80	1-23	1-7	A-3*a*	0.70
12″ avg. thickness—1 yr	1-25	1-8	A-2*c*	0.86	1-27	1-8	A-3*b*	0.77
12″ avg. thickness—ult.	1-26	1-8	A-2*c*	0.90	1-28	1-8	A-3*b*	0.82
3″ slump	A-1	—	A-2*d*	1.02	A-2	—	A-3*c*	1.01
8 bags/cu yd—cem. cont.	A-3	—	A-2*e*	1.00	A-4	—	A-3*d*	1.02
60% fines	A-5	—	A-2*f*	1.02	A-7	—	A-3*e*	1.02
6% air content	A-8	—	A-2*g*	1.00	A-10	—	A-3*f*	1.00

* The table refers to the general procedure, with the first four lines applying to the simplified procedure.

Average values for standard conditions (see Eqs. (1-16), (1-17), and Table 1-4).

$$C_u = C_{7\text{d to ult}} = \underline{2.35}$$

$$(\varepsilon_{sh})_u = (\varepsilon_{sh})_{7\text{d to ult}} = \underline{780 \times 10^{-6} \text{ in/in (cm/cm)}}$$

$$C_{7\text{d to 1 yr}} = \frac{365^{0.60}}{10 + 365^{0.60}} 2.35 = 0.78 \times 2.35 = \underline{1.83}$$

$$(\varepsilon_{sh})_{7\text{d to 1 yr}} = \frac{365}{35 + 365}(780 \times 10^{-6}) = 0.91(780 \times 10^{-6})$$

$$= \underline{710 \times 10^{-6} \text{ in/in (cm/cm)}}$$

$$(\varepsilon_{sh})_{7\text{d to 28d}} = \frac{28}{35 + 28}(780 \times 10^{-6}) = 0.44(780 \times 10^{-6})$$

$$= \underline{343 \times 10^{-6} \text{ in/in (cm/cm)}}$$

$$(\varepsilon_{sh})_{28\text{d to ult}} = (780 - 343)10^{-6} = \underline{437 \times 10^{-6} \text{ in/in (cm/cm)}}$$

$$(\varepsilon_{sh})_{28\text{d to 1 yr}} = (710 - 343)10^{-6} = \underline{367 \times 10^{-6} \text{ in/in (cm/cm)}}$$

Solution for the conditions of the example (see Table A-1)

$$C_{28\text{d to ult}} = (0.84 \times 0.80 \times 0.90 \times 1.02 \times 1.00 \times 1.02 \times 1.00)(2.35)$$

$$= 0.629 \times 2.35 = \underline{1.48} \quad \text{ANS}$$

$$C_{28\text{d to 1 yr}} = 0.629(0.86/0.90)(1.83) = \underline{1.10} \quad \text{ANS}$$

$$(\varepsilon_{sh})_{7\text{d to ult}} = (0.70 \times 0.82 \times 1.01 \times 1.02 \times 1.02 \times 1.00)(780 \times 10^{-6})$$

$$= 0.603(780 \times 10^{-6}) = \underline{470 \times 10^{-6} \text{ in/in (cm/cm)}} \quad \text{ANS}$$

$$(\varepsilon_{sh})_{7\text{d to 1 yr}} = 0.603(0.77/0.82)(710 \times 10^{-6})$$

$$= \underline{402 \times 10^{-6} \text{ in/in (cm/cm)}} \quad \text{ANS}$$

$$(\varepsilon_{sh})_{28\text{d to ult}} = 0.603(437 \times 10^{-6}) = \underline{264 \times 10^{-6} \text{ in/in (cm/cm)}} \qquad \text{ANS}$$

$$(\varepsilon_{sh})_{28\text{d to 1 yr}} = 0.603(0.77/0.82)(367 \times 10^{-6})$$

$$= \underline{208 \times 10^{-6} \text{ in/in (cm/cm)}} \qquad \text{ANS}$$

Numerical Example of Simplified Procedure for Predicting Creep and Shrinkage

For the same problem as that solved by the general procedure, the only effects that apply are: moist-cured concrete, 70 percent ambient relative humidity, age of loading 28 days, shrinkage considered from either 7 days or 28 days, and average thickness of member 12 inches. Referring to the correction factors in Table A-1 for these effects (the first four lines in the table), and the same average values for standard conditions:

$$C_{28\text{d to ult}} = (0.84 \times 0.80 \times 0.90)(2.35) = 0.605 \times 2.35$$

$$= \underline{1.42} \text{ vs } 1.48 \text{ above}$$

$$C_{28\text{d to 1 yr}} = (0.84 \times 0.80 \times 0.86)(1.83) = 0.578 \times 1.83$$

$$= \underline{1.06} \text{ vs } 1.10 \text{ above}$$

$$(\varepsilon_{sh})_{7\text{d to ult}} = (0.70 \times 0.82)(780 \times 10^{-6}) = 0.574(780 \times 10^{-6})$$

$$= \underline{448 \times 10^{-6} \text{ in/in (cm/cm)}} \text{ vs } 470 \times 10^{-6} \text{ in/in (cm/cm) above}$$

$$(\varepsilon_{sh})_{7\text{d to 1 yr}} = (0.70 \times 0.77)(710 \times 10^{-6}) = 0.539(710 \times 10^{-6})$$

$$= \underline{383 \times 10^{-6} \text{ in/in (cm/cm)}} \text{ vs } 402 \times 10^{-6} \text{ in/in (cm/cm) above}$$

$$(\varepsilon_{sh})_{28\text{d to ult}} = (0.70 \times 0.82)(437 \times 10^{-6}) = 0.574(437 \times 10^{-6})$$

$$= \underline{251 \times 10^{-6} \text{ in/in (cm/cm)}} \text{ vs } 264 \times 10^{-6} \text{ in/in (cm/cm) above}$$

$$(\varepsilon_{sh})_{28\text{d to 1 yr}} = (0.70 \times 0.77)(367 \times 10^{-6}) = 0.539(367 \times 10^{-6})$$

$$= \underline{198 \times 10^{-6} \text{ in/in (cm/cm)}} \text{ vs } 208 \times 10^{-6} \text{ in/in (cm/cm) above}$$

The results by the general procedure and the simplified procedure are in satisfactory agreement.

Design Aid and Examples of Both General and Simplified Procedures for Predicting Creep and Shrinkage

All of the correction factors for creep and shrinkage, given by Eqs. (1-20) to (1-30) and (A-1) to (A-10) in the general procedure and by Eqs. (1-20) to (1-30) in the simplified procedure, are either linear or bilinear, except for the creep factor for effect of age of concrete when loaded. These equations have been linearized[53] and represented by a single line, with adjusted scales for each correction factor, in Fig. A-4. The effects of percent fines and cement content on creep and air content on shrinkage are normally negligible and are omitted from Fig. A-4. Also omitted from Fig. A-4 are Eqs. (1-29) and (1-30) for the effect of thickness of large members.

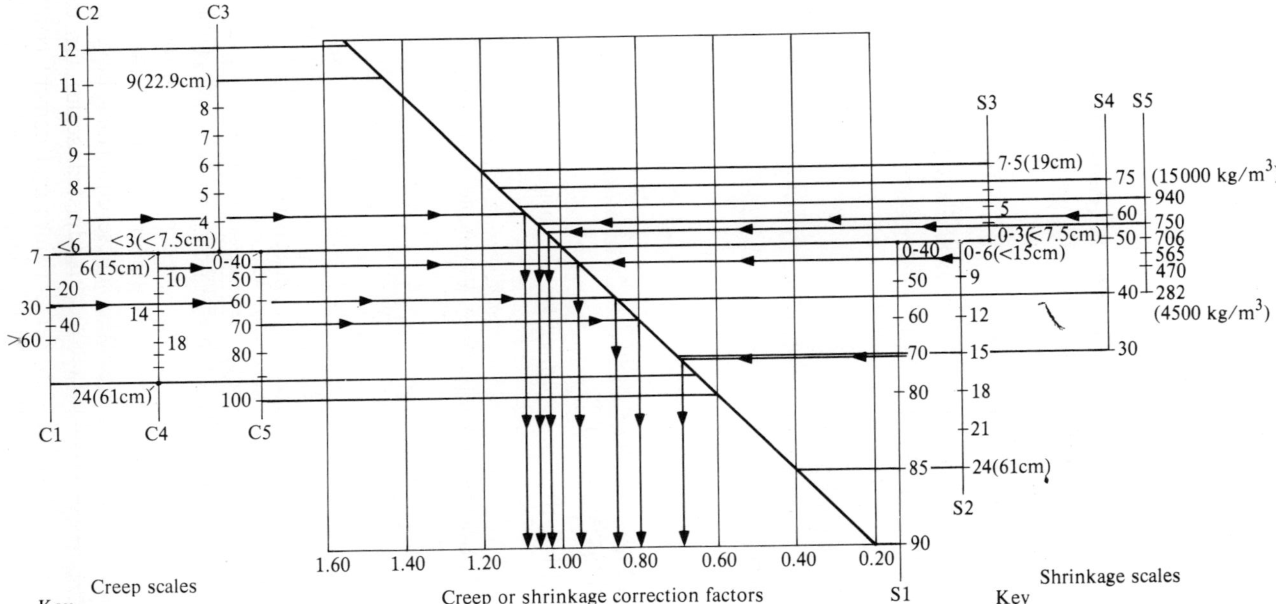

Creep scales

Key
C1 Age at time of loading (d)
C2 Air content (%)
C3 Slump (in.)
C4 Minimum thickness (in.)
C5 Relative humidity (%)

Shrinkage scales

Key
S1 Relative humidity (%)
S2 Minimum thickness (in.)
S3 Slump (in.)
S4 % fines by weight
S5 Cement content (lb/cu ft)

FIGURE A-4
Design aid for determining creep and shrinkage correction factors in the general procedure or the simplified procedure.[53]

For the same example as in the two previous sections, the corresponding correction factors using the design aid in Fig. A-4 are tabulated in Table A-2, and comparisons shown with the results using the equations (from Table A-1). The following comparisons between the two sets of creep and shrinkage answers, using the design aid and using the equations, demonstrates the validity of the design aid.

General procedure—results for previous example using the design aid in Fig. A-4 with factors shown in Table A-2, and comparisons with previous results using Eqs. (1-20) to (1-28) and (A-1) to (A-10)

$$C_{28\text{d to ult}} = (0.85 \times 0.80 \times 0.88 \times 1.00 \times 1.00)(2.35)$$
$$= 0.598 \times 2.35 = \underline{1.41} \text{ vs } 1.48 \text{ using the equations}$$
$$C_{28\text{d to 1 yr}} = 0.598 \times 1.83 = \underline{1.09} \text{ vs } 1.10 \text{ using the equations}$$
$$(\varepsilon_{sh})_{7\text{d to ult}} = (0.68 \times 0.80 \times 1.00 \times 1.02 \times 1.05)(780 \times 10^{-6})$$
$$= 0.583(780 \times 10^{-6})$$
$$= \underline{455 \times 10^{-6} \text{ in/in (cm/cm)}} \text{ vs } 470 \text{ using the equations}$$
$$(\varepsilon_{sh})_{7\text{d to 1 yr}} = 0.583(710 \times 10^{-6})$$
$$= \underline{414 \times 10^{-6} \text{ in/in (cm/cm)}} \text{ vs } 402 \text{ using the equations}$$
$$(\varepsilon_{sh})_{28\text{d to ult}} = 0.583(437 \times 10^{-6})$$
$$= \underline{255 \times 10^{-6} \text{ in/in (cm/cm)}} \text{ vs } 264 \text{ using the equations}$$
$$(\varepsilon_{sh})_{28\text{d to 1 yr}} = 0.583(367 \times 10^{-6})$$
$$= \underline{214 \times 10^{-6} \text{ in/in (cm/cm)}} \text{ vs } 208 \text{ using the equations}$$

*** Table A-2 Tabulation of creep and shrinkage correction factors using the design aid in Fig. A-4**

Condition	Creep corr. factors: Key in Fig. A-4	Creep corr. factors: Factor from Fig. A-4	Creep corr. factors: Factor from Table A-1	Shrinkage corr. factors: Key in Fig. A-4	Shrinkage corr. factors: Factor from Fig. A-4	Shrinkage corr. factors: Factor from Table A-1
28d loading age to ult.	C1	0.85	0.84	—	—	—
70% humidity	C5	0.80	0.80	S1	0.68	0.70
12″ avg. thick.—1 yr or ult.	C4	0.88	0.86–0.90	S2	0.80	0.77–0.82
3″ slump	C3	1.00	1.02	S3	1.00	1.01
8 bags/cu yd or 752 pcf—cement content	—	—	1.00	S5	1.02	1.02
60% fines	—	—	1.02	S4	1.05	1.02
6% air content	C2	1.00	1.00	—	—	1.00

* The table refers to the general procedure, with the first three lines applying to the simplified procedure.

Simplified procedure—results for previous example using the design aid in Fig. A-4 with factors shown in Table A-2, and comparisons with previous results using Eqs. (1-20) to (1-28)

$$
\begin{aligned}
C_{28\text{d to ult}} &= (0.85 \times 0.80 \times 0.88)(2.35) = 0.598 \times 2.35 \\
&= \underline{1.41} \text{ vs } 1.42 \text{ using the equations} \\
C_{28\text{d to 1 yr}} &= 0.598 \times 1.83 = \underline{1.09} \text{ vs } 1.06 \text{ using the equations} \\
(\varepsilon_{sh})_{7\text{d to ult}} &= (0.68 \times 0.80)(780 \times 10^{-6}) = 0.544(780 \times 10^{-6}) \\
&= \underline{424 \times 10^{-6} \text{ in/in (cm/cm)}} \text{ vs } 448 \text{ using the equations} \\
(\varepsilon_{sh})_{7\text{d to 1 yr}} &= 0.544(710 \times 10^{-6}) \\
&= \underline{386 \times 10^{-6} \text{ in/in (cm/cm)}} \text{ vs } 383 \text{ using the equations} \\
(\varepsilon_{sh})_{28\text{d to ult}} &= 0.544(437 \times 10^{-6}) \\
&= \underline{238 \times 10^{-6} \text{ in/in (cm/cm)}} \text{ vs } 251 \text{ using the equations} \\
(\varepsilon_{sh})_{28\text{d to 1 yr}} &= 0.544(367 \times 10^{-6}) \\
&= \underline{200 \times 10^{-6} \text{ in/in (cm/cm)}} \text{ vs } 198 \text{ using the equations}
\end{aligned}
$$

Final Comment on the Starting Times for Creep and Shrinkage Calculations

A significant difference (of the order of 2 to 1) is seen in these examples between shrinkage from age 7 days and from age 28 days. However, the effect on creep of loading ages varying from 28 days (condition in the examples for creep) to 7 days is only to increase the correction factor from 0.84 or 0.85 for 28-day loading to unity for 7-day loading. These time effects should be determined by such construction details as the time of form removal and load application, for example, and should be taken into account in the engineering calculations when possible, especially in the case of shrinkage.

Early Creep–Time Equations[54]

For periods of a few hours to a few days the time-ratio term in Eq. (1-16) requires some modification for accurate prediction, and Eqs. (A-11) and (A-12) appear to be appropriate. The upper limit, average, and lower limit values in each case refer to the standard conditions stated in Chap. 1.

Up to 24 hours after loading

$$
C_t = \frac{t^{0.40}}{2.7 + 0.24t^{0.40}} C_{24\text{hr}} \qquad \begin{aligned} \text{upper limit } C_{24\text{hr}} &= 0.60 \\ \text{average } C_{24\text{hr}} &= 0.40 \\ \text{lower limit } C_{24\text{hr}} &= 0.25 \end{aligned} \qquad \text{(A-11)}
$$

where t is time in hours after loading. 15 data points were used in this determination.

From 1 day to 7 days after loading

$$C_t = \frac{t^{0.50}}{1.32 + 0.50t^{0.50}} C_{7\mathrm{d}} \qquad \begin{aligned} \text{upper limit } C_{7\mathrm{d}} &= 1.00 \\ \text{average } C_{7\mathrm{d}} &= 0.65 \\ \text{lower limit } C_{7\mathrm{d}} &= 0.40 \end{aligned} \tag{A-12}$$

where t is time in days after loading. 21 data points were used in this determination.

From a few days to ultimate time after loading (from Chap. 1 text)

$$C_t = \frac{t^{0.60}}{10 + t^{0.60}} C_{\mathrm{u}} \qquad \begin{aligned} \text{upper limit } C_{\mathrm{u}} &= 4.15 \\ \text{average } C_{\mathrm{u}} &= 2.35 \\ \text{lower limit } C_{\mathrm{u}} &= 1.30 \end{aligned} \tag{1-16}$$

where t is time in days after loading. 479 data points were used in this determination.

Early Shrinkage–Time Equations[76]

For periods of a few hours to a few days, Eq. (1-17) for moist-cured concrete shrinkage requires some modification. Based on a limited amount of data, Eqs. (A-13) and (A-14) for moist-cured shrinkage were found[76] to be appropriate. The upper limit, average, and lower limit values in these two equations refer to the standard conditions stated in Chap. 1.

Moist-cured concrete, Type I (normal) cement (from time of hardening to age 7 days)

$$\begin{aligned} &\text{Upper limit} \quad (\varepsilon_{\mathrm{sh}})_t = (0.083 + 0.125t)(\varepsilon_{\mathrm{sh}})_{7\mathrm{d}} && (\varepsilon_{\mathrm{sh}})_{7\mathrm{d}} = 300 \times 10^{-6} \\ &\text{Average} \quad (\varepsilon_{\mathrm{sh}})_t = (0.15t)(\varepsilon_{\mathrm{sh}})_{7\mathrm{d}} && (\varepsilon_{\mathrm{sh}})_{7\mathrm{d}} = 100 \times 10^{-6} \\ &\text{Lower limit} \quad (\varepsilon_{\mathrm{sh}})_t = (\varepsilon_{\mathrm{sh}})_{7\mathrm{d}} && (\varepsilon_{\mathrm{sh}})_{7\mathrm{d}} = -75 \times 10^{-6} \end{aligned} \tag{A-13}$$

(negative sign means expansion)

Moist-cured concrete, Type III (high-early strength) cement (from time of hardening to age 7 days)

$$(\varepsilon_{\mathrm{sh}})_t = \frac{t}{0.25 + t}(\varepsilon_{\mathrm{sh}})_{7\mathrm{d}} \qquad \begin{aligned} \text{upper limit } (\varepsilon_{\mathrm{sh}})_{7\mathrm{d}} &= 250 \times 10^{-6} \\ \text{average } (\varepsilon_{\mathrm{sh}})_{7\mathrm{d}} &= 175 \times 10^{-6} \\ \text{lower limit } (\varepsilon_{\mathrm{sh}})_{7\mathrm{d}} &= 60 \times 10^{-6} \end{aligned} \tag{A-14}$$

where t is concrete age in days.

The effect of cement type was significant in the case of early shrinkage (or early deformation in the absence of load), apparently because of the difference in the heat of hydration effect (at least in part). Typically there was a period of expansion of the Type I specimens over a period of 2 to 5 days, with a maximum expansion during the period (average for the different specimens) of about 30×10^{-6} in/in (cm/cm). The Type III specimens exhibited little or no expansion. Since the Type III specimens developed the same compressive strength sooner than the Type I specimens, this may account for the lack of expansion of the Type III specimens.

For 15 in. (38 cm) cubic blocks of Type I cement concrete, the average results were predicted by Eq. (A-15).

$$(\varepsilon_{sh})_t = \frac{t}{6.5 + t}(\varepsilon_{sh})_{7d} \qquad (\varepsilon_{sh})_{7d} = 116 \times 10^{-6} \tag{A-15}$$

APPENDIX

B

REVIEW OF DIFFERENT METHODS FOR COMPUTING SHRINKAGE AND CREEP

A number of simple methods for determining shrinkage and creep are summarized in Appendix B for comparison with the procedures in the text and Appendix A. Numerical examples are also included.

Shrinkage Equations

Schorer equation In recognition of the importance of ambient humidity on shrinkage, Schorer's[55] equation (B-1) has been used since 1943 for estimating ultimate shrinkage of hardened concrete for design purposes.

$$(\varepsilon_{sh})_u = 12.5 \times 10^{-6}(90 - H) \tag{B-1}$$

For example, $(\varepsilon_{sh})_u = 250$ microstrain when $H = 70$ percent and 500 microstrain when $H = 50$ percent. Such results can be too low when measured from the end of the curing period, but proved to be relatively useful for general construction in which the pertinent shrinkage strains may begin later (such as following the removal of form supports or the installation of partitions).

Hyperbolic equation One of the most widely used methods for modeling shrinkage versus time has been the hyperbolic equation in the form of Ross's creep equation (B-5) or the author's equations (1-17) and (1-18). As previously discussed in the text and shown in Figs. B-4 to B-7, the early increase and then flat form of the hyperbolic equation lends itself more readily to shrinkage than to creep.

Table B-1 Values of $(\varepsilon_{sh})_t/(\varepsilon_{sh})_u$ by Eq. (B-2) for typical values of v/s and Eqs. (1-17), (1-18) for moist and steam-cured concrete

	Time					
Method used	28 d	3 mth	6 mth	1 yr	5 yr	20 yr
Eq. B-2, $v/s = 2$ in.	0.35	0.61	0.77	0.90	1.00	1.00
Eq. B-2, $v/s = 4$ in.	0.20	0.37	0.52	0.69	0.96	1.00
Eq. B-2, $v/s = 6$ in.	0.14	0.27	0.39	0.54	0.89	1.00
Eq. 1-17,* moist-cured	0.44	0.72	0.84	0.91	0.98	1.00
Eq. 1-18,* steam-cured	0.34	0.62	0.77	0.87	0.97	1.00

* The effect of member size in Eqs. (1-17) and (1-18) is taken into account by means of correction factors for $v/s > 1.5$ in. (3.8 cm). The values in the table refer to $v/s = 1.5$ in. These ratios for Eqs. (1-17) and (1-18) are also shown in Table 1-4.

Exponential equation Using an exponential form of an equation by Lyse,[56] Wallo and Kesler[57] proposed Eq. (B-2) in 1968 with the experimental constants as shown.

$$(\varepsilon_{sh})_t = (\varepsilon_{sh})_u(1 - e^{-0.1t^{0.65}/(v/s)}) \qquad \text{(B-2)}$$

where t is time in days from the day drying was initiated, e is the napierian base, and v/s is the volume–surface ratio in inches.

Values of the ratio $(\varepsilon_{sh})_t/(\varepsilon_{sh})_u$ in Eq. (B-2) are shown in Table B-1 for representative values of v/s. The effect of larger members in delaying shrinkage can be seen. Also shown in Table B-1 are the same ratios for the moist and steam-cured hyperbolic equations (1-17) and (1-18). When using these equations, the effect of member size is accounted for by means of correction factors (Eq. (1-27), (1-28), or (1-30), or Table 1-8 or 1-9). The effect of steam curing in delaying shrinkage can also be seen in Table B-1 or Fig. 1-7.

Creep Equations

Although creep and shrinkage are not independent phenomena, it is convenient to assume that they are additive in analysis, since creep is stress-dependent and shrinkage is not. Thus in this discussion *basic creep* (assumed to be proportional to stress and recoverable) and *drying creep* (assumed to be related to both stress and shrinkage following the time of loading, and not recoverable) will in most cases be considered as *total creep* (total time-dependent deformation minus shrinkage) which is proportional to stress or strain within the usual working stress ranges of concrete structures. Recoverable creep may be treated as delayed elasticity, although creep recovery is partly offset by the increase in the elastic modulus of concrete with time.

Most concrete creep data refer to uniform axial compression. The limited amount of data available for nonuniform stress indicates that somewhat larger creep strains may be found than under uniform stress, since the moisture movement accompanying drying creep is away from the more highly stressed regions toward the more lowly stress regions. Creep under axial tension has been found to be of the same order of magnitude as under axial compression.

Power equation The simple power equation (B-3) was given by Shank[58,59] in 1935 for creep up to one year, since the power equation does not have an ultimate value.

$$c_{sc} = Ct^{1/r} \tag{B-3}$$

where c_{sc} = specific creep or unit creep strain in millionths per psi, t is time after loading in days, $r = 3$ for ordinary strength concrete loaded in air, and $C = 0.500/a^{0.40}$ for Type I cement, a being the age when loaded in days. Values of C are given in Table B-2, including the case of Type III cement. Experimental values of C and r can be obtained from a log–log plot of c_{sc} and t, for which Eq. (B-3) in the form of Eq. (B-4) plots as a straight line, as shown in Fig. B-1.

$$\log c_{sc} = \log C + (1/r) \log t \tag{B-4}$$

For loading periods longer than one year, Table B-3 was recommended.[60]

As an example, for one-year creep using standard Portland cement, $\sigma = 1000$ psi, and loaded at age 28 days (example shown in Fig. B-1):

$$c_{sc} = 0.13(365)^{1/3} = 0.13 \times 7.15 = 0.93 \text{ millionths per psi} \qquad \text{(Eq. (B-3))}$$

and $\varepsilon_{cp} = 0.93(1000)(10^{-6}) = \underline{930 \times 10^{-6} \text{ in/in (cm/cm)}}$

Also, if $E_c = 3.4 \times 10^6$ psi

$$C_{1\text{ yr}} = 0.93 \times 3.4 = \underline{3.2} \qquad \text{(Eq. (1-19))}$$

Table B-2 Values of C in Eq. (B-3) for ordinary-strength concrete, f'_c = 2500 psi to 3750 psi

Age when loaded in air, days	Coefficient C: Standard Portland cement	Coefficient C: High early strength Portland cement
3	—	0.15
7	0.23	0.10
14	0.17	0.07
28	0.13	0.05
60	0.10	0.034

Table B-3 Multiplier for Eq. (B-3) for loading periods greater than 1 year

No. of years loaded	Multiplier
1	1.00
2	1.17
3	1.26
4	1.30
5 or more	1.33*

* The factor 1.33 corresponds to the 20-year data of Troxell *et al.*[22] as shown in Table B-5.

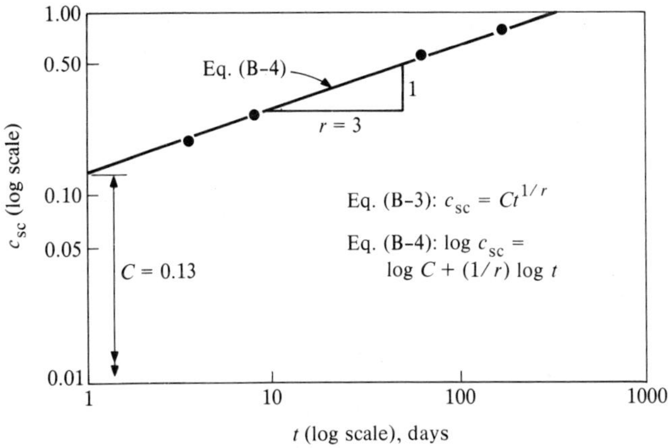

FIGURE B-1
Typical creep constants determined for Shank's equation (B-3).

Shank provided a useful systematic approach, but the method tends to compute values that are too high, probably because it was based on laboratory data (lower humidity than for actual structures) for small specimens (as opposed to actual structures) and the concrete quality of the day.

Hyperbolic equations The hyperbolic equation (B-5) was presented by Ross[20] in 1937.

$$c = \frac{t}{a + bt} \tag{B-5}$$

where $c =$ creep strain in millionths of an in/in or cm/cm, and a and b are constants. For the ultimate value, a is negligible compared to bt and $c = 1/b$. When plotted in the linear form of Eq. (B-6), with t/c as the ordinate and t as the abscissa, the zero intercept determines the value of a and the slope the value of b, as shown for a typical case in Fig. B-2.

$$t/c = a + bt \tag{B-6}$$

As an example, for one-year creep using typical values of $a = 0.05$ and $b = 0.0023$ (examples shown in Fig. B-2):

$$c = 365/(0.05 + 0.0023 \times 365) = 410 \text{ microstrain} \qquad \text{(Eq. (B-5))}$$

and $\quad \varepsilon_{cp} = \underline{410 \times 10^{-6} \text{ in/in (cm/cm)}}$

Also $\quad c_u = \underline{1/b = 1/0.0023 = 435}$

and $\quad (\varepsilon_{cp})_u = \underline{435 \times 10^{-6} \text{ in/in (cm/cm)}}$

Ross also presented the first creep prediction chart for determining the constant b, as a function of age at loading, relative humidity, water–cement ratio, and diameter of specimen (chart also shown in Neville's book,[33] p. 351).

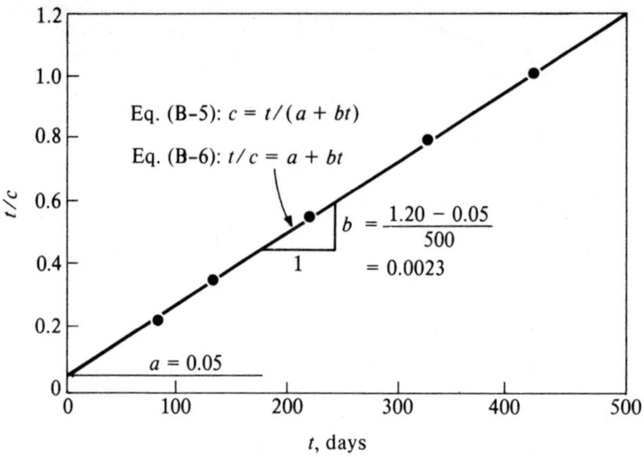

FIGURE B-2
Typical creep constants determined for Ross's equation (B-5).

A similar hyperbolic equation (B-7) was independently developed by Lorman[21] in 1940.

$$c = \frac{mt}{n + t}\sigma \tag{B-7}$$

where m and n are constants. For the ultimate value, $c_u = m\sigma$. Also, when $t = n$ (25 days in the example below), $c = m\sigma/2$, or one-half of the ultimate creep has occurred when $t = n$. When plotted in the linear form of Eq. (B-8), the constants can be determined in the same manner as for Ross's equation, as shown for a typical case in Fig. B-3.

$$t = -n + m\frac{\sigma t}{c} \tag{B-8}$$

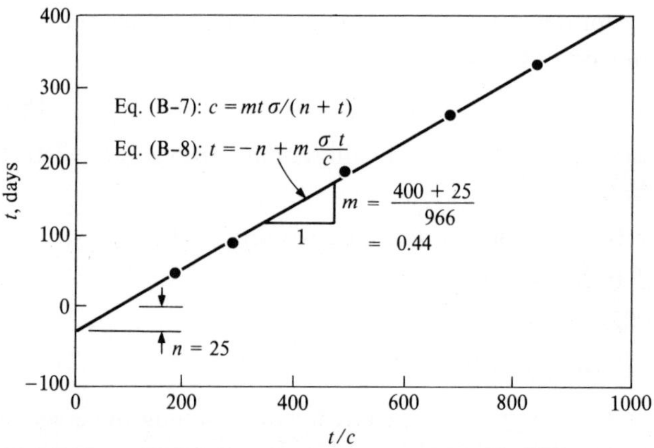

FIGURE B-3
Typical creep constants determined for Lorman's equation (B-7).

As an example, for one-year creep using typical values of $m = 0.44$ and $n = 25$, and $\sigma = 1000$ psi (example shown in Fig. B-3):

$$c = (0.44 \times 365 \times 1000)/(25 + 365) = 412 \text{ microstrain} \qquad \text{(Eq. (B-7))}$$

and $\varepsilon_{cp} = \underline{412 \times 10^{-6} \text{ in/in (cm/cm)}}$

For $E_c = 3.4 \times 10^6$ psi

$$\varepsilon_i = 1000/3.4 \times 10^6 = 294 \times 10^{-6}$$

and $C_{1\,yr} = 412/294 = \underline{1.40}$

Also $c_u = m\sigma = 0.44 \times 1000 = 440$ microstrain

and $(\varepsilon_{cp})_u = \underline{440 \times 10^{-6} \text{ in/in (cm/cm)}}$

$$C_u = 440/294 = \underline{1.50}$$

The hyperbolic expressions of Ross and Lorman have been widely used as they provide a simple means for determining experimental constants and for computing creep deformation. However, this form of equation, while representing concrete shrinkage quite well, tends to flatten out and underestimate long-term creep. The author's modification, Eq. (1-16), was intended to overcome this difficulty and still maintain a simple equation. In fitting a straight line to a set of experimental data, as in Figs. B-1 to B-3, greater weight should be given to the points at later ages.

Logarithmic equation On the basis of an extensive study of creep of concrete used in dams, the Bureau of Reclamation[61] developed the logarithmic equation (B-9) in 1953.

$$c_{sc} = f(K) \log_e (t + 1) \tag{B-9}$$

where K is the age when loaded, $f(K)$ is a function representing the rate of creep with time, and t is time since loading in days. Equation (B-9) plots as a linear function of the logarithm of time under load (actually $t + 1$). The logarithmic equation (also power equation) does not have an ultimate value. Values of $f(K)$ have been found to be:[62] for $K = 7$ days, $f(K) = 0.079$ to 0.096; for $K = 28$ days, $f(K) = 0.041$ to 0.052; and for $K = 90$ days, $f(K) = 0.026$ to 0.030.

For example, when $\sigma = 1000$ psi, $t = 365$ days, $K = 7$ days, and $f(K) =$ say 0.09.

$$c_{sc} = 0.09 \log_e 365 = 0.09 \times 5.90 = 0.53 \text{ millionths per psi}$$

and $\varepsilon_{cp} = 0.53 \times 1000 \times 10^{-6} = \underline{530 \times 10^{-6} \text{ in/in (cm/cm)}}$

For $E_c = 3.4 \times 10^6$ psi,

$$\varepsilon_i = 294 \times 10^{-6} \text{ in/in} \qquad \text{and} \qquad C_{1\,yr} = 530/294 = \underline{1.80}$$

Exponential equation A simple exponential equation (B-10) for the creep coefficient at any time was given by Ulitskii[63] in 1962. The equation is in the form of the time factor in parentheses multiplied by the ultimate creep coefficient. The same expression (that is, same time factor) was used for shrinkage. Ultimate values by Ulitskii are shown in Table B-4.

Table B-4 Ultimate values of creep coefficient and shrinkage strain[63]

Type of concrete	C_u	$(\varepsilon_{sh})_u \times 10^{-6}$
Normal, with large aggregate, normally cured	2.0	300
As above but steam cured	1.8	250
Lightweight concrete, with large aggregate, normally cured	2.5	350
Normal, with small aggregate, normally cured	2.5	500
As above, but heated for "vibro-rolled" placing	2.0	300
Lightweight concrete, with small aggregate	2.5	500
Revibrated concrete	2.5	350
Autoclaved concrete, with small aggregate	1.8	200
Autoclaved silicate concrete (without cement)	1.5	200
Autoclaved silicate form-concrete and gas-concrete (without cement)	2.5	350

$$C_t = (1 - e^{-Bt})C_u \qquad \text{(B-10)}$$

For example, when $B = 0.004$ and $t = 365$ days,

$$C_t/C_u = 1 - 2.72^{-(0.004 \times 365)} = 1 - 0.23 = \underline{0.77}$$

versus 0.78 by the author's equation (1-16) in Table 1-4. Ulitskii's exponential equation (B-10) for creep was also suggested for shrinkage.

As in the case of earlier and more complex exponential expressions for creep by Thomas[64] in 1933, McHenry[51] in 1943, and others, and the shrinkage equation (B-2), this form of equation assumes that creep or shrinkage is proportional to the amount of potential creep or shrinkage remaining. The characteristic time-factor is of the same form in all of these equations.

Total creep in terms of basic creep and drying creep In 1969 Neville, Ward, and Kwei[65] presented a simplified expression for determining total creep, ε_{cp}, in terms of basic creep, ε_{bc}, plus drying creep, ε_{dc}, as given by Eq. (B-11).

$$\begin{aligned} \varepsilon_{cp} &= \varepsilon_{bc} + \varepsilon_{dc} \\ &= \varepsilon_{bc} + Q\varepsilon_{bc}\varepsilon_{sh} \end{aligned} \qquad \text{(B-11)}$$

where Q is a constant, as evaluated in Ref. 65. In this expression both basic creep and drying creep are influenced by the applied stress, with only the drying creep term being a function of shrinkage. For most design purposes this approach is considered by this author to be reasonable and adequate when considering the two aspects of creep. More complex expressions for creep, in terms of basic creep and drying creep, and shrinkage, are given by Wallo and Kesler[57] in terms of volume concentrations of materials and other constants.

Brief Discussion of Shrinkage and Creep Equations

Included in this discussion are power and logarithmic equations, which do not tend toward an ultimate value, and hyperbolic, modified hyperbolic and exponential equations, which do have a finite ultimate value. Only a few representative shrinkage and creep equations are reviewed. It should also be noted that considerably more work has been done on concrete creep (Neville[33]) than on concrete shrinkage.

The average increase in shrinkage and creep with time according to the various forms of the equations are shown in Table B-5 and Figs. B-4 to B-7, as a ratio of the one-year value and also as a ratio of the ultimate value. Where necessary the constants shown with the different equations were used in computing the ratios. Since these equations have all been shown to fit various shrinkage and creep data for shorter periods of time, such comparisons for longer periods (one to ten years) are somewhat revealing.

With regard to Figs. B-4 to B-7, it should be noted firstly that the vertical scale is quite expanded with the origin some distance below the base of each figure; and secondly that different constants could be obtained to more closely fit the particular data shown from Troxell *et al.*, Fig. 1-8, since the constants used did not specifically refer to these data. One should thus study not just the level of the different curves but also their curvature, etc.

In Fig. B-4, which refers to shrinkage as a ratio of the one-year value, it appears that the Wallo and Kesler exponential equation (B-2) and the author's

Table B-5 Average increase in time-dependent concrete strains by various methods*

Method	Time, years					
	1	2	5	10	20	30
As a ratio of the one-year value						
Troxell *et al.* data, Fig. 1-8, shrinkage	1.00	1.05	1.12	1.18	1.27	1.32
Wallo and Kesler exponential eq. (B-2), $v/s = 2''$, shrinkage	1.00	1.08	1.11	1.11	1.11	1.11
Ulitskii exponential eq. (B-10), shrinkage or creep	1.00	1.23	1.30	1.30	1.30	1.30
Author's hyperbolic eq. (1-17), moist-cured, shrinkage	1.00	1.04	1.07	1.09	1.09	1.09
Author's hyperbolic eq. (1-18), steam-cured, shrinkage	1.00	1.07	1.12	1.13	1.14	1.15
Troxell *et al.* data, Fig. 1-8, creep	1.00	1.13	1.21	1.27	1.33	1.36
† Shank power eq. (B-3), creep	1.00	1.26	1.71	2.16	2.72	3.11
Ross hyperbolic eq. (B-5), creep	1.00	1.03	1.05	1.05	1.06	1.06
† Bu. of Reclamation logarithmic eq. (B-9), creep	1.00	1.12	1.28	1.39	1.53	1.58
Author's modified hyperbolic eq. (1-16), creep	1.00	1.08	1.16	1.20	1.23	1.24
As a ratio of the ultimate value						
Troxell *et al.* data, Fig. 1-8, shrinkage	0.75	0.79	0.85	0.89	0.96	1.00
Wallo and Kesler exponential eq. (B-2), $v/s = 2''$, shrinkage	0.90	0.97	1.00	1.00	1.00	1.00
Ulitskii exponential eq. (B-10), shrinkage or creep	0.77	0.95	1.00	1.00	1.00	1.00
Author's hyperbolic eq. (1-17), moist-cured, shrinkage	0.91	0.95	0.98	0.99	1.00	1.00
Author's hyperbolic eq. (1-18), steam-cured, shrinkage	0.87	0.93	0.97	0.99	0.99	1.00
Troxell *et al.* data, Fig. 1-8, creep	0.74	0.83	0.89	0.93	0.98	1.00
† Shank power eq. (B-3), creep	0.32	0.41	0.55	0.69	0.87	1.00
Ross hyperbolic eq. (B-5), creep	0.94	0.97	0.99	0.99	1.00	1.00
† Bu. of Reclamation logarithmic eq. (B-9), creep	0.63	0.71	0.81	0.88	0.97	1.00
Author's modified hyperbolic eq. (1-16), creep	0.78	0.84	0.90	0.93	0.95	0.96

* Where necessary the constants shown herein with the different equations were used in computing the ratios in the table.

† Ultimate values were assumed to be the 30-year values in the power and logarithmic equations.

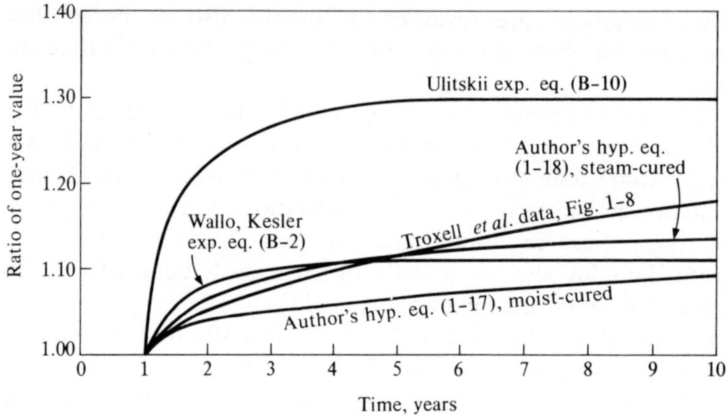

FIGURE B-4
Average increase in shrinkage strains as a ratio of the one-year value, by various methods from Table B-5.

hyperbolic equations (1-17) and (1-18) are in reasonably close agreement with each other and the Troxell *et al.* data, with the hyperbolic forms more closely paralleling the particular data curve.

In Fig. B-5, which refers to shrinkage as a ratio of the ultimate value, the exponential equation (B-2) and the hyperbolic equations (1-17) and (1-18) are all about 15 percent higher than the Troxell *et al.* data curve. This will probably occur for any simple equation devised to fit shrinkage data of a few years when compared to the 30-year data of Troxell *et al.* in terms of the ratio of the

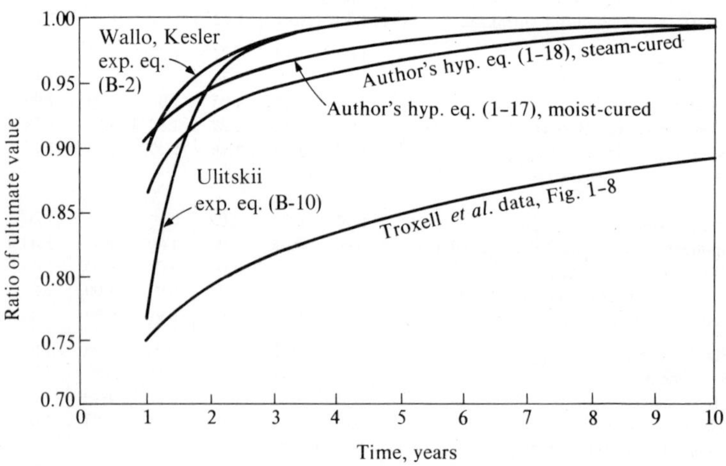

FIGURE B-5
Average increase in shrinkage strains as a ratio of the ultimate value, by various methods from Table B-5.

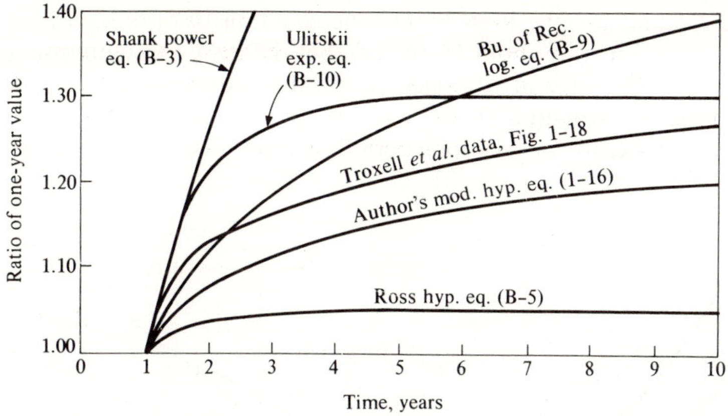

FIGURE B-6
Average increase in creep strains as a ratio of the one-year value, by various methods from Table B-5.

ultimate or 30-year value. In both Figs. B-4 and B-5 for the exponential equation (B-2), a sharp break in the curve at about 2 years followed by a completely flat curve after about 5 years can be seen.

In Fig. B-6, which refers to creep as a ratio of the one-year value, the Ulitskii exponential equation (B-10) and the author's modified hyperbolic equation (1-16) are reasonably close to and bracket the data of Troxell *et al.*, with the modified hyperbolic equation very closely paralleling the data curve.

In Fig. B-7, which refers to creep as a ratio of the ultimate value, the modified hyperbolic equation (1-16) and the Troxell *et al.* data are very close for such an expanded scale, even though the equation was developed independently.

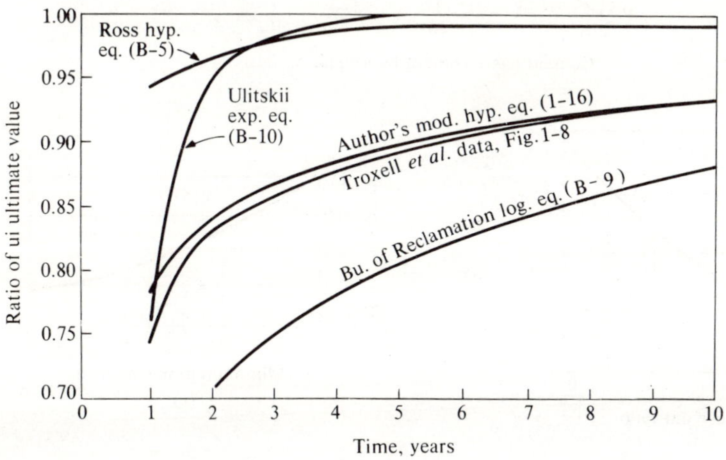

FIGURE B-7
Average increase in creep strains as a ratio of the ultimate value, by various methods from Table B-5.

In both Figs. B-6 and B-7 the Ross hyperbolic equation (B-5) is separated from the particular data curve by about 15 percent, but different constants could improve this agreement. However, at advanced ages both the hyperbolic equation (B-5) and the exponential equation (B-10) tend to flatten out whereas creep continues to increase markedly for very long periods of time. Also the curvatures

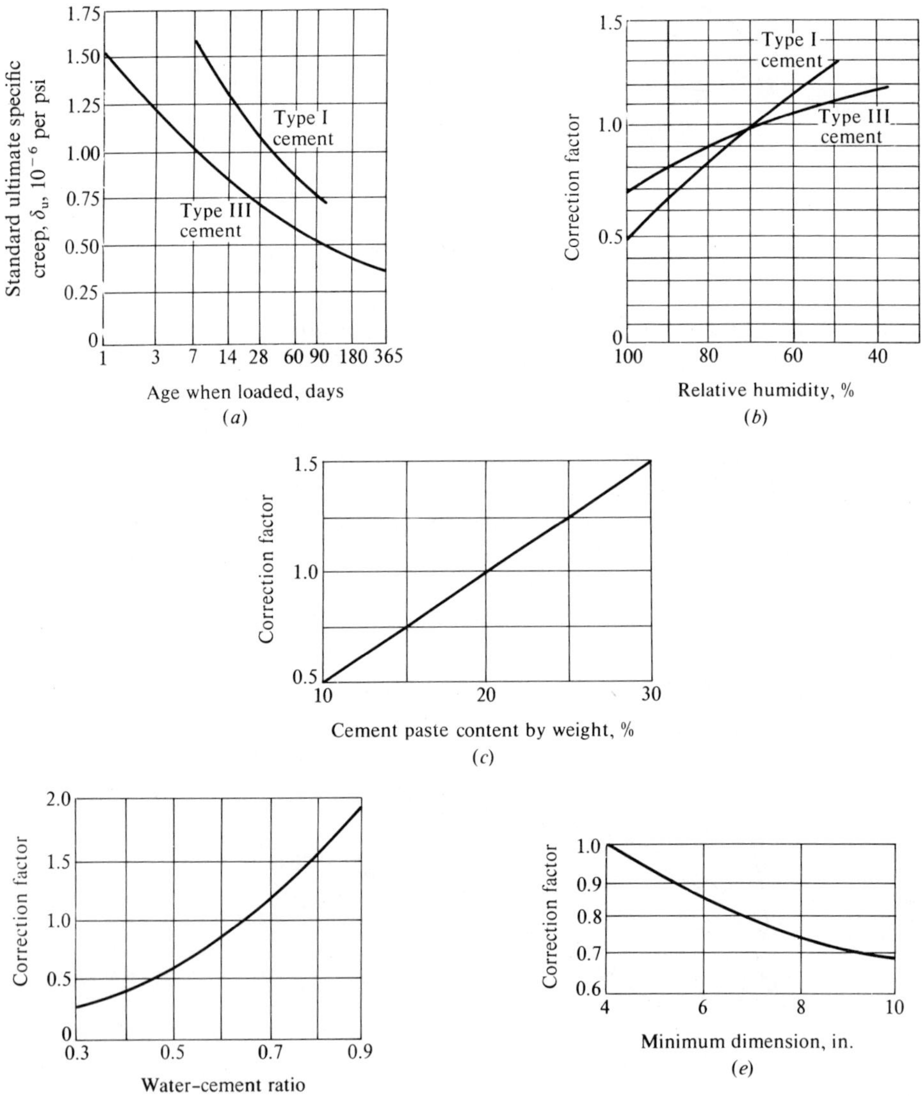

FIGURE B-8
Wagner's standard curve for ultimate specific creep and correction curves. (*From Deutscher Ausschuss für Stahlbeton,*[31] *with permission.*)

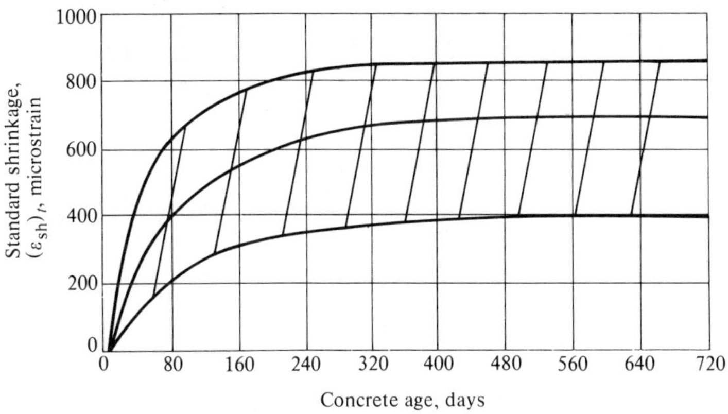

FIGURE B-9
Standard shrinkage curve for a typical lightweight concrete with ranges of variation, by Jones, Hirsch, and Stephenson. (*From Texas Transportation Institute,*[6] *with permission.*)

of these two curves, Eqs. (B-5) and (B-10), are considerably different from that of the data curve, which the author believes is typical. This difficulty appears to be overcome by the modified hyperbolic equation (1-16). The difficulty with the Shank power equation (B-3) and the Bureau of Reclamation logarithmic equation (B-9) can also be seen in Figs. B-6 and B-7.

As noted in the text, the modified hyperbolic equation (1-16) was carefully developed so that as early as 28-day data could be used to predict the entire creep–time curve up to the ultimate value. The same was advocated for shrinkage using the hyperbolic equations (1-17) and (1-18).

Procedures for Predicting Shrinkage and Creep

Creep—Wagner, 1958 Wagner's procedure[31] involves the use of a standard curve for ultimate specific creep of normal-weight concrete, δ_u, as a function of type of cement and age at loading; with correction curves for relative humidity, cement paste content, water–cement ratio, and minimum dimension. These curves are shown in Fig. B-8.

Shrinkage and creep—Jones, Hirsch, and Stephenson, 1959 The procedure of Jones *et al.*[6] refers to standard curves for lightweight concrete for the following standard conditions:

Cement content = 5.7 sacks/cu yd = 536 lb/cu yd
Slump = 2 in. (5.1 cm)
Entrained air = 6%
48% fines less than #4 sieve
Stress level = 1420 psi
3 × 4 × 16 in. (7.6 × 10.2 × 40.6 cm) prism specimens

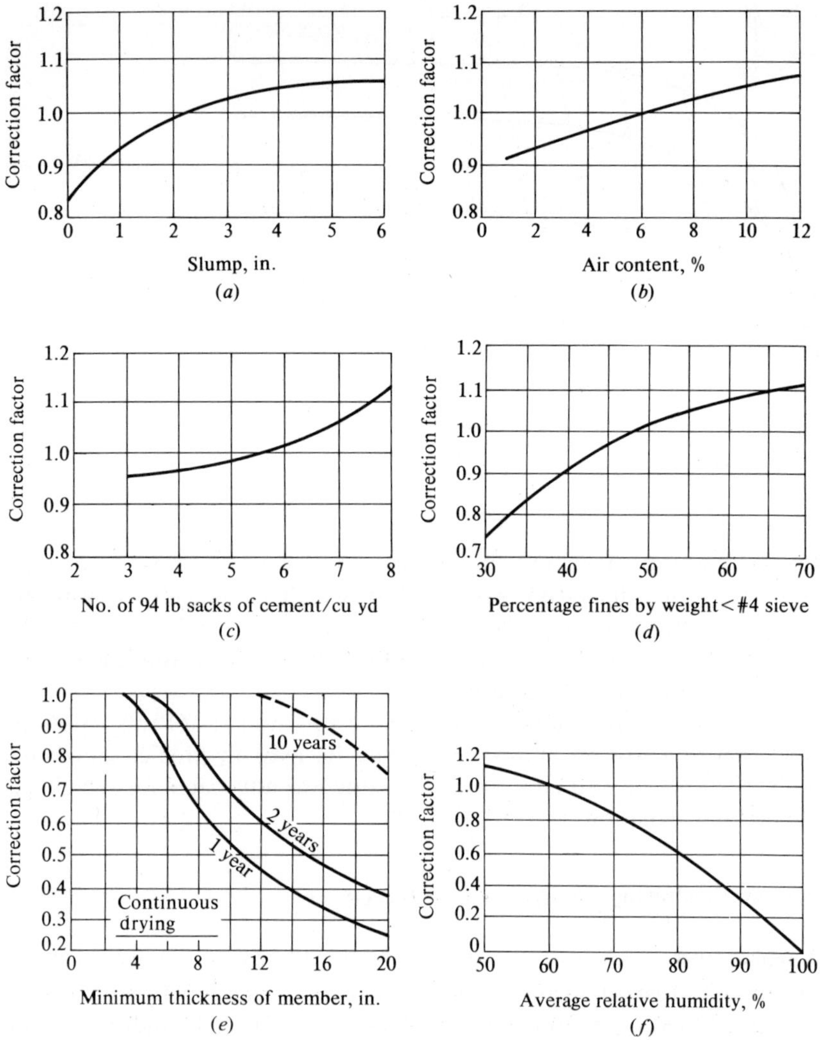

FIGURE B-10
Correction curves of Jones *et al.*, for modifying standard shrinkage curve. (*From Texas Transportation Institute,*[6] *with permission.*)

Relative humidity = 60%
Age when loaded = 14 days for Type I cement and 7 days for Type III cement
Moist cured for 7 days

The standard shrinkage curve is shown in Fig. B-9, with the correction curves for nonstandard conditions given in Fig. B-10. The standard specific creep curve, c_{sc}, is shown in Fig. B-11, with the correction curves for nonstandard conditions given in Fig. B-12.

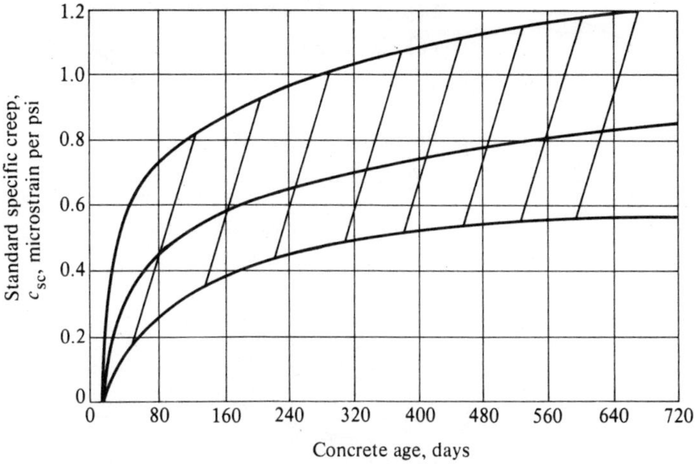

FIGURE B-11
Standard specific creep curve for a typical lightweight concrete with ranges of variation, by Jones *et al.* (*From Texas Transportation Institute,*[6] *with permission.*)

Shrinkage and creep—European Concrete Committee (CEB), 1970 The CEB procedure[32] involves the use of Eqs. (B-12) and (B-13) together with the charts shown in Figs. B-13 and B-14 for computing shrinkage strains and the creep coefficient. These refer to normal-weight concrete, with the suggestion that test results should be obtained in the case of lightweight concrete.

Shrinkage—CEB

$$(\varepsilon_{\text{sh}})_t = (\varepsilon_{\text{sh}})k_{\text{b}}k_{\text{e}}k_t \tag{B-12}$$

where $(\varepsilon_{\text{sh}})$ depends on the environment (Fig. B-13*a* for shrinkage only); k_{b} depends on the composition of the concrete (Fig. B-13*b* for both shrinkage and creep); k_{e} depends on the theoretical thickness of the member (Fig. B-13*c* for shrinkage only); and k_t defines the development of shrinkage and creep as a function of time (Fig. B-13*d*).

The theoretical thickness, e_{m}, is the quotient of the area of the section and the half-perimeter in contact with the atmosphere. If one of the dimensions of the section is very large with respect to the other, e_{m} corresponds almost to the actual thickness. If the dimensions are not constant along the member, an average theoretical thickness can be defined by paying particular attention to those sections in which the stresses are maximum.

An additional factor, k_{p}, which depends on the percentage of reinforcement, is included in Eq. (B-12) by CEB, but is not considered in this discussion. This effect is frequently taken into account in the structural analysis. $k_{\text{p}} = 100/(100 + np)$, where $n = 20$ to include the effect of creep, and $p = 100$ (Area of steel)/(Area of the member).

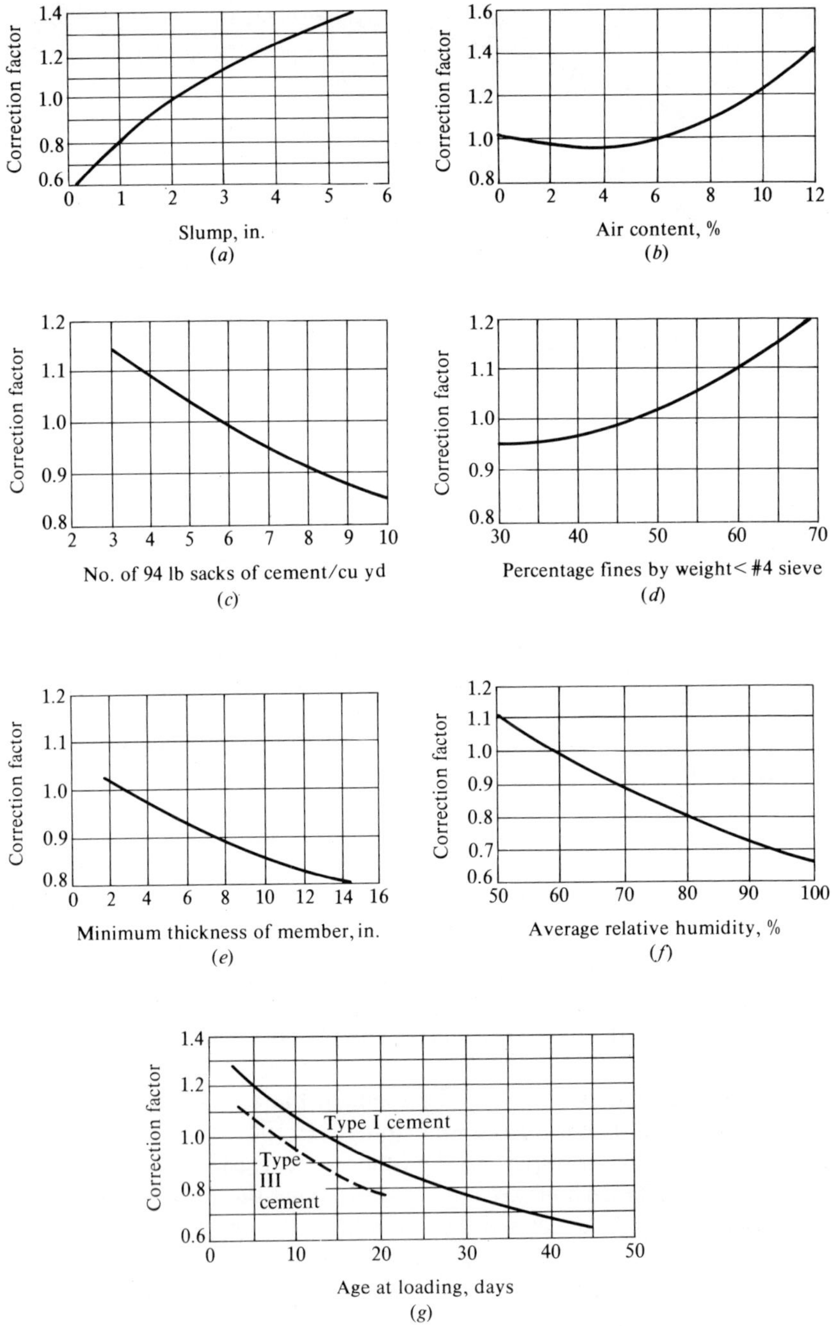

FIGURE B-12
Correction curves of Jones *et al.*, for modifying standard creep curve. (*From Texas Transportation Institute,*[6] *with permission.*)

According to the CEB report, it has been shown experimentally that shrinkage of structural lightweight concrete lies between one and two times that of normal aggregate concrete with the same compressive strength.

Creep coefficient—CEB

$$C_t = Ck_d k_b k_e k_t \tag{B-13}$$

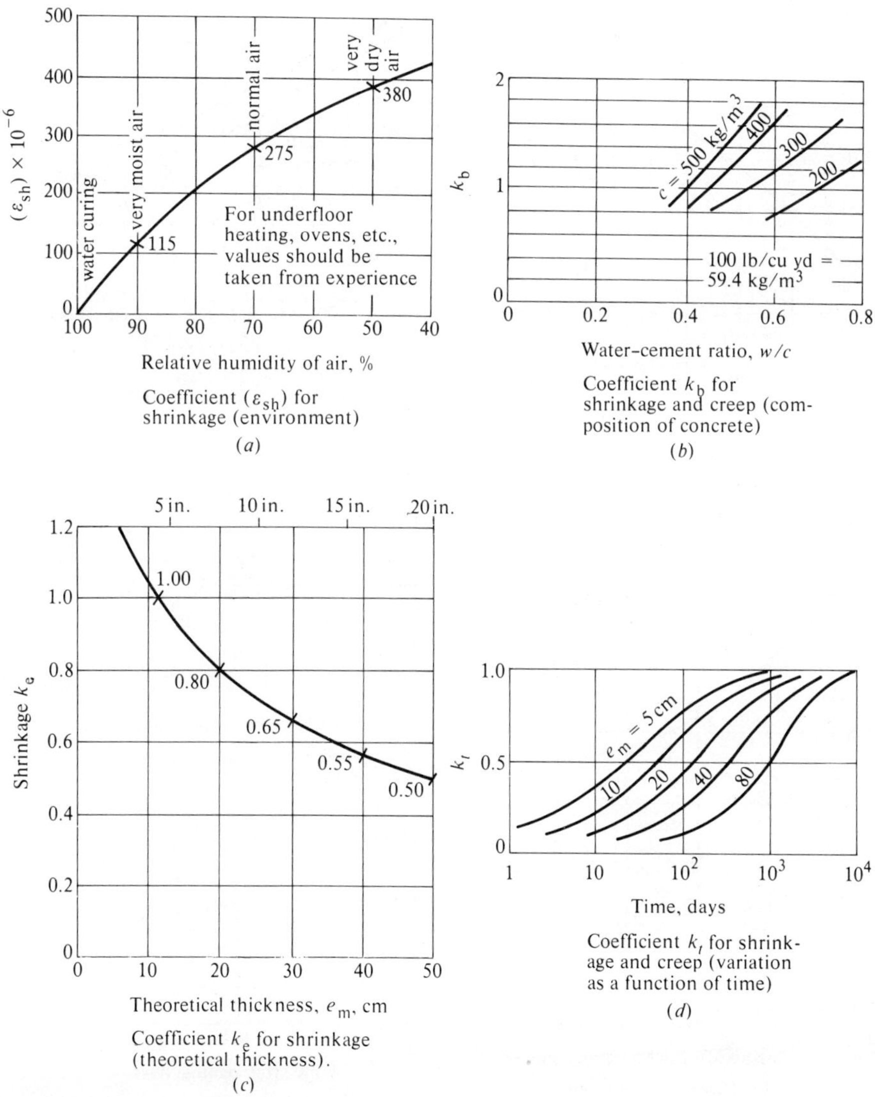

FIGURE B-13
CEB charts for computing concrete shrinkage. (*From Cement and Concrete Association, London,*[32] *with permission.*)

where C depends on the environmental conditions (Fig. B-14*a* for creep only); k_d depends on the hardening of the concrete at the age of loading (Fig. B-14*b* for creep only); k_b depends on the composition of the concrete (Fig. B-14*c* for both creep and shrinkage); k_e depends on the theoretical thickness of the member (Fig. B-14*d* for creep only); and k_t defines the development of creep and shrinkage as a function of time (Fig. B-14*e*).

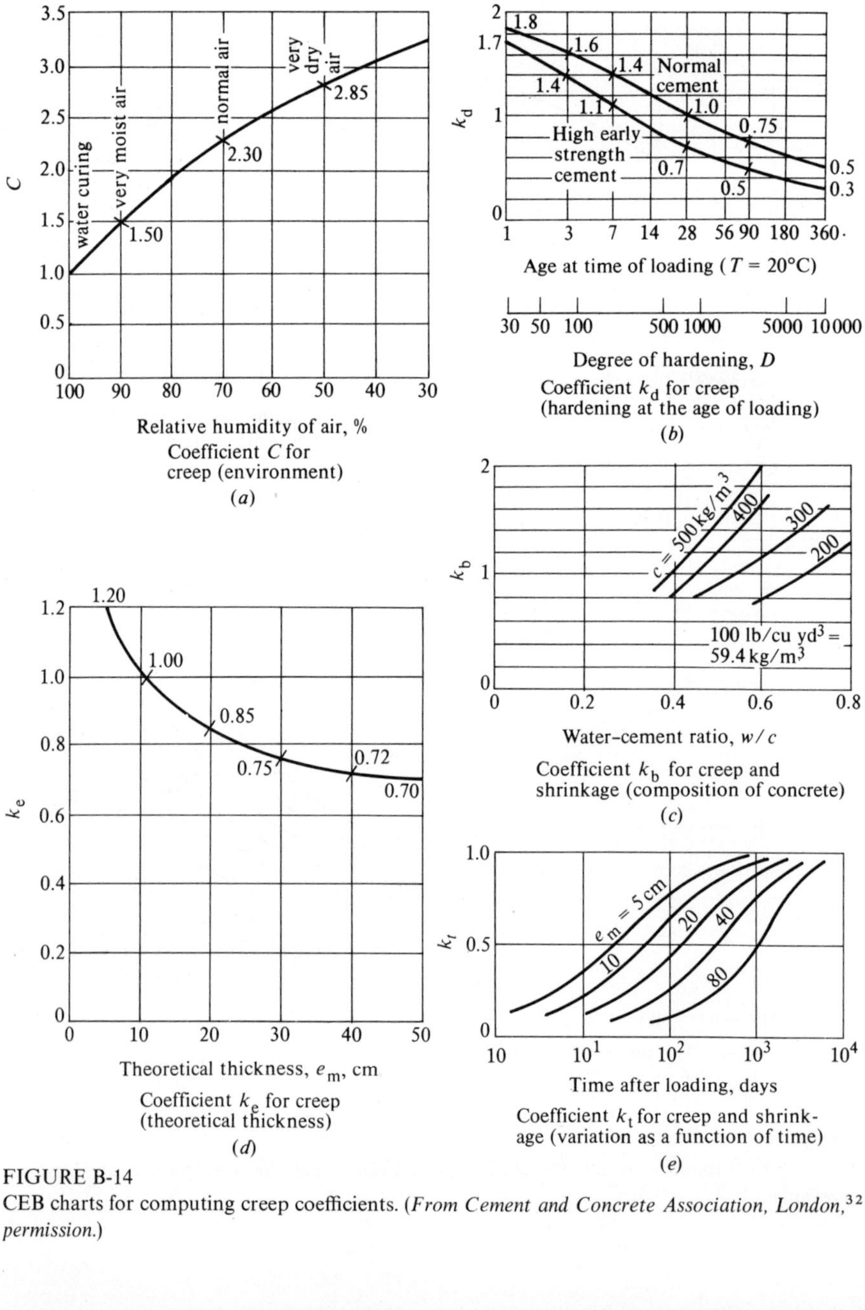

FIGURE B-14
CEB charts for computing creep coefficients. (*From Cement and Concrete Association, London,*[32] *with permission.*)

The values in Fig. B-14*b* for the effect of age at time of loading refer to Portland cement, hardened under normal conditions, i.e., at an average concrete temperature of 20°C (68°F) and with protection against excessive losses of moisture. If the concrete hardens at a temperature other than 20°C, the age at loading is replaced by the corresponding degree of hardening in Fig. B-14*b*:

$$D = \sum \Delta t(T + 10°) \tag{B-14}$$

in which D represents the degree of hardening at the moment of loading, and Δt represents the number of days during which hardening has taken place at T°C.

When creep has a large influence on the limit state under consideration, an increase or a reduction of the order of 15 percent should be considered, so as to cover the most unfavorable case. Superposition of loads should be done on the assumption that the stress applied at the beginning operates until the end of the period under consideration. The same method may be applied for all later stress changes.

The CEB procedure forms a working basis and is valid only for Portland cement concretes of normal quality, hardening under normal conditions and subject to working stresses at the most equal to 40 percent of their rupture stress.

Shrinkage and creep—author and colleagues, 1970 The procedure of the author and colleagues,[1,2,3] which was developed independently of the CEB procedure, is based on a set of standard equations and conditions, with correction factors in the form of equations, or tables, or curves for nonstandard conditions. The method is also recommended in the ACI Committee 209 publication.[9]

Standard equations: (1-16) for creep and (1-17) or (1-18) for shrinkage (ratios in Table 1-4); and standard conditions: ambient relative humidity $\leq$ 40 percent, and average thickness of member (part under consideration) 6 in. (15.2 cm).

Correction factors by general procedure:

1. Age when loaded for creep (Eqs. (1-20), (1-21), Table 1-6, Fig. A-2)
   Age initial shrinkage considered (Eqs. (1-17), (1-18))
2. Humidity for creep (Eq. (1-22), Table 1-7, Fig. A-2)
   Humidity for shrinkage (Eqs. (1-23), (1-24), Table 1-7, Fig. A-3)
3. Size of member for creep (average thickness method—Eqs. (1-25), (1-26), Table 1-8, Fig. A-2) (volume–surface ratio method—Eq. (1-29), Table 1-9)
   Size of member for shrinkage (average thickness method—Eqs. (1-27), (1-28), Table 1-8, Fig. A-3) (volume–surface ratio method—Eq. (1-30), Table 1-9)
4. Slump for creep (Eq. (A-1), Fig. A-2)
   Slump for shrinkage (Eq. (A-2), Fig. A-3)
5. Cement content for creep (Eq. (A-3), Fig. A-2)
   Cement content for shrinkage (Eq. (A-4), Fig. A-3)
6. Percentage fines for creep (Eq. (A-5), Fig. A-2)
   Percentage fines for shrinkage (Eqs. (A-6), (A-7), Fig. A-3)

7. Air content for creep (Eqs. (A-8), (A-9), Fig. A-2)
   Air content for shrinkage (Eq. (A-10), Fig. A-3)

Correction factors by simplified procedure:
1. Age when loaded for creep
   Age initial shrinkage considered
2. Humidity
3. Size of member

(items 1–3: same as above)

Numerical Examples

1. Shrinkage example using methods of Jones *et al.* and general procedure of the author and colleagues

Given: 70 percent ambient relative humidity, thickness of member 10 in. (25.4 cm), 6 bags of cement per cubic yard, 50 percent fines, 6 percent air content, moist-cured, slump 3 in. (7.6 cm).

Required: compute the ultimate shrinkage strain.

Jones, Hirsch, Stephenson—Figs. B-9 and B-10

$$\begin{aligned} &\qquad\quad \textit{B-9} \qquad\quad \textit{B-10a} \quad\; b \qquad\; c \qquad\; d \qquad\; e \qquad\; f \\ (\varepsilon_{sh})_u &= (700 \times 10^{-6}) \times 1.03 \times 0.99 \times 1.02 \times 1.02 \times 1.00 \times 0.83 \\ &= \underline{616 \times 10^{-6} \text{ in/in (cm/cm)}} \end{aligned}$$

(Applies to lightweight concrete.)

Author and colleagues—average $(\varepsilon_{sh})_u$ ***and Fig. A-3***

$$\begin{aligned} &\qquad\quad \text{avg.} \qquad\quad a \qquad\; b \qquad\; c \qquad\; d \qquad\; e \qquad\; f \\ (\varepsilon_{sh})_u &= (780 \times 10^{-6}) \times 0.70 \times 0.88 \times 1.01 \times 0.95 \times 1.00 \times 1.00 \\ &= \underline{461 \times 10^{-6} \text{ in/in (cm/cm)}} \end{aligned}$$

(Applies to normal-weight and lightweight concrete.)

2. Shrinkage example using methods of CEB and the simplified procedure of the author and colleagues

Given: 70 percent ambient relative humidity, shrinkage from both 7 days and 28 days, rectangular section 25 cm (9.8 in.) by 40 cm (15.7 in.), water–cement ratio $w/c = 0.45$, $c = 400 \text{ kg/m}^3$ (673 lb/cu yd), concrete cured for 7 days.

Required: compute the one-year and ultimate shrinkage strain.

CEB—Eq. B-12 and Fig. B-13

$(\varepsilon_{sh}) = 275 \times 10^{-6}$ (Fig. B-13*a*) $\qquad k_b = 1.00$ (*b*)

$e_m = (25 \times 40)/(25 + 40) = 15.4$ cm (6.1 in.) $\qquad k_e = 0.89$ (*c*)

$k_t = 0.35$ at 21 days (age 28 days) (*d*)

$$= 0.75 \text{ at 1 year} \qquad (d)$$
$$= 1.00 \text{ at ultimate} \qquad (d)$$
$$(\varepsilon_{sh})_t = (\varepsilon_{sh}) k_b k_e k_t \qquad \text{(Eq. (B-12))}$$
$$(\varepsilon_{sh})_{7d\text{ to }28d} = (275 \times 10^{-6}) \times 1.00 \times 0.89 \times 0.35 = 86 \times 10^{-6} \text{ cm/cm (in/in)}$$
$$(\varepsilon_{sh})_{7d\text{ to }1\text{ yr}} = (275 \times 10^{-6}) \times 1.00 \times 0.89 \times 0.75 = \underline{184 \text{ microstrain}}$$
$$(\varepsilon_{sh})_{7d\text{ to ult}} = (275 \times 10^{-6}) \times 1.00 \times 0.89 \times 1.00 = \underline{245 \text{ microstrain}}$$
$$(\varepsilon_{sh})_{28d\text{ to }1\text{ yr}} = (184 - 86) = \underline{98 \text{ microstrain}}$$
$$(\varepsilon_{sh})_{28d\text{ to ult}} = (245 - 86) = \underline{159 \text{ microstrain}}$$

Author and colleagues—Eq. (1-17) or Table 1-4 or Fig. 1-7, and avg. $(\varepsilon_{sh})_u$***; Eqs. (1-23), (1-27), (1-28), or Tables 1-7, 1-8, or Fig. A-3***

Average values for standard conditions

$$(\varepsilon_{sh})_{7d\text{ to }28d} = \frac{t}{35 + t}(\varepsilon_{sh})_u \qquad \text{(Eq. (1-17))}$$
$$= \frac{28}{35 + 28}(780 \times 10^{-6}) = 0.44(780 \times 10^{-6})$$
$$= 343 \times 10^{-6} \text{ in/in (cm/cm)}$$
$$(\varepsilon_{sh})_{7d\text{ to }1\text{ yr}} = \frac{365}{35 + 365} 780 = 0.91 \times 780 = 710 \text{ microstrain}$$

The 28-day factor of 0.44 and the 1-year factor of 0.91 can also be obtained from Table 1-4 or Fig. 1-7. It follows that

$$(\varepsilon_{sh})_{28d\text{ to }1\text{ yr}} = (710 - 343) = 367 \text{ microstrain}$$
$$(\varepsilon_{sh})_{28d\text{ to ult}} = (780 - 343) = 437 \text{ microstrain}$$

Correction factors

$$(\text{C.F.})_H = 0.70 \text{ for 70 percent humidity}$$
(Eq. (1-23) or Table 1-7 or Fig. A-3)
$$(\text{C.F.})_T = 0.85 \text{ for member size after 1 year}$$
(Eq. (1-27) or Table 1-8 or Fig. A-3)
$$(\text{C.F.})_T = 0.88 \text{ for member size ultimate value}$$
(Eq. (1-28) or Table 1-8 or Fig. A-3)

Solution for conditions of the problem

$$(\varepsilon_{sh})_{7d\text{ to }1\text{ yr}} = (0.70 \times 0.85)(710) = \underline{422 \text{ microstrain}} \text{ (versus 184 by CEB)}$$
$$(\varepsilon_{sh})_{7d\text{ to ult}} = (0.70 \times 0.88)(780) = \underline{480 \text{ microstrain}} \text{ (versus 245 by CEB)}$$
$$(\varepsilon_{sh})_{28d\text{ to }1\text{ yr}} = (0.70 \times 0.85)(367) = \underline{218 \text{ microstrain}} \text{ (versus 98 by CEB)}$$
$$(\varepsilon_{sh})_{28d\text{ to ult}} = (0.70 \times 0.88)(437) = \underline{269 \text{ microstrain}} \text{ (versus 159 by CEB)}$$

For determining shrinkage that takes place after the installation of partitions in buildings, for example, one might use the shrinkage strain following age 28 days—as computed in Example 2 by the two methods considered.

Comments on shrinkage examples For the conditions of Example 1, $(\varepsilon_{sh})_u = 616$ microstrain and 461 microstrain by the two methods. For the similar conditions of Example 2, $(\varepsilon_{sh})_u = 245$ microstrain and 480 microstrain by the two methods, with similar differences shown for the other results. For these two typical examples, the results by the author *et al.* of 461 and 480 (applies to both normal-weight and lightweight concrete) are intended to be average values based on some 356 data points, while those of Jones *et al.* of 616 (applies only to lightweight concrete) and CEB of 245 (applies only to normal-weight concrete) appear to be somewhat on the high and low side, respectively, of average values. This is based on the conclusion of the author and colleagues[1,2,3] and ACI Committee 209[9] that for nominal design purposes in most cases no consistent distinction seems to be found between shrinkage of normal and lightweight concrete, since the cement gel apparently does most of the shrinking. However, the possibility of exceptional situations in this regard should be guarded against. In this connection, the CEB procedure states that shrinkage of lightweight concrete has been found to be from 1 to 2 times that of normal-weight concrete. It has also been found[4,23] to be less than that of normal-weight concrete.

In studying the different methods for predicting shrinkage, one finds that, although there is not a direct correlation between the various correction factors, the differences in the effect of the correction factors are not large. Rather, it is the choice of the base value of shrinkage that tends to yield significantly different final results (at 70 percent humidity, for example: $(\varepsilon_{sh})_u = 700 \times 0.83 = 581$ microstrain by Jones *et al.*, 275 microstrain by CEB, and $780 \times 0.70 = 546$ microstrain by the author *et al.*). Of course, in all of these methods the base shrinkage value can and should be established by tests for local concrete and conditions, or selected based on local experience, with the correction factors then being applied for varying or nonstandard conditions.

3. Creep example using methods of Wagner, Jones *et al.*, and the general procedure of the author and colleagues

Given: 70 percent ambient relative humidity, age of loading 14 days, thickness of member 10 in. (25.4 cm), Type I cement, 6 bags of cement per cubic yard, cement content by weight 20 percent, water–cement ratio 0.50, 50 percent fines, 6 percent air content, moist-cured, slump 3 in. (7.6 cm).

Required: compute the ultimate creep coefficient.

Wagner—Fig. B-8

$$\delta_u = \overset{a}{(1.30 \times 10^{-6})} \times \overset{b}{1.00} \times \overset{c}{1.00} \times \overset{d}{0.60} \times \overset{e}{0.68} = \underline{0.530 \times 10^{-6} \text{ per psi}}$$

If $E_c = 3.5 \times 10^6$ psi,

$$C_u = \delta_u E_c = 0.530 \times 3.5 = \underline{1.86}$$

(Applies to normal-weight concrete.)

Jones, Hirsch, Stephenson—Figs. B-11 and B-12

$$\begin{array}{l} \qquad\;\; \text{B-11} \quad \text{B-12}a \qquad b \qquad\quad c \qquad\quad d \qquad\quad e \qquad\quad f \qquad\quad g \\ c_{sc} = 0.90 \times 1.14 \times 1.00 \times 0.95 \times 1.00 \times 0.85 \times 0.89 \times 1.00 = \underline{0.737 \text{ per psi}} \end{array}$$

If $E_c = 2.5 \times 10^6$ psi,

$$C_u = 0.737 \times 2.5 = \underline{1.84}$$

(Applies to lightweight concrete.)

Author and colleagues—average C_u and Fig. A-2

$$\begin{array}{l} \qquad\;\; \text{avg.} \qquad a \qquad\quad b \qquad\quad c \qquad\quad d \qquad\quad e \qquad\quad f \qquad\quad g \\ C_u = 2.35 \times 0.93 \times 0.80 \times 0.93 \times 1.03 \times 1.00 \times 1.01 \times 1.00 = \underline{1.70} \end{array}$$

(Applies to normal-weight and lightweight concrete.)

4. Creep example using methods of CEB and the simplified procedure of the author and colleagues

Given: 70 percent ambient relative humidity, age of loading 7 days, rectangular section 25 cm (9.8 in.) by 40 cm (15.7 in.), normal cement, water–cement ratio $w/c = 0.45$, $c = 400$ kg/m^3 (673 lb/cu yd), moist-cured.

Required: compute the one-year and ultimate creep coefficient.

CEB—Eq. B-13 and Fig. B-14

$$C = 2.30 \text{ (Fig. B-14}a) \qquad k_d = 1.40\ (b) \qquad k_b = 1.00\ (c)$$

$$e_m = (25 \times 40)/(25 + 40) = 15.4 \text{ cm (6.1 in.)} \qquad k_e = 0.89\ (d)$$

$$k_t = 0.75 \text{ at 1 year} \qquad k_t = 1.00 \text{ at ultimate } (e)$$

$$C_t = C k_d k_b k_e k_t \qquad \text{(Eq. (B-13))}$$

$$C_{1\,yr} = 2.30 \times 1.40 \times 1.00 \times 0.89 \times 0.75 = \underline{2.15}$$

$$C_{ult} = 2.30 \times 1.40 \times 1.00 \times 0.89 \times 1.00 = \underline{2.87}$$

Author and colleagues—Eq. (1-16) or Table 1-4 or Fig. 1-7, and avg. C_u; Eqs. (1-20), (1-22), (1-25), (1-26), or Tables 1-6, 1-7, 1-8, or Fig. A-2

$$C_t = \frac{t^{0.60}}{10 + t^{0.60}} C_u \qquad \text{Average } C_u = 2.35 \qquad \text{(Eq. (1-16))}$$

$$C_{1\,yr} = \frac{365^{0.60}}{10 + 365^{0.60}} C_u = 0.78 C_u \qquad \text{(Eq. (1-16) or Table 1-4 or Fig. 1-7)}$$

$$(\text{C.F.})_{\text{LA}} = 1.00 \qquad \text{(Eq. (1-20) or Table 1-6 or Fig. A-2)}$$
$$(\text{C.F.})_{\text{H}} = 0.91 \qquad \text{(Eq. (1-22) or Table 1-7 or Fig. A-2)}$$
$$(\text{C.F.})_{\text{T}} = 0.93 \text{ at 1 year (Eq. (1-25) or Table 1-8 or Fig. A-2)}$$
$$= 0.90 \text{ at ultimate (Eq. (1-26) or Table 1-8 or Fig. A-2)}$$

$$C_{1\text{ yr}} = 2.35 \times 0.78 \times 1.00 \times 0.80 \times 0.91 = \underline{1.33} \text{ (versus 2.15 by CEB)}$$
$$C_{\text{u}} = 2.35 \times 1.00 \times 0.80 \times 0.93 = \underline{1.75} \text{ (versus 2.87 by CEB)}$$

Comments on creep examples For the conditions of Example 3 including a loading age of 14 days, $C_{\text{u}} = 1.86$, 1.84, and 1.70 by the three methods. For the similar conditions of Example 4 except a loading age of 7 days, $C_{\text{u}} = 2.87$ and 1.75 by the two methods. For these two typical examples, the results by the author *et al.* of 1.70 and 1.75 (applies to both normal-weight and lightweight concrete) are intended to be average values based on some 479 data points, while those of

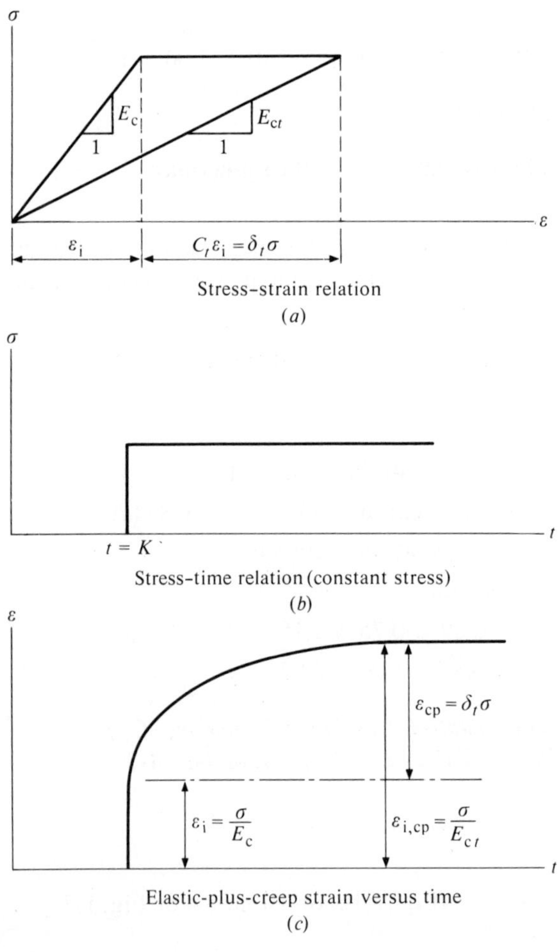

FIGURE B-15
Effective or sustained modulus method.

Wagner of 1.86 (applies only to normal-weight concrete) and Jones *et al.* of 1.84 (applies only to lightweight concrete) are only slightly higher; and that of CEB of 2.87 (applies only to normal-weight concrete) is somewhat higher. It appears to the author that the CEB method may compute somewhat higher than average creep values and somewhat lower than average shrinkage values (Example 2), while the total time-dependent effect may be about average. Thus Rao and Dilger[66] found, for example: the time-dependent parameters based on the CEB recommendations and on Branson's standard curves gave very close results under no stress or constant stress. Under varying stress also, the total deformations according to these two approaches were approximately the same.

As found by the author and colleagues,[1,2,3] ACI Committee 209,[9] Neville,[33] and others, there does not seem to be any fundamental difference between creep (say in the form of the creep coefficient) for normal-weight and lightweight concrete. That is to say, it would appear that the cement gel does most of the creeping with only nominal effects from the aggregate in most cases. However, due to the lower elastic modulus of lightweight concrete and correspondingly higher initial strains for a given load or stress level, the total creep deformation will tend to be greater in lightweight concrete structures under the same general conditions.

In studying the different methods for predicting creep, one finds that, although there is not a direct correlation between the various correction factors, the differences in the effect of the correction factors are not large. Rather, it is the choice of the base value for creep that tends to yield the different final results (such as at 70 percent humidity: $C_u = 2.30$ by CEB versus $2.35 \times 0.80 = 1.88$ by the author *et al.*). Of course, in all of these methods the base creep value can and should be established by tests for local concrete and conditions, or selected based on local experience, with the correction factors then being applied for varying or nonstandard conditions.

Methods of Analyzing Concrete Creep Under Constant and Variable Stress

Effective or sustained modulus method The effective or sustained modulus of elasticity is based on the use of a single strain–time curve under constant stress applied at age $t = K$, as shown in Fig. B-15. It was defined in the text by Eq. (1-46), and here in terms of both specific creep and the creep coefficient by Eq. (B-15).

$$E_{ct} = \frac{\sigma}{\varepsilon_i + \Sigma\varepsilon_{cp}} = \frac{\sigma}{\varepsilon_i + \delta_t\sigma} = \frac{\sigma}{\varepsilon_i(1 + \delta_t E_c)} = \frac{E_c}{1 + C_t} \tag{B-15}$$

since $C_t = \delta_t E_c$ (Eq. (1-19)). It follows that the ordinates to the initial-plus-creep, $\varepsilon_{i,cp}$, versus time curve are those shown in Fig. B-15*c*.

This method is the simplest but least accurate, since it does not take into account the stress history. For example, if creep at any time is assumed to be proportional to stress (or elastic strain) at that time, this method tends to overestimate creep deformation at that time under increasing stress, and underestimate creep deformation at that time under decreasing stress. The effect of age at loading and the elastic modulus at the time of loading also affect these results. On the other hand, if creep is assumed to be proportional to stress (or elastic strain) at the time of initial loading, such as for dead load, creep deformation is underestimated under increasing stress and overestimated under decreasing stress. In most structures concrete stresses tend to decrease gradually with time due to stress redistribution, loss of prestress, etc. This method also has the disadvantage of predicting complete recovery upon unloading. Nevertheless, where applicable the effective modulus concept is so simple that it is the most convenient method of analysis available. For most practical purposes the method is usually of acceptable accuracy, according to Pauw.[67] When it is desired to take into account the effect of variable stress, the age-adjusted effective modulus method provides for some improvement in this approach.

Age-adjusted effective modulus method The age-adjusted effective modulus method developed by Trost[68] and extended by Bazant[69] and others, uses an "aging coefficient," χ, to obtain an "age-adjusted effective modulus," E''_{ct}, as given by Eq. (B-16).

$$E''_{ct} = \frac{E_c}{1 + \chi C_t} \tag{B-16}$$

where
$$\chi = \left[1 - \frac{E_R}{E_c}\right]^{-1} - \frac{1}{C_t}$$

and E_R = a relaxation function = stress at one time caused by a unit strain introduced at another time. χ is always less than 1 and almost always greater than 0.5. Practical values of χ are tabulated in Ref. 69. This method represents a useful refinement of the effective modulus method for variable stress problems.

Rate of creep method The rate of creep method was presented by Glanville[70] in 1930, Dischinger[71,72] in 1937, Ross[50] in 1958, and others. The method is based on the use of a single specific creep–time curve for variable stress problems, as shown in Fig. B-16, and given by Eq. (B-17).

$$\delta_t = Af(t - K) \tag{B-17}$$

where A = a constant, $f(t - K)$ = a function of $(t - K)$, and K = the concrete age when first loaded.

This method assumes that the rate of creep at any age is equal to the stress multiplied by the slope of the specific creep–time curve at that age. This is equivalent to assuming a parallel shift in the initial specific creep–time curve, as shown in Fig. B-16*b*. The creep strain at age t is then determined by integrating Eq. (B-18) over the time interval K to t.

$$\varepsilon_{cp} = A \int_K^t \sigma \frac{df(t - K)}{dt} dt \tag{B-18}$$

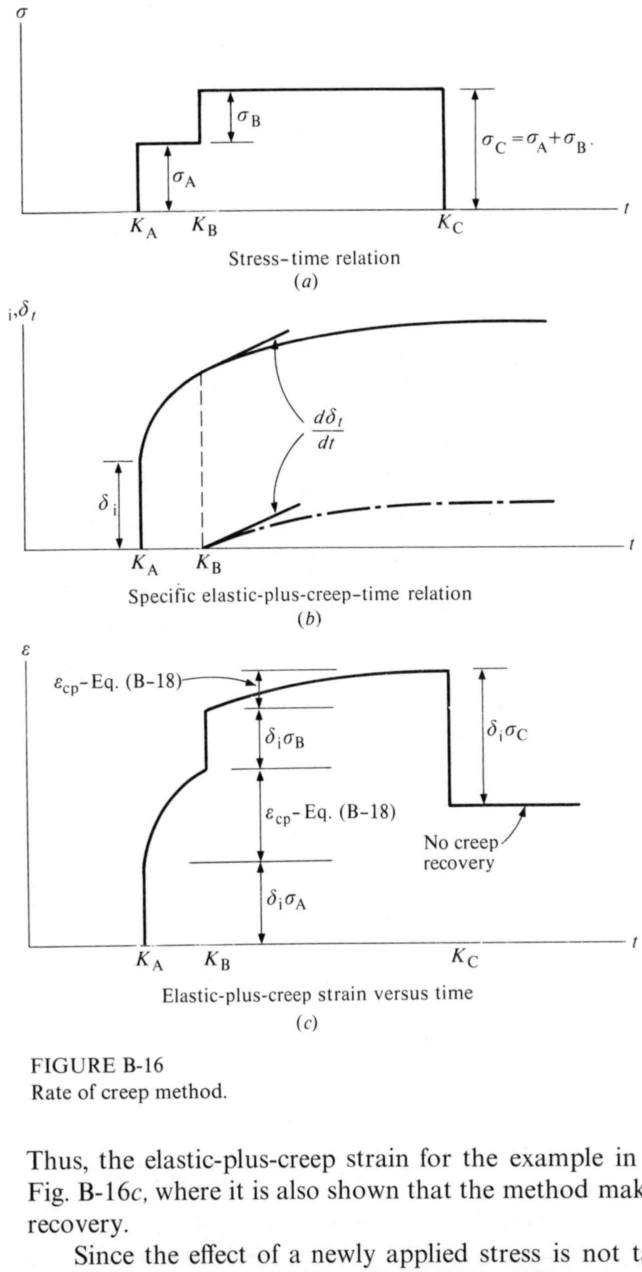

FIGURE B-16
Rate of creep method.

Thus, the elastic-plus-creep strain for the example in Fig. B-16 is described in Fig. B-16*c*, where it is also shown that the method makes no allowance for creep recovery.

Since the effect of a newly applied stress is not taken into account in this method, the rate of creep method underestimates creep under increasing stress and overestimates creep under decreasing stress. According to Neville,[33] the rate of creep method is not more accurate than the effective modulus method and under a gradually decreasing stress the two methods bracket the actual creep behavior.

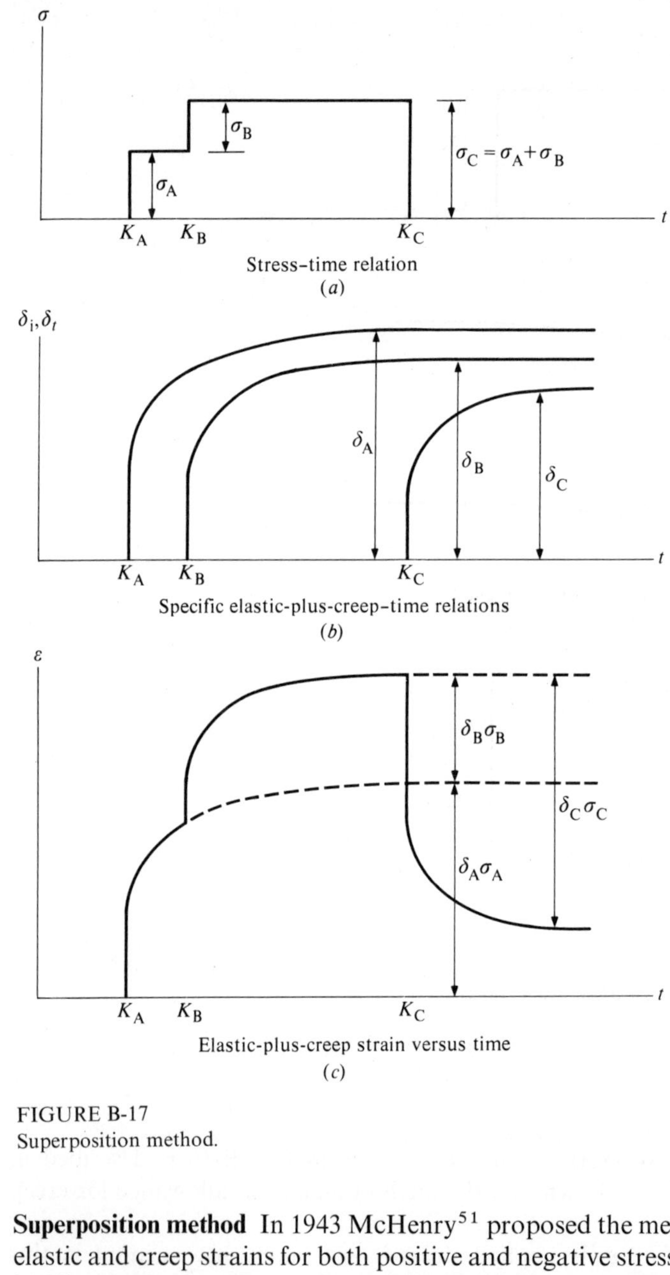

FIGURE B-17
Superposition method.

Superposition method In 1943 McHenry[51] proposed the method of superimposing elastic and creep strains for both positive and negative stress increments, as shown in Fig. B-17. Thus for a given stress–time diagram (Fig. B-17*a*), and with the corresponding specific strain–time information for the particular stress increment ages, K_A, K_B, and K_C (Fig. B-17*b*), the resulting elastic plus creep strain versus time curve is obtained by the principle of superposition (Fig. B-17*c*).

The superposition method tends to overestimate creep recovery, since it assumes that recovery is equivalent to the corresponding creep itself. Hence, under decreasing stress the method tends to underestimate creep strains. The method can be very laborious, requiring computer assistance for many problems. However, while requiring more test information, the superposition method is usually but not always somewhat more accurate than the effective modulus and rate of creep methods.

It may be noted that the various equations of the author and colleagues in Chap. 1 provide a complete set of initial and time-dependent elastic and creep information for the rate of creep and superposition methods.

Other methods Various other methods have been proposed including the rate of flow method by England and Illston,[73] the residual creep limit method of Pauw and Chai,[27] Pauw,[67] the revised rate of creep method of Mossiossian and Gamble,[74] and others. For example, in the rate of flow method it is assumed that the stress-dependent strains of concrete under sustained load can be separated into three major components:

1. Elastic
2. Delayed elastic (recoverable creep)
3. Flow (irrecoverable creep, drying creep, and plastic deformation).

APPENDIX
C

VOLUME–SURFACE RATIOS OF STANDARD PRECAST-PRESTRESSED MEMBERS

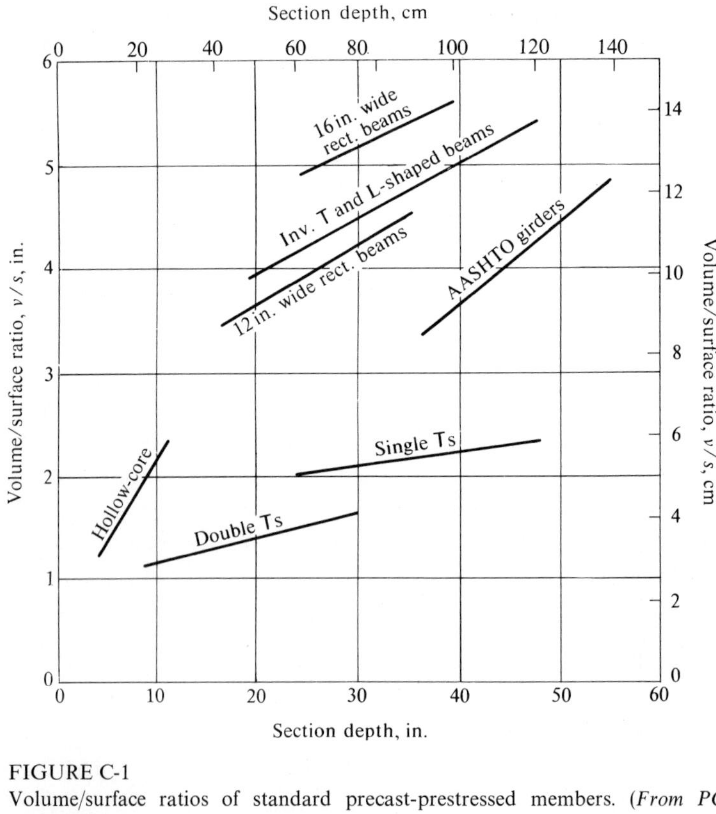

FIGURE C-1
Volume/surface ratios of standard precast-prestressed members. (*From PCI Handbook*,[38] *with permission.*)

CHAPTER
TWO

DIFFERENTIAL SHRINKAGE, CREEP, AND TEMPERATURE IN COMPOSITE BEAMS

2-1 INTRODUCTION

The usual case of differential shrinkage and creep in composite beams, and the one considered in this book, refers to the following condition: for any period following slab casting and bonding of the slab to the precast beam or steel beam, when the shrinkage plus creep strain of the interface fiber (bottom fiber) of the slab would exceed the shrinkage plus creep strain of the interface fiber (top fiber) of the precast beam or steel beam (zero for steel beam) if the two were not bonded together. This condition results in a longitudinal force along the upper part of the beam which produces a downward bending of the composite beam. It may also be noted that the differential shrinkage and creep force is a sustained force that increases with time after slab-to-beam bonding, so that the corresponding creep effects under variable stress due to the force itself must be taken into account.

In analyzing differential shrinkage and creep effects in composite beams, a distinction needs to be made between the cases of *unshored* construction—slab is unstressed under slab dead load, with the precast beam or steel beam supporting the slab plus precast beam (or steel beam) dead load; and *shored* construction—slab is stressed under slab dead load with the composite beam supporting either the slab dead load or slab plus precast beam (or steel beam) dead load, depending normally on the sequence of construction of the shoring and placing of the precast beam or steel beam. Consideration should also be given to whether a precast beam is reinforced or prestressed. These distinctions are taken into account in estimating the differential shrinkage and creep strain or coefficient, with the same procedures used in all cases.

For a given set of conditions, the differential shrinkage and creep effects will be greater for the steel–concrete than the concrete–concrete composite beam, greater for the shored than the unshored concrete–concrete composite beam, and greater for the prestressed than the reinforced precast beam case. The time of slab casting also has an effect on the differential shrinkage and creep coefficient or force, with a later slab casting schedule causing a greater effect. These effects can be seen in the recommended coefficients in Table 2-1.

Two elastic theory methods are presented for computing differential shrinkage and creep stresses and deflections, along with an evaluation of material properties and coefficients for practical application. Numerical examples and comparisons of differential shrinkage stresses and deflections by the two methods (identical results) and for different practical structures are also included. The effects of differential temperature are determined for different cases using the equations presented for differential shrinkage and creep. A simple procedure for computing the effects of differential shrinkage, creep, and temperature in continuous members is also included. Lastly, experimental results of differential shrinkage and creep effects reported in the literature are discussed.

2-2 TWO METHODS OF COMPUTING DIFFERENTIAL SHRINKAGE AND CREEP STRESSES AND DEFLECTIONS

Two simple elastic-theory approaches have been advanced for analyzing differential shrinkage and creep stresses and deflections in composite beams. The distinguishing feature of these is that the slab differential shrinkage and creep force is applied to the slab alone and then to the *composite* beam (section) in one case and to the *separate* slab and beam (sections) in the other case. Hence, these have been referred to[1,2] as the "composite section method" and the "separate section method" respectively.

In the composite section method the differential shrinkage and creep of the slab, after it is bonded to the precast beam, or steel beam, is assumed to be restrained by the composite beam. In this method the slab differential shrinkage and creep force is applied as a concentric tensile force to the slab, and then as an equal eccentric compressive force to the composite beam—*both applied at the slab centroid.* In the separate section method the differential shrinkage and creep of the slab induces an eccentric compressive force at the top of the precast beam, or steel beam, and, in turn, the restraint offered by the precast beam, or steel beam, induces an equal eccentric tensile force at the bottom of the slab—*both applied at the interface.* The section properties and various forces, stresses, and strains can be seen in Figs. 2-1, 2-2 and 2-3. The strains in the precast beam shown in Figs. 2-2*a* and 2-3*a* refer to the case of a prestressed precast beam.

It is demonstrated in Table 2-2 for four rather extreme examples that the two methods lead to identical results. The two methods do, however, provide different insights into the problem.

Subscript notation:

1 cast-in-place slab
2 precast beam or steel beam
B bottom fiber of section
c composite section with transformed slab
I the slab-beam interface
S slab centroid
T top fiber of section

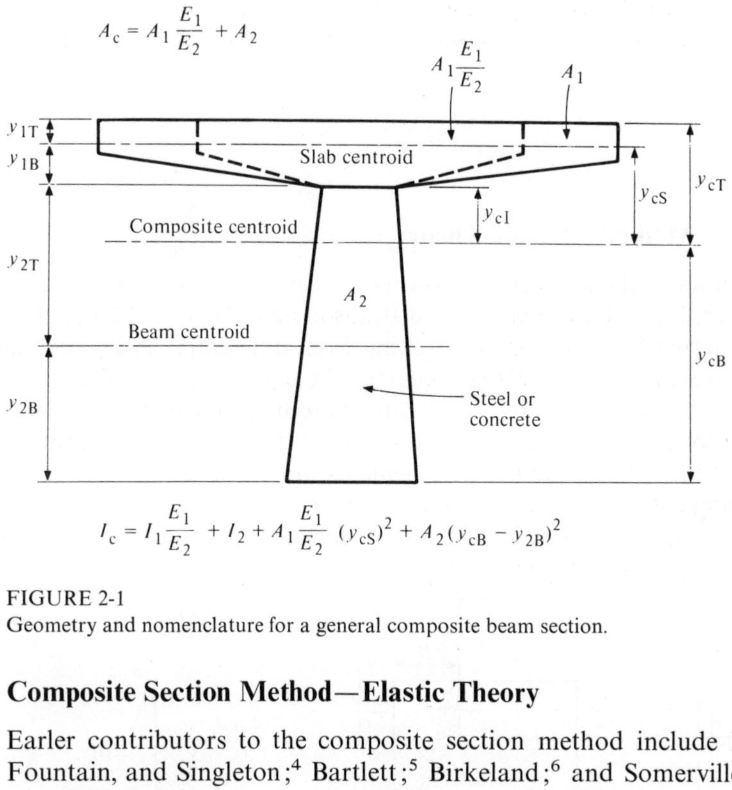

FIGURE 2-1
Geometry and nomenclature for a general composite beam section.

Composite Section Method—Elastic Theory

Earler contributors to the composite section method include Morsch;[3] Viest, Fountain, and Singleton;[4] Bartlett;[5] Birkeland;[6] and Somerville.[7] This method has also recently been recommended by Roll[2] and is used in later chapters of this book.

Referring to Figs. 2-1 and 2-2*a*,

$$Q = DA_1E_1 \tag{2-1}$$

where Q is the slab shrinkage force and D is the differential shrinkage and creep strain or coefficient. Applying this force to the slab and then the composite section, as in Fig. 2-2*b*, yields the following fiber stresses, as shown in Fig. 2-2*c*.

$$\sigma_{1T} = \frac{Q}{A_1} + \left(-\frac{Q}{A_c} - \frac{Qy_{cS}y_{cT}}{I_c} \right) \frac{E_1}{E_2}$$

$$\sigma_{1B} = \frac{Q}{A_1} + \left(-\frac{Q}{A_c} \pm \frac{Qy_{cS}y_{cI}}{I_c} \right) \frac{E_1}{E_2}$$

$$\sigma_{2T} = -\frac{Q}{A_c} \pm \frac{Q y_{cS} y_{cI}}{I_c} \tag{2-2}$$

$$\sigma_{2B} = -\frac{Q}{A_c} + \frac{Q y_{cS} y_{cB}}{I_c}$$

The plus-minus term in the second and third of Eqs. (2-2) is minus, except when the composite centroid is within the slab. These equations follow from elastic theory with the eccentric force, Q, producing the moment, $Q y_{cS}$.

For the constant moment, $Q y_{cS}$, along a simple span, the midspan deflection is given by Eq. (2-3).

$$\Delta = \frac{ML^2}{8EI} = \frac{Q y_{cS} L^2}{8 E_2 I_c} \tag{2-3}$$

Separate Section Method—Elastic Theory

Earlier contributors to the separate section method include Frohlich,[8] Walter,[9] Evans and Parker,[10] Ozell and Diniz[11] and Branson and Ozell.[12,13] Since there is no applied external force or couple resulting from differential shrinkage and creep, the two forces F_1 and F_2 in Fig. 2-3*b* must be equal and collinear. These forces act on their respective sections and contribute both a direct stress and a bending stress. However, both produce a downward deflection that cannot act independently of the other since the two elements are bonded together. Hence the combined moment, $F_1 y_{1B} + F_2 y_{2T}$, must be distributed between the slab and beam sections in proportion to their respective stiffnesses or EI values (Eqs. (2-5)).

Referring to Figs. 2-1 and 2-3,

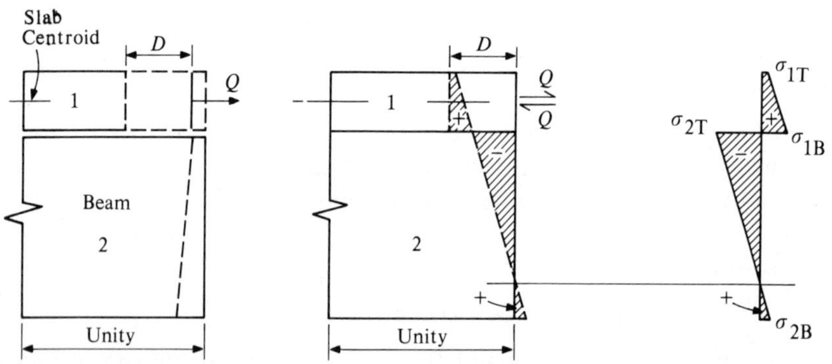

(*a*) Unbonded shrinkage and creep strain of both elements—and slab shrinkage force Q

(*b*) Self-cancelling slab force Q applied as eccentric column load on composite section—and resulting differential shrinkage strains

(*c*) Corresponding differential shrinkage stresses

Time-dependent concrete strains (all from time of slab casting)

FIGURE 2-2
Shrinkage force Q applied to slab and to composite section.[1]

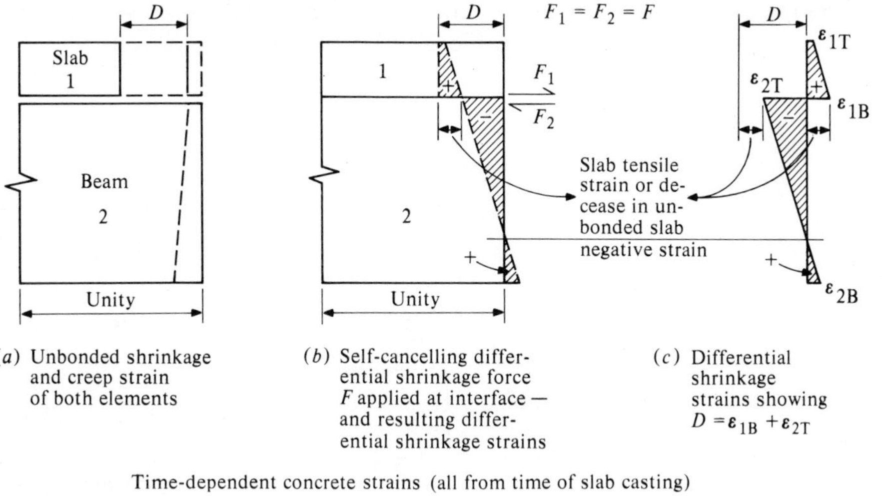

FIGURE 2-3
Shrinkage force F applied to separate sections.[1]

$$\varepsilon_{1B} = \frac{1}{E_1}\left(\frac{F_1}{A_1} + \frac{M_1 y_{1B}}{I_1}\right)$$
$$\varepsilon_{2T} = \frac{1}{E_2}\left(\frac{F_2}{A_2} + \frac{M_2 y_{2T}}{I_2}\right) \tag{2-4}$$

where
$$M_1 = (F_1 y_{1B} + F_2 y_{2T})\ \frac{E_1 I_1}{E_1 I_1 + E_2 I_2}$$
$$M_2 = (F_1 y_{1B} + F_2 y_{2T})\ \frac{E_2 I_2}{E_1 I_1 + E_2 I_2} \tag{2-5}$$

As shown in Figs. 2-3*b* and 2-3*c*,

$$F_1 = F_2 = F \tag{2-6}$$

$$D = \varepsilon_{1B} + \varepsilon_{2T} \tag{2-7}$$

where F is the differential shrinkage and creep force and D is the differential shrinkage and creep strain or coefficient (see Figs. 2-3*a* and 2-3*b*).

Solving Eqs. (2-4) to (2-7),

$$F = \frac{E_1 D}{\dfrac{1}{A_1} + \dfrac{1}{A_2(E_2/E_1)} + \dfrac{(y_{1B} + y_{2T})^2}{I_1 + I_2(E_2/E_1)}}$$

or

$$F = \frac{E_2 D}{\dfrac{1}{A_2} + \dfrac{1}{A_1(E_1/E_2)} + \dfrac{(y_{1B} + y_{2T})^2}{I_2 + I_1(E_1/E_2)}} \tag{2-8}$$

Applying this force to the separate sections, as in Fig. 2-3*b*, yields the following fiber stresses (or strains), as shown in Fig. 2-3*c*.

$$\begin{aligned}\sigma_{1T} &= F/A_1 - M_1 y_{1T}/I_1\\ \sigma_{1B} &= F/A_1 + M_1 y_{1B}/I_1\\ \sigma_{2T} &= -F/A_2 - M_2 y_{2T}/I_2\\ \sigma_{2B} &= -F/A_2 + M_2 y_{2B}/I_2\end{aligned} \tag{2-9}$$

These equations follow from elastic theory with the eccentric forces, F, producing the separate slab and beam moments, M_1 and M_2, respectively.

For the constant moment, $M_1 + M_2$, along a simple span, the midspan deflection is given by Eq. (2-10).

$$\Delta = \frac{ML^2}{8EI} = \frac{F(y_{1B} + y_{2T})L^2}{8(E_1 I_1 + E_2 I_2)} \tag{2-10}$$

where $M = M_1 + M_2 = F(y_{1B} + y_{2T})$, with the EI terms in Eqs. (2-5) cancelling, and $EI = E_1 I_1 + E_2 I_2$.

2-3 EVALUATION OF MATERIAL PROPERTIES AND COEFFICIENTS

Birkeland,[6] Branson,[1] Roll,[2] and others have discussed the need for exercising judgement in the selection of material properties and coefficients to be used in predicting differential shrinkage and creep stresses and deflections. The differential shrinkage and creep force, Q or F, is a sustained force that increases (from zero) at a decreasing rate with time after initial bonding to the precast beam or steel beam, say from the day of slab casting. Hence, this force will be accompanied by a creep strain under the variable sustained stress starting at age 1 day for the slab concrete. A precast concrete beam will also undergo creep strain in addition to elastic strain due to the variable sustained force Q or F applied at age 1 to 3 months say for the precast beam concrete. These creep effects have been referred to as "relaxation" effects.[2,14]

The slab modulus of elasticity also increases (from zero) following initial slab hardening. The differential slab shrinkage, and both the slab and beam moduli of elasticity, are involved directly in computing stresses and deflections in Eqs. (2-1) to (2-10) by the elastic theory.

Evaluation of E and Q or F for Use With Eqs. (2-1) to (2-10)

The following procedure with Eqs. (2-11) to (2-17) approximates the above relaxation effects using the effective modulus method with modified creep coefficients.[1]

General solution

For slab E_c

$$E_1 = (E_c)_t/(1 + \tfrac{2}{5}C_t) \tag{2-11}$$

For precast concrete beam E_c $\quad E_2 = (E_c)_t/(1 + \frac{1}{2}C_t)$ (2-12)

For steel beam E_s $\quad E_2 = E_s$ (2-13)

For Q or F $\quad \begin{Bmatrix} Q_{DS} = \frac{2}{3}\,(Q \text{ in Eq. (2-1)}) \\ F_{DS} = \frac{2}{3}\,(F \text{ in Eq. (2-8)}) \end{Bmatrix}$ (2-14)

Average ultimate values of concrete moduli of elasticity

For slab E_c $\quad E_1 = 0.50(E_c)_{28d}$ (2-15)

For precast concrete beam E_c:

for beam age when slab cast = 1 month $\quad E_2 = 0.55(E_c)_{28d}$ (2-16)

$\geq$ 2 months $\quad E_2 = 0.60(E_c)_{28d}$ (2-17)

where the time-distribution of the sustained force, Q or F, and the age at loading of the slab and beam concretes are taken into account. Values of $C_u = 2.5$, 1.7, and 1.4 were used to obtain Eqs. (2-15) to (2-17), respectively. These were based on the average $C_u = 1.9$ given with Eq. (2-15) for 70 percent relative humidity and loading age 7 days, along with the age when loaded correction factors from Table 1-6. The slab $C_u = 2.5$ was estimated for the case of the force Q or F being applied at age 1 day. By this method, the composite moment of inertia, I_c, would vary slightly from the design I_c, because of the small change in the modular ratio, E_1/E_2.

The following simplified procedure has been found to be satisfactory[15,16] for the somewhat secondary phenomenon of differential shrinkage and creep, and is recommended for routine calculations.

Equations (2-18) to (2-22) recommended for approximate solution for any time after slab concrete age of 28 days, including ultimate values

For slab E_c $\quad E_1 = (E_c)_{28d}$ (2-18)

For precast concrete beam E_c $\quad E_2 = (E_c)_{28d}$ (2-19)

For steel beam E_s $\quad E_2 = E_s$ (2-20)

For Q or F $\quad \left. \begin{matrix} Q_{DS} = (Q \text{ in Eq. (2-1)})/2 \\ F_{DS} = (F \text{ in Eq. (2-8)})/2 \end{matrix} \right\}$ (2-21)

For deflection of concrete–concrete composite beams

$$\Delta_{DS} = \tfrac{4}{3}\,(\Delta \text{ in Eq. (2-3) or (2-10), with (2-18) to (2-21)}) \qquad (2\text{-}22)$$

Values of E_c and E_s are given in Chap. 1.

The modification factor for creep relaxation effects due to differential shrinkage of $\frac{1}{2}$ in Eqs. (2-21) is approximately the same as the relaxation factor of 0.57 used by Roll[2] along with Eqs. (2-1) to (2-3) herein. This factor was obtained using Connolly's[14] equation (2-23) and $C_t = 2.0$.

$$\text{Relaxation factor} = [1 - (1 - e^{-C_t})/C_t] \qquad (2\text{-}23)$$

The factor of $\frac{1}{2}$, along with the evaluation of the coefficient, D, in Table 2-1 which takes into account the additional relatively small effects of creep under gravity loads for different cases, is also consistent with the data of Branson, Meyers, and Kripanarayanan.[15,16] In this work, the differential shrinkage and creep deflection of this magnitude was quite clearly evident for a prestressed composite highway bridge after several months following casting of the deck slab. The factor of $\frac{1}{2}$ is also the same as the previously recommended solution for shrinkage warping[17,18,19] of noncomposite reinforced concrete beams presented in Chap. 3 of this book.

The factor of $\frac{4}{3}$ in Eq. (2-22) takes into account the additional reduced E_2-effect in deflection calculations involving a precast concrete beam.

As illustrated in Tables 2-3 and 2-4 for both steel–concrete and concrete–concrete composite structures, stresses and deflections using Eqs. (2-18) to (2-22) are roughly comparable to the results using Eqs. (2-11) to (2-17).

Evaluation of D for Use With Eqs. (2-1), (2-8), (2-14), (2-21), etc.

The differential shrinkage and creep strain or coefficient, D, at the interface between the composite slab and steel beam or precast concrete beam will be analyzed in two parts. First, the effects due to shrinkage only are determined directly using Eqs. (1-17) and (1-18) from Chap. 1. The combined effects of shrinkage plus creep-under-gravity-loads are then estimated for typical conditions in Table 2-1.

Due to shrinkage only for steel–concrete composite beams

General solution

$$D_t = 1.2 \text{ slab } (\varepsilon_{sh})_t \qquad (2\text{-}24)$$

where slab $(\varepsilon_{sh})_t$ is given by Eq. (1-17) for moist-cured concrete, in which t (also in D_t) is time in days after slab casting less 7 days, as defined in Eq. (1-17); and the factor 1.2, as noted in Chap. 1, is used to convert shrinkage after age 7 days to shrinkage after age 1 day for the slab shrinkage.

Average ultimate value

From Eq. (2-24) $\quad D_u = 1.2(546 \times 10^{-6}) = 655 \times 10^{-6}$ in/in (cm/cm) $\qquad$ (2-25)

Equation (2-25) is computed using the avg. $(\varepsilon_{sh})_u = 546 \times 10^{-6}$ in/in (cm/cm) given with Eq. (1-17) for an average relative humidity of 70 percent.

Due to shrinkage only for concrete–concrete composite beams

General solution

$$D_t = 1.2 \text{ slab } (\varepsilon_{sh})_t - \text{beam } [(\varepsilon_{sh})_t - (\varepsilon_{sh})_{ts}] \qquad (2\text{-}26)$$

In addition to the description of terms with Eq. (2-24) for the slab $(\varepsilon_{sh})_t$ and D_t, the beam $(\varepsilon_{sh})_t$ and $(\varepsilon_{sh})_{ts}$ are given by Eq. (1-17) for moist-cured and Eq. (1-18) for steam-cured concrete, in which

t = age of precast beam concrete (greater than ts) in days, less 7 days for moist-cured and 1–3 days for steam-cured concrete

ts = age of precast beam concrete in days when the slab is cast, less 7 days for moist-cured and 1–3 days for steam-cured concrete

Average ultimate values

From Eq. (2-26)

$$D_u = (1.2 - 1.0 + R_{ts})(\varepsilon_{sh})_u \tag{2-27}$$

where the shrinkage ratio $R_{ts} = ts/(35 + ts)$ for moist-cured concrete from Eq. (1-17) and $R_{ts} = ts/(55 + ts)$ for steam-cured concrete from Eq. (1-18). For 1, 2, and 3-month periods between precast beam and slab casting, ts = 30, 60, 90 (respectively) – (avg. of 1–3 and 7 = 4.5) = 25.5, 55.5, 85.5 days (respectively). Using an average of the R_{ts} values for moist and steam-cured concrete with these values of ts in Eq. (2-27), Eqs. (2-28) to (2-30) are obtained.

For 1-month period between precast beam and slab casting

$$D_u = (1.2 - 1.0 + 0.37)(546 \times 10^{-6}) = 310 \times 10^{-6} \text{ in/in (cm/cm)} \tag{2-28}$$

For 2-month period between precast beam and slab casting

$$D_u = (1.2 - 1.0 + 0.56)(546 \times 10^{-6}) = 415 \times 10^{-6} \text{ in/in (cm/cm)} \tag{2-29}$$

For 3-month period between precast beam and slab casting

$$D_u = (1.2 - 1.0 + 0.66)(546 \times 10^{-6}) = 470 \times 10^{-6} \text{ in/in (cm/cm)} \tag{2-30}$$

Equation (2-29) for the 2-month period is recommended for routine calculations.

For basic concrete shrinkage strains other than 546×10^{-6} in/in (cm/cm) (the average value in Chap. 1 for 70 percent relative humidity and following age 7 days), the results in Eqs. (2-25) and (2-28) to (2-30) are modified by simple proportion. The same is true of all results in Table 2-1.

Due to shrinkage plus creep-under-gravity-loads Only creep effects resulting from sustained gravity loads are considered here. The effect of creep under the differential shrinkage and creep force, Q or F, is taken into account in the previous section.

In the case of unshored construction, the slab is normally unstressed under all dead load, and hence no slab creep strain occurs under gravity load. However, the interface fiber of a precast concrete beam in the unshored case is subjected to stress and hence creep strain following slab casting, due to the slab dead load. This creep strain *tends to reduce the differential shrinkage and creep effect in concrete–concrete composite beams.* In the case of shored construction, since the neutral axis is normally near the slab-to-beam interface, the stresses and creep

Table 2-1 Recommended differential shrinkage and creep strain or coefficient, D, due to shrinkage plus creep-under-gravity-loads—to be used with Eqs. (2-1), (2-8), (2-14), (2-21), etc.*

Beam type	Shored or unshored constr.	Age of precast beam conc. when slab cast	Recommended (average) values of D_t and D_u
Steel–concrete composite beam	Either shored or unshored	—	$D_t = (D_t$ for sh. only—Eq. (2-24)) From Eq. (2-25): $D_u = 655 \times 10^{-6}$ in/in (cm/cm)
Prestressed concrete composite beam	Shored		$D_t = (D_t$ for sh. only—Eq. (2-26)) From Eqs. (2-27) to (2-30):
		1 month	$D_u = 310 \times 10^{-6}$ in/in—Eq. (2-28)
		2 months†	$D_u = 415 \times 10^{-6}$ in/in—Eq. (2-29)
		3 months	$D_u = 470 \times 10^{-6}$ in/in—Eq. (2-30)
	Unshored		$D_t = d_1$ (D_t for sh. only—Eq. (2-26))‡ Using Eqs. (2-27) to (2-30):
		1 month	$D_u = 0.87(310 \times 10^{-6}) = 270 \times 10^{-6}$ in/in
		2 months†	$D_u = 0.90(415 \times 10^{-6}) = 375 \times 10^{-6}$ in/in
		3 months	$D_u = 0.93(470 \times 10^{-6}) = 435 \times 10^{-6}$ in/in
Reinforced concrete composite beam	Shored		$D_t = d_2$ (D_t for sh. only—Eq. (2-26))‡ Using Eqs. (2-27) to (2-30):
		1 month	$D_u = 0.93(310 \times 10^{-6}) = 290 \times 10^{-6}$ in/in
		2 months†	$D_u = 0.95(415 \times 10^{-6}) = 395 \times 10^{-6}$ in/in
		3 months	$D_u = 0.97(470 \times 10^{-6}) = 455 \times 10^{-6}$ in/in
	Unshored		$D_t = d_3$ (D_t for sh. only—Eq. (2-26))‡ Using Eqs. (2-27) to (2-30):
		1 month	$D_u = 0.83(310 \times 10^{-6}) = 260 \times 10^{-6}$ in/in
		2 months†	$D_u = 0.85(415 \times 10^{-6}) = 355 \times 10^{-6}$ in/in
		3 months	$D_u = 0.87(470 \times 10^{-6}) = 410 \times 10^{-6}$ in/in

* The numerical values in the table are based on Eqs. (2-25) and (2-27) to (2-30) for the standard conditions—principally $(\varepsilon_{sh})_u = 546 \times 10^{-6}$ after age 7 days (times 1.2 after 1 day for D.S.), $H = 70$ percent, $h = 6$ in. (15.2 cm). The values are modified by simple proportion for other conditions (see Chaps. 3, 5, examples).

† The 2-month period between precast beam and slab casting is recommended for routine calculations. The recommended values are all underlined.

‡ Estimated effect of creep under sustained gravity loads (primarily dead load) for typical conditions:

Age of precast beam conc. when slab cast	d_1	d_2	d_3
1 month	0.87	0.93	0.83
2 months†	0.90	0.95	0.85
3 months	0.93	0.97	0.87

strains are small in both the slab and precast concrete beam, and can be neglected. Additionally, in the case of reinforced concrete composite beams, the stresses at the interface of the precast concrete beam under its own dead load cause some creep strain for the period following slab casting which *further tends to reduce the differential shrinkage and creep effect in reinforced composite beams.* This latter effect due to precast beam dead load tends to be offset by the effect of prestress in the case of prestressed composite beams.

These various effects are estimated for typical conditions in determining recommended values of the differential shrinkage and creep strain or coefficient, D_t or D_u, presented in Table 2-1. Although the values in the table are only estimated average values, they also illustrate the relative effects for the different cases. The longitudinal slab steel would have the effect of reducing D_t or D_u, depending on the steel percentage, but this effect was not included in the estimated values in Table 2-1 because the principal steel in such slabs is usually in the transverse, not longitudinal, direction.

2-4 NUMERICAL EXAMPLES AND COMPARISON OF DIFFERENTIAL SHRINKAGE AND CREEP STRESSES AND DEFLECTIONS

Comparison of Stresses and Midspan Deflection by the Composite Section and Separate Section Methods

As shown in Table 2-2 for four different cross-sections, computed stresses and deflections by the composite section and separate section methods are identical. It is also shown in Table 2-2 that the slope of the strain distribution diagram or curvature is the same in the slab and beam (plane sections remain plane in the composite beam due to differential shrinkage and creep).

Relative Effects of Larger Slabs, Deeper Beams, etc.

The following observations are made with regard to the computed stresses and deflections for the four rather extreme examples (cross-secs. 1 to 4) in Table 2-2. These comparisons show that it is difficult to form general conclusions as to the relative effects of larger slabs, deeper beams or stems, etc. in determining differential shrinkage and creep effects. For example, note the following:

1. In comparing cross-secs. 1 and 2 having the same size stem, the greater precast beam stresses occur in cross-sec. 1 which has the larger slab. The opposite is true for cross-secs. 3 and 4.

2. In comparing cross-secs. 1 and 3 having the same size slab, the deeper precast beam in cross-sec. 1 results in higher precast beam stresses. The opposite is true for cross-secs. 2 and 4.

3. In comparing cross-secs. 1 and 2 having the same size stem, the larger slab in cross-sec. 1 causes a greater deflection. The opposite is true for cross-secs. 3 and 4.

Table 2-2 Calculated differential shrinkage and creep stresses and deflections for four different cross-sections by the composite section and separate section methods

Section properties and material coefficients

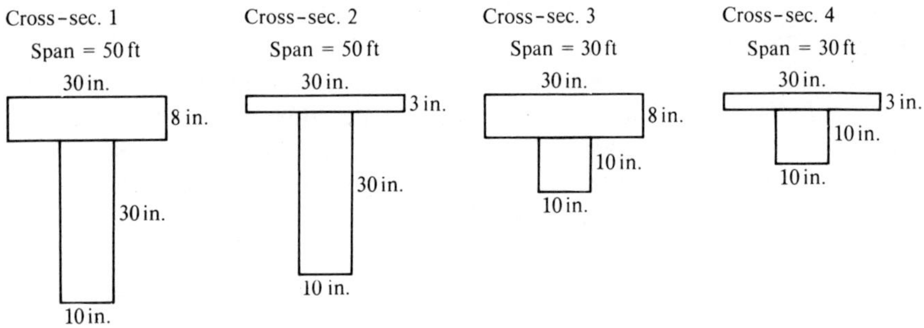

Use for nor. wt. pres. comp. beam (slab $f_c' = 3000$ psi, beam $f_c' = 5000$ psi):

$$E_1 = 0.5 \times 3.3 = 1.7 \times 10^6 \text{ psi} \qquad E_2 = 0.6 \times 3.9 = 2.3 \times 10^6 \text{ psi}$$

$$D = \tfrac{2}{3} \times 375 = 250 \times 10^{-6} \text{ in/in}$$

Sec.	d_1 in.	d_2 in.	A_1 in^2	A_2 in^2	I_1 in^4	I_2 in^4	y_{1T} in.	y_{1B} in.	y_{2T} in.	y_{2B} in.	A_c in^2	I_c in^4	y_{cT} in.	y_{cS} in.	y_c' in.	y_{cB} in.
1	8	30	240	300	1280	22 500	4	4	15	15	477	63 689	15.9	11.9	7.9	22.1
2	3	30	90	300	67.5	22 500	1.5	1.5	15	15	367	37 374	15.0	13.5	12.0	18.0
3	8	10	240	100	1280	833.3	4	4	5	5	277	6959	7.2	3.2	−0.8	10.8
4	3	10	90	100	67.5	833.3	1.5	1.5	5	5	167	2571	5.4	3.9	2.4	7.6

Diff. sh. force, stresses, curvature, and defl. By Eqs. (2-1) to (2-10)

			Stresses, curvature and deflection computed by both composite section method and separate section method						
	D.S. force Q	D.S. force F	* σ_{1T}	* σ_{1B}	* σ_{2T}	* σ_{2B}	† $\frac{\sigma_{1B}-\sigma_{1T}}{d_1}\frac{1}{E_1}$	† $\frac{\sigma_{2B}-\sigma_{2T}}{d_2}\frac{1}{E_2}$	Midspan Δ_{DS}
Sec.	lb	lb	psi	psi	psi	psi	(1/in.) × 10^{-6}	(1/in.) × 10^{-6}	in.
1	102 000	23 597	+42	+155	−366	+208	8.3	8.3	0.37
2	38 250	18 890	+195	+225	−270	+144	6.0	6.9	0.27
3	102 000	9402	−101	+180	−332	+144	20.7	20.7	0.34
4	38 250	7891	+23	+152	−369	+212	25.3	25.3	0.41

(For metric units, see metric conversion factors herein.)

† Curvature or slope of strain distribution diagrams are shown to be the same in the slab and precast beam (plane sections remain plane in the composite beam due to differential shrinkage and creep).

* Positive signs denote tension, negative signs compression.

Table 2-3 Stresses and midspan deflection due to differential shrinkage and creep for a steel–concrete composite building floor system

Design data:[2] shored beam, total dead load = 510 #/ft (60 psf), live load = 578 #/ft (68 psf), 21 wide flange 44, bottom cover plate 5 in. × $\frac{1}{2}$ in., slab thickness = 4 in., simple span = 48 ft, beam spacing = 8 ft 6 in., specified f'_c = 3000 psi, beam steel f_s = 24 000 psi, $E_c = 3.33 \times 10^6$ psi, $E_s = 29 \times 10^6$ psi

Section properties and details (see Fig. 2.1) using Eqs. (2-13) and (2-15)	* Using Eqs. (2-18) and (2-20); same as Roll's[2] solution
$E_1 = 1.67 \times 10^6$ psi	$= 3.33 \times 10^6$ psi
$E_2 = 29 \times 10^6$ psi	$= 29 \times 10^6$ psi
$A_1 = 288$ in^2	$= 288$ in^2
$A_2 = 15.5$ in^2	$= 15.5$ in^2
$A_c = 32.06$ in^2	$= 48.6$ in^2
$d_1 = 4$ in.	$= 4$ in.
$d_2 = 21.16$ in.	$= 21.16$ in.
$y_{2B} = 9.11$ in.	$= 9.11$ in.
$y_{cB} = 16.37$ in.	$= 18.68$ in.
$y_{cI} = 4.79$ in.	$= 2.48$ in.
$y_{cS} = 6.79$ in.	$= 4.48$ in.
$y_{cT} = 8.79$ in.	$= 6.48$ in.
$I_1 = 384$ in^4	$= 384$ in^4
$I_2 = 1077$ in^4	$= 1077$ in^4
$I_c = 2680$ in^4	$= 3200$ in^4

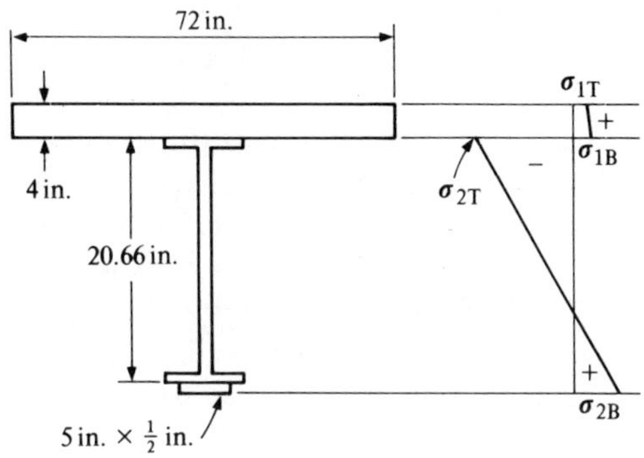

Results using Eqs. (2-1) to (2-3)

Method of estimating time-dependent effects	D in/in	Q_{DS} lb	σ_{1T} psi	σ_{1B} psi	σ_{2T} psi	σ_{2B} psi	Δ_{DS} in.
Using Eqs. (2-13) to (2-15) for life of structure	‡655×10^{-6}	209 300	+83	+320	−7068	+2153	0.64
Using Eqs. (2-18), (2-20) and (2-21) for life of structure*	‡655×10^{-6}	314 000	+20	+223	−7550	+1752	0.63
By Roll's[2] solution for period 14 days to 16 months†	502×10^{-6}	0.57(480) = 274 000	+17	+189	−6580	+1560	0.54

(For metric units, see metric conversion factors herein.)

* Method recommended in this book for approximate solution for any time after slab concrete age of 28 days, including ultimate values.

† Results by last two methods are in same proportion as values of Q. The procedure used in the two methods is identical, except for determining D and Q. The relaxation factor, 0.57, was used in Roll's solution.

‡ Value from Table 2-1.

Comparison of Stresses and Midspan Deflection for a Steel–Concrete Composite Building Floor System and a Prestressed Concrete Composite Bridge Using Eqs. (2-13) to (2-17), Eqs. (2-18) to (2-22), and Roll's Method

Numerical examples of differential shrinkage and creep stresses and deflections using Eqs. (2-1) to (2-3) are shown in Tables 2-3 and 2-4. A rational method[1] for estimating the time-dependent effects (Eqs. (2-13) to (2-17)), an approximate method[1,15,16] using conventional moduli of elasticity and hence conventional section properties along with two relaxation effects (Eqs. (2-18) to (2-22)—*recommended*), and Roll's[2] method are shown to yield generally similar results. One normally unimportant exception to this was the slab top fiber stress for the example in Table 2-3. The recommended method is the same as the one used by Roll, except for the choice of the differential shrinkage and creep coefficient and the estimate of the relaxation coefficient, which are slightly different in the two cases. Suggested values of appropriate differential shrinkage and creep coefficients are given in Table 2-1 for various types of structures.

The results in Tables 2-3 and 2-4 are also similar to, but slightly lower than, those of Birkeland[6] and Zuk.[21] Results in Table 2-4 are shown to be almost exactly the same when the transformed prestressing steel is included in the calculations, as compared to those using the gross section properties.

2-5 DIFFERENTIAL TEMPERATURE

Composite beam stresses and deflections due to differential temperature between the slab and precast concrete beam or steel beam can be estimated using Eqs. (2-1) to (2-10), in which the differential shrinkage and creep strain or coefficient, D, is replaced by D_T in Eq. (2-31).

$$D_T = \alpha \Delta T \tag{2-31}$$

where α is the thermal coefficient of the slab concrete (as 5.5 to 6.0×10^{-6} in. per in. per °F or 9.9 to 10.8 cm per cm per °C, in Chap. 1), and ΔT is the change in temperature in degrees Farenheit.

In this grossly simplified approach, it is assumed that the slab undergoes a uniform temperature change relative to the beam. Hence allowances must be made for nonuniform thermal distribution in estimating the coefficient D_T. Such an approximate solution may be justified in most cases because of the secondary and temporary nature of the problem. More elaborate procedures[21,22] are available for considering nonuniform thermal distribution.

The critical case for differential temperature stresses and deflections will normally be that which causes an additive effect with gravity loads and also differential shrinkage and creep. An example of interest in positive moment regions (including simply supported beams) is that of a sudden cold rain or snow causing a temperature drop of the slab relative to the beam. In negative moment regions the sudden drop in slab temperature has an additive effect with differential shrinkage and creep, but an effect opposite to that of gravity loads and also

Table 2-4 Stresses and midspan deflection due to differential shrinkage and creep for a prestressed concrete composite bridge girder*

Design data:[20] HS 20-44 AASHTO loading, 25 percent impact, unshored Type IV AASHTO girder, slab thickness = 7 in., simple span = 75 ft, girder spacing = 7 ft 8 in., $f'_{ci} = 4000$ psi, $f'_c = 5000$ psi, slab $f'_c = 4000$ psi, girder $E_c = 4.5 \times 10^6$ psi by test, slab $E_c = 3.5 \times 10^6$ psi by test

Section properties and details (see Fig. 2.1) using Eqs. (2-15) and (2-17)	Using Eqs. (2-18) and (2-19) †
$E_1 = 0.50\,(3.5 \times 10^6) = 1.75 \times 10^6$ psi	$= 3.5 \times 10^6$ psi
$E_2 = 0.60\,(4.5 \times 10^6) = 2.70 \times 10^6$ psi	$= 4.5 \times 10^6$ psi
$E_s = (27 \times 10^6)$ psi	$= (27 \times 10^6)$ psi
$A_1 = 644$ in^2 (644)	$= 644$ in^2 (644)
$A_2 = 789$ in^2 (829)	$= 789$ in^2 (811)
$A_c = 1206$ in^2 (1246)	$= 1290$ in^2 (1312)
$y_{2B} = 24.7$ in. (23.7)	$= 24.7$ in. (24.1)
$y_{cB} = 36.1$ in. (35.0)	$= 37.5$ in. (36.8)
$y_{cI} = 17.9$ in. (19.0)	$= 16.5$ in. (17.2)
$y_{cS} = 21.4$ in. (22.5)	$= 20.0$ in. (20.7)
$y_{cT} = 24.9$ in. (26.0)	$= 23.5$ in. (24.2)
$I_1 = 2640$ in^4 (2640)	$= 2640$ in^4 (2640)
$I_2 = 260\,600$ in^4 (277 250)	$= 260\,600$ in^4 (270 070)
$I_c = 554\,200$ in^4 (596 130)	$= 591\,600$ in^4 (618 390)

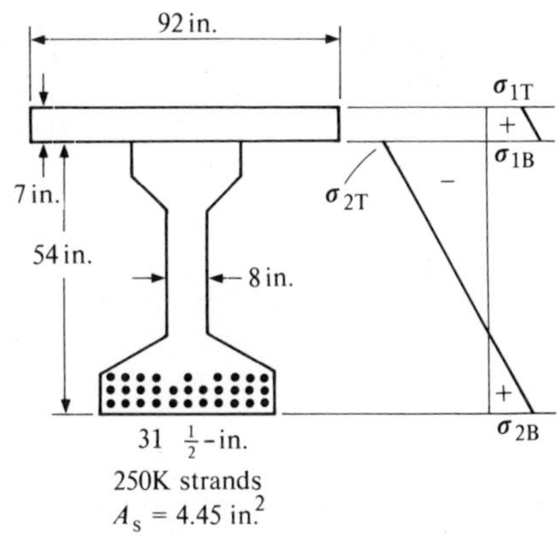

Results using Eqs. (2-1) to (2-3)*

Method of estimating time-dependent effects	D in/in	Q_{DS} lb	σ_{1T} psi	σ_{1B} psi	σ_{2T} psi	σ_{2B} psi	Δ_{DS} in.
Using Eqs. (2-14), (2-15) and (2-17)	‡375×10^{-6} (same)	281 750 (same)	+111 (112)	+160 (160)	−428 (428)	+159 (147)	0.41 (0.40)
Using Eqs. (2-18), (2-19), (2-21) and (2-22)†	‡375×10^{-6} (same)	422 600 (same)	+140 (139)	+218 (216)	−563 (565)	+208 (199)	0.42 (0.42)

(For metric units, see metric conversion factors herein.)

* Values in parentheses are the results when the transformed prestressing steel $(E_s/E_2 - 1)A_s$ is included in the calculation of section properties. Other results are based on the gross section properties.

† Method recommended in this book for approximate solution for any time after slab concrete age of 28 days, including ultimate values.

‡ Value from Table 2-1.

Table 2-5 Stresses and midspan deflection due to differential temperature for concrete–concrete and steel–concrete composite members

Design data for a prestressed concrete composite bridge girder:[20] same as Table 2-4

* Section properties and details (see Fig. 2-1)	* Results using $D_T = 200 \times 10^{-6}$ in/in and Eqs. (2-1) to (2-3)
$E_1 = 3.5 \times 10^6$ psi $E_2 = 4.5 \times 10^6$ psi Section properties, same as second column in Table 2-4, to $I_c = 591\,600$ in^4 (618 390)	$Q = 450\,800$ # (same), midspan $\Delta = 0.34$ in. (0.34)

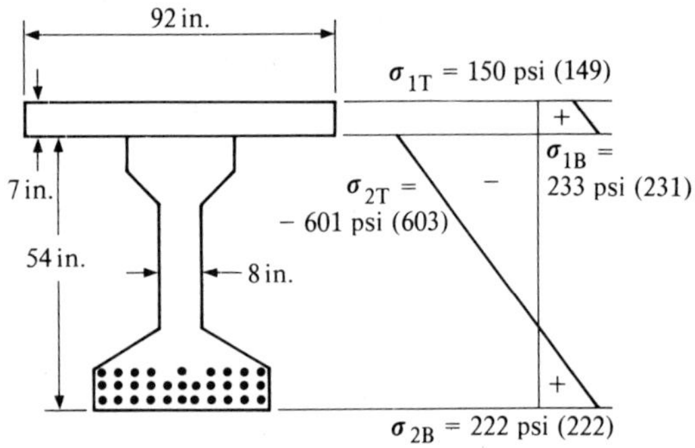

Design data:[4] unshored girder, HS 20-44 AASHTO loading, 28 percent impact, 36 wide flange 150, slab thickness = 6 in., simple span = 55 ft, girder spacing = 6 ft 9 in., $f'_c = 3000$ psi, $f_s = 18\,000$ psi, $E_c = 3.2 \times 10^6$ psi, $E_s = 29 \times 10^6$ psi

Section properties and details (see Fig. 2-1)	Results using $D_T = 200 \times 10^{-6}$ in/in and Eqs. (2-1) to (2-3)
$E_1 = 3.2 \times 10^6$ psi $E_2 = 29 \times 10^6$ psi $A_1 = 432$ in^2 $A_2 = 44.2$ in^2 $A_c = 91.9$ in^2 $y_{2B} = 17.9$ in. $y_{cB} = 28.8$ in. $y_{cI} = 7.0$ in. $y_{cS} = 10.0$ in. $y_{cT} = 13.0$ in. $I_1 = 1296$ in^4 $I_2 = 9030$ in^4 $I_c = 19\,190$ in^4	$Q = 277\,000$ #, midspan $\Delta = 0.27$ in.

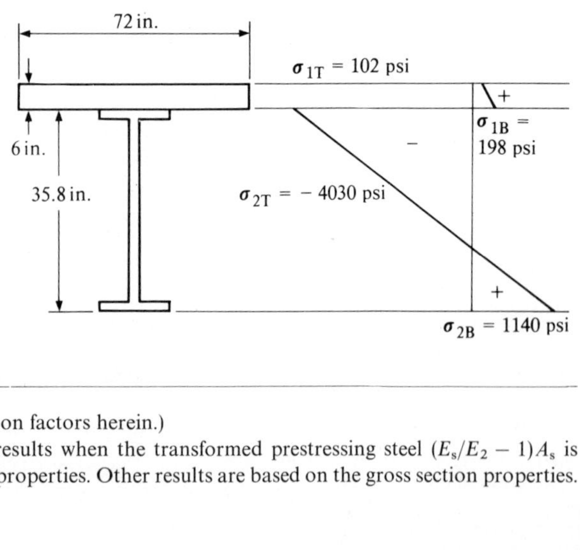

(For metric units, see metric conversion factors herein.)

* Values in parentheses are the results when the transformed prestressing steel $(E_s/E_2 - 1)A_s$ is included in the calculation of section properties. Other results are based on the gross section properties.

warping due to eccentric reinforcement of reinforced concrete members. The reverse is true in all cases here for a sudden hot sun on the slab. See Sec. 2-6 for further discussion of the effect of continuity.

Since a steel beam adjusts to temperature change more rapidly than a concrete beam, as a result of the higher thermal conductivity of the steel and greater bulk of the concrete, the sudden cold rain or snow (or hot sun) would tend to have a greater effect in the concrete–concrete than the steel–concrete composite beam.

Using a maximum temperature drop of 45°F, as provided by AASHTO (see Chap. 1), and assuming this full temperature drop for a slab with no drop in temperature of the precast beam or steel beam, an upper limit value from Eq. (2-31) might be considered to be: $D_T = 6.0 \times 10^{-6} \times 45 = 270 \times 10^6$ in/in. On the basis of a study of other methods and results,[6,21] the author believes that a value of $D_T = 150 \times 10^{-6}$ in/in (for $\Delta T = 25°\text{F}$) to 250×10^{-6} in/in (for $\Delta T = 42°\text{F}$) should provide for a reasonable estimate of temperature stresses and deflections using Eqs. (2-1) to (2-10) in most cases. A value of $D_T = 200 \times 10^{-6}$ in/in (cm/cm) is *recommended* for normal application for both concrete–concrete and steel–concrete composite members.

The procedure and discussion in this section applies equally to the monolithic concrete T-beam, etc.

Numerical examples of differential temperature stresses and deflections are shown in Table 2-5 for a prestressed concrete composite bridge and a steel–concrete composite bridge, both simply supported. The results in the table are similar to, but slightly lower than, those computed by the more elaborate method of Zuk.[21] Results in Table 2-5 are also shown to be essentially the same when the transformed prestressing steel is included in the calculations, as compared to those using the gross section properties.

2-6 EFFECT OF CONTINUITY

Stresses due to differential shrinkage and creep (or temperature) are computed at any section of cantilever, simple, and continuous composite beams using Eqs. (2-2) or (2-9). The Q or F force is computed the same way in all of these cases involving a cast-in-place slab on a precast or steel beam. Also, the moment in these equations, given by Eq. (2-32), is the same in the statically determinate cantilever and simple beam cases (all signs the same as well).

$$M_{\text{DS}} = Qy_{\text{cs}} \qquad \text{or} \qquad F(y_{1\text{B}} + y_{2\text{T}}) \tag{2-32}$$

It should be noted that the two forms of Eq. (2-32) are, in general, not equal, although the two curvature or M/EI expressions in Eq. (2-34) are equal.

In the case of continuous beams this moment must be modified according to the statically indeterminate solution to include "secondary" moments in addition to the "primary" moment computed in Eq. (2-32). These moments and the resulting "total" moments are derived in Figs. 2-4 to 2-6 for 2-span and 3-span continuous beams, with the moment diagrams for cantilever, simple, and 2-span

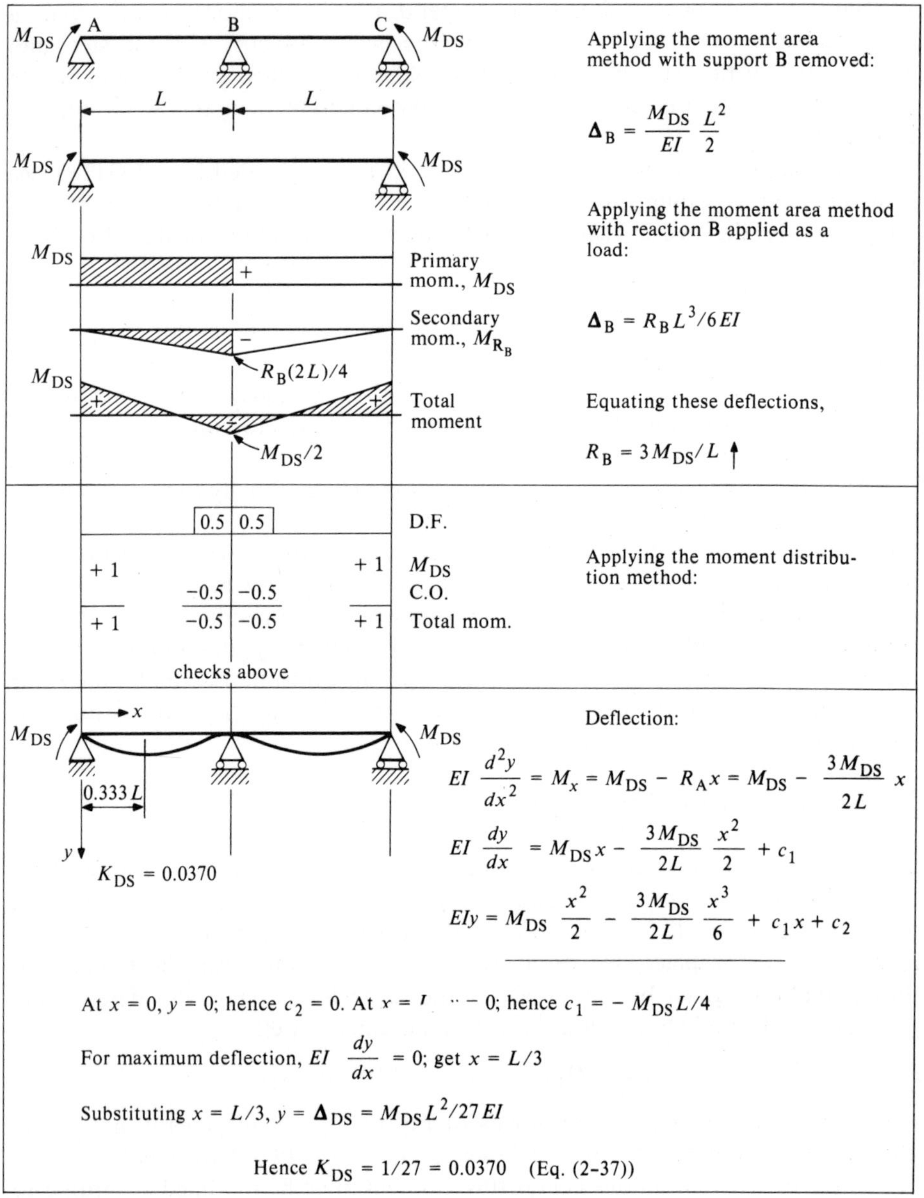

FIGURE 2-4
Moment diagram, deflection equation, and deflection coefficient due to differential shrinkage and creep derived for a 2-span continuous beam.

to 5-span continuous beams summarized in Fig. 2-7. Maximum positive and negative moments are also tabulated in Fig. 2-7.

The maximum deflection due to differential shrinkage and creep (or temperature) is given for cantilever, simple, and continuous beams in the form of Eq. (2-33).

$$\Delta_{DS} = K_{DS}\phi_{DS}L^2 \tag{2-33}$$

where K_{DS} is a deflection coefficient derived in Figs. 2-4 to 2-6 for 2-span and 3-span continuous beams and summarized in Fig. 2-7 for cantilever, simple, and 2-span to 5-span continuous beams by Eqs. (2-35) to (2-40); the curvature, ϕ_{DS}, is given by Eq. (2-34); and L is the span length.

$$\phi_{DS} = M_{DS}/EI = Qy_{cS}/E_2I_c = F(y_{1B} + y_{2T})/(E_1I_1 + E_2I_2) \tag{2-34}$$

The particular form to be used in Eqs. (2-32) and (2-34) depends on the method of analysis selected. Results by the composite section method (first form) and separate section method (second form) are, of course, the same.

Derivation of Moments and Deflections in Continuous Composite Beams Due to Differential Shrinkage and Creep

The approach that is used in Figs. 2-4 to 2-6 to illustrate the calculation of bending moments and deflections in continuous composite beams due to differential shrinkage and creep (or temperature) is the same as is commonly used for computing secondary moments in continuous prestressed beams. With the interior support(s) removed, the downward deflection due to the "primary" moment, M_{DS} (rather than upward camber as is normally the case in continuous prestressed beams), is first determined. The reactions necessary to restore the beam produce the "secondary" moments. The "total" moments are the sum of the primary and secondary moments. Alternately, the moment distribution method is illustrated in Figs. 2-4 and 2-5 for computing the total moments in 2-span and 3-span continuous beams due to differential shrinkage and creep.

After the moments were determined, the deflections were computed in Figs. 2-4 to 2-6 using the double integration, moment area and conjugate beam methods.

The moment diagrams, maximum positive and negative moment, and deflection coefficients due to differential shrinkage and creep are summarized in Fig. 2-7 for cantilever, simple, and 2-span to 5-span continuous beams.

An alternative approach for determining the secondary effects of differential shrinkage and creep when continuity is obtained in composite members by providing reinforcement over the supports (same as above) is that of computing the "hogging" restraint moments at the interior supports.[25,28] For this moment, the British Code[28] uses $M = 0.43DE_1A_1y_{cS}$, which is the same as the moment in Eqs. (2-3) and (2-1) with the factor of 0.43 being used as a reduction coefficient to allow for creep.

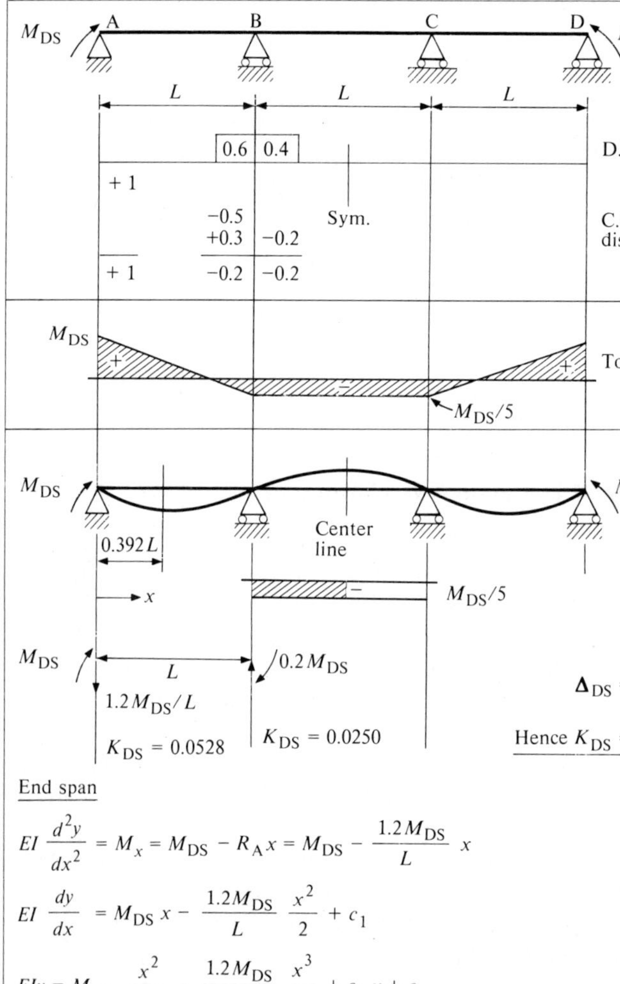

Applying the moment distribution method:

Deflection:

Center span

Applying the moment area method;

$$\Delta_{DS} = \frac{M_{DS}}{5EI}\,\frac{L}{2}\,\frac{L}{4} = M_{DS}L^2/40EI$$

Hence $K_{DS} = 1/40 = 0.0250$

End span

$$EI\,\frac{d^2y}{dx^2} = M_x = M_{DS} - R_A x = M_{DS} - \frac{1.2M_{DS}}{L}\,x$$

$$EI\,\frac{dy}{dx} = M_{DS}\,x - \frac{1.2M_{DS}}{L}\,\frac{x^2}{2} + c_1$$

$$EIy = M_{DS}\,\frac{x^2}{2} - \frac{1.2M_{DS}}{L}\,\frac{x^3}{6} + c_1 x + c_2$$

At $x = 0$, $y = 0$; $c_1 = 0$

At $x = L$, $y = 0$;

$c_2 = -3M_{DS}L/10$

For maximum deflection, $EI\,\frac{dy}{dx} = 0$; get $x = 0.392L$

Substituting $x = 0.392L$, $y = \Delta_{DS} = 0.0528\,M_{DS}L^2/EI$

Hence $K_{DS} = 0.0528$ (Eq. (2-38))

FIGURE 2-5
Moment diagram, deflection equation, and deflection coefficient due to differential shrinkage and creep derived for a 3-span continuous beam.

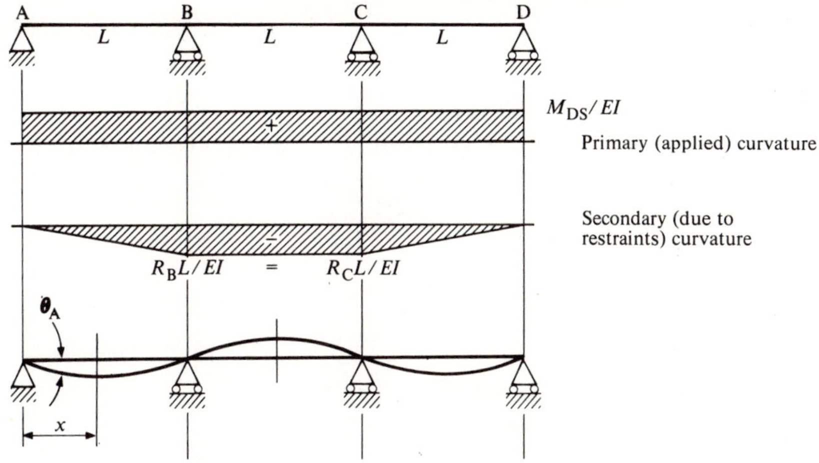

Applying the conjugate beam method with supports B and C removed:

Elastic wt $R_A = (M_{DS}/EI)\,(3L/2) = 3M_{DS}L/2EI$

$\Delta_B = (3M_{DS}L/2EI)L - (M_{DS}/EI)(L)(L/2) = M_{DS}L^2/EI$

Applying the conjugate beam method with R_B and R_C as loads:

Elastic wt $R_A = (R_B L/EI)(L/2 + L/2) = R_B L^2/EI$

$\Delta_B = R_B L^3/EI = (R_B L/EI)(L/2)(L/3) = 5R_B L^3/6EI$

From $\Delta_B = \Delta_B$, $R_B = 6M_{DS}/5L$

End span

$\theta_A = \Sigma$ Elastic wt R_As $= 3M_{DS}L/2EI - R_B L^2/EI$

Also, applying the moment area method, $\theta_A = (M_{DS}/EI)x - (R_B L/EI)(x/L)(x/2)$

From $\theta_A = \theta_A$, $x = 0.392L$

$\Delta_{max} = \Delta_x = (M_{DS}/EI)(x^2/2) - (R_B L/EI)(x/L)(x/2)(2x/3)$

$= (M_{DS}/EI)[x^2/2 - (6/5L)(x^3/3)]$

$= (M_{DS}/EI)[(0.393L)^2/2 - (6/15)(0.393L)^3/L]$

$= 0.0528 M_{DS}L^2/EI$, or $K_{DS} = 0.0528$

Center span

Applying the moment area method:

$\Delta_{℄} = (M_{DS}L/2EI)(L/4) - (R_R L/EI)(L/2)(L/4) = (M_{DS}/EI)[(L^2/8) - (6/5L)(L^3/8)]$

$= (M_{DS}L^2/EI)[(1/8) - (6/40)] = -\,M_{DS}L^2/40EI$, or $K_{DS} = -\,1/40$

This solution checks the solution in Fig. 2-5 for both the end and center span, and confirms that the deflection due to differential shrinkage and creep is negative or upward in certain continuous spans (also see Fig. 2-7)

FIGURE 2-6
Moment diagram, deflection equation, and deflection coefficient due to differential shrinkage and creep for a 3-span continuous beam to check the results in Fig. 2-5.

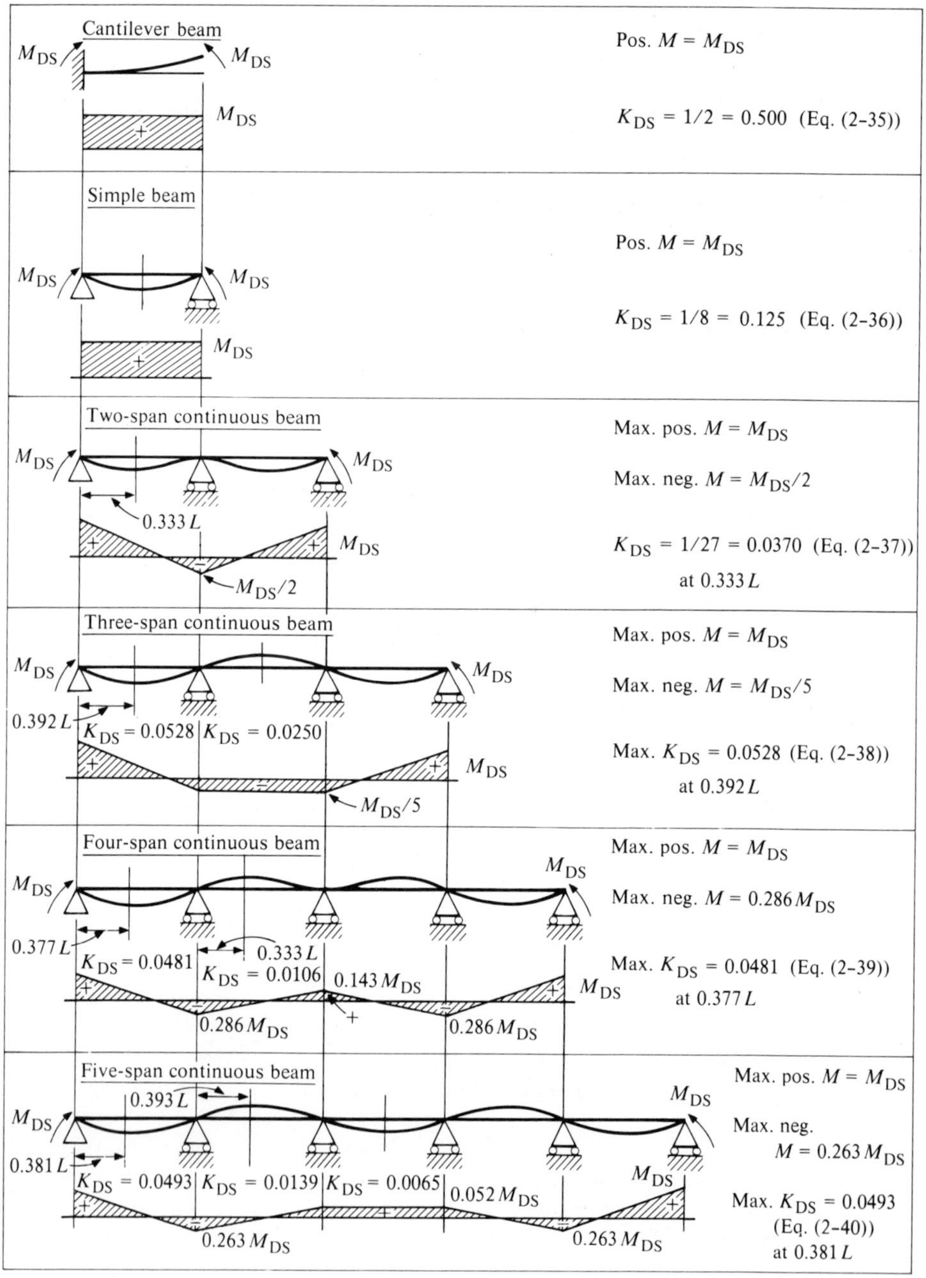

FIGURE 2-7
Summary of moment diagrams and deflection coefficients due to differential shrinkage and creep for cantilever, simple, and 2-span to 5-span (equal spans) continuous beams.

2-7 DIFFERENTIAL SHRINKAGE, CREEP, AND TEMPERATURE EFFECTS IN VERTICAL ELEMENTS OF HIGH-RISE BUILDINGS

Most applications of differential shrinkage, creep, and temperature effects in the past have referred to horizontal structures such as bridges. These effects have usually been neglected in the case of vertical elements in buildings. However, as suggested by Khan and Fintel,[23] such calculations should be considered for high-rise buildings exceeding about 30 stories. In this case the effect of cumulative differential length changes in the vertical elements becomes important.

Nominal calculations of differential length changes in vertical members involve no new procedures other than those for computing axial length changes due to elastic shortening, creep, and shrinkage. The information in Chap. 1 provides the necessary material parameters. For a more detailed discussion of this subject, the reader may consult any number of papers in the literature, including the following papers and their references from ACI Publication SP-27, "Designing for Effects of Creep, Shrinkage, and Temperature in Concrete Structures," 1971: Effects of Column Creep and Shrinkage in Tall Structures: Analysis for Differential Shortening of Columns and Field Observation of Structures, by Mark Fintel and Fazlur R. Khan; Time Dependent Deformations in a 70 Story Structure, by D. W. Pfeifer, D. Morgan, H. G. Russell, and W. G. Corley; and Conceptual Details for Creep, Shrinkage, and Temperature in Ultra High-Rise Buildings, by Fazlur R. Khan and Mark Fintel.

2-8 DISCUSSION OF DIFFERENTIAL SHRINKAGE AND CREEP EXPERIMENTAL RESULTS

Experimental data dealing with the effects of differential shrinkage and creep in composite beams have not been widely reported, although several studies have been published on the subject. Evans and Parker[10] calculated differential shrinkage stresses theoretically (using essentially the separate section method, with estimated differential shrinkage coefficients) and used these results to compare calculated and observed cracking loads for several types of prestressed concrete composite beams. They concluded that differential shrinkage may affect cracking loads as much as 20 percent in normal design.

Birkeland[6] observed time-dependent deflections in two composite prestressed concrete beams. For the time following slab casting he compared the midspan deflection of one of the test beams with the calculated differential shrinkage deflection using the composite section method. As stated in his paper, he combined the effects of shrinkage and creep into the differential shrinkage solution. This required a very large assumed differential shrinkage coefficient of 450×10^{-6} in/in (cm/cm) for 5 days between the casting dates of the two concretes. The portion of the precast beam creep resulting from the prestress force and both stem and slab dead load, for the time following slab casting, were thus combined into the differential shrinkage solution. See Sec. 2-3 herein on evaluation of material properties and coefficients, and Table 2-1 for recommended coefficients.

Branson and Ozell[12,13,24] reported experimental studies of differential shrinkage stresses and deflections using a number of laboratory prestressed composite beams. Differential shrinkage strains plus the creep effects due to differential shrinkage were determined by measuring the direct time-dependent strains on the test composite beams and eliminating other strains through the use of companion noncomposite beams and shrinkage specimens. These results were then used to define the differential shrinkage coefficients for different relative humidities and time periods between the casting of the stem and slab concretes. The coefficients thus determined were used to compare calculated and measured deflections, with generally satisfactory results using the separate section method.

Mattock[25] investigated experimentally the influence of creep and differential shrinkage on the continuity behavior of two half-scale prestressed concrete bridge girders. His data shows a leveling off of the differential shrinkage strain versus time curve at about 245×10^{-6} in/in (cm/cm). The time period between the casting dates for the two concretes was 35 days (taken from graph) and the average relative humidity following the curing period was 50 percent. This value was consistent with the data of Ref. 13 for the same relative humidity and relative casting times, and also consistent with the recommended values herein. Mattock pointed out that creep of the continuous composite beam induces positive restraint moments that can affect cracking, but that shrinkage and creep will not affect the ultimate load capacity of the member.

Roll[2] recently reported an investigation of the stresses and deflections of two typical types of beams used in a small office building designed as a composite steel–concrete structure. Using the composite section method he found excellent correlation between the measured and theoretical deflections after a period of approximately 16 months during which time the structure was subjected to only its own dead load. He indicated that the analyses show the importance of differential shrinkage and creep on the behavior of composite steel–concrete construction. He further concluded that failure to take into account the effects of differential shrinkage and creep can lead to excessive deflections, cracking of the concrete slab, and overstressing of the steel or concrete.

In a study of the time-dependent camber-deflection of a prestressed composite highway bridge, Branson, Meyers, and Kripanarayanan[15,16] found in their analysis of experimental results that the differential shrinkage and creep deflection was quite clearly evident after several months following the casting of the deck slab. They used the composite section method of analysis.

It should be noted that most of the studies described briefly in this section refer to the indirect experimental verification of differential shrinkage and creep effects by measuring strains, deflections, and cracking loads which include all time-dependent effects, not due just to differential shrinkage and creep. Apparently only Branson and Ozell[13] attempted to isolate the differential shrinkage and creep effect from the general time-dependent effects through the use of companion noncomposite beams and shrinkage specimens. But even this procedure results in an indirect evaluation of quite small strains and deflections due only to *differential* shrinkage and creep caused by the differential shrinkage.

In summary, most investigators agree that the effects of differential shrinkage and creep in composite beams can be significant in critical situations involving cracking of the concrete and excessive deflections. Most of the present-day codes and specifications also refer to this effect in certain cases. However, in connection with the determination of cracking loads the author would mention that the extensibility of concrete may practically eliminate any effect that differential shrinkage and creep has on cracking loads in many cases. For example, according to Troxell and Davis,[26] cracking loads up to 2.5 times the instantaneous cracking load have been observed for loads applied over a time span of 2 to 3 months. The increase in concrete strength with time (above the 28-day strength used in design) is also a factor which tends to permit differential shrinkage stresses, but not necessarily deflections, to be neglected in most cases, in the author's opinion. A brief discussion of the extensibility of concrete is presented in Chap. 1.

REFERENCES

1. D. E. Branson, Time-Dependent Effects in Composite Concrete Beams, *ACI Journal,* Proceedings **61,** 2, 213–230, Feb. 1964.
2. Frederic Roll, Effects of Differential Shrinkage and Creep on a Composite Steel–Concrete Structure, in "Designing for Effects of Creep Shrinkage and Temperature in Concrete Structures," 187–214, ACI Publication SP 27-8 (symposium volume), Fall 1971.
3. E. Morsch, "Der Spannbeton Trager," Wittwer, Stuttgart, 1943.
4. I. M. Viest, R. S. Fountain, and R. C. Singleton, "Composite Construction in Steel and Concrete," 1–176, McGraw-Hill Book Co., New York, 1958.
5. R. J. Bartlett, Shrinkage Stresses in Composite Floors, *Concrete and Construction Engineering (London),* **53,** 12, 415–419, Dec. 1958.
6. H. W. Birkeland, Differential Shrinkage in Composite Beams, *ACI Journal,* Proceedings **56,** 11, 1123–1136, May 1960.
7. G. Somerville, An Investigation of the Behavior of Composite Concrete Beams: II. Composite T Beams, *Cement and Concrete Association, London, Technical Report* TRA/371, 1963.
8. H. Frohlich, Theorie der Stahlverbund-Tragwerke, *Der Bauingenieur (Berlin),* **25,** 3, 80–87, Mar. 1950.
9. Helmut Walter, Der Einfluss des Schwindens und Krienchens bei Verbundtragern, *Beton und Stahlbetonbau (Berlin),* **47,** 5, 110–114, May 1952; **47,** 6, 132–137, June 1952.
10. R. H. Evans and A. S. Parker, Behavior of Prestressed Concrete Composite Beams, *ACI Journal,* Proceedings **51,** 9, 861–880, May 1955.
11. A. M. Ozell and J. F. Diniz, Composite Prestressed Concrete Beams Under Repetitive Loading, *PCI Journal,* **2,** 4, 19–27, Mar. 1958.
12. D. E. Branson and A. M. Ozell, A Report on Differential Shrinkage in Composite Prestressed Concrete Beams, *PCI Journal,* **4,** 3, 61–69, Dec. 1959.
13. D. E. Branson and A. M. Ozell, discussion of H. W. Birkeland, Differential Shrinkage in Composite Prestressed Concrete Beams, *ACI Journal,* May 1960; *ACI Journal,* Proceedings **32,** 6, part 2, 1529–1546, Dec. 1960.
14. W. H. Connolly, "Design of Prestressed Concrete Structures," 125, F. W. Dodge Corp., New York, 1960.
15. D. E. Branson, B. L. Meyers, and K. M. Kripanarayanan, Time-Dependent Deformation of

Noncomposite and Composite Prestressed Concrete Structures (paper presented by D.E.B. at the 40th Annual Meeting, Highway Research Board, Washington, D.C., Jan. 1970; based on Iowa State Highway Commission and University of Iowa Research Reports No. 69-1, Feb. 1969, and No. 70-6, Aug. 1970), *National Academy of Sciences—National Academy of Engineering, Highway Research Record,* No. 324, *Symposium on Concrete Deformation,* 15–43, 1970.

16. D. E. Branson and K. M. Kripanarayanan, Loss of Prestress, Camber and Deflection of Non-Composite and Composite Prestressed Concrete Structures (four papers presented by D.E.B. on various aspects at the 6th Congress, Fédération Internationale de la Précontrainte, Prague, June 1970; and at Design Seminars sponsored by the Prestressed Concrete Institute, Chicago, Jan. 1971 and June 1971, and by the California Division of Highways and Ceramic Lightweight Aggregate Association, Sacramento, Mar. 1971), *PCI Journal,* **16,** 5, 22–52, Sept.–Oct. 1971.
17. D. E. Branson, Instantaneous and Time-Dependent Deflections of Simple and Continuous Reinforced Concrete Beams, *Alabama Highway Research Report, Bureau of Public Roads, Report* 7, part 1, 1–78, Aug. 1963.
18. ACI Committee 345, Chairman, D. E. Branson, Deflections of Reinforced Concrete Flexural Members, *ACI Journal,* Proceedings **63,** 6, 637–674, June 1966. Also published in *ACI Manual of Concrete Practice,* part 2: Structural Design, Structural Specifications, Structural Analysis, 1967, 1972.
19. Subcommittee 2, ACI Committee 209, Subcommittee Chairman D. E. Branson, Committee 209 Chairman, J. R. Keeton, Prediction of Creep Shrinkage and Temperature Effects in Concrete Structures (paper presented by D.E.B. at the Annual ACI Convention, New York, Apr. 1970), in "Designing for Effects of Creep, Shrinkage and Temperature in Concrete Structures," 51–93, ACI Publication SP 27-3 (symposium volume), 1971.
20. Prestressed Concrete Institute, "Fundamentals of Prestressed Concrete Design," 2nd edn, 133, Prestressed Concrete Institute, Inc., Chicago, Dec. 1968.
21. William Zuk, Thermal and Shrinkage Stresses in Composite Beams, *ACI Journal,* Proceedings **58,** 3, 327–340, Sept. 1961.
22. B. A. Boley and J. H. Weiner, "Theory of Thermal Stresses," John Wiley & Sons, New York, 1960.
23. Fazlur R. Khan and Mark Fintel, Conceptual Details for Creep, Shrinkage and Temperature in Ultra High-Rise Buildings, in "Designing for Effects of Creep, Shrinkage and Temperature in Concrete Structures," 215–228, ACI Publication SP 27-9 (symposium volume), 1971.
24. D. E. Branson and A. M. Ozell, Camber of Prestressed Concrete Beams, *ACI Journal,* Proceedings **57,** 12, 1549–1574, June 1961.
25. Alan H. Mattock, Precast-Prestressed Concrete Bridges: 5. Creep and Shrinkage Studies, *PCA Journal,* **3,** 2, 32–66, May 1961.
26. G. E. Troxell and H. E. Davis, "Composition and Properties of Concrete," 1–434, McGraw-Hill Book Co., New York, 1956.
27. K. M. Kripanarayanan and D. E. Branson, Some Experimental Studies of Time-Dependent Deflections of Noncomposite and Composite Reinforced Concrete Beams, ACI Publication SP 43-16, *Deflections of Concrete Structures,* 409–419, 1974.
28. British Standards Institution, "Code of Practice for the Structural Use of Concrete, Part 1. Design, Materials, Workmanship," CP 110, 1–155, British Standards Institution, London, 1972.

CHAPTER
THREE

DEFLECTION OF REINFORCED CONCRETE BEAMS AND ONE-WAY SLABS, AND BOTH STEEL AND REINFORCED CONCRETE COMPOSITE BEAMS

3-1 INTRODUCTION

With the present-day use of higher-strength concrete and reinforcing steel, the strength or load-factor method of design, and the resulting shallower sections, the problem of predicting and controlling deflections of reinforced concrete flexural members under service loads has become increasingly important since the 1950s. The first ACI Code provision on deflections, other than load tests, appeared in 1963, with an expanded provision included in the 1971 ACI Code. The AASHTO Highway Bridge Specifications have included a method for calculating deflections since 1957, with a deflection procedure for load-factor design presented in the 1973 edition for the first time. In the case of strength or load factor design, both ACI and AASHTO use the I_{eff} approach derived by the author. Various procedures for computing deflections are studied in this chapter.

In Sec. 3-2 different methods are reviewed for computing initial or short-time deflections. The effects of repeated loading and continuity are also included. Four numerical examples are used to demonstrate the calculations required. These include a simple span rectangular beam, a 3-span continuous one-way slab (sand-lightweight concrete), a continuous T-beam, and a simple span bridge T-beam. Except for the continuous one-way slab, normal-weight concrete is used. Experimental versus computed results are evaluated in Secs. 3-2, 3-6, and Appendix G for short-time deflections.

Procedures for computing deflections due to shrinkage and differential temperature are reviewed in Sec. 3-3, along with a comparison with experimental results. Appropriate solutions for continuous members are also derived. A discussion of possible recommendations for predicting shrinkage warping is presented. The same four numerical examples in Sec. 3-2 are used in Sec. 3-3.

Creep and combined time-dependent deflections are studied in Sec. 3-4 in terms of two basic approaches, referred to as the modified "Δ_i" and increased "n" approaches. Methods available for calculating these deflections are then reviewed in the context of these two approaches. Experimental evaluation is also included in Secs. 3-4 and 3-6. The effect of compression steel and the reduction or k_r-factor are discussed separately, including the results of experimental verification. The same four numerical examples in Sec. 3-2 are used in Sec. 3-4.

Procedures for computing deflections of unshored and shored reinforced concrete and steel–concrete composite beams are outlined in Sec. 3-5, including a limited comparison with experimental results. Numerical examples are used to illustrate the different methods of calculation.

A general discussion of experimental results, practical considerations, and a simplified Code-type calculation procedure for initial and time-dependent deflections of reinforced concrete structures is presented in Sec. 3-6.

The methods of predicting or calculating deflections herein are deterministic in nature. However, the actual behavior is a decidedly probabilistic phenomenon, requiring statistical approaches for a rational analysis.[1,2] Even with the most sophisticated methods of analysis using experimentally determined material properties, the range of variation between measured and computed results for short-time as well as long-time deflections is surprisingly high. Studies[1-4,85] have shown that the coefficient of variation for such deflections is of the order of 15 to 20 percent and higher. In addition, the difficulty is compounded when applied to actual structures and conditions instead of laboratory specimens, since the only property of the concrete usually known at the calculation stage is the specified characteristic compressive strength. Other factors also affect the variability of deflections in actual structures, as discussed in Sec. 3-6.

For these and other reasons, the author has advocated the use of relatively simple, semiempirical, approximate methods based on limiting conditions (such as the I_{eff} and I_{rep} procedures) and/or compatibility of deformations (such as the shrinkage warping procedure) in estimating short-time, creep, shrinkage, and combined time-dependent deformation. These methods developed by the author and his colleagues refer to the following effects: cracking—I_e,[1-24,38,50,52,53] shrinkage warping,[3-5,20,21,25,63] compression steel and movement of the neutral axis—k_r,[3,4,7,8,20,24,26] creep deflections using a simple I_e–k_r–C_t (or when creep and shrinkage calculations are lumped together, I_e–k_r–T) method,[3-6,8,20,24] composite members,[6,27,28,80,81] continuity,[3-6,8,20,24,29,30] and repeated load cycles.[17] Additional information relating to these procedures and similar developments by other authors is also included.

Because of the variability of deflections, it would appear to be not only feasible but essential that relatively simple procedures be used so that engineers will guard against placing undue reliance on the computed or predicted results.

The word "reinforced" has been used in this chapter to refer to nonprestressed concrete structures. Also, the expression "initial or short-time deflection" is preferred to "immediate, instantaneous, or elastic deflection," since there is always some time factor involved in both laboratory or actual structural loading and such deflections are usually neither linear nor completely recoverable.

3-2 INITIAL OR SHORT-TIME DEFLECTION

The principal factors which affect the initial or short-time deflection of reinforced concrete flexural members under service loads are: modulus of elasticity, E_c (Chap. 1); load distribution and support conditions (section on continuity); variable cross-section (section on continuity); load level and degree of cracking along the beam.

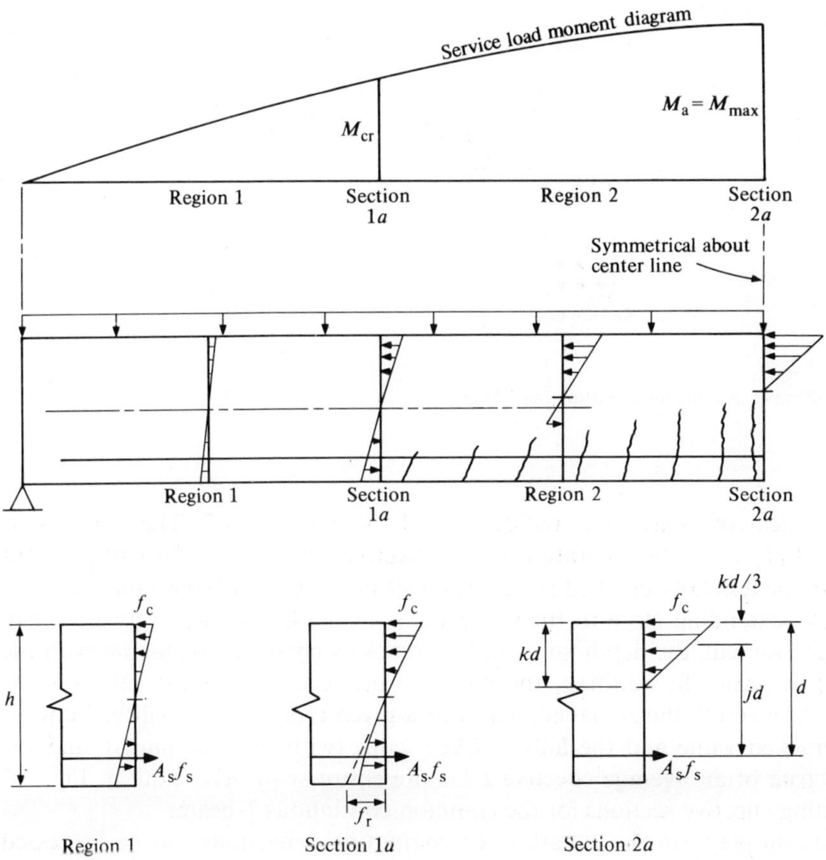

FIGURE 3-1
Various regions of stress and cracking in a reinforced concrete beam carrying service loads.

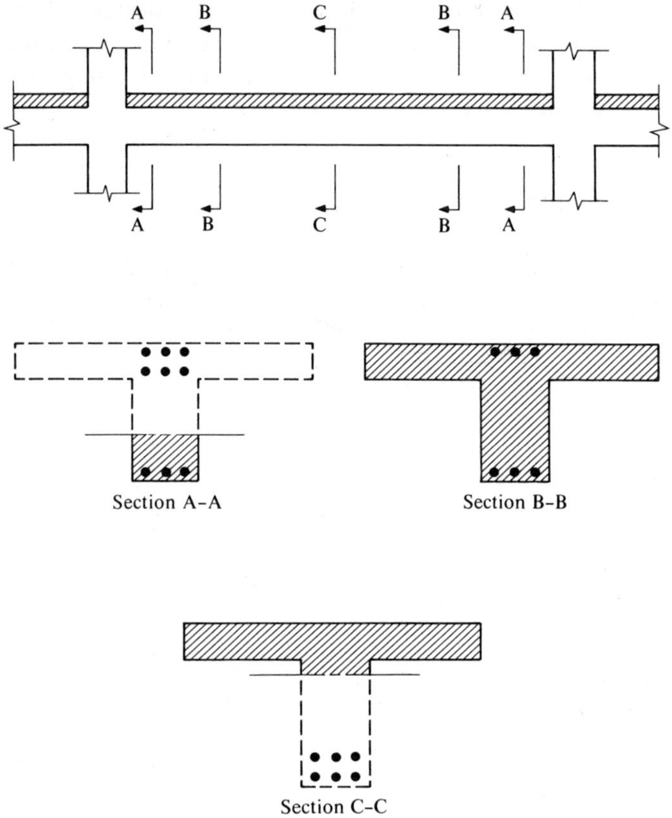

FIGURE 3-2
Limiting effective sections for a continuous T-beam.

The effect of cracking is indicated in Figs. 3-1 and 3-2. The concrete is uncracked (I_g or more accurately I_{ucr} is effective) in regions of small moment and more or less fully cracked (I_{cr} is effective) in regions of large moment, with the cracks extending close to the neutral axis in the latter case. Between small and large moment, the depth and width of cracks vary in some manner with the moment diagram. In addition, the concrete between cracks still carries some tension. As a result, the actual effective I at a given cross-section will be between the uncracked value and the fully cracked value (with these as limits), and the same is true of an average effective I for the entire span. Also note in Fig. 3-2 the limiting effective sections for the common continuous T-beam.

In its simple form the equation for computing deflections can be expressed as Eq. (3-1).

$$\Delta = KM_aL^2/E_cI_e \tag{3-1}$$

where K is a deflection coefficient that depends on the load distribution and support conditions, such as 5/48 for uniformly loaded simple beams, M_a is the maximum moment, such as $wL^2/8$ for uniformly loaded simple beams, and I_e is the average effective moment of inertia. See the notation section for K and M_a values for ideal end conditions and uniform or concentrated loading. Alternative forms of the deflection equation are given by Eqs. (3-37) to (3-42), and derived in Appendix H.

The cracking moment is computed by Eq. (3-2).

$$M_{cr} = f_r I_g / y_t \tag{3-2}$$

where f_r is the modulus of rupture (Chap. 1) and y_t is the distance from the neutral axis to the extreme fiber in tension. For heavily reinforced members and lightweight concrete members (low E_c and hence high modular ratio n), it may be worthwhile to use I_{ucr} instead of I_g.

Methods for Computing Initial or Short-Time Deflections

Maney,[31] 1914

$$\Delta = KM_aL^2/E_cI_{cr} = KL^2(f_s + nf_c)/E_sd \tag{3-3}$$

Equation (3-3) is based on the cracked transformed section throughout the span, in which the curvature M/EI is given by the slope of the strain diagram.

Swain,[32] 1924

$$E_cI_{cr} = E_sA_s(1 - k)jd^2 \tag{3-4}$$

Equation (3-4) is derived from M/E_cI_{cr} = slope of strain diagram $= \varepsilon_s/(1 - k)d$, and $\varepsilon_s = f_s/E_s$, $M = f_sA_sjd$. Equation (3-4) is also based on the cracked section.

ACI Committee 307,[33] 1931 ACI Committee 307 recommended Eq. (3-3) for calculating deflections, which is based on the cracked section. This was a suitably conservative method of the day which computes deflections on the high or safe side.

Murashev,[34] 1940 In order to take into account the participation of tensile concrete in resisting deflections, use the cracked transformed I based on a modular ratio $n = \bar{E}_s/E_c$, where $\bar{E}_s$ is given by Eq. (3-5).

$$\bar{E}_s = E_s/\psi \tag{3-5}$$

where $\psi = 1 - \frac{2}{3}(M_{cr}/M)^2 \leq 1.0$, and M is the moment at service load. This method is based on the consideration that between cracks the steel stress and hence deformation is less than right at the cracks; therefore the average effective steel modulus of elasticity, $\bar{E}_s$, should be greater than the actual steel modulus, E_s, at the cracks. The effect of this procedure is to obtain an effective $I \geq I_{cr}$. A detailed description of Murashev's derivation is given in English by Pregnoff.[35] The method was verified by numerous experiments and incorporated into the U.S.S.R. "Norms and Technical Regulations for Design of Concrete and Reinforced Concrete Structures," NiTU 123-55.

Former PCA,[36] 1947 On the basis of a study of experimental data, PCA recommended in 1947 the use of the gross moment of inertia (neglecting the steel) for computing deflections as long as the extreme fiber stresses in compression stayed within approximately $0.3f'_c$, and as long as the beams were not too shallow or severely cracked.

Dunham,[37] 1953 Another method for taking into account the effect of tensile concrete is given by Eqs. (3-8).

$$\left.\begin{aligned} \Delta &= 0.9\Delta_{cr} && \text{for simple beams} \\ &= 0.8\Delta_{cr} && \text{for one end continuous} \\ &= 0.7\Delta_{cr} && \text{for both ends continuous} \end{aligned}\right\} \tag{3-8}$$

where Δ_{cr} is the computed deflection using I_{cr}. For continuous beams I_{cr} corresponding to the points of maximum positive and negative moment are used as constant Is in each region. As indicated by Eqs. (3-8), the misuse of I_{cr} tends to be more pronounced in continuous beams because a greater length of beam will normally be uncracked or less severely cracked.

AASHTO,[38] 1957 In the 1957, 1961, 1965, 1969, and 1973 editions of the AASHTO Highway Bridge Specifications, the use of I_{cr} for computing deflections is indicated with $n = 8$, regardless of the concrete strength. For load factor design this method has been modified in the 1973 edition, as will be shown.

Former CRSI,[39] 1959 For computing deflections, the gross moment of inertia was recommended.

Yu and Winter,[40] 1960

Method A I_{cr} at midspan is used as a constant value throughout the length of the span for simple spans, and the average I_{cr} for the positive and negative moment regions is used for continuous beams.

Method B To account for the participation of concrete in tension, decrease the deflections in method A by multiplying by ψ_1 from Eq. (3-9).

$$\psi_1 = 1 - M_1/M_a \leq 1.0 \tag{3-9}$$

where $M_1 = 0.1(f'_c)^{2/3}b_w h(h - kd)$. This procedure has the same effect as using a constant $I = I_{cr}/\psi_1 \geq I_{cr}$. The derivation of Eq. (3-9) followed an elastic theory approach with the factor 0.1 having been determined empirically. Comparisons

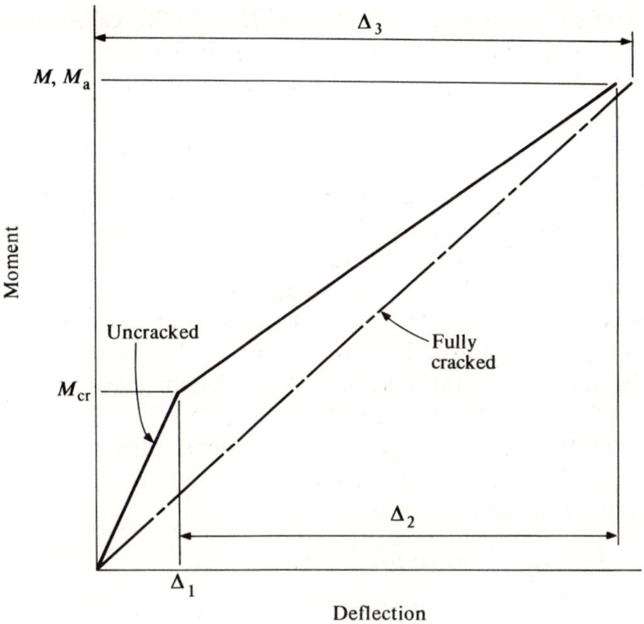

FIGURE 3-3
Bilinear moment–deflection curve similar to the typical experimental curve of Fig. 3-4 and showing the 3 CEB deflection components of Eq. (3-10). The same bilinear approach is used by Beeby's equation (3-22) and the CEB equation (3-24).

with test data for 90 beams indicated that method B provides somewhat better results than method A.

European Concrete Committee (CEB),[41] 1961

$$\Delta = \Delta_1 + \Delta_2 \leq \Delta_3 \tag{3-10}$$

where

$$\Delta_1 = KM_{cr}L^2/E_cI_g \tag{3-11}$$

$$\Delta_2 = \frac{4}{3}\frac{KL^2(M_a - M_{cr})}{E_sA_sd^2(1-2q)(1-2q/3)} \tag{3-12}$$

$$\Delta_3 = \frac{KM_aL^2}{E_sA_sd^2(1-2q)(1-2q/3)} \tag{3-13}$$

$$q = (A_s/bd)(f_y/f'_c) \leq 0.25 \tag{3-14}$$

The beam deflection is thus taken as the smaller of the two values (Fig. 3-3), $(\Delta_1 + \Delta_2)$ and Δ_3, which represents the initial bilinear loading stage and the reloading stage after the beam has cracked, respectively. The denominator of Eq. (3-13) is the equivalent elastic flexural rigidity for a cracked section, as Eq. (3-4), but determined by replacing $(1 - k)$ and j with their approximate equivalents in ultimate strength notation of $(1 - 2q)$ and $(1 - 2q/3)$, respectively. Such a bilinear approach provides for a consideration of loading stages in terms of initial loading before the beam has cracked and reloading after the beam has cracked (Fig. 3-3).

ACI Code (318-63),[42] 1963 Provision 1: when designing by the ultimate strength method, deflections must be checked under working loads by provision 2 when ρ for singly reinforced beams, $(\rho - \rho')$ for doubly reinforced beams, and $(\rho_w - \rho_f)$ for T-beams, exceeds $0.18f'_c/f_y$, or whenever f_y exceeds 40 ksi (2810 kg/cm^2, 276 N/mm^2). This was simply a "red flag" to check deflections at or above the steel ratio of $0.18f'_c/f_y$, which is close to the balanced steel ratio by elastic theory and less than half the balanced steel ratio by ultimate strength theory, since $\rho_{bal} = (0.50, 0.46, 0.43, \text{and } 0.39)f'_c/f_y$ for $f_y = 40$, 50, 60, and 75 ksi, respectively, and $\beta_1 = 0.85$ for $f'_c \leq 4000$ psi (281 kg/cm^2, 27.6 N/mm^2); also to check deflections when f_y is greater than 40 ksi (2810 kg/cm^2, 276 N/mm^2).

Provision 2: when computing deflections, use:

$$I_e = I_g \qquad \text{when } \rho f_y \leq 500 \tag{3-15}$$

$$I_e = I_{cr} \qquad \text{when } \rho f_y > 500 \tag{3-16}$$

From $\rho = 0.18f'_c/f_y$ in provision 1, $\rho f_y = 0.18f'_c = 450$ and 540 when $f'_c = 2500$ psi (176 kg/cm^2, 17.2 N/mm^2) and 3000 psi (211 kg/cm^2, 20.7 N/mm^2), respectively. Hence the Code value of $\rho f_y = 500$ was selected for these and higher-strength concrete as a governing criterion. For continuous spans the moment of inertia may be taken as the average of the values obtained for the positive and negative moment regions.

The rationale for Eqs. (3-15) and (3-16) involves the use of the uncracked I or I_g when the tensile force parameter, ρf_y is small—≤ 500 (which tends to result in a deeper beam because of the greater internal-couple lever-arm required for a given moment, and hence less pronounced cracking effects), and the conservative fully cracked I or I_{cr} when ρf_y is greater than 500 (which tends to result in a shallower beam for a given moment, and hence more pronounced cracking effects).

Author,[5] 1963 One logical form of an expression for the effective I at an individual cracked section that satisfies the limiting conditions ($I_{eff} = I_{ucr}$ when the moment at the individual section $M = M_{cr}$, and I_{eff} approaches I_{cr} when M is very large compared to M_{cr}) was given in Ref. 5 by Eq. (3-17).

$$\text{When } M \geq M_{cr}: \qquad I_{eff} = I_{ucr} - (I_{ucr} - I_{cr})[1 - (M_{cr}/M)^m] \tag{3-17}$$

where m is an unknown power. A precedent for a power function relation relative to the distribution of cracking effects was established by Murashev's expression for ψ in Eq. (3-5) in a different form. However, a considerably different value for the power m was determined in Ref. 5, although initially it was thought that a second degree function might be reasonable, as in the case of Murashev's solution.

Rewriting Eq. (3-17) in the form of Eq. (3-18) and using I_g instead of I_{ucr}, the Newmark numerical procedure[43] together with the test results from some 58 laboratory beams (simple and two-span continuous rectangular beams,[5,44,45] and simple T-beams[40]) were used to empirically determine an appropriate value of $m = 4$ for I_e at individual sections in Eqs. (3-17) and (3-18). The numerical procedure is demonstrated in Appendix D for one of the simple beams and one of the two-span continuous beams from Ref. 5.

$$\text{When } M \geq M_{cr}: \qquad I_e = (M_{cr}/M)^4 I_g + [1 - (M_{cr}/M)^4] I_{cr} \qquad (3\text{-}18)$$

It was further determined that the same type of equation but with the power of 3, or Eq. (3-19), could be used for the average I_e for simple beams (or between inflection points of continuous beams), with an average of values weighted $\frac{2}{3}$ for positive and $\frac{1}{3}$ for negative moment regions being used in the case of the two-span continuous beams. These weighted proportions were later revised for continuous

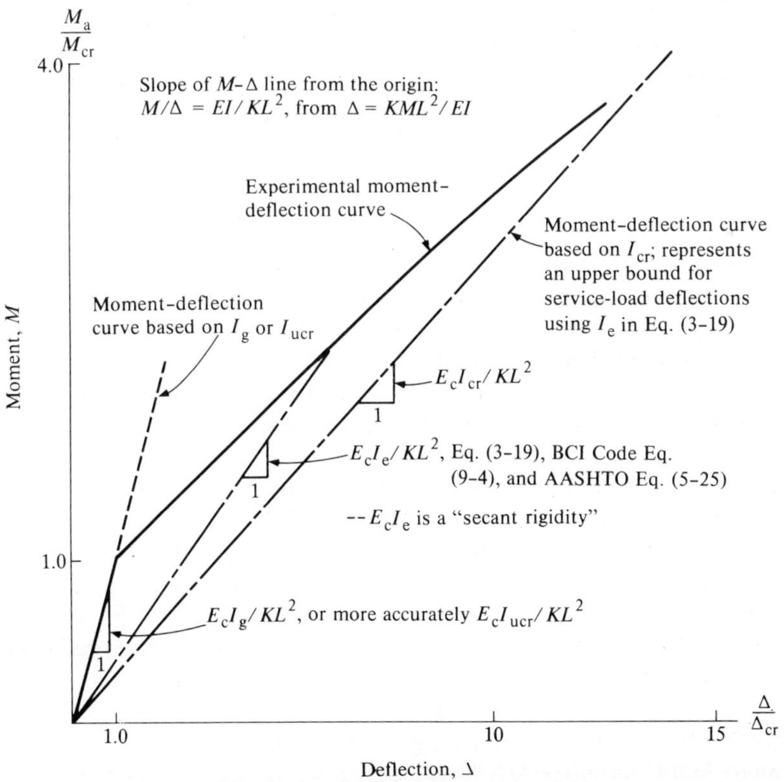

FIGURE 3-4
A typical moment-deflection curve plotted to scale for an experimental reinforced concrete flexural member from Burns and Siess. Also shown are the moments of inertia I_g or I_{ucr}, I_{cr}, and I_e. (*Adapted in part from ASCE Proceedings,*[46] *with permission.*)

beams in Eqs. (3-47b) and (3-49), for example. In the case of heavy concentrated loads it might be desirable to use Eq. (3-18) with a numerical procedure instead of Eq. (3-19), since the development of Eq. (3-19) for the average I_e was based on uniformly distributed loading. When $M_a/M_{cr} \geq$ about 3, I_e is approximately equal to I_{cr} in Eq. (3-19).

The solutions by Eqs. (3-18) and (3-19) differed by a maximum of 3 percent. As one can demonstrate by hand calculations, results by these two equations are not particularly sensitive to the exact powers of 3 and 4, with something like 2.8 or 3.2, and 3.8 or 4.2, for the two equations serving about as well.

$$\text{When } M_a \geq M_{cr}: \quad I_e = (M_{cr}/M_a)^3 I_g + [1 - (M_{cr}/M_a)^3] I_{cr} \tag{3-19}$$

where M_a is the maximum moment at the stage for which the deflection is being computed. When M in Eq. (3-18) or M_a in Eq. (3-19) is less than M_{cr}, $I_e = I_g$. I_e in these two equations provides a transition between the well-defined upper and lower bounds of I_g and I_{cr} (which were used in the 1963 ACI Code), as a function of the level of cracking in the form of M_{cr}/M or M_{cr}/M_a. For different load or moment levels, the deflection is computed in each case using Eq. (3-18) or (3-19) for the total load level being considered, such as dead load and dead plus live load. The incremental deflection, such as due to live load, is then computed as the difference in these values—demonstrated in the examples herein. The various moment levels, moments of inertia, deflections and deflection increments are shown in Fig. 3-4 and 3-5, and in idealized form in Figs. 3-8 (for initial loading and unloading) and 3-19. Also, Fig. 3-15 shows the effect of creep in this idealized sense.

With regard to the effect of repeated loading, since the upper envelope of repeated load–deflection curves approximates the single loading curve for both reinforced and prestressed members, even though increasing residual deflections (due to creep and cracking effects) occur, it appears reasonable to compute short-time deflections using I_e and the residual deflections by an appropriate factor in terms of the total time-dependent or separate creep and shrinkage deflections. This is discussed further in subsequent sections.

I_{ucr} might be more accurately used instead of I_g in Eqs. (3-18) and (3-19), especially for heavily reinforced members and lightweight concrete members (low E_c and hence high modular ratio n). The principal effect of using I_{ucr}, rather than I_g, is in the calculation of an increased value of M_{cr} (due to both I_{ucr} and y_t). Thus if f_r is evaluated correctly, the accuracy in predicting the point of first cracking would be improved using I_{ucr}. If this is the case, then the entire predicted load or moment–deflection curve, such as shown in Fig. G-3, would be improved. The I_e (author's) result in Fig. G-3 used I_g. The effect of using I_{ucr} versus I_g at points well beyond the point of first cracking is small (approaches zero).

According to ACI Committee 435,[102] I_{ucr}/I_g may be as high as 1.30 due to the inclusion of tension steel and 1.15 due to the inclusion of compression steel. In example 1 for a simple beam singly reinforced with 1.98 percent tension steel, $I_{ucr}/I_g = 2658/2287 = 1.16$. Using I_{ucr} instead of I_g, Δ_D was decreased by about 13 percent and Δ_{D+L} was decreased by about 1 percent.

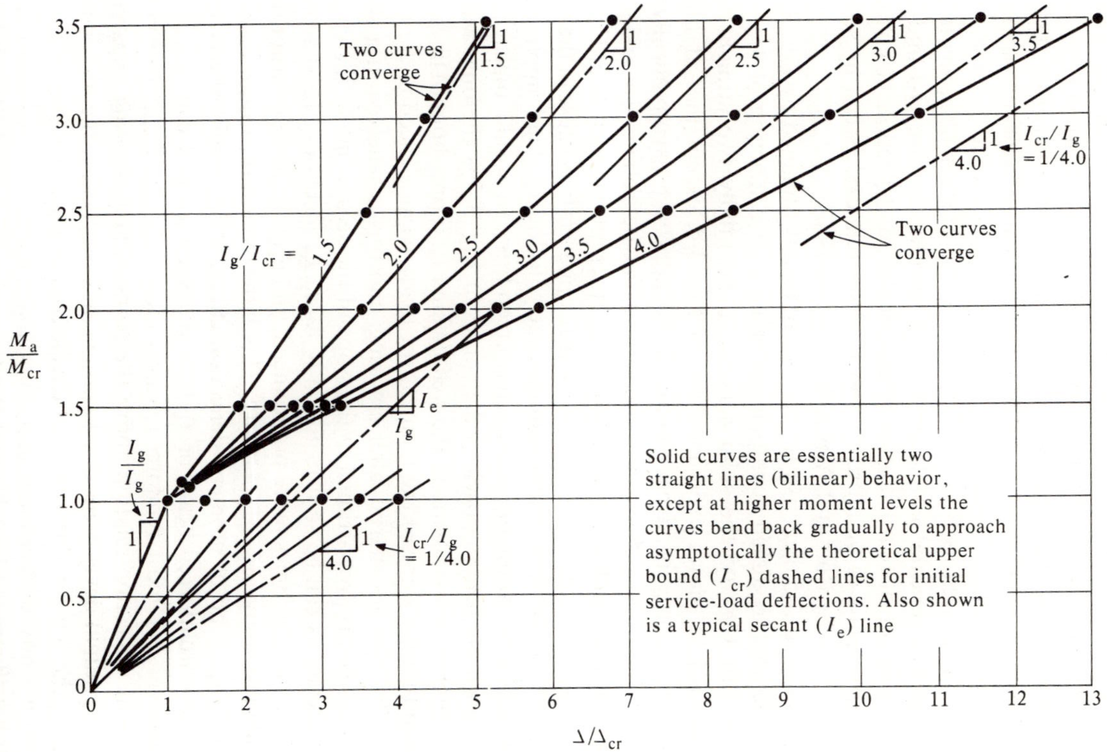

FIGURE 3-5
Generalized moment–deflection curves for various values of I_g/I_{cr} computed using Eq. (3-19) (same as ACI Code equation (9-4) and AASHTO equation (5-25)) for I_e.

Charts presented by Lutz[23] for determining I_g and I_{cr} are shown in Appendix E. He also developed curves for M_{cr} and I_e (Eq. 3-19), although the author believes their direct calculation is easier and preferable. Charts for I_g and I_{cr} are also available in various handbooks.[76,104]

A typical experimental moment (or load) versus deflection curve for a simply supported reinforced concrete flexural member is plotted to scale in Fig. 3-4 from a study by Burns and Siess.[46] With the exception of the slightly curved upper part of the curve, which would appear to be due to creep and other inelastic effects, the marked bilinear nature of the behavior is obvious. The corresponding representations of I_g or I_{ucr}, I_{cr}, and I_e as slopes of lines from the origin are also shown in Fig. 3-4, as well as Fig. 3-5. This essentially bilinear behavior is also shown for other experimental simple beams in Appendix G, Figs. G-1 and G-3. Apparently as a result of redistribution effects, the load–deflection behavior of continuous beams tends to be more "rounded" than bilinear, as can be seen in Fig. 3-7 from a study by Ruiz and Winter.[55]

The curves in Fig. 3-5 demonstrate the type of results obtained by Eq. (3-19), and can also be used to directly compute deflections, as a function of M_a, M_{cr}, I_g/I_{cr}, and Δ_{cr}. In Fig. 3-5 the solid curves are essentially two straight lines (bilinear behavior) as shown experimentally in Fig. 3-4, except that at higher moment levels the curves bend back gradually to approach asymptotically the theoretical upper bound (I_{cr}) dashed lines for service-load deflections. These slightly concave upward curves in Fig. 3-5 are contrary to observations such as in Fig. 3-4 (experimental curve is slightly concave downward), apparently because of the effect of creep which occurs quite rapidly immediately following load application. However, such configuration for a completely creep-free case would appear to be indisputable since the bending back of the curves in Fig. 3-5 is the only way to go from the point of first cracking to the well-defined I_{cr} (upper bound) lines.

It is noted that the slope of the moment (or load) versus deflection curves in Figs. 3-4 and 3-5 changes sharply after cracking, and this is reflected in a sharp decrease in I_e (which is proportional to the slope of the secant line to the curves) following cracking, as can be seen in Fig. 3-6. As I_g/I_{cr} increases, the sharpness of this decrease in I_e becomes more pronounced, and approaches an abrupt drop in I_e for loads just above M_a/M_{cr} = unity. This latter condition would result in a "yield point" type load or moment–deflection curve, which would of course be the case for an unreinforced beam or $I_g/I_{cr} = \infty$. Such a trend can be seen in Fig. 3-5. However, for most reinforced beams there is a significant rise in the load or moment–deflection relation above the point of first cracking apparently because of the gradual development of hairline cracks vertically, the distribution of cracks along the span, and the effect of tensile concrete between cracks.

Based on a study of measured and computed deflections, ACI Committee 435[3] found that the methods of Yu and Winter method B, 1961 CEB, 1963 ACI Code, and the author yielded generally comparable results (within about 20 percent) for predicting short-time deflections, with the methods of Yu and Winter and the author being somewhat more accurate. The same conclusions were drawn by Mirza and Sabnis[11] in connection with other experimental data. They recommended the author's method. Beeby[50] also found satisfactory agreement

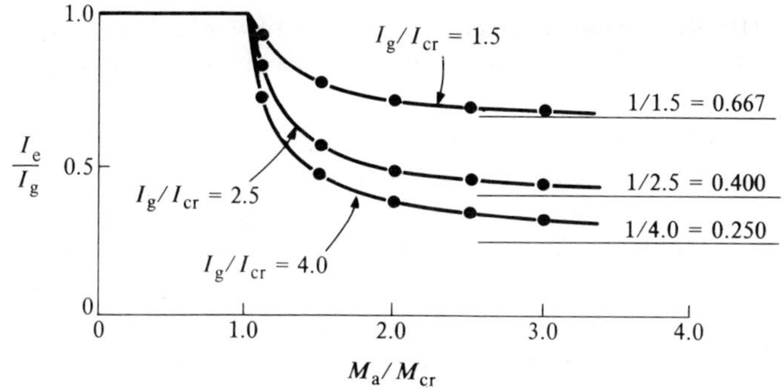

FIGURE 3-6
Generalized effective moment of inertia versus bending moment relation in the cracking range—I_e/I_g versus M_a/M_{cr} curves for different values of I_g/I_{cr}, as computed by Eq. (3-19).

between computed and measured results using the methods of Yu and Winter, the author, and other methods, as shown following Eq. (3-22) and in Appendix G, Fig. G-3. In the usual working load range (say M_{total}/M_{cr} = up to 3 to 4), the author's I_e method shows the best agreement in Fig. G-3, and this result would be improved if I_{ucr} instead of I_g (which would raise the point of first cracking) had been used in Eq. (3-19).

The following sample calculations illustrate the development of Fig. 3-5, in which the first and third columns of the table below are carefully plotted.

For $I_g/I_{cr} = 2.5$ and using Eq. (3-19) for I_e When $M_a = 1.1M_{cr}$, $M_{cr}/M_a = 1/1.1 = 0.909$, and $0.909^3 = 0.751$.

$$\begin{aligned} I_e &= (M_{cr}/M_a)^3 I_g + [1 - (M_{cr}/M_a)^3] I_{cr} \qquad \text{(Eq. (3-19))} \\ &= 0.751 I_g + (1 - 0.751)(I_g/2.5) = 0.851 I_g \end{aligned}$$

It follows that $\Delta/\Delta_{cr} = (M_a/M_{cr})/(I_e/I_g) = 1.1/0.851 = 1.29$

Similarly:

M_a/M_{cr}	I_e/I_g	Δ/Δ_{cr}
(1)	(2)	(1)/(2)
1.0	1.000	1.00
1.1*	0.851*	1.29*
1.5	0.568	2.64
2.0	0.475	4.21
2.5	0.438	5.71
3.0	0.422	7.11
3.5	0.414	8.45
∞	0.400†	∞

* Computed above.
† $I_{cr}/I_g = 1/2.5 = 0.400$.

Bureau of Public Roads (BPR),[47] 1966 The following procedure is recommended by the Bureau of Public Roads for computing deflections of reinforced concrete bridges.

(*a*) General. The usual elastic deflection equations may be used. In general, deflections shall be based on the moment of inertia of the entire superstructure section except railings or any curbs or sidewalks not placed monolithically with the superstructure before falsework is released. For constant depth superstructures, or concrete members, section properties shall be based on the section of maximum positive moment for each continuous or simple span. For superstructures, or concrete members with haunches, or variable moment of inertia, section properties may be the average of the properties at sections of maximum positive and negative moments.

(*b*) Instantaneous deflections caused by dead load (these deflections normally serve in the preparation of dead-load camber diagrams for beams subject to flexure due to dead load only). The moment of inertia to be used in computing instantaneous deflections caused by dead loads shall be determined in accordance with I_g or Eq. (3-20), based upon the working dead load stress in the primary reinforcement.

(*c*) Short-time deflections caused by dead loads after passage of the working live load (these deflections normally serve in the preparation of dead-load camber diagrams for beams subject to flexure due to dead and live loads).

1. The moment of inertia to be used in computing short-time deflections caused by dead loads after passage of the working live load shall be chosen from I_g or Eq. (3-20), based upon the total working load stress in the primary reinforcement equal to the stress due to the working dead load, plus the stress due to the working live load plus impact.
2. The calculation of the working stresses shall be based on the assumption that the superstructure beams or girders act together and have equal deflection under working dead and live load plus impact.
3. The calculation of the stress due to the working live load plus impact shall be based on the assumption that the live loading consists of the number of lane loadings specified in AASHTO, without reduction in load intensity, and is uniformly distributed to the longitudinal beams or girders of the superstructure.

Working load stress in primary reinforcement, f_s	Moments of inertia for calculation of deflections caused by dead loads[48]
$0 < f_s \leq f_y/3$	I_g
$f_y/3 < f_s \leq f_y$	$I_g - \frac{3}{2}(I_g - I_{cr})(f_s/f_y - \frac{1}{3})$ (3-20)

(*d*) Long-time deflections—discussed later in the chapter.

(*e*) Instantaneous deflections caused by live loads.

1. The moment of inertia of the transformed area of the cracked section of the longitudinal deck beams or girders of the superstructure shall be used in computing instantaneous deflections caused by live loads plus impact.
2. The calculation of live-load deflections shall be based on the assumption that the superstructure beams or girders act together and have equal deflection. The live loading shall consist of the number of lane loadings specified in AASHTO, without reduction in load intensity, and shall be considered uniformly distributed to the longitudinal beams or girders.

For simple spans $$\Delta_{L+I} = \frac{KM_{L+I}L^2}{10E_c I_{cr}}$$

and for equal continuous spans $$\Delta_{L+I} = \frac{KM_{L+I}L^2}{12E_c I_{cr}}$$

where M_{L+I} is the maximum positive moment in the end span. An alternative equation is provided for unequal continuous spans.

When	$f_s/f_y = 0.4$	I in Eq. (3-20) $= 0.90I_g + 0.10I_{cr}$
	0.6	$0.60I_g + 0.40I_{cr}$
	0.8	$0.30I_g + 0.70I_{cr}$
	1.0	$0.00I_g + 1.00I_{cr}$
When	$M_a/M_{cr} = 1.1$	I_e in Eq. (3-19) $= 0.75I_g + 0.25I_{cr}$
	1.5	$0.30I_g + 0.70I_{cr}$
	2.0	$0.13I_g + 0.87I_{cr}$
	3.0	$0.04I_g + 0.96I_{cr}$
	4.0	$0.02I_g + 0.98I_{cr}$

Equation (3-20) is based on the indirect determination of I as a function of f_s/f_y, while Eq. (3-19) is based on the direct determination of I_e as a function of the level of cracking—defined by the parameter M_a/M_{cr}. Both equations are in the same form of weighted values of I_g and I_{cr}. In the BPR equation (3-20), the adjusted I is used for dead load only, with I_{cr} used for live load plus impact. However, for determining the dead-load I, f_s is based on the total dead load and live load plus impact, assuming that after passage of the live load the beam is partially cracked with regard to dead-load deflection. This is discussed in the next section, on the effect of repeated load cycles.

Corley and Sozen,[49] 1966

$$\phi = M/E_c I_{cr} = \varepsilon_i/kd \tag{3-21}$$

where ϕ = curvature at an individual section and ε_i = the concrete strain in the extreme fiber in compression = f_c/E_c. Equation (3-21) is to be used when the moment is well above the cracking moment and provided the moment does not cause large inelastic strains. If the applied moment is comparable to the cracking moment, the beam or a portion of it may be considered uncracked. Although no precise rules can be made for determining when to consider a section as fully cracked or partially cracked, the following conditions are offered as a guide:

1. If ρn is greater than about 0.18 and cover of the tension steel is not excessive, consider the member to be fully cracked regardless of the magnitude of the moment.

2. If ρn is less than about 0.08, consider the member to be partially cracked regardless of the magnitude of the moment.

3. If ρn is between 0.18 and 0.08, consider the member to be fully cracked if the moment is more than twice the cracking moment.

4. For T-beams, the value ρn would be calculated using the effective width as that at the extreme fiber in tension.

A numerical procedure is then used to compute deflections. The above section properties may also be used for the entire length of the beam in approximate calculations.

The parameter $\rho n = 0.18$ is roughly the same as $\rho f_y/f'_c = 0.18$, as discussed in the 1963 ACI Code method, Eqs. (3-15) and (3-16).

Beeby,[50] 1968

Method 1 Equation (3-22) represents an empirically modified bilinear equation (see the bilinear moment–deflection diagrams in Figs. 3-3 and 3-4, for example).

$$\phi = \frac{M_{cr}}{E_c I_{ucr}} + \frac{M - M_{cr}}{E'_c I'_c} \qquad (3\text{-}22)$$

where $M_{cr} = f_r I_{ucr}/y_t$, $f_r = 7.5\sqrt{f'_c}$, $E_c = 44\,200\sqrt{f'_c}$, $E'_c = 25\,200\sqrt{f'_c}$ (psi), and I'_c is the I of the cracked transformed section using E'_c. This equation refers to the curvature at an individual section, based on the total moment, M, at the section. A numerical procedure is then used to compute deflections.

Method 2 An alternate and improved empirical method for constructing the bilinear moment–curvature diagram at individual sections is to define three coordinate points as (0, 0), $(M_{cr}, M_{cr}/E_c I_g)$, and $(M_Q, M_Q/E_c I_{cr})$, where M_Q = the ultimate moment resistance of the section based on the strength of the concrete in compression, and M_{cr} is calculated using I_g. This is similar in effect to the author's equation (3-18) in providing an upper bound shown in Fig. 3-4 for deflections based on the fully cracked section, except this is carried all the way to the ultimate moment in Beeby's method 2.

On the basis of an analysis of 133 test beams, Beeby[50] found the mean error by several methods to be (in order of best results), Beeby method 2: 0.000; Yu and Winter: −0.008; Branson: 0.009; CEB, Eq. (3-10): −0.024; transformed section: 0.025; and Beeby method 1: 0.039. An example of the agreement for one beam is shown in Appendix D, Fig. D-3, where it may be noted that the predicted deflections by the various methods are on the high or safe side of experimental results.

European Concrete Committee (CEB),[51] 1968 Equation (3-23) is similar to the 1963 ACI Code approach, with Eq. (3-24) representing a modification of the 1961 CEB bilinear equation (3-10). The $\frac{4}{3}$ factor is retained, but the expression in the denominator of EI is changed to $E_c I_g$ and $E_c I_{cr}$. The $\frac{4}{3}$ factor is used in Eqs. (3-10) and (3-24) to reduce the slope of the moment–deflection curve after cracking (proportional to EI) by multiplying the denominator or $E_c I_{cr}$ by $\frac{3}{4}$. It has more recently been proposed that the $\frac{4}{3}$ factor be changed to $1/0.85 = 1.18$.[52]

$$\text{When} \quad \left.\begin{array}{l}\rho \geq 0.005 \text{ for rectangular beams} \\ \rho \geq 0.001 \text{ for T-beams with } b/b_w \geq 10.0\end{array}\right\} \quad \Delta = \frac{KM_a L^2}{E_c I_{cr}} \tag{3-23}$$

When ρ is less than the above,

$$\Delta = \frac{KL^2}{E_c}\left[\frac{M_{cr}}{I_g} + \frac{4}{3}\frac{M_a - M_{cr}}{I_{cr}}\right] \leq \frac{KM_a L^2}{E_c I_{cr}} \tag{3-24}$$

ACI Code (318-71),[14] 1971 Deflections which occur immediately on application of load shall be computed by the usual methods or formulas for elastic deflections. Unless values are obtained by a more comprehensive analysis, deflections shall be computed using the author's effective moment of inertia equation (3-19) for I_e, which is the Code's equation (9-4). Values of E_c and f_r are given in Chap. 1 herein by the ACI Code and others. For continuous spans, the effective moment of inertia may be taken as the average of the values for critical positive and negative moment sections.

Direct bilinear method The direct bilinear approach is given by Eq. (3-25), which is the same as Eq. (3-24) without the $\frac{4}{3}$ factor.

$$\Delta = \frac{KL^2}{E_c}\left[\frac{M_{cr}}{I_g} + \frac{M_a - M_{cr}}{I_{cr}}\right] \tag{3-25}$$

AASHTO,[38] 1973 As previously noted, the 1973 edition of the AASHTO Highway Bridge Specifications retained the use of I_{cr} for computing deflections, using $n = 8$, regardless of the concrete strength. This apparently applies to live-load plus impact deflections in all cases and dead-load deflections for working stress design. In addition, however, for load factor design the calculation of dead-load deflections is prescribed as follows.

Unless a more comprehensive analysis is made, immediate dead-load deflections upon removal of falsework shall be computed using I_e by Eq. (3-19) (same as ACI Code), which is the AASHTO equation (5-25). Values of E_c are given by Eq. (1-10) for different weight concrete (same as ACI Code), and $f_r = 7.5\sqrt{f'_c}$ (same as ACI Code for normal-weight concrete), without reference to concrete weight. For continuous spans, the effective moment of inertia may be taken as the average of the I_e values for the critical positive and negative moment sections (same as ACI Code).

AASHTO also recommends that the moment of inertia of the entire superstructure sections, except railings or any curbs or sidewalks not placed monolithically with the superstructure before the falsework is released, and the moment of inertia of the full cross-section of the pier or bent, shall be used to determine the elastic properties of the various spans and supports.

ACI Committee 443, concrete bridge design,[53] 1974 For computing immediate deflections of bridges, the ACI Committee 443 recommends the same procedure as the 1971 ACI Code, including Eq. (3-19) for I_e, Eq. (1-10) for E_c, Eq. (1-8) for f_r, and the use of average positive and negative Is for continuous spans.

Fling,[54] 1974 This method is recommended by Fling for simplified computations.

Effect of continuity

$$\Delta = K(5/48)ML^2/EI \tag{3-26}$$

where M is the midspan moment at service loads; for uniform loading $K = 1.0$ for simple beams, 0.74 for beams with one end continuous, 0.60 for beams with both ends continuous; and for a midspan-concentrated load $K = 0.80$ for simple beams, 0.57 for beams with one end continuous, and 0.40 for beams with both ends continuous. Conditions of partial fixity can be interpolated between these K values (but not extrapolated) using Eqs. (3-27) and (3-28).

For uniformly distributed loads $\quad K = 0.4 + 0.6M/M_o \quad$ (3-27)

For a concentrated load at midspan $\quad K = 0.8M/M_o \quad$ (3-28)

where M_o is the total static moment at service loads.

In examples 2 and 3 for computing Δ_D and Δ_{D+L} of both exterior and interior spans, all but one of the results using Eqs. (3-26) and (3-27) were on the low side of the solutions using Eqs. (3-37), (3-39), and (3-40)—by 15, 10, 14, 11, 11, 7, and 12 percent. The one exception was significantly greater than by Eqs. (3-37), (3-39), and (3-40), although this case was extrapolated outside the above limits of $K = 0.60$ to 1.00 for uniformly loaded beams and is included only for illustrative purposes. It should be noted that Eqs. (3-27) and (3-28) are approximate formulas used to interpolate between the K factors for ideal end conditions, and compute somewhat different results from the statical Eqs. (3-40) and (3-41), which are derived in Appendix H.

Estimate of flexural stiffness Table 3-1 is recommended by Fling in Refs. 54 and 104 for the approximate determination of flexural stiffness.

This approach can be useful for certain approximate calculations. However, the author believes the value of $0.6E_cI_g$ in Table 3-1 can be in significant error depending on the effect of I_g/I_{cr}, which is not included as a parameter in the

Table 3-1 Limits of principal tensile steel percentage for the determination of effective flexural rigidity*

	Effective flexural rigidity		
	$E_c I_g$	$0.6 E_c I_g$	$E_c I_{cr}$
Type of member	$M_a < M_{cr}$	$M_{cr} < M_a < 2M_{cr}$	$M_a > 2M_{cr}$
Solid slabs and rectangular beams	$p < 0.4\%$	$0.4\% < p < 0.8\%$	$p > 0.8\%$
T-beams with tension in stem (+mom)	$p_w < 0.6\%$	$0.6\% < p_w < 1.2\%$	$p_w > 1.2\%$
T-beams with tension in flange (−mom)	$p_w < 1.5\%$	$1.5\% < p_w < 3.0\%$	$p_w > 3.0\%$

* The steel percentage refers to that required for the actual moment, and not the actual steel furnished. Also, $p = 100A_s/bd$, and $p_w = 100A_s/b_w d$.

table. For instance, in the case of the rectangular beam in Example 1, $0.6E_c I_g$ is essentially the same as $E_c I_{cr}$, and the dead-load deflection is 0.22 in. (0.56 cm) versus 0.14 in. (0.36 cm) by the I_e method, or a 57 percent differential. This would be compounded when including creep deflection. The other point is that the calculation of I_e is of very little additional trouble to begin with, which also seems to be true of other attempts to simplify an already simple but relatively accurate procedure. In addition, for the T-beam bridge in Example 4, $0.6E_c I_g$ is even less than $E_c I_{cr}$, which is erroneous.

Effect of Repeated Load Cycles

Several methods for computing short-time deflections of cracked reinforced concrete members have directly considered the effect of initial loading and reloading after the member has cracked following the passage of live load. The 1961 and 1968 CEB methods given by Eqs. (3-10) and (3-24) provide for an initial loading bilinear moment–deflection curve and for the fully cracked deflection as an upper limit (Fig. 3-3). The 1966 Bureau of Public Roads method uses I_{cr} for live-load plus impact deflection, and an effective $I = I_g$, or between I_g and I_{cr} by Eq. (3-20), for deflection due to dead load only, in which the effective I is a function of the working load stress in the primary reinforcement under dead load only before passage of live load and under dead plus live plus impact loading after passage of live load. The BPR procedure is illustrated in Example 4.

In repeated load tests of reinforced concrete members by Burns and Siess[46] in 1966, the following typical behavior was found for the beam being described: "When the beam was unloaded at approximately 50% of the yield load, the slope of the repeated loading curve was less than it had been just following first cracking. There was a break in the load–deflection curve on reaching the point to which the beam had previously been loaded; and at this point the flatter slope of the load–deflection *envelope* was resumed just as though the repeated loading had not occurred." For continuous reinforced beams, Ruiz and Winter[55] in 1968 also demonstrated the "envelope effect," as shown typically in Fig. 3-7 for a test beam under 14 load cycles.

Likewise for both reinforced and prestressed beams in repeated load tests, as shown in Chap. 5 herein (with a prestressed beam in Fig. 5-10), Kripanarayanan and the author[17] found in 1972: "An *envelope* of the curved parts of each cycle approximates the single loading curve in the cracking range, which tends to justify the use of I_e for computing deflections, even when a structural member will be subjected to repeated loading in the cracking range." The increased deflection at the top of each load cycle, which is apparently due to both creep and cracking effects that take place during the cycle, can then be taken into account by an appropriate time-dependent factor. In addition, the unloading and reloading back-to-the-previous-maximum-load-point parts of the load–deflection curve were predicted using a modified I, or I_{rep}, as shown in Figs. 5-9 and 5-10, and given by Eq. (5-74). Thus for example I_{rep}, together with I_e, could be used to compute residual deflections due to cracking, or to approximate the entire load–deflection curve under repeated load cycles such as that of Fig. 3-7. In this case, only a single unloading and reloading line is computed for each cycle using I_{rep}, with the effect of the hysteresis loop being neglected.

A practical application of this procedure is shown in Fig. 3-8 which represents an idealized short-time moment versus deflection diagram showing the I_es and I_{rep} for computing dead-load deflection before $(\Delta_D)_1$ and after $(\Delta_D)_2$ passage of repeated live load, and live-load deflection, Δ_L. The procedure for computing these deflections is given by Eqs. (3-29) to (3-36).

I_e method, together with I_{rep}, for computing deflections before and after passage of the design live load (Fig. 3-8)

$$\Delta_{D+L} = \frac{KM_{D+L}L^2}{E_c(I_e)_{D+L}} \quad \text{where } (I_e)_{D+L} = \text{Eq. (3-19) using } M_a = M_{D+L} \tag{3-29}$$

Before passage of live load, and upon first passage of live load

$$(\Delta_D)_1 = \frac{KM_D L^2}{E_{ci}(I_e)_D} \quad \text{where } (I_e)_D = \text{Eq. (3-19) using } M_a = M_D \tag{3-30}$$

$$\Delta_L = \Delta_{D+L} - (\Delta_D)_1 E_{ci}/E_c \tag{3-31}$$

The ratio E_{ci}/E_c is required in Eq. (3-31) because Δ_{D+L} in Eq. (3-29) is a fictitious quantity when different values are used for E_{ci} and E_c.

After passage of live load

$$\Delta_L = \frac{KM_L L^2}{E_c I_{rep}} \quad \text{as shown in Fig. 3-8} \tag{3-32}$$

where the development of I_{rep} (Eq. 5-74) is described in Fig. 5-9, and given here by Eq. (3-33), with $\psi = (M_{ult} - M_{D+L})/(M_{ult} - M_{cr})$.

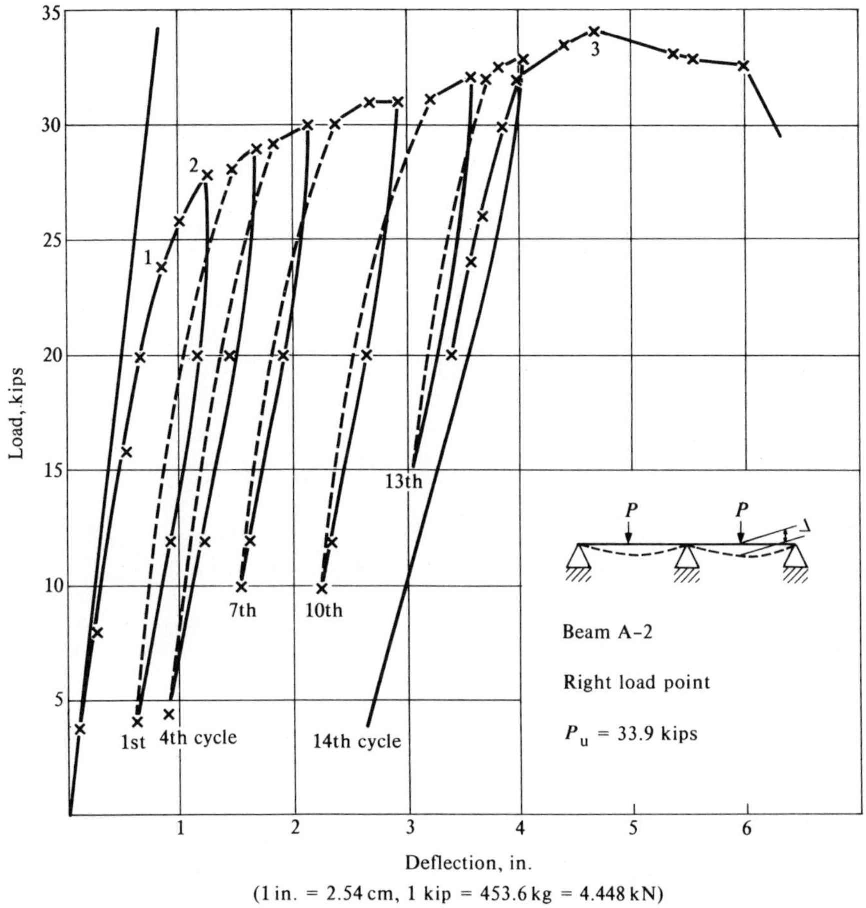

FIGURE 3-7
Load–deflection curve under repeated loading, from Ruiz and Winter. (*From Cornell University Ph.D. Dissertation,*[55] *with permission.*)

$$I_{rep} = \psi(I_e)_{D+L} + (1 - \psi)I_g \tag{3-33}$$

$$(\Delta_D)_2 = [\Delta_{D+L} \text{ in Eq. (3-29)} - \Delta_L \text{ in Eq. (3-32)}](E_c/E_{ci}) \tag{3-34}$$

The primary need to consider dead load separately from live load, and also before and after live-load application, is for the purpose of computing creep deflections. The deflection $(\Delta_D)_2$ represents an approximate shakedown dead-load deflection associated with the worst cracked state under the total load or moment M_{D+L}. The ratio E_c/E_{ci} is included in Eq. (3-34) in order to continue to base the computed dead-load deflection on E_{ci}, as a matter of consistency. I_{rep} refers to the slope of the unloading and reloading line for a given maximum moment (function of $M_{rep} = M_{D+L}$ in the above definition of ψ).

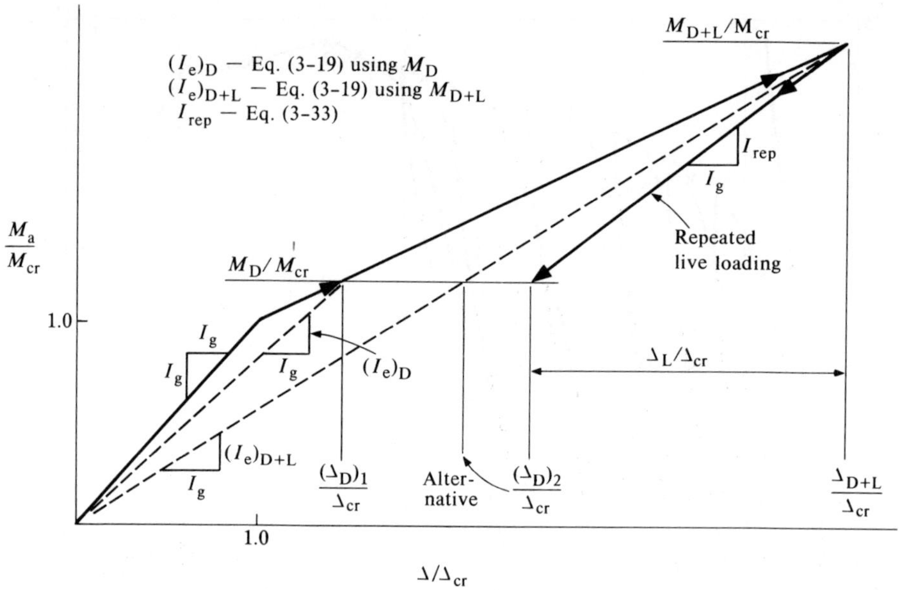

FIGURE 3-8
Idealized short-time moment versus deflection diagram showing I_es and I_{rep} for computing dead-load deflection before $(\Delta_D)_1$ and after $(\Delta_D)_2$ passage of repeated live load, and dead-load plus live-load deflection (Δ_{D+L}). The initial-plus-creep deflections are shown in Fig. 3-15.

As a practical alternative (shown in Fig. 3-8), the additional calculation of I_{rep} can be avoided by assuming I_{rep} = approximately $(I_e)_{D+L}$, in which case Eqs. (3-35) and (3-36) apply.

$$\Delta_L = \frac{KM_L L^2}{E_c(I_e)_{D+L}} \tag{3-35}$$

$$(\Delta_D)_2 = [\Delta_{D+L} \text{ in Eq. (3-29)} - \Delta_L \text{ in Eq. (3-35)}](E_c/E_{ci}) = \frac{KM_D L^2}{E_{ci}(I_e)_{D+L}} \tag{3-36}$$

Equations (3-35) and (3-36) assume the unloading and reloading line follows a complete-recovery secant line, which is the I_e-line.

In the case of early form removal, for example, it may be desirable to use different values of E_{ci} and E_c in Eqs. (3-29) to (3-36). For continuous structures, Eq. (3-37) can of course be used with the appropriate average Is, instead of Eqs. (3-29), (3-30), (3-32), (3-35), and (3-36), as shown in Examples 2 and 3. Calculations using Eqs. (3-29) to (3-36) are illustrated in Example 4.

For other practical problems the unloading and reloading line can also be assumed to follow the I_e line. Thus, in computing deflections where some prior load or moment was larger than the load or moment for which the deflection is being determined, I_e should be based on the maximum load or moment. A typical example of this effect is that of a continuous beam where the maximum negative moments (negative moment envelope) are greater than the negative moments due to the loading pattern producing the maximum deflection in a given span. It is clear then that the moment envelope should be used to determine positive and negative region values of I_e, after which deflections are computed using the loading pattern and moment diagram producing the maximum positive moments and deflections. These calculations are illustrated in Examples 2 and 3 for continuous structures. In connection with such an effect on negative region Is, it should be noted again that the negative region Is have a much smaller effect on deflections than do the positive region Is in continuous structures. The additional effect on deflections of moment redistribution due to cracking, creep, and shrinkage will normally be in the neighborhood of a few percent.

Effect of Continuity

Procedures such as the author's equation (3-18) for I_e at individual sections (demonstrated in Appendix D, Fig. D-2, for a two-span continuous beam, and by Zuraski, Shaikh, and Salmon[30]) and Beeby's equation (3-22) for curvature at individual sections readily lend themselves to statically indeterminate deflection solutions using numerical methods. Another such method based on the use of constant flexural rigidity segments is presented by ACI Committee 435;[29] and others are available. In the absence of a more comprehensive analysis, the following approximate methods are usually satisfactory.

In addition to the simple Eq. (3-1) for ideal end conditions, Eqs. (3-37)[20,56] to (3-42) are derived in Appendices H and I for calculating midspan deflections for simple and continuous beams under uniform or center-concentrated loads in terms of the end and/or midspan moments; and also for cantilever beams. As can be judged in Appendix I for ideal end conditions, the midspan and maximum deflections are essentially the same for interior spans to about 4 and 2 percent different for uniform and center-concentrated loads, respectively, for end spans (as in the fixed-hinged case).

For uniformly distributed loading, beams continuous at one or both ends

$$\Delta_m = \frac{5L^2}{48E_cI_a}[M_m + \tfrac{1}{10}(M_{e1} + M_{e2})] \qquad (3\text{-}37)$$

For a center-concentrated load, beams continuous at one or both ends

$$\Delta_m = \frac{L^2}{48E_cI_a}(4M_m + M_{e1} + M_{e2}) \qquad (3\text{-}38)$$

where M_m in Eqs. (3-37) to (3-41) is the midspan moment (usually positive) and M_{e1}, M_{e2}, are the moments at the ends of the beam span (usually negative or zero).

For uniformly distributed loading or a center-concentrated load, beams continuous at one or both ends

$$\Delta_m = K(5/48)(M_m L^2/E_c I_a) \tag{3-39}$$

For uniform loading $K = 1$ for simple spans, $K = 0.800$ for fixed–hinged beams ($K = 0.738$ when using the maximum moment and computing the maximum deflection), $K = 0.600$ for fixed–fixed beams (example in Appendix H), and for continuous spans (derived in Appendix H),

$$K = 1.20 - 0.20M_o/M_m \tag{3-40}$$

where $M_o = wL^2/8$ = the statistical moment at midspan.

For a center-concentrated load $K = 0.800$ for simple spans, $K = 0.560$ for fixed–hinged beams ($K = 0.5725$ when using the maximum moment and computing the maximum deflection), $K = 0.400$ for fixed–fixed beams (example in Appendix H), and for continuous spans (derived in Appendix H),

$$K = 1.20 - 0.40M_o/M_m \tag{3-41}$$

where $M_o = PL/4$ = the statical moment at midspan.

For uniformly distributed loading or an end-concentrated load, fixed–free (cantilever) beams

$$\Delta_e = K(5/48)(M_e L^2/E_c I_a) \tag{3-42}$$

where M_e and I_a are the moment and moment of inertia at the beam end or support. Since the average I_e in Eq. (3-19) was developed for distributed loading, the author would suggest using that equation for uniform loading and Eq. (3-18) for I_e in the case of an end-concentrated load. For uniform loading, $K = 12/5 = 2.400$. For an end-concentrated load, $K = 16/5 = 3.200$. When beams are cantilevered from other than a fully fixed support, the deflection due to rotation at the support must also be included.

The various moments and deflections mentioned above are summarized in Appendix I for ideal end conditions. For other types of loading, see an appropriate handbook for the K values. As demonstrated in Examples 2 and 3, and Appendix H, Eq. (3-37) or (3-38) and Eq. (3-39) yield the same results, provided the end and midspan moments are statically balanced for a given span.

For beams having a variable flexural rigidity due to changes in cross-section, and/or reinforcement pattern, and/or effects of cracking such as I_e, the following methods for determining an average flexural rigidity, $E_c I_a$, to be used in the above equations, or with other methods such as conjugate beam, have been recommended in the 1973 ACI Committee 435 paper[29] (Eqs. (3-44) to (3-46)) for continuous members.

Cantilever spans Use the flexural rigidity at the face of the support, $E_c I_{sup}$.

$$E_c I_a = E_c I_{sup} \tag{3-43}$$

Continuous spans using midspan value only For spans in which the ratio of end region to midspan flexural rigidity is no larger than 2.0, and, when the moment pattern is between that for a simple span and fixed-ended span, use the midspan flexural rigidity, $E_c I_m$.

$$E_c I_a = E_c I_m \tag{3-44}$$

This method is the simplest and is accurate within ± 25 percent in comparison with the theoretically exact method for $0.4 < M_m/M_o < 1.0$ and $\frac{1}{3} < E_c I/\text{avg. end } EI < 3$. For example, for uniformly loaded prismatic members: $M_m/M_o = 8/24 = 0.33$ for fixed–fixed beams, $M_m/M_o = 9 \times 8/128 = 0.56$ for fixed–hinged beams, $M_m/M_o = 8/16 = 0.50$ and $M_m/M_o = 8/14 = 0.57$ using the ACI Code moment coefficients for interior and end spans, respectively; all of which are within the range of M_m/M_o suggested for Eqs. (3-44) and (3-45).

Continuous spans using a weighted average For spans with flexural rigidity variation more severe than that indicated for the midspan value only case (Eq. (3-44)), or when the negative moment at either end is greater than for a fixed-end span, use the weighted average of end and midspan values according to Eq. (3-45). This procedure maintains reasonable accuracy throughout the entire range of possible moment patterns.

$$E_c I_a = E_c I_m \left[1 - \left(\frac{M_{e1} + M_{e2}}{2M_o} \right)^{\phi} \right] + \frac{E_c I_{e1} + E_c I_{e2}}{2} \left(\frac{M_{e1} + M_{e2}}{2M_o} \right)^{\phi} \tag{3-45}$$

where I_{e1} and I_{e2} are the moments of inertia at the ends of the beam span, and M_o = total statical moment = $wL^2/8$ for uniformly distributed loading, for example.

The advantages of this method are its relative simplicity of use, generally good accuracy and mechanism for increasing accuracy by increasing the value of ϕ. The committee recommends a minimum value of $\phi = 2$. In the range of $0.1 < M_m/M_o < 0.6$, a value of $\phi = 4$ is suggested. In this range with $\phi = 4$ and for $\frac{1}{3} < E_c I_m/\text{avg. end } EI < 3$ the method is accurate within ± 20 percent in comparison with the theoretically exact method.

Continuous spans using a simple average—(ACI 318-71) The 1971 ACI Code suggests the method given by Eq. (3-46) as a simple design approach. However, this method is not recommended by ACI Committee 435, since it was felt that the other procedures were just about as simple and somewhat more accurate.

$$E_c I_a = \tfrac{1}{2} E_c [I_m + \tfrac{1}{2}(I_{e1} + I_{e2})] \tag{3-46}$$

The greater the disparity between the flexural rigidities at the ends and midspan, the lower the accuracy, except when M_m/M_o is about 0.3. At that point, the method is reasonably accurate. This method is accurate within about ± 40 percent in comparison with the theoretically exact method for $0.1 < M_m/M_o < 0.5$ and $\frac{1}{2} < E_c I_m/\text{avg. end } EI < 2$. However, more accurate methods are available, some of which are easier to use.

Author's comments For uniformly loaded prismatic members, Eq. (3-45) reduces to Eqs. (3-47) and (3-48) when $\phi = 2$, 3, and 4 for fixed–fixed and hinged–fixed beams.

Fixed–fixed beams

When $\phi = 2$ $$E_c I_a = E_c[0.56 I_m + 0.22(I_{e1} + I_{e2})] \tag{3-47a}$$

$= 3$ $$E_c I_a = E_c[0.70 I_m + 0.15(I_{e1} + I_{e2})] \tag{3-47b}$$

$= 4$ $$E_c I_a = E_c[0.80 I_m + 0.10(I_{e1} + I_{e2})] \tag{3-47c}$$

Hinged–fixed beams

When $\phi = 2$ $$E_c I_a = E_c[0.75 I_m + 0.125(I_{e1} + I_{e2})] \tag{3-48a}$$

$= 3$ $$E_c I_a = E_c[0.87 I_m + 0.065(I_{e1} + I_{e2})] \tag{3-48b}$$

$= 4$ $$E_c I_a = E_c[0.94 I_m + 0.030(I_{e1} + I_{e2})] \tag{3-48c}$$

Such a relatively heavy weighting of the midspan rigidity can be seen generally from a consideration of the moment area deflection method, and the fact that the statical lever arm for say a fixed–fixed beam is much larger from the support to the positive M/EI-centroid than from the support to the negative M/EI-centroid.

A limited number of comparisons with observed deflections of continuous beams were summarized in the ACI Committee 435 paper[29] for $0.3 < M_m/M_o < 0.65$ and $0.7 < E_c I_m/\text{avg. end } EI < 1.1$. For these comparisons the average discrepancy between calculated and observed results was about 20 percent using the simple average rigidity, about 16 percent using the midspan rigidity, and about 13 percent using the weighted average rigidity with $\phi = 2$ or 4.

Given these limited findings and the approximate nature of the entire approach, it would seem that the midspan rigidity or Eq. (3-44) would be suitable for computing deflections of continuous beams in many cases. However, since the weighted average rigidity or Eq. (3-45) maintains reasonable accuracy throughout the entire range of possible moment patterns, according to ACI Committee 435, this procedure would appear to be somewhat better for routine calculations. Accordingly, for uniformly loaded continuous members the author would suggest using Eqs. (3-74*b*) and (3-49). Both of these equations referred to $\phi = 3$, with Eq. (3-49) having been modified from Eq. (3-48*b*).

Beams with both ends continuous

$$E_c I_a = E_c[0.70 I_m + 0.15(I_{e1} + I_{e2})] \tag{3-47b}$$

Beams with one end continuous

$$E_c I_a = E_c[0.85 I_m + 0.15(I_{\text{cont end}})] \tag{3-49}$$

For a single concentrated load, the use of the midspan-only rigidity or Eq. (3-44) would seem to be appropriate.

The procedures described in this section on continuity refer to short-time deflections under service loads and are usually applied to long-time deflections as well. Moment redistribution effects[57,58] due to cracking (automatically included in numerical solutions such as Appendix D, Fig. D-2, for example), creep and shrinkage are normally considered of a secondary nature and not taken into account in approximate calculations.

Effective Span Length for Deflection Calculations

In the section on deflections the ACI Code (318-71) refers to the span length of members not built integrally with their supports as the clear span plus the depth of the slab or beam, but the span length need not exceed the distance between centers of supports. For monolithic construction the Code refers to the span length in design, and one might assume deflections also, as the clear span. A possible clarification of the Code intention for the span length to be used in deflection calculations is currently (1976) being discussed by ACI Committee 318.

It appears to the author that the *clear span* should be used for *all* deflection calculations, since the edge of the support is the fixed point for determining downward deflections which can occur only in the clear part of the span; that is, assuming any settlement of supports or effects of elastic supports is being excluded. Moments should be determined in the usual way for design, and then only the moment diagram in the clear span used for computing deflections. Of course, the loading pattern for maximum positive moment and deflection in a given span should be used in the deflection calculations, together with the negative region I_es based on the negative moment envelope (maximum negative moments), as previously discussed. The critical negative moment sections are thus located at the face of the supports for deflection purposes. Especially in the case of monolithic construction, but in other cases as well, the clear span is recommended by the author for deflection calculations. As R. S. Fling puts it, he always uses clear spans in deflection calculations because it seems closer to physical reality, and also because computed deflections tend to be greater than those that actually occur. Therefore he is always looking for ways to justify lower computed deflections.

The special cases of the ACI Code moment coefficients are shown in Appendix F, and used in Example 3.

Numerical Examples of Short-Time Deflections

Example 1 Simple span beam, normal-weight concrete; by the cracked-section method including Eqs. (3-3) and (3-4), the gross-section method, Dunham equation (3-8), Yu and Winter equation (3-9), 1963 ACI Code method, author's equation

Table 3-2 Material and section properties, parameters, and conditions for Example 1

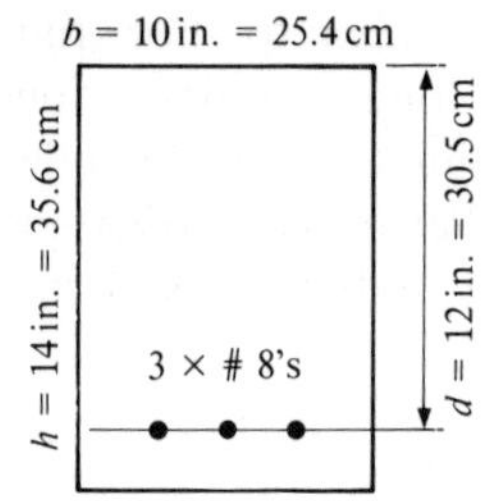

$A_s = 2.37\ \text{in}^2$

Beam designed by the ultimate strength method

Simple span, $L = 20$ ft $= 6.1$ m

Material properties and loading

$f'_c = 4000$ psi, $f_y = 60\,000$ psi

Normal-weight concrete

$E_s = 29 \times 10^3$ ksi (Chap. 1)

$E_c = 3.64 \times 10^3$ ksi (Table 1-3 by ACI Code and by author's equation (1-13))

$n = E_s/E_c = 29/3.64 = 8.0$

$f_r = \underline{496\ \text{psi}}$; use (Table 1-2 by author's equation (1-7), versus $f_r = 475$ psi in Table 1-2 by ACI Code)

$w_{DL} = 0.3$ k/ft (total), $w_{LL} = 1.0$ k/ft

Section properties

$I_g = bh^3/12 = (10)(14)^3/12 = \underline{2287\ \text{in}^4} \qquad \rho = A_s/bd = 2.37/120 = \underline{0.0198}$

$\rho n = (0.0198)(8) = 0.158 \qquad k = \sqrt{2\rho n + (\rho n)^2} - \rho n = 0.426 \qquad kd = 5.11\ \text{in.}$

$I_{cr} = b(kd)^3/3 + nA_s(d - kd)^2 = (10)(5.11)^3/3 + (8)(2.37)(12 - 5.11)^2 = \underline{1345\ \text{in}^4}$

$I_g/I_{cr} = 2287/1345 = 1.70 \qquad (\text{or } I_{cr} = 1331\ \text{in}^4 \text{ by Fig. E-2})$

$$y_o = \frac{(n-1)A_s(d - h/2)}{bh + (n-1)A_s} = \frac{(7)(2.37)(12 - 7)}{(10)(14) + (7)(2.37)} = 0.53\ \text{in.} \qquad y_t = 6.47\ \text{in.}$$

$$\begin{aligned} I_{ucr} &= bh^3/12 + bhy_o^2 + (n-1)A_s(d - h/2 - y_o)^2 \\ &= 2287 + (140)(0.53)^2 + (7)(2.37)(12 - 7 - 0.53)^2 \\ &= 2658\ \text{in}^4 \end{aligned}$$

$I_{ucr}/I_g = 1.16$

Other parameters and conditions

$M_D = w_{DL}L^2/8 = (0.3)(20)^2/8 = 15.0$ ft-k (2075 kg-m, 20.3 kN-m)

$M_L = w_{LL}L^2/8 = (1.0)(20)^2/8 = 50.0$ ft-k (6915 kg-m, 67.8 kN-m)

Average relative humidity = 50 percent   Concrete is moist-cured

Assume 28-day loading age for both dead and live load; hence $E_{ci} = E_c$

(For other metric units, see conversion factors herein.)

(3-19)—same as 1971 ACI Code equation (9-4), 1968 CEB method, direct bilinear equation (3-24), and Fling method.

See Table 3-2 for the material and section properties, parameters and conditions of the problem. The short-time deflection results by the various methods are shown in Table 3-3 along with the ACI Code (318-71) allowable deflections.

Strength design

$$M_u = 1.4(15.0) + 1.7(50.0) = \underline{106\ \text{ft-k}\ (14\,660\ \text{kg-m},\ 144\ \text{kN-m})}$$

$$\omega = \rho f_y/f'_c = (0.0198)(60)/4 = 0.297$$

$$M_u = \phi bd^2 f'_c \omega(1 - 0.59\omega)$$
$$= (0.9)(10)(12)^2(4)(0.297)(1 - 0.59 \times 0.297)/12$$
$$= \underline{106 \text{ ft-k}} \qquad \text{OK}$$

$$a = A_s f_y/(0.85 f'_c b) = (2.37)(60)/(0.85)(4)(10)$$
$$= 4.18 \text{ in. } (10.6 \text{ cm})$$

$$M_u = \phi A_s f_y(d - a/2) = (0.9)(2.37)(60)(12 - 2.09)/12$$
$$= \underline{106 \text{ ft-k}}$$

$$h_{min} = L/16 = (20)(12)/16 = 15 \text{ in. } (38.1 \text{ cm})$$

Hence deflections must be checked according to ACI 318-71 (Chap. 6, Table 6-4 or 6-5 herein). Also, $\rho = 0.18 f'_c/f_y = 0.0120$ and $\rho_{max} = 0.75\rho_b = 0.0214$. Since the actual $\rho = 0.0198$ (Table 3-2), deflections are seen to be potentially critical.

Cracked-section method including Eqs. (3-3) and (3-4)

$$\Delta_D = \frac{KM_D L^2}{E_c I_{cr}} = \frac{(5/48)(15.0)(12)(20)^2(144)}{(3.64 \times 10^3)(1345)} = \frac{1080 \times 10^3}{4896 \times 10^3}$$
$$= \underline{0.221 \text{ in. } (0.56 \text{ cm})} \qquad \text{(Eq. (3-1))}$$

When I is the same for dead load and live load, as in this case, the live-load deflection can be computed directly. However, to form a consistent pattern, the combined dead-load and live-load deflection will be determined in each case first.

$$\Delta_{D+L} = (65.0/15.0)(0.221) = \underline{0.95 \text{ in. } (2.41 \text{ cm})}$$
$$\Delta_L = 0.95 - 0.22 = \underline{0.73 \text{ in. } (1.85 \text{ cm})}$$

Checking with Eq. (3-3),

$$f_s = nM(d - kd)/I_{cr} = (8)(65.0)(12)(12 - 5.11)/1345$$
$$= 31.97 \text{ ksi } (2248 \text{ kg/cm}^2, 220 \text{ N/mm}^2)$$
$$f_c = Mkd/I_{cr} = (65.0)(12)(5.11)/1345$$
$$= 2.96 \text{ ksi } (208 \text{ kg/cm}^2, 20.4 \text{ N/mm}^2)$$

(These steel and concrete stresses would exceed the allowable values for 60 ksi (4219 kg/cm^2, 414 N/mm^2) steel and 4 ksi (281 kg/cm^2, 27.6 N/mm^2) concrete if designed by the working stress method.)

$$\Delta_{D+L} = KL^2(f_s + nf_c)/E_s d = (5/48)(20)^2(144)(31.97 + 8 \times 2.96)/29 \times 10^3)(12)$$
$$= 0.96 \text{ in. } (2.44 \text{ cm}) \qquad \text{(checks within 0.01 in. due to round-off)}$$

(Eq. (3-3))

Checking with Eq. (3-4),

$$E_c I_{cr} = E_s A_s(1 - k)jd^2$$

$$= (29 \times 10^3)(2.37)(1 - 0.427)(1 - 0.142)(12)^2$$
$$= 4874 \times 10^3 \qquad \text{(versus } 4896 \times 10^3 \text{ above—OK)}$$

Gross-section method

$$\Delta_D = \text{cracked-section } \Delta_D/(I_g/I_{cr}) = 0.221/1.70$$
$$= \underline{0.130 \text{ in. } (0.33 \text{ cm})}$$
$$\Delta_{D+L} = (65.0/15.0)(0.130) = 0.56 \text{ in. } (1.42 \text{ cm})$$
$$\Delta_L = 0.56 - 0.13 = \underline{0.43 \text{ in. } (1.09 \text{ cm})}$$

Dunham equation (3-8)

$$\Delta_{D+L} = 0.9\Delta_{cr} = (0.9)(0.95) = \underline{0.86 \text{ in. } (2.18 \text{ cm})}$$
$$\Delta_D = \text{say } \Delta_g = \underline{0.13 \text{ in. } (0.33 \text{ cm})} \qquad \Delta_L = \underline{0.73 \text{ in. } (1.85 \text{ cm})}$$

This simple procedure is probably quite good in many cases. The difficulty is in knowing when it is and when it is not, without checking by other methods.

Yu and Winter Method B, Eq. (3-9) (Method A same as cracked-section)

$$M_1 = 0.1(f_c)^{2/3} b_w h(h - kd) = (0.1)(4000)^{2/3}(10)(14)(14 - 5.11)/12\,000$$
$$= 2.61 \text{ ft-k } (361 \text{ kg-m, } 3.54 \text{ kN-m})$$
$$\psi_1 = 1 - M_1/M_a = (1 - 2.61/65.0) = 0.960 \qquad \text{(Eq. (3-9))}$$
$$M_1 > M_D \qquad \text{and} \qquad M_1 < M_{D+L}$$

Therefore
$$\Delta_D = \Delta_g = 0.13 \text{ in. } (0.33 \text{ cm})$$
$$\Delta_{D+L} = \Delta_{cr}\psi_1 = (0.95)(0.960) = \underline{0.91 \text{ in. } (2.31 \text{ cm})}$$
$$\Delta_L = 0.91 - 0.13 = \underline{0.78 \text{ in. } (1.98 \text{ cm})}$$

The total deflection is slightly less than the fully-cracked deflection for this rather severely cracked beam, as expected.

1963 ACI Code method

$$\rho f_y = (0.0198)(60\,000) = 1188 > 500 \qquad \text{Hence use } I_{cr} \qquad \text{(Eq. (3-16))}$$
$$\Delta_{D+L} = \Delta_{cr} = 0.95 \text{ in. } (2.41 \text{ cm}) \qquad \Delta_D = \text{say } \Delta_g = \underline{0.13 \text{ in. } (0.33 \text{ cm})}$$
$$\Delta_L = \underline{0.82 \text{ in. } (2.08 \text{ cm})}$$

Author's equation (3-19)—same as 1971 ACI Code equation (9-4) and AASHTO equation (5-25)

$$M_{cr} = f_r I_g/y_t = (496)\,(2287)/7 = 162\,050 \text{ in-lb}$$
$$= 13.50 \text{ ft-k } (1867 \text{ kg-m, } 18.3 \text{ kN-m})$$
$$M_D/M_{cr} = 15.0/13.50 = 1.11 \qquad M_{cr}/M_D = 0.900$$

$$(M_{cr}/M_D)^3 = 0.900^3 = 0.73$$

$$M_{D+L}/M_{cr} = 65.0/13.50 = 4.81 \qquad M_{cr}/M_{D+L} = 0.208$$

$$(M_{cr}/M_{D+L})^3 = 0.01$$

$$(I_e)_D = (M_{cr}/M_D)^3 I_g + [1 - (M_{cr}/M_D)^3] I_{cr}$$

$$= (0.73)(2287) + (1 - 0.73)(1345)$$

$$= 2168 \text{ in}^4 \ (90\,230 \text{ cm}^4)$$

$$(I_e)_{D+L} = (0.01)(2287) + (1 - 0.01)(1345) = 1355 \text{ in}^4 \ (56\,400 \text{ cm}^4)$$

$$\Delta_D = \frac{KM_D L^2}{E_c I_e} = \frac{(5/48)(15.0)(12)(20)^2(144)}{(3.64 \times 10^3)(2168)} = \frac{1080 \times 10^3}{7892 \times 10^3}$$

$$= \underline{0.137 \text{ in. } (0.35 \text{ cm})}$$

$$\Delta_{D+L} = (0.137)(65.0/15.0)(2168/1355) = \underline{0.95 \text{ in. } (2.41 \text{ cm})}$$

$$\Delta_L = 0.95 - 0.14 = \underline{0.81 \text{ in. } (2.06 \text{ cm})} \qquad \text{as shown in Fig. 3-19}$$

For this problem with the relatively high steel ratio ($0.0198 = 0.70\rho_b$) and severely cracked state ($M_{D+L}/M_{cr} = 4.8$), the dead-load deflection using I_g and the dead plus live-load deflection using I_{cr} are essentially the same as those using I_e. This is not the case for the one-way slab problem in Example 2. Using I_{ucr},

$$M_{cr} = (496)(2658)/6.47 = 203\,770 \text{ in-lb} = 17.0 \text{ ft-k } (2350 \text{ kg-m, } 23.1 \text{ kN-m})$$

Hence

$$\Delta_D = \Delta_g = 0.13 \text{ in. } (0.33 \text{ cm}) \qquad (\text{versus } 0.14 \text{ in. or } 0.35 \text{ cm above})$$

$$(M_{cr}/M_{D+L}) = (17.0/65.0) = 0.262 \qquad (M_{cr}/M_{D+L})^3 = 0.262^3 = 0.02$$

$$(I_e)_{D+L} = (0.02)(2658) + (0.98)(1345)$$

$$= 1371 \text{ in}^4 \ (57\,060 \text{ cm}^4) \qquad (\text{versus } 1355 \text{ in}^4 \text{ above})$$

$$\Delta_{D+L} = 0.94 \text{ in. } (2.39 \text{ cm}) \qquad (\text{versus } 0.95 \text{ in. or } 2.41 \text{ cm above})$$

1968 CEB method $\rho = 0.0198 > 0.005$ for rectangular beams; hence use Eq. (3-23) and I_{cr} for the cracked-section deflection. The results are the same as by the 1963 ACI Code method.

Direct bilinear equation (3-25)

$$\Delta_{D+L} = \frac{KL^2}{E_c}\left(\frac{M_{cr}}{I_g} + \frac{M_a - M_{cr}}{I_{cr}}\right)$$

$$= \frac{(5/48)(20)^2(144)}{3640}\left[\frac{(13.50)(12)}{2287} + \frac{(65.0 - 13.5)(12)}{1345}\right]$$

$$= \underline{0.87 \text{ in. } (2.21 \text{ cm})} \qquad (\text{Eq. (3-25)})$$

Similarly for dead load with $(M_a - M_{cr}) = (15.0 - 13.5)$,

$$\Delta_D = \underline{0.14 \text{ in. } (0.36 \text{ cm})}$$

which is essentially the same as by the gross-section method and Eq. (3-19) using I_e. $\Delta_{D+L} = 0.87$ in. (2.21 cm) is slightly lower than results by the Yu and Winter method (0.91 in. or 2.31 cm) and the I_e method (0.95 in. or 2.41 cm). Using the CEB bilinear equation (3-25) with the $\frac{4}{3}$ factor, $\Delta_D = 0.15$ in. (0.38 cm) and $\Delta_{D+L} = 1.13$ in. (2.87 cm) except that $\Delta_{cr} = 0.95$ in. (2.41 cm) is given as an upper limit.

Fling equation (3-26) and Table 3-1

$$M_{cr} = 13.50 \text{ ft-k } (1867 \text{ kg-m}, 18.3 \text{ kN-m})$$
$$M_D = 15.0 \text{ ft-k } (2075 \text{ kg-m}, 20.3 \text{ kN-m})$$
$$M_{D+L} = 65.0 \text{ ft-k } (6915 \text{ kg-m}, 67.8 \text{ kN-m})$$

Since design M_u = required M_u = 106 ft-k (14 660 kg-m, 144 kN-m), use p_{D+L} = actual $p = 1.98$ percent (Table 3-2). Also use $p_D = (1.98)(15.0/65.0) = 0.46$ percent. Hence

$$(EI)_D = 0.6E_cI_g = (0.6)(3640)(2287)$$
$$= 4995 \times 10^3 \text{ k-in}^2 \ (14.62 \times 10^6 \text{ kg-cm}^2, 143.3 \times 10^6 \text{ N-cm}^2) \quad \text{(Table 3-1)}$$

$$(EI)_{D+L} = E_cI_{cr} = (3640)(1345)$$
$$= 4896 \times 10^3 \text{ k-in}^2 \ (14.33 \times 10^6 \text{ kg-cm}^2, 140.5 \times 10^6 \text{ N-cm}^2) \quad \text{(Table 3-1)}$$

Table 3-3 Summary of short-time deflection results in inches (cm) by the various methods in Example 1—simple span rectangular beam, normal-weight concrete, shown in Table 3-1

Deflection	Cracked section method	Gross section method	Dunham Eq. (3-8)	Yu and Winter Eq. (3-9)	1963 ACI Code method	Author's Eq. (3-19) (ACI 318-71)	1968 CEB method	Direct bilinear Eq. (3-25)	Fling Eq. (3-26)
†Δ_D	0.22 (0.56)	0.13 (0.33)	0.13* (0.33)	0.13 (0.33)	0.13* (0.33)	0.15 (0.38)	0.13* (0.33)	0.14 (0.36)	0.22 (0.56)
†Δ_{D+L}	0.95 (2.41)	0.56 (1.42)	0.86 (2.18)	0.91 (2.31)	0.95 (2.41)	0.95 (2.41)	0.95 (2.41)	0.87 (2.21)	0.95 (2.41)
†‡$\Delta_L = \Delta_{D+L} - \Delta_D$	0.73 (1.85)	0.43 (1.09)	0.73 (1.85)	0.78 (1.98)	0.82 (2.08)	0.80 (2.03)	0.82 (2.08)	0.73 (1.85)	0.73 (1.85)

* Assumed to be equal to the gross-section deflection.

† The primary differences in the methods are found in computing Δ_{D+L}, except for Δ_D by the cracked-section and Fling methods.

‡ ACI (318-71) allowable deflections based on live load only (Chap. 6 herein). Flat roofs not supporting or attached to nonstructural elements likely to be damaged by large deflections—$\Delta_L \leq L/180 = 240/180 = 1.33$ in. (3.38 cm). Floors not supporting or attached to nonstructural elements likely to be damaged by large deflections—$\Delta_L \leq L/360 = 240/360 = 0.67$ in. (1.70 cm). As determined by all of the methods in the table except the gross-section method, on the basis of the live-load criteria this beam would be limited to use with flat roofs not supporting or attached to nonstructural elements likely to be damaged by large deflections. The allowable deflection for floors is exceeded by say $(0.80 - 0.67) = 0.13$ in. (0.33 cm) only.

$$\Delta_D = K(5/48)(M_D L^2)/(EI)_D$$
$$= (1.00)(5/48)(15.0)(12)(20)^2(144)/4995 \times 10^3$$
$$= \underline{0.216 \text{ in. } (0.55 \text{ cm})} \qquad (\text{Eq. (3-26)})$$

or 0.22 in. (0.56 cm) versus 0.14 in. (0.36 cm) using I_e in Eq. (3-19).

$$\Delta_{D+L} = (0.216)(65.0/15.0)(4995/4896) = \underline{0.95 \text{ in. } (2.41 \text{ cm})}$$
$$\Delta_L = 0.95 - 0.22 = \underline{0.73 \text{ in. } (1.85 \text{ cm})}$$

In this example $0.6E_c I_g$ and $E_c I_{cr}$ were essentially the same. In Example 4, $0.6E_c I_g$ is even less than $E_c I_{cr}$, which is erroneous.

Example 2 Three-span continuous one-way slab, sand-lightweight concrete; by the I_e method, Eq. (3-19), and using Eqs. (3-37), (3-39), (3-47*b*) and (3-49).

See Table 3-4 for the material and section properties, parameters and conditions of the problem, and Table 3-5 for the bending moments and strength design of the slab.

$$h_{min} = L/28.5 = (20)(12)/27.5$$
$$= 8.7 \text{ in. } (21.3 \text{ cm}) \qquad \text{for one end continuous (Chap. 6, Table 6-5)}$$
$$h_{min} = L/33.3 = (20)(12)/32.1$$
$$= 7.5 \text{ in. } (18.3 \text{ cm}) \qquad \text{for both ends continuous (Chap. 6, Table 6-5)}$$

Deflections would have to be checked for the end span by ACI 318-71, for members not supporting or attached to partitions or other construction likely to be damaged by large deflections; and otherwise for both spans. Both end and center span deflections will be computed.

$\rho = 0.0077$, 0.0093, and 0.0049 at sections A–A, B–B, and C–C, respectively, as shown in Table 3-4, versus $\rho = 0.18f_c'/f_y = 0.0180$ and $\rho_{max} = 0.75\rho_b = 0.0371$. The actual steel ratios are sufficiently low that deflections probably are not critical.

The bending moments and average I_e (I_a) for the end span AB and the center span BC are determined in Tables 3-5 and 3-6 (see comment on I_a versus the midspan I_e at bottom of Table 3-6). The bending moments are based on a uniform beam without regard to the variation in $E_c I_e$ due to cracking.

Deflections are computed using Eq. (3-37). In these calculations the positive and negative live-load moments refer to the partial live load producing the maximum positive moment and deflection in each span (Table 3-5), and not to the maximum negative moment envelope, although the latter is used with the negative dead-load moment in computing the negative I_e (Table 3-6).

Table 3-4 Material and section properties, parameters, and conditions for Example 2

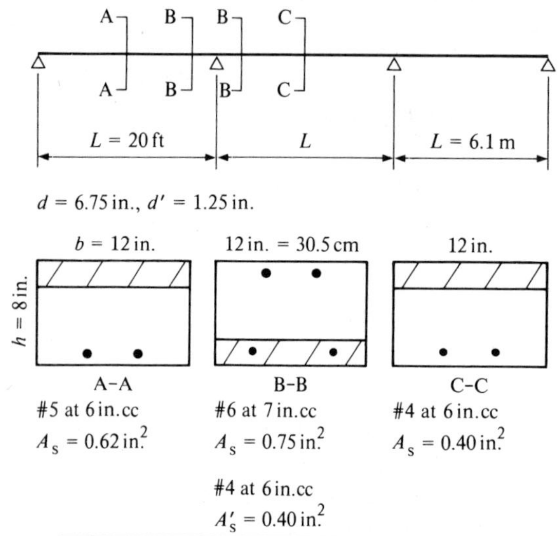

Material properties and loading

$f_c' = 4000$ psi $\quad f_y = 40\,000$ psi

Sand-lightweight concrete (120 pcf)

$E_s = 29 \times 10^3$ ksi (Chap. 1)

$E_c = 2.63 \times 10^3$ ksi (Table 1-3 by ACI Code and by author's equation (1-13))

$n = E_s/E_c = 29/2.63 = 11.0$

$f_r = 450$ psi; use (Table 1-2 by author's equation (1-7), versus $f_r = 405$ psi in Table 1-2 by ACI Code)

$w_{DL} = (8/12)(1)(125)$ [including steel wt.]

$= 83$ #/ft

$w_{LL} = 120$ #/ft

Section properties

$$I_g = bh^3/12 = (12)(8)^3/12 = 512 \text{ in}^4$$

Sec. A–A:

$$\rho = A_s/bd = 0.62/(12)(6.75) = \underline{0.00765} \qquad \rho n = (0.00765)(11) = 0.08415$$

$$k = \sqrt{2\rho n + (\rho n)^2} - \rho n = 0.334 \qquad kd = (0.334)(6.75) = 2.25 \text{ in.}$$

$$I_{cr} = b(kd)^3/3 + nA_s(d - kd)^2 = (12)(2.25)^3/3 + (11)(0.62)(6.75 - 2.25)^2$$

$$= \underline{183.7 \text{ in}^4}$$

$$I_g/I_{cr} = 512/183.7 = 2.79 \qquad (\text{or } I_{cr} = 180.8 \text{ in}^4 \text{ by Fig. E-2})$$

Sec. C–C:

$$\rho = \underline{0.00494} \qquad k = 0.280 \qquad kd = 1.89 \text{ in.} \qquad I_{cr} = \underline{130.9 \text{ in}^4} \qquad I_g/I_{cr} = 3.91$$

Sec. B–B:

$$\rho = A_s/bd = \underline{0.00926} \qquad \rho' = A_s'/bd = \underline{0.00494}$$

$$k = \sqrt{2n(\rho + 2\rho'd'/d) + n^2(\rho + 2\rho')^2} - n(\rho + 2\rho') = 0.326 \qquad kd = 2.20 \text{ in.}$$

$$I_{cr} = b(kd)^3/3 + nA_s(d - kd)^2 + (n - 1)A_s'(kd - d')^2 = \underline{217.0 \text{ in}^2}$$

$$I_g/I_{cr} = 2.36 \qquad (\text{or } I_{cr} = 221 \text{ in}^4 \text{ by Fig. E-2})$$

Other parameters and conditions

Average relative humidity = 70 percent $\quad$ Concrete is moist-cured

Assume 7-day loading age for dead load and Type I cement:

$$f_{ci}' = 0.70(f_c') = 2800 \text{ psi} \qquad (\text{Eq. (1-2)}) \qquad \text{Hence } E_{ci} = \underline{2.42 \times 10^3 \text{ ksi}} \quad (\text{Eq. (1-13)})$$

Assume 2-month loading age for live load, but use 28-day E_c for live load

(For other metric units, see conversion factors herein.)

Table 3-5 Bending moments and strength design of one-way slab in Example 2

Maximum dead-load and live-load moments

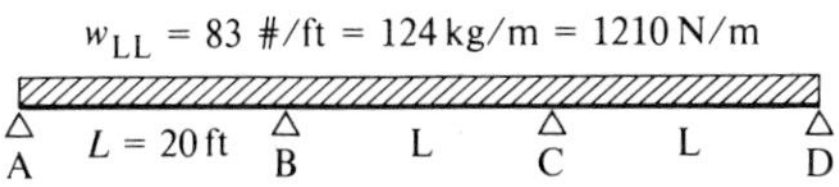

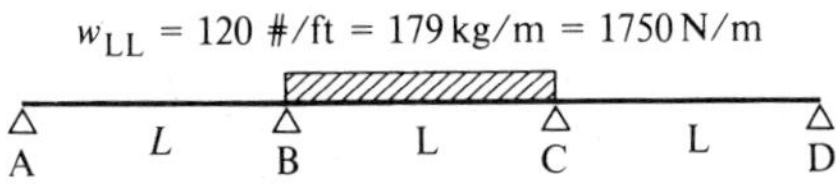

$$\text{Max}\,(M_D)_{AB} = 0.0800wL^2$$
$$= (0.0800)(0.083)(20)^2 = \underline{2.66 \text{ ft-k}}$$
$$℄\ (M_D)_{AB} = 0.0750wL^2 = \underline{2.49 \text{ ft-k}}$$
$$(M_D)_B = 0.1000wL^2 = \underline{3.32 \text{ ft-k}}$$
$$\text{Max}\,(M_D)_{BC} = 0.0250wL^2 = \underline{0.83 \text{ ft-k}}$$

$$(M_L)_B = 0.0500wL^2$$
$$= (0.0500)(0.120)(20)^2 = \underline{2.40 \text{ ft-k}}$$
$$\text{Max}\,(M_L)_{BC} = (0.1250 - 0.0500)wL^2$$
$$= 0.0750wL^2$$
$$= (0.0750)(0.120)(20)^2 = \underline{3.60 \text{ ft-k}}$$

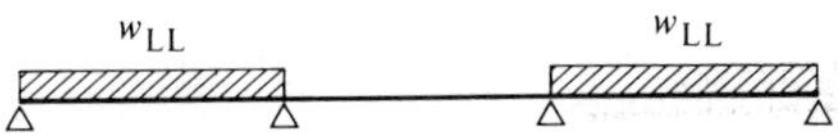

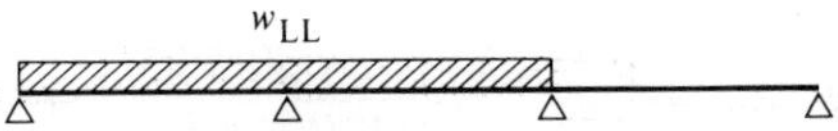

$$(M_L)_B = 0.0500wL^2 = \underline{2.40 \text{ ft-k}}$$
$$\text{Max}\,(M_L)_{AB} = 0.1012wL^2 = \underline{4.86 \text{ ft-k}}$$
$$℄\ (M_L)_{AB} = 0.1000wL^2 = \underline{4.80 \text{ ft-k}}$$

$$(M_L)_B = 0.1167wL^2 = \underline{5.60 \text{ ft-k}}$$

ACI strength design

Sec. A–A, Table 3-4:

$$M_u = 1.4M_D + 1.7M_L = (1.4)(2.66) + (1.7)(4.86) = \underline{11.98 \text{ ft-k}}$$

As in Example 1,

$$M_u = \phi bd^2 f'_c \omega(1 - 0.59\omega) = \underline{11.98 \text{ ft-k}}$$

Checking alternative form,

$$M_u = \phi A_s f_y(d - a/2) = \underline{11.98 \text{ ft-k}}$$

Sec. C–C, Table 3-4:

$$M_u = 1.4M_D + 1.7M_L = (1.4)(0.83) + (1.7)(3.60) = \underline{7.28 \text{ ft-k}}$$
$$M_u = \phi bd^2 f'_c \omega(1 - 0.59\omega) = \underline{7.87 \text{ ft-k}}$$

Checking alternative form,

$$M_u = \phi A_s f_y(d - a/2) = \underline{7.86 \text{ ft-k}}$$

Sec. B–B, Table 3-4:

$$M_u = 1.4M_D + 1.7M_L = (1.4)(3.32) + (1.7)(5.60) = \underline{14.17 \text{ ft-k}}$$

For ultimate load purposes, the compression steel is near the neutral axis, and hence ineffectual. Neglecting the compression steel here (but not in deflection calculations),

$$M_u = \phi bd^2 f'_c \omega(1 - 0.59\omega) = \underline{14.35 \text{ ft-k}}$$

Checking alternative form,

$$M_u = \phi A_s f_y(d - a/2) = \underline{14.36 \text{ ft-k}}$$

Table 3-5 (*continued*)

Moments for deflection calculations

Span AB

$M_D: M_{e1} = 0 \quad M_{e2} = \underline{3.32 \text{ ft-k}} \quad M_m = \underline{2.49 \text{ ft-k}}$

$M_{D+L}: M_{e1} = 0 \quad M_{e2} = 3.32 + 2.40 = \underline{5.72 \text{ ft-k}} \quad M_m = 2.49 + 4.80 = \underline{7.29 \text{ ft-k}}$

Span BC

$M_D: M_{e1} = M_{e2} = \underline{3.32 \text{ ft-k}} \quad M_m = \underline{0.83 \text{ ft-k}}$

$M_{D+L}: M_{e1} = M_{e2} = 3.32 + 2.40 = \underline{5.72 \text{ ft-k}} \quad M_m = 0.83 + 3.60 = \underline{4.43 \text{ ft-k}}$

(1 ft-k = 138.3 kg-m = 1.356 kN-m.)

***End span* AB**

$$\Delta_D = \frac{5L^2}{48E_{ci}I_a}[M_m + \tfrac{1}{10}(M_{e1} + M_{e2})]$$

$$= \frac{(5)(20)^2(144)}{(48)(2420)(512)}[2.49 - 0.10(0 + 3.32)](12)$$

$$= \frac{2.479}{512}(25.92) = \underline{0.13 \text{ in. } (0.33 \text{ cm})} \qquad (\text{Eq. (3-37)})$$

$$\Delta_{D+L} = \frac{5L^2}{48E_cI_a}[M_m + \tfrac{1}{10}(M_{e1} + M_{e2})]$$

$$= \frac{(5)(20)^2(144)}{(48)(2630)(292)}[7.29 - 0.10(0 + 5.72)](12)$$

$$= \frac{2.281}{275}(80.64) = \underline{0.67 \text{ in. } (1.70 \text{ cm})} \qquad (\text{Eq. (3-37)})$$

Δ_{D+L} is a fictitious quantity when different values are used for E_{ci} and E_c. Hence

$$\Delta_L = \Delta_{D+L} - A_D(E_{ci}/E_c) = 0.67 - 0.13(2.42/2.63)$$

$$= \underline{0.55 \text{ in. } (1.40 \text{ cm})} \qquad (\text{Eq. (3-31)})$$

***Center span* BC**

$$\Delta_D = \frac{2.479}{512}[0.83 - (0.10)(3.32)(2)](12) = \frac{2.479}{512}[2.04]$$

$$= \underline{0.01 \text{ in. } (0.03 \text{ cm})} \qquad (\text{Eq. (3-37)})$$

$$\Delta_{D+L} = \frac{2.281}{437}[4.43 - (0.10)(5.72)(2)](12) = \frac{2.281}{437}[39.48]$$

$$= \underline{0.21 \text{ in. } (0.53 \text{ cm})} \qquad (\text{Eq. (3-37)})$$

$$\Delta_L = 0.21 - 0.01(2.42/2.63) = \underline{0.20 \text{ in. } (0.51 \text{ cm})}$$

As in Chap. 6 and Note ‡ of Table 3-3, the allowable deflections based on live load only are: (1) flat roofs not supporting or attached to nonstructural elements likely to be damaged by large deflections

$$\Delta_L = L/180 = 240/180 = 1.33 \text{ in. } (3.38 \text{ cm})$$

and (2) floors not supporting or attached to nonstructural elements likely to be damaged by large deflections

$$\Delta_L = L/360 = 240/360 = 0.67 \text{ in. } (1.70 \text{ cm})$$

The maximum computed live-load deflection is 0.55 in. (1.40 cm). Hence both spans have no limitations so far as live-load deflections are concerned.

Table 3-6 Determination of average values of I_e for Example 2

Span	Load	Section	M_a, Table 3-5 ft-k	* $\dfrac{M_a}{M_{cr}}$	* $\dfrac{M_{cr}}{M_a}$	$\left(\dfrac{M_{cr}}{M_a}\right)^3$	I_{cr}, Table 3-4 in⁴	I_e, Eq. (3-19) in⁴	I_a in⁴
AB	Dead	End	0	—	—	—	—	—	
		Pos.	2.49	0.52	—	—	—	512.0	512†
		Neg.	3.32	0.69	—	—	—	512.0	
	Dead + live	End	0	—	—	—	—	—	
		Pos.	7.29	1.52	0.658	0.285	183.7	277.2§	275¶
		Neg.	8.92⁻ ‡	1.86	0.538	0.156	217.0	263.0	
BC	Dead	Neg.	3.32	0.69	—	—	—	512.0	
		Pos.	0.83	0.17	—	—	—	512.0	512†
		Neg.	3.32	0.69	—	—	—	512.0	
	Dead + live	Neg.	8.92⁻ ‡	1.86	0.538	0.156	217.0	263.0	
		Pos.	4.43	0.92	—	—	—	512.0	437**
		Neg.	8.92⁻ ‡	1.86	0.538	0.156	217.0	263.0	

(For other metric units, see conversion factors herein.)

* $M_{cr} = f_r I_g / y_t = (450)(512)/4 = 57\,600$ in-# $= 4.80$ ft-k (Eq. (3-2)).

† $I_g = 512$ in⁴ (Table 3-4).

‡ $M_a = M_D + M_L = 3.32 + 5.60 = 8.92$ ft-k (Table 3-5).

§ $I_e = (M_{cr}/M_a)^3 I_g + [1 - (M_{cr}/M_a)^3] I_{cr} = (0.285)(512) + (1 - 0.285)(183.7)$
$= 145.9 + 131.3 = \underline{277.2 \text{ in}^4}$ (Eq. (3-19)).

¶ $I_a = [0.85 I_m + 0.15(I_{\text{cont end}})] = [(0.85)(277.2) + (0.15)(263.0)]$
$= 235.6 + 39.5 = \underline{275 \text{ in}^4 \; (11\,400 \text{ cm}^4)}$ (Eq. (3-49)).

** $I_a = [0.70 I_m + 0.15(I_{e1} + I_{e2})] = [(0.70)(512) + (0.15)(263.0)(2)]$
$= 358.4 + 78.9 = \underline{437 \text{ in}^4 \; (18\,190 \text{ cm}^4)}$ (Eq. (3-47*b*)).

Comment

For the end span AB, $I_a = 275$ in⁴ (284 using Eq. (3-48*c*) with $\phi = 4$), which is approximately the same as the midspan $I_e = 277$ in⁴. For the center span BC, $I_a = 437$ in⁴ (462 using Eq. (3-47*c*) with $\phi = 4$), which is 15 percent smaller than the midspan $I_e = I_g = 512$ in⁴. As shown in Example 3, this difference tends to be larger for T-beams.

The above solution differs from ACI 318-71 in that $f_r = 450$ psi (31.6 kg/cm^2, 3.10 N/mm^2) instead of 405 psi (28.5 kg/cm^2, 2.79 N/mm^2) (Table 3-4), and Eqs. (3-47*b*) and (3-49), were used for I_a instead of the simple average of Eq. (3-46). By the ACI Code, $I_a = 320$ in^4 (13 320 cm^4) (versus 275 in^4 or 11 400 cm^4 above), $\Delta_{D+L} = 0.58$ in. (1.47 cm) (versus 0.67 in. or 1.70 cm above) for the end span AB; and $I_a = 383$ in^4 (15 940 cm^4) (versus 437 in^4 or 18 190 cm^4 above), $\Delta_{D+L} =$ 0.24 in. (0.61 cm) (versus 0.21 in. or 0.53 cm above) for the center span BC. The dead-load deflections are the same. The ACI Code method uses a somewhat smaller value of f_r, and applies a greater weight to the negative and smaller weight to the positive moment region values of I_e. The author believes the method used above is more accurate.

Using just the midspan value of I_e, as with Eq. (3-44):

***End span* AB**

$$\Delta_D = \underline{0.13 \text{ in. } (0.33 \text{ cm})}$$

which is the *same as above,* since $I_e = I_g$ in both cases.

$$\Delta_{D+L} = (275/277.2)(0.67) = \underline{0.66 \text{ in. } (1.68 \text{ cm})}$$

versus 0.67 in. (1.70 cm) or *1.5 percent smaller.*

$$\Delta_L = 0.66 - 0.13(2.42/2.63) = \underline{0.54 \text{ in. } (1.37 \text{ cm})}$$

versus 0.55 in. (1.40 cm) or *1.8 percent smaller.*

***Center span* BC**

$$\Delta_D = \underline{0.01 \text{ in. } (0.03 \text{ cm})}$$

which is the *same as above,* since $I_e = I_g$ in both cases.

$$\Delta_{D+L} = (437/512)(0.21) = \underline{0.18 \text{ in. } (0.46 \text{ cm})}$$

versus 0.21 in. (0.53 cm) or *14 percent smaller.*

$$\Delta_L = 0.18 - 0.01(2.42/2.63) = \underline{0.17 \text{ in. } (0.43 \text{ cm})}$$

versus 0.20 in. (0.51 cm) or *15 percent smaller.*

Using the *alternative* procedure shown in Fig. 3-8 and described by Eqs. (3-35) and (3-36), the effect of repeated live load is computed as follows (Is are I_a values in Table 3-6):

***End span* AB**

$$(\Delta_D)_2 = KM_D L^2/E_{ci}(I_e)_{D+L} = (0.13)(512/275)$$
$$= \underline{0.24 \text{ in. } (0.61 \text{ cm})} \qquad (\text{Eq. } (3\text{-}36))$$

$$\Delta_L = KM_L L^2/E_c(I_e)_{D+L} = \frac{2.281}{275}[4.80 - (0.10)(0 + 2.40)](12)$$
$$= \underline{0.45 \text{ in. } (1.14 \text{ cm})} \qquad (\text{Eqs. (3-35), (3-37)})$$

***Center span* BC**

$$(\Delta_D)_2 = (0.01)(512/292) = \underline{0.01 \text{ in. } (0.03 \text{ cm})} \qquad (\text{Eq. (3-36)})$$
$$\Delta_L = \frac{2.281}{437}[3.60 - (0.10)(2.40)(2)](12)$$
$$= \underline{0.20 \text{ in. } (0.51 \text{ cm})} \qquad (\text{Eqs. (3-35), (3-37)})$$

The results for the center span BC are the same as above.

The following solutions by Eq. (3-39) are identical to the previous results using Eq. (3-37), and illustrate the corresponding K factors.

For both end span and center span,

$$(M_o)_D = w_D L^2/8 = 0.083(20)^2/8 = 4.15 \text{ ft-k } (574 \text{ kg-m, } 5.63 \text{ kN-m})$$
$$(M_o)_{D+L} = 0.203(20)^2/8 = 10.15 \text{ ft-k } (1404 \text{ kg-m, } 13.76 \text{ kN-m})$$

***End span* AB**

$$(M)_D = 2.49 \text{ ft-k } (344 \text{ kg-m, } 3.38 \text{ kN-m})$$
$$(M)_{D+L} = 7.29 \text{ ft-k } (1008 \text{ kg-m, } 9.88 \text{ kN-m})$$
$$K_D = 1.20 - 0.20M_o/M_m = 1.20 - (0.20)(4.15/2.49)$$
$$= \underline{0.867} \qquad (\text{Eq. (3-40)})$$
$$\Delta_D = K_D(5/48)(M)_D L^2/E_c I_a = (0.867)(5/48)(2.49)(12)(20)^2(144)/(2420)(512)$$
$$= \underline{0.13 \text{ in. } (0.33 \text{ cm})} \qquad (\text{Eq. (3-39)})$$
$$K_{D+L} = 1.20 - 0.20M_o/M_m = 1.20 - (0.20)(10.15/7.29)$$
$$= \underline{0.922} \qquad (\text{Eq. (3-40)})$$
$$\Delta_{D+L} = K_{D+L}(5/48)(M)_{D+L} L^2/E_c I_a$$
$$= (0.922)(5/48)(7.29)(12)(20)^2(144)/(2630)(275)$$
$$= \underline{0.67 \text{ in. } (1.70 \text{ cm})} \qquad (\text{Eq. (3-39)})$$

***Center span* BC**

$$(M)_D = 0.83 \text{ ft-k } (115 \text{ kg-m, } 1.13 \text{ kN-m})$$
$$(M)_{D+L} = 4.43 \text{ ft-k } (613 \text{ kg-m, } 6.01 \text{ kN-m})$$
$$K_D = 1.20 - 0.20M_o/M_m = 1.20 - (0.20)(4.15/0.83)$$
$$= \underline{0.200} \qquad (\text{Eq. (3-40)})$$
$$\Delta_D = K_D(5/48)(M)_D L^2/E_c I_a = (0.200)(5/48)(0.83)(12)(20)^2(144)/(2420)(512)$$
$$= \underline{0.01 \text{ in. } (0.03 \text{ cm})} \qquad (\text{Eq. (3-39)})$$
$$K_{D+L} = 1.20 - 0.20M_o/M_m = 1.20 - (0.20)(10.15/4.43)$$

$$= \underline{0.742} \qquad \text{(Eq. (3-40))}$$

$$\Delta_{D+L} = K_{D+L}(5/48)(M)_{D+L}L^2/E_c I_a$$
$$= (0.742)(5/48)(4.43)(12)(20)^2(144)/(2630)(437)$$
$$= \underline{0.21 \text{ in. } (0.53 \text{ cm})} \qquad \text{(Eq. (3-39))}$$

The above K values might be compared with the ideal values given with Eq. (3-39) of $K = 0.800$ for fixed–hinged beams (end span) and $K = 0.600$ for fixed–fixed beams (center span).

Using Fling's continuity equations (3-26) and (3-27) with the previous values of I_e and I_a: for both end span and center span.

$$(M_o)_D = w_D L^2/8 = 0.083(20)^2/8 = 4.15 \text{ ft-k } (574 \text{ kg-m}, 5.63 \text{ kN-m})$$
$$(M_o)_{D+L} = 0.203(20)^2/8 = 10.15 \text{ ft-k } (1404 \text{ kg-m}, 13.76 \text{ kN-m})$$

***End span* AB**

$$(M)_D = 2.49 \text{ ft-k } (344 \text{ kg-m}, 3.38 \text{ kN-m})$$
$$(M)_{D+L} = 7.29 \text{ ft-k } (1008 \text{ kg-m}, 9.88 \text{ kN-m})$$
$$K_D = 0.4 + 0.6(M_D/M_o) = 0.4 + 0.6(2.49/4.15) = \underline{0.760} \qquad \text{(Eq. (3-27))}$$

which is slightly greater than $K = 0.74$ as given by Fling for beams with one end continuous when not using Eq. (3-27), and less than $K = 0.867$ by Eq. (3-40).

$$\Delta_D = K_D(5/48)(M)_D L^2/EI$$
$$= (0.760)(5/48)(2.49)(12)(20)^2(144)/(2420)(512) \qquad \text{(Eq. (3-26))}$$
$$= \underline{0.11 \text{ in. } (0.28 \text{ cm})} \qquad \text{versus 0.13 in. (0.33 cm) (15 percent difference) by Eqs. (3-37) and (3-39)}$$

$$K_{D+L} = 0.4 + 0.6(M_{D+L}/M_o) = 0.4 + 0.6(7.29/10.15)$$
$$= \underline{0.831} \qquad \text{(Eq. (3-27))}$$

which is greater than $K = 0.74$ as given by Fling for beams with one end continuous when not using Eq. (3-27), and less than $K = 0.922$ by Eq. (3-40).

$$\Delta_{D+L} = K_{D+L}(5/48)(M)_{D+L}L^2/EI$$
$$= (0.831)(5/48)(7.29)(12)(20)^2(144)/(2630)(275) \qquad \text{(Eq. (3-26))}$$
$$= \underline{0.60 \text{ in. } (1.52 \text{ cm})} \qquad \text{versus 0.67 in. (1.70 cm) (10 percent difference) by Eqs. (3-37) and (3-39)}$$

***Center span* BC**

$$(M)_D = 0.83 \text{ ft-k } (115 \text{ kg-m}, 1.13 \text{ kN-m})$$
$$(M)_{D+L} = 4.43 \text{ ft-k } (613 \text{ kg-m}, 6.01 \text{ kN-m})$$
$$K_D = 0.4 + 0.6(M_D/M_o) = 0.4 + 0.6(0.83/4.15)$$
$$= \underline{0.520} \text{ (an extrapolation)} \qquad \text{(Eq. (3-27))}$$

which is less than $K = 0.60$ as given by Fling for beams with both ends continuous when not using Eq. (3-27), and greater than $K = 0.200$ by Eq. (3-40).

$$\Delta_D = K_D(5/48)(M)_D L^2/EI$$
$$= (0.520)(5/48)(0.83)(12)(20)^2(144)/(2420)(512) \qquad \text{(Eq. (3-26))}$$
$$= \underline{0.02 \text{ in. } (0.05 \text{ cm})} \qquad \text{versus 0.01 in. (0.03 cm) (too small for difference percentage to be significant) by Eqs. (3-37) and (3-39)}$$

$$K_{D+L} = 0.4 + 0.6(M_{D+L}/M_o) = 0.4 + 0.6(4.43/10.15) = \underline{0.662} \qquad \text{(Eq. (3-27))}$$

which is greater than $K = 0.60$ as given by Fling for beams with both ends continuous when not using Eq. (3-27), and less than $K = 0.742$ by Eq. (3-40).

$$\Delta_{D+L} = K_{D+L}(5/48)(M)_{D+L} L^2/EI$$
$$= (0.662)(5/48)(4.43)(12)(20)^2(144)/(2630)(437) \qquad \text{(Eq. (3-26))}$$
$$= \underline{0.18 \text{ in. } (0.46 \text{ cm})} \qquad \text{versus 0.21 in. (0.53 cm) (14 percent difference) by Eqs. (3-37) and (3-39)}$$

Example 3 Continuous T-beam, normal-weight concrete; by the I_e method, Eq. (3-19), and using Eqs. (3-37), (3-39), (3-47*b*), and the ACI Code moment coefficients.

See Table 3-7 for the structural details, material properties, parameters, and conditions of the problem; Table 3-8 for the bending moments and strength design; Table 3-9 for the section properties; and Table 3-10 for the determination of average values of I_e or I_a for each span under dead and dead plus live load. With both ends continuous for exterior and interior spans,

$$h_{min} = L/26.3 = (29)(12)/26.3 = 13.2 \text{ in. } (33.5 \text{ cm}) \qquad \text{(Chap. 6, Table 6-5)}$$

versus

$$h = 25 \text{ in. } (63.5 \text{ cm})$$

Hence, deflections would not have to be checked by ACI 318-71 for members not supporting or attached to partitions or other construction likely to be damaged by large deflections. For other cases deflections must be checked. As shown in Table 3-7, the negative section $\rho - \rho' = 0.0076$ and 0.0239, and the positive section $\rho = 0.0025$, versus $0.75\rho_b = 0.0278$. For such a T-beam, with h_{min} being much lower than the actual h, the computed deflections might be expected to be decidedly on the safe side, which is found to be the case.

For this T-beam built integrally with the supports, the clear span is used for both the design moments and moments used in the deflection calculations (Table 3-8). The transformed compression steel area is given by $(n - 1)A'_s$ for deflection calculations, not $(2n - 1)A'_s$, as would be the case for design by the working stress method (Table 3-7). It is noted that the moment envelope is used for the strength design and for the determination of I_e, while the moment diagram for maximum positive moment and deflection is used in calculating deflections (Table 3-8).

In the negative-moment regions of T-beams, the gross-section moment of inertia, together with the relatively small value of y_t, results in a relatively large

Table 3-7 Structural details, material properties, parameters, and conditions for Example 3

Multi-span T-beam

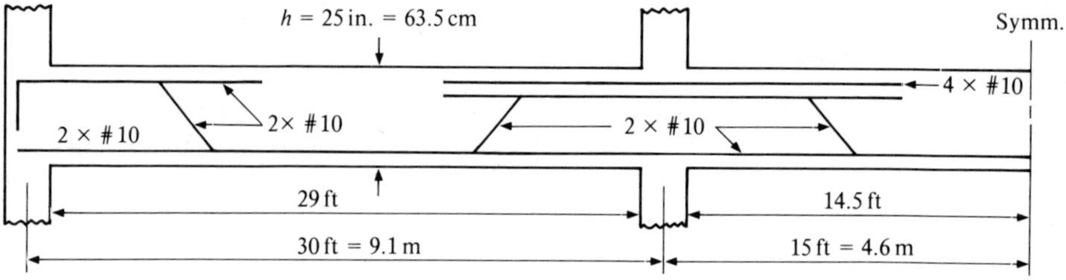

Gross section

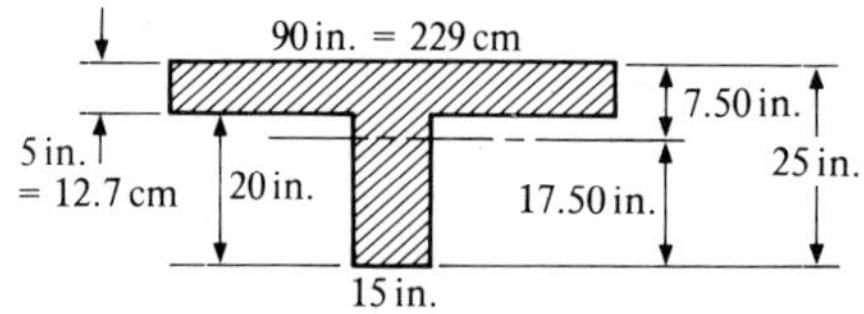

Cracked section—midspan

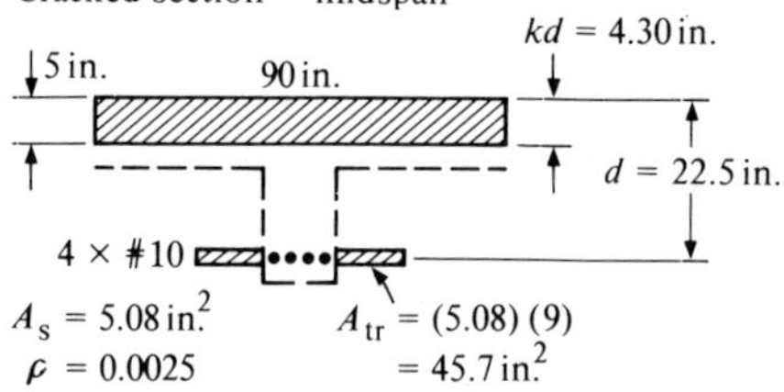

Cracked section—exterior support

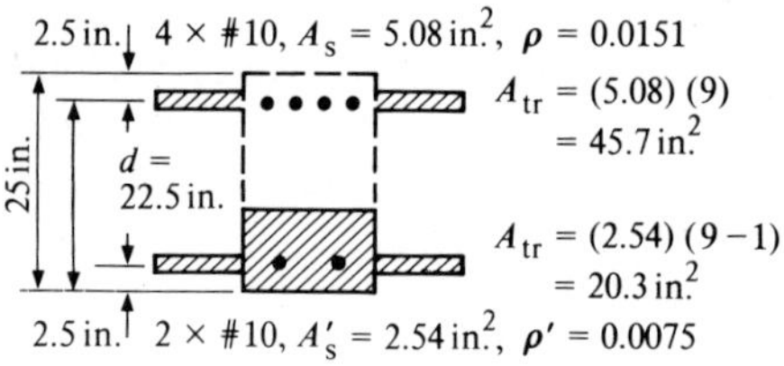

Cracked section—interior support

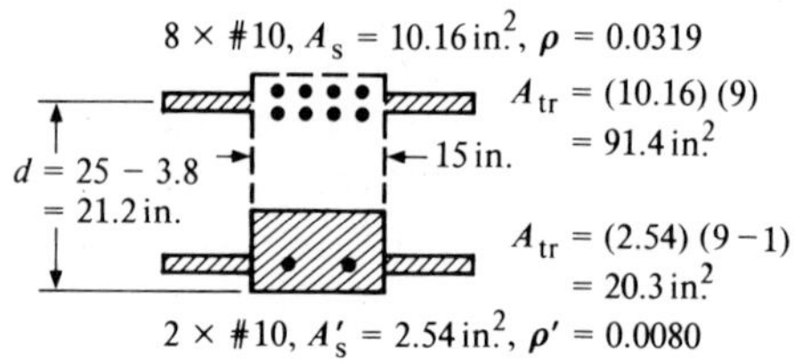

Material properties and loading

$f'_c = 3000$ psi $\quad f_y = 40\,000$ psi $\quad$ Normal-weight concrete (145 pcf)

$E_s = 29 \times 10^3$ ksi (Chap. 1)

$E_c = \underline{3.31 \times 10^3 \text{ ksi; use}}$ (Table 1-3 by author's equation (1-13), versus $E_c = 3.16 \times 10$ ksi in Table 1-3 by ACI Code)

$n = E_s/E_c = 29/3.31 = 8.76$ (ACI Code $n = 9.18$); $\underline{\text{use } n = 9}$

$f_r = \underline{429 \text{ psi; use}}$ (Table 1-2 by author's equation (1-7), versus $f_r = 411$ psi in Table 1-2 by ACI Code)

$w_{DL} = 1.5$ k/ft $\quad w_{LL} = 2.0$ k/ft

Other parameters and conditions

Average relative humidity = 60 percent $\quad$ Concrete is moist-cured

Assume 20-day loading age for dead load and Type I cement:

$f'_{ci} = 0.95(f'_c)_{28d}$ (Eq. (1-2)) $\quad$ Hence assume $E_{ci} = E_c$

Assume 3-month loading age for live load, but use 28-day E_c for live load

(For other metric units, see conversion factors herein.)

Table 3-8 Bending moments and ACI strength design for Example 3

$w_L/w_D = 2.0/1.5 < 3$; hence use ACI Code moment coefficients (Appendix F) for design moment envelope and the determination of I_e in Table 3-10

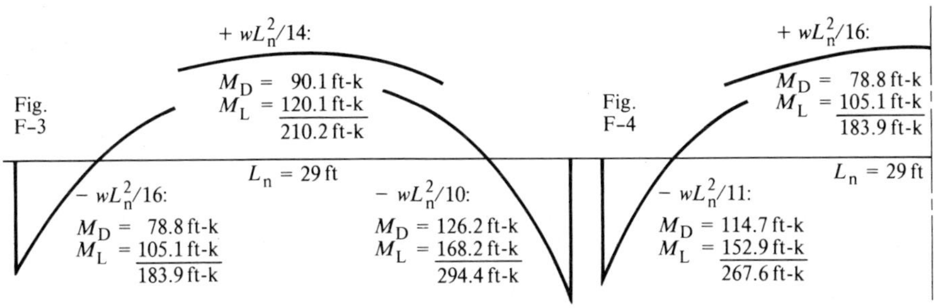

For deflections

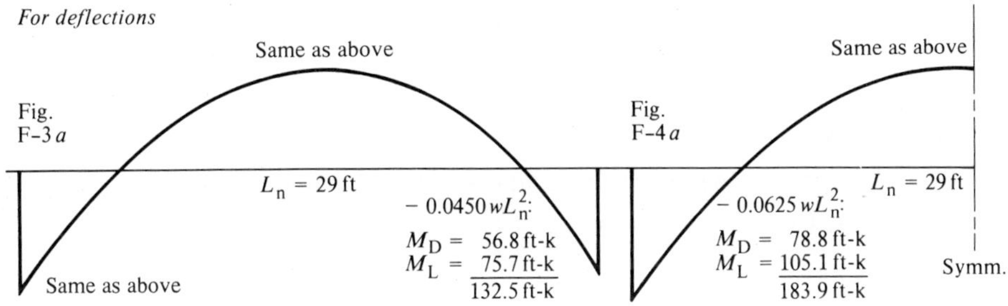

Strength design (Neglecting small compression steel effect)

Exterior support

$$M_u = 1.4M_D + 1.7M_L = 1.4(78.8) + 1.7(105.1) = \underline{289 \text{ ft-k}}$$
$$a = A_s f_y/0.85f'_c b = (5.08)(40)/(0.85)(3)(15) = 5.31 \text{ in.}$$
$$M_u = \phi A_s f_y(d - a/2) = (0.9)(5.08)(40)(22.5 - 2.65)/12 = \underline{303 \text{ ft-k}}$$

Checking:

$$\rho = A_s/bd = 5.08/(15)(22.5) = 0.0151$$
$$\omega = \rho f_y/f'_c = (0.0151)(40)/3 = 0.201$$
$$M_u = \phi b d^2 f'_c \omega(1 - 0.59\omega)$$
$$= (0.09)(15)(22.5)^2(3)(0.201)(1 - 0.59 \times 0.201)/12 = \underline{303 \text{ ft-k}}$$

Interior support

$$M_u = 1.4(126.2) + 1.7(168.2) = \underline{463 \text{ ft-k}}$$
$$a = (10.16)(40)/(0.85)(3)(15) = 10.62 \text{ in.}$$
$$M_u = (0.9)(10.16)(40)(21.2 - 5.31)/12 = \underline{484 \text{ ft-k}}$$

Checking:

$$\rho = 0.0319 \qquad \omega = 0.425$$
$$M_u = (0.9)(15)(21.2)^2(3)(0.425)(1 - 0.59 \times 0.425)/12 = \underline{483 \text{ ft-k}}$$

Exterior midspan

$$M_u = 1.4(90.1) + 1.7(120.1) = \underline{330 \text{ ft-k}}$$
$$\rho = 5.08/(90)(22.5) = 0.0025$$

Table 3-8—continued

$$\omega = (0.0025)(40)/3 = 0.0333$$

$$1.18\omega d/\beta = (1.18)(0.0333)(22.5)/0.85 = 1.04 \text{ in.} < 5 \text{ in.} \qquad \text{hence treat as rectangle}$$

$$a = (5.08)(40)/(0.85)(3)(90) = 0.89 \text{ in.}$$

$$M_u = (0.9)(5.08)(40)(22.5 - 0.44)/12 = \underline{336 \text{ ft-k}}$$

Checking:

$$M_u = (0.9)(90)(22.5)^2(3)(0.0333)(1 - 0.59 \times 0.0333)/12 = \underline{335 \text{ ft-k}}$$

Interior midspan

$$M_u = 1.4(78.8) + 1.7(105.1) = \underline{289 \text{ ft-k}} \qquad \text{Capacity same as exterior midspan}$$

(1 ft-k = 138.3 kg-m = 1.356 kN-m.)

Table 3-9 Section properties for Example 3

For sections in Table 3-7:

Gross section (T-*Beam*)

$$y_b = [(15)(20)(10) + (90)(5)(22.5)]/(300 + 450) = 17.50 \text{ in.}$$

$$I_g = (15)(17.5)^3/3 + (15)(2.5)^3/3 + (90)(5)^3/12 + (450)(5)^2 = \underline{39\,100 \text{ in}^4}$$

Checking by Fig. E-1: $h_f/h = 5/25 = 0.20 \qquad b/b_w = 90/15 = 6.0$

Find $I_g = 0.0275bh^3 = (0.0275)(90)(25)^3 = 38\,700 \text{ in}^4 \qquad \text{OK}$

Cracked section—midspan (same for exterior and interior spans)

$$(90)(kd)^2/2 = (45.7)(22.5 - kd) \qquad kd = 4.30 \text{ in.} < 5 \text{ in. (hence rect. solution)}$$

$$I_{cr} = (90)(4.30)^3/3 + (45.7)(22.5 - 4.30)^2 = \underline{17\,500 \text{ in}^4}$$

Checking by Fig. E-2: h_f/d not used in rect. sol., $\rho n = (0.0025)(9) = 0.0225$

Find $I_{cr} = 0.017bd^3 = (0.017)(90)(22.5)^3 = 17\,400 \text{ in}^4 \qquad \text{OK}$

Cracked section—exterior support

Can be calculated as shown in Table 3-4 for Example 2. However, using Fig. E-2,

$$\rho n + \rho' n' = (0.0151)(9) + (0.0075)(8) = 0.196$$

$$d'/d = 2.5/22.5 = 0.111$$

$$\rho' n'(1 - d'/d) = (0.0075)(8)(0.889) = 0.053$$

Find $I_{cr} = (0.083 - 0.005)bd^3 = (0.078)(15)(22.5)^3 = \underline{13\,300 \text{ in}^4}$

(Not used, Table 3-10)

Cracked section—interior support

Can be calculated as shown in Table 3-4 for Example 2. However, using Fig. E-2,

$$\rho n + \rho' n' = (0.0319)(9) + (0.0080)(8) = 0.351$$

$$d'/d = 2.5/21.2 = 0.118$$

$$\rho' n'(1 - d'/d) = (0.0080)(8)(0.882) = 0.056$$

Find $I_{cr} = (0.129 - 0.006)bd^3 = 0.123(15)(21.2)^3 = \underline{17\,600 \text{ in}^4}$

Additional examples of the use of the charts for determining I_g and I_{cr} are shown in Appendix E.

(1 in. = 2.54 cm, 1 in^4 = 41.62 cm^4.)

Table 3-10 Determination of average values of I_e for Example 3

Span	Load	Section	M_a, Table 3-8 ft-k	$^*\dfrac{M_a}{M_{cr}}$	$^*\dfrac{M_{cr}}{M_a}$	$\left(\dfrac{M_{cr}}{M_a}\right)^3$	I_{cr}, Table 3-9 in^4	I_e, Eq. (3-19) in^4	I_a § in^4
End span	Dead	Ext.	78.8	0.42	—	—	—	39 100†	
		Pos.	90.1	1.13	0.887	0.698	17 500	32 580‡	34 500
		Int.	126.2	0.68	—	—	—	39 100†	
	Dead + live	Ext.	183.9	0.99	—	—	—	39 100†	
		Pos.	210.2	2.63	0.380	0.055	17 500	18 690‡	22 400
		Int.	294.4	1.58	0.633	0.254	17 600	23 060‡	
Interior span	Dead	Neg.	114.7	0.62	—	—	—	39 100†	
		Pos.	78.8	0.99	—	—	—	39 100†	39 100
		Neg.	114.7	0.62	—	—	—	39 100†	
	Dead + live	Neg.	267.6	1.44	0.697	0.339	17 600	24 880‡	
		Pos.	183.9	2.30	0.434	0.082	17 500	19 280‡	21 000
		Neg.	267.6	1.44	0.697	0.339	17 600	24 880‡	

(For other metric units, see conversion factors herein.)

* Neg. $M_{cr} = f_r I_g / y_t = (429)(39\,100)/(7.50)(12\,000) = 186.4$ ft-k (Eq. (3-2))

Pos. $M_{cr} = (429)(39\,100)/(17.50)(12\,000) = 79.9$ ft-k (Eq. (3-2))

† $I_g = 39\,100$ in^4 (1 627 000 cm^4) (Table 3-9)

‡ $I_e = (M_{cr}/M_a)^3 I_g + [1 - (M_{cr}/M_a)^3] I_{cr}$ (Eq. (3-19))

$= (0.698)(39\,100) + [1 - 0.698](17\,500) = 32\,580$ in^4 (1 356 000 cm^4)

$= (0.055)(39\,100) + [1 - 0.055](17\,500) = 18\,690$ in^4 (778 000 cm^4)

$= (0.254)(39\,100) + [1 - 0.254](17\,600) = 23\,060$ in^4 (960 000 cm^4)

$= (0.339)(39\,100) + [1 - 0.339](17\,600) = 24\,880$ in^4 (1 036 000 cm^4)

$= (0.082)(39\,100) + [1 - 0.082](17\,500) = 19\,280$ in^4 (802 000 cm^4)

§ $I_a = [0.70 I_m + 0.15(I_{e1} + I_{e2})]$ (Eq. (3-47*b*))

$= [0.70(32\,580) + 0.15(39\,100 + 39\,100)] = 34\,500$ in^4 (1 436 000 cm^4)

$= [0.70(18\,690) + 0.15(39\,100 + 23\,060)] = 22\,400$ in^4 (932 000 cm^4)

$= [0.70(19\,280) + 0.15(24\,880 + 24\,880)] = 21\,000$ in^4 (874 000 cm^4)

cracking moment. Thus, as can be seen in Table 3-10, the exterior support section remained uncracked under dead load and also dead plus live load in this example. Also, the entire interior span remained uncracked under dead load only (Table 3-10).

It has been mentioned that when M_a/M_{cr} is greater than about 3, I_e may be taken as I_{cr}. For the largest such ratio in Table 3-10, $M_a/M_{cr} = 2.63$, and $I_e = 18\,690$ in^4 (778 000 cm^4), versus $I_{cr} = 17\,500$ in^4 (728 000 cm^4), for example. The average effective *I*s (I_a) were: for dead load only, 34 500 in^4 (1 436 000 cm^4)—end span, 39 100 in^4 (1 627 000 cm^4) (I_g)—interior span; and for dead plus live load, 22 400 in^4 (932 000 cm^4)—end span, 21 000 in^4 (874 000 cm^4)—interior span (Table 3-10).

Exterior (end) span

$$\Delta_D = \frac{5L_n^2}{48E_c I_a}\left[M_m + \tfrac{1}{10}(M_{e1} + M_{e2})\right]$$

$$= \frac{(5)(29)^2(144)}{(48)(3310)(34\,500)}[90.1 - (0.10)(78.8 + 56.8)](12)$$

$$= \frac{3.811}{34\,500}(76.5)(12) = \underline{0.10 \text{ in. } (0.25 \text{ cm})} \qquad \text{(Eq. (3-37))}$$

$$\Delta_{D+L} = \frac{3.811}{22\,400}[210.2 - (0.10)(183.9 + 132.5)](12)$$

$$= \underline{0.36 \text{ in. } (0.91 \text{ cm})} \qquad \text{(Eq. (3-37))}$$

$$\Delta_L = 0.36 - 0.10 = \underline{0.26 \text{ in. } (0.66 \text{ cm})}$$

Interior span

$$\Delta_D = \frac{3.811}{39\,100}[78.8 - (0.10)(78.8 + 78.8)](12)$$

$$= \underline{0.07 \text{ in. } (0.18 \text{ cm})} \qquad \text{(Eq. (3-37))}$$

$$\Delta_{D+L} = \frac{3.811}{21\,000}[183.9 - (0.10)(183.9 + 183.9)](12)$$

$$= \underline{0.32 \text{ in. } (0.81 \text{ cm})} \qquad \text{(Eq. (3-37))}$$

$$\Delta_L = 0.32 - 0.07 = \underline{0.25 \text{ in. } (0.64 \text{ cm})}$$

As shown in Chap. 6, the allowable deflections based on live load only by ACI 318-71 are: (1) flat roofs not supporting or attached to nonstructural elements likely to be damaged by large deflections—$\Delta_L = L/180 \le (29)(12)/180 = 1.93$ in. (4.90 cm); and (2) floors not supporting or attached to nonstructural elements likely to be damaged by large deflections—$\Delta_L \le L/360 = (29)(12)/360 = 0.97$ in. (2.46 cm). The maximum computed live-load deflection is 0.26 in. (0.66 cm). Hence both spans have no limitations so far as live-load deflections are concerned.

The above solution differs from ACI 318-71 in that $f_r = 429$ psi (30.2 kg/cm^2, 2.96 N/mm^2) and $E_c = 3.31 \times 10^3$ ksi (233 000 kg/cm^2, 22.8 kN/mm^2) instead of 411 psi (28.6 kg/cm^2, 2.82 N/mm^2) and 3.16×10^3 ksi (222 000 kg/cm^2, 21.8 kN/mm^2) by ACI (Table 3-7), and Eq. (3-47*b*) was used for I_a instead of the simple average of Eq. (3-46).

Using just the midspan value of I_e, as with Eq. (3-44):

Exterior span

$$\Delta_D = (34\,500/32\,580)(0.10) = \underline{0.11 \text{ in. } (0.28 \text{ cm})}$$

versus 0.10 in. (0.25 cm) or *10 percent greater.*

$$\Delta_{D+L} = (22\,400/18\,690)(0.36) = \underline{0.43 \text{ in. } (1.09 \text{ cm})}$$

versus 0.36 in. (0.91 cm) or *19 percent greater.*

$$\Delta_L = 0.43 - 0.11 = \underline{0.32 \text{ in. } (0.81 \text{ cm})}$$

versus 0.26 in. (0.66 cm) or *23 percent greater.*

Interior span

$$\Delta_D = \underline{0.07 \text{ in. } (0.18 \text{ cm})}$$

(*same, since entire span was uncracked.*)

$$\Delta_{D+L} = (21\,000/19\,280)(0.32) = \underline{0.35 \text{ in. } (0.89 \text{ cm})}$$

versus 0.32 in. (0.81 cm) or *9 percent greater.*

$$\Delta_L = 0.35 - 0.07 = \underline{0.28 \text{ in. } (0.71 \text{ cm})}$$

versus 0.25 in. (0.64 cm) or *12 percent greater.*

As can be seen in Table 3-8 and Appendix F, the maximum positive moments were used in computing deflections with Eq. (3-37). Since this equation refers to the midspan or centerline positive moment, a slight inaccuracy results.

Using the *alternative* procedure shown in Fig. 3-8 and described by Eqs. (3-35) and (3-36), the effect of repeated live load is computed as follows (the values of I_e are the I_a values in Table 3-10):

Exterior span

$$(\Delta_D)_2 = KM_D L^2/E_{ci}(I_e)_{D+L} = \Delta_D(I_e)_D/(I_e)_{D+L}$$

$$= 0.10(34\,500/22\,400) = \underline{0.15 \text{ in. } (0.38 \text{ cm})} \qquad \text{(Eq. (3-36))}$$

$$\Delta_L = KM_L L^2/E_c(I_e)_{D+L} = \frac{3.811}{22\,400}[120.1 - (0.10)(105.1 + 75.7)](12)$$

$$= \underline{0.21 \text{ in. } (0.53 \text{ cm})} \qquad \text{(Eqs. (3-35), (3-37))}$$

Interior span

$$(\Delta_D)_2 = (0.07)(39\,100/21\,000) = \underline{0.13 \text{ in. } (0.33 \text{ cm})} \qquad \text{(Eq. (3-36))}$$

$$\Delta_L = \frac{3.811}{21\,000}[105.1 - (0.10)(105.1 + 105.1)](12)$$

$$= \underline{0.18 \text{ in. } (0.46 \text{ cm})} \qquad \text{(Eqs. (3-35), (3-37))}$$

The following solutions by Eq. (3-39) are identical to the previous results using Eq. (3-37), and illustrate the corresponding K factors.

For both exterior and interior spans,

$$(M_o)_D = w_D L^2/8 = (1.5)(29)^2/8 = 157.7 \text{ ft-k } (21\,800 \text{ kg-m}, 213.8 \text{ kN-m})$$

$$(M_o)_{D+L} = (3.5)(29)^2/8 = 367.9 \text{ ft-k } (50\,900 \text{ kg-m}, 498.9 \text{ kN-m})$$

Exterior span

$$(M)_D = 90.1 \text{ ft-k } (12\,500 \text{ kg-m}, 122.2 \text{ kN-m})$$

$$(M)_{D+L} = 210.2 \text{ ft-k } (29\,100 \text{ kg-m}, 285.0 \text{ kN-m})$$

$$K_D = 1.20 - 0.20 M_o/M_m$$

$$= 1.20 - (0.20)(157.7/90.1) = \underline{0.850} \qquad \text{(Eq. (3-40))}$$

$$\Delta_D = K_D(5/48)(M)_D L^2/E_c I_a$$

$$= (0.850)(5/48)(90.1)(12)(29)^2(144)/(3310)(34\,500)$$

$$= \underline{0.10 \text{ in. } (0.25 \text{ cm})} \qquad \text{(Eq. (3-39))}$$

$$K_{D+L} = 1.20 - 0.20 M_o/M_m = 1.20 - (0.20)(367.9/210.2)$$

$$= \underline{0.850} \qquad \text{(Eq. (3-40))}$$

$$\Delta_{D+L} = K_{D+L}(5/48)(M)_{D+L} L^2/E_c I_a$$

$$= (0.850)(5/48)(210.2)(12)(29)^2(144)/(3310)(22\,400)$$

$$= \underline{0.36 \text{ in. } (0.91 \text{ cm})} \qquad \text{(Eq. (3-39))}$$

Interior span

$$(M)_D = 78.8 \text{ ft-k } (10\,900 \text{ kg-m}, 106.9 \text{ kN-m})$$

$$(M)_{D+L} = 183.9 \text{ ft-k } (25\,400 \text{ kg-m}, 249.4 \text{ kN-m})$$

$$K_D = 1.20 - 0.20 M_o/M_m = 1.20 - (0.20)(157.7/78.8)$$

$$= \underline{0.800} \qquad \text{(Eq. (3-40))}$$

$$\Delta_D = K_D(5/48)(M)_D L^2/E_c I_a$$

$$= (0.800)(5/48)(78.8)(12)(29)^2(144)/(3310)(39\,100)$$

$$= \underline{0.07 \text{ in. } (0.18 \text{ cm})} \qquad \text{(Eq. (3-39))}$$

$$K_{D+L} = 1.20 - 0.20 M_o/M_m = 1.20 - (0.20)(367.9/183.9)$$

$$= \underline{0.800} \qquad \text{(Eq. (3-40))}$$

$$\Delta_{D+L} = K_{D+L}(5/48)(M)_{D+L} L^2/E_c I_a$$

$$= (0.800)(5/48)(183.9)(12)(29)^2(144)/(3310)(21\,000)$$

$$= \underline{0.32 \text{ in. } (0.81 \text{ cm})} \qquad \text{(Eq. (3-39))}$$

The above K values might be compared with the ideal values given with Eq. (3-39) of $K = 0.800$ for fixed–hinged beams (end span) and $K = 0.600$ for fixed–fixed beams (center span).

Using Fling's continuity equations (3-26) and (3-27) with the previous values of I_e and I_a: for both exterior and interior spans,

$$(M_o)_D = w_D L^2/8 = (1.5)(29)^2/8 = 157.7 \text{ ft-k } (21\,800 \text{ kg-m, } 213.8 \text{ kN-m})$$
$$(M_o)_{D+L} = (3.5)(29)^2/8 = 367.9 \text{ ft-k } (50\,900 \text{ kg-m, } 498.9 \text{ kN-m})$$

Exterior span

$$(M)_D = 90.1 \text{ ft-k } (12\,500 \text{ kg-m, } 122.2 \text{ kN-m})$$
$$(M)_{D+L} = 210.2 \text{ ft-k } (29\,100 \text{ kg-m, } 285.0 \text{ kN-m})$$
$$K_D = 0.4 + 0.6(M_D/M_o) = 0.4 + 0.6(90.1/157.7) = \underline{0.743} \qquad \text{(Eq. (3-27))}$$

which is essentially the same as $K = 0.74$ as given by Fling for one end continuous, although both ends are restrained in this exterior span, and less than $K = 0.850$ by Eq. (3-40).

$$\Delta_D = K_D(5/48)(M)_D L^2/EI$$
$$= (0.743)(5/48)(90.1)(12)(29)^2(144)/(3310)(34\,500) \qquad \text{(Eq. (3-26))}$$
$$= \underline{0.089 \text{ in. } (0.23 \text{ cm})} \qquad \text{versus 0.10 in. (0.25 cm) (11 percent difference) by Eqs. (3-37) and (3-39)}$$
$$K_{D+L} = 0.4 + 0.6(M_{D+L}/M_o) = 0.4 + 0.6(210.2/367.9) = \underline{0.743} \qquad \text{(Eq. (3-27))}$$

which is the same as for dead load only above, and less than $K = 0.850$ by Eq. (3-40).

$$\Delta_{D+L} = K(5/48)(M)_{D+L} L^2/EI$$
$$= (0.743)(5/48)(210.2)(12)(29)^2(144)/(3310)(22\,400) \qquad \text{(Eq. (3-26))}$$
$$= \underline{0.32 \text{ in. } (0.81 \text{ cm})} \qquad \text{versus 0.36 in. (0.91 cm) (11 percent difference) by Eqs. (3-37) and (3-39)}$$

Interior span

$$(M)_D = 78.8 \text{ ft-k } (10\,900 \text{ kg-m, } 106.9 \text{ kN-m})$$
$$(M)_{D+L} = 183.9 \text{ ft-k } (25\,400 \text{ kg-m, } 249.4 \text{ kN-m})$$
$$K_D = 0.4 + 0.6(M_D/M_o) = 0.4 + 0.6(78.8/157.7) = \underline{0.700} \qquad \text{(Eq. (3-27))}$$

which is greater than $K = 0.60$ as given by Fling for beams with both ends continuous when not using Eq. (3-27), and less than $K = 0.800$ by Eq. (3-40).

$$\Delta_D = K_D(5/48)(M)_D L^2/EI$$
$$= (0.700)(5/48)(78.8)(12)(29)^2(144)/(3310)(39\,100) \qquad \text{(Eq. (3-26))}$$
$$= \underline{0.065 \text{ in. } (0.17 \text{ cm})} \qquad \text{versus 0.07 in. (0.18 cm) (7 percent difference) by Eqs. (3-37) and (3-39)}$$
$$K_{D+L} - 0.4 + 0.6(M_{D+L}/M_o) = 0.4 + 0.6(183.9/367.9) = \underline{0.700} \qquad \text{(Eq. (3-27))}$$

which is the same as for dead load only above, and less than $K = 0.800$ by Eq. (3-40).

$$\Delta_{D+L} = K(5/48)(M)_{D+L}L^2/EI$$
$$= (0.700)(5/48)(183.9)(12)(29)^2(144)/(3310)(21\,000) \qquad \text{(Eq. (3-26))}$$
$$= \underline{0.28 \text{ in. } (0.71 \text{ cm})} \qquad \text{versus 0.32 in. (0.81 cm) (12 percent difference) by Eqs. (3-37) and (3-39)}$$

Example 4 Interior girder of simple-span T-beam or deck girder bridge, normal-weight concrete; by the I_e method, Eq. (3-19); the BPR method, I_g, Eq. (3-20), I_{cr}; the 1973 AASHTO method, I_e and I_{cr}; and ACI Committee 443 method, I_e.

See Table 3-11 for the structural details, material properties, parameters, and conditions; and Table 3-12 for the loads and moments, AASHTO load factor design, and section properties.

I_e method, Eq. (3-19)

$$M_{cr} = f_r I_g/y_t = (496)(103\,820)/23.74 = 2\,169\,000 \text{ in-lb}$$
$$= 180.8 \text{ ft-k } (25\,000 \text{ kg-m, } 245.2 \text{ kN-m}) \qquad \text{(Eq. (3-2))}$$
$$M_D/M_{cr} = 406.4/180.8 = 2.25 \qquad M_{cr}/M_D = 0.445 \qquad 0.445^3 = 0.088$$
$$(I_e)_D = (M_{cr}/M_D)^3 I_g + [1 - (M_{cr}/M_D)^3] I_{cr} \qquad \text{(Eq. (3-19))}$$
$$= (0.088)(103\,820) + (1 - 0.088)(78\,060)$$
$$= \underline{80\,330 \text{ in}^4 \ (3\,343\,000 \text{ cm}^4)}$$
$$M_{D+L+I}/M_{cr} = 882.7/180.8 = 4.88 > 3 \qquad (\text{hence } I_e = \text{approximately } I_{cr})$$
$$M_{cr}/M_{D+L+I} = 180.8/882.7 = 0.205 \qquad 0.205^3 = 0.0086$$
$$(I_e)_{D+L+I} = (0.0086)(103\,820) + (1 - 0.0086)(78\,060)$$
$$= \underline{78\,280 \text{ in}^4 \ (3\,258\,000 \text{ cm}^4)}$$

This relatively shallow design with high steel content is within the AASHTO minimum depth for long-time deflection, and is marginal by the BPR deflection criteria, as will be shown. For such a design, it may be noted that I_e (80 330 in^4 for dead load and 78 280 in^4 for total load) is close to I_{cr} (78 060 in^4 or 3 249 000 cm^4) for both dead load and total load.

$$\Delta_D = \frac{KM_D L^2}{E_{ci}(I_e)_D} = \frac{(5/48)(406.4)(12)(60)^2(144)}{(3220)(80\,330)}$$
$$= \frac{26\,336 \times 10^4}{25\,866 \times 10^4} = \underline{1.02 \text{ in. } (2.59 \text{ cm})} \qquad \text{(Eq. (3-1))}$$
$$\Delta_{D+L+I} = \frac{KM_{D+L+I}L^2}{E_c(I_e)_{D+L+I}} = \frac{(5/48)(882.7)(12)(60)^2(144)}{(3640)(78\,280)}$$
$$= \frac{57\,200 \times 10^4}{28\,494 \times 10^4} = \underline{2.01 \text{ in. } (5.11 \text{ cm})}$$
$$\Delta_{L+I} = 2.01 - 1.02(E_{ci}/E_c) = 2.01 - 1.02(3.22/3.64)$$
$$= 2.01 - 0.90 = \underline{1.11 \text{ in. } (2.82 \text{ cm})}$$

Table 3-11 Structural details, material properties, parameters, and conditions for Example 4

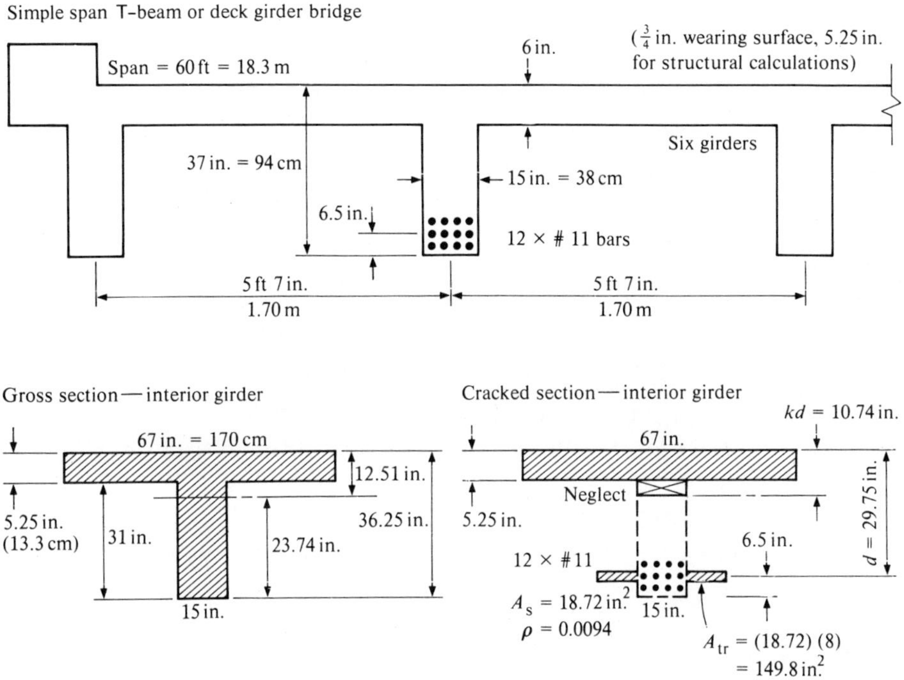

Material properties and conditions

Monolithic T-beam Span center-to-center of bearings = 60 ft 2 lanes

Curb-to-curb = 28 ft 6 in. slab including $\frac{3}{4}$ in. wearing surface LL = HS 20-44

$f'_c = 4000$ psi Grade 40 reinforcement Normal-weight concrete (145 pcf)

$$E_s = 29 \times 10^3 \text{ ksi (AASHTO and Chap. 1)}$$

$E_c = \underline{3.64 \times 10^3 \text{ ksi}}$ (by author's equation (1-13) and AASHTO—same as ACI Code, Table 1-3)

$n = E_s/E_c = 29/3.64 = 7.97$ (use $\underline{n = 8}$ in all calculations)

$$f_r = \underline{496 \text{ psi}} \text{ by author's equation 1-7 (Table 1-2)}$$

$f_r = \underline{475 \text{ psi}}$ by ACI Code (Table 1-2), which is the same as by AASHTO

Average relative humidity = 70 percent Concrete is moist-cured

Assume 7-day loading age for dead load and Type I cement:

$$f'_{ci} = 0.70(f'_c)_{28d} = 2800 \text{ psi} \qquad \text{(Eq. (1-2))}$$

Hence $\qquad E_{ci} = \underline{3.22 \times 10^3 \text{ ksi}} \qquad \text{(Eq. (1-13))}$

Assume 3-month loading age for live load, but use 28-day E_c for live load

(For other metric units, see conversion factors herein.)

Table 3-12 Loads and moments, AASHTO load factor design, and section properties for Example 4

Loads and moments

$$w_{DL} = (150/144)(67 \times 6 + 15 \times 31) = 903 \ \#/\text{ft}$$
$$M_D = (0.903)(60)^2/8 = \underline{406.4 \text{ ft-k}}$$
$$M_L \text{ (absolute max. per axle)} = 806.5 \text{ ft-k (AASHTO Ap. A)}$$
$$M_L \text{ (per interior girder)} = \tfrac{1}{2}(S/6.0)(806.5) = 375.0 \text{ ft-k } (S = 5.58 \text{ ft})$$
$$I = 50/(60 + 125) = 0.27$$
$$M_{L+I} = (1.27)(375.0) = \underline{476.3 \text{ ft-k}} \qquad M_{D+L+I} = \underline{882.7 \text{ ft-k}}$$

AASHTO load factor design

$$M_u = 1.30(M_D + \tfrac{5}{3}M_{L+I}) = 1.30(406.4 + \tfrac{5}{3} \times 476.3) = \underline{1560 \text{ ft-k}}$$
$$A_{sf} = 0.85(b - b')tf_c'/f_y = (0.85)(67 - 15)(5.25)(4/40) = 23.2 \text{ in}^2$$

which is greater than $A_s = 18.72 \text{ in}^2$; hence use rect. beam solution

$$a = A_s f_y/0.85f_c'b = (18.72)(40)/(0.85)(4)(67) = 3.28 \text{ in.}$$
$$M_u = \phi A_s f_y(d - a/2) = (0.9)(18.72)(40)(29.75 - 1.64) = \underline{1579 \text{ ft-k}}$$

Checking: $p = A_s/bd = 18.72/(67)(29.75) = 0.0094$

$$q = pf_y/f_c' = (0.0094)(40/4) = 0.094$$
$$M_u = \phi A_s f_y d(1 - 0.6q) = (0.9)(18.72)(40)(29.75)(1 - 0.6 \times 0.094) = \underline{1577 \text{ ft-k}}$$

Section properties

Gross section—interior girder:

$$y_b = [(15)(31)(15.5) + (67)(5.25)(36.25 - 2.63)]/(465 + 352) = 23.74 \text{ in.}$$
$$I_g = (15)(23.74)^3/3 + (15)(7.26)^3/3 + (67)(5.25)^3/12 + (67)(5.25)(12.51 - 2.63)^2 = \underline{103\,820 \text{ in}^4}$$

Checking by Fig. E-1:

$$h_f/h = 5.25/36.25 = 0.145 \qquad b/b_w = 67/15 = 4.47$$

Find $I_g = 0.0325bh^3 = (0.0325)(67)(36.25)^3 = 103\,700 \text{ in}^4$ OK

Cracked section—interior girder: neglecting web in compression,

$(67)(5.25)(kd - 2.63) = (18.72)(8)(29.75 - kd) \qquad kd = 10.74 \text{ in.} > 5.25 \text{ in.}$

Hence use T-beam solution:

$$I_{cr} = (67)(5.25)^3/12 + (67)(5.25)(10.74 - 2.63)^2 + (18.72)(8)(29.75 - 10.74)^2 = \underline{78\,060 \text{ in}^4}$$

Checking by Fig. E-2,

$$\rho = 0.0094 \qquad \rho n = (0.0094)(8) = 0.0752 \qquad h_f/d = 5.25/29.75 = 0.176$$

Find $I_{cr} = 0.044bd^3 = (0.044)(67)(29.75)^3 = 77\,600 \text{ in}^4$ OK

(For metric units, see conversion factors herein.)

Even though the live-load moment diagram is due to wheel loading and the absolute maximum live-load moment does not occur at midspan, it is deemed appropriate in total load deflection calculations to use Eq. (3-37) for continuous spans and Eq. (3-1), as if uniformly loaded, for simple spans.

I_e method, together with I_{rep}, for computing deflections before and after passage of the design live load (Fig. 3-8, Eqs. (3-29) to (3-36))

$\Delta_{D+L+I} = \underline{2.01\text{ in. }(5.11\text{ cm})}$ (Eq. (3-29)) (same as direct I_e solution)

Before passage of live load, and upon first passage of live load:

$(\Delta_D)_1 = \underline{1.02\text{ in. }(2.59\text{ cm})}$ (Eq. (3-30))

$\Delta_{L+I} = \underline{1.11\text{ in. }(2.82\text{ cm})}$ (Eq. (3-31)) (same as direct I_e solution)

After passage of live load:

$$\psi = (M_{ult} - M_{D+L+I})/(M_{ult} - M_{cr})$$
$$= (1578/0.9 - 882.7)/(1578/0.9 - 180.8) = 0.554$$

$$I_{rep} = \psi(I_e)_{D+L+I} + (1 - \psi)I_g = (0.554)(78\,280) + (1 - 0.554)(103\,820)$$
$$= \underline{89\,670\text{ in}^4\ (3\,732\,000\text{ cm}^4)} \qquad \text{(Eq. (3-33))}$$

$$\Delta_{L+I} = \frac{KM_{L+I}L^2}{E_c I_{rep}} = \frac{(5/48)(476.3)(12)(60)^2(144)}{(3640)(89\,670)}$$
$$= \frac{30\,865 \times 10^4}{32\,640 \times 10^4} \qquad \text{(Eq. (3-32))}$$
$$= \underline{0.95\text{ in. }(2.41\text{ cm})}$$

versus 1.11 in. (2.82 cm) upon first passage of live load

$$(\Delta_D)_2 = [\Delta_{D+L+I}\text{ in Eq. (3-29)} - \Delta_{L+I}\text{ in Eq. (3-32)}](E_c/E_{ci}) \qquad \text{(Eq. (3-34))}$$
$$= (2.01 - 0.95)(3.64/3.22) = \underline{1.20\text{ in. }(3.05\text{ cm})}$$

versus $(\Delta_D)_1 = 1.02$ in. (2.59 cm) before live load

Using the alternative solution for $(\Delta_D)_2$, which avoids the necessity of having to calculate I_{rep},

$$\Delta_{L+I} = \frac{KM_{L+I}L^2}{E_c(I_e)_{D+L+I}} = \frac{(5/48)(476.3)(12)(60)^2(144)}{(3640)(78\,280)}$$
$$= \frac{30\,865 \times 10^4}{28\,494 \times 10^4} \qquad \text{(Eq. (3-35))}$$
$$= \underline{1.08\text{ in. }(2.74\text{ cm})}$$

versus 1.11 in. (2.82 cm) in the direct I_e solution and 0.95 in. (2.41 cm) in the I_{rep} solution

$$(\Delta_D)_2 = [\Delta_{D+L+I}\text{ in Eq. (3-29)} - \Delta_{L+I}\text{ in Eq. (3-35)}](E_c/E_{ci}) \qquad \text{(Eq. (3-36))}$$
$$= (2.01 - 1.08)(3.64/3.22)$$
$$= \underline{1.05\text{ in. }(2.67\text{ cm})}$$

versus 1.20 in. (3.05 cm) in the I_{rep} solution and $(\Delta_D)_1 = 1.02$ in. (2.59 cm) before passage of live load

The results in this alternative solution and in the direct I_e solution were nearly the same only because in this particular example $(I_e)_D = 80\,330$ in^4 (3 343 000 cm^4) and $(I_e)_{D+L+I} = 78\,280$ in^4 (3 258 000 cm^4) were nearly the same.

BPR method (I_g, Eq. (3-20), and I_{cr})

Instantaneous deflections caused by dead load

$$f_s = M_D/A_s\,jd = (406\,400)(12)/(18.72)(29.75 - 10.74/3)$$
$$= 9950 \text{ psi } (70.0 \text{ kg/cm}^2, 6.86 \text{ N/mm}^2)$$

which is less than $f_y/3 = 40\,000/3 = 13\,330$ psi; hence use I_g.

$$\Delta_D = \frac{KM_DL^2}{E_{ci}I_g} = \frac{(5/48)(406.4)(12)(60)^2(144)}{(3220)(103\,820)} = \frac{26\,336 \times 10^4}{33\,430 \times 10^4}$$
$$= \underline{0.79 \text{ in. } (2.01 \text{ cm})}$$

versus 1.02 in. (2.59 cm) by the I_e method, since $M_D/M_{cr} = 2.25$.

Short-time deflections caused by dead loads after passage of the working live load

$$f_s = M_{D+L+I}/A_s\,jd = (882\,700)(12)/(18.72)(29.75 - 10.74/3)$$
$$= 21\,620 \text{ psi } (152.0 \text{ kg/cm}^2, 14.91 \text{ N/mm}^2)$$

$13\,330 < 21\,620 < 40\,000$ ($f_y/3 < f_s \le f_y$); hence use Eq. (3-20).

$$I = I_g - \tfrac{3}{2}(I_g - I_{cr})(f_s/f_y - \tfrac{1}{3})$$
$$= 103\,820 - (1.5)(103\,820 - 78\,060)(21\,620/40\,000 - 0.3333)$$
$$= 103\,820 - 8010 = \underline{95\,810 \text{ in}^4 \text{ } (3\,988\,000 \text{ cm}^4)}$$

versus 80 330 in^4 (3 343 000 cm^4) by the I_e method.

$$\Delta_D = \frac{KM_DL^2}{E_{ci}I} = \frac{26\,336 \times 10^4}{(3220)(95\,810)} = \underline{0.85 \text{ in. } (2.16 \text{ cm})}$$

versus 1.05 in. (2.67 cm) by the I_e method and 1.20 in. (3.05 cm) by the I_{rep} method.

Live load plus impact deflections

$$\Delta_{L+I} = \frac{M_{L+I}L^2}{10E_cI_{cr}} = \frac{(476.3)(12)(60)^2(144)}{(10)(3640)(78\,060)} = \frac{29\,630 \times 10^5}{28\,414 \times 10^5}$$
$$= \underline{1.04 \text{ in. } (2.64 \text{ cm})} \qquad \text{(BPR Eq.)}$$

versus 1.11 in. (2.82 cm) by the I_e method and 0.95 in. (2.41 cm) by the I_{rep} method.

The allowable live-load plus impact deflections by BPR are as follows (Chap. 6): in general (and using $w_{L+I}/w_T = M_{L+I}/M_{D+L+I}$),

$$\Delta_{L+I} \leq \frac{w_{L+I}}{w_T(4.555 - 0.0364L)^2} = \frac{476.3}{(882.7)(4.555 - 0.0364 \times 60)^2}$$
$$= 0.0960 \text{ ft} = 1.15 \text{ in. } (2.92 \text{ cm})$$

versus $\Delta_{L+I} = 1.04$ in. (2.64 cm) OK

For urban areas (which includes pedestrian travel), $L > 42.4$ ft (12.9 m),

$$\Delta_{L+I} \leq w_{L+I}/9w_T = (476.3)/(9)(882.7) = 0.072 \text{ ft} = 0.86 \text{ in. } (2.18 \text{ cm})$$

versus $\Delta_{L+I} = 1.04$ in. (2.64 cm) NG

Additional examples of the BPR method are presented in Ref. 47.

It should be noted that an interior girder (center-to-center portion) together with the AASHTO design wheel distribution was used in these deflection calculations, and not the combined superstructure together with a full truck load in each lane. The computed deflections in the example were higher (on safe side) than they would have been by the latter approach, since the interior girder carried $(\frac{1}{2})(5.58/6.0)(\frac{1}{2})(100) = 23$ percent of two truck loads in the example versus about $(\frac{1}{6})(100) = 17$ percent of the total load carried by each girder in the latter method. An actual comparison would, of course, depend on the effect of the curbs, sidewalks, and exterior girders.

1973 AASHTO method (I_e and I_{cr})—load factor design

$$f_r = 7.5\sqrt{f'_c} = 475 \text{ psi } (33.4 \text{ kg/cm}^2, 3.28 \text{ N/mm}^2) \quad \text{(Table 3-11)}$$
$$M_{cr} = f_r I_g / y_t = (475)(103\,820)/23.74 = 2\,077\,000 \text{ in-\#}$$
$$= 173.1 \text{ ft-k } (23\,940 \text{ kg-m, } 234.7 \text{ kN-m})$$
$$M_D/M_{cr} = 406.4/173.1 = 2.35 \qquad M_{cr}/M_D = 0.426 \qquad 0.426^3 = 0.077$$
$$(I_e)_D = (M_{cr}/M_D)^3 I_g + [1 - (M_{cr}/M_D)^3] I_{cr}$$
$$= (0.077)(103\,820) + (1 - 0.077)(78\,060)$$
$$= \underline{80\,040 \text{ in}^4 \ (3\,331\,000 \text{ cm}^4)}$$

versus 80 330 in^4 (3 343 000 cm^4) (small difference) using the author's value of f_r

$$\Delta_D = \frac{KM_D L^2}{E_{ci}(I_e)_D} = \frac{(5/48)(406.4)(12)(60)^2(144)}{(3220)(80\,040)} = \frac{26\,336 \times 10^4}{25\,773 \times 10^4}$$
$$= \underline{1.02 \text{ in. } (2.59 \text{ cm})}$$

$$\Delta_{D+L+I} = \frac{KM_{D+L+I}L^2}{E_c I_{cr}} = \frac{(5/48)(882.7)(12)(60)^2(144)}{(3640)(78\,060)} = \underline{2.01 \text{ in. } (5.11 \text{ cm})}$$

$$\Delta_{L+I} = 2.01 - 1.02(E_{ci}/E_c) = 2.01 - 1.02(3.22/3.64) = \underline{1.11 \text{ in. } (2.82 \text{ cm})}$$

ACI Committee 443, concrete bridge design

$$f_r = 475 \text{ psi} \qquad \text{(same as AASHTO and ACI)}$$

$$\left.\begin{aligned}(I_e)_D &= \underline{80\,040 \text{ in}^4}\\ (I_e)_{D+L+I} &= \underline{78\,250 \text{ in}^4}\end{aligned}\right\} \text{versus} \left\{\begin{aligned}&80\,330 \text{ in}^4\\ &78\,280 \text{ in}^4\end{aligned}\right. \text{(small difference) using the author's value of } f_r$$

$$\Delta_D = \underline{1.02 \text{ in.}} \qquad \Delta_{D+L+I} = \underline{2.01 \text{ in.}} \qquad \Delta_{L+I} = \underline{1.11 \text{ in.}}$$

These results are the same as by the I_e and AASHTO solutions above. Dead load due to diaphragms was not included in this example.

3-3 DEFLECTION DUE TO SHRINKAGE AND DIFFERENTIAL TEMPERATURE

Concrete shrinkage in both statically determinate and indeterminate reinforced concrete structures induces compressive stresses in the steel which are equilibrated by tensile stresses in the concrete. When the reinforcement is unsymmetrical, the resulting nonuniform strain distribution and accompanying warping cause deflections in the same direction as those caused by the loads for which the structure was designed and reinforced.

The concrete shrinkage stresses can have an effect on cracking although the extent of such an effect is most uncertain. For one thing, much of the shrinkage may occur before the application of the design live load and before cracking. As shown in Table 1-4, 44 percent of the ultimate shrinkage occurs by age 28 days and 72 percent by 3 months, for example. Secondly, even though a section is cracked on the tension side under load, the separate effect of shrinkage warping may be only nominally influenced by such cracks, since the shrinkage shortening on the compression side occurs in a zone that is uncracked by loads, or combined shrinkage forces and loads. To put it another way, shrinkage forces are axial in nature and cause some tension on the compression side of the member, and hence shrinkage forces and loads are not resisted in the same way by a cracked transformed section, for example. Thus it can be seen that a more elaborate analysis than is usually feasible or justified is needed to accurately take into account the combined effects of shrinkage and transverse loads.

In view of these considerations and the fact that shrinkage deflection is only one (although not an insignificant) part of the total deflection, an uncracked section solution or a solution using empirical methods is thought to be preferable in most cases. In fact, as shown later in the chapter, it may be reasonable in many cases to simply lump together the effects of shrinkage and creep by the choice of the time-dependent deflection factor to be used, such as by the ACI Code, the AASHTO Bridge Specifications, etc.

Compression steel has only a minor effect on initial or short-time deflections, as reflected in the uncracked or cracked transformed section moment of inertia. However, for both shrinkage and creep effects, compression steel has the effect of

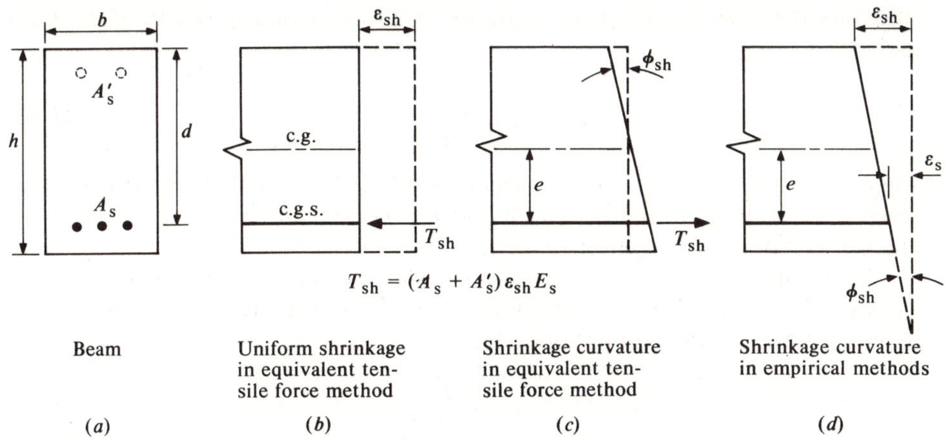

FIGURE 3-9
Shrinkage curvature of a reinforced concrete beam.

significantly reducing deflections. Such reinforcement is advocated by some engineers for no other reason than deflection control, especially in cantilever beams and slabs, and other cases where deflections are frequently critical.

Shrinkage Curvature

Primary interest in analytical methods for predicting shrinkage curvature centers around the basic assumptions, and hence the pertinent variables, involved. The quasi-elastic "tensile force" method of Eqs. (3-50) and (3-51) represents a simple M/EI fictitious-elastic-superposition approach for a nonelastic problem, in which major uncertainties include both E and I: E because of the effect of creep on the increasing sustained shrinkage stresses, and I for the reasons already discussed relating to effects of cracking. The author believes this method can best be used as a check on the somewhat simpler empirical methods, especially for members that are greatly different from those used in defining the empirical constants. The empirical methods relate to the simple but reasonable proposition that the shrinkage curvature or slope of the strain diagram is a direct function of the free shrinkage and steel content, and an inverse function of the depth of the section. The shape of the section is not a factor in the empirical methods (except in defining b), as perhaps it should not be in approximate calculations.

Equivalent tensile force method The equivalent tensile force method has been used in various forms at least since 1936 by Peabody[59] (1936), PCA[36] (1947), Large[60] (1957), Ferguson[21,61] (1958), Pauw and Meyers[74] (1964), and others. In Fig. 3-9*b* an imaginary compressive force, T_{sh}, is applied at the center of the steel in order to permit the beam to shrink uniformly. In Fig. 3-9*c* an equal and opposite tensile force, T_{sh}, is then superimposed, which produces the moment,

$T_{sh}e$, and the corresponding curvature or M/EI expression given by Eqs. (3-50) and (3-51).

$$\phi_{sh} = \frac{T_{sh}e}{E_{ct}I} \tag{3-50}$$

where $T_{sh} = (A_s + A_s')\varepsilon_{sh}E_s$, E_{ct} = sustained modulus of elasticity for the effect of creep (Eqs. (1-46) and (B-15)), and e and I may refer to the gross section, fully cracked section, or some effective section.

In a comparison with experimental data, the author[5] (1963) found (also shown in ACI Committee 435 report[3]) that Eq. (3-51) provided the best agreement by this method, as shown in Table 3-15.

$$\phi_{sh} = \frac{T_{sh}e_g}{\frac{1}{2}E_c I_g} \tag{3-51}$$

where e_g and I_g refer to the gross section and E_c to the 28-day value.

Miller empirical method Miller's empirical method[25] (1958) assumes that the extreme fiber of a beam farthest from the tension steel (method refers to singly reinforced members only) shrinks the same amount as the free shrinkage of the concrete. Following this assumption, the shrinkage curvature of a member is given by Eq, (3-52), as shown in Fig. 3-9*d*.

$$\phi_{sh} = \frac{\varepsilon_{sh} - \varepsilon_s}{d} = \frac{\varepsilon_{sh}}{d}(1 - \varepsilon_s/\varepsilon_{sh}) \tag{3-52}$$

where ε_s is the steel strain due to shrinkage. Miller suggested empirical values of $(\varepsilon_s/\varepsilon_{sh}) = 0.1$ for heavily reinforced members and 0.3 for moderately reinforced members.

Author's empirical method The author's empirical method[5] (1963) for computing shrinkage curvature (also described in ACI Committee 435[3] and 209[4] reports) represents a modification of Miller's method, as shown below by comparison. The steel percentages are expressed as $p = 3$ for 3 percent, and not as steel ratios. Equations (3-53) and (3-54) are applicable to both singly and doubly reinforced members.

$$\phi_{sh} = 0.7\frac{\varepsilon_{sh}}{h}(p - p')^{1/3}\left[\frac{p - p'}{p}\right]^{1/2} \qquad \text{for } p - p' \leq 3.0\% \tag{3-53}$$

where $p = 100A_s/bd$ and $p' = 100A_s'/bd$. The last parameter in Eq. (3-53), $[(p - p')/p]^{1/2}$, was found to be required in order to produce a somewhat smaller curvature for doubly reinforced members than for singly reinforced members when $(p - p')$ for the doubly reinforced case is equal to p for the singly reinforced

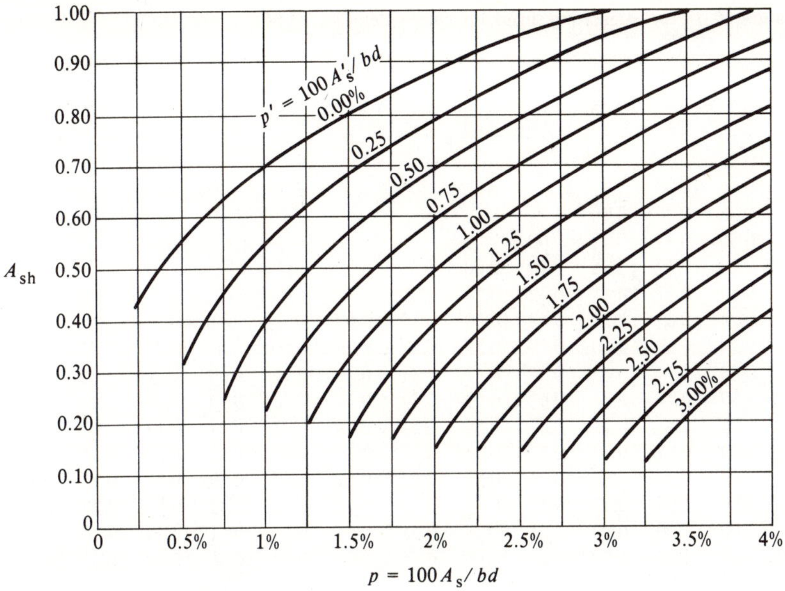

FIGURE 3-10
Values of A_{sh} for calculating shrinkage curvature in Eq. (3-56).

case, other conditions being the same. This modifier becomes unity when $p' = 0$.

$$\phi_{sh} = \varepsilon_{sh}/h \qquad \text{for } p - p' > 3.0\% \tag{3-54}$$

For singly reinforced members, $p' = 0$ and Eq. (3-53) reduces to Eq. (3-55).

$$\phi_{sh} = 0.7\,\frac{\varepsilon_{sh}}{h}\,p^{1/3} \tag{3-55}$$

A single equation (3-56) replaces Eqs. (3-53) to (3-55) for computing shrinkage curvature by the author.

$$\phi_{sh} = A_{sh}\varepsilon_{sh}/h \tag{3-56}$$

where A_{sh} is obtained in Table 3-13 or Fig. 3-10.

This method is also suggested by Neville and Dilger,[10] Mirza and Sabnis,[11] Wang and Salmon,[20] Ferguson,[21] British Standard Code of Practice,[63] and Beeby.[64]

Equations (3-53) to (3-56) were adapted from Miller's approach. For example, his method results in the following expression for singly reinforced members:

$$\begin{aligned} \phi_{sh} &= 0.7\varepsilon_{sh}/d \qquad \text{for moderately reinforced beams} \\ &= 0.9\varepsilon_{sh}/d \qquad \text{for heavily reinforced beams} \end{aligned}$$

Table 3-13 Values of A_{sh} for calculating shrinkage curvature in Eq. (3-56)*

p \ p'	0.00	0.25	0.50	0.75	1.00	1.25	1.50	1.75	2.00	2.25	2.50	2.75	3.00
0.25	0.44	0.00	—	—	—	—	—	—	—	—	—	—	—
0.50	0.56	0.31	0.00	—	—	—	—	—	—	—	—	—	—
0.75	0.64	0.45	0.26	0.00	—	—	—	—	—	—	—	—	—
1.00	0.70	0.55	0.39	0.22	0.00	—	—	—	—	—	—	—	—
1.25	0.75	0.63	0.49	0.35	0.20	0.00	—	—	—	—	—	—	—
1.50	0.80	0.69	0.57	0.45	0.32	0.18	0.00	—	—	—	—	—	—
1.75	0.84	0.74	0.64	0.53	0.42	0.30	0.17	0.00	—	—	—	—	—
2.00	0.88	0.79	0.69	0.60	0.50	0.39	0.28	0.16	0.00	—	—	—	—
2.25	0.92	0.83	0.74	0.65	0.56	0.47	0.37	0.26	0.15	0.00	—	—	—
2.50	0.95	0.87	0.79	0.71	0.62	0.53	0.44	0.35	0.25	0.14	0.00	—	—
2.75	0.98	0.91	0.83	0.75	0.67	0.59	0.51	0.42	0.33	0.24	0.13	0.00	—
3.00	1.00	0.94	0.87	0.79	0.72	0.64	0.57	0.49	0.40	0.32	0.23	0.13	0.00
3.25	1.00	0.97	0.90	0.83	0.76	0.69	0.62	0.54	0.47	0.39	0.31	0.22	0.12
3.50	1.00	1.00	0.94	0.87	0.80	0.74	0.67	0.60	0.52	0.45	0.37	0.29	0.21
3.75	1.00	1.00	1.00	0.90	0.84	0.78	0.71	0.64	0.58	0.51	0.44	0.36	0.28
4.00	1.00	1.00	1.00	1.00	0.88	0.81	0.75	0.69	0.62	0.56	0.49	0.42	0.35

* $p = 100A_s/bd$, $p' = 100A'_s/bd$. When $p' > p$, interchange p' and p to obtain corresponding solution in opposite direction.

The following results are computed by Eq. (3-55), which refers to singly reinforced members:

$$
\begin{aligned}
\phi_{sh} &= 0.56\varepsilon_{sh}/h && \text{when } p = 0.5\% \\
&\ 0.70\varepsilon_{sh}/h && \phantom{\text{when } p = }1.0\% \\
&\ 0.88\varepsilon_{sh}/h && \phantom{\text{when } p = }2.0\% \\
&\ 1.01\varepsilon_{sh}/h && \phantom{\text{when } p = }3.0\%
\end{aligned}
$$

The two methods are seen to yield similar results for the singly reinforced beam. The use of the thickness, h, in Eqs. (3-53) to (3-56), instead of the effective depth, d, was found to provide closer agreement with test data. The difference is small for all but shallow members and for these, the use of h seemed to provide the best fit. It is, of course, assumed that abnormal steel covers are excluded from consideration.

Corley and Sozen empirical method The empirical method of Corley and Sozen[62] (1966) is given by Eq. (3-57).

$$\phi_{sh} = \frac{0.035}{d}(\rho - \rho') \tag{3-57}$$

where the steel ratios are computed as $\rho = A_s/bd$ and $\rho' = A'_s/bd$. The constant 0.035 may be used for sections where the free shrinkage strain is known or where it is of the order of 500 millionths. Where shrinkage is known, the calculated curvature may be multiplied by the ratio of the shrinkage to 500 millionths. This procedure is also recommended by ACI Committee 443[53] (1974) for the Design of Concrete Bridges.

The following comparison is made between Eqs. (3-55) assuming $\varepsilon_{sh} = 500 \times 10^{-6}$, and Eq. (3-57), for singly reinforced members ($p' = \rho' = 0$):

$p = 100A_s/bd$	$\rho = A_s/bd$	ϕ_{sh} in Eq. (3-55)	ϕ_{sh} in Eq. (3-57)
0.5%	0.005	$280 \times 10^{-6}/h$	$175 \times 10^{-6}/d$
1.0%	0.010	$350 \times 10^{-6}/h$	$350 \times 10^{-6}/d$
2.0%	0.020	$440 \times 10^{-6}/h$	$700 \times 10^{-6}/d$
3.0%	0.030	$505 \times 10^{-6}/h$	$1050 \times 10^{-6}/d$

Equations (3-55) and (3-57) are seen to compute comparable results for steel contents of about 1 percent (exactly the same except for the use of h and d), but the linear function of the steel ratio in Eq. (3-57) computes much larger curvatures for high steel ratios (about twice at 3 percent steel) and smaller curvatures for small steel ratios than by Eq. (3-55)—also Eqs. (3-53) and (3-54).

Results by the four methods for computing shrinkage curvature in Eqs. (3-50) to (3-57) are compared with experimental data in Table 3-16 and for four structural examples in Table 3-14. Values of $(\varepsilon_{sh})_t$ or $(\varepsilon_{sh})_u$ may be found in Chap. 1 (particularly in Table 1-10). In the absence of information concerning shrinkage under local conditions, the author suggests the following for routine calculations: $(\varepsilon_{sh})_u = 400 \times 10^{-6}$ in/in (cm/cm).

Curvature Due to Differential Temperature

Since the thermal coefficient of steel and concrete is roughly the same, very little if any curvature of statically determinate reinforced concrete members will result from just ambient temperature changes. Lag effects would usually not have an important effect in such cases. Some exceptions to this are treated in Chap. 2 on differential temperature effects in composite members and other structures, such as those due to a sudden hot sun on the top of a bridge deck. Of course, external restraints in statically indeterminate structures, especially frames, will induce stresses and curvatures due to both temperature changes and shrinkage.

With regard to differential temperatures (such as inside to outside of heated or cooled buildings) applied to reinforced concrete members, the deformation of such members is similar or opposite to that due to shrinkage. For example, the deflection would be downward in the case of roofs of heated buildings (toward the inside for walls) and upward in the case of roofs of air conditioned buildings (toward the outside for walls).

Although the strain distribution resulting from differential temperatures will normally be curved in two directions, for practical purposes in most cases a linear distribution or gradient in the transverse direction can be assumed in approximate deflection calculations. For such cases curvatures due to differential temperatures can be determined using the shrinkage curvature equations in which ϕ_{sh} is replaced by $\alpha\Delta T$, where α is the thermal coefficient of concrete (Eqs. (1-42) to (1-44) for different weight concrete), and ΔT is the differential temperature from one side to the other (in degrees F when using Eqs. (1-42) to (1-44) values for α).

Deflection of Cantilever, Simple, and Continuous Structures

Deflections due to shrinkage (or differential temperature) are computed by Eq. (3-58).

$$\Delta_{sh} = K_{sh}\phi_{sh}L^2 \tag{3-58}$$

where ϕ_{sh} is determined by one or more of Eqs. (3-50) to (3-57), and K_{sh} is the deflection coefficient given by Eqs. (3-59) to (3-63).

Because of the nature of the problem in predicting shrinkage deflections, elaborate procedures are seldom warranted for such calculations. Thus it seems reasonable to compute shrinkage deflections using the deflection coefficients recommended below, and based on the following assumptions.

1. The steel content and eccentricity are constant along the span. For other cases, some practical compromise can be assumed.

2. In the case of continuous beams, shrinkage curvatures are assumed to be the same in the positive and negative regions with the inflection at the quarter-point, for the purpose of the derivations in Figs. 3-11 and 3-12. According to the empirical methods, the positive and negative region curvatures would be precisely the same when the positive p = negative $(p - p')$, which is not unrealistic. In addition, the positive region curvature has a dominative effect on deflections (see Eq. (3-39) and discussion of gravity load deflections of continuous beams), and is recommended in all shrinkage deflection calculations for continuous members using Eq. (3-58). This assumption is on the safe side when negative curvatures are greater than positive curvatures, and vice versa.

3. For end span deflections, $K_{sh} = 0.084$ for two-span beams (derived in Fig. 3-11) and $K_{sh} = 0.090$ for three-span beams (derived in Fig. 3-12). $K_{sh} = 0.090$ is also recommended for the end span of continuous beams with more than three spans.

4. For the center span deflection of a three-span symmetrical continuous beam, $K_{sh} = 0.053$ was derived in Fig. 3-12. For unsymmetrical cases K_{sh} for interior spans may be somewhat larger. Because of this and the approximate nature of such calculations, the author suggests the use of $K_{sh} = 0.065$ for all interior spans or beams continuous at both ends, except in unusual cases. For beams with negative curvatures at the ends that are considerably smaller than the positive curvature, an appropriate value of K_{sh} can be estimated, using the consideration that as the shrinkage curvature diminishes to zero and then to the positive curvature at one end, K_{sh} approaches a value of 0.090 (one end continuous

Cantilever beam

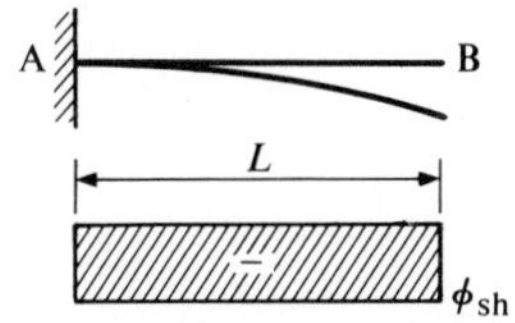

By the moment area method:

$\Delta_B = \phi_{sh} L^2/2$

or $K_{sh} = 1/2 = 0.500$

Simple beam

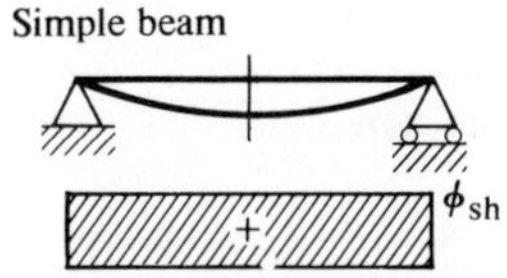

By the moment area method:

$\Delta_{max} = \Delta_{C_L} = (\phi_{sh})(L/2)(L/4) = \phi_{sh} L^2/8$

or $K_{sh} = 1/8 = 0.125$

Two-span continuous beam

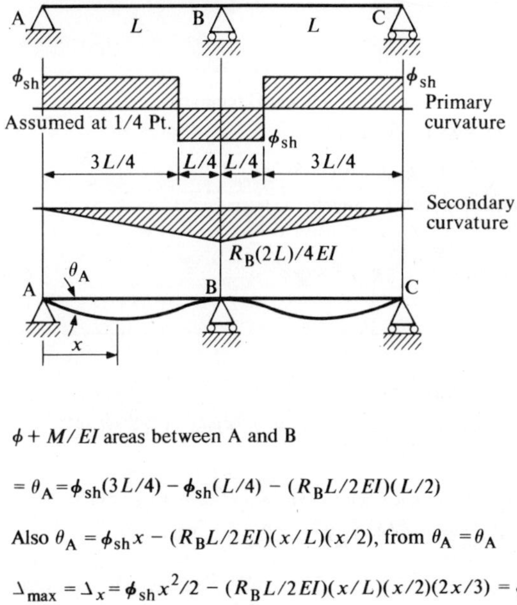

Applying the moment area method with sup. B removed:

$\Delta_B = \phi_{sh}(3L/4)(3L/8)$

$- \phi_{sh}(L/4)(7L/8)$

$= \phi_{sh} L^2/16$

Applying the moment area method with R_B as a load:

$\Delta_B = (R_B L/2EI)(L/2)(2L/3)$

$= R_B L^3/6EI$

Equating these deflections:

$R_B = 3EI\,\phi_{sh}/8L$

$\phi + M/EI$ areas between A and B

$= \theta_A = \phi_{sh}(3L/4) - \phi_{sh}(L/4) - (R_B L/2EI)(L/2)$

Also $\theta_A = \phi_{sh} x - (R_B L/2EI)(x/L)(x/2)$, from $\theta_A = \theta_A$ $\quad \underline{x = 0.42L}$

$$\Delta_{max} = \Delta_x = \phi_{sh} x^2/2 - (R_B L/2EI)(x/L)(x/2)(2x/3) = \phi_{sh} x^2/2 - R_B x^3/6EI$$

$$= \phi_{sh} L^2[(0.42)^2/2 - 3(0.42)^3/(8)(6)] = 0.084\,\phi_{sh} L^2$$

or $\underline{K_{sh} = 0.084}$

FIGURE 3-11
Curvature diagrams, and derived deflection equations and coefficients due to applied shrinkage curvatures for cantilever, simple, and two-span continuous beams (same shrinkage curvature assumed in positive and negative regions).

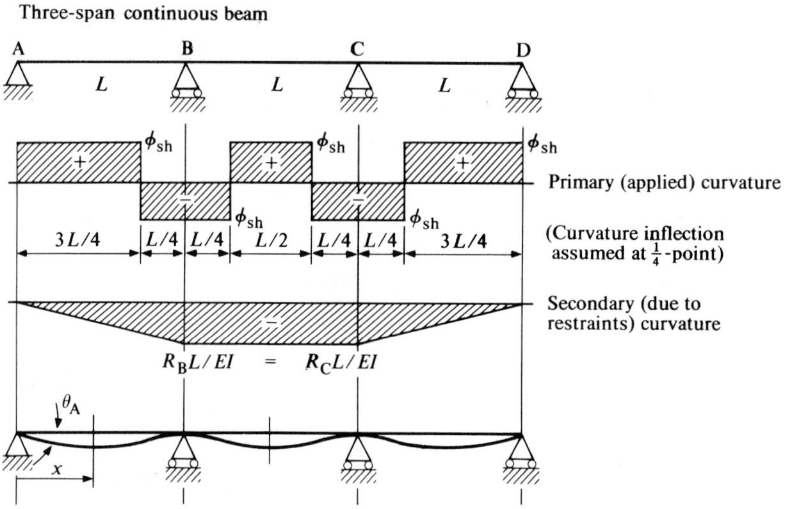

Applying the conjugate beam method with supports B and C removed:

Elastic wt $R_A = \phi_{sh}(3L/4 - L/2 + L/4) = \phi_{sh} L/2$

$\Delta_B = (\phi_{sh} L/2)L - \phi_{sh}(3L/4)(5L/8) + \phi_{sh}(L/4)(L/8) = \phi_{sh} L^2/16$

Applying the conjugate beam method with R_B and R_C as loads:

Elastic wt $R_A = (R_B L/EI)(L/2) = R_B L^2/EI$

$\Delta_B = R_B L^3/EI - (R_B L/EI)(L/2)(L/3) = 5R_B L^3/6EI$

From $\Delta_B = \Delta_B$, $R_B = 3EI\phi_{sh}/40L$

End Span

$\theta_A = \Sigma$ elastic wt R_As $= \phi_{sh} L/2 - R_B L^2/EI$

Also, applying the moment area method, $\theta_A = \phi_{sh} x - (R_B L/EI)(x/L)(x/2)$

From $\theta_A = \theta_A$, $x = 0.43L$

$\Delta_{max} = \Delta_x = \phi_{sh} x^2/2 - (R_B L/EI)(x/L)(x/2)(2x/3) = \phi_{sh} x^2/2 - R_B x^3/3EI$

$= \phi_{sh} L^2[(0.43)^2/2 - (0.43)^3/40] = 0.090\,\phi_{sh} L^2$, or $K_{sh} = 0.090$

Center span

Applying the moment area method:

$\Delta_{C_L} = \phi_{sh}(L/4)(3L/8) - \phi_{sh}(L/4)(L/8) - (R_B L/EI)(L/2)(L/4)$

$= \phi_{sh} L^2[(1/16) - (3/3201)] = 0.053\,\phi_{sh} L^2$, or $K_{sh} = 0.053$

(If the sec. curv. effect is neglected, K_{sh} = 1st term = 1/16 = 0.0625)

FIGURE 3-12
Curvature diagrams, and derived deflection equations and coefficients due to applied shrinkage curvatures for a three-span continuous beam with the same shrinkage curvature assumed in the positive and negative regions.

and one end discontinuous); and as the shrinkage curvature diminishes to zero and then to the positive curvature at both ends, K_{sh} approaches a value of 0.125 (simple beam).

Cantilever beam

$$K_{sh} = \tfrac{1}{2} = 0.500 \qquad \text{(derived in Fig. 3-11)} \qquad (3\text{-}59)$$

Simple beam

$$K_{sh} = \tfrac{1}{8} = 0.125 \qquad \text{(derived in Fig. 3-11)} \qquad (3\text{-}60)$$

Two-span continuous beam

$$K_{sh} = 0.084 \qquad \text{(derived in Fig. 3-11)} \qquad (3\text{-}61)$$

Continuous beam with three or more spans End span (when discontinuous at one end):

$$K_{sh} = 0.090 \qquad \text{(derived in Fig. 3-12; also see Assumption 3)} \qquad (3\text{-}62)$$

Interior spans or spans continuous at both ends:

$$K_{sh} = 0.065 \qquad \text{(see Assumption 4)} \qquad (3\text{-}63)$$

Beam or frame with one or more spans continuous at both ends

$$K_{sh} = 0.065 \qquad \text{(see Assumption 4)} \qquad (3\text{-}63)$$

For computing shrinkage deflections using these procedures for continuous structures, it is recommended that ϕ_{sh} in Eq. (3-58) refer to the positive region curvature in a given span only, as discussed in Assumption 2. Of course, a general deflection analysis could be obtained for any particular shrinkage curvature distribution in lieu of the above approximate solutions.

These procedures are assumed to apply to different weight concrete, as discussed in the section on comparison with test results (also see Example 2 and Observation 5).

Numerical Examples of Shrinkage Deflections

Example 1 Simple span beam, normal weight concrete (see Table 3-2, $p = 1.98\%$, $L = 20$ ft $= 6.1$ m).

Compute the ultimate shrinkage deflection following age 28 days. This might be recommended for the shrinkage deflection that occurs after the installation of partitions, for example.

For $H = 50\%$, $\varepsilon_{sh} = 390 \times 10^{-6}$ in/in (cm/cm) (Table 1-10)

For $b = 10$ in. (25.4 cm), $(\text{C.F.})_T = 0.88$ (Table 1-8)

Hence, $(\varepsilon_{sh})_u = (0.88)(390 \times 10^{-6}) = \underline{343 \times 10^{-6}\ \text{in/in (cm/cm)}}$ (use)

Equivalent tensile force method, Eq. (3-51)

$$T_{sh} = (A_s + A_s')\varepsilon_{sh} E_s = (2.37)(343)(29) = 23\,574 \text{ lb } (10\,690 \text{ kg}, 104\,900 \text{ N})$$

$$\phi_{sh} = \frac{T_{sh} e_g}{\frac{1}{2} E_c I_g} = \frac{(23\,574)(5)}{\frac{1}{2} \times 3.64 \times 10^6 \times 2287}$$

$$= 28.3 \times 10^{-6} \text{ 1/in } (11.1 \times 10^{-6} \text{ 1/cm}) \qquad \text{(Eq. (3-51))}$$

$$\Delta_{sh} = \phi_{sh} L^2/8 = (39.6 \times 10^{-6})(20)^2(144)/8 = \underline{0.20 \text{ in. } (0.51 \text{ cm})} \qquad \text{(Eq. (3-58))}$$

Miller empirical method, Eq. (3-52) Since $p = 1.98\%$, assume $\varepsilon_s/\varepsilon_{sh} = 0.2$ for moderately to heavily reinforced case.

$$\phi_{sh} = (\varepsilon_{sh}/d)(1 - \varepsilon_s/\varepsilon_{sh}) = (343 \times 10^{-6})(0.8)/12$$

$$= 22.9 \times 10^{-6} \text{ 1/in } (9.0 \times 10^{-6} \text{ 1/cm}) \qquad \text{(Eq. (3-52))}$$

$$\Delta_{sh} = \phi_{sh} L^2/8 = (22.9 \times 10^{-6})(20)^2(144)/8 = \underline{0.16 \text{ in. } (0.41 \text{ cm})} \qquad \text{(Eq. (3-58))}$$

Author's empirical method, Eq. (3-55) or (3-56)

$$\phi_{sh} = 0.7(\varepsilon_{sh}/h)p^{1/3} = (0.7)(343 \times 10^{-6})(1.98)^{1/3}/14$$

$$= 21.5 \times 10^{-6} \text{ 1/in } (8.5 \times 10^{-6} \text{ 1/cm}) \qquad \text{(Eq. (3-55))}$$

Or,

$$A_{sh} = 0.88 \qquad \text{(Table 3-13 or Fig. 5-9)}$$

$$\phi_{sh} = A_{sh}\varepsilon_{sh}/h = (0.88)(343 \times 10^{-6})/14 = 21.6 \times 10^{-6} \text{ 1/in} \qquad \text{(Eq. (3-56))}$$

$$\Delta_{sh} = \phi_{sh} L^2/8 = (21.5 \times 10^{-6})(20)^2(144)/8 = \underline{0.15 \text{ in. } (0.38 \text{ cm})} \qquad \text{(Eq. (3-58))}$$

Corley and Sozen empirical method, Eq. (3-57)

$$\phi_{sh} = (\varepsilon_{sh}/500 \times 10^{-6})(0.035/d)(p - p') = (343/500)(0.035/12)(0.0198)$$

$$= 39.6 \times 10^{-6} \text{ 1/in } (15.6 \times 10^{-6} \text{ 1/cm}) \qquad \text{(Eq. (3-57))}$$

$$\Delta_{sh} = \phi_{sh} L^2/8 = (39.6 \times 10^{-6})(20)^2(144)/8 = \underline{0.29 \text{ in. } (0.74 \text{ cm})} \qquad \text{(Eq. (3-58))}$$

The results of four examples by four methods are summarized in Table 3-14.

Example 2 Continuous one-way slab, sand-lightweight concrete (see Table 3-4, $+p = 0.765\%$ and 0.494%, $L = 20$ ft $= 6.1$ m).

Compute the ultimate shrinkage deflection following age 7 days, which may be the time of form removal.

For $H = 70\%$, $\varepsilon_{sh} = 546 \times 10^{-6}$ in/in (cm/cm) (Table 1-10)

For $h = 8$ in. (20.3 cm), $(\text{C.F.})_T = 0.94$ (Table 1-8)

Hence, $(\varepsilon_{sh})_u = (0.94)(546 \times 10^{-6}) = \underline{513 \times 10^{-6} \text{ in/in (cm/cm)}}$ (use)

Equivalent tensile force method, Eq. (3-51)

End span

$$T_{sh} = (A_s + A_s')\varepsilon_{sh}E_s = (0.62)(513)(29)$$
$$= 9224 \text{ lb } (4180 \text{ kg}, 41\,030 \text{ N})$$
$$\phi_{sh} = \frac{T_{sh}e_g}{\frac{1}{2}E_cI_g} = \frac{(9224)(2.75)}{\frac{1}{2} \times 2.63 \times 10^6 \times 512}$$
$$= 37.7 \times 10^{-6} \text{ 1/in } (14.8 \times 10^{-6} \text{ 1/cm}) \qquad (\text{Eq. (3-51)})$$
$$\Delta_{sh} = K_{sh}\phi_{sh}L^2 = (0.090)(37.7 \times 10^{-6})(20)^2(144)$$
$$= \underline{0.20 \text{ in. } (0.51 \text{ cm})} \qquad (\text{Eq. (3-58)})$$

Center span

$$T_{sh} = (A_s + A_s')\varepsilon_{sh}E_s = (0.40)(513)(29)$$
$$= 5951 \text{ lb } (2699 \text{ kg}, 26\,470 \text{ N})$$
$$\phi_{sh} = \frac{T_{sh}e_g}{\frac{1}{2}E_cI_g} = \frac{(5951)(2.75)}{\frac{1}{2} \times 2.63 \times 10^6 \times 512}$$
$$= 24.3 \times 10^{-6} \text{ 1/in } (9.6 \times 10^{-6} \text{ 1/cm}) \qquad (\text{Eq. (3-51)})$$
$$\Delta_{sh} = K_{sh}\phi_{sh}L^2 = (0.065)(24.3 \times 10^{-6})(20)^2(144)$$
$$= \underline{0.09 \text{ in. } (0.23 \text{ cm})} \qquad (\text{Eq. (3-58)})$$

Miller empirical method, Eq. (3-52)

End span Since $p = 0.765\%$, assume $\varepsilon_s/\varepsilon_{sh} = 0.35$ for moderately to lightly reinforced case.

$$\phi_{sh} = (\varepsilon_{sh}/d)(1 - \varepsilon_s/\varepsilon_{sh}) = (513 \times 10^{-6})(0.65)/6.75$$
$$= 49.4 \times 10^{-6} \text{ 1/in } (19.4 \times 10^{-6} \text{ 1/cm}) \qquad (\text{Eq. (3-52)})$$
$$\Delta_{sh} = K_{sh}\phi_{sh}L^2 = (0.090)(49.4 \times 10^{-6})(20)^2(144)$$
$$= \underline{0.26 \text{ in. } (0.66 \text{ cm})} \qquad (\text{Eq. (3-58)})$$

Center span Since $p = 0.494\%$, assume $\varepsilon_s/\varepsilon_{sh} = 0.4$ for lightly reinforced case.

$$\phi_{sh} = (\varepsilon_{sh}/d)(1 - \varepsilon_s/\varepsilon_{sh}) = (513 \times 10^{-6})(0.6)/6.75$$
$$= 45.6 \times 10^{-6} \text{ 1/in } (18.0 \times 10^{-6} \text{ 1/cm}) \qquad (\text{Eq. (3-52)})$$
$$\Delta_{sh} = K_{sh}\phi_{sh}L^2 = (0.065)(45.6 \times 10^{-6})(20)^2(144)$$
$$= \underline{0.17 \text{ in. } (0.43 \text{ cm})} \qquad (\text{Eq. (3-58)})$$

Author's empirical method, Eq. (3-55) or (3-56)

End span

$$\phi_{sh} = 0.7(\varepsilon_{sh}/h)p^{1/3} = (0.7)(513 \times 10^{-6})(0.765)^{1/3}/8$$
$$= 41.1 \times 10^{-6}\ 1/\text{in}\ (16.1 \times 10^{-6}\ 1/\text{cm}) \qquad (\text{Eq. (3-55)})$$

Or,
$$A_{sh} = 0.64 \qquad (\text{Table 3-13 or Fig. 5-9})$$
$$\phi_{sh} = A_{sh}\varepsilon_{sh}/h = (0.64)(513 \times 10^{-6})/8$$
$$= 41.0 \times 10^{-6}\ 1/\text{in} \qquad (\text{Eq. (3-56)})$$
$$\Delta_{sh} = K_{sh}\phi_{sh}L^2 = (0.090)(41.1 \times 10^{-6})(20)^2(144)$$
$$= \underline{0.21\ \text{in.}\ (0.53\ \text{cm})} \qquad (\text{Eq. (3-58)})$$

Center span

$$\phi_{sh} = 0.7(\varepsilon_{sh}/h)p^{1/3} = (0.7)(513 \times 10^{-6})(0.494)^{1/3}/8$$
$$= 35.5 \times 10^{-6}\ 1/\text{in}\ (14.0 \times 10^{-6}\ 1/\text{cm}) \qquad (\text{Eq. (3-55)})$$

Or,
$$A_{sh} = 0.55 \qquad (\text{Table 3-13 or Fig. 5-9})$$
$$\phi_{sh} = A_{sh}\varepsilon_{sh}/h = (0.55)(513 \times 10^{-6})/8$$
$$= 35.3 \times 10^{-6}\ 1/\text{in} \qquad (\text{Eq. (3-56)})$$
$$\Delta_{sh} = K_{sh}\phi_{sh}L^2 = (0.065)(35.5 \times 10^{-6})(20)^2(144)$$
$$= \underline{0.13\ \text{in.}\ (0.33\ \text{cm})} \qquad (\text{Eq. (3-58)})$$

Corley and Sozen empirical method, Eq. (3-57)

End span

$$\phi_{sh} = (\varepsilon_{sh}/500 \times 10^{-6})(0.035/d)(p - p')$$
$$= (513/500)(0.035/6.75)(0.00765)$$
$$= 40.7 \times 10^{-6}\ 1/\text{in}\ (16.0 \times 10^{-6}\ 1/\text{cm}) \qquad (\text{Eq. (3-57)})$$
$$\Delta_{sh} = K_{sh}\phi_{sh}L^2 = (0.090)(40.7 \times 10^{-6})(20)^2(144)$$
$$= \underline{0.21\ \text{in.}\ (0.53\ \text{cm})} \qquad (\text{Eq. (3-58)})$$

Center span

$$\phi_{sh} = (\varepsilon_{sh}/500 \times 10^{-6})(0.035/d)(p - p')$$
$$= (513/500)(0.035/6.75)(0.00494)$$
$$= 26.3 \times 10^{-6}\ 1/\text{in}\ (10.4 \times 10^{-6}\ 1/\text{cm}) \qquad (\text{Eq. (3-57)})$$
$$\Delta_{sh} = K_{sh}\phi_{sh}L^2 = (0.065)(26.3 \times 10^{-6})(20)^2(144)$$
$$= \underline{0.10\ \text{in.}\ (0.25\ \text{cm})} \qquad (\text{Eq. (3-58)})$$

The results of four examples by four methods are summarized in Table 3-14.

Example 3 Continuous T-beam, normal-weight concrete (see Tables 3-7 and 3-9, $+p = 0.25\%$ for all spans, $L_n = 29$ ft $= 8.8$ m).

Compute the ultimate shrinkage deflection following age 20 days.

For $H = 60\%$, $\varepsilon_{sh} = 397 \times 10^{-6}$ in/in (cm/cm) (Table 1-10)

For $t = 5$ in. (12.7 cm), $(\text{C.F.})_T = 1.08$ (Table 1-8)

Hence, $(\varepsilon_{sh})_u = (1.08)(397 \times 10^{-6}) = \underline{429 \times 10^{-6} \text{ in/in (cm/cm)}}$ (use)

In this example all spans, including the end spans, are continuous at both ends, and the positive region section is the same in all spans. Hence the computed shrinkage deflection will be the same for all spans using $K_{sh} = 0.065$ by Eq. (3-63).

Equivalent tensile force method, Eq. (3-51)

$$T_{sh} = (A_s + A'_s)\varepsilon_{sh}E_s = (5.08)(429)(29) = 63\,200 \text{ lb } (28\,670 \text{ kg}, 281\,100 \text{ N})$$

$$\phi_{sh} = \frac{T_{sh}e_g}{\frac{1}{2}E_cI_g} = \frac{(63\,200)(22.5 - 7.5)}{\frac{1}{2} \times 3.31 \times 10^6 \times 39\,100}$$

$$= 14.6 \times 10^{-6} \text{ 1/in } (5.7 \times 10^{-6} \text{ 1/cm}) \qquad \text{(Eq. (3-51))}$$

$$\Delta_{sh} = K_{sh}\phi_{sh}L^2 = (0.065)(14.6 \times 10^{-6})(29)^2(144)$$

$$= \underline{0.11 \text{ in. } (0.28 \text{ cm})} \qquad \text{(Eq. (3-58))}$$

Miller empirical method, Eq. (3-52) Since $p = 0.25\%$, assume $\varepsilon_s/\varepsilon_{sh} = 0.45$ for very lightly reinforced case.

$$\phi_{sh} = (\varepsilon_{sh}/d)(1 - \varepsilon_s/\varepsilon_{sh}) = (429 \times 10^{-6})(0.55)/22.5$$

$$= 10.5 \times 10^{-6} \text{ 1/in } (4.1 \times 10^{-6} \text{ 1/cm}) \qquad \text{(Eq. (3-52))}$$

$$\Delta_{sh} = K_{sh}\phi_{sh}L^2 = (0.065)(10.5 \times 10^{-6})(29)(144)$$

$$= \underline{0.08 \text{ in. } (0.20 \text{ cm})} \qquad \text{(Eq. (3-58))}$$

Author's empirical method, Eqs. (3-55) or (3-56)

$$\phi_{sh} = 0.7(\varepsilon_{sh}/h)p^{1/3} = (0.7)(429 \times 10^{-6})(0.25)^{1/3}/25$$

$$= 7.6 \times 10^{-6} \text{ 1/in } (3.0 \times 10^{-6} \text{ 1/cm}) \qquad \text{(Eq. (3-55))}$$

Or, $A_{sh} = 0.44$ (Table 3-13 or Fig. 5-9)

$$\phi_{sh} = A_{sh}\varepsilon_{sh}/h = (0.44)(429 \times 10^{-6})/25 = 7.6 \times 10^{-6} \text{ 1/in} \qquad \text{(Eq. (3-56))}$$

$$\Delta_{sh} = K_{sh}\phi_{sh}L^2 = (0.065)(7.6 \times 10^{-6})(29)^2(144)$$

$$= \underline{0.06 \text{ in. } (0.15 \text{ cm})} \qquad \text{(Eq. (3-58))}$$

Corley and Sozen empirical method, Eq. (3-57)

$$\phi_{sh} = (\varepsilon_{sh}/500 \times 10^{-6})(0.035/d)(p - p') = (429/500)(0.035/22.5)(0.0025)$$

$$= 3.3 \times 10^{-6} \text{ 1/in } (1.3 \times 10^{-6} \text{ 1/cm}) \qquad \text{(Eq. (3-57))}$$

$$\Delta_{sh} = K_{sh}\phi_{sh}L^2 = (0.065)(3.3 \times 10^{-6})(29)^2(144)$$
$$= \underline{0.03 \text{ in. } (0.08 \text{ cm})} \qquad (\text{Eq. } (3\text{-}58))$$

The results of four examples by four methods are summarized in Table 3-14.

Example 4 Interior girder of simple-span T-beam or deck girder bridge, normal-weight concrete (see Tables 3-11 and 3-12, $p = 0.94\%$, $L = 60$ ft $= 18.3$ m).

Compute the ultimate shrinkage deflection following age 7 days, which may be the time of form removal.

For $H = 70\%$, $\quad \varepsilon_{sh} = 546 \times 10^{-6}$ in/in (cm/cm) $\qquad$ (Table 1-10)

For $t = 6$ in. (15.2 cm), $\quad (\text{C.F.})_T\, 1.00 \qquad$ (Table 1-8)

Hence, $\qquad (\varepsilon_{sh})_u = \underline{546 \times 10^{-6} \text{ in/in (cm/cm)}} \qquad$ (use)

Equivalent tensile force method, Eq. (3-51)

$$T_{sh} = (A_s + A_s')\varepsilon_{sh}E_s = (18.72)(546)(29) = 296\,410 \text{ lb } (134\,500 \text{ kg}, 1\,318\,000 \text{ N})$$
$$\phi_{sh} = \frac{T_{sh}e_g}{\frac{1}{2}E_cI_g} = \frac{(296\,410)(29.75 - 12.51)}{\frac{1}{2} \times 3.64 \times 10^6 \times 103\,820}$$
$$= 27.0 \times 10^{-6} \text{ 1/in } (10.6 \times 10^{-6} \text{ 1/cm}) \qquad (\text{Eq. } (3\text{-}51))$$
$$\Delta_{sh} = \phi_{sh}L^2/8 = (27.0 \times 10^{-6})(60)^2(144) = \underline{1.75 \text{ in. } (4.45 \text{ cm})} \qquad (\text{Eq. } (3\text{-}58))$$

Miller empirical method, Eq. (3-52) Since $p = 0.94\%$, assume $\varepsilon_s/\varepsilon_{sh} = 0.3$ for moderately reinforced case.

$$\phi_{sh} = (\varepsilon_{sh}/d)(1 - \varepsilon_s/\varepsilon_{sh}) = (546 \times 10^{-6})(0.7)/29.75$$
$$= 12.8 \times 10^{-6} \text{ 1/in } (5.0 \times 10^{-6} \text{ 1/cm}) \qquad (\text{Eq. } (3\text{-}52))$$
$$\Delta_{sh} = \phi_{sh}L^2/8 = (12.8 \times 10^{-6})(60)^2(144) = \underline{0.83 \text{ in. } (2.11 \text{ cm})} \qquad (\text{Eq. } (3\text{-}58))$$

Author's empirical method, Eqs. (3-55) or (3-56)

$$\phi_{sh} = 0.7(\varepsilon_{sh}/h)p^{1/3} = (0.7)(546 \times 10^{-6})(0.94)^{1/3}/36.25$$
$$= 10.3 \times 10^{-6} \text{ 1/in } (4.1 \times 10^{-6} \text{ 1/cm}) \qquad (\text{Eq. } (3\text{-}55))$$

Or, $A_{sh} = 0.69$ (Table 3-13 or Fig. 5-9)

$$\phi_{sh} = A_{sh}\varepsilon_{sh}/h = (0.69)(546 \times 10^{-6})/36.25 = 10.4 \times 10^{-6} \text{ 1/in} \qquad (\text{Eq. } (3\text{-}56))$$
$$\Delta_{sh} = \phi_{sh}L^2/8 = (10.3 \times 10^{-6})(60)^2(144) = \underline{0.67 \text{ in. } (1.70 \text{ cm})} \qquad (\text{Eq. } (3\text{-}58))$$

Corley and Sozen empirical method, Eq. (3-57)

$$\phi_{sh} = (\varepsilon_{sh}/500 \times 10^{-6})(0.035/d)(p - p') = (546/500)(0.035/29.75)(0.0094)$$
$$= 12.1 \times 10^{-6} \text{ 1/in } (4.8 \times 10^{-6} \text{ 1/cm}) \qquad (\text{Eq. } (3\text{-}57))$$
$$\Delta_{sh} = \phi_{sh}L^2/8 = (12.1 \times 10^{-6})(60)^2(144) = \underline{0.78 \text{ in. } (1.98 \text{ cm})} \qquad (\text{Eq. } (3\text{-}58))$$

The results of four examples by four methods are summarized in Table 3-14.

Table 3-14 Results of four shrinkage deflection examples by four methods*

Example	Equivalent tensile force method, Eq. (3-51) Δ_{sh}, in. (cm)	Miller empirical method, Eq. (3-52) Δ_{sh}, in. (cm)	Author's empirical method, Eq. (3-55) or (3-56) Δ_{sh}, in. (cm)	Corley and Sozen empirical method, Eq. (3-57) Δ_{sh}, in. (cm)
Ex. 1. Simple span beam, normal-weight concrete, $h = 14$ in. (35.6 cm), $p = 1.98\%$, $L = 20$ ft (6.1 m)	0.20 (0.51)	0.16 (0.41)	0.15 (0.38)	0.29 (0.74)
Ex. 2. 3-span continuous one-way slab, sand-lightweight concrete, $h = 8$ in. (20.3 cm), $+p = 0.765\%$ and 0.494%, $L = 20$ ft (6.1 m)	End span: 0.20 (0.51) Center span: 0.09 (0.23)	End span: 0.26 (0.66) Center span: 0.17 (0.43)	End span: 0.21 (0.53) Center span: 0.13 (0.33)	End span: 0.21 (0.53) Center span: 0.10 (0.25)
Ex. 3. Continuous T-beam, normal-weight concrete, $h = 25$ in. (63.5 cm), $+p = 0.25\%$ for all spans, $L_n = 29$ ft (8.8 m)	All spans: 0.11 (0.28)	All spans: 0.08 (0.20)	All spans: 0.06 (0.15)	All spans: 0.03 (0.08)
Ex. 4. Interior girder of simple span T-beam or deck girder bridge, normal-weight concrete, $h = 36.25$ in. (92.1 cm), $p = 0.94\%$, $L = 60$ ft (18.2 m)	1.75 (4.45)	0.83 (2.11)	0.67 (1.70)	0.78 (1.98)

* Shrinkage curvatures computed by the different methods and equations indicated at the heading of each column. Shrinkage deflections computed by Eqs. (3-58), (3-60), (3-62), and (3-63).

Comparison With Test Results and Discussion

The four methods described by Eqs. (3-51) to (3-57) for computing shrinkage curvatures, and using Eqs. (3-58) and (3-60) for computing deflections of simple beams, are compared with test results in Table 3-16 for 14 beam cases described in Table 3-15. The methods are arranged in the order of increasing accuracy in Table 3-16, with the following results observed.

Using the Corley and Sozen empirical method, Eq. (3-57), computed results agreed with the test data within ± 10 percent in 1 out of 14 or 7 percent of the cases, and within ± 64 percent in all cases.

Table 3-15 Test conditions, concrete properties, and beam details for shrinkage specimens[3,5]

Reference	Beam designation	Relative humidity: Extremes	Relative humidity: Average	f'_c at 28 days	* E_c at 28 days	Age of initial reading	Duration of test	† Section details and properties: b	d	h	e_g	I_g	A_s	A'_s	p	p'	Span L
		%	%	psi	*	days	months	in.	in.	in.	in.	in⁴	in²	in²	%	%	ft
	Series 1, 2																
Miller[25]	3 in.	50	50	3500	3.4	1	3	3.25	2.25	3	0.75	7.3	0.22	0	3.01	0	3.5
	4 in.	50	50	3500	3.4	1	3	3.25	3.25	4	1.25	17.4	0.22	0	2.08	0	3.5
	5 in.	50	50	3500	3.4	1	3	3.25	4.25	5	1.75	33.9	0.22	0	1.60	0	3.5
	6 in.	50	50	3500	3.4	1	3	3.25	5.25	6	2.25	58.5	0.22	0	1.29	0	3.5
	Series 3, 4																
	3 in.	50	50	3500	3.4	1	3	3.25	2.25	3	0.75	7.3	0.11	0	1.50	0	3.5
	4 in.	50	50	3500	3.4	1	3	3.25	3.25	4	1.25	17.4	0.11	0	1.04	0	3.5
	5 in.	50	50	3500	3.4	1	3	3.25	4.25	5	1.75	33.9	0.11	0	0.80	0	3.5
	6 in.	50	50	3500	3.4	1	3	3.25	5.25	6	2.25	58.5	0.11	0	0.65	0	3.5
Washa and Fluck[44]	C2, C5	20–80	50	3290	3.3	14	30	12	4.00	5	0.50	125	0.80	0.40	1.67	0.84	20.8
	D2, D5	20–80	50	3530	3.4	14	30	12	4.00	5	0.50	125	0.80	0.40	1.67	0.84	12.5
	E2, E5	20–80	50	3660	3.5	14	30	12	2.31	3	0.27	27.0	0.44	0.22	1.59	0.80	17.5
	C3, C6	20–80	50	3290	3.3	14	30	12	4.00	5	1.50	125	0.80	0	1.67	0	20.8
	D3, D6	20–80	50	3530	3.4	14	30	12	4.00	5	1.50	125	0.80	0	1.67	0	12.5
	E3, E6	20–80	50	3660	3.5	14	30	12	2.31	3	0.81	27.0	0.44	0	1.59	0	17.5

(For metric units, see conversion factors herein.)

* Computed using $E_c = 57\,700\sqrt{f'_c}$. Shown in the table in psi × 10^6.

† Also $E_s = 29 \times 10^6$ psi.

Using the equivalent tensile force method, Eq. (3-51), computed results agreed with the test data within ±10 percent in 3 out of 14 or 21 percent of the cases, and within ±43 percent in all cases.

Using the Miller empirical method, Eq. (3-52), computed results agreed with the test data within ±10 percent in 6 out of 11 or 55 percent of the cases, and within ±27 percent in all 11 cases (doubly reinforced beams excluded).

Using the author's empirical method, Eqs. (3-53) to (3-56), computed results agreed with the test data within ±10 percent in 12 out of 14 or 86 percent of the cases, and within ±29 percent in all cases.

Note that these percentages were determined as ratios subtracted from 1.00 in Table 3-16.

The last two empirical methods are in somewhat better agreement with the data, as they should be, since their development was based on these particular data.[5] The use of $(E_c/2)I_g$ in the tensile force method, Eq. (3-51), was also based on these data.[5] Also, it should be noted that the scope of the experimental results in Table 3-16 is limited to 14 beam cases that were only 3 in. (7.6 cm) to 6 in. (15.2 cm) deep with p or $(p - p')$ ranging from 0.79 to 3.01 percent. An average of the data for two beams is used for each of the experimental shrinkage warping results in Table 3-16.

Regarding Miller's shrinkage warping data: measurements of warping deflections and shrinkage strains provide the basic physical data, which were reported in the form of curvatures, as shown in Table 3-16. Supports for the specimens were located with overhang portions so as to eliminate dead-load deflection and hence creep deflection due to dead load between the supports. In addition, specimens were placed on their sides and alternated in position to minimize nonuniform drying and other factors that might inhibit free warping.

Regarding the shrinkage warping data of Washa and Fluck[44]: the data were obtained from shorter warp specimens and reported as deflections for the longer beams of the investigation as follows. The warp specimens had the same cross section and reinforcement as the beams they represented but were only 6 ft long. They were supported on their side edges and not loaded. Under these conditions the warp specimens gradually bent into a curved form. Each time readings were taken, the warp specimens were rotated 180° and left to rest on the side which had previously been the upper side. Each warp specimen was fitted with three brass plugs set along the longitudinal center line of the top surface, one at mid-length and the other two a foot from each end. The warp was measured by a dial fitted to a steel bar so arranged that when the ends of the bar contacted the end plugs the dial plunger contacted the center plug. The deflection of the longer beams due to conditions other than loading was then calculated from the warp specimen readings on the assumption that the warp of the longer beams and that of the short warp specimens would cause the two to assume the form of circular arcs of equal radius. Thus, the data were reported in the form of midspan deflections, as shown in Table 3-16.

The following observations by Miller[25] pertain to the concepts of the equivalent tensile force method and the empirical approach:

> Because warping is strictly a geometrical phenomenon due to completely autogenous dimensional changes, it is apparent that conventional concepts of stress and elasticity are not determining factors. Experimental investigations verify this observation and demonstrate that strain is the single decisive factor. Inevitably, strains produce stresses but their relationship is not essential to the analysis of warping. Extensive study of the experimental results clearly demonstrates that warping is no function of f_c' and E_c and further references to these irrelevant items are intentionally omitted, even though a considerable portion of the experimental studies was devoted to them.

The author tends to agree with Miller's observations. This would suggest that essentially no difference would be found for different weight concrete when determining shrinkage deflections, except as the shrinkage of the plain concrete may be different (not so assumed in the method described in the text of Chap. 1).

Another point regarding the empirical methods (first by Miller) is the assumption that the extreme fiber of a beam farthest from the tension steel shrinks the same amount as the free shrinkage of the plain concrete. When the eccentric force (at steel c.g.s.) is outside the kern limit of the cross section (a typical situation, although not in the extreme in most cases), the equivalent tensile force method predicts the extreme fiber of a beam farthest from the tension steel to be greater than the free shrinkage. Ferguson[61] made this point in his discussion of Miller's paper,[25] and correctly so. However, the author found the following in his study[5]:

> The curves of the current investigation indicate that the extreme fiber does shrink more than the free shrinkage of the companion specimens, but not much more. Hence, the effects of the eccentric steel resistance, outside the kern limit of the section, do seem to produce greater-than-free shrinkage strains at the opposite extreme fiber. But Miller's empirical approach (and the other empirical methods) would certainly appear to be a close approximation. Of course, for deeper beams (greater eccentricity) the assumption would tend to be further in error, although in these cases the increased depth may greatly reduce the shrinkage-warping curvature anyway.

Observations and Possible Recommendations

The following observations pertain to the computed shrinkage deflections by the four methods for the four particular examples summarized in Table 3-14, and the 14 experimental beam cases shown in Tables 3-15 and 3-16. It is difficult to determine which methods are the most accurate in general or in specific cases, since comparisons with experimental data that isolate shrinkage effects are limited in scope and difficult to accurately obtain, as discussed in the previous section.

1. The equivalent tensile force method tended to compute shrinkage deflections on the high side in Table 3-14 for the larger members used in Examples 3 and 4, as compared to the three empirical methods (noticeably higher for the bridge girder—1.75 in. or 4.45 cm versus 0.67 in. to 0.83 in. or 1.70 cm to 2.11 cm by the empirical methods). The use of a cracked section moment of inertia instead of I_g, and the use of any greater creep effect by the effective modulus method of Eq. (1-46) or (B-15)—such as $E_c/3$ instead of $E_c/2$ in Eq. (3-50) or (3-51)—would only exaggerate this tendency. However, this tendency to compute high values was

Table 3-16 Computed shrinkage warping compared with test data (test details in Table 3-15)

Reference	Beam designation	Measured free shrinkage at end of test, ε_{sh}	Experimental values*: Average curvature along beam	Experimental values*: Mid-span deflection	Computed curvatures and deflections,† and comparisons: Corley and Sozen Eq. (3-57)‡		Tensile force method, Eq. (3-51)		Assumed in Miller Eq. (3-52), $\varepsilon_s/\varepsilon_{sh}$	Miller Eq. (3-52)		Branson Eqs. (3-53) to (3-56)	
		in/in × 10^{-6}	1/in × 10^{-6}	in.	§	Ratio¶	§	Ratio¶	—	§	Ratio¶	§	Ratio¶
	Series 1, 2												
Miller[25]	3 in.	650	225	—	609	0.37	267	0.84	0.1	260	0.87	219	1.03
	4 in.	650	150	—	291	0.52	175	0.86	0.2	160	0.94	145	1.03
	5 in.	650	108	—	171	0.63	126	0.86	0.3	107	1.01	106	1.02
	6 in.	650	80	—	112	0.71	94	0.85	0.3	87	0.92	83	0.96
	Series 3,4												
	3 in.	550	130	—	257	0.51	113	1.15	0.3	171	0.76	147	0.88
	4 in.	550	88	—	124	0.71	74	1.19	0.3	118	0.75	98	0.90
	5 in.	550	66	—	72.5	0.91	53	1.25	0.3	91	0.73	72	0.92
	6 in.	550	57	—	47.7	1.19	40	1.43	0.4	63	0.90	56	1.02
Washa and Fluck[44]	C2, C5	750	—	0.50	0.85	0.59	0.51	0.98	—	—	—	0.54	0.93
	D2, D5	750	—	0.20	0.31	0.65	0.18	1.11	—	—	—	0.20	1.00
	E2, E5	750	—	0.45	0.99	0.45	0.47	0.96	—	—	—	0.63	0.71
	C3, C6	750	—	1.20	1.71	0.70	1.02	1.18	0.3	1.02	1.18	0.97	1.24
	D3, D6	750	—	0.35	0.62	0.56	0.36	0.97	0.3	0.37	0.95	0.35	1.00
	E3, E6	750	—	1.20	1.99	0.60	0.94	1.28	0.3	1.25	0.96	1.12	1.07

(For metric units, see conversion factors herein.)

* The experimental shrinkage warping data was reported in the form of curvatures by Miller, and in the form of deflections by Washa and Fluck. Each result represents an average of two beams.

† Deflections computed using $\phi_{sh}L^2/8$ for the Washa and Fluck simple-span beams.

‡ Using measured $\varepsilon_{sh}/500 \times 10^{-6}$, as by Eq. (3-57).

§ Computed curvatures in (1/in) × 10^{-6}. Computed deflections in inches.

¶ Ratio of experimental to computed shrinkage curvatures or deflections.

not particularly noticeable in Table 3-16 for the 14 experimental beam cases (relatively small members) or in Table 3-14 for the smaller members used in Examples 1 and 2.

2. For the two highest steel percentages in Table 3-14 of $p = 0.94$ and 1.98 percent, the results by the author's empirical method were on the low side of the results by the other three methods. This is due to the fact that the tensile force method and the Corley and Sozen empirical method are linear functions of the steel content, while the author's method is a function of $p^{1/3}$, for example. However, for the shallow 3 in. and 4 in. (7.6 cm and 10.2 cm) experimental beam cases in Table 3-16, with the highest steel percentages of 3.01 and 2.08 percent (Table 3-15), respectively, the author's method showed the best agreement of the four methods. This was generally true for all of the experimental beam cases (up to 6 in. or 15.2 cm in depth) shown in Table 3-16.

3. The computed results by the empirical method of Corley and Sozen were decidedly higher than the results by the other three methods and the data in Table 3-16 (except for one result). However, this was not the case for the examples shown in Table 3-14, although the tendency can be seen for high steel percentages. The reverse was the case in Table 3-14 for Example 3 with the lowest steel percentage (0.25 percent). As discussed following Eq. (3-57) and as can be seen in Table 3-16, the method of Corley and Sozen tends to compute shrinkage deflections on the high side (about twice as high for 3 percent steel) for higher steel percentages and on the low side for lower steel percentages (lowest of the four methods in Example 3 with $p = 0.25$ percent), as compared to the method of the author.

4. As shown in the previous section on comparison with test results, the computed results by the tensile force method, and the empirical methods of Miller and the author, were compared with the limited experimental results of 14 beam cases in Table 3-16 and found to be roughly comparable in accuracy. These two empirical methods were in somewhat better agreement than the tensile force method (and the empirical method of Corley and Sozen), as they should be, since their development was based on these particular data. The use of $(E_c/2)I_g$ in the tensile force method, Eq. (3-51), was also based on these data.

5. Through the use of E_c, the equivalent tensile force method is the only one of the four methods that distinguishes between different weight concrete (except as it may influence the selection of the shrinkage parameter), since the empirical methods refer to the warped section resulting from free shrinkage and the steel content only. The use of sand-lightweight concrete in Example 2 (Table 3-14) does not seem to have a noticeable effect on the relative computed results by the various methods.

6. Possible recommendations. The empirical method of Miller is expressed only in general terms ("moderately" and "heavily" reinforced beams) and does not apply to doubly reinforced members. With regard to the other three methods, shrinkage curvatures could be computed by the tensile force method and one or both of the empirical methods of Corley and Sozen, and the author, with the quasi-elastic tensile force method used to check the empirical method(s), or vice

versa. An averaging procedure could also be used to estimate results. In special cases, the best estimate could probably be judged from the three solutions in accordance with Observations 1 to 6, the discussion and table following Eq. (3-57), and the results in Tables 3-14 and 3-16. It should be noted that all of these curvature expressions are extremely simple, as was demonstrated in the Examples.

As to the recommendation of one best method of calculation for all cases, this does not seem to be possible at the present state of knowledge. Much more extensive comparisons with experimental data are needed, especially for larger members with moderate to high steel percentages. However, the following may serve as a basis for a qualified recommendation now or in the future:

(*a*) The empirical method of the author, Eqs. (3-53) to (3-56), appears to be a suitable approximate method for slabs (as suggested by Ferguson[21]) and other members as well, although especially the latter is by no means conclusive. This method is in the best agreement with the 14 experimental beam cases in Table 3-16, although the specimens ranged from 3 in. to 6 in. (7.6 cm to 15.2 cm) in depth only, with $p = 1.04$ to 3.01 percent, as shown in Table 3-15. However, there is a question about this method for larger members with moderate to high steel percentages. For example, results in Table 3-14 by this method are on the low side compared to the other methods for Example 1 ($\Delta_{sh} = 0.15$ in. or 0.38 cm versus 0.20 in. or 0.51 cm by the tensile force method and 0.29 in. or 0.74 cm by the method of Corley and Sozen, for a simple beam with $h = 14$ in. or 35.6 cm, $p = 1.98$ percent), and also for Example 4 ($\Delta_{sh} = 0.67$ in. or 1.70 cm versus 0.78 in. or 1.98 cm and 0.83 in. or 2.11 cm by the other two empirical methods and 1.75 in. or 4.45 cm by the tensile force method, for a bridge girder with $h = 36.25$ in. or 92.1 cm, $p = 0.94$ percent). On the other hand, it is noted for smaller members that even for the two experimental beam cases in Table 3-16 with the highest steel percentages ($h = 3$ in. or 7.6 cm, $p = 3.01$ percent and $h = 4$ in. or 10.2 cm, $p = 2.08$ percent), the method of the author was in the best agreement of the four methods. Also, the result seems to be satisfactory by the author's method for the continuous T-beam in Example 3 ($h = 25$ in. or 63.5 cm, $p = 0.25$ percent), which referred to a relatively deep beam with a low steel percentage. The empirical method of the author is suggested by ACI Committees 435[3] and 209,[4] Neville and Dilger,[10] Mirza and Sabnis,[11] Wang and Salmon,[20] Ferguson,[21] the British Standard Code of Practice,[63] and Beeby.[64]

(*b*) As noted in (*a*) above, shrinkage curvatures computed by the equivalent tensile force method, Eq. (3-51), and the empirical method of Corley and Sozen, Eq. (3-57) (also recommended by ACI Committee 443, Concrete Bridge Design), tend to be on the high side of those computed by the empirical method of the author, except for low steel percentages. The high side is the safe side, and hence either of Eqs. (3-51) and (3-57) may be preferable. However, the predictions by these two methods may be *too* high in many cases. The author believes this may well be the case, but suggests that further verification is needed, especially for larger members with moderate to high steel percentages. One difficulty noted with regard to the tensile force method and the method of Corley and Sozen is the following—for the bridge girder in Example 4 ($h = 37$ in. or 94.0 cm, $p = 0.94$

percent), Table 3-14, the computed result by the tensile force method ($\Delta_{sh} = 1.75$ in. or 4.45 cm) is more than twice as high as by the other three methods ($\Delta_{sh} =$ 0.67 in. to 0.83 in. or 1.70 cm to 2.11 cm). Also, for the experimental beam case in Table 3-16 with the highest steel percentage ($h = 3$ in. or 7.6 cm, $p = 3.01$ percent), the computed result by the method of Corley and Sozen ($\phi_{sh} =$ 609×10^{-6} 1/in or 240×10^{-6} 1/cm) is more than twice as high as by the other three methods ($\phi_{sh} = 217, 260, 267 \times 10^{-6}$ 1/in) and the experimental data ($\phi_{sh} = 225 \times 10^{-6}$ 1/in or 89×10^{-6} 1/cm).

3-4 TIME-DEPENDENT DEFLECTION

In most cases the initial deflection of a reinforced concrete beam is substantially altered by the following combined effects.

1. Creep of concrete in compression and tension, including bond creep between concrete and steel The principal effect of creep on beam curvature is that of concrete creep in compression, although the reinforcing steel undergoes some increase in strain (loss of tension stiffening) and stress because of the decreased moment lever arm with time, as shown in Figs. 3-13*b* and *c*. Whereas concrete tensile creep (between cracks in the cracked state) and bond creep may be of interest in precise analyses, their effects (separate k_r-effect > 1) can be included in the combined factor, k_r, for most practical purposes. See Effect 5 (movement of the neutral axis) and Fig. 3-13. This reduction factor, k_r, is defined in Eqs. (3-64) and (3-65), and applied in Eq. (3-74) for creep deflections and Eq. (3-75) for combined time-dependent deflections.

2. Shrinkage of concrete As previously shown in Fig. 3-9, concrete shrinkage causes curvatures and deflections in the same direction as those caused by gravity loading. Such deflections can be determined separately by Eqs. (3-50) to (3-63). Or, since shrinkage and creep deflections are complementary, their combined value may be estimated in approximate calculations with a single time-dependent factor applied to the initial deflection. Such a procedure is used in the ACI Code (318-71) and the AASHTO Highway Bridge Specifications (1973) for Load Factor Design.

3. Formation of new and widening of earlier cracks Laboratory tests reported by Nemirovsky,[65] as described by Pregnoff,[35] showed that the formation of new cracks during sustained loading seems to depend on the development of earlier cracks during the initial loading stage. For example, in beams with a low steel percentage of 0.60 percent and $f_s = 20\,000$ psi (1406 kg/cm^2, 138 N/mm^2), about half of the cracks occurred at initial loading and the remainder during the sustained loading. In beams with the same low steel percentage but $f_s = 30\,000$ psi (2109 kg/cm^2, 207 N/mm^2), almost all cracks occurred during the initial loading. In beams with a higher steel percentage of 2.0 percent, cracks appeared rather early with little additional cracking occurring during the sustained loading. It

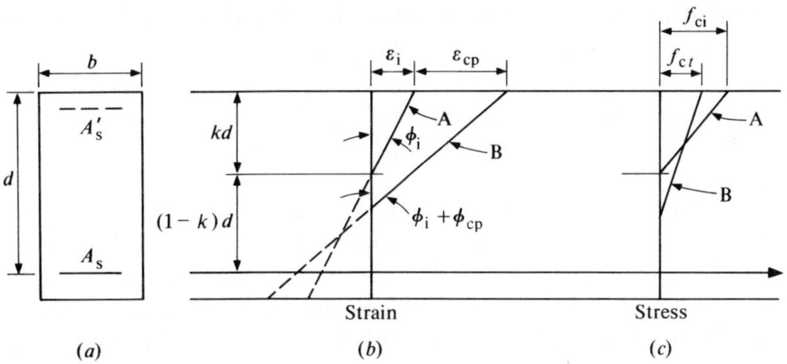

FIGURE 3-13
Stress and strain distribution, and curvature in a reinforced concrete beam due to initial and creep deformation.

would seem that this effect (separate k_r-effect > 1), for most practical purposes, can be taken into account in the determination of the effect of initial cracking (such as by I_e in Eqs. (3-18) or (3-19)) and the combined factor, k_r.

4. Relaxation of tensile stresses in concrete Tensile stresses in the concrete between cracks will be reduced by relaxation, resulting in an increase in curvature and deflection with time. It has been shown[52] that the long-term curvature due to creep of concrete in tension, as a percentage of the total creep curvature, may increase from about 10 percent for high reinforcement percentages to the theoretical value of 50 percent for unreinforced (uncracked) concrete. However, it seems to the author that the important consideration here is the level of initial cracking as determined, for example, by M_a/M_{cr} in Eq. (3-19). When this ratio is greater than about 2.5 to 3.0, the fully cracked section I is essentially the effective I, and hence virtually no concrete in tension is effective for purposes of determining deflections. This would mean that essentially no initial resistance to deflection is provided by concrete in tension, and hence essentially no relaxation effect would exist. An additional point is the fact that the ratio M_a/M_{cr} may be large or small, independent of the steel percentage used. Regardless of the particular level of M_a/M_{cr}, it seems to the author that this effect (separate k_r-effect > 1), for most practical purposes, can also be taken into account in the determination of the effect of initial cracking (such as by I_e in Eqs. (3-18) or (3-19)) and the combined factor, k_r.

5. Movement of the neutral axis The dominative effect of movement of the neutral axis is downward (for positive moment) due to creep, as shown in Fig. 3-13*b* and verified by many experimental observations. This movement results in the overall value of k_r being <1 due to creep, as shown by Eqs. (3-64 and (3-65). Effects 1, 3, and 4 above relating to progressive cracking, creep and relaxation of concrete in tension, and bond creep (all tending to raise the neutral axis and increase time-dependent deflections), have separate k_r-effects > 1. However, the combined value of k_r appears to remain <1 due to creep, relaxation, and cracking. For combined shrinkage and creep, the upper limit of k_r is taken as unity in Eqs. (3-66), (3-68), (3-69), and (3-70) for the case of no-compression steel.

6. Compression steel Compression steel has the effect of significantly reducing both creep and shrinkage deflections, which results in the further reduction in the combined factor, k_r, as shown by Eqs. (3-66) to (3-73). Such reinforcement is advocated by some engineers for no other reason than deflection control, especially in cantilever beams and slabs, and other cases where deflections are frequently critical or may be critical.

7. Effect of repeated load cycles This effect has been discussed earlier in the chapter with reference to Figs. 3-7 and 3-8. For practical purposes, the dead-load deflection (and hence creep deflection) that occurs both before, $(\Delta_D)_1$—Eq. (3-30), and after, $(\Delta_D)_2$—Eq. (3-34) or (3-36), the application or passage of the design live load appears to represent the primary time-dependent deflection consideration under repeated loading in most cases. Equation (3-34) includes residual or recovery effects due to cracking but not to creep, which can be taken into account separately. These time-dependent deflections both before and after live loading are further described by Eqs. (3-80), (3-81), and (3-83).

The effect of repeated loading on the time-dependent response of simply and doubly reinforced beams has been studied by Sinha, Agrawal, Gerstle, and Tulin;[66,67] Miller and Guralnick;[68] and others. Detailed calculations to take this effect into account normally require more information than is usually available, and is beyond the scope of this treatment and of most practical requirements. Also, the authors of these studies indicate that the agreement between experimental and computed results is not particularly good. For example, Miller and Guralnick[68] stated that "neither the equivalent modulus method nor the fictitious axial force method results in satisfactory deflection predictions for a beam subjected to repeated cycles of load."

8. Moment redistribution due to cracking, creep, and shrinkage in statically indeterminate structures This combined effect in statically indeterminate structures causes additional initial and time-dependent deflections that can readily be taken into account by numerical procedures, including the use of computers. An example of such numerical calculations is shown in Fig. D-2 for the effect of redistribution due to cracking in a two-span continuous beam using the Newmark numerical procedure. The combined effect will contribute to the total deflection normally by a few percent.

Curvature Due to Creep

The effect of creep on beam strains, stresses, and curvatures is shown in Fig. 3-13. Line A refers to the initial or elastic strain and stress, and Line B to the initial plus creep strain and stress in Figs. 3-13*b* and *c*. Since the time-dependent change in strain at the level of the steel is small compared to the extreme fiber creep strain in compression, the neutral axis moves downward, causing a similar adjustment in the stress diagram. In order to maintain equilibrium (internal force and couple), the extreme fiber stress must decrease with time, as shown in Fig. 3-13*c*. With respect to Fig. 3-13*b*, it is apparent that the relative increase in curvature due to creep is less than the relative increase in strain due to creep, as shown by Eq. (3-64); and similarly for deflections by Eq. (3-65).

$$\frac{\phi_{\mathrm{cp}}}{\phi_{\mathrm{i}}} = k_{\mathrm{r}} \frac{\varepsilon_{\mathrm{cp}}}{\varepsilon_{\mathrm{i}}} \qquad (3\text{-}64)$$

and similarly for deflections,

$$\frac{\Delta_{\mathrm{cp}}}{\Delta_{\mathrm{i}}} = k_{\mathrm{r}} \frac{\varepsilon_{\mathrm{cp}}}{\varepsilon_{\mathrm{i}}} \qquad (3\text{-}65)$$

where k_{r} is less than 1. Thus the relative increase in the deflection is less than the increase in strain due to creep.

Evaluation of the Combined Factor (For Effects of Creep, Relaxation, and Cracking), k_{r}, Due to Creep Plus Shrinkage and Creep Only, and Including the Primary Effect of Compression Steel

Equation (3-66) was formulated by the author[6] as an approximation of the 1963 ACI Code values, which were multiplied by a time-dependent factor of 2 for computing deflections. This equation is used in the 1971 ACI Code and again multiplied by a factor of 2 for creep plus shrinkage deflections. The lower limit is changed to 0.30 in the 1971 Code.

For creep plus shrinkage deflections

$$k_{\mathrm{r}} = 1.0 - 0.60A'_{\mathrm{s}}/A_{\mathrm{s}} \geq 0.40 \qquad (3\text{-}66)$$

Equation (3-66) was based on the data of Washa and Fluck[44] and Yu and Winter.[40]

Equation (3-67) was suggested by ACI Committee 435[3] for creep deflections only, which provided for $k_{\mathrm{r}} = 0.85$ when $A'_{\mathrm{s}}/A_{\mathrm{s}} = 0$ as indicated in Fig. 3-13*b*, and Eqs. (3-64) and (3-65).

For creep deflections

$$k_{\mathrm{r}} - 0.85 - 0.45A'_{\mathrm{s}}/A_{\mathrm{s}} \geq 0.40 \qquad (3\text{-}67)$$

However, a 1970 study by Hollington[69] showed that for lightly reinforced members, these equations in terms of the steel area ratios can considerably overestimate (on unsafe side) the effect of compression steel on time-dependent deflections. The pertinent conclusion from his study was the following:

The restraint provided by compression reinforcement was found to be considerably less than the calculated value based on ACI Code 318-63 (essentially the same as ACI Code 318-71) or the draft Unified British Code. For members containing a larger percentage of reinforcement, it is likely that the calculated and measured results [using Eq. (3-66)] would be compatible.

On the basis of the combined data of Washa and Fluck, Yu and Winter, and Hollington (shown in Table 3-17 for rectangular, T, and hollow box beams), the author[26] found the following:

[Equation (3-66)] was evaluated and found to be within reasonable limits in most cases for such a grossly simplified approach to this rather complex problem. However, the equation does somewhat overestimate the effect of compression steel for beams with low steel percentages (approximately 1 percent and less) when A'_s/A_s is high (as about 1.0). An alternate method for predicting the effect of compression steel on time-dependent deflections, as a function of the compression steel ratio, ρ', rather than the steel area ratio, A'_s/A_s, was recommended [Eq. (3-70)].

Development of Eqs. (3-70) to (3-73) recommended by the author[26] for estimating k_r Although not directly related, it is interesting to note below that Eq. (3-68) for predicting the effect of nonprestressed tension steel in reducing time-dependent camber of prestressed beams yields approximately the same results as Eq. (3-66) for reducing time-dependent deflection of reinforced beams, except for low values of A'_s/A_s. This equation was theoretically derived by Shaikh and the author using an energy method and experimentally verified, as described by Chap. 5 and Eqs. (5-62) to (5-64). Equation (3-66) was strictly an empirical development.

For creep plus shrinkage deflections

$$k_r = 1/(1 + A'_s/A_s) \tag{3-68}$$

In the theoretical development for prestressed beams, this equation referred to the special case of equal eccentricities for both prestressed and nonprestressed steel.

A'_s/A_s	Eq. (3-66)	Eq. (3-68)
0	1.00	1.00
0.5	0.70 (0.60*)	0.67
1.0	0.40	0.50

* 1963 ACI Code Value. The factors 1.00 and 0.40 are the same in the 1963 and 1971 ACI Codes.

Using this form of equation, it was shown that Eq. (3-69) provided the best fit for the data in Table 3-17, as shown in Table 3-18.

For creep plus shrinkage deflections

$$k_r = 1/(1 + CA'_s/A_s) \qquad C = 0.70 \tag{3-69}$$

Table 3-17 Measured deflections showing the effect of compression steel[26]

Beams	Series	Duration of load, months	* p, %	† Measured deflection ratios				
				$A'_s/A_s = 0$	$A'_s/A_s = 0.5$		$A'_s/A_s = 1.0$	
				‡ Δ_t/Δ_i	‡ Δ_t/Δ_i	Col. 2/1	‡ Δ_t/Δ_i	Col. 4/1
				Col. 1	Col. 2	Col. 3	Col. 4	Col. 5
Washa and Fluck[44]								
Rectangular	A	30	1.63	1.09/0.67 = 1.63	0.65/0.62 = 1.05	0.64	0.40/0.53 = 0.76	0.47
solid beams	B	30	1.67	2.36/1.04 = 2.27	1.58/0.98 = 1.61	0.71	1.09/0.92 = 1.18	0.52
from Ref. 5,	C	30	1.67	3.66/1.88 = 1.95	2.25/1.71 = 1.31	0.67	1.57/1.58 = 0.99	0.51
Table A2-4	D	30	1.67	1.21/0.70 = 1.73	0.74/0.56 = 1.32	0.76	0.62/0.47 = 1.32	0.76
	E	30	1.59	4.80/2.48 = 1.94	2.87/2.20 = 1.30	0.67	2.54/2.34 = 1.09	0.56
Yu and Winter[40] T-beams from Ref. 5, Table A2-4	A, B, C	6	1.01	1.31/1.34 = 0.98	0.99/1.24 = 0.80	0.82	1.05/1.19 = 0.88	0.90§
Hollington[69] Rectangular	Line 14, 17, 20 (Beams 13–21)	26	0.59	0.70/0.46 = 1.52	0.55/0.47 = 1.17	0.77	0.64/0.47 = 1.36	0.89§
						Line 20[A] →	0.53/0.43 = 1.23	0.81
solid and hollow box	Line 35, 38 (Beams 34–39)	26	0.59	1.29/0.57 = 2.26			0.98/0.61 = 1.61	0.71
beams from Table 9	Line 47, 50 (Beams 46–51)	25	0.77	0.56/0.45 = 1.24			0.48/0.48 = 1.00	0.81

* $p = (A_s/bd) \times 100$.

† Each time-dependent deflection was normalized with respect to its initial deflection. Each ratio showing the effect of compression steel content for a given series of tests referred to the same duration of loading, although the duration of loading varied for the different tests. However, this should not materially affect the comparisons from test to test.

‡ Δ_i = initial deflection or short-time deflection; Δ_t = time-dependent deflection or total deflection minus initial deflection.

§ These two points are somewhat inconsistent with col. 3 and the other data in col. 5; they refer to cases in which p is low and A'_s/A_s is high.

Table 3-18 Combined reduction factor, k_r, due to creep plus shrinkage by computational methods and experimental results[26]

Method of determination			Combined reduction factor, k_r, due to creep plus shrinkage		
			* $A'_s/A_s = 0$	* $A'_s/A_s = 0.5$	* $A'_s/A_s = 1.0$
Measured values from Table 3-17, cols. 3 and 5	All data, except as noted		1.00	Range 0.64–0.82 Avg. 0.72	† Range 0.47–0.81 † Avg. 0.64
	Washa and Fluck[44] data, $p = 1.6\%$		1.00	Range 0.64–0.76 Avg. 0.69	Range 0.47–0.76 Avg. 0.56
	Yu and Winter[40] and Hollington[64] data, $p = 0.6$ to 1.0%		1.00	Range 0.77–0.82 Avg. 0.80	† Range 0.71–0.81 † Avg. 0.78
$k_r = 1.0 - 0.60A'_s/A_s$ Eq. (3-66) ACI Code (318-71)			1.00	0.70	0.40
$k_r = 1/(1 + A'_s/A_s)$ Eq. (3-68)			1.00	0.67	0.50
$k_r = 1/(1 + 0.70A'_s/A_s)$ Eq. (3-69)			1.00	0.74	0.59
$k_r = 1/(1 + CA'_s/A_s)$	$\rho = 0.000$	$C = 0.00$	1.00	1.00	1.00
$= 1/(1 + 50\rho')$	0.005	0.25	1.00	0.89	0.80
where $C = 50\rho$	0.006	0.30	1.00	0.87	0.77
$\rho = A_s/bd$	0.010	0.50	1.00	0.80	0.67
$\rho' = A'_s/bd$	0.014	0.70	1.00	0.74	0.59
Eq. (3-70)	0.015	0.75	1.00	0.73	0.57
	0.016	0.80	1.00	0.71	0.56
	0.020	1.00	1.00	0.67	0.50
	0.030	1.20	1.00	0.57	0.40
	0.040	2.00	1.00	0.50	0.33

* Reduction factors are based on, or normalized with respect to, the case of no-compression steel.

† Excluding two apparently inconsistent data points, as noted in Table 3-17.

However, a considerably better fit of the data was found to be that of Eq. (3-70) in terms of the compression steel ratio and not the steel area ratio, as shown in Table 3-18 and Fig. 3-14 (plotted from Table 3-17, cols. 3 and 5).

For creep plus shrinkage deflections—recommended

$$k_r = 1/(1 + CA'_s/A_s) = 1/(1 + 50\rho') \tag{3-70}$$

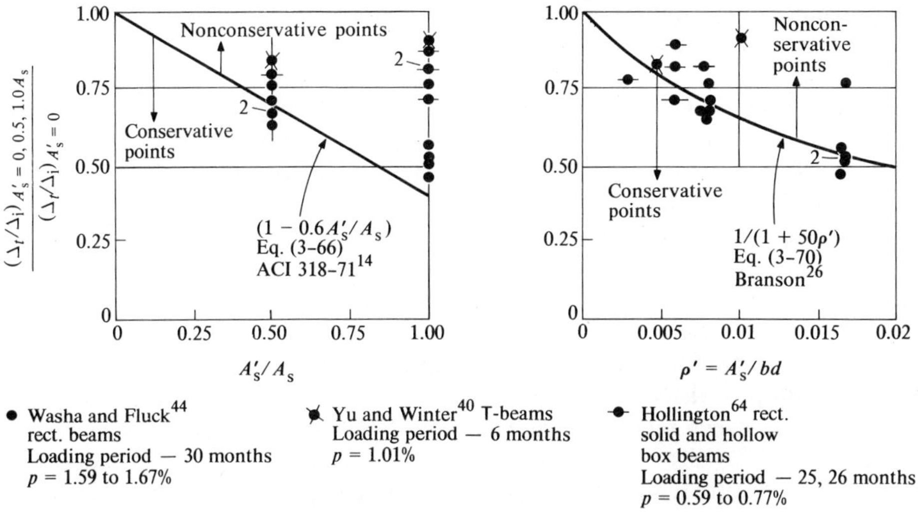

FIGURE 3-14
Comparison between experimental results and computed values of k_r by Eqs. (3-66) and (3-70). The ordinate is used to isolate the compression steel effect.

where $C = 50\rho$, $\rho = A_s/bd$, and $\rho' = A_s'/bd$. No limitation appears to be required for Eq. (3-70), as is the case with Eq. (3-66) below 0.40. Equation (3-71) is suggested for creep deflections only.

For creep deflections—recommended

$$k_r = 0.85/(1 + 50\rho') \tag{3-71}$$

For continuous members—recommended Since both k_r and I_e refer to beam curvature, it is suggested that for continuous members the same positive and negative weighted averages be used for both; such as Eqs. (3-43) to (3-49). Analogously to Eqs. (3-47*b*) and (3-49) for I_a, Eqs. (3-72) and (3-73) are recommended for k_r.

Beams with both ends continuous

$$(k_r)_a = 0.70(k_r)_m + 0.15[(k_r)_{e1} + (k_r)_{e2}] \tag{3-72}$$

Beams with one end continuous

$$(k_r)_a = 0.85(k_r)_m + 0.15(k_r)_{\text{cont end}} \tag{3-73}$$

Observations The following observations are made with regard to the results in Table 3-18 and Eqs. (3-66), (3-69), and (3-70).

1. When $A_s'/A_s = 0.5$, $k_r = 0.70$ by Eq. (3-66) and 0.74 by Eq. (3-69), versus the average experimental values of 0.72 for all the beams, 0.69 for the higher

p-beams, and 0.80 for the lower p-beams. These comparisons might be satisfactory, except possibly for the lower p-beams.

2. When $A'_s/A_s = 1.0$, $k_r = 0.40$ by Eq. (3-66) and 0.59 by Eq. (3-69), versus the average experimental values of 0.64 for all the beams, 0.56 for the higher p-beams, and 0.78 for the lower p-beams. These comparisons are not satisfactory for Eq. (3-66) with any of the p-ranges, and not satisfactory for both Eqs. (3-66) and (3-69) with the lower p-beams.

3. A study of Table 3-18 indicates that Eq. (3-70) represents a considerable improvement over Eqs. (3-66) and (3-69) for the entire range of conditions shown, where the solution for k_r is given in terms of ρ' and not A'_s/A_s.

4. It seems to the author to be much more reasonable to take into account the effect of compression steel in reducing time-dependent deflections in terms of the steel ratio $\rho' = A'_s/bd$, which represents a steel "content" parameter, than in terms of A'_s/A_s, which relates the compression steel only to the tension steel.

Beeby[64,70] has also found that a time-dependent multiplier that includes the effect of compression reinforcement in terms of the compression steel ratio, instead of the ratio of compression steel to tension steel, provides "a much better fit to the data." His expression,[64] $1.6\rho'/(\rho' + 0.00275)$, is in the form of a correction or multiplier for span-depth ratios instead of deflections directly, and this form is also used in the British Standard Code of Practice,[63] Table 6-10 herein (values computed by this equation of Beeby's).

Two Approaches for Computing Initial and Time-Dependent Deflections

Primarily because of major uncertainties regarding elastic, creep, and shrinkage properties of structural concrete and loading histories in advance of construction, and even following construction when the concrete properties are determined for representative specimens, it is normally not feasible to take into account all eight of the separate effects previously discussed. The two principal approaches used by most engineers for computing initial and time-dependent deflections are formulated below. When conditions warrant and the information is available, all of the eight effects can be incorporated in these two approaches. Effect 7 (repeated load cycles) is described by Eqs. (3-80), (3-81), and (3-83) for determining creep deflections.

Modified "Δ_i" approach for creep deflections and creep plus shrinkage deflections

$$\Delta_{cp} = k_r C_t \Delta_i \tag{3-74}$$

$$\Delta_t = \Delta_{sh} + \Delta_{cp} = k_r T \Delta_i \tag{3-75}$$

where k_r is the combined reduction factor given by Eqs. (3-66) to (3-73) or the product of several separate k_r factors, C_t is the creep coefficient, T is the combined creep plus shrinkage (time-dependent) coefficient, and Δ_i is the initial deflection computed by any appropriate method. The author[5] recommended this approach in 1963, using I_e by Eqs. (3-18) or (3-19), and continues to prefer it for a number of reasons, as will be discussed. Equation (3-75) represents the approach used in the Yu and Winter Method D (1960), ACI Code (1963, 1971), Bureau of Public Roads (1966), AASHTO Highway Bridge Specifications for Load Factor Design (1973), and by others.

Values of C_t or C_u and $(\varepsilon_{sh})_t$ or $(\varepsilon_{sh})_u$ (to be used with Eqs. (3-50) to (3-57), for example) may be found in Chap. 1 (particularly in Table 1-10), and values of T_u are suggested in Appendix J, Table J-1. In the absence of information concerning creep and shrinkage under local conditions, the author suggests the following for routine calculations: for beams and one-way slabs—$C_u = 1.6$, $(\varepsilon_{sh})_u = 400 \times 10^{-6}$ in/in (cm/cm), $T_u = 2.5$; and for other slabs (see Chap. 4)—$T_u = 3.0$.

Increased "*n*" approach for initial plus creep deflections and initial plus creep plus shrinkage deflections

$$\Delta_i = \text{elastic deflection using } E_c I_{cr} \text{ and } n = E_s/E_c \text{ in computing } I_{cr} \text{ for fully cracked sections; or using } E_c I_{ucr} \text{ or } E_c I_g \text{ for uncracked sections} \tag{3-76}$$

$$\Delta_i + \Delta_{cp} = \text{elastic + creep deflection using } E_{ct} I_{ct}\text{, the effective modulus} = E_{ct} = E_c/(1 + C_t)\text{, and } n_{ct} = n(1 + C_t) \text{ in computing } I_{ct} \text{ for fully cracked sections; or using } E_{ct} I_{uct} \text{ or } E_{ct} I_g\text{, and } n_{ct} \text{ in computing } I_{uct} \text{ for uncracked sections} \tag{3-77}$$

$$\Delta_i + \Delta_{cp} + \Delta_{sh} = \text{elastic + creep + shrinkage deflection} = \text{same as Eq. (3-77), but using empirically determined values of } E'_{ct} \text{ and } n'_{ct} \text{ instead of } E_{ct} \text{ and } n_{ct}\text{—such as } n'_{ct} = n(1 + T) \tag{3-78}$$

Equations (3-76) to (3-78) refer to the fully cracked and the uncracked sections. For partially cracked sections, Eq. (3-22) by Beeby[50] and an iterative procedure by Pauw[71] may be used. This approach has been used for elastic plus creep deflections by Large[60] (1957); the author[72] (1961); Pauw, Alcock, Meyers[73,74,71] (1962, 1964, 1971); Mayer[75] (1967); Beeby[50] (1968); and others.[52] The approach is used in the Yu and Winter Method C[40] (1960) for elastic plus creep plus shrinkage deflections using E'_{ct} and n'_{ct} from Table 3-19. In the 1957, 1961, 1965, 1969, and 1973 editions of the AASHTO Highway Bridge Specifications this approach is used for elastic design with $n = 8$ for elastic deflections and $n'_{ct} = 30$ for elastic plus creep plus shrinkage deflections.

Shrinkage deflections are computed separately from creep deflections when using Eqs. (3-74) and (3-77). Charts such as Figs. E-1 and E-2 are available for computing section properties using Eqs. (3-76) to (3-78). Additional charts and tables may be found in the ACI Ultimate Strength Design Handbook,[76] Pauw,[71] and Brakel.[77]

Discussion of the modified "Δ_i" and increased "*n*" approaches 1. With an increased *n*-value such as n_{ct} due to creep, which results in an increased effective steel area, the neutral axis is lowered (for a positive-moment singly-reinforced section), as with Fig. 3-13*b*. Accordingly, this approach has a theoretical basis in terms of an elastic procedure.

2. The increased "*n*" method is not very sensitive to the particular creep coefficient to be used, since the larger the value of C_t in Eq. (3-77), the larger the value of I_{cr}, but the smaller the value of E_{ct}, which are offsetting effects. Since prior knowledge of the creep coefficient is uncertain in most cases, this offsetting effect may be said to represent an advantage of the increased "*n*" method, provided the results are sufficiently accurate.

3. It appears that the well-documented effect of compression steel in markedly reducing creep deflections cannot be predicted directly by the increased "*n*" method. That is, the increase in I_{ct} due to compression steel, over the same case without compression steel, does not seem to even approximate the effect of compression steel on deflections. This appears to be a clear example of the inadequacy of a simple quasi-elastic method when applied to a nonelastic problem without taking all pertinent stress and strain compatibility conditions into account. Other examples of this difficulty appear to be the equivalent tensile force method for shrinkage deflections, and the increased "*n*" method itself (see point 4 below). An improvement might be to include or neglect the compression steel in the increased "*n*" solution, and then apply an appropriate reduction factor, k_r, for the effect of compression steel.

4. In the early 1960s the author[72] pursued the increased "*n*" method with great interest, but was forced to abandon it in favor of the modified "Δ_i" method using I_e,[5] after being unable to achieve even remotely satisfactory agreement with experimental deflections. For example, for the beams of Washa and Fluck[44] and Yu and Winter[40] (54 beams with $I_g/I_{cr} = 1.9$ to 3.5 for singly and doubly reinforced rectangular beams and T-beams), the measured time-dependent deflections were of the order of 1 to 2 times the initial deflections (see Table 3-17). The increased "*n*" method requires a creep coefficient of the order of $C_t = 6$ to 8 to predict these initial plus creep deflections, with shrinkage deflections being computed separately by any of the suitable methods available. The measured creep coefficients for the same period as the beam tests were of the order of 3.5 to 5. This seems to show an inconsistency in the direct application of the increased "*n*" method.

It is further noted that empirical 5-year or ultimate values of n'_{ct} by Yu and Winter in Table 3-19 varied from 72.5 to 126 for computing initial plus creep plus shrinkage deflections, which are much higher than the value of 30 used by AASHTO for the same purpose, and much higher than the increased "*n*" approach might rationally suggest. Using empirical values of $k_r T$ and n'_{ct} (both shown in Table 3-19), Yu and Winter[40] found their Method D (modified "Δ_i" approach) to be in better agreement with experimental results than their Method C (increased "*n*" approach). In this regard, for an example problem in Wang and Salmon,[20] p. 535, the time-dependent deflection using $n'_{ct} = 2.8n$ in the increased "*n*" method was about one-third the value by three other methods using the modified "Δ_i" approach. Also, with regard to repeated loading, Miller and Guralnick[68] stated that "neither the equivalent modulus method nor the fictitious axial force method

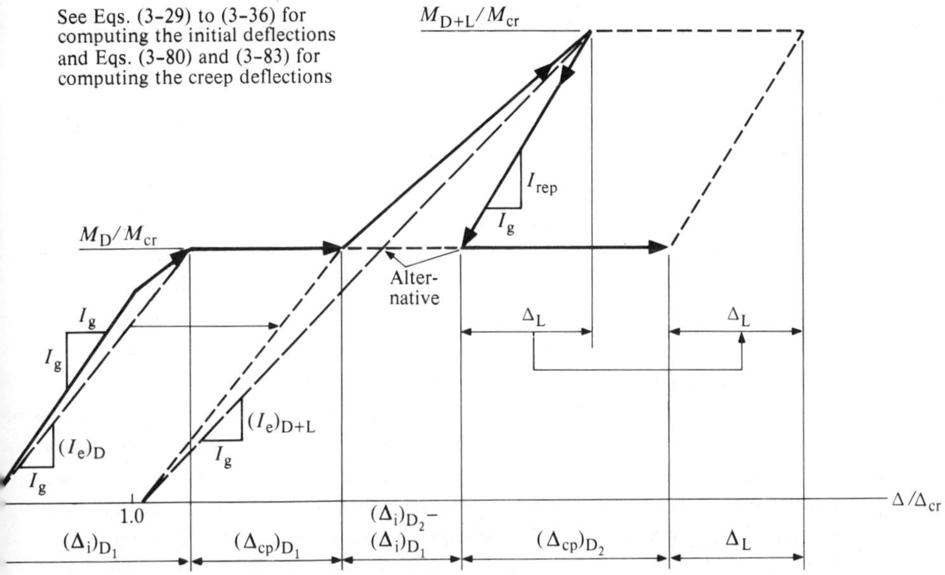

FIGURE 3-15
Idealized moment versus initial plus creep deflection diagram for the case of no sustained live load, and showing the initial and creep deflection increments occurring both before $(\Delta_i)_{D1}$, $(\Delta_{cp})_{D1}$ and after $(\Delta_i)_{D2}$, $(\Delta_{cp})_{D2}$, Δ_L, a single application of live load. The initial or short-time deflections are also shown in Fig. 3-7.

results in satisfactory deflection predictions for a beam subjected to repeated cycles of load." The point is, the author believes, that these relatively simple quasi-elastic methods for predicting creep and shrinkage deflections have significant limitations.

5. A practical inconvenience in the increased "n" approach is the fact that each time a new creep coefficient is to be considered, a new neutral axis location and moment of inertia must be determined, although the use of charts can somewhat offset this disadvantage. On the other hand, such determinations may be cumbersome when considering partially cracked sections for which charts may not be available or feasible. In the modified "Δ_i" approach using I_e, for example, different moments of inertia must be computed for dead load and dead plus live load, but this includes the effect of partial cracking. Then different creep coefficients or incremental creep effects (see point 6 below) can be applied without requiring additional modifications in the section properties.

6. The determination of time-dependent deflections by the modified "Δ_i" approach, as the product of appropriate modifying factors and the initial or short-time deflection, seems to the author to be generally more convenient than the increased "n" approach as an engineering procedure. It also appears to more readily lend itself to the calculation of creep deflections before and after live load application, after the attachment of nonstructural elements, and for partial

sustained live loading, in which the effect of cracking under first dead load and then dead plus live load is taken into account using I_e, for example. See Fig. 3-15 and Eqs. (3-80) to (3-83).

7. The use of I_e in Eqs. (3-18) or (3-19) with the modified "Δ_i" approach appears to be far simpler than the increased "n" approach for including the effect of tensile concrete. Since values of I_g/I_{cr} of 2 to 3 are not uncommon, the author disagrees with those who say that the effect of tensile concrete can be neglected in most cases, and that either I_g for uncracked cases, or I_{cr} for cracked cases, can be used satisfactorily. One cannot be sure without checking that the effect of tensile concrete under first dead load and then dead plus live load can be neglected, when the effective I can be between I_{cr} and $I_g = 2$ to 3 times, or more, I_{cr}. One argument in support of neglecting the effect of tensile concrete is that the increased "n" or effective modulus method implies complete elastic and creep recovery which tends to compensate for the contribution of tensile concrete to beam rigidity. However, such large differences as the factors 2 to 3, or more, suggest that the effect of tensile concrete can be a major consideration. The use of I_e, as shown in Fig. 3-5, represents a simple method of directly and routinely taking into account the effect of tensile concrete in terms of I_g, I_{cr}, and the level of cracking parameter, M/M_{cr}.

8. Primarily because of the variable nature of the deflection problem for concrete structures, the somewhat more empirical approach of the modified "Δ_i" method is not necessarily less reliable than the somewhat more theoretical approach of the increased "n" method. In fact, it has been the experience of the author and others that the reverse is the case, as indicated in point 4 above. It should also be noted that Δ_i in the modified "Δ_i" approach may be said to be the principal parameter in this method, and I_e, when used to compute Δ_i, is determined by an empirical method that provides a transition between the theoretical limits of I_g (or I_{ucr}) and I_{cr}, as shown in Fig. 3-5.

Methods for Computing Time-Dependent Deflections

Former PCA,[36] 1947 (initial plus creep deflection using the modified "Δ_i" approach) The short-time deflection based on the gross section is multiplied by the factor in the hyperbolic equation (3-79) to obtain the total (initial plus creep) deflection under sustained loading.

$$K_t = \frac{t}{20.6 + 0.263t} \tag{3-79}$$

where t = time in days under sustained loading; $K_t = 3.13$ after 1 year, 3.65 after 5 years, and $1/0.263 = 3.80$ ultimately. No distinction was made for beams with compression reinforcement.

Dunham,[37] 1953 (shrinkage and initial plus creep deflection using the modified "Δ_i" approach) Shrinkage deflections (which are considered not likely to be large) may be estimated as 10 to 20 percent of the computed deflection for loading.

What loading was not specified. However, as can be seen for the 4 examples herein (representative range of the ratio of shrinkage to total deflection = 8 to 17 percent), the Dunham percentages may be more nearly correct when applied to the total dead load (including creep) and live load deflection.

Initial plus creep deflections should probably be doubled for loading applied for 1 year, and tripled if the loading is fairly permanent. That is, $k_r C_t = 1$ for 1 year and 2 for fairly permanent loading in Eq. (3-74). See Eq. (3-8) for the initial deflection by this method.

Fully cracked increased "*n*" method—Large,[58] 1957; the author,[72] 1961; Pauw, Alcock, Meyers,[73,74,71] 1962, 1964, 1971; Mayer,[75] 1967; Beeby,[50] 1968; and others[52] (initial plus creep deflection by the increased "*n*" approach) Refers to the increased "*n*" approach directly, using Eq. (3-77) for computing initial plus creep deflections.

AASHTO,[38] 1957 (initial plus creep plus shrinkage deflection using the increased "*n*" approach) In the 1957, 1961, 1965, 1969, and 1973 editions of the AASHTO Highway Bridge Specifications the increased "*n*" approach is specified using $n = 8$ for initial deflections, regardless of the concrete strength, and $n'_{ct} = 30$ for initial plus creep plus shrinkage deflections, as described by Eqs. (3-76) and (3-78). For load factor design the modified "Δ_i" approach is recommended in the 1973 edition, as will be shown.

Yu and Winter,[40] 1960

Method C (initial plus creep plus shrinkage deflection using the increased "n" approach) Empirical values of E'_{ct} and n''_{ct} (using $E_s = 29 \times 10$ psi (2.04 × 10 kg/cm^2, 200 kN/mm^2)) in Table 3-19 are used in the increased "*n*" approach for computing initial plus creep plus shrinkage deflections, as shown in Eq. (3-78). It may be noted in Table 3-19 that most of the values of n'_{ct} (27.9 to 126, or for 28-day loading age—27.9 to 76.3 with ultimate values of 72.5 to 76.3 for $f'_c = 3000$ to 5000 psi or 211 to 352 kg/cm^2 or 20.7 to 34.5 N/mm^2) are much higher than the value of 30 used by AASHTO. This is consistent with the author's observations in point 4 of the previous discussion of the modified "Δ_i" and increased "*n*" approaches.

Method D (creep plus shrinkage deflection using the modified "Δ_i" approach Empirical values of $k_r T$ in Table 3-19 are used in the modified "Δ_i" approach, as shown in Eq. (3-75), for computing creep plus shrinkage deflections. Δ_i refers to Method A, which uses I_{cr}. Hence, when using the $k_r T$ values in Table 3-19 with deflections computed with I_g or I_e (as the 1971 ACI Code), the results may tend to be on the low side.

Based on the long time test results of 68 beams, the simpler Method D was found to be somewhat closer to the experimental values than Method C.

Table 3-19 Values of E'_{ct} and n'_{ct} for Method C and values of $k_r T$ for Method D determined empirically for 68 test beams by Yu and Winter[40]

Duration of loading	Values of * E'_{ct} and † n'_{ct} for Method C									$k_r T$ values for Method D ‡		
	$f'_c = 3000$ psi			$f'_c = 3750$ psi			$f'_c = 5000$ psi					
	Loading age, days			Loading age, days			Loading age, days					
	7	14	28	7	14	28	7	14	28	$A'_s/A_s = 0$	$A'_s/A_s = 0.5$	$A'_s/A_s = 1.0$
1 month	0.60 48.3	0.75 38.7	0.91 31.9	0.63 46.0	0.79 36.7	0.97 29.9	0.66 43.9	0.83 34.9	1.04 27.9	0.58	0.42	0.27
3 months	0.45 64.4	0.56 51.8	0.70 41.4	0.46 63.0	0.58 50.0	0.73 39.7	0.47 61.7	0.61 47.5	0.77 37.7	0.95	0.77	0.55
6 months										1.17	0.95	0.69
9 months										1.31	1.03	0.73
1 yr	0.30 96.7	0.38 76.3	0.48 60.4	0.30 96.7	0.39 74.4	0.50 58.0	0.31 93.5	0.40 72.5	0.51 56.9	1.42	1.08	0.78
1.5 yr										1.54	1.12	0.80
2 yr										1.65	1.15	0.80
2.5 yr										1.72	1.16	0.81
3 yr	0.24 121	0.31 93.5	0.39 74.4	0.24 121	0.31 93.5	0.40 72.5	0.25 116	0.32 90.6	0.42 69.0	1.78	1.18	0.81
4 yr										1.87	1.20	0.82
5 yr	0.23 126	0.29 100	0.38 76.3	0.23 126	0.30 96.7	0.39 74.4	0.24 121	0.31 93.5	0.40 72.5	1.95	1.21	0.82

(Adapted from ACI Journal,[40] with permission.)

* $E'_{ct} \times 10^6$ psi, as 0.60×10^6 psi (42 200 kg/cm^2, 4.14 kN/mm^2), for creep plus shrinkage.

† $n'_{ct} = E_s/E'_{ct} = 29 \times 10^6/E_{ct}$ as $29/0.60 = 48.3$, for creep plus shrinkage.

‡ $k_r T$ values refer to creep plus shrinkage deflections. For ultimate values, the 1963 ACI Code used 2.0, 1.2, and 0.8, respectively. The 1971 ACI Code

Former CRSI,[39] 1961 (creep plus shrinkage deflection using the modified "Δ_i" approach) Additional long-time deflections due to creep and shrinkage are obtained by multiplying the short-time deflection based on the moment of inertia of the gross section by 3.0 when $A'_s = 0$, and by 1.5 for simple beams and 1.0 for continuous beams when $A'_s = A_s$. That is, $k_r T = 3.0$, 1.5, and 1.0 in Eq. (3-75).

It should be noted that PCA and CRSI were sponsors of the 1965 ACI Reinforced Concrete Design Handbook[78] which refers to the ACI Code method for computing initial and time-dependent deflections.

Former European Concrete Committee (CEB),[41] 1961 (creep plus shrinkage deflection using the modified "Δ_i" approach) For long-duration loads, the effects of shrinkage and creep may be included by multiplying by 3 the deflection given by the permanent load. That is, $k_r T = 3$ in Eq. (3-75). A number of corrections are anticipated to permit reduction of deflection when the permanent load is applied on concrete that has already undergone a long process of hardening. The initial deflection is computed by Eq. (3-10).

ACI Code (318-63),[42] 1963 (creep plus shrinkage deflection using the modified "Δ_i" approach) Additional (creep plus shrinkage) ultimate deflections under sustained loading are obtained by multiplying the initial deflection by 2.0 when $A'_s = 0$, 1.2 when $A'_s = 0.5A_s$, and 0.8 when $A'_s = A_s$. That is, $k_r T = 2.0$, 1.2, and 0.8 in Eq. (3-75). The initial deflection is computed by Eq. (3-15) based on the gross section or Eq. (3-16) based on the cracked section, depending on ρf_y. These values of $k_r T$ were the same as the 5-year values of Yu and Winter in Table 3-19, which were based on I_{cr}-deflections. For loading periods less than 5 years, the $k_r T$ values in Table 3-19 were plotted in the commentary of the Code.

With regard to part of the live load being considered as a sustained load, the following is mentioned in the commentary of the Code: in an apartment building, as little as 20 percent of the design live load may fall into this category; in a heavy warehouse, on the other hand, it may be necessary to assume that 80 percent or more of the live load will act as sustained load over extended time intervals.

The British Code Handbook[103] suggests that for normal domestic or office occupancy 25 percent of the imposed load should be considered as permanent and that for structures used for storage at least 75 percent should be considered permanent when the upper limit to the deflection is being assessed.

Author,[5,6,26,79] 1963, 1968, 1971 (initial, shrinkage, and creep deflection calculated separately, with creep deflection using the modified "Δ_i" approach) As shown in Figs. 3-8, 3-15, and 3-19 for initial and creep deflection in idealized form, the author suggests the following procedures for computing the deflection components.

Cantilever beam, uniform loading—Eq. (3-42) for Δ_i, Eq. (3-19) for I_e, and Eq. (3-43) for I_a.

Cantilever beam, end-concentrated load—Eq. (3-42) for Δ_i, Eq. (3-18) for I_e (based on the combined dead and live-load moment), and Eq. (3-43) for I_a.

Simple beam, uniform loading—Eq. (3-39) for Δ_i, and Eq. (3-19) for I_e and I_a.

Simple beam, center-concentrated load—Eq. (3-39) for Δ_i, and Eq. (3-19) for I_e (based on the combined dead and live-load moment), and I_a; except Eq. (3-18) for I_e (based on the combined dead and live-load moment), and I_a in the case of a "heavy" concentrated load.

Continuous beam, uniform loading—Eq. (3-39) for Δ_i, Eq. (3-40) for K, Eq. (3-19) for I_e, and Eq. (3-47*b*) or (3-49) for I_a.

Continuous beam, center-concentrated load—Eq. (3-39) for Δ_i, Eq. (3-41) for K, Eq. (3-19) for I_e (based on the combined dead and live-load moment), and Eq. (3-44) for I_a.

Shrinkage deflection is computed by Eq. (3-58) for Δ_{sh} using Eqs. (3-59) to (3-63) for K_{sh}, Eqs. (3-53) to (3-55) or (3-56) for ϕ_{sh}, and Eqs. (1-17), (1-18) or Table 1-10 for ε_{sh}, with correction factors by Eqs. (1-23) to (1-30) or Tables 1-7 to 1-9. Where information concerning shrinkage under local conditions is lacking, and/or for routine calculations, an average value of $(\varepsilon_{sh})_u = 400 \times 10^{-6}$ in/in (cm/cm) might be used.

Creep deflection is computed by Eq. (3-74) for Δ_{cp} using Eq. (3-71) for k_r, Eqs. (3-72) or (3-73) for $(k_r)_a$, and Eq. (1-16) or Table 1-10 for C_t, with correction factors by Eqs. (1-22) to (1-29) or Tables 1-6 to 1-9. Also see Eqs. (3-80) to (3-84) and Fig. 3-15 for a few special cases of creep deflections. Where information concerning creep under local conditions is lacking, and/or routine calculations, an average value of $C_u = 1.6$ might be used.

For estimating the combined shrinkage plus creep deflection, use Eq. (3-75) for Δ_t, Eq. (3-70) for k_r, Eqs. (3-72) or (3-73) for $(k_r)_a$, and Appendix J, Table J-1 for T_u (and Fig. 1-7 for estimating less than ultimate values). For routine calculations, the author suggests an average value of $T_u = 2.5$ for beams and one-way slabs and 3.0 for other slabs (see Chap. 4).

Equations (1-2) to (1-5) are used for concrete strengths at different ages (used to determine f_r by Eq. (1-7) or Table 1-2 and E_c by Eq. (1-13) or Table 1-3). For loading periods and shrinkage periods less than to ultimate, Eqs. (1-16) to (1-18) or Fig. 1-7 may be used for creep and shrinkage (also in Table 1-10 for shrinkage following a given age) increments. A few special cases are described below by Eqs. (3-80) to (3-84).

The short-time dead-load deflection that takes place both before $(\Delta_D)_1$ and after $(\Delta_D)_2$ application of the design live load and the short-time live-load deflection are shown in Fig. 3-8 and computed by Eqs. (3-29) to (3-36). Equations (3-80) to (3-84) are used for computing dead-load creep deflection before and after live-load application or passage, partial live-load creep deflection, and dead-load creep deflection that occurs after the attachment of nonstructural elements. An idealized moment versus initial plus creep deflection diagram for the case of no sustained live load is shown in Fig. 3-15.

Dead-load creep deflection before application or passage of live load

$$(\Delta_{cp})_{D1} = k_r \alpha C_u (\Delta_i)_{D1} \tag{3-80}$$

where αC_u is the part of the total creep that takes place before live-load application,

Table 3-20 Representative values of α and α_1

t, days*	α, α_1	t, days*	α, α_1
7	0.24	45	0.49
14	0.33	60	0.54
21	0.38	90	0.60
30	0.44	120	0.64

* For determining α in Eqs. (3-80) to (3-83), t is the time in days between dead-load and live-load application. For determining α_1 in Eq. (3-81), t is the time in days between dead-load application and the attachment of nonstructural elements.

or $\alpha = C_t/C_u = t^{0.60}/(10 + t^{0.60})$ in Eq. (1-16), t is the time in days between dead-load and live-load application, k_r is given by Eq. (3-71), and $(\Delta_i)_{D1}$ is computed by Eq. (3-30). Equation (1-16) can also be used for any combination of sustained loading periods, including intermittent loading. Representative values of α are given in Table 3-20.

Dead-load creep deflection after the attachment of nonstructural elements but before application or passage of live load

$$(\Delta_{cp})_{D1} = k_r(1 - \alpha_1)\alpha C_u(\Delta_i)_{D1} \tag{3-81}$$

where α_1 also $= C_t/C_u$ in Eq. (1-16), and t is the time in days between dead-load application and the attachment of nonstructural elements. Equation (3-81) is an alternate to Eq. (3-80). Representative values of α_1 are also given in Table 3-20.

Dead-load creep deflection after the attachment of nonstructural elements without regard to the effect of live load

$$(\Delta_{cp})_D = k_r(1 - \alpha_1)C_u(\Delta_i)_D \tag{3-82}$$

where $(\Delta_i)_D$ is computed by Eq. (3-30).

Dead-load creep deflection after application or passage of live load

$$(\Delta_{cp})_{D2} = k_r(1 - \alpha)C_u(\Delta_i)_{D2} \tag{3-83}$$

where $(\Delta_i)_{D2}$ is computed by Eq. (3-34) or (3-36).

Partial live-load creep deflection

$$(\Delta_{cp})_L = k_r C_u(\Delta_L)_s \tag{3-84}$$

where $(\Delta_L)_s = 0.30\Delta_L$ when 30 percent of the live load is assumed to be sustained, for example, C_u is the ultimate creep coefficient in which the concrete age when the live load is applied is taken into account, and Δ_L is computed by Eqs. (3-31) or (3-32) or (3-35). When all of the live load or superimposed load is sustained, of course, $(\Delta_L)_s = \Delta_L$.

These procedures provide a complete set of simple formulas for computing average initial, shrinkage, and creep deflections for any loading age or shrinkage period, period of loading including intermittent loading, member size and other effects. Specific cases of creep deflections before and after live load application, after the attachment of nonstructural elements, and for partial sustained live loading are described. Charts and tables are also presented in Chap. 1 for the appropriate elastic, cracking, creep, and shrinkage parameters, including correction factors. These procedures are also illustrated in the examples herein.

BPR,[47] 1966 (creep plus shrinkage deflection using the modified "Δ_i" approach) The following procedure is recommended by the Bureau of Public Roads for computing deflections of reinforced concrete bridges: the short-time dead-load deflection caused by dead loads, creep, and shrinkage. That is, values of $k_r T$ in computed by Eq. (3-20) and/or the procedure described in conjunction with Eq. (3-20), shall be multiplied by a factor from Table 3-21 to obtain the long-time deflection caused by dead loads, creep and shrinkage. That is, values of $k_r T$ in Eq. (3-75) are given in Table 3-21. The long-time deflection, thus calculated, might be expected to occur over a period of about three years.

Corley and Sozen,[49] 1966 (initial, shrinkage, and creep deflection calculated separately, with creep deflection using the modified "Δ_i" approach) The initial curvature (and hence deflection) is computed by Eq. (3-21). Shrinkage deflection is computed by Eq. (3-57).

Creep deflection is computed by Eq. (3-74) for Δ_{cp} using Eqs. (3-85) or (3-86) for $k_r C_u$. These equations were based on a study of the available data in the literature along with a consideration of the strain diagram in Fig. 3-12 and Eqs. (3-64), (3-65).

For rectangular sections: $$k_r C_u = k(3 - \rho'/\rho) \tag{3-85}$$

For flanged sections with $kd > 2t/3$: $$k_r C_u = 2k(3 - \rho'/\rho) \tag{3-86}$$

where t = the thickness of the top flange, and k = the ratio of the depth of the compression block of the cracked transformed section to the effective depth.

Table 3-21 Values of $k_r T$ in Eq. (3-75) recommended by BPR[47]

Humidity condition	$A'_s = 0$	$A'_s = 0.5A_s$	$A'_s = A_s$
Climate of high humidity	2.5	1.8	1.5
Climate of average humidity	3.0	2.2	1.8
Climate of low humidity	3.5	2.5	2.0

As discussed previously herein, more recent work by the author[26] and Beeby[64] has found this effect (k_r effect) to be more appropriately expressed as a function of ρ' rather than ρ'/ρ, although Eqs. (3-85) and (3-86) are given in terms of k (no relation to k_r) and ρ'/ρ.

ACI Code (318-71),[14] 1971 (creep plus shrinkage deflection using the modified "Δ_i" approach) The initial deflection is determined using I_e in Eq. (3-19). Additional (creep plus shrinkage) ultimate deflections under sustained loading are obtained by Eq. (3-75), using $T = 2$ and k_r by Eq. (3-66). For loading periods less than 5 years, the $k_r T$ values in Table 3-19 are plotted in the commentary of the Code.

The $k_r T$ values used in this method were based on the results by Yu and Winter in Table 3-19, which were experimental values based on computed deflections using I_{cr}. Thus, when I_g or I_e are used in the Code method, the computed results may tend to be on the low side. For this and other reasons related to average shrinkage and creep effects, the author believes this will usually be the case with regard to long-time or total deflections (as the examples tend to bear out compared to other solutions—see Table 3-22, note §, and Table 3-23). No provision is made for the potentially significant effect of loading age in the Code procedure.

AASHTO,[38] 1973

Working stress design (initial plus shrinkage plus creep deflection using the increased "n" approach) See description of method for AASHTO, 1957 to 1973 Editions herein.

Load factor design (creep plus shrinkage deflection using the modified "Δ_i" approach) Initial deflection under dead load is determined using I_e in Eq. (3-19). Creep plus shrinkage deflections are found by Eq. (3-75), using values of $k_r T$ from Table 3-21 (same as by BPR). The long-time deflections thus calculated might be expected to occur over a period of about three years.

European Concrete Committee (CEB),[52] 1973 In their 1973 CEB draft,[52] nine procedures by the following authors were reviewed in detail: Arga e Lima and Monteiro (Portugal), Beeby (England), Brakel (Holland), Branson (USA), Leonhardt and Kayatz (Germany), Lewicki and Szymanski (Poland), Mayer (Germany), Monnier (Holland), and Pauw (USA). These methods include both the modified "Δ_i" and increased "n" approaches, and modifications thereof.

ACI Committee 443, Concrete Bridge Design,[53] 1974 (creep plus shrinkage deflection and alternately creep and shrinkage deflection separately using the modified "Δ_i" approach) The initial deflection is computed using I_e in Eq. (3-19). Creep plus shrinkage ultimate deflections under sustained loading are obtained by Eq. (3-75), using $T = 2$ and k_r by Eq. (3-66). This is the same method as that of the ACI Code (318-71). Alternately, creep and shrinkage deflections may be computed separately using the method of Corley and Sozen for creep (Eqs. (3-85) and (3-86)) and for shrinkage (Eq. (3-57)).

Numerical Examples of Time-Dependent Deflections

Example 1 Simple span rectangular beam, normal-weight concrete, by various methods.

See Table 3-2 for the material and section properties, parameters and conditions of the problem. Compute the ultimate time-dependent deflection following age 28 days. Assume the loading age is also 28 days and that only the dead load is to be considered as a sustained load. The results by the various methods are summarized in Table 3-22 along with the ratios of the different deflection components to the total deflection.

For $H = 50\%$ and loading age = 28 days, $C_u = 1.86$ (Table 1-10)

For $b = 10$ in. (25.4 cm), $(\text{C.F.})_T = 0.93$ (Table 1-8)

Hence, $C_u = (0.93)(1.86) = \underline{1.73}$ (use)

Cracked-section increased "n" method (initial plus creep deflection by Eq. (3-77))

$$E_{ct} = E_c/(1 + C_u) = 3.64 \times 10^3/(1 + 1.73)$$
$$= \underline{1.33 \times 10^3 \text{ ksi } (93\,500 \text{ kg/cm}^2, 9.2 \text{ kN/mm}^2)} \quad \text{(Eq. (1-46) or (B-15))}$$
$$n_{ct} = n(1 + C_u) = 8(1 + 1.73) = 21.8 \quad \text{(Eq. (1-47))}$$
$$\rho n_{ct} = (0.0198)(21.8) = 0.432$$
$$k = \sqrt{2\rho n_{ct} + (\rho n_{ct})^2} - \rho n_{ct}$$
$$= \sqrt{(2)(0.432) + 0.432^2} - 0.432 = 0.593$$
$$kd = (0.593)(12) = 7.12 \text{ in. } (18.1 \text{ cm})$$
$$I_{ct} = b(kd)^3/3 + n_{ct}A_s(d - kd)^2 = (10)(7.12)^3/3 + (21.8)(2.37)(12 - 7.12)^2$$
$$= \underline{2433 \text{ in}^4 \ (101\,300 \text{ cm}^4)}$$

Checking by using Fig. E-2 with $\rho n = 0.432$ and $\rho' n' = 0$,

$$I_{ct} = 0.141bd^3 \quad (0.141)(10)(12)^3 = 2436 \text{ in}^4 \ (101\,400 \text{ cm}^4) \quad \text{OK}$$
$$(\Delta_i + \Delta_{cp}) = \frac{KM_D L^2}{E_{ct} I_{ct}} = \frac{(5/48)(15.0)(12)(20)^2(144)}{(1.33 \times 10^3)(2433)} = \frac{1080 \times 10^3}{3236 \times 10^3}$$
$$= \underline{0.33 \text{ in. } (0.84 \text{ cm})}$$
$$(\Delta_i)_D = \underline{0.22 \text{ in. } (0.56 \text{ cm})} \quad \text{(prev. calc.)}$$

Hence $(\Delta_{cp})_D = 0.33 - 0.22 = \underline{0.11 \text{ in. } (0.28 \text{ cm})}$

As previously discussed, this method tends to compute creep deflections on the low side (also see note ‡, Table 3-22). However, the combined initial plus creep deflections by this method may be more nearly correct.

Yu and Winter

Method C (initial plus creep plus shrinkage deflection using the increased "n" approach) For $f'_c = 4000$ psi and loading age = 28 days,

$$E'_{ct} = \underline{0.39 \times 10^3 \text{ ksi } (27\,400 \text{ kg/cm}^2, 2.7 \text{ kN/mm}^2)}$$

$$n'_{ct} = 74.4 \quad \text{(Table 3-19)}$$

$$b(kd)^2/2 = n'_{ct}A_s(d - kd) \qquad 5(kd)^2 = (74.4)(2.37)(12 - kd)$$

$$kd = 9.45 \text{ in. } (24.0 \text{ cm})$$

$$I'_{ct} = b(kd)^3/3 + n'_{ct}A_s(d - kd)^2$$

$$= (10)(9.45)^3/3 + (74.4)(2.37)(12 - 9.45)^2$$

$$= \underline{3960 \text{ in}^4 \ (164\,800 \text{ cm}^4)}$$

$$(\Delta_i + \Delta_{sh} + \Delta_{cp})_D = \frac{KM_DL^2}{E'_{ct}I'_{ct}} = \frac{1080 \times 10^3}{(0.39 \times 10^3)(3960)} = \frac{1080 \times 10^3}{1544 \times 10^3}$$

$$= \underline{0.70 \text{ in. } (1.78 \text{ cm})} \qquad \text{(Eq. (3-78))}$$

$$(\Delta_i)_D = \underline{0.22 \text{ in. } (0.56 \text{ cm})} \qquad \text{(prev. calc. for cracked section)},$$

Hence,

$$(\Delta_{sh} + \Delta_{cp})_D = \Delta_t = 0.70 - 0.22 = \underline{0.48 \text{ in. } (1.22 \text{ cm})}$$

Method D (creep plus shrinkage deflection using the modified "Δ_i" approach)

$$(\Delta_i)_D = \underline{0.22 \text{ in. } (0.56 \text{ cm})} \qquad \text{(prev. calc. for cracked section)}$$

$$k_r T = 1.95 \quad \text{(Table 3-19)}$$

$$(\Delta_{sh} + \Delta_{cp})_D = \Delta_t = k_r T(\Delta_i)_D = (1.95)(0.22)$$

$$= \underline{0.43 \text{ in. } (1.09 \text{ cm})} \qquad \text{(Eq. (3-75))}$$

In this method the $k_r T$ values, and hence Δ_t, are based on the cracked-section $(\Delta_i)_D$.

ACI Code (318-63) (creep plus shrinkage deflection using the modified "Δ_i" approach)

$$(\Delta_i)_D = \underline{0.13 \text{ in. } (0.33 \text{ cm})} \qquad \text{(prev. calc.)} \qquad k_r T = 2.0$$

$$(\Delta_{sh} + \Delta_{cp})_D = \Delta_t = k_r T(\Delta_i)_D = (2.0)(0.13)$$

$$= \underline{0.26 \text{ in. } (0.66 \text{ cm})} \qquad \text{(Eq. (3-75))}$$

Author (initial, shrinkage, and creep deflection calculated separately, with creep deflection using the modified "Δ_i" approach)

$$(\Delta_i)_D = \underline{0.15 \text{ in. } (0.38 \text{ cm})} \qquad (\Delta_{sh})_u = \underline{0.15 \text{ in. } (0.38 \text{ cm})} \qquad \text{(prev. calc.)}$$

$$k_r = 0.85 \qquad \text{(Eq. (3-71))}$$

$$(\Delta_{cp})_D = k_r C_u(\Delta_i)_D = (0.85)(1.73)(0.15) = \underline{0.22 \text{ in. } (0.56 \text{ cm})} \qquad \text{(Eq. (3-74))}$$

The following illustrates the alternate (simpler) method for including the effect of repeated live load on the dead-load creep deflection, assuming the live load is applied 1 month after dead load ($\alpha = 0.44$ in Table 3-20):

$$(\Delta_i)_{D1} = 0.15 \text{ in. } (0.38 \text{ cm}) \qquad \text{(Eq. (3-30) and prev. calc.)}$$
$$(\Delta_{cp})_{D1} = k_r \alpha C_u (\Delta_i)_{D1} = (0.85)(0.44)(1.73)(0.15)$$
$$= 0.19 \text{ in. } (0.25 \text{ cm}) \qquad \text{(Eq. (3-80))}$$
$$(\Delta_i)_{D2} = (\Delta_i)_{D1}(I_e)_D/(I_e)_{D+L} = (0.15)(2033/1355)$$
$$= 0.22 \text{ in. } (0.56 \text{ cm}) \qquad \text{(Eq. (3-36))}$$

Hence, for the purpose of computing the dead-load creep deflection, the full design live load has the effect of increasing the initial or short-time dead-load deflection from 0.15 in. (0.38 cm) to 0.22 in. (0.56 cm).

$$(\Delta_{cp})_{D2} = k_r(1 - \alpha)C_u(\Delta_i)_{D2} = (0.85)(1 - 0.44)(1.73)(0.22)$$
$$= 0.18 \text{ in. } (0.46 \text{ cm}) \qquad \text{(Eq. (3-83))}$$
$$(\Delta_{cp})_D = (\Delta_{cp})_{D1} + (\Delta_{cp})_{D2} = 0.10 + 0.18 = \underline{0.28 \text{ in. } (0.71 \text{ cm})}$$

versus 0.22 in. (0.56 cm) or 27 percent greater than above when the effect of repeated live load is neglected. These various deflection components are shown in Figs. 3-8 and 3-15. Also, when taking into account the effect of repeated live loading,

$$\Delta_L = \frac{KM_L L^2}{E_c(I_e)_{D+L}} = \frac{(5/48)(50.0)(12)(20)^2(144)}{(3.64 \times 10^3)(1335)} = \underline{0.74 \text{ in. } (1.88 \text{ cm})} \qquad \text{(Eq. (3-35))}$$

versus 0.80 in. (2.03 cm) or 9 percent less than before under a single live load application.

ACI Code (318-71) (creep plus shrinkage deflection using the modified "Δ_i" approach)

$$(\Delta_i)_D = \underline{0.15 \text{ in. } (0.38 \text{ cm})} \qquad \text{(prev. calc.)}$$
$$k_r = 1.0 \text{ (Eq. (3-66))} \qquad T = 2.0$$
$$(\Delta_{sh} + \Delta_{cp})_D = \Delta_t = k_r T(\Delta_i)_D = (1.0)(2.0)(0.15)$$
$$= \underline{0.30 \text{ in. } (0.76 \text{ cm})} \qquad \text{(Eq. (3-75))}$$

As shown in Table 3-22, shrinkage plus creep deflections by this method (as well as by the 1963 ACI Code) tend to be on the low (unsafe) side, since the $k_r T$ values used were obtained from Yu and Winter's empirical values in Table 3-19, which were derived on the basis that the initial deflections referred to the fully cracked section. In addition, when considering typical conditions of member size, humidity, loading age, and average creep and shrinkage levels, the author believes the use of $T = 2.5$ would more nearly approximate an average value in such Code calculations for combined creep and shrinkage deflections.

Summary of deflections for Example 1 The different deflection components and total deflection by the various methods are summarized in Table 3-22 for Example 1. Several observations regarding these methods are made in the footnotes to the table.

The following allowable deflections from the ACI Code (318-71) refer to the time-dependent plus live-load deflections (see note ‡ to Table 3-3) for a comparison with the allowable initial or live-load deflection).

Roof or floor construction supporting or attached to nonstructural elements *likely* to be damaged by large deflections—$\Delta_t + \Delta_L \leq L/480 = 240/480 = 0.50$ in. (1.27 cm).

Roof or floor construction supporting or attached to nonstructural elements *not likely* to be damaged by large deflections—$\Delta_t + \Delta_L \leq L/240 = 240/240 = 1.00$ in. (2.54 cm).

It is seen in Table 3-22 that Max $\Delta_t + \Delta_L = \Delta_{\text{Total}} - (\Delta_i)_D > 1.00$ in. (2.54 cm). Hence, as determined by all of the methods summarized in the table, this beam does not satisfy either of the above criteria for roof or floor construction supporting or attached to nonstructural elements. However, as shown in note ‡ of Table 3-3, the beam is satisfactory for use with flat roofs (but not floors) not supporting or attached to nonstructural elements likely to be damaged by large deflections.

Example 2 Three-span continuous one-way slab, sand-lightweight concrete; by the author's method, and ACI Code (318-71). See Table 3-4 for the material and section properties, parameters and conditions of the problem. Compute the ultimate time-dependent deflection following age 7 days, which may be the time of form removal. Assume the loading age is also 7 days for dead load, and that 30 percent of the live load is to be considered as a sustained load applied at age 2 months ($\alpha = 0.52$ from Table 3-20 for $t = 60 - 7 = 53$ days). The results are summarized in Table 3-23.

For $H = 70\%$ and loading age = 7 days,

$$C_u = 1.88 \text{ for dead load} \qquad \text{(Table 1-10)}$$

For $H = 70\%$ and loading age = 60 days,

$$C_u = 1.45 \text{ for sustained live load} \qquad \text{(Table 1-10)}$$

For $h = 8$ in. (20.3 cm),

$$(\text{C.F.})_T = 0.96 \qquad \text{(Table 1-8)}$$

Hence,

$$C_u = (0.96)(1.88) = \underline{1.80} \qquad \text{(use for dead load)}$$

$$C_u = (0.96)(1.45) = \underline{1.39} \qquad \text{(use for sustained live load)}$$

Author's method

End span AB

Pos. $k_r = 0.85$

Neg. $k_r = 0.85/(1 + 50\rho') = 0.85/(1 + 50 \times 0.00494) = 0.68 \qquad$ (Eq. (3-71))

Table 3-22 Summary of deflections and deflection ratios for Example 1* (1 in. = 2.54 cm.)

	$(\Delta_i)_D$ in.	Δ_{sh} in.	$(\Delta_{cp})_D$ in.	Δ_t † in.	$(\Delta_i)_D + \Delta_t$ in.	Δ_L in.	Total = $(\Delta_i)_D + \Delta_t + \Delta_L$ in.	$\frac{(\Delta_i)_D}{\Delta_{Tot}}$ %	$\frac{\Delta_{sh}}{\Delta_{Tot}}$ %	$\frac{(\Delta_{cp})_D}{\Delta_{Tot}}$ %	$\frac{\Delta_t}{\Delta_{Tot}}$ %	$\frac{(\Delta_i)_D + \Delta_t}{\Delta_{Tot}}$ %	$\frac{\Delta_L}{\Delta_{Tot}}$ %
Cracked section for Δ_L, cracked-section increased "*n*" method for $(\Delta_i)_D + (\Delta_{cp})_D$, and tensile force method for Δ_{sh}	0.22	0.20‡	0.11‡	0.31	0.53	0.73	1.26	17	16‡	9‡	25	42	58
Yu and Winter Method B for Δ_L, and Method C for $(\Delta_i + \Delta_{sh} + \Delta_{cp})_D$	—	—	—	—	0.70	0.78	1.48	—	—	—	—	47	53
Yu and Winter Method B for Δ_L and $(\Delta_i)_D$, and Method D (also refers to Method A) for $(\Delta_{sh} + \Delta_{cp})_D$	0.13	—	—	0.43	0.56	0.78	1.34	10	—	—	32	43	58
ACI Code (318-63)	0.13	—	—	0.26	0.39	0.82	1.21§	11	—	—	21	32	68
Method of the author	0.15	0.15	0.22	0.37	0.52	0.80	1.32	11	11	17	28	39	61
Method of the author including the effect of live load on $(\Delta_i)_D$, Δ_{cp}, Δ_L	0.22	0.15	0.28	0.43	0.65	0.74	1.39¶	16	11	20	31	47	53
ACI Code (318-71)	0.15	—	—	0.30	0.45	0.80	1.25§	12	—	—	24	36	64

* The initial deflections are summarized in Table 3-3, and the shrinkage deflections in Table 3-14. † $\Delta_t = \Delta_{sh} + (\Delta_{cp})_D$.

‡ As previously discussed, the author believes this procedure in general tends to overestimate shrinkage deflection (16 percent of total in this example) and underestimate creep deflection (9 percent of total in this example). However, with $(\Delta_i)_D$ also high, the total deflection is consistent with other results above.

§ The results by the 1963 and 1971 ACI Codes are on the low (unsafe) side of the other results. As previously noted, the author believes this tendency is

$$(k_r)_a = 0.85(k_r)_m + 0.15(k_r)_{\text{cont end}} \qquad (\text{Eq. (3-73)})$$
$$= (0.85)(0.85) + (0.15)(0.68) = \underline{0.82}$$

which is nearly the same as the Pos. k_r in this case.

$$(\Delta_i)_D = \underline{0.13 \text{ in. } (0.33 \text{ cm})} \qquad \Delta_{sh} = \underline{0.21 \text{ in. } (0.53 \text{ cm})}$$
$$\Delta_L = \underline{0.55 \text{ in. } (1.40 \text{ cm})} \qquad (\text{prev. calc.})$$
$$(\Delta_{cp})_D = k_r C_u (\Delta_i)_D = (0.82)(1.80)(0.13)$$
$$= \underline{0.19 \text{ in. } (0.48 \text{ cm})} \qquad (\text{Eq. (3-74)})$$
$$(\Delta_{cp})_L = k_r C_u (\Delta_L)_s = (0.82)(1.39)(0.30 \times 0.55)$$
$$= \underline{0.19 \text{ in. } (0.48 \text{ cm})} \qquad (\text{Eq. (3-84)})$$
$$\Delta_{cp} = (\Delta_{cp})_D + (\Delta_{cp})_L = 0.19 + 0.19 = \underline{0.38 \text{ in. } (0.97 \text{ cm})}$$
$$\Delta_{\text{Total}} = (\Delta_i)_D + \Delta_{sh} + \Delta_{cp} + \Delta_L = 0.13 + 0.21 + 0.38 + 0.55$$
$$= \underline{1.27 \text{ in. } (3.23 \text{ cm})}$$

For checking the ACI Code allowable deflections,

$$\Delta_t = \Delta_{sh} + \Delta_{cp} = 0.21 + 0.38 = 0.59 \text{ in. } (1.50 \text{ cm}),$$

and $\qquad \Delta_t + \Delta_L = 0.59 + 0.55 = \underline{1.14 \text{ in. } (2.90 \text{ cm})}$

Or, $\qquad \Delta_t + \Delta_L = \Delta_{\text{Total}} - (\Delta_i)_D = 1.27 - 0.13 = 1.14$ in.

The following calculations refer to the alternative (simpler) method for including the effect of sustained and repeated live load on elastic and creep deflections—$(\Delta_i)_{D2}$, Δ_{cp}, Δ_L. See Figs. 3-8 and 3-15 for a description of the various deflection components.

Initial dead-load deflection before live load $= (\Delta_i)_{D1} = (\Delta_i)_D$
$= \underline{0.13 \text{ in. } (0.33 \text{ cm})}$ (prev. calc.)

Initial dead-load deflection after live load $= (\Delta_i)_{D2}$
$= \underline{0.24 \text{ in. } (0.61 \text{ cm})}$ (prev. calc.)

Creep dead-load deflection before live load $= (\Delta_{cp})_{D1} = k_r \alpha C_u (\Delta_i)_{D1}$
$= (0.82)(0.52)(1.80)(0.13)$
$= \underline{0.10 \text{ in. } (0.25 \text{ cm})}$ (Eq. (3-80))

Creep dead-load deflection after live load $= (\Delta_{cp})_{D2} = k_r (1 - \alpha) C_u (\Delta_i)_{D2}$
$= (0.82)(1 - 0.52)(1.80)(0.24)$
$= \underline{0.17 \text{ in. } (0.43 \text{ cm})}$

Partial live-load creep deflection when $\Delta_L = 0.55$ in. (1.40 cm) (prev. calc.)
$= (\Delta_{cp})_L = k_r C_u (\Delta_L)_s$
$= (0.82)(1.39)(0.30 \times 0.55)$
$= \underline{0.19 \text{ in. } (0.48 \text{ cm})}$ (Eq. (3-84))

Hence, $\qquad \Delta_{cp} = (\Delta_{cp})_{D1} + (\Delta_{cp})_{D2} + (\Delta_{cp})_L$

$$= 0.10 + 0.17 + 0.19$$
$$= \underline{0.46 \text{ in. } (1.17 \text{ cm})}$$
$$\Delta_{\text{Total}} = (\Delta_i)_{D2} + \Delta_{sh} + \Delta_{cp} + \Delta_L$$
$$= 0.24 + 0.21 + 0.46 + 0.45$$
$$= \underline{1.36 \text{ in. } (3.45 \text{ cm})}$$

which is 0.09 in. (0.23 cm) or 7 percent greater than above when the effect of repeated live load is neglected. Note from previous calculation that $\Delta_L = 0.45$ in. (1.14 cm) after repeated live loading.

For checking the ACI Code allowable deflections,

$$\Delta_t = \Delta_{sh} + \Delta_{cp} = 0.21 + 0.46 = 0.67 \text{ in. } (1.70 \text{ cm}),$$

and $\Delta_t + \Delta_L = 0.67 + 0.45 = \underline{1.12 \text{ in. } (2.48 \text{ cm})}$

Or, $\Delta_t + \Delta_L = \Delta_{\text{Total}} - (\Delta_i)_{D2} = 1.36 - 0.24 = 1.12$ in.

Center span BC

$$\text{Pos. } k_r = 0.85 \qquad \text{Neg. } k_r = 0.85/(1 + 50 \times 0.00494) = 0.68 \qquad (\text{Eq. (3-71)})$$
$$(k_r)_a = 0.70(k_r)_m + 0.15[(k_r)_{e1} + (k_r)_{e2}] \qquad (\text{Eq. (3-72)})$$
$$= [(0.70)(0.85) + (0.15)(0.68 + 0.68)] = \underline{0.80}$$
$$(\Delta_i)_D = \underline{0.01 \text{ in. } (0.03 \text{ cm})} \qquad \Delta_{sh} = \underline{0.13 \text{ in. } (0.33 \text{ cm})}$$
$$\Delta_L = \underline{0.20 \text{ in. } (0.51 \text{ cm})} \qquad (\text{prev. calc.})$$
$$(\Delta_{cp})_D = k_r C_u (\Delta_i)_D = (0.80)(1.80)(0.01) = \underline{0.01 \text{ in. } (0.03 \text{ cm})} \qquad (\text{Eq. (3-74)})$$
$$(\Delta_{cp})_L = k_r C_u (\Delta_L)_s = (0.80)(1.39)(0.30 \times 0.20)$$
$$= \underline{0.07 \text{ in. } (0.18 \text{ cm})} \qquad (\text{Eq. (3-84)})$$
$$\Delta_{cp} = (\Delta_{cp})_D + (\Delta_{cp})_L = 0.01 + 0.07 = \underline{0.08 \text{ in. } (0.20 \text{ cm})}$$
$$\Delta_{\text{Total}} = (\Delta_i)_D + \Delta_{sh} + \Delta_{cp} + \Delta_L = 0.01 + 0.13 + 0.08 + 0.20$$
$$= \underline{0.42 \text{ in. } (1.07 \text{ cm})}$$

For checking the ACI Code allowable deflections,

$$\Delta_t = \Delta_{sh} + \Delta_{cp} = 0.13 + 0.08 = 0.21 \text{ in. } (0.53 \text{ cm}),$$

and $\Delta_t + \Delta_L = 0.21 + 0.20 = \underline{0.41 \text{ in. } (1.04 \text{ cm})}$

Or, $\Delta_t + \Delta_L = \Delta_{\text{Total}} - (\Delta_i)_D = 0.42 - 0.01 = 0.41$ in.

Since the initial dead-load deflection for this interior span is very small (0.01 or 0.03 cm), the effect of live load on the dead-load deflection is to increase the creep deflection and the total deflection by only 0.01 in. (or 13 percent for Δ_{cp} and 2 percent for Δ_{Total}). Thus for this case,

$$\Delta_t + \Delta_L = \underline{0.42 \text{ in. } (1.07 \text{ cm})}$$

Allowable time-dependent plus live-load deflections (ACI Code 318-71) Roof or floor construction supporting or attached to nonstructural elements *likely* to be damaged by large deflections—$\Delta_t + \Delta_L \leq L/480 = 240/480 = 0.50$ in. (1.27 cm).

Roof or floor construction supporting or attached to nonstructural elements *not likely* to be damaged by large deflections—$\Delta_t + \Delta_L \leq L/240 = 240/240 = 1.00$ in. (2.54 cm).

By previous calculations for live-load deflections, both spans could be used for flat roofs or floors *not* supporting or attached to nonstructural elements likely to be damaged by large deflections, according to the ACI Code (318-71). The above calculations for time-dependent plus live-load deflections indicate that the center span has no limitations at all (0.41 in. or 0.42 in. < 0.50 in. and 1.00 in.). However, the end span (where 1.14 in. or 1.12 in. > 0.50 in. and 1.00 in.), and presumably the design itself, does not satisfy either of the above ACI Code criteria for roof or floor construction supporting or attached to nonstructural elements. On the other hand, if the nonstructural elements are to be installed some time after the removal of forms (after dead-load application), the calculation of Δ_t should take this into account (which alters Δ_{cp}), in which case the latter of the two time-dependent ACI Code criteria may be satisfied in this problem.

For example, assume that partitions are to be installed one month after dead-load application. Use $\alpha_1 = 0.44$ from Table 3-20. For the end span,

$$(\Delta_{cp})_D = k_r(1 - \alpha_1)C_u(\Delta_i)_D = (0.82)(1 - 0.44)(1.80)(0.13)$$
$$= \underline{0.11 \text{ in. } (0.28 \text{ cm})} \qquad \text{(Eq. (3-82))}$$

which reduces the previous calculations of $(\Delta_{cp})_D$, Δ_{cp}, or Δ_t by 0.08 in. (0.20 cm). This would be sufficient to satisfy the latter of the two time-dependent ACI Code criteria when using the first of the two procedures above, for example.

ACI Code (318-71)

Pos. $k_r = 1.00$ Neg. $k_r = 1.00 - 0.60(A'_s/A_s) = 1.00 - (0.60)(0.40/0.75)$

$$= 0.68 \qquad \text{(Eq. (3-66))}$$

Avg. $k_r = (1.00 + 0.68)/2 = 0.84$ for both end span and center span values

Hence, for both spans

$$k_r T = (0.84)(2) = \underline{1.68}$$

End span AB

$$(\Delta_i)_D = \underline{0.13 \text{ in. } (0.33 \text{ cm})} \qquad \Delta_{D+L} = 0.58 \text{ in. } (1.47 \text{ cm}) \qquad \text{(prev. calc.)}$$
$$\Delta_L = 0.58 - 0.13(2.42/2.63) = \underline{0.46 \text{ in. } (1.17 \text{ cm})}$$
$$\Delta_t = k_r T(\Delta_i)_D + k_r T(\Delta_L)_s = (1.68)(0.13) + (1.68)(0.30 \times 0.46)$$

$$= \underline{0.45 \text{ in. } (1.14 \text{ cm})}$$

$$\Delta_{\text{Total}} = (\Delta_i)_D + \Delta_t + \Delta_L = 0.13 + 0.45 + 0.46 = \underline{1.04 \text{ in. } (2.64 \text{ cm})}$$

versus 1.27 in. (3.23 cm) or 1.36 in. (3.45 cm) by the author's procedures.

$$\Delta_t + \Delta_L = 0.45 + 0.46 = \underline{0.91 \text{ in. } (2.31 \text{ cm})}$$

versus 1.14 in. (2.90 cm) or 1.12 in. (2.84 cm) by the author's procedures.

Center span BC

$$(\Delta_i)_D = \underline{0.01 \text{ in. } (0.03 \text{ cm})} \qquad \Delta_{D+L} = 0.24 \text{ in. } (0.61 \text{ cm}) \qquad \text{(prev. calc.)}$$

$$\Delta_L = 0.24 - 0.01(2.42/2.63) = \underline{0.23 \text{ in. } (0.58 \text{ cm})}$$

$$\Delta_t = k_r T(\Delta_i)_D + k_r T(\Delta_L)_s = (1.68)(0.01) + (1.68)(0.30 \times 0.23)$$

$$= \underline{0.14 \text{ in. } (0.36 \text{ cm})}$$

$$\Delta_{\text{Total}} = (\Delta_i)_D + \Delta_t + \Delta_L = 0.01 + 0.14 + 0.23 = \underline{0.38 \text{ in. } (0.97 \text{ cm})}$$

versus 0.42 in. (1.07 cm) or 0.43 in. (1.09 cm) by the author's procedures.

$$\Delta_t + \Delta_L = 0.14 + 0.23 = \underline{0.37 \text{ in. } (0.94 \text{ cm})}$$

versus 0.41 in. (1.04 cm) or 0.42 in. (1.07 cm) by the author's procedures.

As previously discussed, the results by the ACI Code method tend to be on the low side compared to other methods such as above (especially for this relatively early starting time for shrinkage and dead-load creep of age 7 days). The ACI Code does not provide for a distinction to be made based on loading age.

Allowable time-dependent plus live-load deflections Since $\Delta_t + \Delta_L = 0.91$ in. (2.31 cm) for the end span (greater than 0.50 in. or 1.27 cm but less than 1.00 in. or 2.54 cm allowable values) and 0.37 in. (0.94 cm) for the center span by the ACI Code method of calculation, the end span, and presumably the design itself, may be used for roof or floor construction supporting or attached to nonstructural elements *not likely* to be damaged by large deflections, but may not be used for roof or floor construction supporting or attached to nonstructural elements *likely* to be damaged by large deflections.

Example 3 Continuous T-beam, normal-weight concrete, by the author's method.

See Table 3-7 for the material and section properties, parameters and conditions of the problem. Compute the ultimate time-dependent deflection following age 20 days. Assume the loading age is also 20 days for dead load, and that 40 percent of the live load is to be considered as a sustained load applied at age 3 months ($\alpha = 0.56$ in Table 3-20 for $t = 90 - 20 = 70$ days).

For $H = 60\%$ and loading age = 20 days,

$$C_u = 1.78 \text{ for dead load} \qquad \text{(Table 1-10)}$$

For $H = 60\%$ and loading age = 90 days,

$$C_u = 1.51 \text{ for sustained live load} \qquad \text{(Table 1-10)}$$

Since the positive region (where the T-beam flange is in compression and assumed to be the primary creep element) has the dominative effect on deflections, the size correction factor will be based on $t = 5$ in. (12.7 cm).

For $t = 5$ in., $(\text{C.F.})_T = 1.04$ (Table 1-8)

Hence,

$$C_u = (1.04)(1.78) = \underline{1.85} \qquad \text{(use for dead load)}$$
$$C_u = (1.04)(1.51) = \underline{1.57} \qquad \text{(use for sustained live load)}$$

Exterior (end) span

$$\text{Pos. } k_r = 0.85$$

Since ρ' is about the same at both ends, use

$$\text{Neg. } k_r = 0.85/(1 + 50\rho') = 0.85/(1 + 50 \times 0.0080) = 0.61 \qquad \text{(Eq. (3-71))}$$

For an average k_r, use Eq. (3-72) for both the exterior and interior spans (exterior span also continuous at both ends).

$$(k_r)_a = 0.70(k_r)_m + 0.15[(k_r)_{e1} + (k_r)_{e2}] \qquad \text{(Eq. (3-72))}$$
$$= (0.70)(0.85) + (0.15)(0.61 + 0.61) = \underline{0.78}$$

for both exterior and interior spans.

$$(\Delta_i)_D = \underline{0.10 \text{ in. } (0.25 \text{ cm})} \qquad \Delta_{sh} = \underline{0.06 \text{ in. } (0.15 \text{ cm})}$$
$$\Delta_L = \underline{0.26 \text{ in. } (0.66 \text{ cm})} \qquad \text{(prev. calc.)}$$
$$(\Delta_{cp})_D = k_r C_u (\Delta_i)_D = (0.78)(1.85)(0.10) = \underline{0.14 \text{ in. } (0.36 \text{ cm})} \qquad \text{(Eq. (3-74))}$$
$$(\Delta_{cp})_L = k_r C_u (\Delta_L)_s = (0.78)(1.57)(0.40 \times 0.26)$$
$$= \underline{0.13 \text{ in. } (0.33 \text{ cm})} \qquad \text{(Eq. (3-84))}$$
$$\Delta_{cp} = (\Delta_{cp})_D + (\Delta_{cp})_L = 0.14 + 0.13 = \underline{0.27 \text{ in. } (0.69 \text{ cm})}$$
$$\Delta_{Total} = (\Delta_i)_D + \Delta_{sh} + \Delta_{cp} + \Delta_L = 0.10 + 0.06 + 0.27 + 0.26$$
$$= \underline{0.69 \text{ in. } (1.75 \text{ cm})}$$

For checking the ACI Code allowable deflections,

$$\Delta_t = \Delta_{sh} + \Delta_{cp} = 0.06 + 0.27 = 0.33 \text{ in. } (0.84 \text{ cm}),$$

and

$$\Delta_t + \Delta_L = 0.33 + 0.26 = \underline{0.59 \text{ in. } (1.50 \text{ cm})}$$

Or,

$$\Delta_t + \Delta_L = \Delta_{Total} - (\Delta_i)_D = 0.69 - 0.10 = 0.59 \text{ in.}$$

The following calculations refer to the alternative (simpler) method for including the effect of sustained and repeated live load on elastic and creep deflections—$(\Delta_i)_{D2}$, Δ_{cp}, Δ_L. See Example 2 and Figs. 3-8 and 3-15 for a description of the various deflection components.

$$(\Delta_i)_{D1} = (\Delta_i)_D = \underline{0.10 \text{ in.}} \qquad (\Delta_i)_{D2} = \underline{0.15 \text{ in.}} \qquad \Delta_{sh} = \underline{0.06 \text{ in.}}$$
$$\Delta_L = \underline{0.26 \text{ in.}}, \underline{0.21 \text{ in.}} \qquad \text{(prev. calc.)}$$
$$(\Delta_{cp})_{D1} = k_r \alpha C_u (\Delta_i)_{D1} = (0.78)(0.56)(1.85)(0.10)$$
$$= \underline{0.08 \text{ in. } (0.20 \text{ cm})} \qquad \text{(Eq. (3-80))}$$
$$(\Delta_{cp})_{D2} = k_r(1-\alpha)C_u(\Delta_i)_{D2} = (0.78)(1-0.56)(1.85)(0.15)$$
$$= \underline{0.10 \text{ in. } (0.25 \text{ cm})} \qquad \text{(Eq. (3-83))}$$
$$(\Delta_{cp})_L = k_r C_u (\Delta_L)_s = (0.78)(1.57)(0.40 \times 0.26)$$
$$= \underline{0.13 \text{ in. } (0.33 \text{ cm})} \qquad \text{(Eq. (3-84))}$$

Hence,
$$\Delta_{cp} = (\Delta_{cp})_{D1} + (\Delta_{cp})_{D2} + (\Delta_{cp})_L = 0.08 + 0.10 + 0.13$$
$$= \underline{0.31 \text{ in. } (0.79 \text{ cm})}$$
$$\Delta_{Total} = (\Delta_i)_{D2} + \Delta_{sh} + \Delta_{cp} + \Delta_L = 0.15 + 0.06 + 0.31 + 0.21$$
$$= \underline{0.73 \text{ in. } (1.85 \text{ cm})}$$

which is 0.04 in. (0.10 cm) or 6 percent greater than above when the effect of repeated live load is neglected. Note from previous calculation that $\Delta_L = 0.26$ in. (0.66 cm) for sustained loading and $\Delta_L = 0.21$ in. (0.53 cm) for repeated loading.

For checking the ACI Code allowable deflections,

$$\Delta_t = \Delta_{sh} + \Delta_{cp} = 0.06 + 0.31 = 0.37 \text{ in. } (0.94 \text{ cm}),$$

and
$$\Delta_t + \Delta_L = 0.37 + 0.21 = \underline{0.58 \text{ in. } (1.47 \text{ cm})}$$

Or,
$$\Delta_t + \Delta_L = \Delta_{Total} - (\Delta_i)_{D2} = 0.73 - 0.15 = 0.58 \text{ in.}$$

Interior span

$$\text{Avg. } k_r = \underline{0.78} \qquad \text{(same as for exterior span, as noted above)}$$
$$(\Delta_i)_D = \underline{0.07 \text{ in. } (0.18 \text{ cm})} \qquad \Delta_{sh} = \underline{0.06 \text{ in. } (0.15 \text{ cm})}$$
$$\Delta_L = \underline{0.25 \text{ in. } (0.64 \text{ cm})} \qquad \text{(prev. calc.)}$$
$$(\Delta_{cp})_D = k_r C_u (\Delta_i)_D = (0.78)(1.85)(0.07) = \underline{0.10 \text{ in. } (0.25 \text{ cm})} \qquad \text{(Eq. (3-74))}$$
$$(\Delta_{cp})_L = k_r C_u (\Delta_L)_s = (0.78)(1.57)(0.40 \times 0.25)$$
$$= \underline{0.12 \text{ in. } (0.30 \text{ cm})} \qquad \text{(Eq. (3-84))}$$
$$\Delta_{cp} = (\Delta_{cp})_D + (\Delta_{cp})_L = 0.10 + 0.12 = \underline{0.22 \text{ in. } (0.56 \text{ cm})}$$
$$\Delta_{Total} = (\Delta_i)_D + \Delta_{sh} + \Delta_{cp} + \Delta_L = 0.07 + 0.06 + 0.22 + 0.25$$
$$= \underline{0.60 \text{ in. } (1.52 \text{ cm})}$$

For checking the ACI Code allowable deflections,

$$\Delta_t = \Delta_{sh} + \Delta_{cp} = 0.06 + 0.22 = 0.28 \text{ in. } (0.71 \text{ cm}),$$

and
$$\Delta_t + \Delta_L = 0.28 + 0.25 = \underline{0.53 \text{ in. } (1.35 \text{ cm})}$$

Or,
$$\Delta_t + \Delta_L = \Delta_{Total} - (\Delta_i)_D = 0.60 - 0.07 = 0.53 \text{ in.}$$

Including the effect of live load on dead-load creep deflections:

$$(\Delta_i)_{D1} = (\Delta_i)_D = \underline{0.07 \text{ in.}} \qquad (\Delta_i)_{D2} = \underline{0.13 \text{ in.}} \qquad \Delta_{sh} = \underline{0.06 \text{ in.}}$$

$$\Delta_L = \underline{0.25 \text{ in.}}, \underline{0.18 \text{ in.}} \qquad \text{(prev. calc.)}$$

$$(\Delta_{cp})_{D1} = k_r \alpha C_u (\Delta_i)_{D1} = (0.78)(0.56)(1.85)(0.07)$$
$$= \underline{0.06 \text{ in. } (0.15 \text{ cm})} \qquad \text{(Eq. (3-80))}$$

$$(\Delta_{cp})_{D2} = k_r(1-\alpha)C_u(\Delta_i)_{D2} = (0.78)(1-0.56)(1.85)(0.13)$$
$$= \underline{0.08 \text{ in. } (0.20 \text{ cm})} \qquad \text{(Eq. (3-83))}$$

$$(\Delta_{cp})_L = k_r C_u (\Delta_L)_s = (0.78)(1.57)(0.40 \times 0.25)$$
$$= \underline{0.12 \text{ in. } (0.30 \text{ cm})} \qquad \text{(Eq. (3-84))}$$

Hence, $\Delta_{cp} = 0.06 + 0.08 + 0.12 = \underline{0.26 \text{ in. } (0.66 \text{ cm})}$

$$\Delta_{Total} = (\Delta_i)_{D2} + \Delta_{sh} + \Delta_{cp} + \Delta_L = 0.13 + 0.06 + 0.26 + 0.18$$
$$= \underline{0.63 \text{ in. } (1.60 \text{ cm})}$$

which is 0.03 in. (0.08 cm) or 5 percent greater than above when the effect of repeated live load is neglected. Note from previous calculation that $\Delta_L = 0.25$ in. (0.64 cm) for sustained loading and $\Delta_L = 0.18$ in. (0.46 cm) for repeated loading.

For checking the ACI Code allowable deflections,

$$\Delta_t = \Delta_{sh} + \Delta_{cp} = 0.06 + 0.26 = 0.32 \text{ in. } (0.81 \text{ cm}),$$

and $\Delta_t + \Delta_L = 0.32 + 0.18 = \underline{0.50 \text{ in. } (1.27 \text{ cm})}$

Or, $\Delta_t + \Delta_L = \Delta_{Total} - (\Delta_i)_{D2} = 0.63 - 0.13 = 0.50$ in.

Allowable time-dependent plus live-load deflections (ACI Code 318-17) Roof or floor construction supporting or attached to nonstructural elements *likely* to be damaged by large deflections—$\Delta_t + \Delta_L \le L/480 = L_n/480 = 29 \times 12/480 = 0.73$ in. (1.85 cm).

Roof or floor construction supporting or attached to nonstructural elements *not likely* to be damaged by large deflections—$\Delta_t + \Delta_L \le L/240 = L_n/240 = 29 \times 12/240 = 1.45$ in. (3.68 cm).

Since 0.59 in. (1.50 cm) or 0.58 in. (1.47 cm) for the exterior span and 0.53 in. (1.35 cm) or 0.50 in. (1.27 cm) for the interior span are both less than the allowable values of 0.73 in. (1.85 cm) and 1.45 in. (3.68 cm), the T-beam has no deflection limitations according to the ACI Code (318-71) based on the time-dependent plus live-load deflection criteria. Previous calculations also showed no deflection limitations based on live load only.

The approximate deflections by the ACI Code calculation method are as follows:

$$\text{Avg. } k_r = (1.00 + 0.61)/2 = 0.805 \qquad k_r T = (0.805)(2) = \underline{1.61}$$

Exterior span

$$(\Delta_i)_D = \underline{0.10 \text{ in. } (0.25 \text{ cm})} \qquad \Delta_L = \underline{0.26 \text{ in. } (0.66 \text{ cm})}$$

(Assumed to be the same as above, since the ACI Code uses the same method for short-time deflections although with slightly different constants and average values for continuous members.)

$$\Delta_t = k_r T(\Delta_i)_D + k_r T(\Delta_L)_s = (1.61)(0.10) + (1.61)(0.40 \times 0.26)$$
$$= \underline{0.33 \text{ in. } (0.84 \text{ cm})}$$
$$\Delta_{\text{Total}} = (\Delta_i)_D + \Delta_t + \Delta_L = 0.10 + 0.33 + 0.26 = \underline{0.69 \text{ in. } (1.75 \text{ cm})}$$

versus 0.69 in. (1.75 cm) or 0.73 in. (1.85 cm) by the author's method.

Interior span

$$(\Delta_i)_D = \underline{0.07 \text{ in. } (0.18 \text{ cm})} \qquad \Delta_L = \underline{0.25 \text{ in. } (0.64 \text{ cm})} \qquad k_r T = \underline{1.61}$$
$$\Delta_t = k_r T(\Delta_i)_D + k_r T(\Delta_L)_s = (1.61)(0.07) + (1.61)(0.40 \times 0.25)$$
$$= \underline{0.27 \text{ in. } (0.69 \text{ cm})}$$
$$\Delta_{\text{Total}} = (\Delta_i)_D + \Delta_t + \Delta_L = 0.07 + 0.27 + 0.25 = 0.59 \text{ in. } (1.50 \text{ cm})$$

versus 0.60 in. (1.52 cm) and 0.63 in. (1.60 cm) by the author's method.

In this example with the somewhat late beginning age for shrinkage and dead-load creep of 20 days, the results by the ACI Code method and the method above for calculating creep and shrinkage deflections separately were essentially the same. The ACI Code does not provide for a distinction to be made based on the effect of loading age.

Example 4 Interior girder of simple-span T-beam or deck girder bridge, normal-weight concrete; by the author, BPR, AASHTO, and ACI Committee 443—Concrete Bridge Design. See Table 3-11 for the structural details, material properties, parameters, and conditions of the problem. Compute the ultimate time-dependent deflection following age 7 days, which may be the time of form removal. Assume the loading age is also 7 days for dead load and that the live load is applied at age 3 months ($\alpha = 0.59$ from Table 3-20 for $t = 90 - 7 = 83$ days). The results by the various methods are summarized in Table 3-24.

For $H = 70\%$ and loading age = 7 days, $C_u = 1.88$ (Table 1-10)

Since the T-beam flange is in compression and assumed to be the primary creep element, the size correction factor will be based on $t = 6$ in. (15.2 cm) (including wearing surface).

For $t = 6$ in. (15.2 cm), $(\text{C.F.})_T = 1.00$ (Table 1-8)

Hence, $C_u = (1.00)(1.88) = \underline{1.88}$ (use)

Author

$$(\Delta_i)_D = \underline{1.02 \text{ in. } (2.59 \text{ cm})} \qquad \Delta_{sh} = \underline{0.67 \text{ in. } (1.70 \text{ cm})}$$
$$\Delta_{L+I} = \underline{1.11 \text{ in. } (2.82 \text{ cm})} \qquad \text{(prev. calc.)}$$
$$k_r = 0.85 \qquad \text{(Eq. (3-71))}$$

Table 3-23 Summary of deflections and deflection ratios for Examples 2 and 3

Method	Span	$(\Delta_i)_D$ in.	Δ_{sh} in.	Δ_{cp} in.	Δ_t ‡ in.	Δ_L in.	Δ_{Total} in.	$\frac{(\Delta_i)_D}{\Delta_{Tot}}$ %	$\frac{\Delta_{sh}}{\Delta_{Tot}}$ %	$\frac{\Delta_{cp}}{\Delta_{Tot}}$ %	$\frac{\Delta_t}{\Delta_{Tot}}$ %	$\frac{\Delta_L}{\Delta_{Tot}}$ %
*Example 2**												
Author	End	0.13	0.21	0.38	0.59	0.55	1.27	10	17	30	47	43
	Center	0.01	0.13	0.08	0.21	0.20	0.42	2	31	19	50	48
Author, including effect of LL on $(\Delta_i)_D$, Δ_{cp}, Δ_L	End	0.24	0.21	0.46	0.67	0.45	1.36	18	15	34	49	33
	Center	0.01	0.13	0.09	0.22	0.20	0.43	2	30	21	51	47
ACI Code (318-71)	End	0.13	—	—	0.45	0.46	1.04	13	—	—	43	44
	Center	0.01	—	—	0.14	0.23	0.38	3	—	—	37	60
Example 3†												
Author	Ext.	0.10	0.06	0.27	0.33	0.26	0.69	14	9	39	48	38
	Int.	0.07	0.06	0.22	0.28	0.25	0.60	12	10	37	47	42
Author, including effect of LL on $(\Delta_i)_D$, Δ_{cp}, Δ_L	Ext.	0.15	0.06	0.31	0.37	0.21	0.73	20	8	43	51	29
	Int.	0.13	0.06	0.26	0.32	0.18	0.63	21	10	41	51	28
ACI Code (318-71)	Ext.	0.10	—	—	0.33	0.26	0.69	14	—	—	48	38
	Int.	0.07	—	—	0.27	0.25	0.59	12	—	—	46	42

(1 in. = 2.54 cm.)

* 30 percent of the live load was considered as a sustained load.

† 40 percent of the live load was considered as a sustained load.

‡ $\Delta_t = \Delta_{sh} + \Delta_{cp}$.

$$(\Delta_{cp})_D = k_r C_u (\Delta_i)_D = (0.85)(1.88)(1.02) = \underline{1.63 \text{ in. } (4.14 \text{ cm})} \qquad \text{(Eq. (3-74))}$$

$$\Delta_{Total} = (\Delta_i)_D + \Delta_{sh} + (\Delta_{cp})_D + \Delta_{L+I} = 1.02 + 0.67 + 1.63 + 1.11$$
$$= \underline{4.43 \text{ in. } (11.25 \text{ cm})}$$

The following includes the complete set of calculations using the I_{rep} method for including the effect of live load on dead-load elastic and creep deflection (the various deflection components are shown in Figs. 3-8 and 3-15):

$$\text{Initial dead-load deflection before live load} = (\Delta_i)_{D1} = (\Delta_i)_D$$
$$= \underline{1.02 \text{ in. } (2.59 \text{ cm})} \qquad \text{(prev. calc.)}$$
$$\text{Initial dead-load deflection after live load} = (\Delta_i)_{D2}$$
$$= \underline{1.20 \text{ in. } (3.05 \text{ cm})} \qquad \text{(prev. calc.)}$$
$$\text{Creep dead-load deflection before live load} = (\Delta_{cp})_{D1} = k_r \alpha C_u (\Delta_i)_{D1}$$
$$= (0.85)(0.59)(1.88)(1.02)$$
$$= \underline{0.96 \text{ in. } (2.44 \text{ cm})} \qquad \text{(Eq. (3-80))}$$
$$\text{Creep dead-load deflection after live load} = (\Delta_{cp})_{D2} = k_r(1-\alpha)C_u(\Delta_i)_{D2}$$
$$= (0.85)(1 - 0.59)(1.88)(1.20)$$
$$= \underline{0.79 \text{ in. } (2.01 \text{ cm})} \qquad \text{(Eq. (3-83))}$$

Hence,

$$\Delta_{cp} = (\Delta_{cp})_{D1} + (\Delta_{cp})_{D2} = 0.96 + 0.79 = \underline{1.75 \text{ in. } (4.45 \text{ cm})}$$
$$\Delta_{Total} = (\Delta_i)_{D2} + \Delta_{sh} + \Delta_{cp} + \Delta_{L+I} = 1.20 + 0.67 + 1.75 + 0.95$$
$$= \underline{4.57 \text{ in. } (11.61 \text{ cm})}$$

which is 0.14 in. (0.36 cm) or 3 percent greater than above when the effect of repeated live load on $(\Delta_i)_D$, Δ_{cp}, and Δ_{L+I} is neglected.

Bureau of Public Roads (BPR) For climate of average humidity and $A'_s = 0$, select from Table 3-21, $k_r T = 3.0$.

Before live load, $(\Delta_i)_{D1} = \underline{0.79 \text{ in. } (2.01 \text{ cm})}$ (prev. calc.)

After live load, $(\Delta_i)_{D2} = \underline{0.85 \text{ in. } (2.16 \text{ cm})}$

Also, $\Delta_{L+I} = \underline{1.04 \text{ in. } (2.64 \text{ cm})}$ (prev. calc.)

The Bureau of Public Roads method does not explain how to proportion the time-dependent factor with each of the two dead-load deflections. However, since they are not greatly different, use an average or $(\Delta_i)_D = \underline{0.82 \text{ in. } (2.08 \text{ cm})}$.

$$\Delta_t = k_r T (\Delta_i)_D = (3.0)(0.82) = \underline{2.46 \text{ in. } (6.25 \text{ cm})} \qquad \text{(Eq. (3-75))}$$

Hence, $\Delta_{Total} = \Delta_{D2} + \Delta_t + \Delta_{L+I} = 0.85 + 2.46 + 1.04 = \underline{4.35 \text{ in. } (11.05 \text{ cm})}$

AASHTO

Working stress design Using $n = 8$ for all concrete strengths and the fully cracked section,

$$I_{cr} = 78\,060 \text{ in}^4 \ (3\,249\,000 \text{ cm}^4) \qquad \text{(Table 3-12)}$$

$$\Delta_{D+L+I} = \underline{2.01 \text{ in. } (5.11 \text{ cm})} \qquad \text{(prev. calc.)}$$

$$\Delta_D = 2.01 M_D/M_{D+L+I} = (2.01)(406.4/882.7) = \underline{0.93 \text{ in. } (2.36 \text{ cm})}$$

Using $n'_{ct} = 30$ for initial plus shrinkage plus creep dead-load deflection,

$$E'_{ct} = E_s/n'_{ct} = 29 \times 10^3/30$$

$$= 0.97 \times 10^3 \text{ ksi } (68\,200 \text{ kg/cm}^2, 6.7 \text{ kN/mm}^2)$$

$$\rho n'_{ct} = (0.0094)(30) = 0.282$$

$$k = \sqrt{2\rho n'_{ct} + (\rho n'_{ct})^2} - \rho n'_{ct} = 0.520$$

$$kd = (0.520)(29.75) = 15.47 \text{ in. } (39.29 \text{ cm})$$

$$I'_{ct} = b(kd)^3/3 + n'_{ct} A_s (d - kd)^2$$

$$= (67)(15.47)^3/3 + (30)(18.72)(29.75 - 15.47)^2$$

$$= \underline{197\,210 \text{ in}^4 \ (8\,208\,000 \text{ cm}^4)},$$

Checking by using Fig. B-2 with $\rho n = 0.282$, $\rho' n' = 0$,

$$I'_{ct} = 0.112 bd^3 = (0.112)(67)(29.75)^3$$

$$= 197\,600 \text{ in}^4 \ (8\,224\,000 \text{ cm}^4) \qquad \text{OK}$$

$$(\Delta_i + \Delta_{sh} + \Delta_{cp})_D = \frac{K M_D L^2}{E'_{ct} I'_{ct}} = \frac{(5/48)(406.4)(12)(60)^2(144)}{(0.97 \times 10^3)(197\,210)}$$

$$= \underline{1.38 \text{ in. } (3.51 \text{ cm})} \qquad \text{(Eq. (3-78))}$$

$$\Delta_{Total} = \Delta_{D+L+I} - \Delta_D + (\Delta_i + \Delta_{sh} + \Delta_{cp})_D$$

$$= 2.01 - 0.93 + 1.38 = \underline{2.46 \text{ in. } (6.25 \text{ cm})}$$

which is considerably less than the results by the other methods, as shown in Table 3-24, and as previously discussed for this increased "n" approach.

Load factor design For climate of average humidity and $A'_s = 0$, select from Table 3-21, $k_r T = 3.0$.

$$\Delta_D = \underline{1.02 \text{ in. } (2.59 \text{ cm})}$$

$$\Delta_{L+I} = \underline{1.11 \text{ in. } (2.82 \text{ cm})} \qquad \text{(prev. calc.)}$$

$$\Delta_t = k_r T (\Delta_i)_D = (3.0)(1.02) = \underline{3.06 \text{ in. } (7.77 \text{ cm})} \qquad \text{(Eq. (3-75))}$$

Hence, $\Delta_{Total} = \Delta_D + \Delta_t + \Delta_{L+I} = 1.02 + 3.06 + 1.11 = \underline{5.19 \text{ in. } (13.18 \text{ cm})}$

ACI Committee 443, Concrete Bridge Design Using I_e and the same values of f_r and E_c as ACI and AASHTO,

$$(\Delta_i)_D = \underline{1.02 \text{ in. } (2.59 \text{ cm})} \qquad \Delta_{L+I} = \underline{1.11 \text{ in. } (2.82 \text{ cm})} \qquad \text{(prev. calc.)}$$

For $A'_s = 0$, $k_r = 1.00$ (Eq. (3-66)), $T = 2$; hence $k_r T = \underline{2.00}$.

$$\Delta_t = k_r T (\Delta_i)_D = (2.00)(1.02) = \underline{2.04 \text{ in. } (5.18 \text{ cm})} \qquad \text{(Eq. (3-75))}$$

$$\Delta_{Total} = (\Delta_i)_D + \Delta_t + \Delta_{L+I} = 1.02 + 2.04 + 1.11 = \underline{4.17 \text{ in. } (10.59 \text{ cm})}$$

Table 3-24 Summary of deflections and deflection ratios for Example 4

Method	$(\Delta_i)_D$ in.	Δ_{sh} in.	Δ_{cp} in.	Δ_t in.	Δ_{L+I} in.	* Δ_{Total} in.	$\frac{(\Delta_i)_D}{\Delta_{Tot}}$ %	$\frac{\Delta_{sh}}{\Delta_{Tot}}$ %	$\frac{\Delta_{cp}}{\Delta_{Tot}}$ %	$\frac{\Delta_t}{\Delta_{Tot}}$ %	$\frac{\Delta_{L+I}}{\Delta_{Tot}}$ %
Author	1.02	0.67	1.63	2.30	1.11	4.43	23	15	37	52	25
Author, including effect of live load on $(\Delta_i)_D$, Δ_{cp}, Δ_L	1.20	0.67	1.75	2.42	0.95	4.57	26	15	38	53	21
Bureau of Public Roads	0.85	—	—	2.46	1.04	4.35	20	—	—	56	24
AASHTO Working Stress Design method	0.93	—	—	0.45†	1.08	2.46†	38	—	—	18	44
AASHTO Load Factor Design method	1.02	—	—	3.06‡	1.11	5.19‡	20	—	—	59	21
ACI Committee 443—concrete bridge design	1.02	—	—	2.04	1.11	4.17	24	—	—	49	27
ACI Committee 443—alternative method	1.02	0.78	2.21	2.99‡	1.11	5.12‡	20	15	43	58	22

(1 in. = 2.54 cm.)

* It should be noted that part of the total deflection is usually offset by cambering.

† These results are considerably less than the results by the other methods, as previously discussed for the increased "*n*" method.

‡ These results are on the high side compared to the results by the other methods.

Alternately, creep and shrinkage deflections may be computed separately using the method of Corley and Sozen for creep (Eqs. (3-85) and (3-86)) and for shrinkage (Eq. (3-57)).

$$(\Delta_i)_D = \underline{1.02 \text{ in. } (2.59 \text{ cm})}$$

and

$$\Delta_{L+I} = \underline{1.11 \text{ in. } (2.82 \text{ cm})} \qquad \text{(same as above)}$$

$$\Delta_{sh} = \underline{0.78 \text{ in. } (1.98 \text{ cm})} \qquad \text{(prev. calc.)}$$

$$kd = 10.74 \text{ in. } (27.28 \text{ cm}) > 2t/3 = (2)(5.25)/3$$

$$= 3.50 \text{ in. } (8.89 \text{ cm}) \qquad \text{OK}$$

$$k = 10.74/29.75 = 0.361$$

Hence,

$$k_r C_u = 2k(3 - \rho'/\rho) = 6k = (6)(0.361) = \underline{2.17} \qquad \text{(Eq. (3-85))}$$

$$\Delta_{cp} = k_r C_u (\Delta_i)_D = (2.17)(1.02) = \underline{2.21 \text{ in. } (5.61 \text{ cm})}$$

$$\Delta_{\text{Total}} = (\Delta_i)_D + \Delta_{sh} + \Delta_{cp} + \Delta_{L+I} = 1.02 + 0.78 + 2.21 + 1.11$$

$$= \underline{5.12 \text{ in. } (13.00 \text{ cm})}$$

Summary of the deflection ratios for the four examples The ratios of the different deflection components to the total deflection for the four examples by the various calculation methods are shown in Tables 3-22, 3-23, and 3-24—26 different solutions. These examples include different loading ages and up to 40 percent of the design live load assumed to act as a sustained load. The following observations are made.

Ratio of dead-load to total deflection—for Examples 1, 2, and 3, the ratio varies from 10 to 21 percent, except 2 to 3 percent for the center span of the 3-span continuous one-way slab in Example 2 in which the short-time dead-load deflection was very small. For the highway bridge girder in Example 4, the ratio varies from 20 to 26 percent, except 38 percent by the AASHTO Working Stress Design method using I_{cr} with $n = 8$.

Ratio of shrinkage to total deflection—for all four examples, the ratio varies from 8 to 17 percent, except 30 to 31 percent for the center span of the 3-span continuous one-way slab in Example 2 in which the deflection due to load is small.

Ratio of creep to total deflection—for all four examples, the ratio varies from 17 to 43 percent, except 9 percent by the direct use of the increased "n" method for the rectangular beam in Example 1. The wide range in the creep deflection ratios is due to the effect of different loading ages and percentages of the live load acting as a sustained load.

Ratio of time-dependent (shrinkage plus creep) to total deflection—for all four examples, the ratio varies from 28 to 59 percent, except 18 percent by the AASHTO Working Stress Design method for Example 4 using I'_{ct} with $n'_{ct} = 30$, and 21 percent by the ACI Code (318-63), 23 percent by the ACI Code (318-71), and 25 percent by the increased "n" method for the rectangular beam in Example 1.

Ratio of live load to total deflection—for Examples 1, 2, and 3, the ratio varies from 28 to 68 percent. For the highway bridge girder in Example 4, the ratio of live load plus impact to total deflection varies from 21 to 27 percent, except 44 percent by the AASHTO Working Stress Design method using I_{cr} with $n = 8$. The reason for this high percentage is the low total deflection by this method due to the low initial dead-load plus time-dependent deflection using I'_{ct} with $n'_{ct} = 30$.

The following is a summary of the representative ranges from above:

Ratio of dead-load to total deflection—10 to 21 percent for Examples 1 to 3
20 to 26 percent for the bridge in Example 4
Ratio of shrinkage to total deflection—8 to 17 percent for all four examples
Ratio of creep to total deflection—17 to 43 percent for all four examples
Ratio of time-dependent to total deflection—28 to 59 percent for all four examples
Ratio of live-load to total deflection—28 to 68 percent for Examples 1 to 3
Ratio of live-load plus impact to total deflection—21 to 27 percent for the bridge in Example 4.

3-5 DEFLECTION OF REINFORCED CONCRETE AND STEEL–CONCRETE COMPOSITE BEAMS

Procedures for computing the deflection of composite beams consisting of a precast reinforced concrete beam with composite slab, and steel beam with composite slab are described in this section.[4,6,10,80–84] Subscripts 1 and 2 are used to refer to the slab (or effect of the slab such as under slab dead load) and precast beam or steel beam, respectively.

Reinforced Concrete Composite Beams (ordinary reinforced precast beam)

The deflection of reinforced concrete composite beams using unshored and shored construction is computed by Eqs. (3-87), (3-91), and (3-93) at any time t, and Eqs. (3-90), (3-92), (3-94), (3-95), and (3-96) for ultimate values.

Unshored construction

$$\begin{aligned}\Delta_{i,t} = {} & \overbrace{(\Delta_i)_2}^{(1)} + \overbrace{k_r C_{s2}(\Delta_i)_2}^{(2)} + \overbrace{k_r(C_{t2} - C_{s2})(\Delta_i)_2 I_2/I_c}^{(3)} + \overbrace{K_{sh}(\phi_{sh})_s L^2}^{(4)} \\ & + \overbrace{K_{sh}[(\phi_{sh})_t - (\phi_{sh})_s]L^2 I_2/I_c}^{(5)} + \overbrace{(\Delta_i)_1}^{(6)} + \overbrace{k_r C_{t1}(\Delta_i)_1 I_{2s}/I_c}^{(7)} \\ & + \overbrace{\Delta_{DS}}^{(8)} + \overbrace{\Delta_L}^{(9)} + \overbrace{(\Delta_{cp})_L}^{(10)} \end{aligned} \tag{3-87}$$

It can be seen by inspection that Eq. (3-87) for composite beams reduces to the basic equation for noncomposite reinforced beams, Eq. (3-88), when $I_2/I_c = 1$, $C_{t2} = C_t$, subscript 2 = D, and terms (6), (7), and (8)—effects of composite slab—are deleted.

$$\Delta_{i,t} = (\Delta_i)_D + k_r C_t (\Delta_i)_D + K_{sh}(\phi_{sh})_t L^2 + \Delta_L + (\Delta_{cp})_L \tag{3-88}$$

In Eq. (3-87), use the same $k_r = 0.85/(1 + 50\rho')$ by Eq. (3-71), in terms (2), (3), and (7), where ρ' is the precast beam compression steel ratio.

Term (1) is the initial dead-load deflection of the precast beam. For example, in its simplest form for ideal end conditions, $(\Delta_i)_2 = KM_2L^2/E_{ci}I_2$ by Eq. (3-1). For simple and continuous beams, the midspan deflection may be computed by Eqs. (3-37), (3-38), or (3-39). For computing I_2 (using I_e in Eq. (3-19)), M_a refers to the precast beam dead load and M_{cr} to the precast beam.

Term (2) is the dead-load creep deflection of the precast beam up to the time of slab casting. C_{s2} is the creep coefficient of the precast beam concrete at the time of slab casting.

Term (3) is the creep deflection of the composite beam for any period following slab casting due to the precast beam dead load. C_{t2} is the creep coefficient of the precast beam concrete at any time after slab casting. I_2 is the same as in term (1). I_c refers to the composite section in which the gross-section I and the 28-day E_c are recommended for most calculations. The expression I_2/I_c modifies the initial stress (strain) and accounts for the effect of the composite section in restraining additional creep curvature (strain) after slab casting.

Term (4) is the deflection due to shrinkage warping of the precast beam up to the time of slab casting, as defined by Eq. (3-58). $(\phi_{sh})_s$ is the shrinkage curvature up to the time of slab casting, such as by Eqs. (3-53) to (3-55), or by Eq. (3-56)—$(\phi_{sh})_s = A_{sh}(\varepsilon_{sh})_s/h$, in which $(\varepsilon_{sh})_s$ is the shrinkage strain of the precast beam concrete up to the time of slab casting, and A_{sh} is obtained from Table 3-13 or Fig. 3-10.

Term (5) is the shrinkage deflection of the composite beam for any period following slab casting due to the shrinkage of the precast beam concrete, but not including the effect of differential shrinkage and creep which is included in term (8). $(\phi_{sh})_t$ is the shrinkage curvature for any period following slab casting, such as by Eq. (3-56)—$(\phi_{sh})_t = A_{sh}(\varepsilon_{sh})_t/h$, in which $(\varepsilon_{sh})_t$ is the shrinkage strain of the precast beam concrete at any time after slab casting. I_2 and I_c are the same as in term (3).

Term (6) is the initial deflection of the precast beam under slab dead load. For example, in its simplest form for ideal end conditions, $(\Delta_i)_1 = [(\Delta_i)_1 + (\Delta_i)_2] - (\Delta_i)_2(E_{ci}/E_c)$, where $[(\Delta_i)_1 + (\Delta_i)_2] = K(M_1 + M_2)L^2/E_cI_{2s}$ by Eq. (3-1). For simple and continuous beams, the midspan deflection may be computed by Eqs. (3-37), (3-38), or (3-39). For computing I_{2s} ($I_{2s} = I_e$ in Eq. (3-19), for example), M_a refers to the precast beam plus slab dead load and M_{cr} to the precast beam. When diaphragms are cast with the slab or prior to slab casting, add $(\Delta_i)_{1D}$ to $(\Delta_i)_1$. For two symmetrical diaphragms,

$$(\Delta_i)_{1D} = \frac{M_{1D}}{E_c I_{2s}}\left(\frac{L^2}{8} - \frac{a^2}{6}\right) \tag{3-89}$$

where M_{1D} is the moment between diaphragms and a is $L/4$, $L/3$, etc., for diaphragms at the quarter points, third points, etc., respectively. When diaphragms are used, M_a should include the diaphragm dead load in computing I_{2s} for both term (6) and Eq. (3-89).

Term (7) is the creep deflection of the composite beam due to slab dead load. C_{t1} is the creep coefficient for the slab loading, where the age of the precast beam concrete at the time of slab casting is considered. I_{2s} is the same as in term (6) and I_c the same as in term (3).

Term (8) is the deflection due to differential shrinkage and creep. For simple spans, $\Delta_{DS} = \frac{4}{3}Q_{DS}y_{cS}L^2/8E_cI_c$ by Eq. (2-22). I_c is the same as in term (3). See Chap. 2, Eqs. (2-18) to (2-21), etc., for Q_{DS}, and Fig. 2-1 for the definition of y_{cS}. Also see Chap. 2 for the effect of continuity on differential shrinkage and creep, and the selection of appropriate differential shrinkage and creep coefficients in Table 2-1.

Term (9) is the live-load deflection of the composite beam. For example, in its simplest form for ideal end conditions, $\Delta_L = KM_LL^2/E_c(I_c)_{D+L}$ by Eq. (3-1). For simple and continuous beams, the midspan deflection may be computed by Eqs. (3-37), (3-38), or (3-39). For computing $(I_c)_{D+L}$, ($(I_c)_{D+L} = I_e$ in Eq. (3-19), for example), M_a refers to the total precast beam plus slab plus diaphragm, partitions, ceilings, etc., dead load and live load, and M_{cr} refers to the composite beam. This solution for Δ_L is an approximation that is similar to Eq. (3-35). When the precast beam is severely cracked prior to composite action, the solution becomes less accurate because $(I_c)_{D+L}$ refers to the cracked composite section under total load. However, this condition is usually not the case. Also included in M_L are the moments due to diaphragms, partitions, ceilings, etc., placed after composite action.

Term (10) is the partial live-load (also including loads due to diaphragms, partitions, ceilings, etc., placed after composite action) creep deflection for any permanent live load. By Eq. (3-84), $(\Delta_{cp})_L = (M_P/M_L)(k_r)_sC_L\Delta_L$, where M_P is the moment due to the permanent load supported by the composite beam, $(k_r)_s$ refers to the slab compression steel in Eq. (3-71), and C_L is the slab creep coefficient based on the slab age when the permanent load is applied.

$$\Delta_u = \overbrace{(\Delta_i)_2}^{(1)} + \overbrace{(\alpha_s)_{cp}k_rC_u(\Delta_1)_2}^{(2)} + \overbrace{[1-(\alpha_s)_{cp}]k_rC_u(\Delta_i)_2I_2/I_c}^{(3)} + \overbrace{K_{sh}(\alpha_s)_{sh}(\phi_{sh})_uL^2}^{(4)}$$

$$+ \overbrace{K_{sh}[1-(\alpha_s)_{sh}](\phi_{sh})_uL^2I_2/I_c}^{(5)} + \overbrace{(\Delta_i)_1}^{(6)} + \overbrace{\beta_sk_rC_u(\Delta_i)_1I_{2s}/I_c}^{(7)}$$

$$+ \overbrace{\Delta_{DS}}^{(8)} + \overbrace{\Delta_L}^{(9)} + \overbrace{(\Delta_{cp})_L}^{(10)} \qquad (3\text{-}90)$$

Equation (3-90) for ultimate values corresponds term by term to Eq. (3-87) for values at any time t.

Shored construction, assuming the precast beam supports its own dead load and the composite beam supports the slab dead load and live load

$$\Delta_{i,t} = \text{Eq. (3-87), with terms (6), (7), and (8) modified as follows} \qquad (3\text{-}91)$$

Term (6) is the initial deflection of the composite beam under slab dead load. $(\Delta_i)_1 = KM_1L^2/E_cI_c$, by Eqs. (3-1), (3-37), (3-38), or (3-39) for simple and continuous beams. The gross-section I_c using 28-day E_c values for both concretes is recommended for most cases.

Term (7) is the creep deflection of the composite beam under slab dead load. $(\Delta_{cp})_1 = (k_r)_s C_{s1}(\Delta_i)_1$. The composite-section effect is already included in $(\Delta_i)_1$ in term (6). $(k_r)_s$ refers to the slab compression steel in Eq. (3-71), and C_{s1} is the slab creep coefficient based on the slab age when the shores are struck.

Term (8) may be adjusted to reflect only the differential shrinkage and creep deflection that occurs after the shores are struck, by using a reduced differential shrinkage and creep coefficient, $D_t = [1 - (\alpha_s)_{sh}$ from Table 3-25 based on t = time between slab casting and removal of shores] $\times$ [D_t based on the differential shrinkage and creep following slab casting, as described in Chap. 2 and Table 2-1 for the shored case].

$$\Delta_u = \text{Eq. (3-90), with the corresponding modifications described in Eq. (3-91) applied to ultimate values} \qquad (3\text{-}92)$$

Shored construction, assuming the composite beam supports all dead and live load

$$\Delta_{i,t} = \text{Eq. (3-91), with terms (1), (2), (3), and (4) modified as follows} \qquad (3\text{-}93)$$

Term (1) is the initial deflection of the composite beam under precast beam dead load. $(\Delta_i)_2 = KM_2L^2/E_cI_c$.

Terms (2) and (3) are replaced by $(\Delta_{cp})_2 = (k_r)_s C_{s1}(\Delta_i)_2$. This term is now the creep deflection of the composite beam under precast beam dead load.

There is some question about term (4), which is the shrinkage warping up to the time of slab casting, since the precast beam is now shored or supported. It may be reasonable to exclude the effect of term (4) completely.

$$\Delta_u = \text{Eq. (3-92), with the corresponding modifications described in Eq. (3-93) applied to ultimate values.} \qquad (3\text{-}94)$$

Equations (3-87) and (3-90) to (3-94), could be greatly shortened by combining terms, but are kept in the form of separate terms in order to show the separate effects or contributions that take place. The equations for the shored case assume more or less continuous shoring. Special consideration would be required for one or two-point shoring, for example.

The parameters $(\alpha_s)_{cp}$, $(\alpha_s)_{sh}$, and β_s are defined and representative values tabulated in Table 3-25.

Table 3-25 Representative values of $(\alpha_s)_{cp}$, $(\alpha_s)_{sh}$, and β_s *

† *For* $(\alpha_s)_{cp}$ *and* β_s t = Time in days between precast beam dead load application and slab casting for $(\alpha_s)_{cp}$, and concrete age for β_s	$(\alpha_s)_{cp}$	Moist cured β_s	Steam cured β_s
2	0.13	—	—
4	0.19	—	0.99
7	0.24	1.00	0.97
10	0.28	0.95	0.90
20	0.38	0.87	0.85
28	0.42	0.84	0.83
60	0.54	0.77	0.76
90	0.60	0.74	0.74
120	0.64	0.71	0.71

For $(\alpha_s)_{sh}$ ‡ t_s = age of precast beam concrete when the slab is cast in days

§ t_p	7	10	20	28	60	90	120
4	0.05, —	0.09, —	0.21, —	0.28, —	0.47, —	0.57, —	0.63, —
7	—	0.04, 0.08	0.16, 0.27	0.23, 0.38	0.42, 0.60	0.52, 0.70	0.59, 0.76
10	—	—	0.12, 0.19	0.19, 0.30	0.38, 0.52	0.48, 0.62	0.54, 0.69
20	—	—	—	0.07, 0.10	0.26, 0.33	0.36, 0.43	0.43, 0.49
28	—	—	—	—	0.19, 0.23	0.29, 0.33	0.36, 0.39
60	—	—	—	—	—	0.10, 0.10	0.17, 0.16
90	—	—	—	—	—	—	0.07, 0.06

* The *underlined* values for β_s and $(\alpha_s)_{sh}$ refer to steam cured precast beam concrete (usual case), and the other values to moist cured concrete. The values of $(\alpha_s)_{cp}$ apply for both moist and steam cured concrete. The values in the table apply to different weight concrete.

† $(\alpha_s)_{cp}$ is the ratio of the precast beam creep that takes place up to the time of slab casting to the total creep. From Eq. (1-16) for both moist and steam cured concrete: $(\alpha_s)_{cp} = C_t/C_u = t^{0.60}/(10 + t^{0.60})$, where t = time in days between precast beam dead load application and slab casting.

β_s is the creep correction factor based on the age of the precast beam concrete when the slab is cast, $t = t_{LA}$. From Eq. (1-20) for M.C.: $\beta_s = 1.25 t_{LA}^{-0.118}$. From Eq. (1-21) for S.C.: $\beta_s = 1.13 t_{LA}^{-0.095}$.

‡ $(\alpha_s)_{sh}$ is the ratio of the precast beam shrinkage that takes place between the time of dead load application and slab casting to the ultimate shrinkage. Using Eq. (1-18) for S.C.: $(\alpha_s)_{sh} = [(t_s - 1)/(55 + t_s - 1)] - [(t_p - 1)/(55 + t_p - 1)]$. Using Eq. (1-17) for M.C.: $(\alpha_s)_{sh} = [(t_s - 7)/(35 + t_s - 7)] - [(t_p - 7)/(35 + t_p - 7)]$.

§ t_p = age of precast beam concrete when the precast beam dead load moment is applied in days.

Grossly approximate method for unshored reinforced concrete composite beams

$$\Delta_u = \Delta_i + (k_r T)\Delta_i(I_{2s}/I_c) + \Delta_L + (\Delta_{cp})_L \tag{3-95}$$

where $\Delta_i = KM_D L^2/E_c I_{2s}$, M_D = slab plus precast beam dead load moment, $(k_r T) = 4.0/(1 + 50\rho')$, ρ' refers to the precast beam compression steel ratio in Eq. (3-71), the constant 4.0 takes into account the combined effects of precast beam creep, precast beam shrinkage and differential shrinkage and creep of the composite beam, $I_{2s} = I_c$ in Eq. (3-19) for the precast beam under slab plus precast beam dead load, I_c is the gross-section composite I, and Δ_L, $(\Delta_{cp})_L$, are computed as in terms (9) and (10) of Eq. (3-87).

Grossly approximate method for shored reinforced concrete composite beams

$$\Delta_u = \Delta_i + (k_r)_s T\Delta_i + \Delta_L \tag{3-96}$$

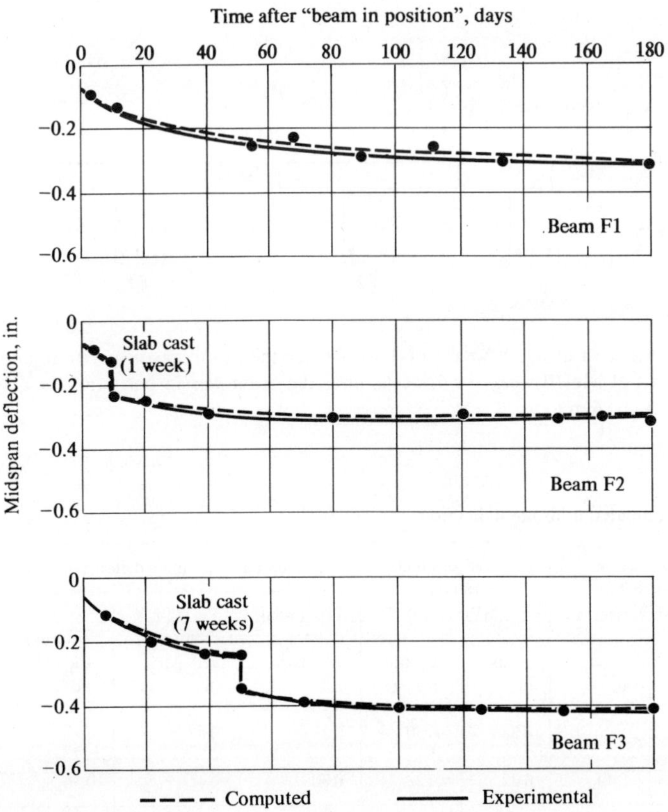

FIGURE 3-16
Computed and experimental midspan deflection of beams of Group F (one noncomposite and two composite beams)[81] (1 in = 2.54 cm).

where $\Delta_i = KM_D L^2/E_c(I_c)_{D+L}$, M_D = total sustained load including all dead load and any permanent live load, $(k_r)_s T = 4.0/(1 + 50\rho')$, ρ' refers to the compression steel ratio of the slab steel in Eq. (3-71), $(I_c)_{D+L}$ is defined in term (9) of Eq. (3-87), and $\Delta_L = K(M_L - M_P)L^2/E_c(I_c)_{D+L}$, M_L = total live load, M_P = permanent live load.

Experimental observations From a study by Kripanarayanan and the author,[81] measured and computed deflections (using Eqs. (3-87) and (3-88)) for one noncomposite and two composite beams (described in Table 3-26) are shown in Fig. 3-16 to be in satisfactory agreement. The results are also tabulated in Table 3-27 in which both experimental material parameters and the general or average material parameters from Chap. 1 are used. These results include the use of Eq. (3-90) for ultimate values. The term by term contributions to the ultimate computed deflections are presented in Table 3-28.

It was observed that the total results were sensitive to the stiffness, creep, and shrinkage characteristics of the concrete. The effect of the 4-week (age 4 weeks

Table 3-26 Details of laboratory specimens[81]

Beam stems are 6 in. by 8 in., sand-lightweight concrete Slabs are 21 in. by 3 in., normal-weight concrete Beam steel = 3 × #4 bars Spans = 15 ft	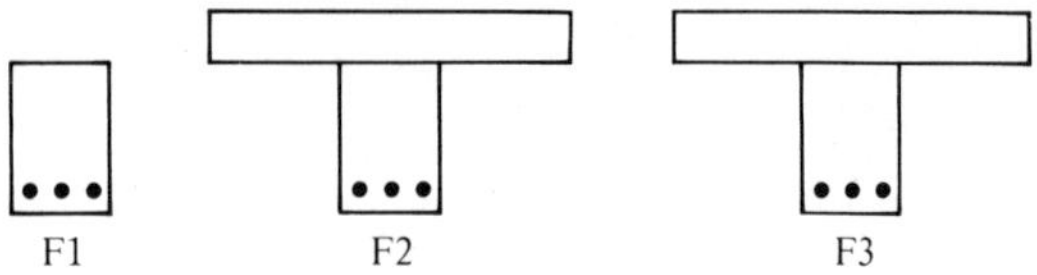

The slab for beam F2 was cast at age 4 weeks or 1 week after the beam was in position, and the slab for beam F3 was cast at age 10 weeks or 7 weeks after the beam was in position. (6 in. = 15.2 cm, 8 in. = 20.3 cm, 15 ft = 4.6 m.)

Table 3-27 Measured and computed midspan deflections

Beam	Computed initial deflection in.	Deflection just before slab cast: Meas. in.	Deflection just before slab cast: Comp.* in.	Deflection at 180 days (Fig. 3-14): Meas. in.	Deflection at 180 days (Fig. 3-14): Comp.* in.	Computed ultimate deflection: Eqs. (3-88), (3-90) with exp. para. in.	Computed ultimate deflection: Eqs. (3-88), (3-90) with gen. para. in.	Computed ultimate deflection: Approx. Eq. (3.95) in.
F1	0.07	—	—	0.34	0.30	0.38	0.47	—
F2	0.07	0.13	0.14	0.32	0.28	0.30	0.45	0.36
F3	0.07	0.25	0.24	0.45	0.40	0.43	0.59	0.36

(1 in. = 2.54 cm.)

* Using Eqs. (3-87) and (3-88), with experimental material parameters.

Table 3-28 Term-by-term results for the ultimate midspan deflections

Terms*	Initial defl. due to beam D.L. (1)	Creep defl. up to slab cast (2)	Creep defl. after slab cast (3)	Shrink. defl. up to slab cast (4)	Shrink. defl. after slab cast (5)	Elastic defl. due to slab D.L. (6)	Creep defl. due to slab D.L. (7)	Defl. due to diff. shrink. and creep (8)	Total
Beam F1, Eq. (3-88)	0.07†	0.09†	—	0.22†	—	—	—	—	0.38
Beam F2, Eq. (3-90)	0.07	0.02	0.01	0.02	0.04	0.10	0.02	0.02	0.30
Beam F3, Eq. (3-90)	0.07	0.05	0.01	0.11	0.02	0.10	0.02	0.05	0.43

All deflections are in inches. The experimental material parameters were used in the calculations. (1 in. = 2.54 cm.)

* Terms in Eq. (3-90).

† Terms (1), (2), (3), respectively, in Eq. (3-88).

or 1 week after the beam was in position) and 10-week (age 10 weeks or 7 weeks after the beam was in position) slab-casting schedules had a very noticeable effect on deflections (Fig. 3-16 and Table 3-27). An earlier slab tends to reduce total creep and shrinkage deformation (causing downward deflection) by forming an earlier composite section, and also reduces differential shrinkage deformation (also causing downward deflection). When considering a 4-week slab as compared to the 10-week slab, the ultimate deflection was 0.13 in. (3.30 mm) or 30 percent less for the 4-week case. These results serve to point out the relatively large beneficial effect of casting the slab as early as possible. In comparing the noncomposite beam deflection with the deflection of the composite beams, the ultimate deflection of the noncomposite beam was about 0.08 in. (2.03 mm) or 27 percent greater than the 4-week case, but about 0.05 in. (1.27 mm) or 12 percent less than the 10-week case. As can be seen in Table 3-28, the significantly larger creep and shrinkage deflection occurring before the slab was cast accounted for most of the additional deflection for the 10-week case, compared to the 4-week case. The estimated ultimate deflections shown in Table 3-28 by the approximate method for the composite beams (Eq. (3-95)) were also satisfactory. It is noted that the expression for k_r and the approximate method herein are slightly different from that of Ref. 81.

Numerical example of unshored reinforced concrete composite beam deflection See Table 3-29 for the material properties, loading, parameters, and conditions, and Table 3-30 for the moments, strength design and section properties of the simply supported unshored composite beam deflection example. Compute the

Table 3-29 Material properties, loading, parameters, and conditions for reinforced concrete composite beam deflection example

Simply supported unshored composite beam

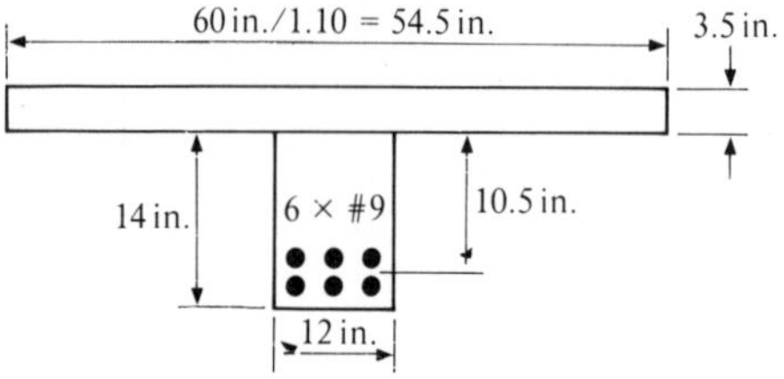

$A_s = 6.0\ \text{in}^2 \qquad d_c = 14$ in.

Neglect any compression steel effect in precast and composite beams.

Hence $\quad \rho' = 0$

and $\quad k_r = 0.85$ in Eq. (3-71)

Beam spacing = 5 ft $\qquad$ Span = 30 ft

Material properties and loading

All concrete—normal-weight

Slab $f'_c = 3000$ psi $\qquad$ Beam $f'_c = 4000$ psi

Steel $f_y = 40\,000$ psi

From Table 1-3 by Eq. (1-13),

$$\text{Slab } E_c = E_1 = \underline{3.31 \times 10^3 \text{ ksi}}$$

$$\text{Beam } E_c = E_2 = \underline{3.64 \times 10^3 \text{ ksi}}$$

$$E_2/E_1 = \underline{1.10}$$

From Table 1-2 by Eq. (1-7),

$$f_r = \underline{496 \text{ psi}}$$

$$E_s = 29 \times 10^3 \text{ ksi}$$

$$n = E_s/E_2 = 29/3.64 = 7.97 \qquad \underline{\text{Use } n = 8}$$

$$(w_{DL})_{\text{Slab}} = (60)(3.5)(150)/144 = 218\ \#/\text{ft}$$

$$(w_{DL})_{\text{Beam}} = (12)(14)(150)/144 = 175\ \#/\text{ft}$$

$$w_{LL} = 800\ \#/\text{ft}$$

Assume no sustained LL

Other parameters and conditions

Avg. rel. humidity = 60 percent $\qquad$ Precast beam steam-cured and slab moist-cured.

Assume the precast beam is in place at age 28 days and the slab is cast at age 2 months (1 month following beam in place). Use 28-day E_c throughout.

For steam-cured precast beam concrete, $H = 60$ percent, loading age = 28 days, and shrinkage following age 28 days:

$C_u = 1.70 \qquad (\varepsilon_{sh})_u = 414 \times 10^{-6}$ in/in $\qquad$ (Table 1-10)

For $b = 12$ in., $\quad$ Creep $(\text{C.F.})_T = 0.90 \qquad$ Shrinkage $(\text{C.F.})_T = 0.82 \qquad$ (Table 1-8)

Hence, $\quad C_u = (0.90)(1.70) = 1.53 \qquad$ (use)

$(\varepsilon_{sh})_u = (0.82)(414 \times 10^{-6}) = \underline{339 \times 10^{-6} \text{ in/in}} \qquad$ (use)

For moist-cured slab concrete in differential shrinkage and creep calc.:

$K_{sh} = 1/8 \qquad$ (Eq. (3-60))

From Table 3-25,

$(\alpha_s)_{cp} = 0.42 \qquad (\alpha_s)_{sh} = 0.19 \qquad \beta_s = 0.83$

From Table 2-1 for 6 in. slab and 70 percent humidity,

$D_u = 355 \times 10^{-6}$ in/in

For 3.5 in. slab,

$(\text{C.F.})_T = 1.21 \qquad$ (Table 1-8)

Shrinkage $(\text{C.F.})_H = 0.70$ at 70 percent humidity and 0.80 at 60 percent humidity $\qquad$ (Table 1-7)

Hence, $\quad D_u = (355 \times 10^{-6})(0.80/0.70)(1.21) = \underline{491 \times 10^{-6} \text{ in/in}} \qquad$ (use)

(For metric units, see conversion factors herein.)

Table 3-30 Moments, strength design and section properties for reinforced concrete composite beam deflection example

Moments for computing deflections

$$M_{\text{Slab}} = M_1 = (0.218)(30)^2/8 = \underline{24.5 \text{ ft-k}}$$
$$M_{\text{Beam}} = M_2 = (0.175)(30)^2/8 = \underline{19.7 \text{ ft-k}} \qquad M_1 + M_2 = 24.5 + 19.7 = \underline{44.2 \text{ ft-k}}$$
$$M_L = (0.800)(30)^2/8 = \underline{90.0 \text{ ft-k}}$$
$$M_1 + M_2 + M_L = 24.5 + 19.7 + 90.0 = \underline{134.2 \text{ ft-k}}$$

ACI Code strength design

Precast beam:

$$M_u = 1.4M_D = 1.4(M_1 + M_2) = (1.4)(44.2) = \underline{61.9 \text{ ft-k}}$$
$$a = A_s f_y/0.85f'_c b = (6.0)(40)/(0.85 \times 4 \times 12) = 5.88 \text{ in.}$$
$$M_u = \phi A_s f_y(d - a/2) = (0.9)(6.0)(40)(10.5 - 2.94)/12 = \underline{136 \text{ ft-k}} \qquad \text{OK}$$

Composite beam:

$$M_u = 1.4M_D + 1.7M_L = 61.9 + 1.7(90.0) = 61.9 + 153.0 = \underline{215 \text{ ft-k}}$$
$$a = A_s f_y/0.85f'_c b = (6.0)(40)/(0.85 \times 3 \times 60) = 1.56 \text{ in.}$$
$$M_u = \phi A_s f_y(d - a/2) = (0.9)(6.0)(40)(14 - 0.78)/12 = \underline{238 \text{ ft-k}} \qquad \text{OK}$$

Basic section properties

Precast beam:

$$I_g = (12)(14)^3/12 = \underline{2744 \text{ in}^4} \qquad \rho = 6.0/(12 \times 10.5) = 0.0476$$
$$\rho n = (0.0476)(8) = 0.381 \qquad \rho' n' = 0$$

From Fig. E-2,

$$I_{cr} = 0.133bd^3 = (0.133)(12)(10.5)^3 = \underline{1970 \text{ in}^4}$$
$$I_g/I_{cr} = 2744/1970 = 1.39$$

Composite beam:

$$y_t = [(14)(12)(7) + (54.5)(3.5)(15.75)]/[(14)(12) + (54.5)(3.5)] = \underline{11.65 \text{ in.}}$$
$$b/b_w = 54.5/12 = 4.54 \qquad h_f/h = 3.5/17.5 = 0.20$$

From Fig. E-1,

$$I_c = I_g = 0.034bh^3 = (0.034)(54.5)(17.5)^3 = \underline{9930 \text{ in}^4}$$
$$\rho = 6.0/(54.5 \times 14) = 0.0079 \qquad \rho n = (0.0079)(8) = 0.0632$$
$$h_f d = 3.5/14 = 0.250$$

From Fig. E-2,

$$I_{cr} = 0.040bd^3 = (0.040)(54.5)(14)^3 = \underline{5980 \text{ in}^4}$$
$$I_g/I_{cr} = 9930/5980 = 1.66$$

Effective section properties

Precast beam:

$$M_{cr} = f_r I_g/y_t = (496)(2744)/7 = 194\,400 \text{ in.}\# = \underline{16.2 \text{ ft-k}}$$
$$M_2/M_{cr} = 19.7/16.2 = 1.22 \qquad M_{cr}/M_2 = 16.2/19.7 = 0.822 \qquad 0.822^3 = 0.555$$
$$I_2 = (M_{cr}/M_2)^3 I_g + [1 - (M_{cr}/M_2)^3]I_{cr}$$
$$= (0.555)(2744) + (0.445)(1970) = \underline{2400 \text{ in}^4} \qquad \text{(Eq. (3-19))}$$

Table 3-30—continued

$(M_1 + M_2)/M_{cr} = 44.2/16.2 = 2.73 \qquad M_{cr}/(M_1 + M_2) = 0.367 \qquad 0.367^3 = 0.049$

$I_{2s} = (0.049)(2744) + (0.951)(1970) = \underline{2007 \text{ in}^4}$ (Eq. (3-19))

Composite beam:

$M_{cr} = f_r I_g/y_t = (496)(9930)/11.65 = 422\,700 \text{ in.}\# = \underline{35.2 \text{ ft-k}}$

$(M_1 + M_2 + M_L)/M_{cr} = 134.2/35.2 = 3.81 \qquad M_{cr}/(M_1 + M_2 + M_L) = 0.262 \qquad 0.262^3 = 0.018$

$(I_c)_{D+L} = (0.018)(9930) + (0.982)(5980) = \underline{6051 \text{ in}^4}$

or essentially the same as I_{cr}

(For metric units, see conversion factors herein.)

term by term and ultimate deflection following age 28 days. The results are summarized in Table 3-31 along with the ACI Code allowable deflections.

From ACI Code (318-71) and Chap. 6, precast beam deflections need not be checked when $h_{min} = (L/16)(0.8$ for 40 ksi steel, or 2810 kg/cm^2 or 276 N/mm$^2) = (360/16)(0.8) = 18$ in. (45.7 cm), and composite beam deflections need not be checked when $h_{min} = 18$ in. Since precast beam $h = 14$ in. (35.6 cm) and composite beam $h = 17.5$ in. (44.5 cm), all critical deflections must be computed.

Solution by Eq. (3-90), as defined term by term in Eq. (3-87) Initial dead-load deflection of the precast beam—term (1):

$$(\Delta_i)_2 = \frac{KM_2L^2}{E_{ci}I_2} = \frac{(5/48)(19.7)(12)(30)^2(144)}{(3640)(2400)} = \frac{3191 \times 10^3}{8736 \times 10^3} = \underline{0.37 \text{ in. } (0.94 \text{ cm})}$$

Dead-load creep deflection of the precast beam up to the time of slab casting—term (2):

$$(\alpha_s)_{cp} k_r C_u (\Delta_i)_2 = (0.42)(0.85)(1.53)(0.37) = \underline{0.20 \text{ in. } (0.51 \text{ cm})}$$

Creep deflection of the composite beam following slab casting due to the precast beam dead load—term (3):

$$[1 - (\alpha_s)_{cp}] k_r C_u (\Delta_i)_2 (I_2/I_c) = (0.58)(0.85)(1.53)(0.37)(2400)/9930$$
$$= \underline{0.07 \text{ in. } (0.18 \text{ cm})}$$

Deflection due to shrinkage warping of the precast beam up to the time of slab casting—term (4):

$$\rho = 0.0476 \qquad \rho' = 0$$

From Fig. 3-9 and Eq. (3-56),

$$A_{sh} = 1.00$$
$$(\phi_{sh})_u = A_{sh}(\varepsilon_{sh})_u/h = (1.00)(339 \times 10^{-6})/14$$

$$= 24.2 \times 10^{-6}\ 1/\text{in}\ (9.5 \times 10^{-6}\ 1/\text{cm})$$

$$K_{sh} = 1/8 \qquad (\text{Eq. (3-60)})$$

$$K_{sh}(\alpha_s)_{sh}(\varepsilon_{sh})_u L^2 = (1/8)(0.19)(24.2 \times 10^{-6})(30)^2(144) = \underline{0.07\ \text{in.}\ (0.18\ \text{cm})}$$

Shrinkage deflection of the composite beam following slab casting due to the shrinkage of the precast beam concrete—term (5):

$$K_{sh}[1 - (\alpha_s)_{sh}](\phi_{sh})_u L^2(I_2/I_c) = (1/8)(0.81)(24.2 \times 10^{-6})(30)^2(144)(2400)/9930$$

$$= \underline{0.08\ \text{in.}\ (0.20\ \text{cm})}$$

Initial deflection of the precast beam under slab dead load—term (6):

$$(\Delta_i)_1 + (\Delta_i)_2 = \frac{K(M_1 + M_2)L^2}{E_c I_{2s}} = \frac{(5/48)(44.2)(12)(30)^2(144)}{(3640)(2007)}$$

$$= \frac{7161 \times 10^3}{7305 \times 10^3} = 0.98\ \text{in.}\ (2.49\ \text{cm})$$

$$(\Delta_i)_1 = [(\Delta_i)_1 + (\Delta_i)_2] - (\Delta_i)_2 = 0.98 - 0.37 = \underline{0.61\ \text{in.}\ (1.55\ \text{cm})}$$

Table 3-31 Term-by-term deflections in inches (cm) for reinforced concrete composite beam example by Eq. (3-90), and allowable values*

Initial D.L. defl. of beam	D.L. creep defl. up to slab cast	D.L. creep defl. after slab cast	Shrink. defl. up to slab cast	Shrink. defl. after slab cast	Elastic defl. due to slab D.L.	Creep defl. due to slab D.L.	Defl. due to diff. shrinkage and creep	Live load defl.	Total†
(1)	(2)	(3)	(4)	(5)	(6)	(7)	(8)	(9)	Δ_u
0.37 (0.94)	0.20 (0.51)	0.07 (0.18)	0.07 (0.18)	0.08 (0.20)	0.61 (1.55)	0.13 (0.33)	0.42 (1.07)	0.66 (1.68)	2.61 (6.63)

* Comparison with ACI Code (318-71) allowable deflections (see Chap. 6 herein):

Flat roofs not supporting or attached to nonstructural elements likely to be damaged by large deflections—$\Delta_L = L/180 = 360/180 = 2.00$ in. (5.08 cm).

Floors not supporting or attached to nonstructural elements likely to be damaged by large deflections—$\Delta_L = L/360 = 360/360 = 1.00$ in. (2.54 cm).

$\Delta_L = 0.66$ in. (1.68 cm); hence the composite beam has no deflection limitations based on live load.

Roof or floor construction supporting or attached to nonstructural elements *likely* to be damaged by large deflections—$\Delta_t + \Delta_L = L/480 = 360/480 = 0.75$ in. (1.91 cm).

Roof or floor construction supporting or attached to nonstructural elements *not likely* to be damaged by large deflections—$\Delta_t + \Delta_L = L/240 = 360/240 = 1.50$ in. (3.81 cm).

For the purpose of checking the ACI Code allowable deflections, $\Delta_t + \Delta_L$ logically includes only the deflection that takes place after the slab has hardened, or terms (3) + (5) + (7) + (8) + (9).

Thus, $\Delta_t + \Delta_L = 0.07 + 0.08 + 0.13 + 0.42 + 0.66 = 1.36$ in. (3.45 cm); hence the beam is limited to roof or floor construction supporting or attached to nonstructural elements *not likely* to be damaged by large deflections.

† It should be noted that the total deflection of 2.61 in. (6.63 cm) or $\Delta_u/L = 2.61/360 = 1/138$ is quite large and may warrant cambering or shoring during construction. With the use of shoring, the relatively large term (6) is significantly reduced, the relatively small term (7) may be increased or reduced, and term (8) is increased because of the larger differential shrinkage and creep coefficient (Table 2-1).

Creep deflection of the composite beam due to slab dead load—term (7):

$$\beta_s k_r C_u(\Delta_i)_1(I_{2s}/I_c) = (0.83)(0.85)(1.53)(0.61)(2007)/9930 = \underline{0.13\text{ in. }(0.33\text{ cm})}$$

Deflection due to differential shrinkage and creep—term (8):

$$Q_{DS} = (Q \text{ in Eq. (2-1)})/2 = (DA_1E_1)/2 \qquad \text{(Eq. (2-21))}$$
$$= (491 \times 10^{-6})(60 \times 3.5)(3.31 \times 10^6)/2 = 170\,600\text{ lb }(77\,400\text{ kg}, 759\,000\text{ N})$$
$$y_{cS} = 17.5 - y_t - 3.5/2 = 17.5 - 11.65 - 1.75 = 4.10\text{ in. }(10.41\text{ cm}) \qquad \text{(Fig. 2-1)}$$
$$\Delta_{DS} = \frac{(4/3)Q_{DS}\,y_{cS}L^2}{eE_cI_c} = \frac{(4/3)(170.6)(4.10)(30)^2(144)}{(8)(3640)(9930)}$$
$$= \frac{1209 \times 10^5}{2892 \times 10^5} = \underline{0.42\text{ in. }(1.07\text{ cm})} \qquad \text{(Eq. (2-22))}$$

Live-load deflection of the composite beam—term (9):

$$\Delta_L = \frac{KM_LL^2}{E_c(I_c)_{D+L}} = \frac{(5/48)(90.0)(12)(30)^2(144)}{(3640)(6051)} = \frac{1458 \times 10^4}{2203 \times 10^4} = \underline{0.66\text{ in. }(1.68\text{ cm})}$$

Total ultimate deflection (summarized in Table 3-31)—Eq. (3-90):

$$\Delta_u = 0.37 + 0.20 + 0.07 + 0.07 + 0.08 + 0.61 + 0.13 + 0.42 + 0.66$$
$$= \underline{2.61\text{ in. }(6.63\text{ cm})}$$

Solution by grossly approximate method, Eq. (3-93)

$$\Delta_i = (\Delta_i)_1 + (\Delta_i)_2 = 0.98\text{ in. }(2.49\text{ cm}) \qquad \text{in term (6) above}$$
$$\Delta_L = 0.66\text{ in. }(1.68\text{ cm}) \qquad \text{in term (9) above}$$
$$\Delta_u = \Delta_i + (k_rT)\Delta_i(I_{2s}/I_c) + \Delta_L = 0.98 + (4.0)(0.98)(2007/9930) + 0.66$$
$$= 0.98 + 0.79 + 0.66 = \underline{2.43\text{ in. }(6.17\text{ cm})}$$

versus 2.61 in. (6.63 cm) by the more general solution above.

Steel–Concrete Composite Beams

The deflection of steel–concrete composite beams using unshored and shored construction can be computed by Eqs. (3-97) to (3-101).

Unshored construction

$$\begin{matrix} & (1) & & (2) & & (3) & & (4) & & (5) \\ \Delta_{i,t} = & \Delta_{s2} & + & \Delta_{s1} & + & \Delta_{DS} & + & \Delta_L & + & (\Delta_{cp})_L \end{matrix} \qquad (3\text{-}97)$$

where:

Term (1) is the dead-load deflection of the steel beam. $\Delta_{s2} = KM_2L^2/E_sI_s$, by Eq. (3-1), (3-37), (3-38), or (3-39).

Term (2) is the deflection of the steel beam under slab dead load. $\Delta_{s1} = KM_1L^2/E_sI_s$, by Eq. (3-1), (3-37), (3-38), or (3-39).

Term (3) is the deflection due to shrinkage of the slab concrete. $\Delta_{DS} = (Q_{DS}y_{cs}L^2)/(8E_sI_c)$ by Eqs. (2-3), (2-20), and (2-21), where $Q_{DS} = (Q \text{ in Eq. (2-1)})/2 = (DA_1E_1)/2$. See Chap. 2 for the effect of continuity on differential shrinkage, the selection of an appropriate shrinkage coefficient in Table 2-1, the use of I_c based on the 28-day E_c for the slab concrete with these equations, and Fig. 2-1 for the definition of y_{cs}.

Term (4) is the short-time live-load deflection. $\Delta_L = KM_LL^2/E_sI_c$, where I_c is computed using the 28-day E_c for the slab concrete. Also included in M_L are the moments due to diaphragms, partitions, ceilings, etc., placed after composite action.

Term (5) is the partial live-load creep deflection for any permanent live load (plus diaphragm, partition, ceiling, etc., dead load) supported by the composite beam. $(\Delta_{cp})_L = KM_PL^2/E_s(I_c)_{cp}$, where M_P refers to the permanent live-load and dead-load moment supported by the composite beam, $(I_c)_{cp}$ is computed using an increased n-value, or $n_{ct} = nC_L$—not $n(1 + C_L)$, since the elastic deflection component is already included in term (4)—and C_L is the creep coefficient of the slab concrete based on the concrete age when the permanent live load, etc., is applied. Also, n and E_c may refer to the 28-day modulus of elasticity of the slab concrete in most cases.

Alternatively, term (5) may be estimated more simply by $(\Delta_{cp})_L = (0.70)(M_P/M_L)C_L\Delta_L$, where M_P is the moment due to the permanent load supported by the composite beam and C_L is the slab creep coefficient based on the slab age when the permanent live load, etc., is applied. The factor 0.70 is analogous to $k_r = 0.85$ in Eq. (3-71) for reinforced concrete beams without compression steel as shown in Fig. 3-13 and Eqs. (3-64) and (3-65). The author prefers this alternative method which avoids the necessity of having to compute $(I_c)_{cp}$.

The two forms of term (5) compute essentially the same results in the example of a shored composite beam.

Shored construction, assuming the steel beam supports its own dead load and the composite beam supports the slab dead load and live load

$$\Delta_{i,t} = \text{same as Eq. (3-97), except as follows.} \qquad (3\text{-}98)$$

Term (2) is the deflection of the composite beam under slab dead load. $\Delta_{s1} = KM_1L^2/E_sI_c$, where I_c is computed using E_c for the slab concrete, based on the slab age when the shores are struck. For most purposes, the 28-day E_c can be used, in which case I_c is the same as in terms (3) and (4) of Eq. (3-97). When using the same value of I_c, this term can be included in the calculations by simply adding M_1 to M_L in term (4) of Eq. (3-97), although the separate calculation of Δ_{s1} is needed in the alternative solution of term (6) below.

Term (3) may be adjusted to reflect only the shrinkage deflection that occurs after the shores are struck, as explained in term (8) of Eq. (3-91).

An additional term (6) is required for the creep deflection of the composite beam under slab dead load: $(\Delta_{cp})_{s1} = KM_1L^2/E_s(I_c)_{cp}$, where $(I_c)_{cp}$ is computed using an increased n-value, $n_{ct} = nC_{s1}$, and C_{s1} is the creep coefficient of the slab concrete based on the concrete age when the shores are struck and n may refer to the 28-day E_c of the slab concrete in most cases.

Alternatively, term (6) may be estimated more simply by $(\Delta_{cp})_{s1} = (0.70)C_{s1}\Delta_{s1}$. See term (5) of Eq. (3-97) for comments.

When the difference between the effect of slab age when the shores are struck and when other permanent loads are applied to the composite section can be neglected (that is, when $C_{s1} = C_L$), term (6) can be deleted and the effect of this term included by adding M_1 to both M_L and M_P in term (5) of Eq. (3-97).

Shored construction, assuming the composite beam supports all dead and live load

$$\Delta_{i,t} = \text{same as Eq. (3-97), except as follows.} \tag{3-99}$$

Term (1) + (2) is the deflection of the composite beam under slab plus steel beam dead load. $(\Delta_{s1} + \Delta_{s2}) = K(M_1 + M_2)L^2/E_sI_c$. See term (2) of Eq. (3-98) for comments concerning E_c and I_c.

Term (3) may be adjusted to reflect only the shrinkage deflection that occurs after the shores are struck, as explained in term (8) of Eq. (3-91).

An additional term (6) is required for the creep deflection of the composite beam under slab plus steel beam dead load, as in term (6) of Eq. (3-98), with $(M_1 + M_2)$ replacing M_1. The alternative solution and the solution when $C_{s1} = C_L$ in term (6) of Eq. (3-98) also apply in this case with $(M_1 + M_2)$ replacing M_1.

Ultimate deflections are also computed by Eqs. (3-97) to (3-99). These equations could be shortened by combining terms, but are kept in the form of separate terms in order to show the separate effects or contributions that take place.

Approximate AISC method for unshored and shored steel–concrete composite beams Equation (3-100) is recommended in the American Institute of Steel Construction Handbook[82] for uniformly loaded simple-span beams.

$$\Delta = \frac{(5)(12M)(144L^2)}{(48)(E = 29\,000)(I = Sy)} = \frac{ML^2}{160Sy} \tag{3-100}$$

where Δ = deflection, in.; M = moment, ft-kips; S = section modulus, in^3; y = distance from bottom of steel section to neutral axis, in.; and L = span length, ft.

For unshored beams:

Dead-load deflection	$M = M_D$	$S = S_s$	$y = y_{bs}$
Live-load deflection (short-term)	$M = M_L$	$S = S_{tr}$	$y = y_b$

For shored beams:

Dead-load deflection	$M = M_D$	$S = S_{tr}$	$y = y_b$
Live-load deflection (short-term)	$M = M_L$	$S = S_{tr}$	$y = y_b$

where S_s and y_{bs} refer to the steel beam with or without cover plates, and S_{tr} and y_b to the composite beam. For the shored case, it is seen that the method assumes the composite beam supports all dead and live load.

If it is desired to consider long-term creep deflection, S_{tr} and y_b should be based on an n-value double that shown in the tables (for 3000 psi concrete, $n = 9$). The method does not include the effect of shrinkage deflection, unless it is assumed to be included in the modular ratio, $2n$.

The allowable deflections for the examples in the AISC Handbook are 1 and $\frac{1}{2}$ in. for dead-load deflection and $L/360$ for live-load deflection by this method, with long-term creep deflection not considered significant in the examples.

Approximate AISI method for steel–concrete composite beams fully shored during construction Equation (3-101) by Hooper[83] is recommended by the American Iron and Steel Institute for computing deflections of shored simple-span steel–concrete composite beams carrying uniformly distributed loads.

$$\Delta = \underset{(1)}{\Delta_{sh}} + \underset{(2)}{\Delta_{short}} + \underset{(3)}{\Delta_{long}}$$

$$\Delta = \frac{L^2}{100 y_b}\left[\underset{(1)}{0.33} + \frac{\overset{(2)}{(f_b)_{short}} + \overset{(3)}{1.40(f_b)_{long}}}{20}\right] \tag{3-101}$$

where:

Term (1) is the shrinkage deflection based on a unit stress between 450 and 600 psi of the concrete area assumed to act in the composite beam. $\Delta_{sh} = (L^2/100y_b)[(f_b)_{sh}/20]1.2 = K_{sh}L^2/100y_b = 0.33L^2/100y_b$, where y_b is the distance from the composite centroid to the bottom fiber.

Term (2) is the short-term deflection usually assumed to be based on half of the live load with the other half included in term (3). $\Delta_{short} = (L^2/100y_b)[(f_b)_{short}/20]$, where $(f_b)_{short} = 50\%\ M_{short}/S_b = 0.50M_L/S_b$ in ksi, and S_b = composite section modulus for the bottom fiber.

Term (3) is the long-term deflection, including the effect of creep, under permanent load, usually assumed to be based on the total dead load plus half of the live load. $\Delta_{long} = K_{cp}(L^2/100y_b)[(f_b)_{long}/20] = 1.40(L^2/100y_b)[(f_b)_{long}/20]$, where $(f_b)_{long} = (M_D + 0.50M_L)/S_b$ in ksi, and $K_{cp} = I_{short}/I_{long} = 1.40$ in which I_{short} and I_{long} are the composite-section Is using n (based on E_c for 3000 psi concrete) and n_{ct} (based on $E_{ct} = E_c/3$), respectively.

In Eq. (3-101) the basic expression, $\Delta = (L^2/100y_b)(f_b/20)$, is derived as follows for uniformly distributed loading and simple-span beams:

$$\Delta = 5wl^4/384EI \qquad M = wl^2/8 \qquad f_b = My_b/I$$

Hence, $$\Delta = 5Ml^2/48EI = (5/48)(f_b/E)(l^2/y_b)$$

Letting $l = 12L$ and $E = 30 \times 10^3$ ksi,

$$\Delta = (5/48)(f_b/30 \times 10^3)(144L^2/y_b) = (L^2/100y_b)(f_b/20).$$

with Δ—in., L—ft, y_b—in., f_b—ksi.

Numerical example of shored steel–concrete composite beam deflection See Table 3-32 for the material properties, loading, moments, parameters, and conditions, and Table 3-33 for the design of the member and section properties of a simply supported shored composite beam deflection example. Compute the ultimate deflection by the general Eqs. (3-98), (3-99), the approximate AISC Eq. (3-100), and the approximate AISI Eq. (3-101).

From AISC Handbook,

$$\text{Depth/span} = F_y/800 = 36/800 = 1/22.2$$

or

$$\text{Depth} = \text{span}/22.2 = (35)(12)/22.2 = 18.9 \text{ in. } (48.01 \text{ cm})$$

Since composite depth = 17.68 in. (44.91 cm) (less than 18.9 in.), deflections may be critical, and are intended to be checked in any case.

Solution by Eq. (3-98), assuming the steel beam supports its own dead load and the composite beam supports the slab dead load and live load Initial dead-load deflection of the steel beam—term (1):

$$\Delta_{s2} = \frac{KM_2L^2}{E_sI_s} = \frac{(5/48)(6.58)(12)(35)^2(144)}{(29\,000)(429)} = \frac{145.1 \times 10^4}{1244 \times 10^4} = \underline{0.12 \text{ in. } (0.30 \text{ cm})}$$

Deflection of the composite beam under slab dead load—term (2):

$$\Delta_{s1} = \frac{KM_1L^2}{E_sI_c} = \frac{(5/48)(45.94)(12)(35)^2(144)}{(29\,000)(1180)} = \frac{1013 \times 10^4}{3422 \times 10^4} = \underline{0.30 \text{ in. } (0.76 \text{ cm})}$$

Deflection due to shrinkage of the slab concrete—term (3):

For $t = 10$ days, $(\alpha_s)_{sh} = t/(35 + t) = 10/45 = 0.22$ (Table 3-25)

Hence,

$$D_u = (1 - 0.22)(766 \times 10^6 \text{ from Table 3-32}) = 597 \times 10^{-6} \text{ in/in (cm/cm)}$$

$$Q_{DS} = (Q \text{ in Eq. (2-1)})/2 = (DA_1E_1)/2 = (597 \times 10^{-6})(72 \times 4)(3.22 \times 10^6)$$
$$= 277\,000 \text{ (126 000 kg, 1 232 000 N)}$$

(using a 72 in. × 4 in., rather than a 7.5 ft × 4 in. slab)

$$y_{cs} = d + 2 - y_b = 13.68 + 2 - 13.18 = 2.50 \text{ in.} \quad \text{(Fig. 2-1)}$$

$$\Delta_{DS} = \frac{Q_{DS}\,y_{cs}L^2}{8E_sI_c} = \frac{(277)(2.50)(35)^2(144)}{(8)(29\,000)(1180)} = \frac{1222 \times 10^5}{2738 \times 10^5} = \underline{0.45 \text{ in. } (1.14 \text{ cm})}$$

Short-time live-load (plus ceiling and partition dead-load) deflection of the composite beam—term (4):

$$\Delta_L = \frac{KM_LL^2}{E_sI_c} = \frac{(5/48)(\overset{126.31}{91.86 + 34.45})(12)(35)^2(144)}{(29\,000)(1180)}$$

$$= \frac{2785 \times 10^4}{3422 \times 10^4} = \underline{0.81 \text{ in. } (2.06 \text{ cm})}$$

Table 3-32 Material properties, loading, moments, parameters, and conditions for steel–concrete composite beam deflection example

Simply supported shored composite beam

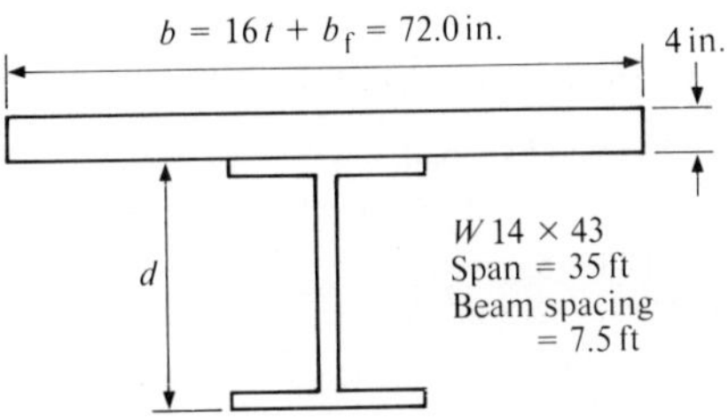

Steel: $d = 13.68$ in. $A_s = 12.6\ \text{in}^2$

$y_{bs} = 6.84$ in. $S_s = 62.7\ \text{in}^3$

$I_s = 429\ \text{in}^4$

Composite: $y_b = 13.18$ in.

$S_b = S_{tr} = 89.4\ \text{in}^3$

$I_c = I_{tr} = 1180\ \text{in}^4$

Material properties, loading, moments

$F_y = 36$ ksi $\quad E_s = 29\,000$ ksi

Normal-weight concrete slab

$f'_c = 3000$ psi $\quad n = 9$

$E_c = 29\,000/9 = 3220$ ksi

$w_{slab} = (72)(4)(150)/144 = 300\ \#/\text{ft}$

$w_{steel} = 43\ \#/\text{ft}$

$w_{ceiling+partition} = (30\ \text{psf})7.5 = 225\ \#/\text{ft}$

$w_{LL} = (80\ \text{psf})7.5 = 600\ \#/\text{ft}$

$M_{slab} = (300)(35)^2/8 = 45\,940$ ft-#

$M_{steel} = (43)(35)^2/8 = 6\,580$ ft-#

$M_{c+p} = (225)(35)^2/8 = 34\,450$ ft-#

$M_{LL} = (600)(35)^2/8 = \underline{91\,860}$ ft-#

$M_{Total} = 178\,830$ ft-#

Other parameters and conditions

Avg. rel. humidity = 70 percent, slab moist-cured

For shrinkage calculation:

$$K_{sh} = 1/8 \qquad \text{(Eq. (3-60))}$$

From Table 2-1 for 6 in. slab and 70 percent humidity,

$$D_u = 655 \times 10^{-6}\ \text{in/in}$$

For 4 in. slab, $(\text{C.F.})_T = 1.17$ (Table 1-8)

Hence, $D_u = (655 \times 10^{-6})(1.17) = \underline{766 \times 10^{-6}\ \text{in/in}}$ (use)

Assume shores are struck at slab age of 10 days,

$$C_u = 1.79 \qquad \text{(Table 1-10)}$$

For 4 in. slab,

$$\text{Creep } (\text{C.F.})_T = 1.11 \qquad \text{(Table 1-8)}$$

Hence for slab, partition and ceiling load,

$$C_{s1} = (1.79)(1.11) = \underline{2.00} \qquad \text{(use)}$$

For sustained live load applied at slab age 2 months (assumed as 30 percent of the full live load),

$$C_u = 1.45 \qquad \text{(Table 1-10)}$$

Hence, for sustained live load,

$$C_L = (1.45)(1.11) = \underline{1.61} \qquad \text{(use)}$$

(1 in. = 2.54 cm, 1 ft = 0.305 m. For other metric units, see conversion factors herein.)

Table 3-33 Member design and section properties for steel–concrete composite beam deflection example

Design—shored steel–concrete composite beam

$$\text{Req. } S_{tr} = M_{\text{Total}}/F_b = (178.8)(12)/24 = 89.4\ \text{in}^3$$

From AISC Handbook for 4 in. slab, select W 14 × 43, $S_{tr} = 89.4\ \text{in}^3$.

Section properties	
Composite I, $n_{ct} = 9(2) = 18$ (when shores struck, also by AISC method)	Composite I, $n_{ct} = 9(1.61) = 14.49$ (for sustained live load)

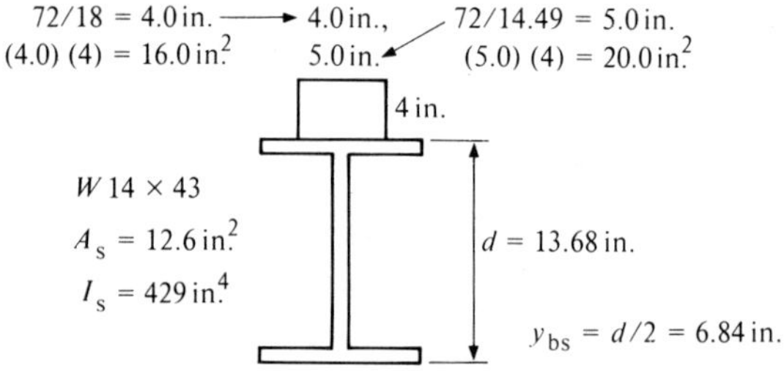

$y_b = \dfrac{(12.6)(6.84) + (16.0)(15.68)}{12.6 + 16.0} = 11.79$ in.	$y_b = \dfrac{(12.6)(6.84) + (20.0)(15.68)}{12.6 + 20.0} = 12.26$ in.
$(I_c)_{cp} = I_{tr} = 429 + (12.6)(11.79 - 6.84)^2 + (4.0)(4)^3/12 + (16.0)(15.68 - 11.79)^2 = 1001\ \text{in}^4$	$(I_c)_{cp} = I_{tr} = 429 + (12.6)(12.26 - 6.84)^2 + (5.0)(4)^3/12 + (20.0)(15.68 - 12.26)^2 = 1060\ \text{in}^4$

Note: $I_s = \underline{429\ \text{in}^4}$. Using $nC_{s1} = 9 \times 2 = 18$, $(I_c)_{cp} = \underline{1001\ \text{in}^4}$.
Using $nC_L = 9 \times 1.61 = 14.49$, $(I_c)_{cp} = \underline{1060\ \text{in}^4}$.
And using $n = 9$, $I_{tr} = \underline{1180\ \text{in}^4}$ (from AISC Handbook)

(For metric units, see conversion factors herein.)

Partial live-load (plus ceiling and partition dead-load) creep deflection of the composite beam—term (5):

$$(\Delta_{cp})_L = \frac{KM_PL^2}{E_s(I_c)_{cp}} = \frac{(5/48)[\overset{62.01}{(0.30)(91.86) + 34.45}](12)(35)^2(144)}{(29\,000)(1060)}$$

$$= \frac{1367 \times 10^4}{3074 \times 10^4} = \underline{0.44\ \text{in. } (1.12\ \text{cm})}$$

Using the alternative solution for term (5):

$$(\Delta_{cp})_L = (0.70)(M_P/M_L)C_L\Delta_L = (0.70)(62.01/126.31)(1.61)(0.81)$$
$$= \underline{0.45 \text{ in. } (1.14 \text{ cm})}$$

which is essentially the same as the more general solution above, which requires the additional calculation of $(I_c)_{cp}$ in Table 3-33.

Creep deflection of the composite beam under slab dead load—term (6):

$$(\Delta_{cp})_{s1} = \frac{KM_1L^2}{E_s(I_c)_{cp}} = \frac{(5/48)(45.94)(12)(35)^2(144)}{(29\,000)(1001)} = \frac{1013 \times 10^4}{2903 \times 10^4} = \underline{0.35 \text{ in. } (0.89 \text{ cm})}$$

Using the alternative solution for term (6):

$$(\Delta_{cp})_{s1} = (0.70)C_{s1}\Delta_{s1} = (0.70)(2.00)(0.30) = \underline{0.42 \text{ in. } (1.07 \text{ cm})}$$

versus 0.35 in. (0.89 cm) by the more general solution above, which requires the additional calculation of $(I_c)_{cp}$ in Table 3-33.

$$\begin{array}{ccccccc} & (1) & (2) & (3) & (4) & (5) & (6) \\ \Delta_u = & 0.12 + & 0.30 + & 0.45 + & 0.81 + & 0.44 + & 0.35 = \underline{2.47 \text{ in. } (6.27 \text{ cm})} \end{array}$$

$$\text{Total live-load deflection} = 0.81(91.86/126.31) + 0.44[(0.30)(91.86)/62.01]$$
$$= \underline{0.78 \text{ in. } (1.98 \text{ cm})}$$

which is less than $L/360 = (35)(12)/360 = 1.17$ in. (2.97 cm).

$$\text{Total dead-load deflection} = 2.47 - 0.78 = \underline{1.69 \text{ in. } (4.29 \text{ cm})}$$

which is slightly greater than the suggested AISC limit of 1.5 in. (3.8 cm). However, the beam was designed as a shored beam.

Solution by Eq. (3-99), assuming the composite beam supports all dead and live load Deflection of the composite beam under slab plus steel beam dead load—terms (1) + (2):

$$(\Delta_{s1} + \Delta_{s2}) = \frac{K(M_1 + M_2)L^2}{E_sI_c} = \frac{(5/48)(45.94 + 6.58)(12)(35)^2(144)}{(29\,000)(1180)}$$

$$= \frac{1158 \times 10^4}{3422 \times 10^4} = \underline{0.34 \text{ in. } (0.86 \text{ cm})}$$

Deflection due to shrinkage of the slab concrete—term (3):

$$\Delta_{DS} = \underline{0.45 \text{ in. } (1.14 \text{ cm})} \qquad \text{(same as prev. sol. by Eq. (3-98))}$$

Short-time live-load (plus ceiling and partition dead-load) deflection of the composite beam—term (4):

$$\Delta_L = \underline{0.81 \text{ in. } (2.06 \text{ cm})} \qquad \text{(same as prev. sol. by Eq. (3-98))}$$

Partial live-load (plus ceiling and partition dead-load) creep deflection of the composite beam—term (5):

$$(\Delta_{cp})_L = \underline{0.44 \text{ in. } (1.12 \text{ cm})} \qquad \text{(same as prev. sol. by Eq. (3-98))}$$

Creep deflection of the composite beam under slab plus steel beam dead load—term (6):

$$(\Delta_{cp})_{s1+s2} = (\text{prev. sol. by Eq. (3-98)})(M_1 + M_2)/M_1$$
$$= (0.35)(45.94 + 6.58)/45.94 = \underline{0.40 \text{ in. } (1.02 \text{ cm})}$$

$$\begin{aligned} & \quad\ (1) + (2) \quad (3) \qquad (4) \qquad (5) \qquad (6) \\ \Delta_u &= 0.34 + 0.45 + 0.81 + 0.44 + 0.40 \\ &= \underline{2.44 \text{ in. } (6.20 \text{ cm})} \\ & \quad \text{versus } 2.47 \text{ in. } (6.27 \text{ cm}) \text{ by Eq. (3-98)} \end{aligned}$$

The distinction between the solutions by Eq. (3-98) and (3-99) is seen to be negligible for this example, since the steel beam dead load was small compared to the other loads, which is typical.

If the beam had been unshored when the slab was cast:

$$\Delta_{s1} = 0.30(I_c/I_s) = 0.30(1180/429) = \underline{0.83 \text{ in. } (2.11 \text{ cm})} \qquad \text{in term (2), Eq. (3-97)}$$

$$\Delta_{DS} = 0.45 \text{ (unshored } D/\text{shored } D) = 0.45(355/395)$$
$$= \underline{0.41 \text{ in. } (1.04 \text{ cm})} \text{ in term (3), Eq. (3-97)} \qquad \text{(Table 2-1)}$$

There is no creep deflection under slab dead load, since the slab is unstressed under its own dead load in the unshored case. Hence,

$$\begin{aligned} & \quad\ (1) \qquad (2) \qquad (3) \qquad (4) \qquad (5) \\ \Delta_u &= 0.12 + 0.83 + 0.41 + 0.81 + 0.44 = \underline{2.61 \text{ in. } (6.63 \text{ cm})} \qquad \text{in Eq. (3-97)} \end{aligned}$$

versus 2.47 in. (6.27 cm) and 2.44 in. (6.20 cm) for the two shored cases above.

The deflections for the two shored cases and the unshored case are summarized in Table 3-34.

Solution by approximate AISC Eq. (3-100) for shored steel–concrete composite beams

$$\Delta = \frac{ML^2}{160Sy} = \frac{M_{\text{Total}}L^2}{160S_{\text{tr}}y_b} + \frac{(M_{\text{Total}} - M_L + 0.30M_L)}{160[S_{\text{tr}}y_b = (I_c)_{cp}]_{n_{ct}=18}} \qquad \text{(Eq. (3-100))}$$

$$= \frac{(178.83)(35)^2}{(160)(89.4)(13.18)} + \frac{\overset{114.53}{(178.83 - 0.70 \times 91.86)}(35)^2}{(160)(1001)}$$

$$= \frac{21.91 \times 10^4}{18.85 \times 10^4} + \frac{14.03 \times 10^4}{16.02 \times 10^4}$$

$$= 1.16 + 0.88 = \underline{2.04 \text{ in. } (5.18 \text{ cm})}$$

Table 3-34 Term by term deflections in inches for the steel–concrete composite beam example by Eqs. (3-97), (3-98), and (3-99)

Type of construction	Defl. of steel beam under steel D.L.	Defl. of steel beam under slab D.L.	Defl. of comp. beam under slab D.L.	Defl. of comp. under slab + steel D.L.	Defl. of comp. beam due to slab shrink.	Defl. of comp. beam under L.L. + partition + ceiling D.L.	Creep defl. under L.L. + partition + ceiling D.L.	Creep defl. of comp. beam under slab D.L.	Creep defl. of comp. beam under slab + steel D.L.	Total Δ_u
	(1)	(2)	(2)	(1) + (2)	(3)	(4)	(5)	(6)	(6)	
Unshored, Eq. (3-97)	0.12	0.83	—	—	0.41	0.81	0.44	—	—	2.61
Shored, Eq. (3-98)	0.12	—	0.30	—	0.45	0.81	0.44	0.35	—	2.47
Shored, Eq. (3-99)	—	—	—	0.34	0.45	0.81	0.44	—	0.40	2.44

Total live-load deflection in terms (4) and (5) = 0.78 in. in all three cases. This is less than the AISC allowable deflection of $L/360 = (35)(12)/360 = 1.17$ in.

Total dead-load deflection is slightly greater than suggested AISC value of 1.5 in.:

Unshored, Eq. (3-97) Total dead-load deflection = 2.61 − 0.78 = 1.83 in.

Shored, Eq (3-98) Total dead-load deflection = 2.47 − 0.78 = 1.69 in.

Shored, Eq (3-99) Total dead-load deflection = 2.44 − 0.78 = 1.66 in.

(1 in. = 2.54 cm.)

versus 2.47 in. (6.27 cm) by Eq. (3-98) and 2.44 in. (6.20 cm) by Eq. (3-99). The second term represents the long-term creep deflection. However, it appears that Eq. (3-100) does not include the effect of slab shrinkage.

$$\text{Total live-load deflection} = 1.16(91.86/178.83) + 0.88[0.30(91.86)/114.53]$$
$$= 0.60 + 0.21 = \underline{0.81 \text{ in. } (2.06 \text{ cm})}$$

which is less than $L/360 = (35)(12)/360 = 1.17$ in. (2.97 cm).

$$\text{Total dead-load deflection} = 2.04 - 0.81 = \underline{1.23 \text{ in. } (3.12 \text{ cm})}$$

which is less than the suggested AISC limit of 1.5 in. (3.8 cm).

If the beam had been unshored, the solution by the approximate AISC Eq. (3-100) is as follows:

$$\Delta = \frac{ML^2}{160Sy} = \frac{(M_{\text{steel}} + M_{\text{slab}})L^2}{160S_s y_{bs}} + \frac{(M_{c+p} + M_L)L^2}{160S_{tr} y_b} + \frac{(M_{c+p} + 0.30M_L)L^2}{160[(I_c)_{cp}]_{n_{ct}=18}}$$

126.31

$$= \frac{(6.58 + 45.94)(35)^2}{(160)(62.7)(6.84)} + \frac{(34.45 + 91.86)(35)^2}{(160)(89.4)(13.18)} + \frac{(34.45 + 0.30 \times 91.86)(35)^2}{(160)(1001)}$$

$$= \frac{6.434 \times 10^4}{6.862 \times 10^4} + \frac{15.47 \times 10^4}{18.85 \times 10^4} + \frac{7.596 \times 10^4}{16.02 \times 10^4}$$

$$= 0.94 + 0.82 + 0.47 = \underline{2.23 \text{ in. } (5.66 \text{ cm})}$$

versus 2.61 in. (6.63 cm) by Eq. (3-97).

$$\text{Total live-load deflection} = 0.82(91.86/126.31) + 0.21 \text{ in. (above)}$$
$$= 0.60 + 0.21 = \underline{0.81 \text{ in. } (2.06 \text{ cm})} \qquad \text{(same as above)}$$

which is less than $L/360 = 1.17$ in. (2.97 cm).

$$\text{Total dead-load deflection} = 2.23 - 0.81 = \underline{1.52 \text{ in. } (3.86 \text{ cm})}$$

which is essentially the same as the suggested AISC limit of 1.5 in. (3.8 cm).

Solution by approximate AISI Eq. (3-101) for shored steel–concrete composite beams

$$(f_b)_{\text{short}} = 0.50M_L/S_b = (0.50)(91.86)(12)/89.4$$
$$= 6.17 \text{ ksi } (434 \text{ kg/cm}^2, 42.5 \text{ N/mm}^2)$$
$$(f_b)_{\text{long}} = (M_{\text{Total}} - 0.50M_L)/S_b = (178.83 - 0.50 \times 91.86)(12)/89.4$$
$$= 17.84 \text{ ksi } (1254 \text{ kg/cm}^2, 123 \text{ N/mm}^2)$$

Checking the design,

$$(f_b)_{\text{Total}} = 6.17 + 17.84$$
$$= 24.0 \text{ ksi } (1687 \text{ kg/cm}^2, 165 \text{ N/mm}^2)$$

$$= F_b = 24 \text{ ksi} \qquad \text{OK}$$

$$\Delta = \frac{L^2}{100 y_b}\left[0.33 + \frac{(f_b)_{\text{short}} + 1.40(f_b)_{\text{long}}}{20}\right] \qquad \text{(Eq. (3-101))}$$

$$= \frac{35^2}{(100)(13.18)}\left[0.33 + \frac{6.17 + (1.40)(17.84)}{20}\right]$$

$$= 0.929[1.89] = \underline{1.76 \text{ in. } (4.47 \text{ cm})}$$

versus 2.47 in. (6.27 cm) by Eq. (3-98), 2.44 in. (6.20 cm) by Eq. (3-99), and 2.04 in. (5.18 cm) by the approximate AISC equation (3-100).

3-6 EXPERIMENTAL RESULTS, PRACTICAL CONSIDERATIONS, AND A SIMPLIFIED CALCULATION PROCEDURE

Discussion of Experimental Results

A discussion of experimental and computed results has already been included in Sec. 3-2 for the effect of short-time loading, repeated load cycles, and continuity; Sec. 3-3 for shrinkage deflections; Sec. 3-4 for the effect of creep, creep plus shrinkage, and compression steel on time-dependent deflections; and Sec. 3-5 for the deflection of reinforced concrete composite beams. This section will attempt to present an overview of the situation with regard to the nature of the deflection problem and the correlation that may be expected between computed and actual initial and time-dependent deflections. Many experimental studies on the subject have been reported through the years and only a few that relate to the material in this chapter will be mentioned.

Typical of the agreement between computed and measured deflections is that observed in a 1966 ACI Committee 435[3] study of both short-time and long-time deflections, as shown in Figs. 3-17 and 3-18. With the exception of the gross-section methods using earlier time-dependent factors, the agreement between computed and measured deflections for these 40 test results was in the range of plus-minus 20 percent in most, but not all, cases. Also, as Beeby[85] states, the CEB[86] says its method will predict deflections within plus-minus 20 percent, and it is probable that the methods of the British Code and ACI Code are of the same order of accuracy. However, it is clear that such plus-minus percentages are not reliable without extensive qualification, because of the randomness of material properties and variations in construction practices, loading histories, etc. On the other hand, such percentages may be of some value as a rough index of the agreement between computed and measured results by various procedures when material properties and conditions are reasonably closely approximated. It should also be noted that the plus-minus 20 percent in Figs. 3-17 and 3-18 represents a nominal limit band, and that most results are better than these limits.

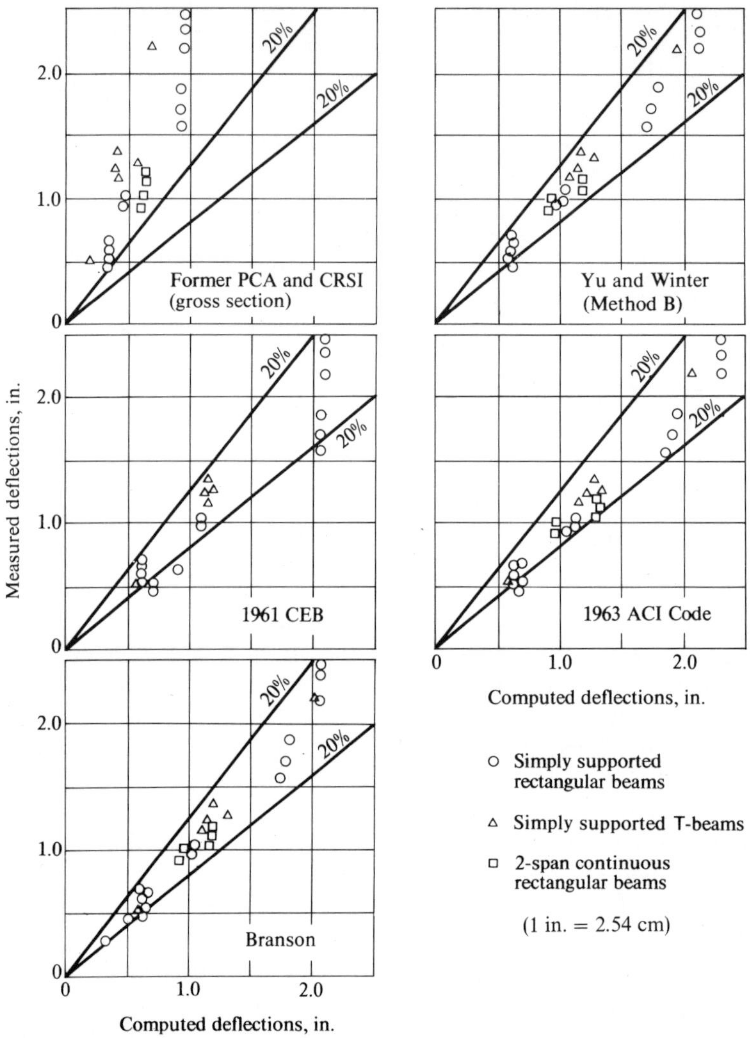

FIGURE 3-17
Comparison of computed and measured short-time deflections from ACI Committee 435. (*From ACI Journal*,[3] *with permission.*)

A more recent statistical study of short-time simply-supported beam deflections by ACI Committee 435[1] in 1972 found for controlled laboratory conditions that there is approximately a 90 percent chance that the deflection of a particular beam will be within the range of 20 percent less than to 30 percent more than the calculated value when using the 1971 ACI Code (I_e) method. The Committee also found that this was only a slight improvement over the 1963 ACI Code method which showed that for the 90 percent interval the range of

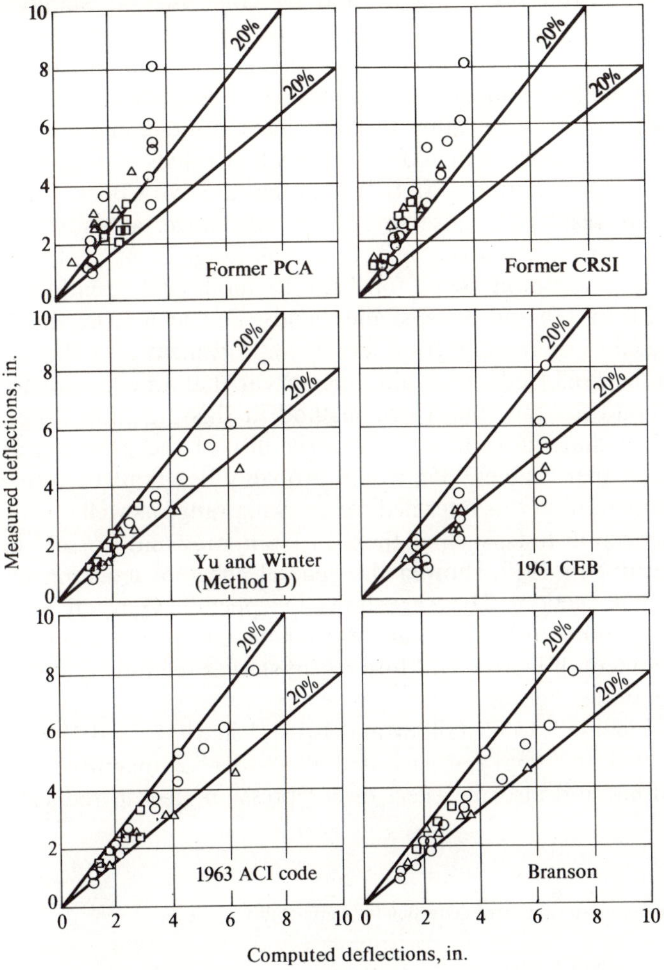

FIGURE 3-18
Comparison of computed and measured long-time deflections from ACI Committee 435. (*From ACI Journal,*[3] *with permission.*)

agreement was 30 percent less than to 30 percent more than the calculated value. The probability study by the Committee is summarized in Table 3-35, and an earlier study by the Committee (reported in Ref. 6) of mean values is shown in Table 3-36. From Table 3-35, the 90 percent probability figures above seem to apply to singly reinforced beams only, with the doubly reinforced beams having a lower probability. Table 3-35 can be used to evaluate the minimum expected variability of deflections calculated by the two methods. These studies were based primarily on the work of Professor Benjamin and his colleagues at Stanford University.

As was pointed out by the author[2] in his discussion of the Committee 435 paper, the comparison of the 1963 and 1971 ACI Codes could be misleading, since most of the experimental beams were rather severely cracked under the applied loads (the vast majority of the more than 100 test beams had values of M_{max}/M_{cr} greater than 2.0, and many greater than 2.5). It follows that the two Codes would compute about the same results (I_e close to I_{cr}) for this range, as shown in Table 3-37. But this will normally not be the case for the range of M_{max}/M_{cr} = say 1.1 to 2.0, which is a common range for light to moderately reinforced members (I_g/I_{cr} = say 2.0 to 3.0) and for sustained loads. For such cases the I_e procedure can be expected to represent a considerable improvement over the use of either I_g or I_{cr}, as in the 1963 Code. It would thus appear that an even greater variability would be found for the 1963 Code method in this cracking range. However, it is recognized that the characteristic variability of the problem, as found in the Committee paper (Table 3-35), would probably be about the same for the 1971 ACI Code method in this intermediate cracking range as well.

Early short-time deflection tests, such as those of Humphrey and Losse[87] in 1911 and Bach and Graf[88] in 1917, showed the general effect of cracking on reinforced concrete beam deflections. These are shown in Appendix G, where such representations are comparable to Figs. 3-4 and 3-5 for the effect of cracking in terms of the effective moment of inertia, I_e. More recent studies relating to short-time deflections are mentioned below.

Tests by Faber[89,90] in 1927, 1932; Gilkey and Ernst[91] in 1935; and Washa and Fluck[44,45,92] in 1948 to 1956 demonstrated the order of magnitude of time-dependent deflections, and also the effect of compression steel in reducing

Table 3-35 Ranges of ratio of actual to calculated deflection for *t* probability

	ACI 318-63			ACI 318-71		
Beam	$t = 0.6$	$t = 0.8$	$t = 0.9$	$t = 0.6$	$t = 0.8$	$t = 0.9$
Short-term with top steel	0.82 to 1.21	0.73 to 1.32	0.65 to 1.41	0.87 to 1.31	0.78 to 1.31	0.71 to 1.38
Short-term without top steel	0.86 to 1.04	0.81 to 1.09	0.78 to 1.13	0.91 to 1.11	0.86 to 1.17	0.82 to 1.25
Long-term with top steel	0.93 to 1.22	0.84 to 1.30	0.78 to 1.36	0.93 to 1.23	0.89 to 1.32	0.89 to 1.40
Long-term without top steel	0.79 to 1.02	0.73 to 1.07	0.69 to 1.22	1.00 to 1.29	0.89 to 1.29	0.81 to 1.29

(From Subcommittee 2, ACI Committee 435 Report, *ACI Journal*,[1] with permission.)

Table 3-36 Statistical study results*[6]

	Mean value of $\dfrac{\text{meas. defl.}}{\text{calc. defl.}}$	
	ACI 318-63	ACI 318-71
Short-time with top steel	1.02	1.04
Short-time without top steel	0.95	1.01
Long-time with top steel	1.07	1.03
Long-time without top steel	0.90	1.02

* These results are based on 107 tests for short-time deflections and 30 tests for long-time deflections, including T-beams and continuous beams.

Table 3-37 Relation between M_a/M_{cr} and I_e, I_g, I_{cr} in Eq. (3-19)*

When $M_a/M_{cr} = 1.1$,	$I_e = 0.75I_g$	$+ 0.25I_{cr}$
1.5	0.30	0.70
2.0	0.13	0.87
2.5	0.06	0.94
3.0	0.04	0.96

* It is seen that when M_a/M_{cr} is greater than about 2.0 to 2.5, I_e is close to I_{cr}.

When $M_a/M_{cr} = 2.0$, $I_g/I_{cr} = 1.5: I_e = 1.07I_{cr}$
$= 2.0: \quad = 1.13$
$= 3.0: \quad = 1.26$

When $M_a/M_{cr} = 2.5$, $I_g/I_{cr} = 1.5: I_e = 1.03I_{cr}$
$= 2.0: \quad = 1.06$
$= 3.0: \quad = 1.12$

such deflections. The experimental results of Washa and Fluck[44,45] in 1952, 1956, Miller[25] in 1958, and Yu and Winter[40] in 1960 are studied in some detail in Secs. 3-3 and 3-4 as they relate to various methods of analysis. One point that can be observed in these studies is the fact that for most beams the ratio of time-dependent to initial deflection was of the order of 1 to 2, even though the ratio of creep (exclusive of shrinkage) to initial strain was of the order of 2 to 5. It was thus clear that the reduction factor k_r in Eqs. (3-64) and (3-65), and the effect of compression steel as by Eqs. (3-66) to (3-73), were the principal reasons for such smaller deflection ratios, as compared to strain ratios. Miller's data referred primarily to shrinkage warping.

As can be seen in Sec. 3-2 for short-time deflections and Sec. 3-4 for long-term deflections, the study of Yu and Winter is especially noteworthy. Two short-time (Methods A, B, and Eq. (3-9)) and two long-time (Methods C, D, and Table 3-19) calculation methods were compared with some 175 deflection tests from the project itself and from the literature at the time.

The study of Hajnal-Konyi[93] presented important time-dependent deflection information, principally as influenced by different span–depth ratios and steel strengths.

The author's[5] data in 1963, in addition to that of Washa and Fluck, Miller, and Yu and Winter, were used in the development of I_e for short-time deflections (Eqs. (3-18), (3-19), and Figs. 3-4, 3-5, 3-6), shrinkage warping equations (3-53) to (3-56)—and Table 3-13, Fig. 3-10, and creep deflection equation (3-74). Several shrinkage warping methods were compared with experimental results (Table 3-16). Some 66 test specimens were involved in this study.

Corley and Sozen[49] in 1966 studied some 75 test results (both initial and time-dependent) from the literature and including their own, for the purpose of describing simplified procedures for short-time deflections (Eq. (3-21)), shrinkage warping (Eq. (3-57)), and creep deflections (Eqs. (3-85) and (3-86)).

The effect of repeated loading on short-time and long-time deflections was investigated by Sinha, Agrawal, Gerstle, and Tulin[66,67] in 1964, 1965; Maldague[48] in 1965; Burns and Siess[46] in 1966; Miller and Guralnick[68,94] in 1967, 1970; Ruiz and Winter[55] in 1968; and Kripanarayanan and Branson[17] in 1972; as discussed in Secs. 3-2 and 3-4, and Figs. 3-7, 3-8, and 3-15. The conclusion drawn by the author is that the effect of repeated loading can usually be neglected, at least in simplified calculations.

As previously discussed in Sec. 3-2, Beeby[50] in 1968 compared the computed short-time deflection by the methods of the transformed section, Yu and Winter, Branson, CEB, and two of his own (Methods 1, 2, and Eq. (3-22)) with some 133 test results. An example of the agreement is shown in Fig. G-3.

The study by Hollington[69] in 1970 involved deflection data for representative precast concrete floor components, and showed among other things that the effect of compression steel in reducing time-dependent deflections in lightly reinforced members was not as large as the British and ACI Codes indicate. This led to the development of Eqs. (3-70) and (3-71) by the author[26] for computing k_r.

The k_r and compression steel reduction effects discussed above can also be seen in the results by Pauw[95] in 1971 for a continuous box-girder bridge, in which the measured time-dependent (total minus initial) deflections, including shrinkage warping, were only of the order of 2 to 3 (average about 2.4) times the initial deflections, while the creep coefficient or ratio of creep strain (exclusive of shrinkage) to initial strain was between 2.5 and 3.0.

In a 1973 paper Heiman[96] compared, calculated, and measured deflections of lightweight concrete beams under construction, service, and test loads. The calculations were made using a live load of 100 psf (488 kg/m^2, 4.8 kN/m^2) for short-term and 15 psf (73 kg/m^2, 718 N/m^2) for long-term deflection. The following typical results for the total deflection (including 5-year long-term values) were presented by Heiman using three calculation methods (also summarized by Clarke, Neville, and Houghton-Evans in Ref. 97): by Branson, 0.46 in. (11.7 mm); by Neville,[98] 0.41 in. (10.4 mm); and by the 1971 ACI Code, 0.35 in. (8.9 mm). The measured value was 0.52 in. (13.2 mm). According to Heiman, the principal reason for the discrepancy was that the calculation of long-term deflection using a live load of 15 psf did not allow for the heavy construction loading applied soon after casting, which causes cracking and high initial creep.

In another study of measured and computed deflections of four buildings in 1974, Heiman[99] found "reasonable agreement" between the measured values and those calculated by Branson's method using separate initial, shrinkage, and creep values.

In a 1974 paper Kripanarayanan and Branson[81] compared, computed, and measured initial and time-dependent deflections for two reinforced concrete composite beams, as shown in Fig. 3-16.

Practical Considerations

The previous discussion has referred primarily to laboratory beams. The problem of predicting deflections of actual structures appears to be greatly compounded by a number of factors, principally the following:

1. The only property of the concrete normally known to the designer at the calculation stage is the specified characteristic strength, as distinguished from the actual strength, and the elastic and cracking properties are nominally related to the actual strength. Also knowledge of time-dependent properties for a particular concrete under local conditions is either unknown or a variable quantity between rather broad limits in many cases (such as Figs. 1-9 to 1-11).

2. The load history in actual structures is a major uncertain factor in many cases, as discussed by Ferguson.[21] For example, the time of form removal, casting of a composite slab, and the installation of partitions, roofing, etc., can have an important effect on creep deflections. Also of importance is the effect of construction loads due to equipment and the stacking of materials, and the time of application of the full design live load, on beam cracking, which in turn can affect the time-dependent (both creep and shrinkage) deflections as well.

3. Other effects such as the interaction of partitions, shear walls, cladding, etc., and the location of construction joints, also influence the deflection of actual structures, but such effects are seldom taken into account in calculations.

However, the author wonders how much all of these factors affect the problem on the average in most cases, especially if the control of deflections by means of computed and allowable values is viewed in terms of "characteristic deflections" as opposed to "predicted deflections." In this context one tends to think in terms of other simpler means of controlling deflections, such as stiffness criteria and span–depth ratios (see Chap. 6). On the other hand, it would seem that some form of a reasonably simple provision for checking deflections would be important from both a physical and intuitive point of view, even though this may not be required for all members. By whatever means, the importance of deflection control is widely recognized as a result of the wide variety of serviceability problems known to the profession, some of which have been cataloged by Plewes[100] and by Mayer and Reusch.[101]

With regard to point 1 on material properties, the type of information given in Chap. 1, for example, appears to represent a consistent and logical approach for estimating material properties in advance of construction. Using such a procedure, the principal variations between computed and actual properties are found in the time-dependent properties, and these discrepancies can be reduced as more creep and shrinkage data are obtained and cataloged for local concretes and conditions. A more simplified Code-type procedure for estimating time-dependent effects, such as the 1971 ACI Code, simply refers to a more grossly approximate or characteristic result.

With regard to point 2 on load histories and the effects of cracking and sustained loading, the author found in writing the earlier parts of this chapter and in the example problems that these effects do not seem to greatly affect the total deflection. For example, the effect of repeated loading complicates the theoretical problem. However, since the envelope of repeated load versus deflection tends to approximate the single loading curve (Figs. 3-7 and 5-10), the effect of cracking due to the application and removal of construction loads and live loads on the total deflection does not appear to be overly significant in most cases. As shown in Fig. 3-8, upon the removal of the superimposed load, the residual short-time dead-load deflection is increased. But when the beam is reloaded, the total deflection is again about the same. In addition, in connection with Fig. 3-15 and the example problems, the effect of cracking due to say live loading and unloading on sustained load creep deflection also does not seem to markedly alter the total deflection. Also, as discussed in Sec. 3-3, it is believed that the effect of cracking can usually be ignored in estimating shrinkage deflections. It thus seems to the author that the nominal creep and shrinkage, or time-dependent, factor used in deflection calculations can be thought of as including these various interrelated time-dependent and cracking effects, provided the overall effect of cracking is taken into account such as by the use of an effective moment of inertia, I_e. Using I_{cr} for all loads appears to significantly overestimate deflections in most cases.

Simplified Calculation Procedure

With the previously discussed limitations in mind, together with the earlier observations and examples in the chapter, the following procedure is thought to represent an appropriate simplified Code-type approach for calculating deflections:

1. For cantilever, simple or continuous beams with uniform loading or a single concentrated load, use Eqs. (3-39) to (3-42) to compute Δ_i. Equations (3-37) and (3-38) will also compute the same results as Eq. (3-39) for continuous beams. In the case of cantilevers, the deflection due to the rotation at the support must also be included.

2. For different load levels, the deflection should be computed in each case using Eq. (3-19) for I_e based on the total load level being considered, such as dead load or dead plus live load. The incremental deflection, such as for live load, is then computed as the difference in these values, as shown in Fig. 3-19. However, the following is suggested for most cases to simplify calculations: use I_e in Eq.

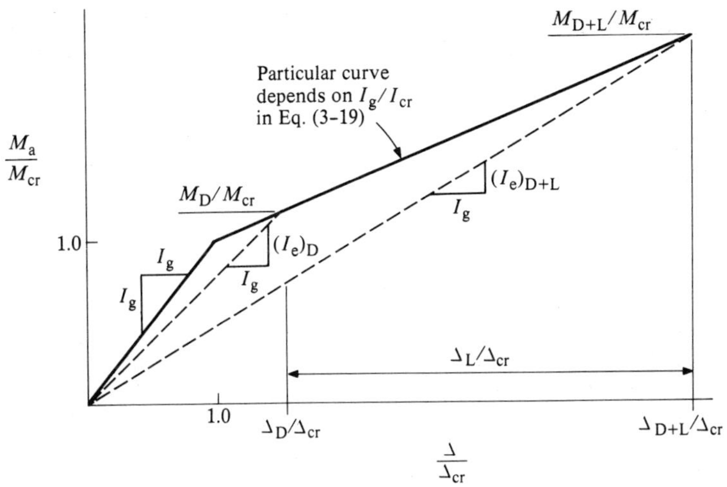

FIGURE 3-19
Idealized short-time moment versus deflection diagram using Eq. (3-19).

(3-19) for the initial deflection due to all sustained loads, including beam, composite slab, partition, roofing, etc., dead load; and partial live load. It will probably be satisfactory in most cases to use only one value of I_e based on the total sustained load. Also use I_{cr} for live-load deflection in which a separate calculation should normally be satisfactory (even though the correct procedure is to determine live-load deflection as the total minus the dead-load deflection). This is the procedure used in the 1973 AASHTO Highway Bridge Specifications for computing deflections using I_e and I_{cr}, in the case of load factor design.

3. Use the 28-day concrete strength, f'_c, to compute f_r by Eq. (1-7) or Table 1-2 and E_c by Eq. (1-13) or Table 1-3.

4. For creep deflections, use Eq. (3-74) for Δ_{cp}, with k_r given by Eq. (3-71). Also use $C_u = 1.6$, or obtain C_u from Chap. 1, Table 1-10, with size, humidity, and age when loaded correction factors.

5. For shrinkage deflections, use Eq. (3-58) for Δ_{sh}, using Eqs. (3-59) to (3-63) for K_{sh}, Eq. (3-56) for ϕ_{sh}, and Table 3-13 or Fig. 3-10 for A_{sh}. Also use $(\varepsilon_{sh})_u = 400 \times 10^{-6}$ in/in (cm/cm), or obtain $(\varepsilon_{sh})_u$ from Chap. 1, Table 1-10, with size, humidity, and age of initial shrinkage correction factors.

6. For combined time-dependent deflections, use Eq. (3-75) for Δ_t, with k_r given by Eq. (3-70). Also use $T_u = 2.5$ for beams and one-way slabs, and $T_u = 3.0$ for other slabs (see Chap. 4), or obtain T_u from Appendix J, Table J-1.

7. Calculate I_e, k_r (based on ρ'), and A_{sh} (based on ρ and ρ' in computing ϕ_{sh}) at the support section for cantilevers and the midspan section for simple and continuous prismatic beams. The use of these values is considered satisfactory in approximate calculations primarily because these rigidities or curvature properties have the dominant effect on deflections, as shown by ACI Committee 435[29] in the case of continuous beams.

8. Use the clear span for all deflection calculations after the moments have been determined using the same span lengths as used in design. That is, consider only the part of the moment diagram in the clear span as contributing to deflections, as discussed in Sec. 3-2.

This procedure was recommended by ACI Committee 435[105] for the 1977 ACI Code and Commentary.

REFERENCES

1. Subcommittee 2, ACI Committee 435, Subcommittee Chairman, J. R. Benjamin, Committee 435 Chairman, B. L. Meyers, Variability of Deflections of Simply Supported Reinforced Concrete Beams, *ACI Journal,* Proceedings **69,** 1, 29–35, Jan. 1972.
2. D. E. Branson, discussion of Ref. 1, in *ACI Journal,* Proceedings **69,** 7, 449–450, July 1972.
3. ACI Committee 435, Chairman, D. E. Branson, Deflections of Reinforced Concrete Flexural Members, *ACI Journal,* Proceedings **63,** 6, part 1, 637–674, June 1966. Republished in *ACI Manual of Concrete Practice,* Part 2: Structural Design, Structural Specifications, Structural Analysis, 1967, 1972.
4. Subcommittee 2, ACI Committee 209, Subcommittee Chairman, D. E. Branson, Committee 209 Chairman, J. R. Keeton, Prediction of Creep, Shrinkage and Temperature Effects in Concrete Structures, in "Designing for Effects of Creep, Shrinkage and Temperature in Concrete Structures," 51–93, ACI Publication SP 27-3 (Symposium Volume), 1971.
5. D. E. Branson, Instantaneous and Time-Dependent Deflections of Simple and Continuous Reinforced Concrete Beams, *HPR Publication* 7, Part 1, 1–78, Alabama Highway Department, Bureau of Public Roads, Aug. 1963.
6. D. E. Branson, Design Procedures for Computing Deflections, *ACI Journal,* Proceedings **65,** 9, 730–742, Sept. 1968.
7. A. F. Shaikh and D. E. Branson, Non-Tensioned Steel in Prestressed Concrete Beams, *PCI Journal,* **15,** 1, 14–36, Feb. 1970.
8. D. E. Branson, discussion of ACI Committee 318, Proposed Revision of ACI 318-63: Building Code Requirements for Reinforced Concrete, *ACI Journal,* Proceedings **67,** 9, 692–695, Sept. 1970.
9. E. G. Nawy and J. G. Potyondy, Moment Rotation, Cracking and Deflection of Spirally Bound, Pretensioned Prestressed Concrete Beams, *Engineering Research Bulletin* 51, 1–97, Rutgers University, 1970.
10. A. M. Neville and W. Dilger, "Creep of Concrete: Plain, Reinforced, and Prestressed," 1–622, North-Holland Publishing Company, Amsterdam, 1970.
11. M. S. Mirza and G. M. Sabnis, Deflections of One-Way Slabs and Beams, *Proceedings of Symposium, ACI Canadian Capital Chapter, Montreal, Canada,* 53–87, Oct. 1971.
12. E. G. Nawy and J. G. Potyondy, authors' closure and discussion by D. E. Branson of Flexural Cracking Behavior of Pretensioned Prestressed Concrete I- and T-beams, *ACI Journal,* May 1971; discussion and closure in *ACI Journal,* Proceedings **68,** 11, 873–877, Nov. 1971.
13. D. E. Branson, discussion of J. C. Jofriet and G. M. McNeice, Finite Element Analysis of Reinforced Concrete Slabs, *Journal of the Structural Division, ASCE,* Mar. 1971; in *Journal of the Structural Division, ASCE,* **97,** ST 12, Proceedings Paper 8557, 2907–2911, Dec. 1971.
14. American Concrete Institute, "Building Code Requirements for Reinforced Concrete," 1–78, and "Commentary," 1–96, American Concrete Institute, Detroit, Michigan, 1971.
15. Prestressed Concrete Institute, "PCI Design Handbook," 1-1 to 12-8, Prestressed Concrete Institute, Chicago, Illinois, 1971.
16. B. Forsyth, "Unified Design of Reinforced Concrete Members," McGraw-Hill Book Co., New York, 1971.
17. K. M. Kripanarayanan and D. E. Branson, Short-Time Deflections of Beams Under Single and Repeated Load Cycles, *ACI Journal,* Proceedings **69,** 2, 110–117, Feb. 1972.

18. R. F. Denkins, "Serviceability Requirements—Control of Deflections," Chap. 6, Notes on ACI 318-71, Building Code Requirements with Design Applications, 6-1 to 6-16, Portland Cement Association, 1972.
19. G. Winter and A. H. Nilson, "Design of Concrete Structures," 8th edn, 1–615, McGraw-Hill Book Co., New York, 1972.
20. C. K. Wang and C. G. Salmon, "Reinforced Concrete Design," 2nd edn, 1–934, Intext Press, New York, 1973.
21. P. M. Ferguson, "Reinforced Concrete Fundamentals," 3rd edn, 1–750, John Wiley & Sons, New York, 1973.
22. W. W. Bennett and N. Veerasubramanian, Behavior of Non-rectangular Beams With Limited Prestress After Flexural Cracking, *ACI Journal,* Proceedings **69,** 9, 533–542, Sept. 1972. Discussion by D. E. Branson and C. I. Chen in *ACI Journal,* Mar. 1973.
23. L. A. Lutz, Graphical Evaluation of the Effective Moment of Inertia for Deflection, *ACI Journal,* Proceedings **70,** 3, 207–213, Mar. 1973.
24. C. G. Salmon, A. F. Shaikh, and M. S. Mirza, Computation of Deflections for Beams and One-Way Slabs, *ACI Committee 435 Symposium Paper,* ACI Convention, San Francisco, Apr. 1974.
25. A. Miller, Warping of Reinforced Concrete Due to Shrinkage, *ACI Journal,* Proceedings **54,** 11, 939–950, May 1958.
26. D. E. Branson, Compression Steel Effect on Long-Time Deflections, *ACI Journal,* Proceedings **68,** 8, 555–559, Aug. 1971.
27. D. E. Branson, Calculation of Deflections of Concrete–Concrete Composite Beams, unpublished draft of report to Subcommittee 3, ACI Committee 435, 1–14, July 1967.
28. K. M. Kripanarayanan and D. E. Branson, Some Experimental Studies of Time-Dependent Deflections of Noncomposite and Composite Reinforced Concrete Beams, *ACI Committee 435 Symposium Paper,* ACI Convention, San Francisco, Apr. 1974.
29. Subcommittee 7, ACI Committee 435, Subcommittee Chairmen, R. S. Fling and A. F. Shaikh, Committee 435 Chairmen, B. L. Meyers and G. M. Sabnis, Deflections of Continuous Concrete Beams, *ACI Journal,* Proceedings **70,** 12, 781–787, Dec. 1973.
30. J. D. Zuraski, A. F. Shaikh, and C. G. Salmon, Calculation of Instantaneous Deflections of Reinforced Concrete Beams, *ACI Committee 435 Symposium Paper,* ACI Convention, San Francisco, Apr. 1974.
31. G. A. Maney, Relation Between Deformations and Deflections in Reinforced Concrete Beams, *ASTM Proceedings,* **14,** part II, 1914.
32. G. F. Swain, "Structural Engineering, Strength of Materials," McGraw-Hill Book Co., New York, 1924.
33. ACI Committee 307, Chairman, T. D. Myrlea, Deflection of Reinforced Concrete Members: Progress Report, *ACI Journal,* Proceedings **27,** 351, 1931.
34. V. E. Murashev, Theory of Appearance and Opening of Cracks, Computation of Rigidity of Reinforced Concrete Members, *Stroitelnaya Promishlenost* (*Moscow*), 11, 1940.
35. M. V. Pregnoff, Deflection of Concrete Beams, unpublished draft of report to ACI Committee 435, 1–92, Feb. 1962.
36. Portland Cement Association, Deflection of Reinforced Concrete Members, *Bulletin* ST-70, PCA, 1947.
37. C. W. Dunham, "The Theory and Practice of Reinforced Concrete," 3rd edn, 1–499, McGraw-Hill Book Co., New York, 1953.
38. AASHTO, "Standard Specifications for Highway Bridges," 341 National Press Building, Washington, D.C., 20004, 1957, 1961, 1965, 1969, and 1973.
39. CRSI, "Design Handbook," Concrete Reinforcing Steel Institute, Chicago, revised 1959, 1961.
40. W. W. Yu and G. Winter, Instantaneous and Long-Time Deflections of Reinforced Concrete Beams Under Working Loads, *ACI Journal,* Proceedings **57,** 1, 29–50, July 1960.
41. F. Levi, Work of the European Concrete Committee, *ACI Journal,* Proceedings **57,** 9, 1041–1070, Mar. 1961.
42. American Concrete Institute, "Building Code Requirements for Reinforced Concrete," and Commentary, American Concrete Institute, Detroit, Michigan, 1963.

43. N. M. Newmark, Numerical Procedure for Computing Deflections, Moments, and Buckling Loads, *ASCE Transactions,* **1–8,** 1161–1234, 1943.
44. G. W. Washa and P. G. Fluck, The Effect of Compressive Reinforcement on the Plastic Flow of Reinforced Concrete Beams, *ACI Journal,* Proceedings **49,** 8, 89–108, Oct. 1952.
45. G. W. Washa and P. G. Fluck, Plastic Flow (Creep) of Reinforced Concrete Continuous Beams, *ACI Journal,* Proceedings **52,** 5, 549–561, Jan. 1956.
46. N. H. Burns and C. P. Siess, Repeated and Reverse Loading in Reinforced Concrete, *ASCE Proceedings,* **92,** ST5, 65–78, Oct. 1966.
47. Bureau of Public Roads, "Reinforced Concrete Bridge Members, Strength and Serviceability Criteria, Ultimate Design," 1–81, U.S. Department of Commerce, Bureau of Public Roads, Aug. 1966.
48. J. C. Maldague, Détermination Expérimental des Lois Moments—Courbures des Poutres en Béton Armé, 517–628, *Annales de l'Institut Technique du Bâtiment et des Travaux Publics,* 9 Rue La Perouse, Paris 16, May 1965.
49. W. G. Corley and M. A. Sozen, Time-Dependent Deflections of Reinforced Concrete Beams, *ACI Journal,* Proceedings **63,** 3, 373–386, Mar. 1966.
50. A. W. Beeby, Short-Term Deformations of Reinforced Concrete Members, *Cement and Concrete Association, Technical Report* TRA 408, London, Mar. 1968.
51. Commission IV, CEB, New 1968 Version Worked Out by Commission IVb, "Deformations," *Portland Cement Association, Foreign Literature Study* 547, 1968.
52. I. J. Brakel, Calculation and Limitation of Deflections, *Comité Européen du Béton, Bulletin d'Information* 90, 1-1 to 1-31, Apr. 1973.
53. ACI Committee 443, Chairman, W. G. Corley, Analysis and Design of Reinforced Concrete Bridge Structures, *ACI Journal,* Proceedings **71,** 4, 171–200, Apr. 1974.
54. R. S. Fling, Simplified Deflection Computations, *ACI Publication SP-43, Deflections of Concrete Structures,* 205–223, 1974.
55. W. M. Ruiz and G. Winter, Reinforced Concrete Beams Under Repeated Loads, *ASCE Proceedings,* **95,** ST6, 1189–1211, June 1969. Also Ph.D. dissertation, Cornell University, Sept. 1968.
56. J. J. Hogan, Explanation of Control of Deflections—Section 909, *Portland Cement Association, Bulletins* 63S-3-1 to 63S-3-14, Skokie, Ill., 1963.
57. D. P. Ray and A. H. Nilson, Tests of Continuous Concrete Beams Designed for Moment Redistribution, research report, Cornell University, 1962.
58. B. J. Dave, "Moment Redistribution Effects in Continuous Reinforced Concrete Beams Resulting From Time-Dependent Causes," Ph.D. Thesis, University of Iowa, 1968.
59. Dean Peabody, "The Design of Reinforced Concrete Structures," 2nd edn, 1–532, John Wiley & Sons, New York, 1936.
60. G. E. Large, "Basic Reinforced Concrete Design: Elastic and Creep," 2nd edn, 1–527, Ronald Press Co., New York, 1957.
61. P. M. Ferguson, discussion of A. L. Miller, Warping of Reinforced Concrete Due to Shrinkage, May 1958; in *ACI Journal,* Proceedings **54,** 6, 1393–1402, Dec. 1958.
62. W. G. Corley and M. A. Sozen, Time-Dependent Deflections of Reinforced Concrete Beams, *ACI Journal,* Proceedings **63,** 3, 373–386, Mar. 1966.
63. British Standards Institution, "Code of Practice for The Structural Use of Concrete, Part 1. Design, Materials, Workmanship," CP 110, 1–155, British Standards Institution, 2 Park Street, London, 1972.
64. A. W. Beeby, Modified Proposals for Controlling Deflections by Means of Ratios of Span to Effective Depth, *Cement and Concrete Association, Technical Report* SBN 7210 0816 X, London, 1–19, Apr. 1971.
65. I. M. Nemirovsky, "The Rigidity of Reinforced Concrete Construction Under Long Time Loading," Government Publication on Construction, Architecture and Structural Materials, Moscow, 1958.
66. B. P. Sinha, K. H. Gerstle, and L. G. Tulin, Response of Singly Reinforced Beams to Cyclic Loading, *ACI Journal,* Proceedings **61,** 8, 1021–1038, Aug. 1964.

67. G. L. Agrawal, L. G. Tulin, and K. H. Gerstle, Response of Doubly Reinforced Concrete Beams to Cyclic Loading, *ACI Journal,* Proceedings **62,** 7, 823–835, July 1965.
68. C. A. Miller and S. A. Guralnick, Reinforced Concrete Beams Subjected to Repeated Loads, *ASCE Proceedings,* **93,** ST 5, 67–84, Oct. 1967.
69. M. R. Hollington, A Series of Long-Term Tests to Investigate the Deflection of a Representative Precast Concrete Floor Component, *Cement and Concrete Association, London, Technical Report* TRA 442, 1–43, Apr. 1970.
70. A. W. Beeby, A Note on the Influence of Compression Steel on the Long-Term Deformations of Flexural Members, unpublished report, May 1973.
71. A. Pauw, "Time-Dependent Deformations of Concrete," 1–275, Missouri State Highway Department, Federal Highway Administration, University of Missouri, Sept. 1971.
72. D. E. Branson, Designing for Time-Dependent Effects in Reinforced and Prestressed Concrete Beams, *ASCE, Alabama Section Meeting, Tuscaloosa,* 1–17, Apr. 1961.
73. D. G. Alcock and A. Pauw, Controlled-Deflection Design Method for Reinforced Concrete Beams and Slabs, *ACI Journal,* Proceedings **59,** 5, 645–658, May 1962.
74. A. Pauw and B. L. Meyers, Effect of Creep and Shrinkage on the Behavior of Reinforced Concrete Members, *ACI Publication SP-9,* 129–158, 1964.
75. H. Mayer, Die Berechnung der Durchbiegung von Stahlbetonbauteilen, *Deutscher Ausschluss für Stahlbeton,* **194,** 1967.
76. ACI Committee 340, Chairman, F. Kramrisch, "Design Handbook—In Accordance with the Strength Design Method of ACI 318-71," vol. 1, 2nd edn, 2nd printing, 1–403, ACI Publication SP-17(73), 1974.
77. I. J. Brakel, The Moment of Inertia of the Transformed Reinforced Concrete Cross-Section, When the Neutral Axis is Not Coincident With the Centre of Gravity, *Comité Européen du Béton, Bulletin d'Information* 90, 7-1 to 7-6, Apr. 1973.
78. ACI Committee 317, Chairman, W. E. Kunze, "Reinforced Concrete Design Handbook—Working Stress Method," 3rd edn, 1–271, ACI Publication SP-3, 1965.
79. D. E. Branson and M. L. Christiason, Time-Dependent Concrete Properties Related to Design—Strength and Elastic Properties, Creep and Shrinkage (based on M.S. thesis, University of Iowa, Jan. 1970), in "Designing for Effects of Creep, Shrinkage and Temperature in Concrete Structures," 257–277, ACI Publication SP 27-13 (symposium volume), 1971.
80. D. E. Branson, Time-Dependent Effects in Composite Concrete Beams, *ACI Journal,* Proceedings **61,** 2, 213–230, Feb. 1964.
81. K. M. Kripanarayanan and D. E. Branson, Some Experimental Studies of Time-Dependent Deflections of Noncomposite and Composite Reinforced Concrete Beams, *ACI Publication SP 43-16, Deflections of Concrete Structures,* 409–419, 1974.
82. American Institute of Steel Construction, "Manual of Steel Construction," 7th edn, 1-1 to 7-12, American Institute of Steel Construction, New York, 1970.
83. Ira Hooper, Rapid Deflection Calculations, *American Iron and Steel Institute, New York, Contemporary Steel Design,* **2,** 1, 1–8.
84. I. M. Viest, R. S. Fountain, and R. C. Singleton, "Composite Construction in Steel and Concrete," 1–176, McGraw-Hill Book Co., New York, 1958.
85. A. W. Beeby, A Note on an Aspect of the Variability of Deflections, *Cement and Concrete Association, paper for publication,* 1–16, Apr. 1974.
86. Comité Européen du Béton—Fédération Internationale de la Pré-Contrainte, "International Recommendations for the Design and Construction of Concrete Structures," 1–80, Cement and Concrete Association, London, June 1970.
87. R. L. Humphrey and L. H. Losse, The Strength of Reinforced Concrete Beams, Results of Tests of 333 Beams, *U.S. Bureau of Standards, Technological Paper* 2, 1911.
88. C. Bach and O. Graf, Tests on Reinforced Concrete Beams to Obtain Relationship Between Angle Change and Bending Moment, *Deutscher Ausschuss für Eisenbeton,* **38,** 1917.
89. O. Faber, Plastic Yield, Shrinkage, and Other Problems of Concrete and Their Effect on Design, *Abstracts in Engineering,* **124,** 661, 1927.
90. O. Faber, Elasticity, Plasticity and Shrinkage, *Preliminary Publication of the First Congress of*

International Association for Bridge and Structural Engineering, Paris, 565–576, 1932.

91. H. J. Gilkey and G. C. Ernst, Report of Project Committee on the Use of High Elastic Limit Steel as Reinforcement for Concrete, Sustained Loading Tests on Slender Concrete Beams Reinforced with High Elastic Limit Steel, *Proceedings, Highway Research Board,* 81, 1935.
92. George W. Washa, Plastic Flow of Thin Reinforced Concrete Slabs, *ACI Journal,* Proceedings **44,** 3, 237–260, Nov. 1947.
93. K. Hajnal-Konyi, Tests on Beams with Sustained Loading, *Magazine of Concrete Research, Cement and Concrete Association,* **15,** 43, 3–14, Mar. 1963.
94. C. A. Miller and S. A. Guralnick, Creep Deflection of Reinforced Concrete Beams, *Proceedings, ASCE, Structural Division,* **96,** ST 12, 2625–2638, Dec. 1970.
95. Adrian Pauw, Time Dependent Deflections of a Box Girder Bridge, in "Designing for Effects of Creep, Shrinkage and Temperature in Concrete Structures," 141–158, ACI Publication SP 27-6 (symposium volume), 1971.
96. J. L. Heiman, Long-Term Deformations in the Tower Building, Australia Square, Sydney, *ACI Journal,* Proceedings **70,** 4, 277–284, Apr. 1973.
97. C. V. Clarke, A. M. Neville, and W. Houghton-Evans, Deflection—Problems and Treatment in Various Countries, *ACI Publication SP-43, Deflections of Concrete Structures,* 129–178, 1974.
98. A. M. Neville, "Creep of Concrete: Plain, Reinforced, and Prestressed," 1–622, North-Holland Publishing Company, Amsterdam, 1970.
99. J. L. Heiman, A Comparison of Measured and Calculated Deflections of Flexural Members in Four Reinforced Concrete Buildings, *ACI Publication SP-43, Deflections of Concrete Structures,* 515–545, 1974.
100. W. G. Plewes, Deflections—Is There a Problem? *Proceedings of Symposium, ACI Canadian Capital Chapter, Montreal, Canada,* 5–14, Oct. 1971.
101. H. Mayer and H. Reusch, "Building Damage Caused by Deflection of Reinforced Concrete Building Components," *Deutscher Ausschuss für Stahlbeton,* 193, 1967; trans. J. H. Rainer, National Research Council of Canada, *Technical Translation* 1412, 1–115, Ottawa, 1970.
102. Subcommittee 1, ACI Committee 435, Subcommittee Chairman, R. S. Fling, Committee 435 Chairman, D. E. Branson, Allowable Deflections, *ACI Journal,* Proceedings **65,** 6, 433–444, June 1968.
103. Cement and Concrete Association, "Handbook on the Unified Code for Structural Concrete," 1–153, Cement and Concrete Association, London, 1972.
104. R. S. Fling, Deflections, Chap. 2 in M. Fintel (ed.), "Handbook of Concrete Engineering," 44–54, Van Nostrand Reinhold Co., New York, 1974.
105. Building Code Subcommittee, ACI Committee 435, Subcommittee Chairman, D. E. Branson, Committee 435 Chairman, G. M. Sabnis, Revised ACI Code and Commentary Provision on Deflections, Section 9.5, *ACI Committee 435 Report,* submitted to ACI Committee 318, 1–26, Dec. 1975.

FIGURE D-1
Example of Newmark[43] numerical solution for computing deflections of simple beams using the fourth power effective moment of inertia at individual sections, $I_e(I_{eff})$, by Eq. (3-18).[5]

APPENDIX

D

EXAMPLES OF NEWMARK NUMERICAL SOLUTION FOR COMPUTING DEFLECTIONS USING I_{eff} AT INDIVIDUAL SECTIONS

Simple beam

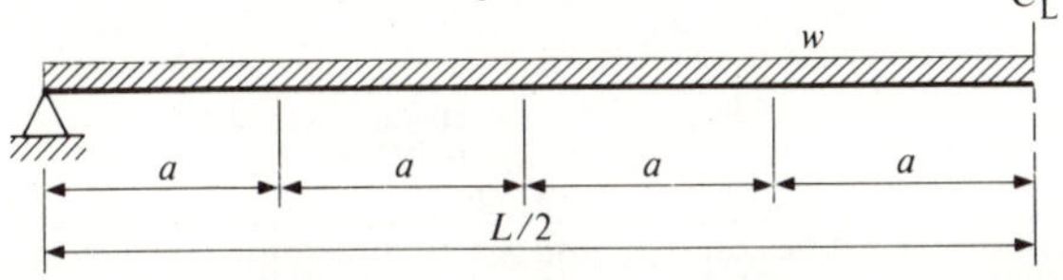

Description											Multiplier
Moment	M	0		0.4375		0.7500		0.9375		1.0000	$wL^2/8$
Curvature	ϕ	0		0.4375		0.7500		0.9375		1.0000	$wL^2/8EI$
*Equiv. conc. ϕ	Φ			5.1250		8.8750		11.1250		11.8750	$wL^2a/96EI$
Average slope	θ		31.0580		25.9330		17.0625		5.9375		$wL^2a/96EI$
Deflection	y	0		31.0580		56.9910		74.0535	↑ Start	79.9910	$wL^2a^2/96EI$

Exact answer: $\Delta = \dfrac{5wL^4}{384EI} = 80\,\dfrac{wL^2(L/8)^2}{96EI}$

(*a*) Example of ideal solution for constant *EI* beam under uniform load

	M	0		0.4375		0.7500		0.9375		1.0000	16 400 in. #
†	M_{cr}/M			1+ use I_g		0.732		0.586		0.549	—
‡	I_{eff}			41.7		25.0		21.0		20.3	in.^{4}
	ϕ			0.0105		0.0300		0.0446		0.0493	$16\,400/E_c$
*	Φ			0.135		0.355		0.525		0.582	$16\,400a/12E_c$
	θ		1.306		1.171		0.816		0.291		$16\,400a/12E_c$
	y	0		1.306		2.477		3.293	↑ Start	3.584	$16\,400a^2/12E_c$

$$\Delta = 3.584\,\frac{(16\,400)(9/8)^2(12)^2}{(12)(4.4)(10)^6} = 0.203\text{ in.}$$

versus 0.206 in. computed by the more approximate third power equation (3-19), and 0.153 in. determined experimentally. This example demonstrates the worst agreement between computed and measured deflections for the 58 experimental beams[5,44,45] (34 separate evaluations after averaging). The best agreement is demonstrated in Fig. D-2.

$f'_c = 5130$ psi, $E_c = 4.4 \times 10^6$ psi, $L = 9$ ft, $I_g = 41.7$ in.^{4}, $I_{cr} = 18.2$ in.^{4}

(*b*) Example of solution in which cracking is considered

* $\Phi_1 = (a/12)(\phi_0 + 10\phi_1 + \phi_2)$, as $\Phi_{5.125} = (a/12)[0 + (10)(0.4375) + 0.7500] = (a/12)(5.1250)$

† $M_{cr} = f_r I_g/(h/2), f_r = 7.5\sqrt{f'_c}$

‡ $I_{\text{eff}} = (M_{cr}/M)^4 I_g + [1 - (M_{cr}/M)^4] I_{cr}$ (Eq. (3-18))

Two-span continuous beam

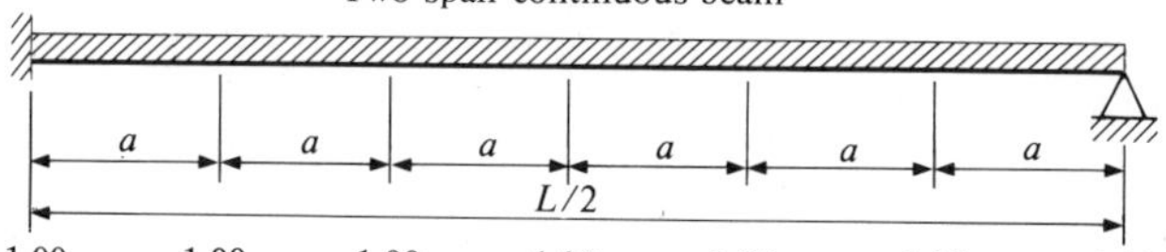

Description		0		1		2		3		4		5		6	Multiplier
Total distr. load		1.00		1.00		1.00		1.00		1.00		1.00		1.00	135.2#/ft
†Equiv. conc. load	Q	—		1.00		1.00		1.00		1.00		1.00		—	$135.2a$
*Assumed avg. sh.	V		3.25		2.25		1.25		Start 0.25		−0.75		−1.75		$135.2a$
← Follow procedure below →															
Trial deflection	y_t	0		0.878		2.254		3.366		3.863		3.757		*3.180	$135.2a^4/12E_c$

*The shear distribution must be adjusted until the required linear deflection corrections are zero (or small), in which case the deflection curve has been determined. Also, the effects of redistribution due to cracking in statically indeterminate beams are included in this procedure. The following is the final cycle:

Description		0		1		2		3		4		5		6	Multiplier
From above	Q	—		1.00		1.00		1.00	Start	1.00		1.00		—	$135.2a$
*New	V		3.16		2.16		1.16		0.16		−0.84		−1.84		$135.2a$
Moment	M	−3.96		−0.80		1.36		2.52		2.68		1.84		0	$135.2a^2$
	‡ M_{cr}/M	0.623		1+ use Ig		1+		0.979		0.920		1+		—	—
	§ I_{eff}	21.7		41.7		41.7		39.8		35.0		41.7		—	in.^{4}
Curvature	ϕ	0.1825		0.0192		−0.0326		−0.0633		−0.0766		−0.0441		0	$135.2a^2/E_c$
¶ Equiv. conc. ϕ, Φ		0.713		0.342		−0.370		−0.742		−0.873		−0.518		—	$135.2a^3/12E_c$
Average slope	θ	0	−0.713		−1.055		−0.685		0.057		0.930		1.448		$135.2a^3/12E_c$
Trial defl.	y_t	0		0.713		1.768		2.453		2.396		1.466		0.018	$135.2a^4/12E_c$
Linear correction		0		−0.003		−0.006		−0.009		−0.012		−0.015		−0.018	$135.2a^4/12E_c$
Deflection	y	0		0.710		1.762		2.444		2.384		1.451		0	$135.2a^4/12E_c$

$$\Delta = 2.444\,\frac{(135.2)(9/6)^4(12)^3}{(12)(4.4)(10)^6} = 0.0548\text{ in.}$$ versus 0.0550 in. computed by the more approximate third power equation (3-19), with an average of values weighted ($\frac{2}{3}$) for positive and ($\frac{1}{3}$) for negative moment regions, and 0.056 in. determined experimentally. This example demonstrates the best agreement between computed and measured deflections for the 58 experimental beams[5,44,45] (34 separate evaluations after averaging). The worst agreement is demonstrated in Fig. D-1.

$f'_c = 5130$ psi, $E_c = 4.4 \times 10^6$ psi, $L = 9$ ft, $I_g = 41.7$ in.^{4}, neg. $I_{cr} = 16.4$ in.^{4}, pos. $I_{cr} = 9.2$ in.^{4}

(For metric units, see conversion factors herein.)

†Values of Q can also be easily computed for other than uniform loading.

‡$M_{cr} = f_r I_g/(h/2),\ f_r = 7.5\sqrt{f'_c}$

§ $I_e = I_{eff} = (M_{cr}/M)^4 I_g + [1 - (M_{cr}/M)^4] I_{cr}$ (Eq. (3-18))

¶ $\Phi_0 = \frac{a}{12}(\frac{7}{2}\phi_0 + \frac{6}{2}\phi_1 - \frac{1}{2}\phi_2)$, as $\Phi_{0.713} = \frac{a}{12}[\frac{7}{2}(0.1825 + \frac{6}{2}(0.0192) - \frac{1}{2}(-0.0326)] = \frac{a}{12}(0.713)$

$\Phi_3 = \frac{a}{12}(\phi_2 + 10\phi_3 + \phi_4)$, as $\Phi_{0.742} = \frac{a}{12}[0.0326 + (10)(0.0633) + 0.0766] = \frac{a}{12}(0.742)$

FIGURE D-2
Example of Newmark[43] numerical solution for computing deflections of two-span continuous beams using the fourth power effective moment of inertia at individual sections, $I_e(I_{eff})$, by Eq. (3-18). Effects of moment redistribution due to cracking are incorporated in the numerical solution.[5]

APPENDIX

E

CHARTS AND EXAMPLES FOR DETERMINING THE GROSS MOMENT OF INERTIA, I_g, AND THE CRACKED TRANSFORMED MOMENT OF INERTIA, I_{cr}

The charts shown in Figs. E-1 and E-2 for I_g and I_{cr}, and the following examples, were presented by Lutz.[23] He also developed curves for M_{cr} and I_e, although this author believes their direct calculation is easier and preferable. Charts for I_g and I_{cr} are also available in Refs. 76 and 104 and other handbooks.

Example 1

Determine I_g and I_{cr} for a rectangular beam, $b = 14$ in. (35.6 cm), $h = 24$ in. (61.0 cm), $d = 21.5$ in. (54.6 cm), three #10 bars, $A_s = 3.81$ in^2 (24.6 cm^2), $n = 8$.

$$b/b_w = 1$$

From Fig. E-1, find $\quad I_g = \underline{0.0833bh^3}$

$$\rho = 3.81/(14)(21.5) = 0.01266 \qquad \rho n = (0.01266)(8) = 0.1013$$

$$\rho n + \rho' n' = 0.1013 \qquad \rho' n'(1 - d'/d) = 0 \qquad \text{(no-compression steel)}$$

From Fig. E-2, find $\quad I_{cr} = \underline{0.057bd^3}$

Example 2

Determine I_g and I_{cr} for a T-beam, $b = 50$ in. (127 cm), $b_w = 12$ in. (30.5 cm), $h = 20$ in. (50.8 cm), $h_f = 4$ in. (10.2 cm), $d = 17.5$ in. (44.5 cm), three #11 bars, $A_s = 4.68$ in^2 (30.2 cm^2), $n = 9$, $kd' = 4.70$ in. > 4 in. (hence T-beam solution).

$$b/b_w = 50/12 = 4.17 \qquad h_f/h = 4/20 = 0.20$$

From Fig. E-1, find $\quad I_g = \underline{0.0355bh^3}$

$$\rho = 4.68/(50)(17.5) = 0.00535 \qquad \rho n = (0.00535)(9) = 0.0481$$

$$h_f/d = 4/17.5 = 0.229$$

Using h_f/d and ρn in Fig. E-2, find $\quad I_{cr} = \underline{0.032bd^3}$

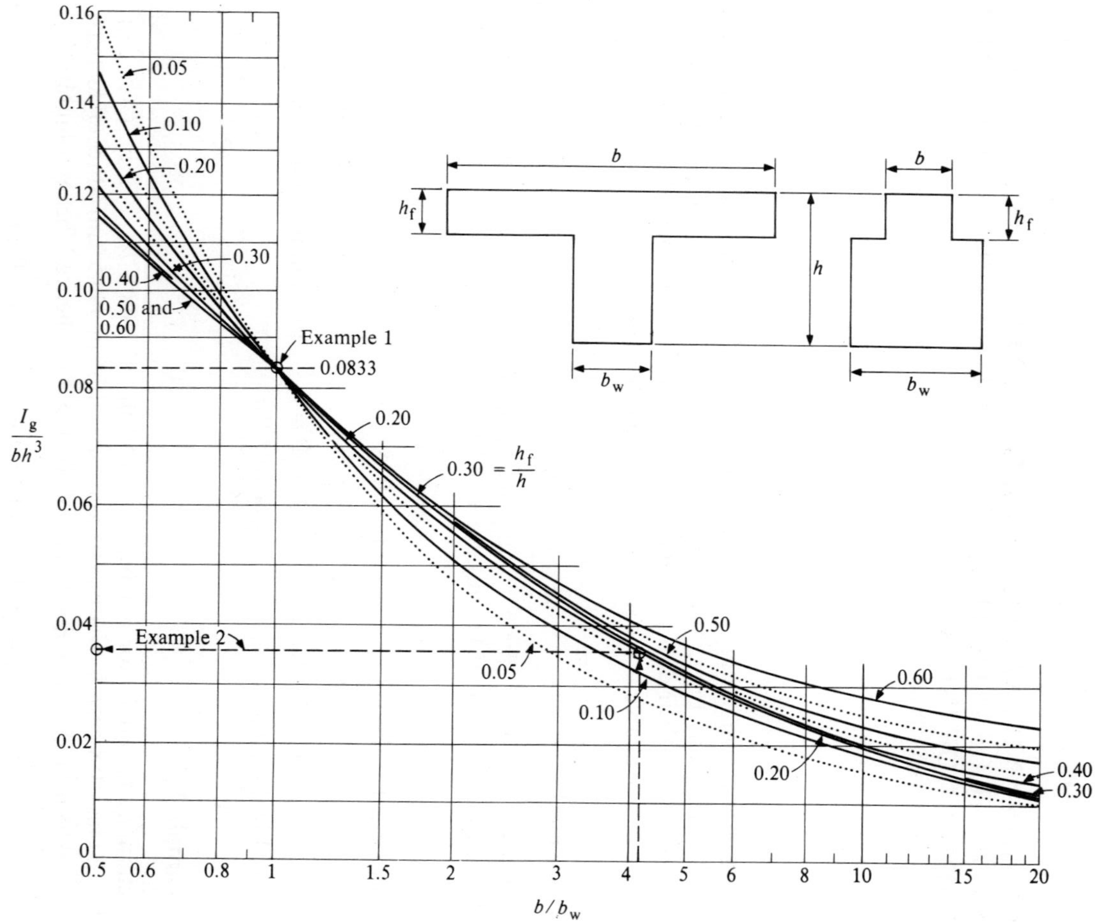
0.05
0.10
0.20
0.30
0.40
0.50 and
0.60
Example 1
0.0833
0.20
0.30 = $\frac{h_f}{h}$
0.50
0.60
0.40
0.30
0.20
0.10
0.05
Example 2
$\frac{I_g}{bh^3}$
b/b_w
0.16
0.14
0.12
0.10
0.08
0.06
0.04
0.02
0
0.5
0.6
0.8
1
1.5
2
3
4
5
6
8
10
15
20
b
h_f
b_w
h

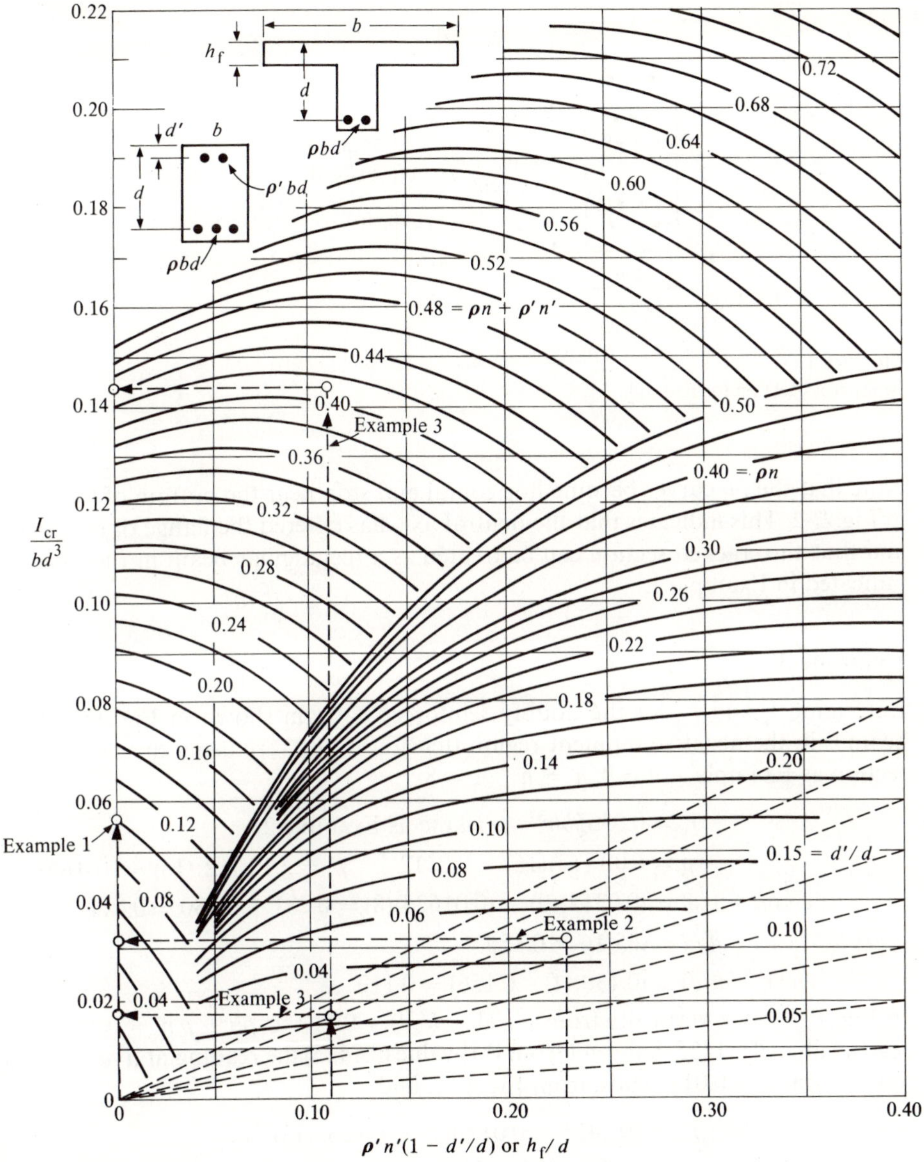

FIGURE E-2
Cracked section moment of inertia, I_{cr}, by Lutz. (*From ACI Journal,*[23] *with permission.*)

FIGURE E-1
Gross section moment of inertia, I_g, by Lutz. (*From ACI Journal,*[23] *with permission.*)

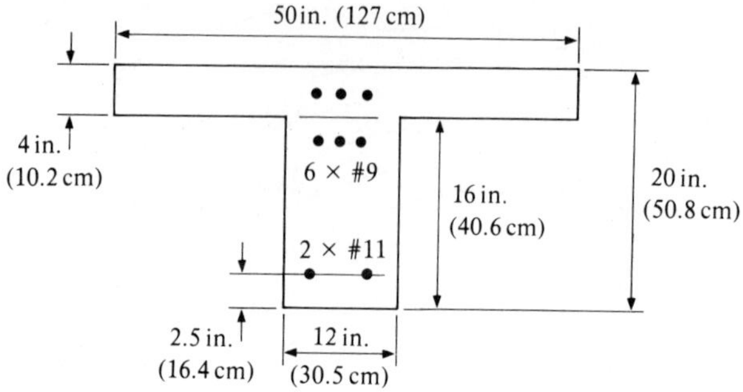

FIGURE E-3
Beam cross-section for Example 3.

Note that the ρn curves become horizontal and stop near the bottom of the chart in Fig. E-2. This indicates that the neutral axis has entered the flange of the beam, and thus the cracked section can be treated as a rectangular beam in the manner indicated in Example 1.

Example 3

Determine I_g and I_{cr} for the doubly reinforced T-beam shown in Fig. E-3. This beam is in the negative moment region (tension in the top) using the same gross section as Example 2. $n = 9$, $n' = n - 1 = 8$.

$$I_g = \underline{0.0355bh^3} \qquad \text{(same as Example 2)}$$

$$\rho = 6.00/(12)(16) = 0.0312 \qquad \rho' = 3.12/(12)(16) = 0.01625$$

$$\rho n + \rho' n' = (0.0312)(9) + (0.01625)(8) = 0.281 + 0.130 = 0.411$$

$$d'/d = 2.5/16 = 0.156$$

$$\rho' n'(1 - d'/d) = (0.130)(1 - 0.156) = 0.110$$

In Fig. E-2 move vertically from $\rho' n'(1 - d'/d) = 0.110$ to $\rho n + \rho' n' = 0.411$, and read a value of 0.1435. Subtract from this value (0.1435) the reading at $d'/d = 0.156$ (dashed lines) of 0.017. Thus, from Fig. E-2, find

$$I_{cr} = (0.1435 - 0.017)bd^3 = 0.1265(12)(16)^3$$

$$= \underline{6220 \text{ in}^4 \ (259\,000 \text{ cm}^4)}$$

Additional examples, in which comparisons between theoretical and chart values are shown, may be seen in Table 3-9.

APPENDIX
F

ACI CODE (ACI 318-71) MOMENT COEFFICIENTS

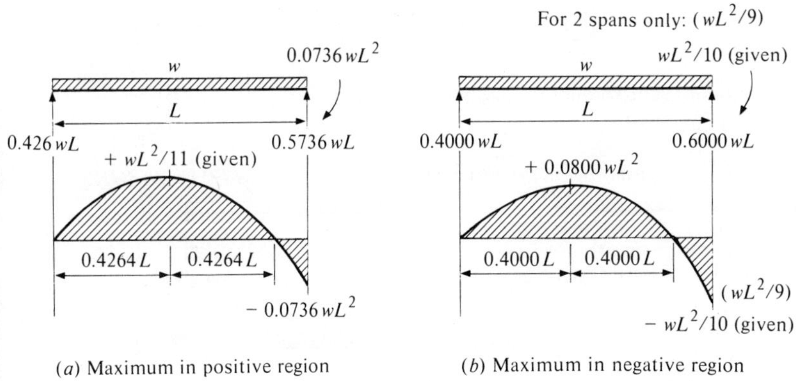

FIGURE F-1
Exterior span with discontinuous end unrestrained.

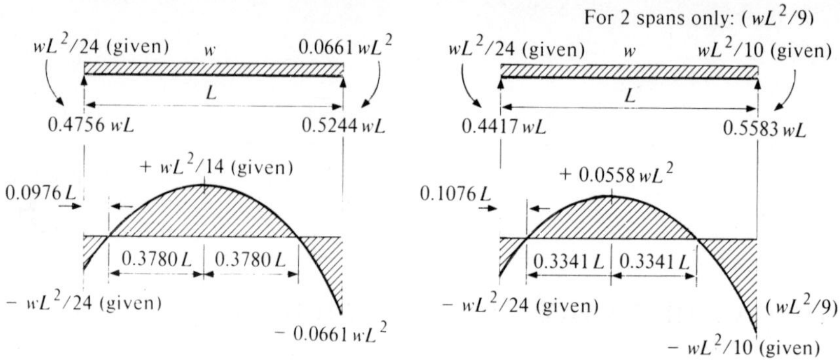

(*a*) Maximum in positive region (*b*) Maximum in negative region

FIGURE F-2
Exterior span with exterior support built integrally with spandrel beam or girder.

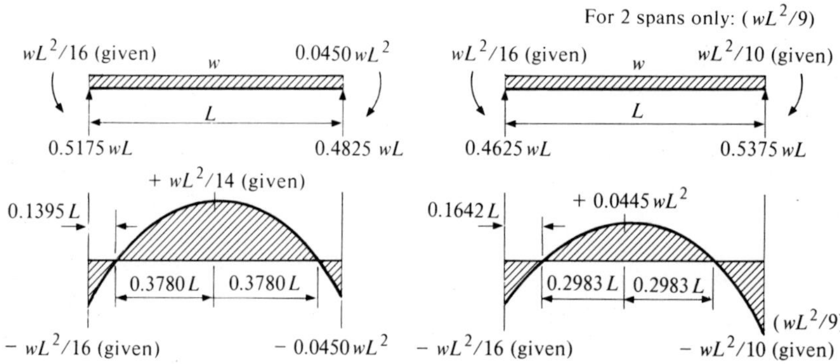

(*a*) Maximum in positive region (*b*) Maximum in negative region

FIGURE F-3
Exterior span with exterior support built integrally with column.

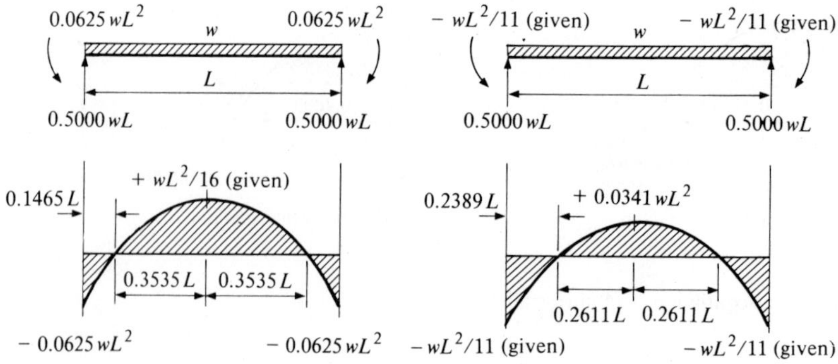

(*a*) Maximum in positive region (*b*) Maximum in negative region

FIGURE F-4
Interior span.

APPENDIX
G

EXPERIMENTAL RESULTS FOR INITIAL OR SHORT-TIME DEFLECTIONS

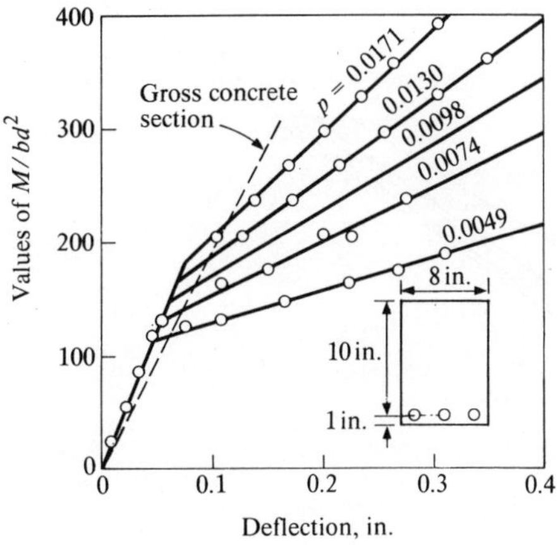

FIGURE G-1
Deflection tests by Humphrey and Losse[87] (1 in. = 2.54 cm).

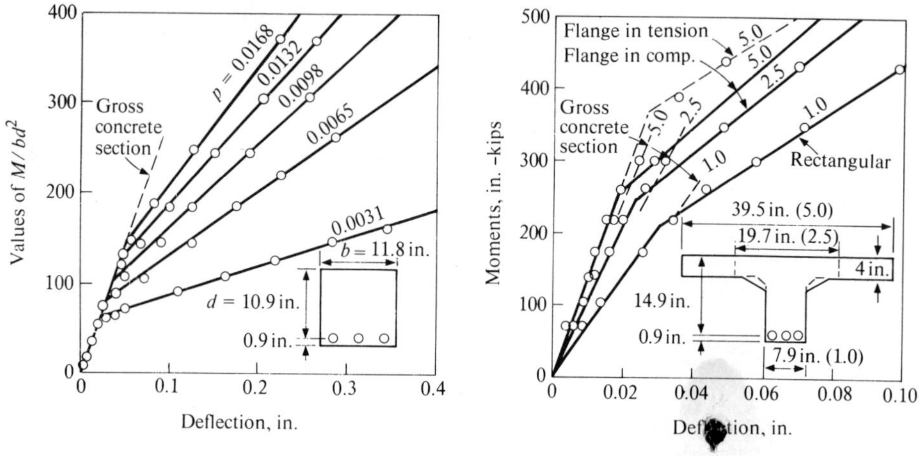

FIGURE G-2
Deflection tests by Bach and Graf[88] (1 in. = 2.54 cm).

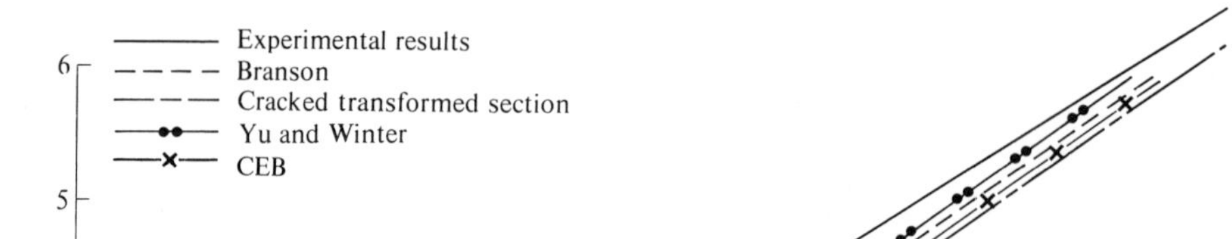

FIGURE G-3
Experimental load–deflection curve for beam with 0.93 percent reinforcement compared with deflections calculated by various methods—typical results as presented by Beeby. (*From Cement and Concrete Association,*[50] *with permission.*)

APPENDIX
H

DERIVATION OF MIDSPAN DEFLECTIONS FOR CONTINUOUS BEAMS WITH UNIFORM LOADING OR A CENTER CONCENTRATED LOAD

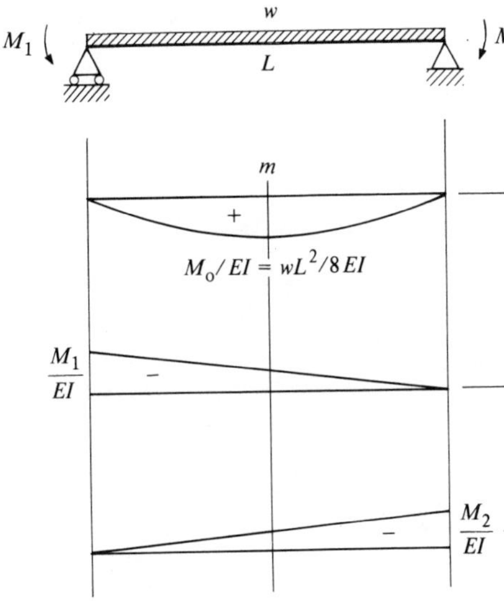

Using the conjugate beam method:

$$\Delta_m = \frac{5wL^4}{384EI} = \frac{5M_oL^2}{48EI}$$

$$\Delta_m = \frac{1}{EI}\left[\frac{M_1 L}{6}\frac{L}{2} - \frac{M_1}{2}\frac{L}{4}\frac{L}{6}\right]$$

$$= M_1 L^2/16EI$$

Similarly, $\Delta_m = M_2L^2/16EI$

FIGURE H-1
M/EI diagrams and midspan deflections for uniformly loaded continuous beams.

From Fig. H-1,

$$\Delta_m = \frac{5M_oL^2}{48EI} - (M_1 + M_2)\frac{L^2}{16EI} = \frac{L^2}{48EI}[5M_o - 3(M_1 + M_2)]$$

Since the net midspan moment $M_m = M_o - (1/2)(M_1 + M_2)$,

$$\Delta_m = \frac{L^2}{48EI}[5M_m + (5/2)(M_1 + M_2) - 3(M_1 + M_2)]$$

$$= \frac{5L^2}{48EI}[M_m - (1/10)(M_1 + M_2)]$$

which is Eq. (3-37) in the text.

Since $M_m = M_o - (1/2)(M_1 + M_2)$, $M_1 + M_2 = 2(M_o - M_m)$. Hence,

$$\Delta_m = \frac{5L^2}{48EI}[M_m - (1/5)(M_o - M_m)] = \frac{5L^2}{48EI}[1.20M_m - 0.20M_o]$$

$$= \frac{5L^2}{48EI}M_m[1.20 - 0.20M_o/M_m] = K(5/48)M_mL^2/EI$$

where $K = 1.20 - 0.20M_o/M_m$. This is Eq. (3-40) in the text.

Example of beam fixed at both ends:

$$M_o = wL^2/8 \qquad M_m = wL^2/24$$
$$K = 1.20 - 0.20(24/8) = 0.60$$
$$\Delta_m = (0.60)(5/48)(wL^2/24)L^2/EI = wL^4/384EI$$

From Fig. H-2,

$$\Delta_m = \frac{M_oL^2}{12EI} - (M_1 + M_2)\frac{L^2}{16EI} = \frac{L^2}{48EI}[4M_o - 3(M_1 + M_2)]$$

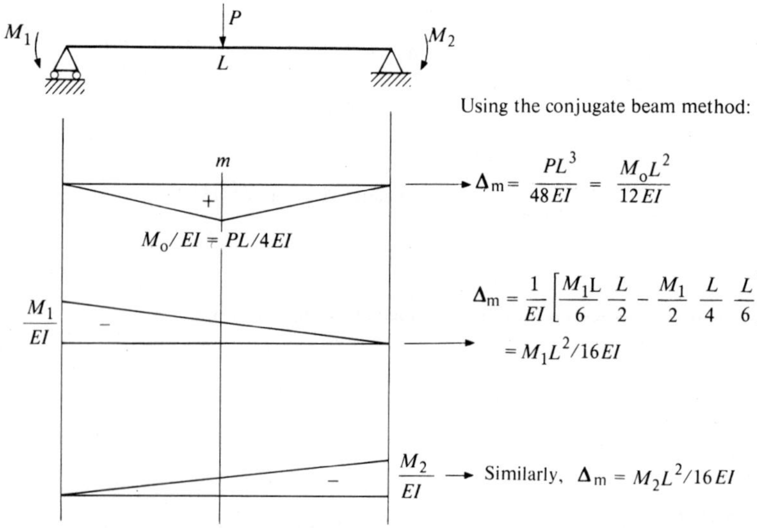

FIGURE H-2.
M/EI diagrams and midspan deflections for continuous beams with a center-concentrated load.

Since the net midspan moment, $M_m = M_o - (1/2)(M_1 + M_2)$,

$$\Delta_m = \frac{L^2}{48EI}[4M_m + (4/2)(M_1 + M_2) - 3(M_1 + M_2)]$$

$$= \frac{L^2}{48EI}[4M_m - (M_1 + M_2)]$$

which is Eq. (3-38) in the text.

Since $M_m = M_o - (1/2)(M_1 + M_2)$, $M_1 + M_2 = 2(M_o - M_m)$. Hence,

$$\Delta_m = \frac{L^2}{48EI}[4M_m - 2(M_o - M_m)] = \frac{L^2}{48EI}[6M_m - 2M_o]$$

$$= \frac{5L^2}{48EI}M_m[1.20 - 0.40M_o/M_m] = K(5/48)M_m L^2/EI$$

where $K = 1.20 - 0.40M_o/M_m$. This is Eq. (3-41) in the text.

Example of beam fixed at both ends:

$$M_o = PL/4 \qquad M_m = PL/8$$

$$K = 1.20 - 0.40(8/4) = 0.40$$

$$\Delta_m = (0.40)(5/48)(PL/8)L^2/EI = PL^3/192EI$$

APPENDIX
I

MIDSPAN MOMENTS, MAXIMUM POSITIVE MOMENTS, END MOMENTS; AND MIDSPAN DEFLECTIONS, MAXIMUM DEFLECTIONS FOR IDEAL CONDITIONS

Cantilever beam, uniform load

$$M_e = wL^2/2 = 0.500\,wL^2$$

$$\Delta_{max} = \frac{wL^4}{8EI} = \frac{M_eL^2}{4EI} = 0.1250\,wL^4/EI$$

Cantilever beam, end-concentrated load

$$M_e = PL = 1.0000\,PL$$

$$\Delta_{max} = \frac{PL^3}{3EI} = \frac{M_eL^2}{3EI} = 0.3333\,PL^3/EI$$

Simple beam, uniform load

$$^{+}M_{max} = M_m = wL^2/8 = 0.1250\,wL^2$$

$$\Delta_{max.} = \Delta_m = \frac{5\,wL^4}{384\,EI} = \frac{5\,M_mL^2}{48\,EI} = 0.01302\,wL^4/EI$$

Simple beam, center-concentrated load

$$^{+}M_{max} = M_m = PL/4 = 0.2500\,PL$$

$$\Delta_{max} = \Delta_m = \frac{PL^3}{48\,EI} = \frac{M_mL^2}{12\,EI} = 0.02083\,PL^3/EI$$

Fixed-hinged beam, uniform load

$$M_e = wL^2/8 = 0.1250\,wL^2$$

$$M_m = wL^2/16 = 0.06250\,wL^2$$

$$^{+}M_{max} = 9\,wL^2/128 = 0.07031\,wL^2$$

$$\Delta_m = \frac{wL^4}{192\,EI} = \frac{M_mL^2}{12\,EI} = 0.005208\,wL^4/EI$$

$$\Delta_{max} = \frac{wL^4}{185\,EI} = 0.005405\,wL^4/EI$$

Fixed-hinged beam, center-concentrated load

$$M_e = 3\,PL/16 = 0.1875\,PL$$

$$^{+}M_{max} = M_m = 5\,PL/32 = 0.1563\,PL$$

$$\Delta_m = \frac{7\,PL^3}{768\,EI} = \frac{56\,M_mL^2}{960\,EI} = 0.009115\ PL^3/EI$$

$$\Delta_{max} = \frac{5\,PL^3}{240\,EI} = 0.009317\,PL^3/EI$$

Fixed-fixed beam, uniform load

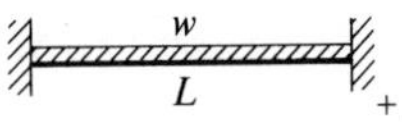

$$M_e = wL^2/12 = 0.08333\,wL^2$$

$$^+M_{max} = M_m = wL^2/24 = 0.04167\,wL^2$$

$$\Delta_{max} = \Delta_m = \frac{wL^4}{384\,EI} = \frac{M_m L^2}{16\,EI} = 0.002604\,wL^4/EI$$

Fixed-fixed beam, center-concentrated load

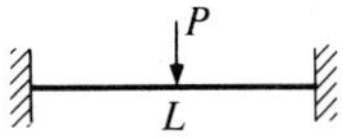

$$M_e = {}^+M_{max} = M_m = PL/8 = 0.1250\,PL$$

$$\Delta_{max} = \Delta_m = \frac{PL^3}{192\,EI} = \frac{M_m L^2}{24\,EI} = 0.005208\,PL^3/EI$$

FIGURE I-1

Midspan moments (M_m), maximum positive moments ($^+M_{max}$), end moments (M_e); and midspan deflections (Δ_m), maximum deflections (Δ_{max}) for ideal conditions.

APPENDIX

J

MULTIPLIERS FOR ADDITIONAL LONG-TIME DEFLECTIONS

Table J-1 Multipler, T_u*, for additional long-time deflections due to creep and shrinkage for both normal-weight and lightweight concrete members of common types, sizes, and composition[6]

	Average relative humidity, age when loaded								
	100%			70%			50%†		
Concrete strength, f'_c, at age 28 days	≤7d	14d	≥28d	≤7d	14d	≥28d	≤7d	14d	≥28d
2500 to 4000 psi (176 to 281 kg/cm^2) (17.2 to 27.6 N/mm^2)	2.0	1.5	1.0	3.0	2.0	1.5	4.0	3.0	2.0
>4000 psi (>281 kg/cm^2) (>27.6 N/mm^2)	1.5	1.0	0.7	2.5	1.8	1.2	3.5	2.5	1.5

* For the period: 1 month or less — 3 months — 1 year — 5 years or more
use: 25% — 50% — 75% — 100% of table values

† The 50% humidity values may normally be used for lower relative humidities, such as in heated buildings for example.

CHAPTER

FOUR

DEFLECTION OF TWO-WAY REINFORCED CONCRETE SLAB SYSTEMS

4-1 INTRODUCTION

The analytical determination of the deflection of two-way reinforced concrete slab systems is complicated by the three-dimensional nature of the problem; the effect of the flexural and torsional rigidities of supporting beam and/or slab elements; the stiffening effect of columns, drop panels, and column capitals; the aspect ratio and continuity of the panels; cracking of the concrete; and the time-dependent concrete deformation. As a result of these complexities, in addition to the usual uncertainties regarding loading conditions and material properties, great accuracy in the calculations is normally not warranted. The procedures described in this chapter for computing initial and time-dependent slab deflections range from simple estimates to reasonably accurate approximate methods.

Review of Approaches for Computing Deflections

A number of classical and contemporary methods for calculating slab deflections were reviewed in some detail by ACI Committee 435[1] in 1974. The following is a brief summary of the principal approaches that have been used for computing short-time deflections of slabs.

Classical methods of elasticity The elastic solution of slab or plate deflections has been presented by Timoshenko and Woinowsky-Krieger[2] and others, primarily for ideal boundary conditions and full panel loading. Such results can be useful in many cases for uncracked slabs and also when estimating the effect of cracking on the slab flexural rigidity. Various modifications of the theoretical deflection coefficients have also been suggested.[3-6]

Crossing beam analogy In 1904 Marsh[7] proposed a simple method of crossing beams (grid) with a uniform load divided between the short-direction (A) and the long-direction (B) beams according to the proportion given by Eq. (4-1).

$$\begin{aligned} &\text{Short-dir. load} = qr_s \qquad \text{where} \quad r_s = \frac{B^4}{A^4 + B^4} \\ &\text{Long-dir. load} = (1 - r_s)q \end{aligned} \tag{4-1}$$

However, this method did not take into account the torsional moments in the slab, nor the effect of yielding supports. Marcus[8] modified the method to approximate the effect of the torsional moments. A variation of the method by the PCA[4] that somewhat improves the results for interior panel flat plates is to consider a beam of unit width with a span equal to the diagonal length of the slab.

Analogous gridwork method The analogous gridwork method presented by Ewell, Okubo, and Abrams[9] in 1952 is similar to the crossing beam method but includes the effect of torsional moments directly. The slab is replaced by a small number of intersecting beams (such as three or four per panel in both directions) having the equivalent flexural and torsional stiffnesses of the slab. An analytical solution is then obtained for the deflection of the grid system using a displacement method of analysis in most cases.

Statics ratio method The statics ratio method represents another variation of the crossing beam analogy as suggested by Furr[10] in 1959. It makes use of an assumed deflection configuration and statically equivalent concentrated loads at the beam strip intersection points. The assumed deflection curve is then modified successively until the requirements of statics are met.

Wide beam and equivalent frame methods In this approach a system of wide beams or equivalent frames is used to determine the midpanel deflection as the sum of the midspan column line deflection in one direction and the deflection at midspan of the middle line between columns in the other direction. The equivalent stiffnesses and distribution factors for such a system can be determined by the method of the ACI Code (318-71)[11] for moment design, for example. Contributors to this procedure from 1948 to date include Peabody,[12] Corley,[13,15] Sozen,[3,13,14] Siess,[3,13,14] Vanderbilt,[3,5] Jirsa,[15] Fling,[16] Nawy,[17] Nilson and Walters,[18] and others.

Finite difference and finite element methods The finite difference and finite element methods have been used for the general analysis of slab systems[19,20,21] and in the evaluation of slab deflections.[3,5,22] Such numerical solutions are extremely valuable especially for complex problems and for experimental verification.

A number of procedures for computing initial and time-dependent deflections of slabs without and with beams, drop panels, and column capitals (flat plates, flat slabs, two-way slabs) is presented herein, together with numerical examples and comparisons between methods. The procedure for calculating deflections of flat slabs may also be applied to waffle slabs. The effect of partial cracking on deflections is included. A discussion of experimental results for both laboratory and actual structures is also presented.

4-2 CALCULATION PROCEDURES

Short-Time Deflections

Method of deflection coefficients The deflection of plates or slabs may be computed by Eqs. (4-2) to (4-5) using deflection coefficients. This approach is probably the simplest, and may be useful in many cases for estimating deflections. When the deflection coefficients are accurately defined for the conditions of a particular structure, the computed deflections are, of course, of comparable accuracy.

$$\Delta = K_s q L^4/D \tag{4-2}$$

where K_s = slab deflection coefficient, q = uniformily distributed loading, L = longer span, D = flexural rigidity = $Eh^3/12(1 - v^2)$, h = slab thickness and v = Poisson's ratio.

Assuming $v = 0.2$ for concrete, Eq. (4-2) reduces to Eq. (4-3).

$$\Delta = 11.5 K_s q L^4/E_c h^3 \tag{4-3}$$

Neglecting the Poisson ratio effect ($v = 0$), Eq. (4-2) reduces to Eq. (4-4).

$$\Delta = 12 K_s q L^4/E_c h^3 \tag{4-4}$$

Equation (4-4) can also be written in the form of Eq. (4-5) for a unit width of slab.

$$\Delta = K_s q L^4/E_c I_e \tag{4-5}$$

where $I_e = I_g$ when the slab is uncracked and less than I_g when the slab is partially cracked. Equations (4-4) and (4-5) are conservative compared to Eq. (4-3).

The following solution of Eqs. (4-4) and (4-5) demonstrates one set of appropriate units for each term in these equations. Let $K_s = 0.005$, $q = 100$ psf (488 kg/m², 4788 N/m²), $L = 20$ ft (6.1 m), $E_c = 3.0 \times 10^6$ psi (211 000 kg/cm², 20.7 kN/mm²), and $h = 6$ in. (15.2 cm). By Eq. (4-4),

$$\Delta = \frac{12K_s q L^4}{E_c h^3} = \frac{(12)(0.005)(100 \text{ psf})(20 \text{ ft})^4(144 \text{ in}^2/\text{ft}^2)}{(3.0 \times 10^6 \text{ psi})(6 \text{ in.})^3}$$
$$= 0.21 \text{ in. } (0.53 \text{ cm})$$

By Eq. (4-5) for a unit width of slab,

$$\Delta = \frac{K_s q L^4}{E_c I_e} = \frac{(0.005)(100\ \text{lb/ft})(20\ \text{ft})^4(1728\ \text{in}^3/\text{ft}^3)}{(3.0 \times 10^6\ \text{psi})(216\ \text{in}^4)}$$

$$= 0.21\ \text{in.}\ (0.53\ \text{cm})$$

where $I_e = I_g = (12\ \text{in.})(6\ \text{in.})^3/12 = 216\ \text{in}^4\ (8990\ \text{cm}^4)$.

Theoretical coefficients from the method of elasticity The theoretical values of the slab deflection coefficients for ideal boundary conditions in Tables 4-1 and 4-2 were adapted from Timoshenko and Woinowsky-Krieger[2] and Szilard[23] when using the long span, L, in the deflection equation. These coefficients can often be useful in making simple estimates of deflections, similar to the following.

1. In Table 4-1 the case for all sides simply supported may approximate the conditions of a corner panel (maximum panel deflection) of a slab with beams (two-way slab), in which the effect of yielding supports (which would increase the panel deflection) tends to be offset by the effect of continuity on two sides (which would decrease the panel deflection). Example: for $L/S = 1.0$ and uniform loading; approximate K_s for a two-way slab corner panel $= 0.0041$.

2. In Table 4-1 the case of the single simply supported flat plate panel represents an absolute upper bound for deflections of a corner panel (maximum panel deflection) of a flat plate or flat slab (with drop panels and/or column capitals), since the restraining effect of the two continuous sides is neglected. Example: for $L/S = 1.0$ and uniform loading; absolute upper bound K_s for a flat plate or flat slab corner panel $= 0.0263$.

3. In Table 4-1 the case of all sides fixed represents an absolute lower bound for deflections of an interior panel for all types of slab systems. Example: for $L/S = 1.0$ and uniform loading; absolute lower bound K_s for an interior panel for all types of slab systems $= 0.0013$.

4. In Table 4-1 the case of the flat plate interior panel represents either the ideal conditions for an interior flat plate panel (without beams, drop panels or column capitals), or an upper bound for deflections of an interior panel with any stiffening elements (flat slab or two-way slab). Example: for $L/S = 1.0$ and uniform loading; ideal K_s for a flat plate interior panel = upper bound K_s for a flat slab or two-way slab interior panel $= 0.0058$.

5. In Table 4-2 the case of a single square panel supported elastically on all four sides represents an absolute upper bound for deflections of a square corner panel (maximum panel deflection) for all types of slab systems, since the restraining effect of the two continuous sides and the stiffening effect of any drop panels or column capitals are neglected. Note that the flat plate condition is included when $EI/LD = 0$. Examples: for $L/S = 1.0$, uniform loading, and $EI/LD = 0$; absolute upper bound K_s for a flat plate or flat slab corner panel $= 0.0257$. For $L/S = 1.0$, uniform loading, and $EI/LD = 4.0$, which corresponds to a typical case of $h = 6$ in. (15.2 cm), $L = 20$ ft (6.1 m), beams 14 in. (35.6 cm) by 25 in. (63.5 cm); absolute upper bound K_s for a two-way slab corner panel $= 0.0055$.

Table 4-1 Theoretical deflection coefficients, K_s, for ideal boundary conditions[2,23]

L/S*	Uniform loading, all sides simply supported, K_s for max Δ in Eqs. (4-2) to (4-5)	Uniform loading, all sides fixed, K_s for max Δ in Eqs. (4-2) to (4-5)	Uniform loading, flat plate,† K_s for max Δ in Eqs. (4-2) to (4-5)		Center-concentrated load, all sides simply supported, max $\Delta = K_s PL^2/D$	Center-concentrated load, all sides fixed, max $\Delta = K_s PL^2/D$
			Single isolated panel	Interior panel without checkerboard loading		
1.0	0.00406	0.001260	0.0263	0.00581	0.01160	0.00560
1.1	0.00331	0.001025	0.0222	0.00487	0.01045	—
1.2	0.00279	0.000850	0.0189	0.00428	0.00940	0.00449
1.3	0.00223	0.000669	0.0173	0.00387	—	—
1.4	0.001835	0.000539	0.0162	0.00358	0.00757	0.00353
1.5	0.001525	0.000435	0.0155	0.00337	—	—
1.6	0.001266	0.000351	—	—	0.00613	0.00278
1.7	0.001057	0.000285	—	—	—	—
1.8	0.000887	0.000233	—	—	0.00500	0.00222
1.9	0.000747	0.000191	—	—	—	—
2.0	0.000633	0.000159	0.0140	0.00292	0.00413	0.00181
3.0	0.000151	—	—	—	0.00188	—

* L = long span, S = short span.

† Flat plate refers to slab supported only by columns without drop panels and capitals.

Table 4-2 Theoretical deflection coefficients, K_s, for uniformly loaded single square panel supported elastically on all four sides[2]

EI/LD*	∞	100	50	25	10	5
K_s for max Δ in Eqs. (4-2) to (4-5)	0.00406	0.00412	0.00418	0.00429	0.00464	0.00519
EI/LD*	4	3	2	1	0.5	0
K_s for max Δ in Eqs. (4-2) to (4-5)	0.00546	0.00588	0.00668	0.00873	0.01174	0.0257

* Examples: (1) when $h = 4$ in. (10.2 cm), $L = 16$ ft = 192 in. (4.9 m),

For beams 8 in. (20.3 cm) by 12 in. (30.5 cm) $EI/LD = 1.1$
For beams 14 in. (35.6 cm) by 24 in. (61.0 cm) $EI/LD = 15.1$

(2) When $h = 8$ in. (20.3 cm), $L = 30$ ft = 360 in. (9.1 m),

For beams 10 in. (25.4 cm) by 14 in. (35.6 cm) $EI/LD = 0.14$
For beams 16 in. (40.6 cm) by 30 in. (76.2 cm) $EI/LD = 2.25$

(using $\nu = 0.2$).

Coefficients from the crossing beam analogy Approximate values of the deflection coefficient, K_s, can also be determined for two-way slabs using the crossing beam analogy.[7,8] This procedure neglects the torsional stiffness of the slab (which would decrease the panel deflection if included) and also neglects the deflection of the supporting beams (which would increase the panel deflection if included).

Examples: for two-way square slab panels, uniformly loaded—interior panel $K_s = \frac{1}{2}(1/384$ – fixed–fixed beam case$) = 0.0013$. Side panel $K_s =$ average of 0.0013 and $\frac{1}{2}(0.0054$ – hinged–fixed beam case$) = \frac{1}{2}(0.0013 + 0.0027) = 0.0020$. Corner panel $K_s = 0.0027$. For other than square panels, use Eq. (4-1), and an average of the deflections computed in the two directions.

Comments: for the corner panel of two-way square slabs, $K_s = 0.0027$ here, compared with 0.0041, which was the approximate value in Example 1 of Table 4-1, and with 0.0055, which was the absolute upper bound value in Example 5 of Table 4-2 when $EI/LD = 4.0$ (a typical value).

For the interior panel of two-way square slabs, $K_s = 0.0013$ here, compared with the same value in Example 3 of Table 4-1, which was a lower bound value.

Slab deflections computed by this method usually tend to be on the low side, primarily because of the use of ideal end restraint conditions.

Coefficients obtained using the diagonal beam strip analogy The deflection coefficients, K_s, in Table 4-3 were suggested by PCA[4] for use in Eq. (4-4) for interior panel flat plates, based on the use of a diagonal beam strip analogy. The PCA coefficients are divided by 12 here to apply to Eq. (4-4), and the long span, L, not the diagonal length, is used in the equation and in Ref. 4. It may be seen that the coefficients in Table 4-3 are roughly the same as the theoretical values in Table 4-1 for the interior flat plate panel: when $L/S = 1.0$, $K_s = 0.0053$ versus 0.0058 in Table 4-1; and when $L/S = 2.0$, $K_s = 0.0027$ versus 0.0029 in Table 4-1. This method is seen to have a rational basis, since as L/S becomes large, the coefficient approaches the one-way fixed–fixed coefficient of $1/384 = 0.0026$. The coefficients used by CRSI[24] (divided by 12) are also the same as those in Table 4-3.

Coefficients from finite difference solutions The deflection coefficients, K_s, in Table 4-4 were derived by Vanderbilt, Sozen, and Siess[3] for interior panels using the finite difference method. The table refers principally to two-way slabs (slabs with supporting beams), but also to flat plates (slabs without supporting beams) when $EI_L/SD = EI_S/LD = 0$. These coefficients are applicable to interior panels in Eqs. (4-2) to (4-5) for the panel deflections, and in the usual elastic beam

Table 4-3 Deflection coefficients, K_s, for interior flat plate panels, based on the diagonal beam strip analogy

L/S	1.0	1.1	1.2	1.3	1.4	1.5	2.0
K_s	0.00528	0.00443	0.00389	0.00353	0.00326	0.00307	0.00266

S/L	EI_L/SD	EI_S/LD	Center of panel			Midspan of long beam			Midspan of short beam		
			$c/L = 0.0$	0.1	0.2	0.0	0.1	0.2	0.0	0.1	0.2
	$I_S = I_L$										
1.0	0.0	0.0	0.00581	0.00441	0.00289	0.00435	0.00304	0.00173	Same as long beam		
	0.2	0.2	0.00438	0.00340	0.00240	0.00299	0.00207	0.00122			
	0.25	0.25	0.00415	0.00324	0.00233	0.00277	0.00192	0.00114			
	0.5	0.5	0.00331	0.00271	0.00205	0.00198	0.00141	0.00085			
	1.0	1.0	0.00260	0.00222	0.00179	0.00130	0.00092	0.00056			
	2.0	2.0	0.00206	0.00184	0.00158	0.00077	0.00054	0.00033			
	2.5	2.5	0.00196	0.00174	0.00153	0.00065	0.00045	0.00028			
	4.0	4.0	0.00174	0.00159	0.00144	0.00043	0.00030	0.00018			
	5.0	5.0	0.00162	0.00154	0.00141	0.00035	0.00024	0.00015			
0.8	0.0	0.0	0.00420	0.00301	0.00189	0.00378	0.00262	0.00155	0.00230	0.00131	0.00057
	0.2	0.16	0.00316	0.00237	0.00159	0.00271	0.00192	0.00116	0.00149	0.00088	0.00040
	0.5	0.4	0.00246	0.00191	0.00136	0.00195	0.00138	0.00085	0.00099	0.00059	0.00028
	1.0	0.8	0.00191	0.00154	0.00177	0.00134	0.00095	0.00058	0.00063	0.00038	0.00018
	2.0	1.6	0.00147	0.00124	0.00100	0.00083	0.00058	0.00036	0.00036	0.00022	0.00011
	4.0	3.2	0.00116	0.00103	0.00089	0.00048	0.00033	0.00020	0.00019	0.00012	0.00006
0.6	0.0	0.0	0.00327	0.00234	0.00143	0.00321	0.00228	0.00137	0.00099	0.00040	0.00008
	0.2	0.12	0.00256	0.00189	0.00119	0.00246	0.00178	0.00108	0.00063	0.00027	0.00006
	0.5	0.3	0.00201	0.00150	0.00098	0.00187	0.00134	0.00082	0.00040	0.00017	0.00004
	1.0	0.6	0.00153	0.00116	0.00079	0.00135	0.00096	0.00059	0.00025	0.00011	0.00003
	2.0	1.2	0.00110	0.00085	0.00061	0.00087	0.00061	0.00037	0.00013	0.00006	0.00002
	4.0	2.4	0.00077	0.00063	0.00048	0.00051	0.00035	0.00022	0.00007	0.00003	0.00001
0.4	0.0	0.0	0.00284	0.00204	—	0.00284	0.00204	—	0.00031	0.00004	—
	0.2	0.08	0.00231	0.00166	—	0.00230	0.00165	—	0.00020	0.00003	—
	0.5	0.2	0.00183	0.00131	—	0.00181	0.00128	—	0.00012	0.00002	—
	1.0	0.4	0.00137	0.00098	—	0.00134	0.00094	—	0.00007	0.00001	—
	2.0	0.8	0.00093	0.00066	—	0.00088	0.00061	—	0.00004	0.00001	—
	4.0	1.6	0.00059	0.00042	—	0.00053	0.00036	—	0.00002	0.00000	—

Table 4-4—continued

			Center of panel			Midspan of long beam			Midspan of short beam		
S/L	EI_L/SD	EI_S/LD	$c/L = 0.0$	0.1	0.2	0.0	0.1	0.2	0.0	0.1	0.2
	$I_S = (S/L)I_L$										
1.0	Same as above										
0.8	0.0	0.0	0.00405	0.00301	0.00189	0.00365	0.00262	0.00155	0.00218	0.00131	0.00057
	0.2	0.128	0.00321	0.00240	0.00160	0.00274	0.00193	0.00116	0.00157	0.00093	0.00043
	0.5	0.32	0.00251	0.00193	0.00137	0.00198	0.00139	0.00085	0.00108	0.00065	0.00031
	1.0	0.64	0.00195	0.00156	0.00117	0.00136	0.00095	0.00059	0.00072	0.00043	0.00021
	2.0	1.28	0.00149	0.00125	0.00101	0.00084	0.00058	0.00036	0.00042	0.00026	0.00013
	4.0	2.56	0.00118	0.00104	0.00090	0.00048	0.00033	0.00020	0.00023	0.00014	0.00007
0.6	0.0	0.0	0.00327	0.00234	0.00143	0.00321	0.00228	0.00137	0.00099	0.00040	0.00008
	0.2	0.072	0.00260	0.00190	0.00119	0.00250	0.00178	0.00108	0.00070	0.00030	0.00007
	0.5	0.18	0.00204	0.00151	0.00098	0.00190	0.00135	0.00082	0.00049	0.00022	0.00006
	1.0	0.36	0.00156	0.00116	0.00079	0.00137	0.00096	0.00059	0.00032	0.00015	0.00004
	2.0	0.72	0.00111	0.00085	0.00061	0.00088	0.00061	0.00037	0.00019	0.00009	0.00003
	4.0	1.44	0.00078	0.00063	0.00049	0.00051	0.00035	0.00022	0.00010	0.00005	0.00002
0.4	0.0	0.0	0.00284	0.00205	—	0.00284	0.00204	—	0.00031	0.00004	—
	0.5	0.08	0.00185	0.00131	—	0.00182	0.00128	—	0.00016	0.00003	—
	1.0	0.16	0.00139	0.00098	—	0.00135	0.00094	—	0.00011	0.00002	—
	2.0	0.32	0.00094	0.00066	—	0.00089	0.00061	—	0.00006	0.00001	—
	4.0	0.64	0.00059	0.00042	—	0.00053	0.00036	—	0.00003	0.00001	—

$I_S = (S/L)^2 I_L$

1.0	Same as above										
0.8	0.0	0.0	0.00420	0.00301	0.00189	0.00378	0.00262	0.00155	0.00230	0.00131	0.00057
	0.2	0.1024	0.00325	0.00242	0.00161	0.00277	0.00194	0.00116	0.00164	0.00097	0.00047
	0.5	0.256	0.00255	0.00196	0.00138	0.00201	0.00140	0.00085	0.00117	0.00071	0.00034
	1.0	0.512	0.00199	0.00158	0.00118	0.00138	0.00096	0.00059	0.00080	0.00049	0.00024
	2.0	1.024	0.00152	0.00126	0.00102	0.00085	0.00059	0.00036	0.00049	0.00030	0.00016
	4.0	2.048	0.00119	0.00105	0.00090	0.00048	0.00033	0.00020	0.00028	0.00017	0.00009
0.6	0.0	0.0	0.00327	0.00234	0.00143	0.00321	0.00228	0.00137	0.00099	0.00040	0.00008
	0.2	0.0432	0.00263	0.00190	0.00119	0.00252	0.00179	0.00108	0.00074	0.00032	0.00008
	0.5	0.108	0.00208	0.00151	0.00098	0.00192	0.00135	0.00082	0.00056	0.00025	0.00007
	1.0	0.216	0.00158	0.00117	0.00079	0.00138	0.00096	0.00059	0.00040	0.00019	0.00005
	2.0	0.432	0.00113	0.00086	0.00061	0.00089	0.00061	0.00037	0.00026	0.00012	0.00004
	4.0	0.864	0.00079	0.00063	0.00048	0.00052	0.00035	0.00022	0.00015	0.00007	0.00002
0.4	0.0	0.0	0.00284	0.00205	—	0.00284	0.00204	—	0.00031	0.00004	—
	0.2	0.0128	0.00234	0.00166	—	0.00233	0.00165	—	0.00025	0.00003	—
	0.5	0.032	0.00186	0.00131	—	0.00183	0.00128	—	0.00019	0.00003	—
	1.0	0.064	0.00140	0.00098	—	0.00136	0.00094	—	0.00014	0.00003	—
	2.0	0.128	0.00094	0.00066	—	0.00089	0.00061	—	0.00010	0.00002	—
	4.0	0.256	0.00059	0.00042	—	0.00053	0.00036	—	0.00006	0.00002	—

Coefficients in this table apply directly to flat plates and two-way slabs with relatively shallow supporting beams. For stiffer beams, coefficients may be obtained from charts in Ref. 5, or estimated between values in this table and the case of rigid supports. (From *ASCE Proceedings* paper[3] by Vanderbilt, Sozen, and Siess, with permission.)

* L = long span, S = short span, c = width of support, and I_L and I_S are the moments of inertia of the beams in the long and short directions, respectively.

equations for the beam deflections. The panel deflection coefficients refer to the total panel deflection, which is not to be added to the beam deflection in addition. All panels were uniformly loaded. In this paper deflection coefficients were also derived using a frame analysis method for one-quarter scale model test structures including a flat plate, two flat slabs, and two two-way slabs. Both uncracked and fully cracked values were shown for corner, side, and center panels.

Equivalent frame method An approximate method for calculating deflections of two-way reinforced concrete floor systems that uses the same assumptions and terminology incorporated in the 1971 ACI Code[11] for moment analysis was proposed by Nilson and Walters.[18] The procedure is based on the equivalent frame method and applies equally well to the direct design method of the ACI Code, in which moment coefficients are used in lieu of a moment analysis. The method is essentially the same for flat plates, flat slabs, and two-way slabs, waffle slabs, and combination slabs, with the various stiffnesses, equivalent stiffnesses, and distribution factors also being determined as in the ACI Code for moment design.

Following the moment analysis and design by either the equivalent frame or direct design method, the midpanel deflection is computed as the sum of the deflection at midspan of the column strip or column line in one direction, and the deflection at midspan of the middle strip or middle line between columns in the other direction, as shown in Fig. 4-1. For rectangular panels, or other panels that have different characteristics in the two directions, an average of the two deflections thus determined is used. The column strip in both directions is defined as the width on each side of the column centerline equal to one-fourth the smaller of the two panel dimensions. The middle strip is the strip bounded by two column strips. An alternate method is to use the "ersatz" frame consisting of column strips equal to 0.4 of the span (0.2 on each side) and middle strips equal to 0.6 of the span (dividing line is approximate inflection line).

The midspan deflection of the equivalent frame, column strip, or middle strip is computed as the sum of three parts: that of the panel assumed to be fixed at both ends, plus that due to the known rotation at each of the two support lines. First, the midspan fixed-end deflection of the full width equivalent frame is computed by Eq. (4-6).

$$\text{Fixed } \Delta_{\text{frame}} = wL^4/384E_c I_{\text{frame}} \tag{4-6}$$

where w is the distributed load per panel width (psf $\times L_2$) and L is the span center-to-center of columns. Other basic deflection formulas would, of course, apply for concentrated loads. To include the effect of different positive and negative moment region Is, especially when using drop panels and/or I_e (effect of cracking based on the moment level, as by Eq. (3-19)), a weighted average of the positive and negative moments of inertia according to Eq. (3-48) is recommended for slabs as for beams. For computing deflections, it would appear that the stiffening effect of the columns, as determined by the ACI Code magnification factor,

$(1 - c_2/L_2)^2$, might normally be neglected. Typical sections for computing the *I*s of an interior frame are shown in Fig. 4-2 for flat plates and two-way slabs.

The deflection in Eq. (4-6) provides the basis for calculating the fixed-end deflection of the column and middle strips in Eq. (4-7), according to the relative lateral moment distribution factors (*LDF*) and stiffnesses (*EI*) of the strips—or the *M*/*EI* ratio of the strips to that of the full panel width.

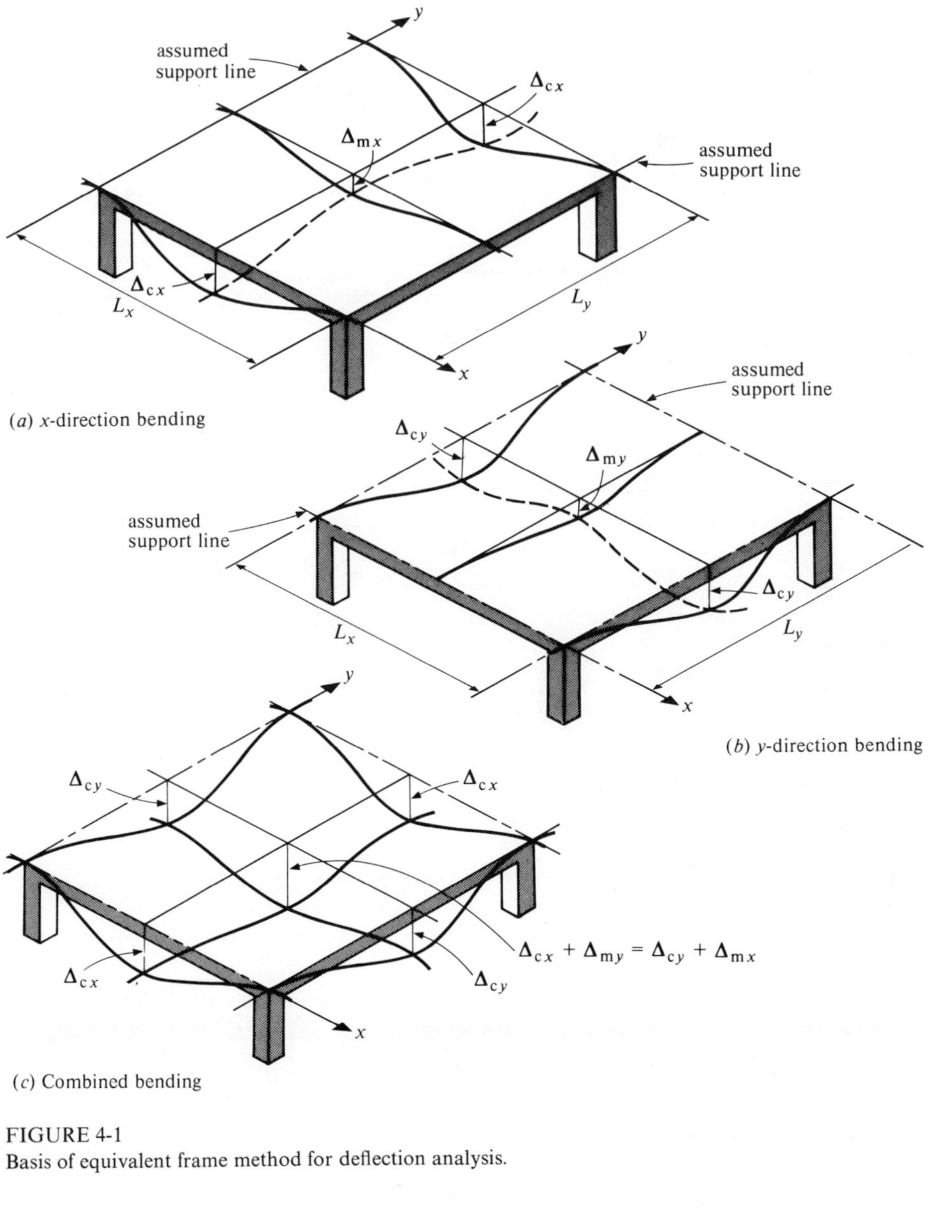

FIGURE 4-1
Basis of equivalent frame method for deflection analysis.

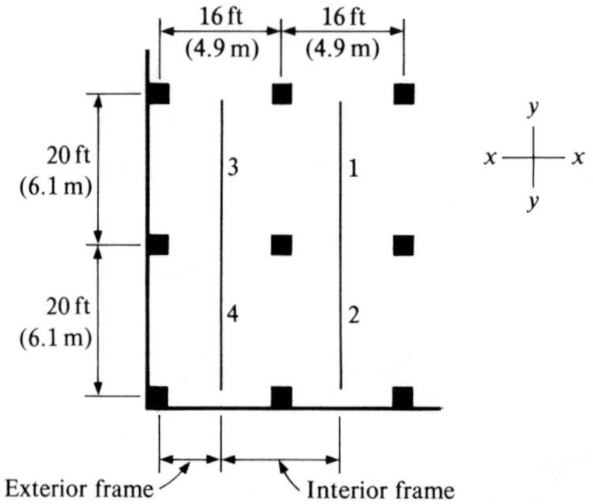

Interior equivalent frame sections—example $L_1 = 20\text{ ft} = 240\text{ in.}$ (6.1 m), $L_2 = 16\text{ ft} = 192\text{ in.}$ (4.9 m), $h = 7\text{ in.}$ (17.8 cm):

Flat plate

	Section	
Frame:	$L_2 = 192$ in., $h = 7$ in.	$I_g = (192)(7)^3/12 = 5488\text{ in.}^4$ (228 400 cm^4)
Column strip:	96 in., $h = 7$ in.; $L_1/2$ or $L_2/2$, whichever is smaller	$I_g = (96)(7)^3/12 = 2744\text{ in.}^4$ (114 200 cm^4)
Middle strip:	48 in., 48 in., $h = 7$ in.	$I_g = (2)(48)(7)^3/12 = 2744\text{ in.}^4$ (114 200 cm^4)

Two-way slab

	Section	
Frame:	$L_2 = 192$ in.	For I_g, use design aid, such as Fig. E-1, Chap. 3
Column strip:	96 in.	For I_g, use design aid, such as Fig. E-1, Chap. 3
Middle strip:	Same as middle strip for flat plate	

These same sections would also be used in computing I_e.

FIGURE 4-2
Exterior and interior equivalent frames, with examples of interior frame, column strip, and middle strip sections for a flat plate and two-way slab.

$$\text{Fixed } \Delta_{c,m} = (LDF)_{c,m}(\text{Fixed } \Delta_{\text{frame}}) \frac{(EI)_{\text{frame}}}{(EI)_{c,m}} \tag{4-7}$$

where $(LDF)_{c,m} = M_{c,m}/M_{\text{frame}}$. See Table 4-5 for values of LDF.

If the ends of the columns at the floor above and below are assumed fixed (usual case for an equivalent frame analysis), the rotation of the column at the floor in question is equal to the net applied moment divided by the stiffness of the equivalent column, as given by Eq. (4-8) for the frame, column strip, or middle strip.

$$\text{End } \theta_c = \text{End } \theta_m = \text{End } \theta_{\text{frame}} = (M_{\text{net}})_{\text{frame}}/K_{ec} \tag{4-8}$$

where K_{ec} refers to the equivalent column stiffness, as defined in Fig. 4-3. The various stiffnesses of the equivalent frame used in the moment design are also shown in Fig. 4-3 and defined in detail in Ref. 25, pp. 16–2 to 16–17, for example.

In the direct design method, moments are based on the clear span and should theoretically be adjusted to obtain moments and rotations at the column

Table 4-5 Lateral distribution factors, *LDF*, for column and middle strips, using an average of the positive and negative section values

From ACI Code (318-71), Sec. 13.3.4, and using Eqs. (Q), p. 15–13, PCA, Ref. 26, for column strip percentages:

Exterior negative	$100 - 10\beta_t + 12\beta_t(\alpha_1 L_1/L_1)(1 - L_2/L_1)$
Interior negative	$75 + 30(\alpha_1 L_2/L_1)(1 - L_2/L_1)$
Positive	$60 + 30(\alpha_1 L_2/L_1)(1.5 - L_2/L_1)$

except when $\alpha_1 L_2/L_1 > 1$ (typical two-way slab case), use $\alpha_1 L_2/L_1 = 1$.

α_1 is ratio of flexural stiffness of beam section to the flexural stiffness of a width of slab bounded laterally by the center line of the adjacent panel, if any, on each side of the beam ($= E_{cb}I_b/E_{cs}I_s$).

β_t is ratio of torsional stiffness of edge beam section to the flexural stiffness of a width of slab equal to the span length of the beam, center-to-center of supports ($= E_{cb}C/2E_{cs}I_s$).

Example when $\alpha_1 = \beta_t = 0$, which might be assumed for a flat plate with no beams (for panels from Fig. 4-2).

Interior panel, both directions, $(LDF)_c = \dfrac{60\text{ (pos.)} + 75\text{ (neg.)}}{2} = 67.5\%$

$(LDF)_m = 100 - 67.5 = 32.5\%$

Side panel, 2, (y–y), $(LDF)_c = \dfrac{60 + (100 + 75)/2}{2} = 73.8\%$

$(LDF)_m = 100 - 73.8 = 26.2\%$

(x–x),	$(LDF)_c = 67.5\%$	$(LDF)_m = 32.5\%$
Side panel, 3, (y–y),	$(LDF)_c = 67.5\%$	$(LDF)_m = 32.5\%$
(x–x),	$(LDF)_c = 73.8\%$	$(LDF)_m = 26.2\%$
Corner panel, both directions,	$(LDF)_c = 73.8\%$	$(LDF)_m = 26.2\%$

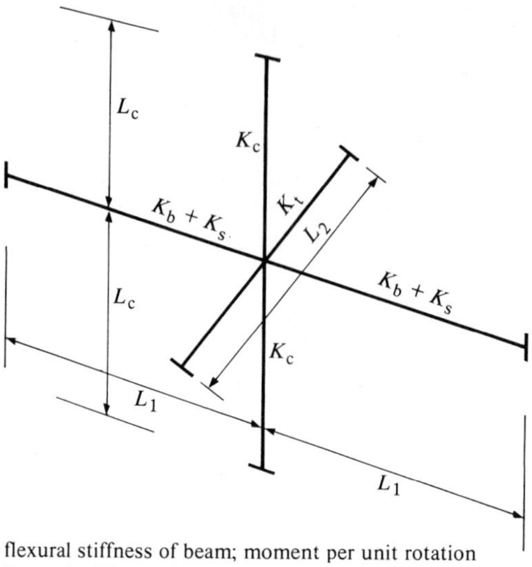

K_b = flexural stiffness of beam; moment per unit rotation
K_c = flexural stiffness of column; moment per unit rotation
K_{ec} = flexural stiffness of an equivalent column, consisting of the two columns and the torsional elements on each side (or one side) of the column, according to ACI Code equation (13-5)—$1/K_{ec} = 1/\Sigma K_c + 1/K_t$, moment per unit rotation
K_s = flexural stiffness of slab; moment per unit rotation
K_t = torsional stiffness of torsional element; moment per unit rotation

FIGURE 4-3
Equivalent frame showing the stiffnesses of the beams (if any), columns, equivalent column, slab, and torsional elements.

centerlines. However, the use of moments at the face of the supports should cause little error, and might even be preferable when used to compute I_e. Also, as a practical matter, only the exterior column rotation need be considered in most cases when using the direct design moment coefficients with equal adjacent spans. When pattern loading is used, such as in the more general equivalent frame moment analysis, and when adjacent spans are unequal, this would normally not be the case.

The center deflection of a beam having an end rotation of θ radians, with the far end fixed, is easily derived as Eq. (4-9).

$$\Delta_\theta = \theta L/8 \tag{4-9}$$

When using I_e, for consistency use Eq. (4-10) instead of Eq. (4-9).

$$\Delta_\theta = (\theta L/8)(I_g/I_e)_{\text{frame}} \tag{4-10}$$

The combined midspan deflection of a column or middle strip is thus found by Eq. (4-11).

$$\Delta_{c,m} = \text{Fixed } \Delta_{c,m} + (\Delta_{\theta 1})_{c,m} + (\Delta_{\theta 2})_{c,m} \tag{4-11}$$

where $\Delta_{\theta 1}$ and $\Delta_{\theta 2}$ refer to the midspan deflection due to end rotations at both ends. The final midpanel deflection is computed by Eq. (4-12), as shown in Fig. 4-1.

$$\Delta = \Delta_{cy} + \Delta_{mx} = \Delta_{cx} + \Delta_{my} \tag{4-12}$$

An average of the two results in Eq. (4-12) is used when they are not the same, which is usually the case for rectangular panels, for example.

It should be noted that the loads and moments used in computing deflections refer to working loads, not factored loads, and that the moments are not subject to redistribution (as in the case of the strength design moments) which is based on the concept of limit analysis.

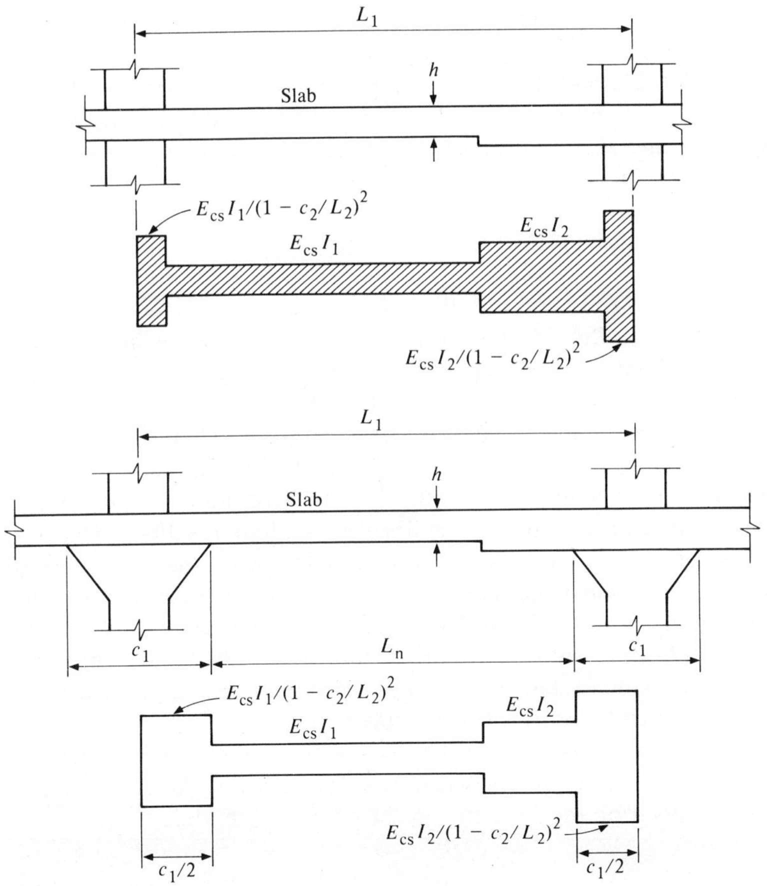

FIGURE 4-4
Equivalent slab–beam stiffness diagrams for slabs without and with drop panels and column capitals.

Effect of drop panels, column capitals, supporting beams, pattern loading, and wall loads The method of deflection coefficients usually applies directly to flat plates and slabs with rigid side supports or supporting beams, while neglecting the effect of drop panels, column capitals, and pattern loading. The coefficients in Table 4-4, however, do include the effect of size of supporting beams (also included in Table 4-2 for a single square panel), columns, or capitals, but not the effect of drop panels or pattern loading.

In the case of the equivalent frame method the stiffening effect of supporting beams (demonstrated herein) and drop panels is included directly in the stiffness calculations. Also, the stiffening effect of columns and capitals is included when using the ACI Code magnification factor, $(1 - c_2/L_2)^2$, in the slab-beam stiffness calculations, as shown in Fig. 4-4. Such an effect may normally be neglected for flat plates (as in the example herein), but should be included when column capitals are used. This might require a modification of the basic deflection formula (Eq. (4-6)) for this nonuniform member, or in approximate calculations just the use of a weighted average of positive and negative region EIs.

It is noted in Table 4-4 for a flat plate or slab interior panel with $S/L = 1.0$ and $I_S = I_L = 0$, for example, that $K_s = 0.00581$, 0.00441, and 0.00289 when $c/L = 0.0$, 0.1, and 0.2, respectively. Thus the deflection range is about 2 to 1 for these values of c/L, with the latter coefficient being $289/441 = 0.66$ of the middle one (possible indication of a comparison between deflections of slabs without and with capitals). Values of c/L may range from 0.02 to 0.33, as shown in Ref. 25.

In the equivalent frame method the effect of pattern loading is included in the frame moment analysis (as Ref. 26, p. 16–18), and implicitly when using the direct design moment coefficients. The effect of both checkerboard and strip pattern loading on moments and deflections was studied in Refs. 3, 27, and 28. For example, the deflection of an interior panel under checkerboard loading was found to vary less than 14 percent from that with all panels loaded. Also, solutions obtained for checkerboard loading consisting of a uniform load on the interior and corner panels indicated that pattern loading seldom resulted in critical deflections, and where it did, the deflections were little greater than those for all panels loaded. Since dead load (which contributes to both short-time and long-time deflection) is frequently of the order of half the live load and acts on all panels, this is further reason for not laboriously considering various patterns of live load when calculating approximate slab deflections, even though increased deflections due to cracking under dead plus live loading may take place.

The effect of line loads such as wall loads acting along panel edges, or occasionally across the panels themselves, can be included directly in the equivalent frame method when determining moments in the column or middle strips.

Effect of cracking The effect of cracking can be included directly in the moment of inertia calculations in the equivalent frame method (previously described) and in the method of deflection coefficients, as shown in the examples herein.

The following procedure was suggested by Matsui and Branson[29] (and used by ACI Committee 209[6]) as a method that compared favorably with a number of test results for determining an approximate value of I_e in Eq. (4-5) to be used with deflection coefficients for partially cracked flat plates and two-way slabs: for combined dead and live loads, use I_e as the average of the positive and negative moment region values, as computed in Eq. (3-19), in the following manner—for flat plates, both positive and negative values for the long direction column strip; and for two-way slabs, both positive and negative values for the short direction middle strip. These values were based on the determination of moments by the 1963 ACI Code. Scanlon[22] subsequently found that somewhat better agreement was obtained using the average I_e of both column and middle strips. A numerical average of the panel values of I_e (column plus middle strip values, with each one determined by a weighted average of 70 percent positive and 15 percent, 15 percent, negative) in the two directions is used with deflection coefficients in the examples herein for both flat plates and two-way slabs.

The approximate interior panel deflection coefficients suggested by ACI Committee 209[6] to be used with I_e referred to the theoretical values in Table 4-1 with all four sides fixed for two-way slabs, and the values in Table 4-3 for flat plates (same as by PCA[4] and CRSI,[24] except differing by a factor of 12 due to units).

When using the equivalent frame method, the following section properties for slabs without and with supporting beams may apply in typical cases, as was found to be the case for the examples herein.

Slabs without beams: for all dead-load deflections—I_g; and for dead plus live-load deflections—I_g for the middle strips in both directions and I_e for the column strips in both directions.

Slabs with beams: for all dead-load deflections—I_g; and for dead plus live-load deflections—I_g for the column strips in both directions and I_e for the middle strips in both directions.

Where they differ in positive and negative moment regions, a weighted average of 70 percent positive and 15 percent, 15 percent, negative values of I is used in each case. The above procedure appears to be consistent with the tensile cracking patterns shown in Fig. 4-7, for example, which are along column lines for the flat plate[30] and flat slab[31] at design load (with the addition of middle strip cracking for the flat slab) and diagonal for the two-way slab[32] at overloading. The two-way slab was uncracked at design load. The principal exception to the above Is might be for slabs without beams when the middle strip positive moment (midpanel moment) in one or both directions exceeds the cracking moment (not the case in the example herein, but was the case for the flat slab in Fig. 4-7), in which case partial cracking may have a pronounced effect on deflections and should be taken into account by using I_e instead of I_g for such middle strips. In addition, when shallow supporting beams are used in two-way slabs so that the column strip cracking moment is exceeded in one or both directions (not the case in the example herein), especially in the positive region, it might be necessary to use I_e instead of I_g for such column strips.

Long-Time Deflections

Only a limited amount of work has been reported on the time-dependent deflection of slabs using precise methods.[22] The following simplified approach (with modifications) for computing creep and shrinkage, or combined time-dependent deflections of slabs (using the author's procedure for beams in Chap. 3) has been applied by Matsui and Branson,[29] Taylor,[33] Scanlon,[22] Heiman,[34] and Rangan.[35] Use Eqs. (3-56) and (3-58) for shrinkage deflection, Eq. (3-74) for creep deflection, and Eq. (3-75) for combined time-dependent deflection, in which Eqs. (3-70) and (3-71) may be used for the effect of compression steel, k_r. However, as was noted in Chap. 3, the primary effect of compression steel in reducing time-dependent deflection is in the positive moment regions, and the author suggests that this effect may normally be neglected for slabs, except when the compression steel percentages in the positive regions are relatively large. Thus, the creep or combined time-dependent deflection may normally be taken as C_t or T times the short-time deflection, respectively. See Table 1-10 for values of C_u and $(\varepsilon_{sh})_u$.

As described in the last section of this chapter, the limited data that have been reported on time-dependent slab deflections indicate a somewhat higher ratio of time-dependent to initial deflection than may have been used in structural calculations in the past, such as $T_u = 2$ in the 1971 ACI Code. Such behavior for slabs may be due to the biaxial nature of the deformation and the fact that slabs are usually thinner than beams. For the purpose of comparing computed deflections with the usual standard allowable deflections (Chap. 6 herein), the author believes that values of $C_u = 2.0$ and $T_u = 3.0$ (versus $C_u = 1.6$ and $T_u = 2.5$ for beams herein) might be reasonable for slab deflection calculations in most practical cases. Also see Chap. 3, Appendix J, Table J-1, for other values of T_u.

4-3 NUMERICAL EXAMPLES

Solid Slab System Without Beams (Flat Plate)

The following check of the ACI Code, Sec. 13.3.1, is made for the purpose of using the direct design method in the strength design and calculation of service load moments for the flat plate example, which are shown in detail in Tables 4-6 and 4-8.

There are at least three spans.
$L_1/L_2 = 20/16$, which is less than 2.
Successive span lengths are equal.
Columns are not offset.
Live load/dead load = 100/108, which is less than 3.
There are no beams.
Therefore the direct design method applies.

Table 4-6 Strength design of solid slab system without beams (flat plate) by the ACI Code (318-71)

Geometry and material properties

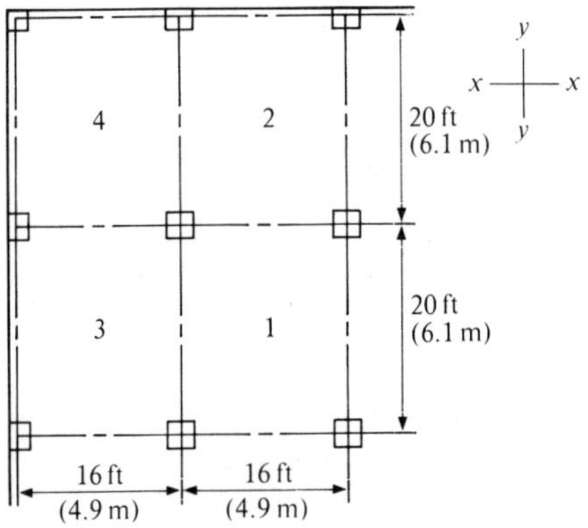

Square columns, $c = 18$ in. (45.7 cm)

$L_c = 10$ ft (3.05 m)

y–y, $L_1 = 20$ ft (6.1 m), $L_2 = 16$ ft (4.9 m), $L_n = 18.5$ ft (5.6 m), width of CS and MS = 8 ft (2.4 m)

x–x, $L_1 = 16$ ft (4.9 m), $L_2 = 20$ ft (6.1 m), $L_n = 14.5$ ft (4.4 m), width of CS = 8 ft (2.4 m), width of MS = 12 ft (3.7 m)

All concrete: $f'_c = 4000$ psi (281 kg/cm^2, 27.6 N/mm^2)

$w = 145$ pcf (2320 kg/m^3, 22.8 kN/m^3)

All steel: $f_y = 50\,000$ psi (3520 kg/cm^2, 345 N/mm^2)

Preliminary slab thickness (ACI 318-71)

From Table 6-6, Eq. (6-3), which is Code Eq. (9-8), applies to flat pls.

For long direction (y–y), $h = L_n(800 + 0.005f_y)/36\,000$

$= (18.5)(12)(800 + 250)/36\,000 = 6.48$ in. (16.5 cm)

plus 10% for slab without beam = 7.13 in. (18.1 cm)

Use $h = 7$ in. (17.8 cm) and check deflections.

Loads

Live load = 100 psf (488 kg/m^2, 4788 N/m^2)

Superimposed dead load = 20 psf (98 kg/m^2, 958 N/m^2)

Slab dead load = (7/12)(150) = 88 psf (430 kg/m^2, 4210 N/m^2)

Total service load = 208 psf (1015 kg/m^2, 5210 N/m^2)

Strength design load = 1.4(108) + 1.7(100) = 321 psf (1570 kg/m^2, 15 370 N/m^2)

Preliminary shear check

Two-way action—

Full panel load = $V_u = wL_1L_2$

$= (0.321)(20)(16) = 103$ kips (46 700 kg, 458 000 N)

Estimate avg. $d = 5.5$ in. (14.0 cm)

$b_0 = 2(2c + 2d) = 2(36 + 11) = 94$ in. (239 cm)

$v_u = V_u/\phi b_0 d = 103\,000/(0.85 \times 94 \times 6)$

$= 215$ psi (15.1 kg/cm^2, 1.48 N/mm^2) $< 4\sqrt{f'_c}$

$= 4\sqrt{4000} = 253$ psi (17.8 kg/cm^2, 1.74 N/mm^2) OK

Table 4-6 (*continued*)

One-way action—

$$V_u = wL_2L_n/2 = (0.321)(16)(18.5)/2 = 47.5 \text{ kips}$$

$$v_u = V_u/\phi bd = (47\,500)/(0.85 \times 16 \times 12 \times 6) = 49 \text{ psi}$$

$$< 2\sqrt{f_c'} = 2\sqrt{4000} = 126 \text{ psi } (8.9 \text{ kg/cm}^2, 0.87 \text{ N/mm}^2) \qquad \text{OK}$$

Hence, $h = 7$ in. (17.8 cm) is OK for shear initially.

Long direction ($y - y$) *section properties and design moments—interior frame (separate calculations omitted for the exterior half-panel frame).*

$$K_b = 0 \text{ (no beams)} \qquad K_s = 4E_{cs}I_s/L_1 \qquad \text{where } I_s = L_2h^3/12$$

$$= 4E_c(16 \times 12)(7^3)/(12 \times 20 \times 12) = 91.5E_c$$

$$K_c = 4E_{cc}I_c/L_c = 4E_c(18^4)/(12 \times 10 \times 12) = 145.8E_c$$

$$\Sigma K_c = 2(145.8E_c) = 291.6E_c \text{ for columns above and below floor}$$

The torsional element on each side of the column is 18 in. × 7 in.

$$C = \Sigma(1 - 0.63x/y)(x^3y/3) \qquad \text{(ACI Code Eq. (13-7))}$$

$$= (1 - 0.63\tfrac{7}{18})\frac{7^3 \times 18}{3} = 1554$$

$$K_t = \Sigma 9E_{cs}C/L_2(1 - c_2/L_2)^3 \qquad \text{(ACI Code Eq. (13-6))}$$

$$c_2/L_2 = 1.5/16 = 0.0938$$

$$K_t = (2)(9)(E_c)(1554)/(16 \times 12)(1 - 0.0938)^3 = 195.7E_c$$

$$1/K_{ec} = 1/\Sigma K_c + 1/K_t \qquad \text{(ACI Code Eq. (13-5))}$$

$$= 1/291.6E_c + 1/195.7E_c = 0.008539/E_c$$

$$K_{ec} = E_c/0.008539 = 117.1E_c$$

$$\alpha_{ec} = K_{ec}/\Sigma(K_s + K_b) = 117.1E_c/91.5E_c = 1.280$$

Mom. distr. ratios—interior span—neg. = 0.65, pos. = 0.35
Mom. distr. ratios—exterior span—

$$\text{Ext. neg.} = 0.65/(1 + 1/\alpha_{ec}) = 0.65/(1 + 1/1.280) = 0.365$$

$$\text{Pos.} = 0.63 - 0.28/(1 + 1/\alpha_{ec}) = 0.63 - 0.28/(1 + 1/1.280) = 0.473$$

$$\text{Int. neg.} = 0.75 - 0.10/(1 + 1/\alpha_{ec}) = 0.75 - 0.10/(1 + 1/1.280) = 0.694$$

Col. strip mom. ratios (for flat plate, assume $\alpha_1 = \beta_t = 0$), Table 4-5,

$$\text{Ext. neg. CS} = 1.00 - 0.10\beta_t + 0.12\beta_t(\alpha_1L_2/L_1)(1 - L_2/L_1) = 1.00$$

$$\text{Pos. CS} = 0.60 + 0.30(\alpha_1L_2/L_1)(1.5 - L_2/L_1) = 0.60$$

$$\text{Int. neg. CS} = 0.75 + 0.30(\alpha_1L_2/L_1)(1 - L_2/L_1) = 0.75$$

Summary of design moments—long direction

$$M_0 = wL_2L_n^2/8 = (0.321)(16)(18.5)^2/8 = 219.8 \text{ ft-k } (20\,400 \text{ kgm}, 298\,000 \text{ Nm})$$

Table 4-6 *(continued)*

	Exterior span			Interior span		
	Ext. neg.	Pos.	Int. neg.	Neg.	Pos.	Neg.
1. Mom. distr. ratios (above)	0.365	0.473	0.694	0.650	0.350	0.650
2. Col. strip mom. ratios (above)	1.00	0.60	0.75	0.75	0.60	0.75
3. Panel mom. = M_0 (line 1), ft-k	80.2	104.0	152.5	142.9	76.9	142.9
4. Panel mom. w 10% redistr., ft-k	80.2	111.7	137.2	137.2	82.6	137.2
5. CS mom. = line 2 × line 4, ft-k	80.2	67.0	102.9	102.9	49.6	102.9
6. CS mom. = line 2 × line 4, Nm × 10^3	108.8	90.9	139.5	139.5	67.3	139.5
7. MS mom. = line 4 − line 5, ft-k	0	44.7	34.3	34.3	33.0	34.3
8. MS mom. = line 4 − line 5, Nm × 10^3	0	60.6	46.5	46.5	44.7	46.5

Short direction (x–x) section properties and design moments—interior frame (separate calculations omitted for the exterior half-panel frame)

$$K_b = 0 \text{ (no beams)} \qquad K_s = 4E_{cs}I_s/L_2 \qquad \text{where } I_s = L_1h^3/12$$

$$= 4E_c(20 \times 12)(7^3)/(12 \times 16 \times 12) = 142.9E_c$$

$$\Sigma K_c = 291.6E_c \qquad \text{and} \qquad C = 1554 \qquad \text{(same as for long direction)}$$

$$c_2/L_2 = 1.5/20 = 0.075 \qquad K_t = 18E_c(1554)/(20 \times 12)(1 - 0.075)^3 = 147.3E_c$$

$$1/K_{ec} = 1/\Sigma K_c + 1/K_t \qquad \text{(ACI Code Eq. (13-5))}$$

$$= 1/291.6E_c + 1/147.3E_c = 0.01022/E_c$$

$$K_{ec} = E_c/0.01022 = 97.9E_c$$

$$\alpha_{ec} = K_{ec}/\Sigma(K_s + K_b) = 97.9E_c/142.9E_c = 0.685$$

Mom. distr. ratios—interior span—neg. = 0.65, pos. = 0.35
Mom. distr. ratios—exterior span—

$$\text{Ext. neg.} = 0.65/(1 + 1/\alpha_{ec}) = 0.65/(1 + 1/0.685) = 0.264$$

$$\text{Pos.} = 0.63 - 0.28/(1 + 1/\alpha_{ec}) = 0.63 - 0.28/(1 + 1/0.685) = 0.516$$

$$\text{Int. neg.} = 0.75 - 0.10/(1 + 1/\alpha_{ec}) = 0.75 - 0.10/(1 + 1/0.685) = 0.709$$

Col. strip mom. ratios (same as for long direction)

Summary of design moments—short direction

$$M_0 = wL_2L_n^2/8 = (0.321)(20)(14.5)^2/8 = 168.7 \text{ ft-k } (23\,300 \text{ kgm, } 228\,800 \text{ Nm})$$

Table 4-6 *(continued)*

	Exterior span			Interior span		
	Ext. neg.	Pos.	Int. neg.	Neg.	Pos.	Neg.
1. Mom. distr. ratios (above)	0.264	0.516	0.709	0.650	0.350	0.650
2. Col. strip mom. ratios (above)	1.00	0.60	0.75	0.75	0.60	0.75
3. Panel mom. = M_0 (line 1), ft-k	44.5	87.0	119.6	109.7	59.0	109.7
4. Panel mom. *w* 10% redistr., ft-k	44.5	93.0	107.6	107.6	61.1	107.6
5. CS mom. = line 2 × line 4, ft-k	44.5	55.8	80.7	80.7	36.7	80.7
6. CS mom. = line 2 × line 4, Nm × 10^3	60.3	75.7	109.4	109.4	49.8	109.4
7. MS mom. = line 4 – line 5, ft-k	0	37.2	26.9	26.9	24.4	26.9
8. MS mom. = line 4 – line 5, Nm × 10^3	0	50.4	36.5	36.5	33.1	36.5

For the exterior columns and ACI Code Sec. 11.13.2, 13.2.4, the combined *moment transfer shear and punching shear* was found to be 348 psi, versus the allowable of 253 psi. Hence, stirrups or shearhead reinforcement would be required for the exterior columns to carry the difference in these shear stresses. Such calculations may be found in Ref. 26, pp. 15–24, 15–25, 17–4 to 17–9, for example.

For using #5 bars with $h = 7$ in. (17.8 cm):

Long direction (y–y), $d = 7 - 0.75 - 0.31 = 5.94$ in. (15.1 cm)

Short direction (x–x), $d = 7 - 0.75 - 0.94 = 5.31$ in. (13.5 cm)

Design of reinforcement—long direction (y–y)

Column strip, $b = 96$ in. (244 cm), $d = 5.94$ in. (15.1 cm)

Value	Ext. neg.	Ext. pos.	Int. neg.	Int. pos.
1. Design M_u, ft-k	80.2	67.0	102.9	49.6
2. Design M_u, Nm × 10^3	108.8	90.9	139.5	67.3
3. $\bar{M}_u = M_u/\phi = M_u/0.90$, ft-k	89.1	74.4	114.3	55.1
4. * $R_u = \bar{M}_u/bd^2$, psi	316	264	405	195
5. †‡ Req. ρ, %	0.70	0.60	0.90	0.40
6. Req. $A_s = \rho bd$, in^2	3.99	3.42	5.13	2.28
7. Req. N of #5s = $A_s/0.31$	12.9	11.0	16.5	7.4
8. Use N =	13	11	17	8
9. Actual $A_s = 0.31N$, in^2	4.03	3.41	5.27	2.48
10. Actual $A_s = 0.31N$, cm^2	26.0	22.0	34.0	16.0
11. Actual ρ = line 9/bd, %	0.71	0.60	0.92	0.43
12. Actual spacing = b/line 8, in.	7.4	8.7	5.7	12.0
13. Actual spacing = b/line 8, cm	18.8	22.1	14.5	30.5

For Footnotes see p. 302.

Table 4-6 *(continued)*

Middle strip, $b = 96$ in. (244 cm), $d = 5.94$ in. (15.1 cm)				
Value	Ext. neg.	Ext. pos.	Int. neg.	Int. pos.
1. Design M_u, ft-k	—	44.7	34.3	33.0
2. Design M_u, Nm × 10^3	—	60.6	46.5	44.7
3. $\bar{M}_u = M_u/\phi = M_u/0.90$, ft-k	—	49.7	38.1	36.7
4. * $R_u = \bar{M}_u/bd^2$, psi	—	176	135	130
5. †‡ Req. ρ, %	—	0.40‡	0.40‡	0.40‡
6. Req. $A_s = \rho bd$, in^2	—	2.28	2.28	2.28
7. Req. N of #5s = $A_s/0.31$	—	7.4	7.4	7.4
8. Use N =	8	8	8	8
9. Actual $A_s = 0.31N$, in^2	2.48	2.48	2.48	2.48
10. Actual $A_s = 0.31N$, cm^2	16.0	16.0	16.0	16.0
11. Actual ρ = line 9/bd, %	0.43	0.43	0.43	0.43
12. Actual spacing = b/line 8, in.	12.0	12.0	12.0	12.0
13. Actual spacing = b/line 8, cm	30.5	30.5	30.5	30.5

Design of reinforcement—short direction (x–x)

Column strip, $b = 96$ in. (244 cm), $d = 5.31$ in. (13.5 cm)				
Value	Ext. neg.	Ext. pos.	Int. neg.	Int. pos.
1. Design M_u, ft-k	44.5	55.8	80.7	36.7
2. Design M_u, Nm × 10^3	60.3	75.7	109.4	49.8
3. $\bar{M}_u = M_u/\phi = M_u/0.90$, ft-k	49.4	62.0	89.7	40.8
4. * $R_u = \bar{M}_u/bd^2$, psi	219	275	398	181
5. †‡ Req. ρ, %	0.45	0.64	0.85	0.40‡
6. Req. $A_s = \rho bd$, in^2	2.29	3.26	4.33	2.04
7. Req. N of #5s = $A_s/0.31$	7.4	10.5	14.0	6.9‡
8. Use N =	8	11	14	7
9. Actual $A_s = 0.31N$, in^2	2.48	3.41	4.34	2.17
10. Actual $A_s = 0.31N$, cm^2	16.0	22.0	28.0	14.0
11. Actual ρ = line 9/bd, %	0.49	0.67	0.85	0.43
12. Actual spacing = b/line 8, in.	12.0	8.7	6.9	13.7
13. Actual spacing = b/line 8, cm	30.5	22.1	17.5	34.8

Table 4-6 *(continued)*

Middle strip, $b = 144$ in. (366 cm), $d = 5.31$ in. (13.5 cm)				
Value	Ext. neg.	Ext. pos.	Int. neg.	Int. pos.
1. Design M_u, ft-k	—	37.2	26.9	24.4
2. Design M_u, Nm × 10^3	—	50.4	36.5	33.1
3. $\bar{M}_u = M_u/\phi = M_u/0.90$, ft-k	—	41.3	29.9	27.1
4. * $R_u = \bar{M}_u/bd^2$, psi	—	122	88	80
5. †‡ Req. ρ, %	—	0.40‡	0.40‡	0.40‡
6. Req. $A_s = \rho bd$, in^2	—	3.06	3.06	3.06
7. Req. N of #5s = $A_s/0.31$	—	10.3‡	10.3‡	10.3‡
8. Use N =	11	11	11	11
9. Actual $A_s = 0.31N$, in^2	3.41	3.41	3.41	3.41
10. Actual $A_s = 0.31N$, cm^2	22.0	22.0	22.0	22.0
11. Actual ρ = line 9/bd, %	0.45	0.45	0.45	0.45
12. Actual spacing = b/line 8, in.	13.1	13.1	13.1	13.1
13. Actual spacing = b/line 8, cm	33.3	33.3	33.3	33.3

* $R_u = \bar{M}_u/bd^2 = (89\,100)(12)/(96)(5.94)^2 = 316$ psi (22.2 kg/cm^2, 2.18 N/mm^2)

† $\rho = (1/m)\left(1 - \sqrt{1 - 2mR_u/f_y}\right) \qquad m = f_y/0.85f'_c$

or use design aid, as Ref. 39, page 47, for example.

‡ ACI 10.5.1, $\rho_{\min} = 200/f_y = 200/50\,000 = 0.0040$ or 0.40%

ACI 7.13, min. $A_s = 0.002bh = (0.002)(96)(7) = 1.34$ in^2 (also 2.02 in^2)

ACI 13.5.1, min. $N = b/2h = 96/14 = 6.9$ (also 10.3 when $b = 144$ in.).

Separate calculations for the exterior half-panel frame are omitted in this example. Also, shear reinforcement is required at the exterior columns, as shown in Table 4-6. If shearhead sections are used for this purpose, their effect on stiffnesses and deflections is not included in the calculations herein.

Deflections computed by several different methods are tabulated in Table 4-11.

Equivalent frame method using the direct design procedure and I_g Except for the calculation of K_{ec} and α_{ec} in Table 4-6, all of the necessary deflection calculations are included in the following example, with the various deflection components shown in Fig. 4-1. Computed results are compared with the allowable deflections from the ACI Code (see Chap. 6 herein, Table 6-15), and summarized in Table 4-7.

Section properties and fixed frame deflections for all panels As shown in Fig. 4-2,

Frame (y–y), $\quad I_g = (192)(7)^3/12 = 5488$ in^4 (228 400 cm^4)

(x–x), $\quad I_g = (240)(7)^3/12 = 6860$ in^4 (285 500 cm^4)

Column strip (y–y), $\quad I_g = (96)(7)^3/12 = 2744$ in^4 (114 200 cm^4)

(x–x), $\quad I_g = (96)(7)^3/12 = 2744$ in^4 (114 200 cm^4)

Middle strip (y–y), $\quad I_g = (96)(7)^3/12 = 2744$ in^4 (114 200 cm^4)

(x–x), $\quad I_g = (144)(7)^3/12 = 4116$ in^4 (171 300 cm^4)

For $f'_c = 4000$ psi (281 kg/cm², 27.6 N/mm²) and normal-weight concrete,

$$E_c = 3.64 \times 10^6 \text{ psi } (256\,000 \text{ kg/cm}^2, 25.1 \text{ kN/mm}^2) \qquad \text{(Table 1-3)}$$

$$\text{Fixed } \Delta_{\text{frame}} = wL^4/384E_cI_{\text{frame}} \qquad \text{(Eq. (4-6))}$$

$$\text{Fixed } (\Delta_{\text{frame},y})_{\text{D,D+L}} = (108 \text{ or } 208)(16)(20)^4(1728)/(384)(3.64 \times 10^6)(5388)$$

$$\text{Fixed } (\Delta_{\text{frame},y})_{\text{D}} = 0.0623 \text{ in. } (0.1582 \text{ cm})$$

$$\text{Fixed } (\Delta_{\text{frame},y})_{\text{D+L}} = 0.1200 \text{ in. } (0.3048 \text{ cm})$$

$$\text{Fixed } (\Delta_{\text{frame},x})_{\text{D,D+L}} = (108 \text{ or } 208)(20)(16)^4(1728)/(384)(3.64 \times 10^6)(6860)$$

$$\text{Fixed } (\Delta_{\text{frame},x})_{\text{D}} = 0.0255 \text{ in. } (0.0648 \text{ cm})$$

$$\text{Fixed } (\Delta_{\text{frame},x})_{\text{D+L}} = 0.0491 \text{ in. } (0.1247 \text{ cm})$$

Interior panel deflections

$$\text{Fixed } \Delta_{\text{c,m}} = (LDF)_{\text{c,m}} (\text{Fixed } \Delta_{\text{frame}}) \frac{I_{\text{frame}}}{I_{\text{c,m}}} \qquad \text{(Eq. (4-7))}$$

For both y–y and x–x,

$$(LDF)_c = 67.5\% \qquad (LDF)_m = 32.5\% \text{ (Table 4-5)}$$

$$\text{Fixed } (\Delta_{cy})_{\text{D}} = (0.675)(0.0623)(5488/2744) = 0.084 \text{ in. } (0.213 \text{ cm})$$

$$\text{Fixed } (\Delta_{cy})_{\text{D+L}} = (0.675)(0.1200)(5488/2744) = 0.162 \text{ in. } (0.411 \text{ cm})$$

$$\text{Fixed } (\Delta_{cy})_{\text{L}} = 0.162 - 0.084 = 0.078 \text{ in. } (0.198 \text{ cm})$$

$$\text{Fixed } (\Delta_{my})_{\text{D}} = (0.325)(0.0623)(5488/2744) = 0.040 \text{ in. } (0.102 \text{ cm})$$

$$\text{Fixed } (\Delta_{my})_{\text{D+L}} = (0.0325)(0.1200)(5488/2744) = 0.078 \text{ in. } (0.198 \text{ cm})$$

$$\text{Fixed } (\Delta_{my})_{\text{L}} = 0.078 - 0.040 = 0.038 \text{ in. } (0.097 \text{ cm})$$

$$\text{Fixed } (\Delta_{cx})_{\text{D}} = (0.675)(0.0255)(6860/2744) = 0.043 \text{ in. } (0.109 \text{ cm})$$

$$\text{Fixed } (\Delta_{cx})_{\text{D+L}} = (0.675)(0.0491)(6860/2744) = 0.083 \text{ in. } (0.211 \text{ cm})$$

$$\text{Fixed } (\Delta_{cx})_{\text{L}} = 0.083 - 0.043 = 0.040 \text{ in. } (0.102 \text{ cm})$$

$$\text{Fixed } (\Delta_{mx})_{\text{D}} = (0.325)(0.0255)(6860/4116) = 0.014 \text{ in. } (0.036 \text{ cm})$$

$$\text{Fixed } (\Delta_{mx})_{\text{D+L}} = (0.325)(0.0491)(6860/4116) = 0.027 \text{ in. } (0.069 \text{ cm})$$

$$\text{Fixed } (\Delta_{mx})_{\text{L}} = 0.027 - 0.014 = 0.013 \text{ in. } (0.033 \text{ cm})$$

For interior panels, assume M_{net} in Eq. (4-8) is zero for equal adjacent spans. Hence end θs = 0 (Eq. (4-8)) and Δ_θs = 0 (Eq. (4-9)). Thus,

$$\Delta_{\text{c,m}} = \text{Fixed } \Delta_{\text{c,m}} \text{ (Eq. (4-11))} \quad \text{and} \quad \Delta = \Delta_{cy} + \Delta_{mx} = \Delta_{cx} + \Delta_{my} \quad \text{(Eq. (4-12))}$$

$(\Delta_{cy})_{\text{D}} + (\Delta_{mx})_{\text{D}} = 0.084 + 0.014 = 0.098$ in.

$(\Delta_{cx})_{\text{D}} + (\Delta_{my})_{\text{D}} = 0.043 + 0.040 = 0.083$ in. Avg. $\Delta_{\text{D}} = 0.091$ in. (0.231 cm)

$(\Delta_{cy})_{\text{L}} + (\Delta_{mx})_{\text{L}} = 0.078 + 0.013 = 0.091$ in.

$(\Delta_{cx})_L + (\Delta_{my})_L = 0.040 + 0.038 = 0.078$ in. Avg. $\Delta_L = 0.085$ in. (0.216 cm)

$$\Delta_t = 3.0\Delta_D = (3.0)(0.091) = 0.273 \text{ in. } (0.693 \text{ cm})$$

$$\Delta_L = 0.09 \text{ in.} < L/180 = (16)(12)/180 = 1.07 \text{ in.} \quad \text{OK}$$
$$< L/360 = 192/360 = 0.53 \text{ in.} \quad \text{OK}$$

$$\Delta_t + \Delta_L = 0.273 + 0.085 = 0.36 \text{ in.} < L/240 = 192/240 = 0.80 \text{ in.} \quad \text{OK}$$
$$< L/480 = 192/480 = 0.40 \text{ in.} \quad \text{OK}$$

Corner and side panel exterior column rotations, and deflections

y–y, $K_{ec} = 117.1E_c = (117.1)(3.64 \times 10^6)$
$= 426.2 \times 10^6$ lb-in^2 (1247×10^6 kg-cm^2, $12\,230 \times 10^6$ N-cm^2) (Table 4-6)

$\alpha_{ec} = 1.280$ (Table 4-6)

Service load $M_0 = w_{D+L}L_2L_n^2/8 = (0.208)(16)(18.5)^2/8 = 142.4$ ft-k

$(M_{net})_{D+L} = 0.65M_0/(1 + 1/\alpha_{ec}) = (0.65)(142.4)/(1 + 1/1.280)$
$= 52.0$ ft-k (7190 kgm, 70 500 Nm) (ACI Code, 13.3.3.3)

$(M_{net})_D = (108/208)(52.0) = 27.0$ ft-k (3730 kgm, 36 600 Nm)

End $\theta_{D+L} = (M_{net})_{frame}/K_{ec} = (52.0)(12\,000)/426.2 \times 10^6$
$= 1.464 \times 10^{-3}$ rad (Eq. (4-8))

End $\theta_D = (27.0)(12\,000)/426.2 \times 10^6 = 0.760 \times 10^{-3}$ rad (Eq. (4-8))

$(\Delta_\theta)_{D+L} = \theta_{D+L}L/8 = (1.464 \times 10^{-3})(20 \times 12)/8 = 0.044$ in. (0.112 cm)

$(\Delta_\theta)_D = (0.760 \times 10^{-3})(20 \times 12)/8 = 0.023$ in. (0.058 cm) (Eq. (4-9))

$(\Delta_\theta)_L = 0.044 - 0.023 = 0.021$ in. (0.053 cm)

x–x, $K_{ec} = 97.9E_c = (97.9)(3.64 \times 10^6)$
$= 356.4 \times 10^6$ lb-in^2 (1043×10^6 kg-cm^2, $10\,230 \times 10^6$ N-cm^2) (Table 4-6)

$\alpha_{ec} = 0.685$ (Table 4-6)

Service load $M_0 = w_{D+L}L_2L_n^2/8 = (0.208)(20)(14.5)^2/8 = 109.3$ ft-k

$(M_{net})_{D+L} = 0.65M_0/(1 + 1/\alpha_{ec}) = (0.65)(109.3)/(1 + 1/0.685)$
$= 28.9$ ft-k (4000 kgm, 39 200 Nm)

$(M_{net})_D = (108/208)(28.9) = 15.0$ ft-k (2070 kgm, 20 300 Nm)

End $\theta_{D+L} = (M_{net})_{frame}/K_{ec} = (28.9)(12\,000)/356.4 \times 10^6$
$= 0.973 \times 10^{-3}$ rad (Eq. (4-8))

End $\theta_D = (15.0)(12\,000)/356.4 \times 10^6 = 0.505 \times 10^{-3}$ rad (Eq. (4-8))

$(\Delta_\theta)_{D+L} = \theta_{D+L}L/8 = (0.973 \times 10^{-3})(16 \times 12)/8 = 0.023$ in. (0.058 cm)

$(\Delta_\theta)_D = (0.505 \times 10^{-3})(16 \times 12)/8 = 0.012$ in. (0.030 cm) (Eq. (4-9))

$(\Delta_\theta)_L = 0.023 - 0.012 = 0.011$ in. (0.028 cm)

The rotations of all interior columns are neglected for this slab with equal adjacent spans, as a practical matter.

Corner panel deflections

$$\text{Fixed } \Delta_{c,m} = (LDF)_{c,m}\,(\text{Fixed } \Delta_{frame})\,\frac{I_{frame}}{I_{c,m}} \qquad \text{(Eq. (4-7))}$$

For both y–y and x–x,

$$(LDF)_c = 73.8\% \qquad (LDF)_m = 26.2\% \text{ (Table 4-5)}$$

$$\text{Fixed } (\Delta_{cy})_D = (0.738)(0.0623)(5488/2744) = 0.092 \text{ in. } (0.234 \text{ cm})$$

$$\text{Fixed } (\Delta_{cy})_{D+L} = (0.738)(0.1200)(5488/2744) = 0.177 \text{ in. } (0.452 \text{ cm})$$

$$\text{Fixed } (\Delta_{cy})_L = 0.177 - 0.092 = 0.085 \text{ in. } (0.216 \text{ cm})$$

$$\text{Fixed } (\Delta_{my})_D = (0.262)(0.0623)(5488/2744) = 0.033 \text{ in. } (0.084 \text{ cm})$$

$$\text{Fixed } (\Delta_{my})_{D+L} = (0.262)(0.1200)(5488/2744) = 0.063 \text{ in. } (0.160 \text{ cm})$$

$$\text{Fixed } (\Delta_{my})_L = 0.063 - 0.033 = 0.030 \text{ in. } (0.076 \text{ cm})$$

$$\text{Fixed } (\Delta_{cx})_D = (0.738)(0.0255)(6860/2744) = 0.047 \text{ in. } (0.119 \text{ cm})$$

$$\text{Fixed } (\Delta_{cx})_{D+L} = (0.738)(0.0491)(6860/2744) = 0.091 \text{ in. } (0.231 \text{ cm})$$

$$\text{Fixed } (\Delta_{cx})_L = 0.091 - 0.047 = 0.044 \text{ in. } (0.112 \text{ cm})$$

$$\text{Fixed } (\Delta_{mx})_D = (0.262)(0.0255)(6860/4116) = 0.011 \text{ in. } (0.028 \text{ cm})$$

$$\text{Fixed } (\Delta_{mx})_{D+L} = (0.262)(0.0491)(6860/4116) = 0.021 \text{ in. } (0.053 \text{ cm})$$

$$\text{Fixed } (\Delta_{mx})_L = 0.021 - 0.011 = 0.010 \text{ in. } (0.025 \text{ cm})$$

$$\Delta_{c,m} = \text{Fixed } \Delta_{c,m} + (\Delta_{\theta 1})_{c,m} + 0 \qquad \text{(Eq. (4-11))}$$

$$(\Delta_{cy})_D = 0.092 + 0.023 = 0.115 \text{ in. } (0.292 \text{ cm})$$

$$(\Delta_{my})_D = 0.033 + 0.023 = 0.056 \text{ in. } (0.142 \text{ cm})$$

$$(\Delta_{cx})_D = 0.047 + 0.012 = 0.059 \text{ in. } (0.150 \text{ cm})$$

$$(\Delta_{mx})_D = 0.011 + 0.012 = 0.023 \text{ in. } (0.058 \text{ cm})$$

$$(\Delta_{cy})_L = 0.085 + 0.021 = 0.106 \text{ in. } (0.269 \text{ cm})$$

$$(\Delta_{my})_L = 0.030 + 0.021 = 0.051 \text{ in. } (0.130 \text{ cm})$$

$$(\Delta_{cx})_L = 0.044 + 0.011 = 0.055 \text{ in. } (0.140 \text{ cm})$$

$$(\Delta_{mx})_L = 0.010 + 0.011 = 0.021 \text{ in. } (0.053 \text{ cm})$$

$$\Delta = \Delta_{cy} + \Delta_{mx} = \Delta_{cx} + \Delta_{my} \qquad \text{(Eq. (4-12))}$$

$$(\Delta_{cy})_D + (\Delta_{mx})_D = 0.115 + 0.023 = 0.138 \text{ in.}$$

$$(\Delta_{cx})_D + (\Delta_{my})_D = 0.059 + 0.056 = 0.115 \text{ in.} \qquad \text{Avg. } \Delta_D = 0.127 \text{ in. } (0.323 \text{ cm})$$

$$(\Delta_{cy})_L + (\Delta_{mx})_L = 0.106 + 0.021 + 0.127 \text{ in.}$$

$$(\Delta_{cx})_L + (\Delta_{my})_L = 0.055 + 0.051 = 0.107 \text{ in.} \qquad \text{Avg. } \Delta_L = 0.117 \text{ in. } (0.297 \text{ cm})$$

$$\Delta_t = 3.0\Delta_D = (3.0)(0.127) = 0.381 \text{ in. } (0.968 \text{ cm})$$

$$\Delta_L = 0.12 \text{ in.} < L/180 = (16)(12)/180 = 1.07 \text{ in.} \quad \text{OK}$$
$$< L/360 = 192/360 = 0.53 \text{ in.} \quad \text{OK}$$
$$\Delta_t + \Delta_L = 0.381 + 0.117 = 0.50 \text{ in.} < L/240 = 192/240 = 0.80 \text{ in.} \quad \text{OK}$$
$$> L/480 = 192/480 = 0.40 \text{ in.} \quad \text{Not OK}$$

The $L/480$ criterion is exceeded when using the shorter span as L, but would be met when using the longer span of $L = 20$ ft (6.1 m). However, any effect of cracking was excluded in this example.

Side panel 2 deflections

$$(\Delta_{cy})_D = 0.115 \text{ in. } (0.292 \text{ cm})\text{—same as corner panel}$$
$$(\Delta_{my})_D = 0.056 \text{ in. } (0.142 \text{ cm})\text{—same as corner panel}$$
$$(\Delta_{cy})_L = 0.106 \text{ in. } (0.269 \text{ cm})\text{—same as corner panel}$$
$$(\Delta_{my})_L = 0.051 \text{ in. } (0.130 \text{ cm})\text{—same as corner panel}$$

$$(\Delta_{cx})_D = 0.043 \text{ in. } (0.109 \text{ cm})\text{—same as interior panel}$$
$$(\Delta_{mx})_D = 0.014 \text{ in. } (0.036 \text{ cm})\text{—same as interior panel}$$
$$(\Delta_{cx})_L = 0.040 \text{ in. } (0.102 \text{ cm})\text{—same as interior panel}$$
$$(\Delta_{mx})_L = 0.013 \text{ in. } (0.033 \text{ cm})\text{—same as interior panel}$$

$$\Delta = \Delta_{cy} + \Delta_{mx} = \Delta_{cx} + \Delta_{my} \qquad \text{(Eq. (4-12))}$$
$$(\Delta_{cy})_D + (\Delta_{mx})_D = 0.115 + 0.014 = 0.129 \text{ in.}$$
$$(\Delta_{cx})_D + (\Delta_{my})_D = 0.043 + 0.056 = 0.099 \text{ in.} \qquad \text{Avg. } \Delta_D = 0.114 \text{ in. } (0.290 \text{ cm})$$
$$(\Delta_{cy})_L + (\Delta_{mx})_L = 0.106 + 0.013 = 0.119 \text{ in.}$$
$$(\Delta_{cx})_L + (\Delta_{my})_L = 0.040 + 0.051 = 0.091 \text{ in.} \qquad \text{Avg. } \Delta_L = 0.105 \text{ in. } (0.267 \text{ cm})$$
$$\Delta_t = 3.0\Delta_D = (3.0)(0.114) = 0.342 \text{ in. } (0.869 \text{ cm})$$

$$\Delta_L = 0.11 \text{ in.} < L/180 = (16)(12)/180 = 1.07 \text{ in.} \quad \text{OK}$$
$$< L/360 = 192/360 = 0.53 \text{ in.} \quad \text{OK}$$
$$\Delta_t + \Delta_L = 0.342 + 0.105 = 0.45 \text{ in.} < L/240 = 192/240 = 0.80 \text{ in.} \quad \text{OK}$$
$$> L/480 = 192/240 = 0.40 \text{ in.} \quad \text{Not OK}$$

Side panel 3 deflections

$$(\Delta_{cy})_D = 0.084 \text{ in. } (0.213 \text{ cm})\text{—same as interior panel}$$
$$(\Delta_{my})_D = 0.040 \text{ in. } (0.102 \text{ cm})\text{—same as interior panel}$$
$$(\Delta_{cy})_L = 0.078 \text{ in. } (0.198 \text{ cm})\text{—same as interior panel}$$
$$(\Delta_{my})_L = 0.038 \text{ in. } (0.097 \text{ cm})\text{—same as interior panel}$$

$$(\Delta_{cx})_D = 0.059 \text{ in. } (0.150 \text{ cm})\text{—same as corner panel}$$
$$(\Delta_{mx})_D = 0.023 \text{ in. } (0.058 \text{ cm})\text{—same as corner panel}$$

$(\Delta_{cx})_L = 0.055$ in. (0.140 cm)—same as corner panel

$(\Delta_{mx})_L = 0.021$ in. (0.053 cm)—same as corner panel

$$\Delta = \Delta_{cy} + \Delta_{mx} = \Delta_{cx} + \Delta_{my} \quad \text{(Eq. (4-12))}$$

$(\Delta_{cy})_D + (\Delta_{mx})_D = 0.084 + 0.023 = 0.107$ in.

$(\Delta_{cx})_D + (\Delta_{my})_D = 0.059 + 0.040 = 0.099$ in. Avg. $\Delta_D = 0.103$ in. (0.262 cm)

$(\Delta_{cy})_L + (\Delta_{mx})_L = 0.078 + 0.021 = 0.099$ in.

$(\Delta_{cx})_L + (\Delta_{my})_L = 0.055 + 0.038 = 0.093$ in. Avg. $\Delta_L = 0.096$ in. (0.244 cm)

$\Delta_t = 3.0\Delta_D = (3.0)(0.103) = 0.309$ in. (0.785 cm)

$\Delta_L = 0.10$ in. $< L/180 = (16)(12)/180 = 1.07$ in. OK

$< L/360 = 192/360 = 0.53$ in. OK

$\Delta_t + \Delta_L = 0.309 + 0.096 = 0.41$ in. $< L/240 = 192/240 = 0.80$ in. OK

$< L/480 = 192/480 = 0.40$ in. OK, close enough

Equivalent frame method using the direct design procedure and I_e The deflections of the flat plate designed in Table 4-6 are computed in this example using I_e. Based on the service load moments and section properties determined in Table 4-8 and summarized in Table 4-9, the following section properties are used: for all dead load-deflections—I_g; and for dead plus live-load deflections—I_g for the middle strips in both directions and I_e for the column strips in both directions (using a weighted average of 70 percent positive and 15 percent, 15 percent, negative values).

The various deflection components are shown in Fig. 4-1. Computed results are compared with the allowable deflections from the ACI Code (see Chap. 6 herein, Table 6-15), and summarized in Table 4-10.

Although a number of the calculations are the same as for the previous example, all calculations are shown in this example for completeness.

Interior panel deflections

$$\text{Fixed } \Delta_{\text{frame}} = wL^4/384E_cI_{\text{frame}} \quad \text{(Eq. (4-6))}$$

$$\text{Fixed } (\Delta_{\text{frame},y})_{D,D+L} = (108 \text{ or } 208)(16)(20)^4(1728)/(384)(3.64 \times 10^6)(5488 \text{ or } 4992)$$

$$\text{Fixed } (\Delta_{\text{frame},y})_D = 0.0623 \text{ in. } (0.1582 \text{ cm})$$

$$\text{Fixed } (\Delta_{\text{frame},y})_{D+L} = 0.1319 \text{ in. } (0.3350 \text{ cm})$$

$$\text{Fixed } (\Delta_{\text{frame},x})_{D,D+L} = (108 \text{ or } 208)(20)(16)^4(1728)/(384)(3.64 \times 10^6)(6860 \text{ or } 6363)$$

$$\text{Fixed } (\Delta_{\text{frame},x})_D = 0.0255 \text{ in. } (0.0648 \text{ cm})$$

$$\text{Fixed } (\Delta_{\text{frame},x})_{D+L} = 0.0530 \text{ in. } (0.1346 \text{ cm})$$

Table 4-7 Summary of flat plate deflections by the equivalent frame method using I_g

20 ft (6.1 m)	*Corner panel 4* $\Delta_D = 0.13$ in. (0.32 cm) * $\Delta_t = 0.38$ in. (0.97 cm) $\Delta_L = 0.12$ in. (0.30 cm) $\Delta_t + \Delta_L = 0.50$ in. (1.27 cm)	*Side panel 2* $\Delta_D = 0.11$ in. (0.29 cm) * $\Delta_t = 0.34$ in. (0.87 cm) $\Delta_L = 0.11$ in. (0.27 cm) $\Delta_t + \Delta_L = 0.45$ in. (1.14 cm)
20 ft (6.1 m)	*Side panel 3* $\Delta_D = 0.10$ in. (0.26 cm) * $\Delta_t = 0.31$ in. (0.79 cm) $\Delta_L = 0.10$ in. (0.24 cm) $\Delta_t + \Delta_L = 0.41$ in. (1.03 cm)	† *Interior panel 1* $\Delta_D = 0.09$ in. (0.23 cm) * $\Delta_t = 0.27$ in. (0.69 cm) $\Delta_L = 0.09$ in. (0.22 cm) $\Delta_t + \Delta_L = 0.36$ in. (0.90 cm)
	16 ft (4.9 m)	16 ft (4.9 m)

* $\Delta_t = 3.0\Delta_D$.

† In the solution herein, no distinction is made between interior panel 1 and the other interior panels.

Comparison with ACI Code Allowable Deflections:

$\Delta_L < L/180 = (16)(12)/180 = 1.07$ in. (2.72 cm)	OK all panels
$< L/360 = 192/360 = 0.53$ in. (1.35 cm)	OK all panels
$\Delta_t + \Delta_L < L/240 = 192/240 = 0.80$ in. (2.03 cm)	OK all panels
$< L/480 = 192/480 = 0.40$ in. (1.02 cm)	OK all panels except the corner and side panel 2 using $L = 16$ ft. When using $L = 20$ ft (6.1 m), all panels would satisfy the ACI Code criteria. However, any effect of cracking was excluded in this example. Also, $T = 3.0$ was used instead of the ACI Code $T = 2$.

The deflections are the smallest in the interior panel and the largest in the corner panel. The deflections of side panel 2 are larger than those of side panel 3 because the exterior end of side panel 2 is rotating in the long direction while the exterior end of side panel 3 is rotating in the short direction.

$$\text{Fixed } \Delta_{c,m} = (LDF)_{c,m} (\text{Fixed } \Delta_{\text{frame}}) \frac{I_{\text{frame}}}{I_{c,m}} \qquad \text{(Eq. (4-7))}$$

For both y–y and x–x

$$(LDF)_c = 67.5\% \qquad (LDF)_m = 32.5\% \text{ (Table 4-5)}$$

$$\text{Fixed } (\Delta_{cy})_D = (0.675)(0.0623)(5488/2744) = 0.084 \text{ in. } (0.213 \text{ cm})$$

$$\text{Fixed } (\Delta_{cy})_{D+L} = (0.675)(0.1319)(4992/2248) = 0.198 \text{ in. } (0.503 \text{ cm})$$

$$\text{Fixed } (\Delta_{cy})_L = 0.198 - 0.084 = 0.114 \text{ in. } (0.290 \text{ cm})$$

Table 4-8 Service load moments and section properties for deflection calculations of solid slab system without beams (flat plate)

For $f_c' = 4000$ psi (281 kg/cm², 27.6 N/mm²) and normal-weight concrete,

$f_r = \underline{496 \text{ psi}}$ (34.0 kg/cm², 3.42 N/mm²) (Table 1-2)

$E_c = \underline{3.64 \times 10^6 \text{ psi}}$ (256 000 kg/cm², 25.1 kN/mm²) (Table 1-3)

$n = E_s/E_c = 29/3.64 = \underline{8.0}$

Long direction (y–y)

Service load moments

$M_0 = w_{D+L} L_2 L_n^2/8 = (0.208)(16)(18.5)^2/8 = 142.4$ ft-k (19 700 kgm, 193 100 Nm)

	Exterior span			Int. span	
	Ext. neg.	Pos.	Int. neg.	Neg.	Pos.
1. Mom. distr. ratios (from Table 4-6)	0.365	0.473	0.694	0.65	0.35
2. $(M_a)_{D+L} = M_0$ (line 1), ft-k	52.0	67.4	98.8	92.6	49.8
3. $(M_a)_D = (108/208)$ (line 2), ft-k	27.0	35.0	51.3	48.1	25.9
4. Col. strip mom. ratios (from Table 4-6)	1.00	0.60	0.75	0.75	0.60
5. CS $(M_a)_{D+L}$ = (line 2) (line 4), ft-k	52.0	40.4	74.1	69.5	29.9
6. MS $(M_a)_{D+L}$ = (line 2) – (line 5), ft-k	0	27.0	24.7	23.1	19.9
7. CS $(M_a)_D$ = (line 3) (line 4), ft-k	27.0	21.0	38.5	36.1	15.5
8. MS $(M_a)_D$ = (line 3) – (line 7), ft-k	0	14.0	12.8	12.0	10.4

For both column and middle strips,

$$I_g = (96)(7)^3/12 = 2744 \text{ in}^4 \ (114\,200 \text{ cm}^4)$$

$$M_{cr} = f_r I_g/y_t = (0.496)(2744)/(3.5 \times 12) = 32.4 \text{ ft-k} \ (4480 \text{ kgm}, 43\,900 \text{ Nm})$$

$(M_a)_{D+L} < M_{cr}$ at all middle strip locations. Hence use I_g for all middle strip deflection calculations. Use I_e for column strip under $D + L$.

$(M_a)_D > M_{cr}$ only at the interior columns (38.5, 36.1 versus 32.4 ft-k), and the moments are smaller in the short direction. Hence, for practical purposes, assume the slab to be uncracked under dead load.

Column strip I_e, $b = 96$ in. (244 cm), $d = 5.94$ in. (15.1 cm)

	Exterior span			Int. span	
	Ext. neg.	Pos.	Int. neg.	Neg.	Pos.
9. ρn = (Actual ρ, Table 4-6) (8.0)	0.057	0.048	0.074	0.074	0.034
10. I_{cr}/bd^3 (Fig. B-2, Chap. 3)	0.038	0.031	0.045	0.045	0.025
11. $I_{cr} = (96)(5.94)^3$ (line 10), in⁴	765	624	905	905	503
12. I_g/I_{cr} = 2744/(line 11)	3.6	4.4	3.0	3.0	5.4
13. $M_{cr}/(M_a)_{D+L}$ = 32.4/(line 5)	0.623	0.802	0.437	0.466	>1
14. I_e, Eq. (3-19), in⁴*	1244	1718	1058	1091	2744
15. $I_e \times 10^{-3}$, cm⁴	51.8	71.5	44.0	45.4	114.2

* $I_e = (M_{cr}/M_a)^3 I_g + [1 - (M_{cr}/M_a)^3] I_{cr}$

Table 4-8 (*continued*)

Short direction (x–x)

Service load moments

$$M_0 = w_{D+L}L_2L_n^2/8 = (0.208)(20)(14.5)^2/8 = 109.3 \text{ ft-k } (15\,100 \text{ kgm, } 148\,200 \text{ Nm})$$

	Exterior span			Int. span	
	Ext. neg.	Pos.	Int. neg.	Neg.	Pos.
1. Mom. distr. ratios (from Table 4-6)	0.264	0.516	0.709	0.65	0.35
2. $(M_a)_{D+L} = M_0$ (line 1), ft-k	28.9	56.4	77.5	71.0	38.3
3. $(M_a)_D = (108/208)$ (line 2), ft-k	15.0	29.3	40.2	36.9	19.9
4. Col. strip mom. ratios (from Table 4-6)	1.00	0.60	0.75	0.75	0.60
5. CS $(M_a)_{D+L}$ = (line 2) (line 4), ft-k	28.9	33.8	58.1	53.3	23.0
6. MS $(M_a)_{D+L}$ = (line 2) − (line 5), ft-k	0	22.6	19.4	17.7	15.3
7. CS $(M_a)_D$ = (line 3) (line 4), ft-k	15.0	17.5	30.2	27.7	11.9
8. MS $(M_a)_D$ = (line 3) − (line 7), ft-k	0	11.8	10.0	9.2	8.0

For column strip, $I_g = 2744 \text{ in}^4$, $M_{cr} = 32.4$ ft-k (same as long dir.,)

For middle strip, $I_g = (144)(7)^3/12 = 4116 \text{ in}^4$ (171 000 cm^4)

$$M_{cr} = f_r I_g/y_t = (0.496)(4116)/(3.5 \times 12) = 48.6 \text{ ft-k } (6720 \text{ kgm, } 65\,900 \text{ Nm})$$

M_{cr} is exceeded only by CS $(M_a)_{D+L}$. Hence compute I_e only for the column strip under dead plus live load.

Column strip I_e, $b = 96$ in. (244 cm), $d = 5.31$ in. (13.5 cm)

	Exterior span			Int. span	
	Ext. neg.	Pos.	Int. neg.	Neg.	Pos.
9. ρn = (Actual ρ, Table 4-6) (8.0)	0.039	0.054	0.068	0.068	0.034
10. I_{cr}/bd^3 (Fig. B-2, Chap. 3)	0.027	0.036	0.042	0.042	0.025
11. $I_{cr} = (96)(5.31)^3$ (line 10), in^4	388	517	604	604	359
12. I_g/I_{cr} = 2744/(line 11)	7.0	5.3	4.5	4.5	7.6
13. $M_{cr}/(M_a)_{D+L}$ = 32.4/(line 5)	>1	0.959	0.558	0.608	>1
14. I_e, Eq. (3-19), in^4 *	2744	2481	976	1085	2744
15. $I_e \times 10^{-3}$, cm^4	114.2	103.3	40.6	45.2	114.2

$$\text{Fixed } (\Delta_{my})_D = (0.325)(0.0623)(5488/2744) = 0.040 \text{ in. } (0.102 \text{ cm})$$

$$\text{Fixed } (\Delta_{my})_{D+L} = (0.325)(0.1319)(4992/2744) = 0.078 \text{ in. } (0.198 \text{ cm})$$

$$\text{Fixed } (\Delta_{my})_L = 0.078 - 0.040 = 0.038 \text{ in. } (0.097 \text{ cm})$$

$$\text{Fixed } (\Delta_{cx})_D = (0.675)(0.0255)(6860/2744) = 0.043 \text{ in. } (0.109 \text{ cm})$$

$$\text{Fixed } (\Delta_{cx})_{D+L} = (0.675)(0.0530)(6363/2247) = 0.101 \text{ in. } (0.257 \text{ cm})$$

$$\text{Fixed } (\Delta_{cx})_L = 0.101 - 0.043 = 0.058 \text{ in. } (0.147 \text{ cm})$$

Table 4-9 Summary of service load moments and section properties from Table 4-8

Summary of service load moments and section properties
Use I_g for all dead load deflection calculations.

Long direction $(y\text{–}y)$—$K_{ec} = \underline{117.1E_c}$, $(I_g)_{c,m} = \underline{2744\ \text{in}^4}$ (114 200 cm⁴),
$(I_g)_{\text{frame}} = (2)(2744) = \underline{5488\ \text{in}^4}$ (228 400 cm⁴).

Column strip: M_D, ft-k	−27.0	21.0	−38.5	−36.1	15.5	−36.1
M_{D+L}, ft-k	−52.0	40.4	−74.1	−69.5	29.9	−69.5
$(I_e)_{D+L}$, in⁴	1244	1718	1058	1091	2744*	1091
Middle strip: M_D, ft-k	0	14.0	−12.8	−12.0	10.3	−12.0
M_{D+L}, ft-k	0	27.0	−24.7	−23.1	19.9	−23.1
$(I_e)_{D+L}$, in⁴	2744*	2744*	2744*	2744*	2744*	2744*

Panel 1 and 3 (interior panels in long direction)—$(I_e)_{D+L}$

$$(I_e)_c = 0.70I_m + 0.15(I_{e1} + I_{e2})$$
$$= (0.70)(2744) + (0.30)(1091) = \underline{2248\ \text{in}^4}\ (93\,560\ \text{cm}^4)$$
$$(I_e)_m = \underline{2744\ \text{in}^4}\ (114\,200\ \text{cm}^4)$$
$$(I_e)_{\text{frame}} = (I_e)_c + (I_e)_m = 2248 + 2744 = \underline{4992\ \text{in}^4}\ (207\,800\ \text{cm}^4)$$

Panel 2 and 4 (exterior panels in long direction)—$(I_e)_{D+L}$

$$(I_e)_c = (0.70)(1718) + 0.15(1058 + 1244) = \underline{1548\ \text{in}^4}\ (64\,400\ \text{cm}^4)$$
$$(I_e)_m = \underline{2744\ \text{in}^4}\ (114\,200\ \text{cm}^4)$$
$$(I_e)_{\text{frame}} = 1548 + 2744 = \underline{4292\ \text{in}^4}\ (178\,600\ \text{cm}^4)$$

Short direction $(x\text{–}x)$—$K_{ec} = \underline{97.9E_c}$, $(I_g)_c = \underline{2744\ \text{in}^4}$ (114 200 cm⁴),
$(I_g)_m = \underline{4116\ \text{in}^4}$ (171 300 cm⁴), $(I_g)_{\text{frame}} = 2744 + 4116 = \underline{6860\ \text{in}^4}$ (285 500 cm⁴)

Column strip: M_D, ft-k	−15.0	17.5	−30.2	−27.7	11.9	−27.7
M_{D+L}, ft-k	−28.9	33.8	−58.1	−53.3	23.0	−53.3
$(I_e)_{D+L}$, in⁴	2744*	2481	976	1085	2744*	1085
Middle strip: M_D, ft-k	0	11.7	−10.1	−9.2	7.9	−9.2
M_{D+L}, ft-k	0	22.6	−19.4	17.7	15.3	−17.7
$(I_e)_{D+L}$, in⁴	4116*	4116*	4116*	4116*	4116*	4116*

Panel 1 and 2 (interior panels in short direction)—$(I_e)_{D+L}$

$$(I_e)_c = 0.70(2744) + 0.30(1085) = \underline{2247\ \text{in}^4}\ (93\,500\ \text{cm}^4)$$
$$(I_e)_m = \underline{4116\ \text{in}^4}\ (171\,300\ \text{cm}^4)$$
$$(I_e)_{\text{frame}} = 2247 + 4116 = \underline{6363\ \text{in}^4}\ (264\,800\ \text{cm}^4)$$

Panel 3 and 4 (exterior panels in short direction)—$(I_e)_{D+L}$

$$(I_e)_c = 0.70(2481) + 0.15(976 + 2744) = \underline{2295\ \text{in}^4}\ (95\,500\ \text{cm}^4)$$
$$(I_e)_m = \underline{4116\ \text{in}^4}\ (171\,300\ \text{cm}^4)$$
$$(I_e)_{\text{frame}} = 2295 + 4116 = \underline{6411\ \text{in}^4}\ (266\,800\ \text{cm}^4)$$

* Denotes I_g.

$$\text{Fixed } (\Delta_{mx})_D = (0.325)(0.0255)(6860/4116) = 0.014 \text{ in. } (0.036 \text{ cm})$$
$$\text{Fixed } (\Delta_{mx})_{D+L} = (0.325)(0.0530)(6363/4116) = 0.027 \text{ in. } (0.069 \text{ cm})$$
$$\text{Fixed } (\Delta_{mx})_L = 0.027 - 0.014 = 0.013 \text{ in. } (0.033 \text{ cm})$$

For interior panels, assume M_{net} in Eq. (4-8) is zero for equal adjacent spans. Hence end θs $= 0$ (Eq. (4-8)) and Δ_θs $= 0$ (Eq. (4-10)). Thus,

$$\Delta_{c,m} = \text{Fixed } \Delta_{c,m} \qquad (\text{Eq. (4-11)})$$

and

$$\Delta = \Delta_{cy} + \Delta_{mx} = \Delta_{cx} + \Delta_{my} \qquad (\text{Eq. (4-12)})$$

$$(\Delta_{cy})_D + (\Delta_{mx})_D = 0.084 + 0.014 = 0.098 \text{ in.}$$
$$(\Delta_{cx})_D + (\Delta_{my})_D = 0.043 + 0.040 = 0.083 \text{ in.} \qquad \text{Avg. } \Delta_D = 0.091 \text{ in. } (0.231 \text{ cm})$$
$$(\Delta_{cy})_L + (\Delta_{mx})_L = 0.114 + 0.013 = 0.127 \text{ in.}$$
$$(\Delta_{cx})_L + (\Delta_{my})_L = 0.058 + 0.038 = 0.096 \text{ in.} \qquad \text{Avg. } \Delta_L = 0.112 \text{ in. } (0.284 \text{ cm})$$
$$\Delta_t = 3.0\Delta_D = (3.0)(0.091) = 0.273 \text{ in. } (0.693 \text{ cm})$$

$$\Delta_L = 0.11 \text{ in.} < L/180 = (16)(12)/180 = 1.07 \text{ in.} \qquad \text{OK}$$
$$< L/360 = 192/360 = 0.53 \text{ in.} \qquad \text{OK}$$
$$\Delta_t + \Delta_L = 0.273 + 0.112 = 0.39 \text{ in.} < L/240 = 192/240 = 0.80 \text{ in.} \qquad \text{OK}$$
$$< L/480 = 192/480 = 0.40 \text{ in.} \qquad \text{OK}$$

Corner and side panel exterior column rotations, and deflections *y–y*, interior frame used with corner panel 4 and side panel 2:

$$K_{ec} = 117.1E_c = (117.1)(3.64 \times 10^6) = 426.2 \times 10^6 \text{ lb-in}^2 \quad (\text{Table 4-6})$$
$$\alpha_{ec} = 1.280 \qquad (\text{Table 4-6})$$
$$\text{Service load } M_0 = w_{D+L}L_2L_n^2/8 = (0.208)(16)(18.5)^2/8 = 142.4 \text{ ft-k}$$
$$(M_{net})_{D+L} = 0.65M_0/(1 + 1/\alpha_{ec}) = (0.65)(142.4)/(1 + 1/1.280)$$
$$= 52.0 \text{ ft-k } (7190 \text{ kgm}, 70\,500 \text{ Nm}) \qquad (\text{ACI Code, 13.3.3.3})$$
$$(M_{net})_D = (108/208)(52.0) = 27.0 \text{ ft-k } (3730 \text{ kgm}, 36\,600 \text{ Nm})$$
$$\text{End } \theta_{D+L} = (M_{net})_{frame}/K_{ec} = (52.0)(12\,000)/426.2 \times 10^6$$
$$= 1.464 \times 10^{-3} \text{ rad} \qquad (\text{Eq. (4-8)})$$
$$\text{End } \theta_D = (27.0)(12\,000)/426.2 \times 10^6 = 0.760 \times 10^{-3} \text{ rad} \qquad (\text{Eq. (4-8)})$$
$$(\Delta_\theta)_{D+L} = (\theta_{D+L}L/8)(I_g/I_e)_{frame} \qquad (\text{Eq. (4-10)})$$
$$= (1.464 \times 10^{-3})(20 \times 12/8)(5488/4292) = 0.056 \text{ in. } (0.142 \text{ cm})$$
$$(\Delta_\theta)_D = (0.760 \times 10^{-3})(20 \times 12/8)(5488/5488) = 0.023 \text{ in. } (0.058 \text{ cm})$$
$$(\Delta_\theta)_L = 0.056 - 0.023 = 0.033 \text{ in. } (0.084 \text{ cm})$$

x–x, interior frame used with corner panel 4 and side panel 3:

$$K_{ec} = 97.9E_c = (97.9)(3.64 \times 10^6) = 356.4 \times 10^6 \text{ lb-in}^2 \qquad (\text{Table 4-6})$$
$$\alpha_{ec} = 0.685 \qquad (\text{Table 4-6})$$
$$\text{Service load } M_0 = w_{D+L}L_2L_n^2/8 = (0.208)(20)(14.5)^2/8 = 109.3 \text{ ft-k}$$

$$(M_{\text{net}})_{\text{D+L}} = 0.65M_0/(1 + 1/\alpha_{\text{ec}}) = (0.65)(109.3)/(1 + 1/0.685)$$
$$= 28.9 \text{ ft-k } (4000 \text{ kgm}, 39\,200 \text{ Nm})$$
$$(M_{\text{net}})_{\text{D}} = (108/208)(28.9) = 15.0 \text{ ft-k } (2070 \text{ kgm}, 20\,300 \text{ Nm})$$
$$\text{End } \theta_{\text{D+L}} = (M_{\text{net}})_{\text{frame}}/K_{\text{ec}} = (28.9)(12\,000)/356.4 \times 10^6$$
$$= 0.973 \times 10^{-3} \text{ rad} \qquad (\text{Eq. (4-8)})$$
$$\text{End } \theta_{\text{D}} = (15.0)(12\,000)/356.4 \times 10^6 = 0.505 \times 10^{-3} \text{ rad} \qquad (\text{Eq. (4-8)})$$
$$(\Delta_\theta)_{\text{D+L}} = (\theta_{\text{D+L}}L/8)(I_g/I_e)_{\text{frame}} \qquad (\text{Eq. (4-10)})$$
$$= (0.973 \times 10^{-3})(16 \times 12/8)(6860/6411) = 0.025 \text{ in. } (0.064 \text{ cm})$$
$$(\Delta_\theta)_{\text{D}} = (0.505 \times 10^{-3})(16 \times 12/8)(6860/6860) = 0.012 \text{ in. } (0.030 \text{ cm})$$
$$(\Delta_\theta)_{\text{L}} = 0.025 - 0.012 = 0.013 \text{ in. } (0.033 \text{ cm})$$

The rotations of all interior columns are neglected for this slab with equal adjacent spans, as a practical matter.

Corner panel deflections

$$\text{Fixed } \Delta_{\text{frame}} = wL^4/384E_cI_{\text{frame}} \qquad (\text{Eq. (4-6)})$$
$$\text{Fixed } (\Delta_{\text{frame},y})_{\text{D,D+L}} = (108 \text{ or } 208)(16)(20)^4(1728)/(384)(3.64 \times 10^6)(5488 \text{ or } 4292)$$
$$\text{Fixed } (\Delta_{\text{frame},y})_{\text{D}} = 0.0623 \text{ in. } (0.1582 \text{ cm})$$
$$\text{Fixed } (\Delta_{\text{frame},y})_{\text{D+L}} = 0.1639 \text{ in. } (0.4163 \text{ cm})$$

$$\text{Fixed } (\Delta_{\text{frame},x})_{\text{D,D+L}} = (108 \text{ or } 208)(20)(16)^4(1728)/(384)(3.64 \times 10^6)(6860 \text{ or } 6411)$$
$$\text{Fixed } (\Delta_{\text{frame},x})_{\text{D}} = 0.0255 \text{ in. } (0.0648 \text{ cm})$$
$$\text{Fixed } (\Delta_{\text{frame},x})_{\text{D+L}} = 0.0526 \text{ in. } (0.1336 \text{ cm})$$

$$\text{Fixed } \Delta_{\text{c,m}} = (LDF)_{\text{c,m}} (\text{Fixed } \Delta_{\text{frame}}) \frac{I_{\text{frame}}}{I_{\text{c,m}}} \qquad (\text{Eq. (4-7)})$$

For both y–y and x–x,

$$(LDF)_c = 73.8\% \qquad (LDF)_m = 26.2\% \text{ (Table 4-5)}$$
$$\text{Fixed } (\Delta_{cy})_{\text{D}} = (0.738)(0.0623)(5488/2744) = 0.092 \text{ in. } (0.234 \text{ cm})$$
$$\text{Fixed } (\Delta_{cy})_{\text{D+L}} = (0.738)(0.1639)(4292/1548) = 0.335 \text{ in. } (0.851 \text{ cm})$$
$$\text{Fixed } (\Delta_{cy})_{\text{L}} = 0.335 - 0.092 = 0.243 \text{ in. } (0.617 \text{ cm})$$

$$\text{Fixed } (\Delta_{my})_{\text{D}} = (0.262)(0.0623)(5488/2744) = 0.033 \text{ in. } (0.084 \text{ cm})$$
$$\text{Fixed } (\Delta_{my})_{\text{D+L}} = (0.262)(0.1639)(4292/2744) = 0.067 \text{ in. } (0.170 \text{ cm})$$
$$\text{Fixed } (\Delta_{my})_{\text{L}} = 0.067 - 0.033 = 0.034 \text{ in. } (0.086 \text{ cm})$$

$$\text{Fixed } (\Delta_{cx})_{\text{D}} = (0.738)(0.0255)(6860/2744) = 0.047 \text{ in. } (0.119 \text{ cm})$$
$$\text{Fixed } (\Delta_{cx})_{\text{D+L}} = (0.738)(0.0526)(6411/2295) = 0.108 \text{ in. } (0.274 \text{ cm})$$
$$\text{Fixed } (\Delta_{cx})_{\text{L}} = 0.108 = 0.047 = 0.061 \text{ in. } (0.155 \text{ cm})$$

$$\text{Fixed } (\Delta_{mx})_{\text{D}} = (0.262)(0.0255)(6860/4116) = 0.011 \text{ in. } (0.028 \text{ cm})$$

$$\text{Fixed } (\Delta_{mx})_{D+L} = (0.262)(0.0526)(6411/4116) = 0.021 \text{ in. } (0.053 \text{ cm})$$
$$\text{Fixed } (\Delta_{mx})_{L} = 0.021 - 0.011 = 0.010 \text{ in. } (0.025 \text{ cm})$$

$$\Delta_{c,m} = \text{Fixed } \Delta_{c,m} + (\Delta_{\theta 1})_{c,m} + 0 \qquad \text{(Eq. (4-11))}$$
$$(\Delta_{cy})_D = 0.092 + 0.023 = 0.115 \text{ in. } (0.292 \text{ cm})$$
$$(\Delta_{my})_D = 0.033 + 0.023 = 0.056 \text{ in. } (0.142 \text{ cm})$$
$$(\Delta_{cx})_D = 0.047 + 0.012 = 0.059 \text{ in. } (0.150 \text{ cm})$$
$$(\Delta_{mx})_D = 0.011 + 0.012 = 0.023 \text{ in. } (0.058 \text{ cm})$$

$$(\Delta_{cy})_L = 0.243 + 0.033 = 0.276 \text{ in. } (0.701 \text{ cm})$$
$$(\Delta_{my})_L = 0.034 + 0.033 = 0.067 \text{ in. } (0.170 \text{ cm})$$
$$(\Delta_{cx})_L = 0.061 + 0.013 = 0.074 \text{ in. } (0.188 \text{ cm})$$
$$(\Delta_{mx})_L = 0.010 + 0.013 = 0.023 \text{ in. } (0.058 \text{ cm})$$

$$\Delta = \Delta_{cy} + \Delta_{mx} = \Delta_{cx} + \Delta_{my} \qquad \text{(Eq. (4-12))}$$
$$(\Delta_{cy})_D + (\Delta_{mx})_D = 0.115 + 0.023 = 0.138 \text{ in.}$$
$$(\Delta_{cx})_D + (\Delta_{my})_D = 0.059 + 0.056 = 0.115 \text{ in.} \qquad \text{Avg. } \Delta_D = 0.127 \text{ in. } (0.323 \text{ cm})$$
$$(\Delta_{cy})_L + (\Delta_{mx})_L = 0.276 + 0.023 = 0.299 \text{ in.}$$
$$(\Delta_{cx})_L + (\Delta_{my})_L = 0.074 + 0.067 = 0.141 \text{ in.} \qquad \text{Avg. } \Delta_L = 0.220 \text{ in. } (0.559 \text{ cm})$$
$$\Delta_t = 3.0\Delta_D = (3.0)(0.127) = 0.381 \text{ in. } (0.968 \text{ cm})$$

$$\Delta_L = 0.22 \text{ in.} < L/180 = (16)(12)/180 = 1.07 \text{ in.} \qquad \text{OK}$$
$$< L/360 = 192/360 = 0.53 \text{ in.} \qquad \text{OK}$$
$$\Delta_t + \Delta_L = 0.381 + 0.220 = 0.60 \text{ in.} < L/240 = 192/240 = 0.80 \text{ in.} \qquad \text{OK}$$
$$> L/480 = 192/480 = 0.40 \text{ in.} \qquad \text{Not OK}$$

The $L/480$ criterion is not satisfied, even when the longer span is used—$L/480 = (20)(12)/480 = 0.50$ in. Hence, for this condition to apply in the structure (Chap. 6, Table 6-15), the corner panel must be cambered by the difference. It is noted that a value of $T = 3.0$ was used, instead of the ACI Code value of 2.0

Side panel 2 deflections

$$(\Delta_{cy})_D = 0.115 \text{ in. } (0.292 \text{ cm})\text{—same as corner panel}$$
$$(\Delta_{my})_D = 0.056 \text{ in. } (0.142 \text{ cm})\text{—same as corner panel}$$
$$(\Delta_{cy})_L = 0.276 \text{ in. } (0.701 \text{ cm})\text{—same as corner panel}$$
$$(\Delta_{my})_L = 0.067 \text{ in. } (0.170 \text{ cm})\text{—same as corner panel}$$

$$(\Delta_{cx})_D = 0.043 \text{ in. } (0.109 \text{ cm})\text{—same as interior panel}$$
$$(\Delta_{mx})_D = 0.014 \text{ in. } (0.036 \text{ cm})\text{—same as interior panel}$$
$$(\Delta_{cx})_L = 0.058 \text{ in. } (0.147 \text{ cm})\text{—same as interior panel}$$
$$(\Delta_{mx})_L = 0.013 \text{ in. } (0.033 \text{ cm})\text{—same as interior panel}$$

$$\Delta = \Delta_{cy} + \Delta_{mx} = \Delta_{cx} + \Delta_{my} \qquad \text{(Eq. (4-12))}$$

$(\Delta_{cy})_D + (\Delta_{mx})_D = 0.115 + 0.014 = 0.129$ in.

$(\Delta_{cx})_D + (\Delta_{my})_D = 0.043 + 0.056 = 0.099$ in. Avg. $\Delta_D = 0.114$ in. (0.290 cm)

$(\Delta_{cy})_L + (\Delta_{mx})_L = 0.276 + 0.013 = 0.289$ in.,

$(\Delta_{cx})_L + (\Delta_{my})_L = 0.058 + 0.067 = 0.125$ in. Avg. $\Delta_L = 0.207$ in. (0.526 cm)

$$\Delta_t = 3.0\Delta_D = (3.0)(0.114) = 0.342 \text{ in. } (0.869 \text{ cm})$$

$\Delta_L = 0.21$ in. $< L/180 = (16)(12)/180 = 1.07$ in. OK

$< L/360 = 192/360 = 0.53$ in. OK

$\Delta_t + \Delta_L = 0.342 + 0.207 = 0.55$ in. $< L/240 = 192/240 = 0.80$ in. OK

$> L/480 = 192/480 = 0.40$ in. Not OK

See comments for corner panel with regard to the $L/480$ criterion.

Side panel 3 deflections

$(\Delta_{cy})_D = 0.084$ in. (0.213 cm)—same as interior panel

$(\Delta_{my})_D = 0.040$ in. (0.102 cm)—same as interior panel

$(\Delta_{cy})_L = 0.114$ in. (0.290 cm)—same as interior panel

$(\Delta_{my})_L = 0.038$ in. (0.097 cm)—same as interior panel

$(\Delta_{cx})_D = 0.059$ in. (0.150 cm)—same as corner panel

$(\Delta_{mx})_D = 0.023$ in. (0.058 cm)—same as corner panel

$(\Delta_{cx})_L = 0.074$ in. (0.188 cm)—same as corner panel

$(\Delta_{mx})_L = 0.023$ in. (0.058 cm)—same as corner panel

$$\Delta = \Delta_{cy} + \Delta_{mx} = \Delta_{cx} + \Delta_{my} \qquad \text{(Eq. (4-12))}$$

$(\Delta_{cy})_D + (\Delta_{mx})_D = 0.084 + 0.023 = 0.107$ in.

$(\Delta_{cx})_D + (\Delta_{my})_D = 0.059 + 0.040 = 0.099$ in. Avg. $\Delta_D = 0.103$ in. (0.262 cm)

$(\Delta_{cy})_L + (\Delta_{mx})_L = 0.114 + 0.023 = 0.137$ in.

$(\Delta_{cx})_L + (\Delta_{my})_L = 0.074 + 0.038 = 0.112$ in. Avg. $\Delta_L = 0.125$ in. (0.318 cm)

$$\Delta_t = 3.0\Delta_D = (3.0)(0.103) = 0.309 \text{ in. } (0.785 \text{ cm})$$

$\Delta_L = 0.13$ in. $< L/180 = (16)(12)/180 = 1.07$ in. OK

$< L/360 = 192/360 = 0.53$ in. OK

$\Delta_t + \Delta_L = 0.309 + 0.125 = 0.43$ in. $< L/240 = 0.80$ in. OK

$\simeq L/480 = 0.40$ in. Not OK

Effect of uncracked and partially cracked slabs using the method of deflection coefficients

Using I_g, $I_g = (12)(7)^3/12 = 343$ in^4 per ft width (14 300 cm^4)

Using I_e, I_e = an average effective I for the panel under consideration

Table 4-10 Summary of flat plate deflections by the equivalent frame method using I_e*

20 ft (6.1 m)	*Corner panel 4* $\Delta_D = 0.13$ in. (0.32 cm) † $\Delta_t = 0.38$ in. (0.97 cm) $\Delta_L = 0.22$ in. (0.56 cm) $\Delta_t + \Delta_L = 0.60$ in. (1.52 cm)	*Side panel 2* $\Delta_D = 0.11$ in. (0.29 cm) † $\Delta_t = 0.34$ in. (0.87 cm) $\Delta_L = 0.21$ in. (0.53 cm) $\Delta_t + \Delta_L = 0.55$ in. (1.39 cm)
20 ft (6.1 m)	*Side panel 3* $\Delta_D = 0.10$ in. (0.26 cm) † $\Delta_t = 0.31$ in. (0.79 cm) $\Delta_L = 0.13$ in. (0.32 cm) $\Delta_t + \Delta_L = 0.43$ in. (1.10 cm)	‡ *Interior panel 1* $\Delta_D = 0.09$ in. (0.23 cm) † $\Delta_t = 0.27$ in. (0.69 cm) $\Delta_L = 0.11$ in. (0.28 cm) $\Delta_t + \Delta_L = 0.39$ in. (0.98 cm)
	16 ft (4.9 m)	16 ft (4.9 m)

* As summarized in Table 4-9, the following section properties were used: for all dead-load calculations—I_g; and for dead plus live-load calculations—I_g for the middle strips and I_e for the column strips (using a weighted average of 70 percent positive and 15 percent, 15 percent, negative values).

† $\Delta_t = 3.0\Delta_D$.

‡ In the solution herein, no distinction is made between interior panel 1 and the other interior panels.

Comparison with ACI Code allowable deflections:

$\Delta_L < L/180 = (15)(12)/180 = 1.07$ in. (2.72 cm)	OK all panels
$< L/360 = 192/360 = 0.53$ in. (1.35 cm)	OK all panels
$\Delta_t + \Delta_L < L/240 = 192/240 = 0.80$ in. (2.03 cm)	OK all panels
$< L/480 = 192/480 = 0.40$ in. (1.02 cm)	Not OK, except for interior panels

The deflections are smallest in the interior panel and the largest in the corner panel. The deflections of side panel 2 are larger than those of side panel 3 because the exterior end of side panel 2 is rotating in the long direction while the exterior end of side panel 3 is rotating in the short direction.

= as suggested herein—an average of the panel (column + middle strip) I_e in the two directions, where I_e is computed in each direction for the column and middle strips as the weighted average of 70 percent positive and 15 percent, 15 percent, negative region values of I_e

In this example $I_e = I_g$ for all dead load calculations; $I_e = I_g$ for all dead plus live-load calculations for the middle strips in both directions; and $I_e < I_g$ for all dead plus live-load calculations for the column strips in both directions. This combination would probably be satisfactory for most flat plate and flat slab problems.

As computed in Table 4-8 and summarized in Table 4-9, $I_e = 4992$ in^4 and 6363 in^4 in the two directions for an interior panel under dead plus live loading. Hence for an interior panel of the flat plate designed in Table 4-6,

$$(I_e)_D = I_g = 343 \text{ in}^4 \text{ per ft width } (14\,300 \text{ cm}^4)$$

$$\qquad\qquad 312 \qquad\qquad 318$$

$$(I_e)_{D+L} = \tfrac{1}{2}[(4992/16) + (6363/20)] = 315 \text{ in}^4 \text{ per ft width } (13\,100 \text{ cm}^4)$$

Method of deflection coefficients using the theoretical coefficients from the method of elasticity[2,23] For the flat plate designed in Table 4-6, $L/S = 20/16 = 1.25$.

From Table 4-1: $K_s = 0.00408$—interior panel without checkerboard loading
$K_s = 0.0181$—single isolated panel

Interior panel without checkerboard loading using I_g and I_e

$$\Delta_{D+L} = K_s q L^4 / E_c I_g \qquad \text{(Eq. (4-5))}$$

$$= (0.00408)(208)(20)^4(12)^3/(3.64 \times 10^6)(343) = 0.188 \text{ in. } (0.478 \text{ cm})$$

$$\Delta_D = (108/208)(0.188) = 0.10 \text{ in. } (0.25 \text{ cm})$$

versus 0.09 in. (0.20 cm) to 0.11 in. (0.28 cm) by the other methods.

$$\Delta_L = 0.19 - 0.10 = 0.09 \text{ in. } (0.23 \text{ cm})$$

versus 0.08 in. (0.20 cm) to 0.11 in. (0.28 cm) by the other methods.

Using $(I_e)_{D+L}$, $\Delta_{D+L} = (0.188)(343/315) = 0.21$ in. (0.53 cm)

$$\Delta_L = 0.21 - 0.10 = 0.11 \text{ in. } (0.28 \text{ cm})$$

versus 0.10 in. (0.25 cm) to 0.13 in. (0.33 cm) by the other methods.

Single isolated panel using I_g

$$\Delta_{D+L} = K_s q L^4 / E_c I_g \qquad \text{(Eq. (4-5))}$$

$$= (0.0181)(208)(20)^4(12)^3/(3.64 \times 10^6)(343) = 0.834 \text{ in. } (2.12 \text{ cm})$$

$$\Delta_D = (108/208)(0.834) = 0.43 \text{ in. } (1.09 \text{ cm})$$

versus 0.13 in. (0.32 cm) by the equivalent frame method shown in Table 4-7 for a corner panel using I_g.

$$\Delta_L = 0.83 - 0.43 = 0.40 \text{ in. } (1.02 \text{ cm})$$

versus 0.12 in. (0.30 cm) by the equivalent frame method shown in Table 4-7 for a corner panel using I_g.

Deflections using the theoretical coefficients for a single isolated flat plate panel may serve as a gross upper bound for all types of slab panels.

Method of deflection coefficients using the coefficients obtained in the diagonal beam strip analogy for an interior panel (PCA,[4] ACI Committee 209[6] and CRSI[24]), using I_g and I_e For the flat plate designed in Table 4-6, $L/S = 20/16 = 1.25$.

From Table 4-3: $K_s = 0.00371$

$$\Delta_{D+L} = K_s q L^4 / E_c I_g \qquad \text{(Eq. (4-5))}$$

$$= (0.00371)(208)(20)^4(12)^3/(3.64 \times 10^6)(343) = 0.171 \text{ in. } (0.434 \text{ cm})$$

$$\Delta_D = (108/208)(0.171) = 0.09 \text{ in. } (0.23 \text{ cm})$$

versus 0.09 in. (0.23 cm) to 0.11 in. (0.28 cm) by the other methods.

$$\Delta_L = 0.17 - 0.09 = 0.08 \text{ in. } (0.20 \text{ cm})$$

versus 0.09 in. (0.22 cm) to 0.11 in. (0.28 cm) by the other methods.

Using $(I_e)_{D+L}$, $\quad \Delta_{D+L} = (0.171)(343/315) = 0.19$ in. (0.48 cm)

$$\Delta_L = 0.19 - 0.09 = 0.10 \text{ in. } (0.25 \text{ cm})$$

versus 0.11 in. (0.28 cm) to 0.13 in. (0.33 cm) by the other methods.

Method of deflection coefficients using the coefficients from the finite difference solutions for an interior panel by Vanderbilt, Sozen, and Siess,[3] using I_g and I_e For the flat plate designed in Table 4-6,

$$S/L = 16/20 = 0.8 \qquad c/L = 1.5/20 = 0.075$$

$$I_S = I_L = 0 \qquad EI_S/LD = EI_L/SD = 0$$

From Table 4-4: $\quad K_s = 0.00476$

$$\Delta_{D+L} = K_s q L^4 / E_c I_g \qquad \text{(Eq. (4-5))}$$

$$= (0.00476)(208)(20)^4(12)^3/(3.64 \times 10^6)(343) = 0.219 \text{ in. } (0.556 \text{ cm})$$

$$\Delta_D = (108/208)(0.219) = 0.11 \text{ in. } (0.28 \text{ cm})$$

versus 0.09 in. (0.20 cm) to 0.10 in. (0.25 cm) by the other methods.

$$\Delta_L = 0.22 - 0.11 = 0.11 \text{ in. } (0.28 \text{ cm})$$

versus 0.08 in. (0.20 cm) to 0.09 in. (0.22 cm) by the other methods.

Using $(I_e)_{D+L}$, $\quad \Delta_{D+L} = (0.219)(343/315) = 0.24$ in. (0.61 cm)

$$\Delta_L = 0.24 - 0.11 = 0.13 \text{ in. } (0.33 \text{ cm})$$

versus 0.10 in. (0.25 cm) to 0.11 in. (0.28 cm) by the other methods.

Summary of deflection results for the flat plate As can be seen in Table 4-11, the results by all of the methods for the interior panel are in close agreement. Only the equivalent frame method provides a means of computing deflections for other than interior panels, as shown in Tables 4-7 and 4-10. These computed deflections were rounded to practical values.

Solid Slab System With Beams (Two-Way Slab)

The solution for interior, side, and corner panel deflections of two-way slabs follows the procedure for flat plates, with the effect of the beams included directly in the stiffness calculations. The determination of these stiffnesses (shown in Fig. 4-3) is described in detail in Ref. 26, Chap. 15 and 16, and in Ref. 39, Chap. 17,

Table 4-11 Summary of deflections for the solid slab system without beams (flat plate) by the different methods

Method	Using I_g or I_e	Δ_D in.	Δ_D cm	Δ_t* in.	Δ_t* cm	Δ_L in.	Δ_L cm	$\Delta_t + \Delta_L$ in.	$\Delta_t + \Delta_L$ cm
Interior panels									
Equivalent frame method	Using I_g	0.09	0.23	0.27	0.69	0.09	0.22	0.36	0.90
	Using I_e	0.09	0.23	0.27	0.69	0.11	0.28	0.39	0.98
Method of deflection coefficients from the theory of elasticity	Using I_g	0.10	0.25	0.30	0.76	0.09	0.23	0.39	0.99
	Using I_e	0.10	0.25	0.30	0.76	0.11	0.28	0.41	1.04
Method of deflection coefficients from the diagonal beam strip analogy	Using I_g	0.09	0.23	0.27	0.69	0.08	0.20	0.35	0.89
	Using I_e	0.09	0.23	0.27	0.69	0.10	0.25	0.37	0.94
Method of deflection coefficients from the finite difference solutions	Using I_g	0.11	0.28	0.33	0.84	0.11	0.28	0.44	1.12
	Using I_e	0.11	0.28	0.33	0.84	0.13	0.33	0.46	1.17
Side panel 2									
Equivalent frame method	Using I_g	0.11	0.29	0.34	0.87	0.11	0.27	0.45	1.14
	Using I_e	0.11	0.29	0.34	0.87	0.21	0.53	0.55	1.39
Side panel 3									
Equivalent frame method	Using I_g	0.10	0.26	0.31	0.79	0.10	0.24	0.41	1.03
	Using I_e	0.10	0.26	0.31	0.79	0.13	0.32	0.43	1.10
Corner panels									
Equivalent frame method	Using I_g	0.13	0.32	0.38	0.97	0.12	0.30	0.50	1.27
	Using I_e	0.13	0.32	0.38	0.97	0.22	0.56	0.60	1.52

* $\Delta_t = 3.0\Delta_D$.

Comparison with ACI Code allowable deflections:

$\Delta_L < L/180 = (16)(12)/180 = 1.07$ in. (2.72 cm) OK, all panels, all methods

$< L/360 = 192/360 = 0.53$ in. (1.35 cm) OK, all panels, all methods

$\Delta_t + \Delta_L < L/240 = 192/240 = 0.80$ in. (2.03 cm) OK, all panels, all methods

$< L/480 = 192/480 = 0.40$ in. (1.02 cm) Not OK, corner panels and side panel 2; Marginal, interior panels and side panel 3

The results by all of the methods for the interior panel are in reasonably close agreement, especially when comparing results using the same *I*.

for example. For illustrative purposes, the detailed calculations will be shown in this problem only for the interior panel.

For determining the lateral distribution factors (*LDF*) for two-way slabs, $\alpha_1 = E_{cb}I_b/E_{cs}I_s$ is required. However, when $\alpha_1 L_2/L_1 > 1.0$, it is taken as unity—which is usually the case. Since adjacent spans are equal, it is not necessary to calculate K_{ec} for an interior panel by the procedure herein. The principal additional calculations needed for unequal adjacent spans, pattern loading, or side and corner panels are for K_{ec} and the end rotation and deflection in Eqs. (4-8) to (4-10).

Referring to the sample two-way slab from Ref. 39, the slab design details and gross-section properties are shown in Table 4-12. The following loads were used:

$$\text{Service live load} = 120 \text{ psf } (586 \text{ kg/m}^2, 5750 \text{ N/m}^2)$$
$$\text{Service dead load} = (6.5/12)(150) = 81 \text{ psf } (395 \text{ kg/m}^2, 3880 \text{ N/m}^2)$$
$$\text{Total service load} = 201 \text{ psf } (981 \text{ kg/m}^2, 9620 \text{ N/m}^2)$$

Deflections are computed by several different methods and compared with the allowable deflections from the ACI Code (see Chap. 6 herein, Table 6-15), as summarized in Table 4-15. The various deflection components are shown in Fig. 4-1. Computed deflections by the equivalent frame method using I_g and I_e are also shown in Table 4-14 for the interior, side, and corner panels.

Equivalent frame method using the direct design procedure and I_g As shown in Table 4-12, the full frame I_g for both directions is about 12 percent higher than the column strip I_g plus the middle strip I_g, because of the effect of the different T-beams. Also, when using the effective flange width for the T-beams, $(x\text{–}x)$ $(I_b)_g = 45\,000$ in^4 (1 870 000 cm^4) and $(y\text{–}y)$ $(I_b)_g = 24\,000$ in^4 (999 000 cm^4)—calculations not shown, which are 19 to 28 percent lower than the column strip values in the two directions. Thus, the appropriate effective section properties to be used are not at all clear. As a matter of consistency in approximate deflection calculations by this method, the author suggests the use of the following values:

$$(x\text{–}x)\ I_{\text{frame}} = \text{CS } I + \text{MS } I = 55\,300 + 2750 = \underline{58\,050 \text{ in}^4} \ (2\,416\,000 \text{ cm}^4)$$
$$(y\text{–}y)\ I_{\text{frame}} = 33\,200 + 4120 = \underline{37\,320 \text{ in}^4} \ (1\,553\,000 \text{ cm}^4)$$

The above values of I_b are used to demonstrate the calculation of α_1 and *LDF* for both directions, as follows (ACI Code, 318-71):

Long direction $(x\text{–}x)$, $I_s(240)(6.5)^3/12 = 5490$ in^4 (228 000 cm^4)

$$\alpha_1 = E_{cb}I_b/E_{cs}I_s = 45\,000/5490 = 8.20$$

Since $\alpha_1 L_2/L_1$ is (much) greater than 1.0, use $\alpha_1 L_2/L_1 = 1$. From Table 4-5,

$$\text{Int. neg. } LDF = 75 + 30(\alpha_1 L_2/L_1)(1 - L_2/L_1) = 75 + 30(1)(1 - 20/25) = 81.0\%$$
$$\text{Int. pos. } LDF = 60 + 30(\alpha_1 L_2/L_1)(1.5 - L_2/L_1) = 60 + 30(1)(1.5 - 20/25) = 81.0\%$$

Table 4-12 Design details for solid slab system with beams (two-way slab) and gross-section properties

Geometry and material properties

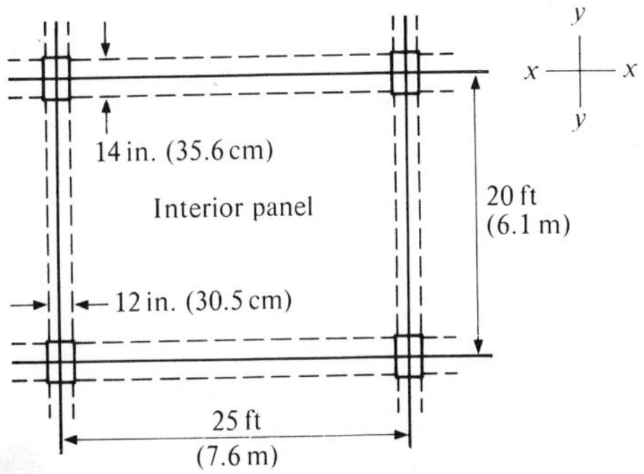

Square columns, $c = 15$ in. (38.1 cm)
$L_c = 12$ ft (3.7 m)

Long beams, 14 in. (35.6 cm) by 28 in. (71.1 cm) overall

Short beams, 12 in. (30.5 cm) by 24 in. (61.0 cm) overall

Slab thickness $= h = 6.5$ in. (16.5 cm)

x–x, $L_1 = 25$ ft (7.6 m)
$L_2 = 20$ ft (6.1 m)
$L_n = 24$ ft (7.3 m)
Width of CS, MS = 10 ft (3.0 m)

y–y, $L_1 = 20$ ft (6.1 m)
$L_2 = 25$ ft (7.6 m)
$L_n = 18.83$ ft (5.7 m)
Width of CS = 10 ft (3.0 m)
Width of MS = 15 ft (4.6 m)

Concrete: $f'_c = 3000$ psi (211 kg/cm^2, 20.7 N/mm^2)

$w = 145$ pcf (2320 kg/m^3, 22.8 kN/m^3)

Steel: $f_y = 40\,000$ psi (2810 kg/cm^2, 276 N/mm^2)

Gross-section moments of inertia

Full frame section and column strip beam-slab

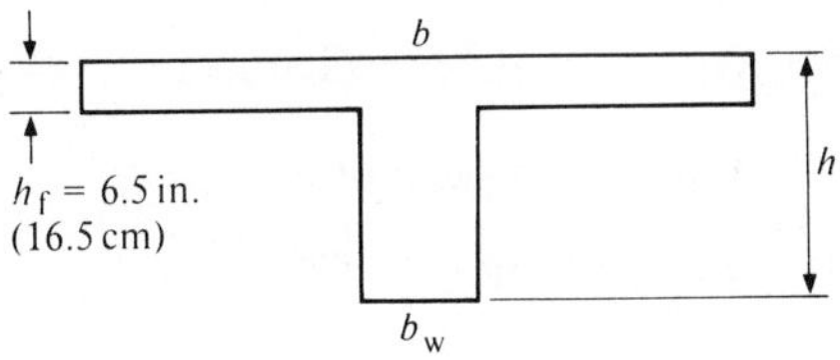

Middle strip slab

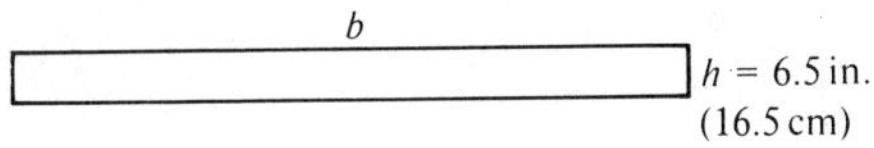

Long direction (x–x)

Full frame section:

$$b/b_w = 240/14 = 17.1$$

$$h_f/h = 6.5/28 = 0.23$$

From Fig. B-1, Chap. 3,

$$I_g/bh^3 = 0.013$$

and

$$I_g = (0.013)(240)(28)^3 = 65\,900 \text{ in}^4 \ (2\,740\,000 \text{ cm}^4)$$

Short direction (y–y)

Full frame section:

$$b/b_w = 300/12 = 25.0$$

$$h_f/h = 6.5/24 = 0.27$$

From Fig. B-1, Chap. 3,

$$I_g/bh^3 = 0.010$$

and

$$I_g = (0.010)(300)(24)^3 = 41\,500 \text{ in}^4 \ (1\,727\,000 \text{ cm}^4)$$

Table 4-12 *(continued)*

Column strip beam-slab: $b/b_w = 120/28 = 8.6$ $h_f/h = 6.5/28 = 0.23$ From Fig. B-1, Chap. 3, $I_g/bh^3 = 0.021$ and $I_g = (0.021)(120)(28)^3$ $= 55\,300$ in^4 (2 300 000 cm^4) Middle strip slab: $I_g = bh^3/12 = (120)(6.5)^3/12$ $= 2750$ in^4 (114 500 cm^4)	Column strip beam-slab: $b/b_w = 120/12 = 10.0$ $h_f/h = 6.5/24 = 0.27$ From Fig. B-1, Chap. 3, $I_g/bh^3 = 0.020$ and $I_g = (0.020)(120)(24)^3$ $= 33\,200$ in^4 (1 138 000 cm^4) Middle strip slab: $I_g = bh^3/12 = (180)(6.5)^3/12$ $= 4120$ in^4 (171 000 cm^4)

For interior panel, Avg. $(LDF)_c = \underline{0.810}$ $\quad (LDF)_m = 1 - 0.810 = \underline{0.190}$

Short direction (y–y), $I_s = (300)(6.5)^3/12 = 6870$ in^4 (286 000 cm^4)

$$\alpha_1 = E_{cb}I_b/E_{cs}I_s = 24\,000/6870 = 3.49$$

Since $\alpha_1 L_2/L_1$ is (much) greater than 1.0, use $\alpha_1 L_2/L_1 = 1$. From Table 4-5,

$$\text{Int. neg. } LDF = 75 + 30(1)(1 - 25/20) = 67.5\%$$
$$\text{Int. pos. } LDF = 60 + 30(1)(1.5 - 25/20) = 67.5\%$$

For interior panel,

$$\text{Avg. } (LDF)_c = 0.675 \qquad (LDF)_m = 1 - 0.675 = \underline{0.325} \quad \text{(Table 4-5)}$$

For $f_c' = 3000$ psi (211 kg/cm^2, 20.7 N/mm^2) and normal-weight concrete,

$$E_c = 3.31 \times 10^6 \text{ psi } (233\,000 \text{ kg/cm}^2, 22.8 \text{ kN/mm}^2) \qquad \text{(Table 1-3)}$$

Interior panel deflections

$$\text{Fixed } \Delta_{\text{frame}} = wL^4/384E_c I_{\text{frame}} \qquad \text{(Eq. (4-6))}$$
$$\text{Fixed } (\Delta_{\text{frame},x})_{\text{D,D+L}} = (81 \text{ or } 201)(20)(25)^4(1728)/(384)(3.31 \times 10^6)(58\,050)$$
$$\text{Fixed } (\Delta_{\text{frame},x})_{\text{D}} = 0.0148 \text{ in. } (0.0376 \text{ cm})$$
$$\text{Fixed } (\Delta_{\text{frame},x})_{\text{D+L}} = 0.0368 \text{ in. } (0.0935 \text{ cm})$$

$$\text{Fixed } (\Delta_{\text{frame},y})_{\text{D,D+L}} = (81 \text{ or } 201)(25)(20)^4(1728)/(384)(3.31 \times 10^6)(37\,320)$$
$$\text{Fixed } (\Delta_{\text{frame},y})_{\text{D}} = 0.0118 \text{ in. } (0.0300 \text{ cm})$$
$$\text{Fixed } (\Delta_{\text{frame},y})_{\text{D+L}} = 0.0293 \text{ in. } (0.0744 \text{ cm})$$

$$\text{Fixed } \Delta_{\text{c,m}} = (LDF)_{\text{c,m}}(\text{Fixed } \Delta_{\text{frame}}) \frac{I_{\text{frame}}}{I_{\text{c,m}}} \qquad \text{(Eq. (4-7))}$$

$$\text{Fixed } (\Delta_{cx})_D = (0.810)(0.0148)(58\,050/55\,300) = 0.013 \text{ in. } (0.033 \text{ cm})$$
$$\text{Fixed } (\Delta_{cx})_{D+L} = (0.810)(0.0368)(58\,050/55\,300) = 0.031 \text{ in. } (0.079 \text{ cm})$$
$$\text{Fixed } (\Delta_{cx})_L = 0.031 - 0.013 = 0.018 \text{ in. } (0.046 \text{ cm})$$

$$\text{Fixed } (\Delta_{mx})_D = (0.190)(0.0148)(58\,050/2750) = 0.059 \text{ in. } (0.150 \text{ cm})$$
$$\text{Fixed } (\Delta_{mx})_{D+L} = (0.190)(0.0368)(58\,050/2750) = 0.148 \text{ in. } (0.376 \text{ cm})$$
$$\text{Fixed } (\Delta_{mx})_L = 0.148 - 0.059 = 0.089 \text{ in. } (0.226 \text{ cm})$$

$$\text{Fixed } (\Delta_{cy})_D = (0.675)(0.0118)(37\,320/33\,200) = 0.009 \text{ in. } (0.023 \text{ cm})$$
$$\text{Fixed } (\Delta_{cy})_{D+L} = (0.675)(0.0293)(37\,320/33\,200) = 0.022 \text{ in. } (0.056 \text{ cm})$$
$$\text{Fixed } (\Delta_{cy})_L = 0.022 - 0.009 = 0.013 \text{ in. } (0.033 \text{ cm})$$

$$\text{Fixed } (\Delta_{my})_D = (0.325)(0.0118)(37\,320/4120) = 0.035 \text{ in. } (0.089 \text{ cm})$$
$$\text{Fixed } (\Delta_{my})_{D+L} = (0.325)(0.0293)(37\,320/4120) = 0.086 \text{ in. } (0.218 \text{ cm})$$
$$\text{Fixed } (\Delta_{my})_L = 0.086 - 0.035 = 0.051 \text{ in. } (0.130 \text{ cm})$$

For interior panels, assume M_{net} in Eq. (4-8) is zero for equal adjacent spans. Hence End θs = 0 (Eq. (4-8)) and Δ_θs = 0 (Eq. (4-9)). Thus,

$$\Delta_{c,m} = \text{Fixed } \Delta_{c,m} \qquad \text{(Eq. (4-11))}$$

and

$$\Delta = \Delta_{cy} + \Delta_{mx} = \Delta_{cx} + \Delta_{my} \qquad \text{(Eq. (4-12))}$$

$$(\Delta_{cy})_D + (\Delta_{mx})_D = 0.009 + 0.059 = 0.068 \text{ in.}$$
$$(\Delta_{cx})_D + (\Delta_{my})_D = 0.013 + 0.035 = 0.048 \text{ in.} \qquad \text{Avg. } \Delta_D = 0.058 \text{ in. } (0.147 \text{ cm})$$
$$(\Delta_{cy})_L + (\Delta_{mx})_L = 0.013 + 0.089 = 0.102 \text{ in.}$$
$$(\Delta_{cx})_L + (\Delta_{my})_L = 0.018 + 0.051 = 0.069 \text{ in.} \qquad \text{Avg. } \Delta_L = 0.086 \text{ in. } (0.218 \text{ cm})$$

It is noted that the middle strip deflection increment in the long direction (Δ_{mx}) is considerably larger than the other deflection increments for both dead load and live load.

$$\Delta_t = 3.0\Delta_D = (3.0)(0.058) = 0.174 \text{ in. } (0.425 \text{ cm})$$

$$\Delta_L = 0.09 \text{ in.} < L/180 = (20)(12)/180 = 1.33 \text{ in.} \qquad \text{OK}$$
$$< L/360 = 240/360 = 0.67 \text{ in.} \qquad \text{OK}$$
$$\Delta_t + \Delta_L = 0.174 + 0.086 = 0.26 \text{ in.} < L/240 = 240/240 = 1.00 \text{ in.} \qquad \text{OK}$$
$$< L/480 = 240/480 = 0.50 \text{ in.} \qquad \text{OK}$$

If half of the live load is assumed to be a sustained load, for example,

$$\Delta_t + \Delta_L = (3.0)(0.058 + 0.086/2) + 0.086 = 0.39 \text{ in. } (0.99 \text{ cm})$$

which is still less than 0.50 in.—OK.

The above results for an interior panel are summarized in Table 4-14, along with the deflections of side and corner panels (calculations not shown herein).

Equivalent frame method using the direct design procedure and I_e The deflections of the two-way slab shown in Table 4-12 are computed in this example for an interior panel using I_e. Based on the service load moments and section properties determined in Table 4-13, the following section properties are used: for all dead-load deflection calculations—I_g; and for dead plus live-load deflection calculations—I_g for the column strips in both directions and I_e for the middle strips in both directions (using a weighted average of 70 percent positive and 15 percent, 15 percent, negative values).

Interior panel deflections

$$\text{Fixed } \Delta_{\text{frame}} = wL^4/384E_c I_{\text{frame}} \qquad (\text{Eq. (4-6)})$$

$$\text{Fixed } (\Delta_{\text{frame},x})_{\text{D,D+L}} = (81 \text{ or } 201)(20)(25)^4(1728)/(384)(3.31 \times 10^6)(58\,050 \text{ or } 57\,800)$$

$$\text{Fixed } (\Delta_{\text{frame},x})_{\text{D}} = 0.0148 \text{ in. } (0.0376 \text{ cm})$$

$$\text{Fixed } (\Delta_{\text{frame},x})_{\text{D+L}} = 0.0369 \text{ in. } (0.0937 \text{ cm})$$

$$\text{Fixed } (\Delta_{\text{frame},y})_{\text{D,D+L}} = (81 \text{ or } 201)(25)(20)^4(1728)/(384)(3.31 \times 10^6)(37\,320 \text{ or } 37\,200)$$

$$\text{Fixed } (\Delta_{\text{frame},y})_{\text{D}} = 0.0118 \text{ in. } (0.0300 \text{ cm})$$

$$\text{Fixed } (\Delta_{\text{frame},y})_{\text{D+L}} = 0.0294 \text{ in. } (0.0747 \text{ cm})$$

$$\text{Fixed } \Delta_{\text{c,m}} = (LDF)_{\text{c,m}}(\text{Fixed } \Delta_{\text{frame}}) \frac{I_{\text{frame}}}{I_{\text{c,m}}} \qquad (\text{Eq. (4-7)})$$

$$\text{Fixed } (\Delta_{\text{c}x})_{\text{D}} = (0.810)(0.0148)(58\,050/55\,300) = 0.013 \text{ in. } (0.033 \text{ cm})$$

$$\text{Fixed } (\Delta_{\text{c}x})_{\text{D+L}} = (0.810)(0.0369)(57\,800/55\,300) = 0.031 \text{ in. } (0.079 \text{ cm})$$

$$\text{Fixed } (\Delta_{\text{c}x})_{\text{L}} = 0.031 - 0.013 = 0.018 \text{ in. } (0.046 \text{ cm})$$

$$\text{Fixed } (\Delta_{\text{m}x})_{\text{D}} = (0.190)(0.0148)(58\,050/2750) = 0.059 \text{ in. } (0.150 \text{ cm})$$

$$\text{Fixed } (\Delta_{\text{m}x})_{\text{D+L}} = (0.190)(0.0369)(57\,800/2500) = 0.162 \text{ in. } (0.411 \text{ cm})$$

$$\text{Fixed } (\Delta_{\text{m}x})_{\text{L}} = 0.162 - 0.059 = 0.103 \text{ in. } (0.262 \text{ cm})$$

$$\text{Fixed } (\Delta_{\text{c}y})_{\text{D}} = (0.675)(0.0118)(37\,320/33\,200) = 0.009 \text{ in. } (0.023 \text{ cm})$$

$$\text{Fixed } (\Delta_{\text{c}y})_{\text{D+L}} = (0.675)(0.0294)(37\,200/33\,200) = 0.022 \text{ in. } (0.056 \text{ cm})$$

$$\text{Fixed } (\Delta_{\text{c}y})_{\text{L}} = 0.022 - 0.009 = 0.013 \text{ in. } (0.033 \text{ cm})$$

$$\text{Fixed } (\Delta_{\text{m}y})_{\text{D}} = (0.325)(0.0118)(37\,320/4120) = 0.035 \text{ in. } (0.089 \text{ cm})$$

$$\text{Fixed } (\Delta_{\text{m}y})_{\text{D+L}} = (0.325)(0.0294)(37\,200/4010) = 0.089 \text{ in. } (0.226 \text{ cm})$$

$$\text{Fixed } (\Delta_{\text{m}y})_{\text{L}} = 0.089 - 0.035 = 0.054 \text{ in. } (0.137 \text{ cm})$$

For interior panels, assume M_{net} in Eq. (4-8) is zero for equal adjacent spans. Hence End θs $= 0$ (Eq. (4-8)), and Δ_θs $= 0$ (Eq. (4-9)). Thus,

$$\Delta_{\text{c,m}} = \text{Fixed } \Delta_{\text{c,m}} \qquad (\text{Eq. (4-11)})$$

Table 4-13 Service load moments and effective section properties for deflection calculations of solid slab system with beams (two-way slab)

For $f_c' = 3000$ psi (211 kg/cm², 20.7 N/mm²) and normal-weight concrete,

$$f_r = \underline{429 \text{ psi}} \; (30.2 \text{ kg/cm}^2, 2.96 \text{ N/mm}^2) \qquad \text{(Table 1-2)}$$

$$E_c = \underline{3.31 \times 10^6 \text{ psi}} \; (233\,000 \text{ kg/cm}^2, 22.8 \text{ kN/mm}^2) \qquad \text{(Table 1-3)}$$

$$n = E_s/E_c = 29/3.31 = \underline{8.8}$$

	(x–x)		(y–y)	
	Neg.	Pos.	Neg.	Pos.
1. Mom. distr. ratios, int. panel (ACI Code, 318-71)	0.65	0.35	0.65	0.35
2. * $(M_a)_{D+L} = M_0$ (line 1), ft-k	188.2	101.3	144.8	77.9
3. $(M_a)_D = (81/201)$ (line 2), ft-k	75.8	40.8	58.4	31.4
4. † CS $(M_a)_{D+L} = (LDF)_c$ (line 2), ft-k	152.4	82.1	97.7	52.6
5. † MS $(M_a)_{D+L} = (LDF)_m$ (line 2), ft-k	35.8	19.2	47.1	25.3
6. † CS $(M_a)_D = (LDF)_c$ (line 3), ft-k	61.4	33.0	39.4	21.2
7. † MS $(M_a)_D = (LDF)_m$ (line 3), ft-k	14.4	7.8	19.0	10.2

* (x–x), Service $M_0 = w_{D+L} L_2 L_n^2/8 = (0.201)(20)(24)^2/8$

$= 289.5$ ft-k (40 000 kgm, 393 000 Nm)

(y–y), Service $M_0 = (0.201)(25)(18.83)^2/8$

$= 222.7$ ft-k (30 800 kgm, 302 000 Nm)

† From previous example, interior panel:

$(x\text{–}x)-(LDF)_c = 0.810 \qquad (LDF)_m = 0.190$

$(y\text{–}y)-(LDF)_c = 0.675 \qquad (LDF)_m = 0.325$

Check for cracking in column strip beam-slab—sections from Table 4-12

Long direction (x–x)

$$\text{Bot } y_t = \frac{(301)(10.75) + (780)(24.75)}{\underset{301}{(14)(21.5)} + \underset{780}{(120)(6.5)}}$$

$= 20.85$ in. (53.0 cm)

Bot $M_{cr} = f_r I_g / y_t$

$= (0.429)(55\,300)/(20.85)(12)$

$= 82.8$ ft-k

> CS Pos. $(M_a)_{D+L} = 82.1$ ft-k, above

Top $M_{cr} = (0.429)(55\,300)/(7.15)(12)$

$= 276.6$ ft-k

> CS Neg. $(M_a)_{D+L} = 152.4$ ft-k, above

Short direction (y–y)

$$\text{Bot } y_t = \frac{(210)(8.75) + (780)(20.75)}{\underset{210}{(12)(17.5)} + \underset{780}{(120)(6.5)}}$$

$= 18.21$ in. (46.3 cm)

Bot $M_{cr} = (0.429)(33\,200)/(18.21)(12)$

$= 65.2$ ft-k

> CS Pos. $(M_a)_{D+L} = 52.6$ ft-k, above

Top $M_{cr} = (0.429)(33\,200)/(5.79)(12)$

$= 204.9$ ft-k

> CS Neg. $(M_a)_{D+L} = 97.7$ ft-k, above

Hence the column strip is uncracked in both directions in both negative and positive moment regions under dead plus live loading (maximum loading).

Table 4-13 *(continued)*

Check for cracking in middle strip slab—sections from Table 4-12

Long direction $(x–x)$	*Short direction* $(y–y)$
Bot. and Top $M_{cr} = f_r I_g / y_t$ $= (0.429)(2750)/(3.25)(12)$ $= 30.3$ ft-k > MS Pos. $(M_a)_{D+L} = 19.2$ ft-k, above (hence uncracked) < MS Neg. $(M_a)_{D+L} = 35.8$ ft-k, above (hence partially cracked)	Bot. and Top $M_{cr} = (0.429)(4120)/(3.25)(12)$ $= 45.3$ ft-k > MS Pos. $(M_a)_{D+L} = 25.3$ ft-k, above (hence uncracked) < MS Neg. $(M_a)_{D+L} = 47.1$ ft-k, above (hence partially cracked)

Summary of effective moments of inertia—for dead-load deflections, consider all parts of the slab to be uncracked (by inspection above); for dead plus live-load deflections, consider the column strips to be uncracked in both directions, and the middle strips to be uncracked in the positive moment regions and partially cracked in the negative moment regions in both directions.

Calculation of MS Neg. I_e, *MS Avg.* I_e, *and* I_{frame}—sections from Table 4-12

The following properties are determined as shown in Tables 4-6 and 4-8 (calculations not included herein).

Long direction $(x–x)$	*Short direction* $(y–y)$
Design $M_0 = 457.9$ ft-k MS Neg. $M_{D+L} = 56.6$ ft-k (76 700 Nm) $d = 5.44$ in. (13.8 cm) 13 No. 5 bars $A_s = 4.03\ \text{in}^2\ (26.0\ \text{cm}^2)$ $\rho = 0.62\%$ $I_{cr}/bd^3 = 0.035$ $I_{cr} = (0.035)(120)(5.44)^3$ $= 676\ \text{in}^4$ (28 100 cm^4 $I_g/I_{cr} = 2750/676 = 4.1$	Design $M_0 = 352.4$ ft-k MS Neg. $M_{D+L} = 74.4$ ft-k (100 900 Nm) $d = 4.81$ in. (12.2 cm) 24 No. 5 bars $A_s = 7.44\ \text{in}^2\ (48.0\ \text{cm}^2)$ $\rho = 0.86\%$ $I_{cr}/bd^3 = 0.046$ $I_{cr} = (0.046)(180)(4.81)^3$ $= 921\ \text{in}^4\ (38\,300\ \text{cm}^4)$ $I_g/I_{cr} = 4120/921 = 4.5$

$$I_e = (M_{cr}/M_a)^3 I_g + [1 - (M_{cr}/M_a)^3] I_{cr} \qquad \text{(Eq. (3-19))}$$

$M_{cr}/(M_a)_{D+L} = 30.3/35.8 = 0.846$ $(0.846)^3 = 0.605$ MS Neg. $I_e = (0.605)(2750) + (1 - 0.605)(676)$ $= 1931\ \text{in}^4\ (804\,000\ \text{cm}^4)$ MS Avg. $I_e = (0.70)(2750) + (0.30)(1931)$ $= \underline{2500\ \text{in}^4}\ (104\,000\ \text{cm}^4)$ $I_{frame} = 55\,300 + 2500$ $= \underline{57\,800\ \text{in}^4}\ (2\,405\,000\ \text{cm}^4)$	$M_{cr}/(M_a)_{D+L} = 45.3/47.1 = 0.962$ $(0.962)^3 = 0.890$ MS Neg. $I_e = (0.890)(4120) + (1 - 0.890)(921)$ $= 3768\ \text{in}^4\ (1\,568\,000\ \text{cm}^4)$ MS Avg. $I_e = (0.70)(4120) + (0.30)(3768)$ $= \underline{4010\ \text{in}^4}\ (167\,000\ \text{cm}^4)$ $I_{frame} = 33\,200 + 4010$ $= \underline{37\,200\ \text{in}^4}\ (1\,548\,000\ \text{cm}^4)$

and $\Delta = \Delta_{cy} + \Delta_{mx} = \Delta_{cx} + \Delta_{my}$ (Eq. (4-12))

$(\Delta_{cy})_D + (\Delta_{mx})_D = 0.009 + 0.059 = 0.068$ in.

$(\Delta_{cx})_D + (\Delta_{my})_D = 0.013 + 0.035 = 0.048$ in. Avg. $\Delta_D = 0.058$ in. (0.147 cm)

$(\Delta_{cy})_L + (\Delta_{mx})_L = 0.013 + 0.103 = 0.116$ in.

$(\Delta_{cx})_L + (\Delta_{my})_L = 0.018 + 0.054 = 0.072$ in. Avg. $\Delta_L = 0.094$ in. (0.239 cm)

versus Avg. $\Delta_D = 0.058$ in. (0.147 cm) and Avg. $\Delta_L = 0.086$ in. (0.218 cm) in the previous solution by the same method using I_g.

$$\Delta_t = 3.0\Delta_D = (3.0)(0.058) = 0.174 \text{ in. } (0.425 \text{ cm})$$

$\Delta_L = 0.09$ in. $< L/180 = (20)(12)/180 = 1.33$ in. OK

$< L/360 = 240/360 = 0.67$ in. OK

$\Delta_t + \Delta_L = 0.174 + 0.094 = 0.27$ in. $< L/240 = 240/240 = 1.00$ in. OK

$< L/480 = 240/480 = 0.50$ in. OK

If half of the live load is assumed to be a sustained load, for example, the critical deflection can be approximated as $\Delta_t + \Delta_L = (3.0)(0.058 + 0.094/2) + 0.094 = 0.41$ in. (1.04 cm), which is still less than 0.50 in. However, when cracking is more severe, a separate deflection calculation using I_e under dead plus sustained live load would be required.

The above results for an interior panel are summarized in Table 4-14, along with the deflections of side and corner panels (calculations not shown herein).

Effect of uncracked and partially cracked two-way slabs using the method of deflection coefficients The author suggests that an average of the middle strip moments of inertia per unit width in the two directions be used with the method of deflection coefficients for two-way slabs. As computed in Table 4-13, $I = 2500$ in^4 and 4010 in^4 in the two directions for the middle strips of the interior panel under dead plus live loading. The slab was uncracked under dead loading. Thus, in this example,

$$(I_e)_D = (12)(6.5)^3/12 = 275 \text{ in}^4 \text{ per ft width } (11\,400 \text{ cm}^4)$$

$$(I_e)_{D+L} = \tfrac{1}{2}[\underset{250}{(2500/10)} + \underset{269}{(4010/15)}] = \underline{259 \text{ in}^4} \text{ per ft width } (10\,800 \text{ cm}^4)$$

Method of deflection coefficients using the theoretical coefficients from the method of elasticity,[2,23] and I_g, I_e For the two-way slab shown in Table 4-12, $L/S = 25/20 = 1.25$.

From Table 4-1:

$K_s = 0.000760$ for the case of all sides fixed

$\Delta_{D+L} = K_s q L^4/E_c I_g$ (Eq. (4-5))

Table 4-14 Summary of two-way slab deflections by the equivalent frame method using I_g and I_e.* The calculations are shown herein only for the interior panel

Using I_g		
20 ft (6.1 m)	*Corner panel 4* $\Delta_D = 0.07$ in. (0.18 cm) † $\Delta_t = 0.21$ in. (0.53 cm) $\Delta_L = 0.10$ in. (0.25 cm) $\Delta_t + \Delta_L = 0.31$ in. (0.79 cm)	*Side panel 2* $\Delta_D = 0.06$ in. (0.16 cm) † $\Delta_t = 0.18$ in. (0.46 cm) $\Delta_L = 0.09$ in. (0.24 cm) $\Delta_t + \Delta_L = 0.27$ in. (0.69 cm)
20 ft (6.1 m)	*Side panel 3* $\Delta_D = 0.07$ in. (0.17 cm) † $\Delta_t = 0.20$ in. (0.51 cm) $\Delta_L = 0.10$ in. (0.25 cm) $\Delta_t + \Delta_L = 0.30$ in. (0.76 cm)	‡ *Interior panel 1* $\Delta_D = 0.06$ in. (0.15 cm) † $\Delta_t = 0.17$ in. (0.43 cm) $\Delta_L = 0.09$ in. (0.22 cm) $\Delta_t + \Delta_L = 0.26$ in. (0.66 cm)
	25 ft (7.6 m)	25 ft (7.6 m)
*Using I_e**		
20 ft (6.1 m)	*Corner panel 4* $\Delta_D = 0.07$ in. (0.18 cm) † $\Delta_t = 0.21$ in. (0.53 cm) $\Delta_L = 0.15$ in. (0.38 cm) $\Delta_t + \Delta_L = 0.36$ in. (0.91 cm)	*Side panel 2* $\Delta_D = 0.06$ in. (0.16 cm) † $\Delta_t = 0.18$ in. (0.46 cm) $\Delta_L = 0.13$ in. (0.33 cm) $\Delta_t + \Delta_L = 0.31$ in. (0.79 cm)
20 ft (6.1 m)	*Side panel 3* $\Delta_D = 0.07$ in. (0.17 cm) † $\Delta_t = 0.20$ in. (0.51 cm) $\Delta_L = 0.14$ in. (0.36 cm) $\Delta_t + \Delta_L = 0.34$ in. (0.86 cm)	‡ *Interior panel 1* $\Delta_D = 0.06$ in. (0.15 cm) † $\Delta_t = 0.17$ in. (0.43 cm) $\Delta_L = 0.09$ in. (0.22 cm) $\Delta_t + \Delta_L = 0.27$ in. (0.69 cm)
	25 ft (7.6 m)	25 ft (7.6 m)

* As shown in Table 4-13, the following section properties were used: for all dead-load calculations—I_g; and for dead plus live-load calculations—I_g for the column strips and I_e for the middle strips (using a weighted average of 70 percent positive and 15 percent, 15 percent, negative values).

† $\Delta_t = 3.0\Delta_D$.

‡ In the solution herein, no distinction is made between interior panel 1 and other interior panels.

Comparison with ACI Code allowable deflections: As shown in the footnote of Table 4-15, computed deflections are within the ACI Code allowable values for all panels by all methods.

The deflections are smallest in the interior panel and the largest in the corner panel. The deflections of side panel 3 are larger than those of side panel 2 because the exterior end of side panel 3 is rotating in the long direction while the exterior end of side panel 2 is rotating in the short direction.

$$= (0.000760)(201)(25)^4(12)^3/(3.31 \times 10^6)(275) = 0.114 \text{ in. } (0.290 \text{ cm})$$
$$\Delta_D = (81/201)(0.114) = 0.05 \text{ in. } (0.13 \text{ cm})$$

versus 0.05 in. (0.13 cm) and 0.06 in. (0.15 cm) by the other two methods.

$$\Delta_L = 0.11 - 0.05 = 0.06 \text{ in. } (0.15 \text{ cm})$$

versus 0.06 in. (0.15 cm) and 0.09 in. (0.22 cm) by the other two methods.

Using $(I_e)_{D+L}$, $\Delta_{D+L} = (0.114)(275)/259 = 0.12$ in. (0.30 cm)

$$\Delta_L = 0.12 - 0.05 = 0.07 \text{ in. } (0.18 \text{ cm})$$

versus 0.07 in. (0.18 cm) and 0.09 in. (0.22 cm) by the other two methods.

Method of deflection coefficients using the coefficients from the finite difference solutions for an interior panel by Vanderbilt, Sozen, and Siess,[3] using I_g and I_e For the two-way slab shown in Table 4-12, $S/L = 20/25 = 0.8$, $c/L = 1/25 = 0.04$, $I_L = 45\,000$ in^4 and $I_S = 24\,000$ in^4—using the T-beam *I*s with the effective flange width from the previous problem (it is not particularly critical which *I*s are used).

$$EI_L/SD = (I_L/S)(12 \times 0.96/h^3) = (45\,000/20 \times 12)(12 \times 0.96/6.5^3) = 7.9$$
$$EI_S/LD = (24\,000/25 \times 12)(12 \times 0.96/6.5^3) = 3.4$$
$$I_S = (24/45)I_L = 0.53I_L$$

In Table 4-4, $I_S = (S/L)I_L = 0.8I_L$ or $I_S = (S/L)^2 I_L = 0.64I_L$

Use the latter case. Hence, from Table 4-4, $K_s = 0.00070$ to 0.00080 (extrapolated). Use $K_s = 0.00075$.

$$\Delta_{D+L} = K_s q L^4/E_c I_g \qquad \text{(Eq. (4-5))}$$
$$= (0.00075)(201)(25)^4(12)^3/(3.31 \times 10^6)(275) = 0.112 \text{ in. } (0.284 \text{ cm})$$
$$\Delta_D = (81/201)(0.112) = 0.05 \text{ in. } (0.10 \text{ cm})$$

versus 0.05 in. (0.13 cm) and 0.06 in. (0.15 cm) by the other two methods.

$$\Delta_L = 0.11 - 0.05 = 0.06 \text{ in. } (0.15 \text{ cm})$$

versus 0.06 in. (0.15 cm) and 0.09 in. (0.22 cm) by the other two methods.

Using $(I_e)_{D+L}$, $\Delta_{D+L} = (0.112)(275/259) = 0.12$ in. (0.30 cm)

$$\Delta_L = 0.12 - 0.05 = 0.07 \text{ in. } (0.18 \text{ cm})$$

versus 0.07 in. (0.18 cm) and 0.09 in. (0.22 cm) by the other two methods.

Summary of deflection results for the two-way slab As can be seen in Table 4-15, the results by all of the methods for the interior panel are in close agreement. Only the equivalent frame method provides a means of computing deflections for other than interior panels, as shown in Table 4-14. These computed deflections were rounded to practical values.

Table 4-15 Summary of deflections for the solid slab system with beams (two-way slab) by the different methods

Method	Using I_g or I_e	Δ_D in.	Δ_D cm	Δ_t* in.	Δ_t* cm	Δ_L in.	Δ_L cm	$\Delta_t + \Delta_L$ in.	$\Delta_t + \Delta_L$ cm
Interior panels									
Equivalent frame method	Using I_g	0.06	0.15	0.17	0.43	0.09	0.22	0.26	0.66
	Using I_e	0.06	0.15	0.17	0.43	0.09	0.22	0.27	0.69
Method of deflection coefficients from the theory of elasticity	Using I_g	0.05	0.13	0.15	0.38	0.06	0.15	0.21	0.53
	Using I_e	0.05	0.13	0.15	0.38	0.07	0.18	0.22	0.56
Method of deflection coefficients from the finite difference solutions	Using I_g	0.05	0.13	0.15	0.38	0.06	0.15	0.21	0.53
	Using I_e	0.05	0.13	0.15	0.38	0.07	0.18	0.22	0.56
Side panel 2									
Equivalent frame method	Using I_g	0.06	0.15	0.18	0.46	0.09	0.22	0.27	0.69
	Using I_e	0.06	0.15	0.18	0.46	0.13	0.33	0.31	0.79
Side panel 3									
Equivalent frame method	Using I_g	0.07	0.18	0.20	0.51	0.10	0.25	0.30	0.76
	Using I_e	0.07	0.18	0.20	0.51	0.14	0.36	0.34	0.86
Corner panels									
Equivalent frame method	Using I_g	0.07	0.18	0.21	0.53	0.10	0.25	0.31	0.79
	Using I_e	0.07	0.18	0.21	0.53	0.15	0.38	0.36	0.91

* $\Delta_t = 3.0\Delta_D$.

Comparison with ACI Code allowable deflections:

$\Delta_L < L/180 = (20)(12)/180 = 1.33$ in. (3.38 cm) OK, all panels, all methods

$< L/360 = 240/360 = 0.67$ in. (1.70 cm) OK, all panels, all methods

$\Delta_t + \Delta_L < L/240 = 240/240 = 1.00$ in. (2.54 cm) OK, all panels, all methods

$< L/480 = 240/480 = 0.50$ in. (1.27 cm) OK, all panels, all methods

The results by all of the methods for the interior panel are in reasonably close agreement, especially when comparing results using the same I.

Observations from the Deflection Examples

1. As can be seen in Table 4-11 for the flat plate and Table 4-15 for the two-way slab, the results by all of the methods for the interior panel were in close agreement, especially when comparing results using the same I. Only the equivalent frame method provided a means of computing deflections for the side and corner

panels, as shown in Tables 4-7, 4-10 for the flat plate and Table 4-14 for the two-way slab.

2. As shown in Tables 4-7, 4-10, and 4-14 for both the flat plate and two-way slab, the deflections are the smallest in the interior panel, and the largest in the corner panel, as expected. Also, for both slabs the side panel deflections were largest in the panels whose exterior ends were rotating in the long direction (side panel 2 for the flat plate and 3 for the two-way slab).

3. As can be seen in Table 4-11 for the flat plate and Table 4-15 for the two-way slab, the effect of cracking (using I_e instead of I_g) on Δ_L and $\Delta_t + \Delta_L$ varied from small in the interior panel to relatively large in the corner panel. The following table illustrates.

Flat plate	*Interior panel*	*Corner panel*	*Ratio*
Using I_g (Table 4-7)	$\Delta_L = 0.09$ in. (0.22 cm) $\Delta_t + \Delta_L = 0.36$ in. (0.90 cm)	0.12 in. (0.30 cm) 0.50 in. (1.27 cm)	1.3 1.4
Using I_e (Table 4-10)	$\Delta_L = 0.11$ in. (0.28 cm) $\Delta_t + \Delta_L = 0.39$ in. (0.98 cm)	0.22 in. (0.56 cm) 0.60 in. (1.52 cm)	2.0 1.5
Two-way slab			
Using I_g (Table 4-14)	$\Delta_L = 0.09$ in. (0.22 cm) $\Delta_t + \Delta_L = 0.26$ in. (0.66 cm)	0.10 in. (0.25 cm) 0.31 in. (0.79 cm)	1.1 1.2
Using I_e (Table 4-14)	$\Delta_L = 0.09$ in. (0.22 cm) $\Delta_t + \Delta_L = 0.27$ in. (0.69 cm)	0.15 in. (0.38 cm) 0.36 in. (0.91 cm)	1.7 1.3

As shown by these ratios, the corner panel deflections were from 10 to 100 percent greater than the interior panel deflections.

4. The following section properties and cracking locations were found in Table 4-8 for the flat plate and Table 4-13 for the two-way slab.

Flat plate: for all dead-load deflections—I_g (even though the slab was slightly cracked at the interior columns in the long but not the short direction under dead load); and for dead plus live-load deflections—I_g for the middle strips in both directions (slab was uncracked at all locations) and I_e for the column strips in both directions (slab was partially cracked at all positive and negative locations in the long direction, except the interior panel positive regions, and partially cracked only at the interior negative locations in the short direction).

Two-way slab: for all dead-load deflections—I_g (slab and beams were uncracked at all locations); and for dead plus live-load deflections—I_g for the column strips in both directions (slab and beams are uncracked at all locations) and I_e for the middle strips in both directions (slab was uncracked in the positive region at all locations and partially cracked in the negative regions of all panels in both directions).

Where they differ in positive and negative moment regions, a weighted average of 70 percent positive and 15 percent, 15 percent, negative values of I was used in each case. The above procedure appears to be consistent with the tensile cracking patterns found in a series of tests at the University of Illinois and the Portland Cement Association, as shown in Fig. 4-7 for example, which are along column lines for the flat plate[30] and flat slab[31] at design load (with the addition of middle strip cracking for the flat slab) and diagonal for the two-way slab[32] at overloading. The two-way slab and beams remained uncracked under service loading.

5. When compared with the ACI Code allowable deflections, as shown in Tables 4-11 for the flat plate and 4-15 for the two-way slab, all panels were satisfactory by all methods, except in the case of the flat plate side and corner panels and the severest limiting deflection of $L/480$ (Table 4-11 using the smallest dimension of $L = 16$ ft or 4.9 m, and $T = 3.0$, rather than say the largest dimension of $L = 20$ ft or 6.1 m, and $T = 2$).

4-4 DISCUSSION OF EXPERIMENTAL RESULTS

Sozen and Siess[14] have presented a historical review of early analytical and experimental research on slab systems by Lord[40] (1910), Nichols[41] (1914), Westergaard and Slater[42] (1921), Di Stasio and van Buren[43] (1936), and others. This early work, which was accompanied by considerable controversy (especially with regard to flat slabs), led to completely different approaches for the design of slabs without beams (flat plates, flat slabs) and slabs with beams (two-way slabs). Not until the 1971 ACI Code, for example, was a unified design procedure

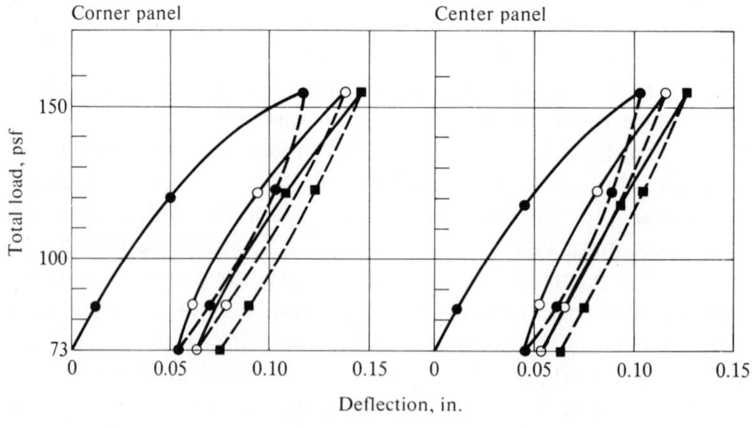

FIGURE 4-5
Midpanel deflections of the 45 ft (13.7 m) square nine-panel (each panel 15 ft by 15 ft or 4.6 m by 4.6 m) flat plate from Ref. 30. The design load was 156 psf (762 kg/m^2, 7470 N/m^2). (*From ACI Journal paper by Guralnick and La Fraugh, with permission.*)

provided for the different types of slabs; a highly desirable objective since many slab systems are a combination of the different basic types. The discussion in this section will focus on deflections and research conducted since the early 1960s.

Short-Time Deflections

A comprehensive research program on multiple-panel floor slabs was begun[44] at the University of Illinois in the 1950s and continued at that institution and the Portland Cement Association in the 1960s. Of principal interest with regard to deflections are Refs. 3, 14, 30–32, 45–47, by Sozen, Siess, Guralnick, La Fraugh, Vanderbilt, Jirsa, Hatcher and Gamble. All of these papers contain useful information on load–deflection behavior and cracking patterns, with three of the papers specifically discussed as follows.

Guralnick and LaFraugh[30] tested a 45 ft (13.7 m) square normal-weight reinforced concrete flat plate structure consisting of nine 15 ft by 15 ft (4.6 m by 4.6 m) panels arranged three by three. The slab was approximately 5.25 in. (13.3 cm) thick with square supporting columns and spandrel beams also used. The design concrete strength was 3000 psi (210 kg/cm^2, 20.7 N/mm^2). This laboratory structure was a three-quarter scale model of an actual structure.

The load–deflection curves for the different panels under repeated loading into the cracking range (which occured at the design load for this structure) were similar to those found in the literature and in Chap. 3 herein for beams. On first loading the load–deflection curves exhibit markedly inelastic behavior. However, subsequent unloading and reloading produces almost linear behavior, as

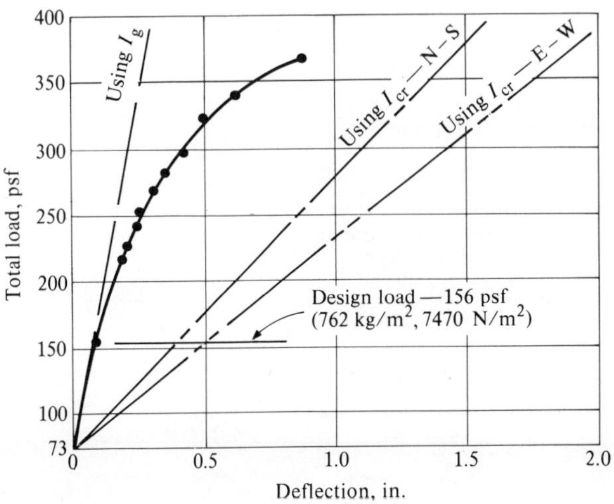

FIGURE 4-6
Midpanel load–deflection curve for the center panel of the 45 ft (13.7 m) square nine-panel flat plate. (*From ACI Journal paper*[30] *by Guralnick and La Fraugh, with permission.*)

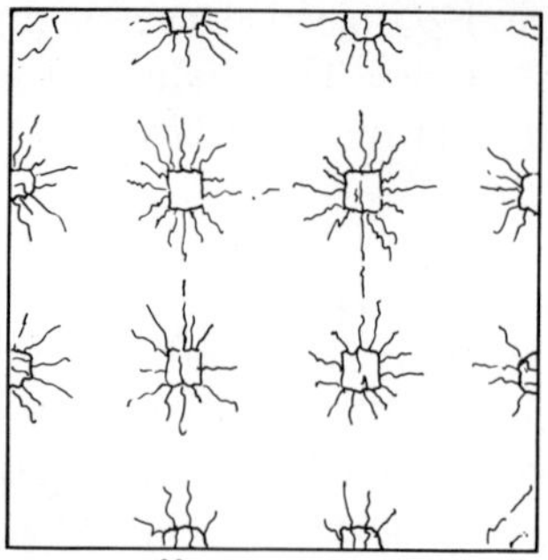

Flat Plate,[30],45 ft (13.7 m) square
By Guralnick and La Fraugh

Top surface at design load

Cracking of the bottom was very slight at design load.

Subsequent cracking when subjected to overloading was along column lines for the top and along lines midway between columns for the bottom.

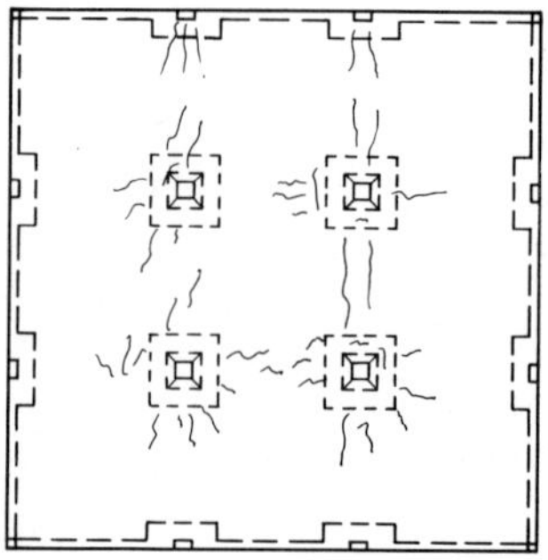

Flat Slab,[31],15 ft (4.6 m) square
By Hatcher, Sozen, and Siess

Top surface at design load

Cracking of the bottom was fairly extensive along lines midway between columns (except the center panel) at design load.

Subsequent cracking when subjected to overloading was similar to the flat plate above.

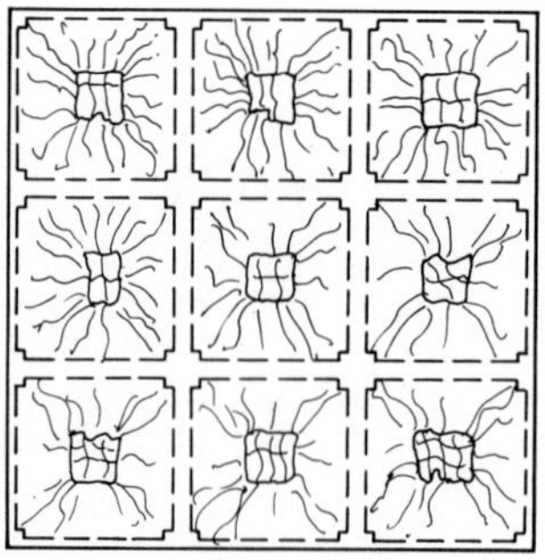

Two-way slab,[32],15 ft (4.6 m) square
By Gamble, Sozen, and Siess

Bottom surface at failure

No cracks were found at design load.

When subjected to overloading diagonal cracks developed in the bottom of the slab and near the centers of some of the end spans of the interior beams.

FIGURE 4-7
Cracking patterns for the conditions stated with each figure. (*From ACI Journal*[30] *and ASCE Proceedings,*[31,32] *with permission.*)

shown in Fig. 4-5 for a corner panel and the center panel. The side panel deflections were in between the corner and center panel deflections. Tensile cracking of the top of the slab at the design load is shown in Fig. 4-7, while cracking of the bottom of the slab was very slight at the service load level. Subsequent tensile cracking due to overloading was along column lines for the top of the slab and along lines midway between columns for the bottom of the slab.

Center panel deflections were calculated using Eq. (4-5), which neglects Poisson's ratio, and $K_s = 0.00581$ (coefficient in Table 4-1 for an interior panel without checkerboard loading). Three different values for I were used (I_g, and I_{cr} in both directions—difference due to different effective depths). These deflections are shown in Fig. 4-6, where it is seen that the uncracked value (I_g) accurately predicted the experimental result up to the service load of 156 psf (762 kg/m^2, 7470 N/m^2) for this center panel. Under service loading the ratio of the initial corner panel deflection (average 0.105 in., 0.27 cm) to the center panel deflection (0.10 in., 0.25 cm) was 1.1. This low ratio was apparently due to the stiffening effect of the spandrel beams.

Hatcher, Sozen, and Siess[31] tested a square normal-weight reinforced concrete flat slab consisting of nine 5 ft by 5 ft (1.5 m by 1.5 m) panels. The slab was 1.75 in. (4.4 cm) thick with drop panels, column capitals, and spandrel beams also used. The concrete strength at age 78 days was 2760 psi (194 kg/cm^2, 19.0 N/mm^2). This laboratory structure was a one-quarter scale model of a prototype designed for a live load of 200 psf (976 kg/m^2, 9580 N/m^2) and a dead load of 85 psf (415 kg/m^2, 4070 N/m^2).

The load–deflection behavior of the flat slab was similar to that of the flat plate[30] previously described. Tensile cracking of the top of the slab at the design load of 290 psf (1416 kg/m^2, 13 890 N/m^2) was along column lines, as shown in Fig. 4-7, the same as for the flat plate. At this load level bottom cracks were fairly extensive along lines midway between columns for this flat slab (except for the interior panel), which was the case for the flat plate[30] only when subjected to overloading. With regard to this difference in behavior, it should be remembered that the flat slab was a one-quarter scale model while the flat plate was three-quarter scale. Under service loading the ratio of the initial corner panel deflection (average 0.20 in., 0.51 cm) to the center panel deflection (0.07 in., 0.18 cm) was 2.9, which is the largest such ratio (the other maximum ratio was 2.0) found in this chapter for various types of slabs and methods of computing and measuring deflections.

Gamble, Sozen, and Siess[32] tested a square normal-weight reinforced concrete two-way slab consisting of nine 5 ft by 5 ft (1.5 m by 1.5 m) panels. The slab was 1.5 in. (3.8 cm) thick with relatively deep supporting beams used for all panels as follows: exterior beams—3 in. by 4.25 in. (7.6 cm by 10.8 cm), interior beams—3 in. by 5 in. (7.6 cm by 12.7 cm). The concrete strength at age 76 days was 3420 psi (240 kg/cm^2, 23.6 N/mm^2). This laboratory structure was a one-quarter scale model of a prototype designed for a live load of 70 psf (342 kg/m^2, 3350 N/m^2) and a dead load of 75 psf (366 kg/m^2, 3590 N/m^2).

The load–deflection behavior of the two-way slab was similar to that of the

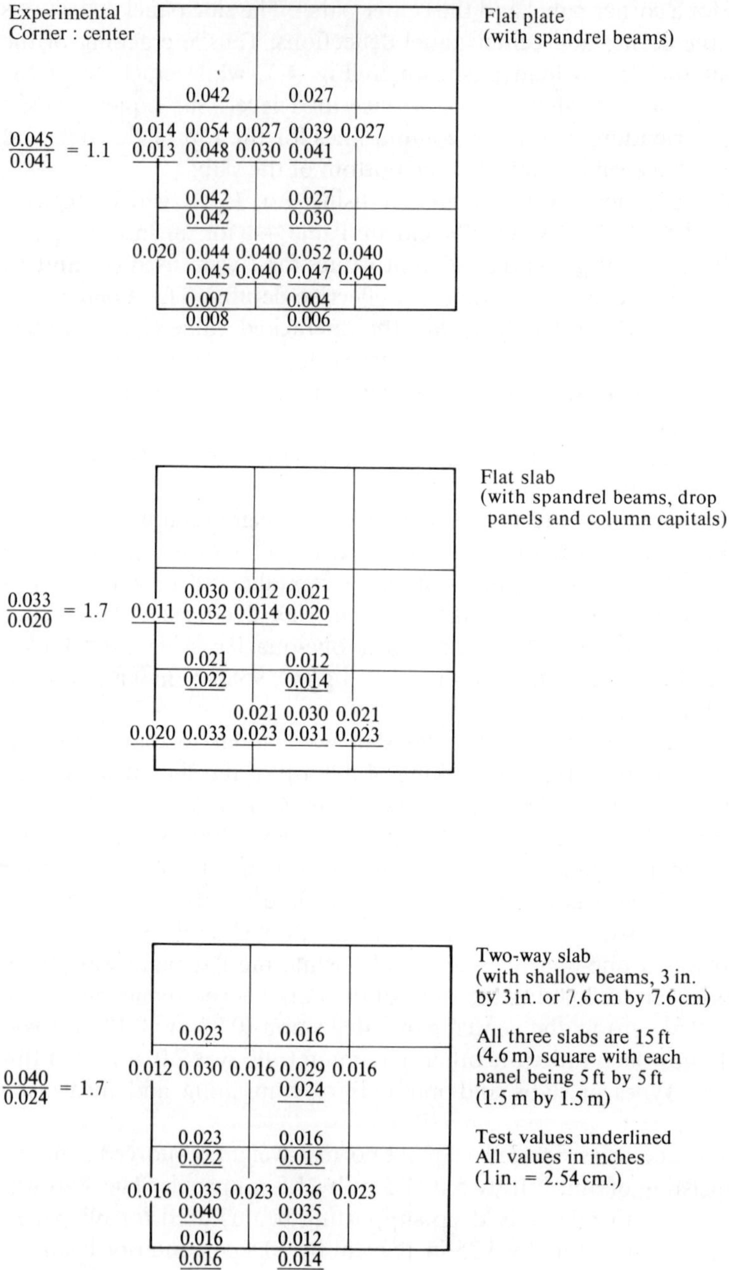

FIGURE 4-8
Computed deflections by the equivalent frame method[18] and measured deflections from University of Illinois tests.[3,45] (*From ACI Journal paper*[18] *by Nilson and Walters, with permission.*)

flat plate[30] and flat slab[31] previously described. No cracks were found in the test structure at the design load of 145 psf (708 kg/m^2, 6940 N/m^2). When subjected to overloading the principal cracking occurred diagonally at the bottom surface, as shown in Fig. 4-7. Under service loading the ratio of the initial corner panel deflection (average 0.043 in., 0.11 cm) to the center panel deflection (0.027 in., 0.07 cm) was 1.6.

In the flat plate, flat slab, and two-way slab tests described above, the envelope of each load–deflection curve under repeated loading and unloading into the cracking range tends to approximate a single loading curve into the cracking range, which is consistent with the behavior described in Chap. 3 for reinforced beams and Chap. 5 for cracked prestressed beams.

Nilson and Walters[18] used a flat plate, a flat slab, and a two-way slab structure from the University of Illinois tests[3,45] (all three were one-quarter scale models consisting of nine panels—each panel 5 ft by 5 ft or 1.5 m by 1.5 m) to compare computed versus measured deflections, as shown in Fig. 4-8. The equivalent frame method described herein was used in the calculations. It may be seen in Fig. 4-8 that very good agreement was obtained, with the maximum difference being about 20 percent, and the typical value being much closer. For these experimental results the ratio of corner to center panel deflection was 1.1 for the flat plate, 1.7 for the flat slab, and 1.7 for the two-way slab. Spandrel beams were used with all three of the slab systems.

In 1967 Magura and Corley[48] reported the results of tests to destruction of a multipanel waffle roof slab at the New York World's Fair. The test structure was a box-like building measuring approximately 180 ft (54.9 m) by 120 ft (36.6 m) in plan and 20 ft (6.1 m) in height. A 2 ft (61.0 cm) thick waffle slab supported on columns about 30 ft (9.1 m) on center formed the roof of the "Rathskeller."

The load–deflection curves were found to be linear until the load exceeded dead plus about 1.5 live load. Average measured deflections under service loading varied from 0.24 in. (0.61 cm) for the interior panels (versus 0.27 in., 0.69 cm, computed) to 0.35 in. (0.89 cm) for the outside panels (versus 0.28 in., 0.71 cm, computed). The computed deflections using an equivalent frame analysis (with the "ersatz" frame) were in reasonably good agreement with the measured values. Magura and Corley concluded that, although the frame analysis is cumbersome, it appears to be a reliable method for predicting deflections. All measured deflections were within the ACI Code allowable values for roof slabs.

In 1970 Simmonds[49] presented the results of an analytical and experimental study of a flat plate test structure supported on columns elongated in plan. The test structure was a one-third scale model consisting of nine panels, each 6 ft by 6 ft (1.8 m by 1.8 m). The thickness of the slab was 2 in. (5.1 cm). No edge beams were used. Calculated deflections for the interior panel using the coefficient method compared reasonably well with observed values. Simmonds concluded that deflections can be computed for the panel midpoints using the same procedures as for slabs supported on square columns if the span in the direction of the column elongation is appropriately reduced.

Short-Time and Long-Time Deflections

In 1961 Blakey[36] described an experimental lightweight concrete structure whose ratio of long-time to short-time deflection at the center of an interior panel was about 7 after 200 days. Using a 3.5 in. (8.9 cm) thick experimental lightweight flat plate spanning three bays of 9 ft (2.7 m) in one direction and three bays of 12 ft (3.7 m) in the other, with cantilevers of 4.5 ft (1.4 m) in the latter direction, Blakey[37] reported in 1963 that the measured time-dependent dead-load deflection at the center after 8 months was 9 times (when neglecting differential settlement) the initial elastic deflection. He attributed about half of this to progressive cracking and bond slip, and half to creep. The span–depth ratios were 41 in one direction and 31 in the other, which may not be too high for some two-way floor systems, but apparently were too high for this lightweight concrete flat plate structure. In 1969 Taylor[38] found for a reinforced concrete flat plate structure built at North Sydney, Australia, that the measured deflections at the center of the interior panels over a period of 2.5 years were 6.5 to 10 times the initial deflections. These ratios or multipliers are abnormally high compared to other measured values that have been reported. Taylor[33] mentions, for example, that both creep and shrinkage of the Sydney concrete were greater than that usually found in the U.S.A.

In 1967, 1968 Shukla, Gupta, and Branson[50] measured initial and time-dependent deflections of a normal-weight two-way slab consisting of nine panels, each 6 ft by 6 ft (1.8 m by 1.8 m) with relatively deep beams. The thickness of the slab was 1.5 in. (3.8 cm). The structure was loaded with sand bags at age 30 days, with the loading maintained for 500 days. The average relative humidity was 50 percent. Measured deflections of the slab indicated an ultimate ratio of time-dependent to initial deflections of about 5. The author's method for beams, which was given earlier in this chapter for slabs, was later used by Matsui and Branson[29] to compute shrinkage warping and creep deflections. The computed results compared favorably with the measured deflections.

In 1970 Taylor[33] used the measured long-time deflection results obtained from his North Sydney flat plate structure[38] to compare with computed results based on initial elastic deflections and code multipliers. The calculated values were derived from a combination of the initial slab deflection procedure of Vanderbilt, Sozen, and Siess[3] and the author's[51,52] procedure for determining long-time deflections of beams. Taylor found that the usual code multipliers were inadequate and that a multiplier greater than 6 was needed for combined time-dependent deflections. He concludes that either the separate creep and shrinkage or the combined time-dependent methods may be used with a reasonable degree of accuracy. He also points out that these methods account for the change from an uncracked to a cracked section.

In 1971 Scanlon[22] reported a study of the initial and time-dependent deflections of slabs using both a finite element method and the author's simplified methods for cracking, creep, shrinkage, and combined time-dependent results. From a limited comparison of results obtained by the finite element analysis with those from the simplified procedures, he concluded that reasonable results may be obtained using the simplified procedures, but that the procedures are not sufficiently developed to reliably predict long-time deflections of slabs, including the effect of compression steel. However, he noted that for many practical situations, the agreement obtained between the computer results and the simplified procedures would be considered adequate.

In 1974 Heiman[34] presented a comparison of measured and calculated deflections of flexural members in four reinforced concrete buildings. One consisted of beam and slab floor construction and a flat plate roof, another of flat slab construction, a third of beam and slab construction, and the fourth was a flat plate structure. He calculated deflections using a combination of the method of Vanderbilt, Sozen, and Siess[3] and the author's I_{eff} method for short-time deflections, Neville's[53] method and the author's method for shrinkage deflections, the author's method for creep deflections, and the multiplier method for combined time-dependent deflections.

Heiman found that the ACI Code method (using $T = 2$) tended to underpredict deflections and that more accurate results were obtained when the long-time deflections caused by creep and shrinkage were calculated separately. He found reasonable agreement between the measured values and those calculated by the author's creep and shrinkage methods, together with creep and shrinkage parameters from the author and Christiason[54] and ACI Committee 209.[6] For the four buildings, the computed deflections thus obtained differed from the measured values by 17, 14, 13, and 7 percent.

Heiman also observed that large short-time deflections under heavy construction loads can have an important influence on subsequent long-time deformations. Where service loads are small, shrinkage can sometimes play an important part in the long-time deflections, and in addition to causing warping may also result in extensive cracking and loss of stiffness.

In 1975 Rangan[35] presented a method for predicting deflections of flat plates and slabs that consisted of a modified I_{eff} method for short-time deflections and the same multiplier method for creep and shrinkage deflections previously described. An equivalent frame analysis was also used. Computed results were compared with measured values for three existing floors with reasonably good agreement. The computed deflections differed from the measured values by 7, 17, and 24 percent.

Thus the studies described above tend to substantiate the procedures described herein for computing separate creep and shrinkage deflection or combined time-dependent deflection of slab systems, at least for most practical purposes.

One slight modification of the calculation procedures herein was found to be appropriate in a 1976 computer study of experimental and computed short-time deflections by Kripanarayanan and the author.[55] The study showed that more accurate results were obtained when using a weighted average of (0.70 Pos. I_e + 0.30 Neg. I_e) for two-way slabs (same as herein), but using (0.50 Pos. I_e + 0.50 Neg. I_e) for flat plates and flat slabs.

Additional experimental research on the deflection of slab systems not specifically referred to above includes studies of Ockleston,[56,57,58] Rosenthal,[59] PCA,[60] Smith and Simmonds,[61] Beresford,[62] Taylor and Heiman,[63] and Heiman and Taylor.[64]

REFERENCES

1. Subcommittee 5, ACI Committee 435, Subcommittee Chairman, A. H. Nilson, Committee 435 Chairman, G. M. Sabnis, State-of-the-Art Report, Deflection of Two Way Reinforced Concrete Floor Systems, *ACI Publication SP 43-3, Deflections of Concrete Structures,* 55–81, 1974.
2. S. Timoshenko and S. Woinowsky-Krieger, "Theory of Plates and Shells," McGraw-Hill Book Co., New York, 1959.
3. M. D. Vanderbilt, M. A. Sozen, and C. P. Siess, Deflections of Multiple-Panel Reinforced Concrete Floor Slabs, *Proceedings, ASCE,* **91,** ST4, 77–101, Aug. 1965.
4. Portland Cement Association, Deflections of Flat Plate Floors, *Portland Cement Association, Concrete Report* XS6504, 1–3, Aug. 1965.
5. M. D. Vanderbilt, Deflection Calculations: Two-way Slabs, *Proc. Symposium on Deflections of Structural Concrete, Montreal,* 113–140, 1971.
6. Subcommittee 2, ACI Committee 209, Subcommittee Chairman, D. E. Branson, Committee 209 Chairman, J. R. Keeton, Prediction of Creep, Shrinkage and Temperature Effects in Concrete Structures, in "Designing for Effects of Creep, Shrinkage and Temperature in Concrete Structures," 51–93, ACI Publication SP 27-3, 1971.
7. C. F. Marsh, "Reinforced Concrete," 1–283, D. Van Nostrand Co., New York, 1904.
8. H. Marcus, "Die Theorie elasticher Gewebe und ihre Anwendung auf die Berechnung biegsamer Platten," Julius Springer, Berlin, 1924.
9. W. W. Ewell, S. Okubo, and J. I. Abrams, Deflections in Gridworks and Slabs, *Transactions, ASCE,* **117,** 869, 1952.
10. W. L. Furr, Numerical Method for Approximate Analysis of Building Slabs, *ACI Journal,* **31,** 6, 511, Dec. 1959.
11. American Concrete Institute, "Building Code Requirements for Reinforced Concrete," 1–78, and Commentary, 1–96, American Concrete Institute, Detroit, Michigan, 1971.
12. D. Peabody, Continuous Frame Analysis of Flat Slabs, *Journal of Boston Society of Civil Engineers,* **35,** 1, Jan. 1948.
13. W. G. Corley, M. A. Sozen, and C. P. Siess, The Equivalent Frame Analysis for Reinforced Concrete Slabs, *Dept. of Civil Engineering, University of Illinois, Urbana, Structural Research Series* No. 218, June 1961.
14. M. A. Sozen and C. P. Siess, Investigation of Multiple-Panel Reinforced Concrete Floor Slabs: Design Methods—Their Evolution and Comparison, *ACI Journal,* Proceedings **60,** 8, 999–1028, Aug. 1963.
15. W. G. Corley and J. O. Jirsa, Equivalent Frame Analysis for Slab Design, *ACI Journal,* Proceedings **67,** 11, 875–884, Nov. 1970.
16. R. S. Fling, Approximate Method for Computing Slab Deflections, private communication, R. S. Fling and Partners, Columbus, Ohio, 1972.
17. E. G. Nawy, Deflection Control in Two-Way Slab and Plate Systems, *Seminar on Serviceability for Concrete Structures, Joint Conference, PCA–ACI–ASCE, and the University of Miami,* Jan. 1975.
18. A. H. Nilson and D. B. Walters, Deflection of Two-Way Floor Systems by the Equivalent Frame Method, *ACI Journal,* **72,** 5, 210–218, May 1975.
19. A. C. Scordelis, Finite Element Analysis of Reinforced Concrete Structures, *Proceedings, Conference on Finite Element Methods in Civil Engineering, McGill University,* 71, 1972.
20. L. G. Selna, Time Dependent Behavior of Reinforced Concrete Structures, *Dept. of Civil*

Engineering, University of California at Berkeley, Structures and Materials Research Report 67-19, 1967.

21. A. A. R. Youssef, J. O. McCutcheon, A. A. Mufti, and M. S. Mirza, Inelastic Analysis of Plates and Slabs Using Finite Elements, *McGill University, Montreal, Structural Concrete Series* No. 71-1, Mar. 1971.
22. A. Scanlon, Time Dependent Deflections of Reinforced Concrete Slabs, *University of Alberta, Canada, Structural Engineering Report* 38, 1–174, Dec. 1971.
23. R. Szilard, "Theory of Analysis of Plates, Classical and Numerical Methods," 1–724, Prentice Hall, Englewood Cliffs, New Jersey, 1974.
24. Concrete Reinforcing Steel Institute, "CRSI Handbook," 2nd Edn, 1–1 to 15–53, Concrete Reinforcing Steel Institute, Chicago, Illinois, 1972, 1975.
25. A. Zweig, Design Aids for the Direct Design Method of Flat Slabs, *ACI Journal,* Proceedings **70,** 4, 285–299, Apr. 1973.
26. Portland Cement Association, Notes on ACI 318-71, Building Code Requirements with Design Applications, 1–1 to 26–27, Portland Cement Association, Skokie, Illinois, 1972.
27. S. H. Simmonds, "Effects of Column Stiffness on the Moments in Two-Way Floor Slabs," thesis, Univ. of Illinois, Urbana, 1962.
28. J. O. Jirsa, M. A. Sozen, and C. P. Siess, Pattern Loadings on Reinforced Concrete Floor Slabs, *Proceedings, ASCE,* **95,** ST6, 1117–1137, June 1969.
29. K. Matsui and D. E. Branson, Initial and Time-Dependent Deflections of Reinforced Concrete Flat Plates and Two-Way Slabs, interim research report, University of Iowa, 1970.
30. S. A. Guralnick and R. W. La Fraugh, Laboratory Study of a 45-Foot Square Flat Plate Structure, *ACI Journal,* Proceedings **60,** 9, 1107–1185, Sept. 1963.
31. D. S. Hatcher, M. A. Sozen, and C. P. Siess, Test of a Reinforced Concrete Flat Slab, *Proceedings, ASCE,* **95,** ST6, 1051–1072, June 1969.
32. W. L. Gamble, M. A. Sozen, and C. P. Siess, Test of a Two-Way Reinforced Concrete Floor Slab, *Proceedings, ASCE,* **95,** ST6, 1073–1096, June 1969.
33. P. J. Taylor, Long-term Deflection Calculation Methods for Flat Plates, *Construction Review (Sydney),* **43,** 2, 68–74, May 1970.
34. J. L. Heiman, A Comparison of Measured and Calculated Deflections of Flexural Members in Four Reinforced Concrete Buildings, *ACI Publication SP 43-21, Deflections of Concrete Structures,* 515–545, 1974.
35. B. V. Rangan, Prediction of Long-Term Deflections of Flat Plates and Slabs, manuscript in progress, *ACI Journal,* 1975.
36. F. A. Blakey, Deformations of an Experimental Lightweight Flat Plate Structure, *Civil Engineering Transactions, I. E. Aust.,* **C.E.3,** 1, Mar. 1961.
37. F. A. Blakey, Australian Experiments with Flat Plates, *ACI Journal,* Proceedings **60,** 4, 515–525, Apr. 1963.
38. P. J. Taylor, Initial and Long-Term Deflections of a Reinforced Concrete Flat Plate Structure, *Inst. of Eng. Aust., Conference on the Deformation of Concrete and Concrete Structures, Brisbane, Paper* 2794, Sept. 1969.
39. C. K. Wang and C. G. Salmon, "Reinforced Concrete Design," 2nd edn, 1–934, Intext Educational Publishers, New York, 1973.
40. A. R. Lord, A Test of a Flat Slab Floor in a Reinforced Concrete Building, *National Association of Cement Users (ACI),* **7,** 156–179, 1910.
41. J. R. Nichols, Statical Limitations Upon the Steel Requirements in Reinforced Concrete Flat Slab Floors, *Transactions, ASCE,* **77,** 1670–1736, 1914.
42. H. M. Westergaard and W. A. Slater, Moments and Stresses in Slabs, *Proceedings, ACI,* **17,** 415–538, 1921.
43. J. Di Stasio and M. P. van Buren, Slabs Supported on Four Sides, *ACI Journal,* Proceedings **32,** 3, 350–364, Jan.–Feb. 1936.
44. Research on Multiple-Panel Reinforced Concrete Floor Systems, *ACI Journal,* Proceedings **56,** 2, newsletter, 11–13, Aug. 1959.
45. M. D. Vanderbilt, M. A. Sozen, and C. P. Siess, Deflections of Reinforced Concrete Floor Slabs,

University of Illinois, Urbana, Civil Engineering Studies, Structural Research Series 263, Apr. 1963.

46. J. O. Jirsa, M. A. Sozen, and C. P. Siess, Test of a Flat Slab Reinforced with Welded Wire Fabric, *Proceedings, ASCE*, **92,** ST3, 199–224, June 1966.
47. M. D. Vanderbilt, M. A. Sozen, and C. P. Siess, Test of a Modified Reinforced Concrete Two-Way Slab, *Proceedings, ASCE*, **95,** ST6, 1097–1116, June 1969.
48. D. D. Magura and W. G. Corley, Tests to Destruction of a Multi-Panel Waffle Slab Structure, 1964–65 New York World's Fair, *Portland Cement Association, PCA R/D Ser.* 1302, 1–62, June 1967.
49. S. H. Simmonds, Flat Slabs Supported on Columns Elongated in Plan, *ACI Journal,* Proceedings **67,** 12, 967–975, Dec. 1970.
50. M. Shukla, S. P. Gupta, and D. E. Branson, Deflection of Two-Way Slabs, interim research reports, University of Iowa, 1967, 1968.
51. D. E. Branson, Design Procedures for Computing Deflections, *ACI Journal,* Proceedings **65,** 9, 730–742, Sept. 1968.
52. ACI Committee 435, Chairman, D. E. Branson, Deflections of Reinforced Concrete Flexural Members, *ACI Journal,* Proceedings **63,** 6, 637–674, June 1966.
53. A. M. Neville and W. Dilger, "Creep of Concrete: Plain, Reinforced, and Prestressed," 1–622, North-Holland Publishing Company, Amsterdam, 1970.
54. D. E. Branson and M. L. Christiason, Time Dependent Concrete Properties Related to Design—Strength and Elastic Properties, Creep and Shrinkage, in "Designing for Effects of Creep, Shrinkage and Temperature in Concrete Structures," 257–277, ACI Publication SP 27-13, 1971.
55. K. M. Kripanarayanan and D. E. Branson, Short-Time Deflections of Two-Way Slab Systems, accepted for ACI Journal, 1976.
56. A. J. Ockleston, I. R. Scott, and A. C. Liebenburg, General Report on Tests on the Old Dental Hospital, Johannesburg; The Determination of the Properties of Materials from the Old Dental Hospital; The Effect of Floors and Walls on the Behavior of Reinforcement Concrete Frameworks Subject to Horizontal Loading; Load Tests on Stairways of a Reinforced Concrete Building in Johannesburg, *The Concrete Association, South Africa, Paper* 1–4, 1956; *Structural Engineer (London)*, **33,** 10, 305–323, Oct. 1955.
57. A. J. Ockleston, Loading Tests on a Large Reinforced Concrete Slab Spanning in One Direction, *The Concrete Association, South Africa, Paper* 5, 1957.
58. A. J. Ockleston, Loading Tests on a Large Reinforced Concrete Slab Spanning in Two Directions, *The Concrete Association, South Africa, Paper* 6, 1958; *Structural Engineer (London)*, **33,** 10, 305–323, Oct. 1955.
59. I. Rosenthal, Experimental Investigation of Flat Plate Floors, *ACI Journal,* Proceedings **56,** 2, 153–166, Aug. 1959.
60. Portland Cement Association, Deflections of Flat Plate Floors of Apartment Buildings, *Portland Cement Association, Concrete Report* CR008.01D (XS 6504), Jan. 1965.
61. A. E. Smith and S. H. Simmonds, Test of Flat Plate Supported on Columns Elongated in Plan, *University of Alberta, Structural Engineering Report* 21, 1–72, May 1969.
62. F. D. Beresford, Experimental Lightweight Flat Plate Structures, Part V: Deformation Under Lateral Load, *Constructional Review, Sydney,* **35,** 12, 17–23, 1962.
63. P. J. Taylor and J. L. Heiman, A Long-term Investigation of Deflections of a Flat-slab Structure, *Fourth Australian Building Research Congress, Sydney,* 27–31, Aug. 1970.
64. J. L. Heiman and P. J. Taylor, Long-term Deflections of a Reinforced Flat Plate, *Architectural Science Review, Sydney,* **15,** 2, 25–29, June 1972.

CHAPTER

FIVE

LOSS OF PRESTRESS, CAMBER, AND DEFLECTION OF NONCOMPOSITE AND COMPOSITE PRESTRESSED CONCRETE MEMBERS

5-1 INTRODUCTION

Loss of prestress equations using the approximate step-function method, the relaxation method, the AASHTO method, and the PCI Handbook method; and camber equations using the approximate step-function method recommended by ACI Committees 435 and 209 are presented in this chapter for noncomposite and both unshored and shored composite prestressed members. The effects of nonprestressed steel and sustained loads other than a composite slab or topping applied sometime after prestressing are also included in the equations. In addition to specific equations for ultimate (in time) values, continuous time equations are given which, along with the continuous material parameter equations in Chap. 1, facilitate the calculation of values at any time t. These equations also readily lend themselves to computer solution. Design simplifications are outlined along with simplified equations for computing loss of prestress, stress in nonprestressed steel, and camber for normal-weight, sand-lightweight, and all-lightweight concrete. The effect of partial continuity involving a continuous composite deck slab on prestress loss or gain, and camber or deflection is also evaluated.

Numerical examples of loss of prestress, concrete and steel stresses, and camber and deflection calculations are presented for a noncomposite double-T of normal-weight concrete, a noncomposite double-T of sand-lightweight concrete, a noncomposite single-T of sand-lightweight concrete both without and with nonprestressed tension steel and with additional sustained loading, and an unshored composite AASHTO Type IV girder of normal-weight concrete. Both stress and deflection results in each example are discussed and compared with appropriate limiting values—ACI Code (318-71) and AASHTO (1973).

The effect of nonprestressed tension steel in reducing time-dependent creep, shrinkage and camber is reviewed both analytically and experimentally, as is the short-time deflection of cracked prestressed beams. Both the bilinear method and the I_e or I-effective approach used in the 1971 ACI Code for reinforced members and the 1971 PCI Handbook for satisfying the 1971 ACI Code provision on deflections of partially prestressed members are included. Alternative methods for calculating I_e are shown for both noncomposite and composite members. The effect of repeated load cycles into the cracking range is also described.

A number of analytical and experimental studies since 1960 which relate to various methods of calculation are discussed. Both laboratory specimens and actual structural members are included in the experimental studies. Typical ranges of variation between experimental and calculated values are shown, along with pertinent conclusions and recommendations made by different investigators.

5-2 LOSS OF PRESTRESS

While the loss of prestress has essentially no effect on the ultimate capacity of prestressed members (unless the tendons are unbonded), significant overprediction or underprediction of losses can adversely affect such serviceability aspects as camber, deflection, cracking and connection performance. According to the ACI Code (318-71) the following contributions to loss of prestress shall be considered:

Slip at anchorage (The PCI Handbook[1] states that seating losses are usually very small on long beds, but strand deflection losses can be large. It suggests that tensioning procedures should not cause strand tension to vary from the theoretical more than 5 percent. Slippage losses are normally taken into account by overtensioning.)
Elastic shortening of concrete (normally not compensated for by overtensioning)
Creep of concrete
Shrinkage of concrete
Relaxation of steel stress (see Chap. 1)
Frictional loss due to intended or unintended curvature in the tendons (post-tensioned)

For deflected strands, the effect of losses in prestress at the deflecting devices should also be taken into account. The elastic and creep losses usually vary along the length of the member, but it will normally suffice to calculate values at midspan.

Four procedures for computing the loss of prestress are outlined and compared. These are referred to as the approximate step-function method of the author and colleagues,[2-10] the relaxation method of Trost,[11,12] the AASHTO method,[13,14] and the PCI Handbook method.[1] These procedures are formulated primarily for pretensioned members with appropriate modifications indicated for post-tensioned members. A detailed study of the first two more basic approaches may also be found in Ref. 15.

Methods of Calculation

The equations in this section refer to steel stress, such as the initial plus time-dependent loss of prestress, $(\Delta f_s)_{i,t}$, and nonprestressed steel stress, $(f_{ns})_{i,t}$. For computing the loss of prestress force, the loss of prestress in percent of say the initial tensioning stress, and the force in nonprestressed steel, use Eqs. (5-1), (5-2), and (5-3), respectively.

$$(\Delta P)_{i,t} = (\Delta f_s)_{i,t} A_{ps} \tag{5-1}$$

$$PL_{i,t} = (\Delta f_s)_{i,t}(100)/f_{si} \tag{5-2}$$

$$(F_{ns})_{i,t} = (f_{ns})_{i,t} A_{ns} \tag{5-3}$$

where A_{ps} and A_{ns} refer to the areas of prestressed steel and nonprestressed tension steel, respectively. Use A'_{ns} instead of A_{ns} in Eq. (5-3) for compression steel. The choice of expressing the loss of prestress in units of stress (as psi) or percent of some specified stress, such as the initial tensioning stress, is a matter of preference, since the loss in psi must also refer to a specified stress in order to define the force after loss.

It may be noted for example, that prestressing steel percentages ($100A_{ps}/A_g$ or $100A_{ps}/bd$) vary from about 0.30 to 0.80 percent for typical prestressed beams, and tension steel percentages (A_s/bd) vary from about 0.70 to 2.50 percent for typical reinforced beams. The prestressing steel ratio tends to apply more appropriately when expressed as A_{ps}/A_g at working loads and A_{ps}/bd after cracking and at ultimate load.

Approximate step-function method Based on the step-function approach for computing camber as derived by the author and Ozell,[2,3] and used by ACI Committees 435[9] and 209,[10] Eqs. (5-4) to (5-26) are recommended for computing the initial and time-dependent loss of prestress and stress in nonprestressed steel at any time including ultimate values (subscript $t = \mathrm{u}$). These equations are thought to represent a reasonable compromise between complexity and reality that also analytically describes the physical behavior due to the different effects. This approach also lends itself to a relatively simple procedure in the case of composite beams. Typical loss of prestress versus time curves are shown in Fig. 5-1 for both noncomposite and composite beams.

Noncomposite beams with prestressed steel only (containing no nonprestressed tension steel) The loss of prestress at any time including ultimate values is computed by Eq. (5-4).

$$(\Delta f_s)_{i,t} = \overbrace{(nf_{ci})}^{(1)} + \overbrace{(nf_{ci})C_t(1 - \Delta P_t/2P_o)}^{(2)} + \overbrace{(\varepsilon_{sh})_t E_s/(1 + b_{11})}^{(3)} + \overbrace{(f_{sr})_t}^{(4)} \tag{5-4}$$

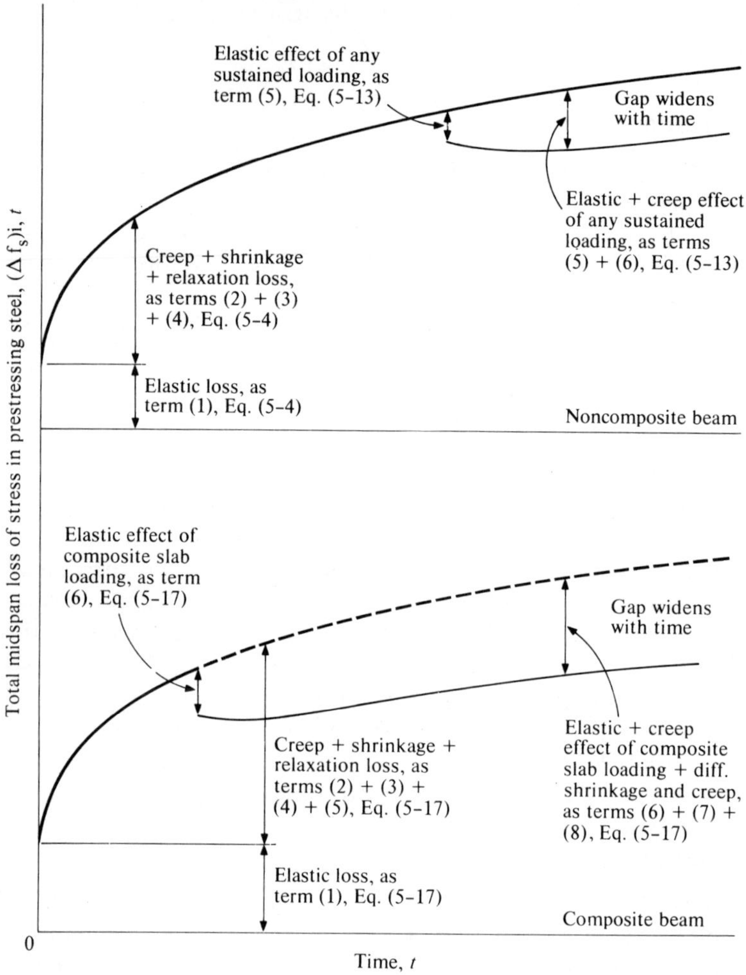

FIGURE 5-1
Typical loss of stress in prestressing steel versus time curves for noncomposite and composite beams.

where:

Term (1) is the prestress loss due to elastic shortening. $f_{ci} = P_i/A_t + P_i e^2/I_t - M_D e/I_t$, and n is the modular ratio at the time of prestressing. Frequently P_o, A_g, and I_g are used as a close approximation instead of P_i, A_t, and I_t, where $P_o = P_i -$ (about $2)(P_o/A_g E_{ci})(A_{ps} E_s) = P_i/(1 + 2n\rho_{ps})$. P_i is the tensioning or jacking force and P_o is the prestress force after transfer. Only the first two terms for f_{ci} apply at the ends of simple beams, for example. See Table 5-1 for values of n. Special allowances should be made for post-tensioned members where tendons are stressed successively, such as reducing the elastic loss to one-half as recommended by the AASHTO–PCI Joint Committee,[13] and using the net section properties. For continuous post-tensioned members, the effect of secondary moments due to prestressing should also be considered, although this effect normally results in a gain in prestress, not a loss. Frictional losses must also be computed in the case of draped post-tensioned tendons.

Term (2) is the prestress loss due to concrete creep. See Chap. 1 for values of the creep coefficient, C_t, including Table 1-10 for ultimate values, C_u. As explained in the footnote, Table 1-10 contains average values for different loading ages, with a correction factor for the effect of member size required from Table 1-8 or 1-9, or the corresponding equations. The expression $C_t(1 - \Delta P_t/2P_o)$ is used to approximate the creep effect resulting from the variable stress history under the decreasing prestress force. See Table 5-2 for approximate values of $\Delta P_t/P_o$ (in the form of $\Delta P_s/P_o$ and $\Delta P_u/P_o$) for this somewhat secondary effect at 3 weeks to 1 month, 2 to 3 months, and ultimate values. $\Delta P_t/P_o$ is defined as the total loss minus the initial elastic loss (that is, the time-dependent loss) divided by the prestress force after elastic loss.

Term (3) is the prestress loss due to concrete shrinkage. See Chap. 1 for values of the shrinkage strain, $(\varepsilon_{sh})_t$, including Table 1-10 for ultimate values, $(\varepsilon_{sh})_u$. As explained in the footnote, Table 1-10 contains average values for different starting ages, with a correction factor for the effect of member size required from Table 1-8 or 1-9, or the corresponding equations. The somewhat secondary effect of the parameter b_{11} takes into account the reduction in the free shrinkage due to the stiffening effect of the steel,[16] $n(\rho_{ps})(1 + e^2/r^2)$, and creep effect, $(1 + \eta C_t)$. See term (3) of Eq. (5-27), and the corresponding derivation. Also see the section on design simplifications for a recommended approximate value of b_{11}.

$$b_{11} = n(\rho_{ps})(1 + e^2/r^2)(1 + \eta C_t) \tag{5-5}$$

Term (4) is the prestress loss due to steel relaxation. See Chap. 1, Eqs. (1-31) to (1-41), and discussion pertaining to the reduction in the constant-length relaxation of steel stress in a prestressed concrete structure. Representative values of Eqs. (1-33) to (1-40) are shown in Tables 1-11 to 1-13, as PL_r in percent of initial tensioning stress, f_{si}. Hence, use in Eq. (5-4),

$$(f_{sr})_t = (PL_r)(f_{si})/100$$

When $f_{si}/f_y = 0.65$ to 0.80 (a typical range), *the author prefers to use Eqs. (1-39) and (1-40) for most design calculations,* as noted in Chap. 1. These two equations are thought to represent a degree of refinement (note uncertainties due to effects of prestretching, temperature, decreasing rather than constant-length behavior, etc.) that is more consistent with the approximate nature of one's ability to predict creep and shrinkage losses, for example. As can be seen in Tables 1-11 to 1-13, these two equations may compute relaxation losses that are less than the constant-length relaxation, as represented by Eqs. (1-31) to (1-34). Since the decrease in strain or stress from causes other than relaxation is usually of the order of 10 to 30 percent, this decreasing stress may reduce the usual applied stress ratios to a value approaching the 0.55 factor used in Eqs. (1-31) to (1-34). The practice of prestretching is recommended with these procedures.

In beams containing both prestressed steel and compression steel (top steel in a simple beam, for example), the compression steel will normally have a negligible effect on the loss of prestress and concrete tensile stresses in the precompressed tensile zone. However, this can be taken into account by Eq. (5-6) for the effect on the prestressed steel, and by Eq. (5-10) for the effect on the compression steel, as modified by the paragraph following Eq. (5-11).

Noncomposite beams containing both prestressed and nonprestressed steel The loss of stress in the prestressing steel is given by Eq. (5-6) for beams containing prestressed steel and nonprestressed tension steel (refers to bottom steel in a simple beam, for example).

$$(\Delta f_s)_{i,t} = \overbrace{(nf_{ci})}^{(1)} + \overbrace{(nf_{ci})(k_r C_t)(1 - \Delta P_t/2P_o)}^{(2)} + \overbrace{k_r(\varepsilon_{sh})_t E_s/(1 + b_{11})}^{(3)} + \overbrace{(f_{sr})_t}^{(4)} \tag{5-6}$$

where:

Equation (5-6) is the same as Eq. (5-4), except the creep and shrinkage loss terms (terms (2) and (3)) are multiplied by the reduction factor k_r to take into account the effect of the nonprestressed tension steel in reducing the creep and shrinkage strains. A very slight decrease in the relaxation of steel stress would also accompany this reduction in the loss of prestress due to creep and shrinkage. Using the method of Eq. (5-27), k_r is given by Eq. (5-7).

$$k_r = 1/(1 + b_{12}) \tag{5-7}$$

where

$$b_{12} = n(\rho_{ns})(1 + e_{ps}e_{ns}/r^2)(1 + \eta C_t) \tag{5-8}$$

For approximate calculations in most cases, the author suggests the simpler Eq. (5-9), which is a modification of Eqs. (5-57) to (5-59).

$$k_r = 1/(1 + A_{ns}/2A_{ps}) \qquad \text{where } A_{ns}/A_{ps} \le 1 \tag{5-9}$$

For beams with prestressed steel and nonprestressed tension steel, the compressive stress in the nonprestressed tension steel is given by Eq. (5-10).

$$(f_{ns})_{i,t} = \overbrace{(nf_{ci})}^{(1)} + \overbrace{(nf_{ci})C_t/(1 + b_{22})}^{(2)} + \overbrace{(\varepsilon_{sh})_t E_s/(1 + b_{22})}^{(3)} \tag{5-10}$$

where:

Term (1) is the compression due to elastic shortening. $f_{ci} = P_i/A_t + P_i e_{ps} e_{ns}/I_t - M_D e_{ns}/I_t$, and n is the modular ratio of the nonprestressed tension steel at the time of prestressing. Frequently P_o, A_g, and I_g are used as a close approximation instead of P_i, A_t, and I_t. See term (1) of Eq. (5-4) for further discussion.

Term (2) is the compression due to concrete creep, where b_{22} is given by Eq. (5-11).

$$b_{22} = n(\rho_{ns})(1 + e_{ns}^2/r^2)(1 + \eta C_t) \tag{5-11}$$

Term (3) is the compression due to concrete shrinkage, where b_{22} is given by Eq. (5-11).

In the case of beams containing prestressed steel and compression steel, Eqs. (5-6) to (5-8), (5-10), (5-11), but not (5-9), also apply, except that e_{ns} is replaced by $(-e'_{ns})$. However, the effect of compression steel on the loss of prestress and concrete stresses in the precompressed tension zone is normally very small, since the concrete stress at the level of the compression steel is usually small. This is, of course, not the case in the doubly reinforced nonprestressed concrete beam.

When a prestressed beam contains both nonprestressed tension steel and compression steel in addition to prestressed steel, as a rough approximation the correction factors, k_r or $(1 + b)$, can be superimposed; that is, the appropriate terms are multiplied by the correction factors due to both nonprestressed steels.

For beams containing both prestressed and nonprestressed steel, it is important to note that the calculation of concrete stresses must take into account not only the force in the prestressing steel (tension in the steel producing concrete compression) but also the force in the nonprestressed steel (in the case of nonprestressed tension steel, for example, compression in the steel producing concrete tension). Thus, in the case of beams with prestressed steel and nonprestressed tension steel, concrete stresses on any fiber, y, are calculated by Eq. (5-12)

$$f_c = P_{ps}/A_g \mp P_{ps}e_{ps}y/I - F_{ns}/A_g \pm F_{ns}e_{ns}y/I \pm My/I \tag{5-12}$$

using the gross-section properties and (+) for concrete compression.

Effect of sustained loads other than a composite slab or topping applied some time after the transfer of prestress When a sustained load other than a composite slab or topping (such as a girder supporting floor beams) is applied some time after the transfer of prestress, the loss of prestress is computed by Eq. (5-13), the stress in nonprestressed tension steel by Eq. (5-14).

$$(\Delta f_s)_{i,t} = \overbrace{(\Delta f_s)_{i,t} \text{ in Eq. (5-4) or (5-6)}}^{\text{Terms (1, 2, 3, 4)}} - \overbrace{(mf_{cs})}^{(5)} - \overbrace{(mf_{cs})(k_r C_{ts})/(1 + b_{11})}^{(6)} \tag{5-13}$$

where:

Term (5) is the elastic prestress gain due to the superimposed sustained load. $f_{cs} = M_S e/I_t$, or approximately, $f_{cs} = M_S e/I_g$, and m is the modular ratio at the time of load application. See Table 5-1 for representative values of m. M_S is the moment due to the superimposed load.

Term (6) is the prestress gain due to concrete creep under the superimposed sustained load. C_{ts} is the creep coefficient, where the age of the beam concrete at the time of load application is taken into account. For ultimate values, term (6) becomes

$$(mf_{cs})(\beta_s k_r C_u)/(1 + b_{11})_u$$

See Table 5-3 for representative values of β_s, and Eq. (5-5) for b_{11}.

$$(f_{ns})_{i,t} = \overbrace{(f_{ns})_{i,t} \text{ in Eq. (5-10)}}^{\text{Terms (1, 2, 3)}} - \overbrace{(mf_{cs})}^{(4)} - \overbrace{(mf_{cs})C_{ts}/(1 + b_{22})}^{(5)} \tag{5-14}$$

where:

Term (4) is the same as term (5) of Eq. (5-13), except that e is replaced by e_{ns} in computing f_{cs}.

Term (5) is the same as term (6) of Eq. (5-13), except that e is replaced by e_{ns} in computing f_{cs}, and b_{22} (Eq. (5-11)) replaces b_{11}.

General equations at ultimate time for noncomposite beams containing both prestressed steel and nonprestressed tension steel The loss of prestress in the prestressed steel, and compression in the nonprestressed tension steel, at ultimate time are given by Eqs. (5-15) and (5-16), respectively, for the case of noncomposite beams containing both types of steel. These equations correspond to Eqs. (5-13) and (5-14), respectively, for any time, t. Equations (5-15) and (5-16) also reduce to the correct equations when the nonprestressed tension steel and/or the subsequent sustained load application are eliminated. That is, Eqs. (5-15) and (5-16) are general equations for the ultimate values.

$$(\Delta f_s)_u = \overbrace{(nf_{ci})}^{(1)} + \overbrace{(nf_{ci})(k_r C_u)(1 - \Delta P_u/2P_o)}^{(2)} + \overbrace{k_r(\varepsilon_{sh})_u E_s/(1 + b_{11})_u}^{(3)}$$

$$+ \overbrace{(f_{sr})_u}^{(4)} - \overbrace{(mf_{cs})}^{(5)} - \overbrace{(mf_{cs})(\beta_s k_r C_u)/(1 + b_{11})_u}^{(6)} \tag{5-15}$$

$$(f_{ns})_u = \overbrace{(nf_{ci})}^{(1)} + \overbrace{(nf_{ci})C_u/(1 + b_{22})_u}^{(2)} + \overbrace{(\varepsilon_{sh})_u E_s/(1 + b_{22})_u}^{(3)}$$

$$- \overbrace{(mf_{cs})}^{(4)} - \overbrace{(mf_{cs})(\beta_s C_u)/(1 + b_{22})_u}^{(5)} \tag{5-16}$$

Following an evaluation of the stresses, and hence forces, in the prestressed steel and the nonprestressed tension steel, concrete stresses are computed using Eq. (5-12).

When prestressed steel only is used, the loss of prestress is given by Eq. (5-15), with $k_r = 1$.

The term β_s is defined and tabulated in Table 5-3.

Columns The preceding equations for computing the stress in prestressed and nonprestressed steel also apply to columns, with appropriate modifications made for steel eccentricities, bending moments, etc. For example, in the case of symmetrical prestressed and nonprestressed steel, and total sustained axial load

F_{col} but neglecting the effect of transverse bending moments (which normally will have a small effect), the loss of prestress is given by Eq. (5-6) with the following modifications:

$$f_{ci} = (P_i + F_{col})/A_t \qquad \text{or} \qquad f_{ci} = (P_o/A_g) + (F_{col})/A_t$$

$$\text{or} \qquad f_{ci} = (P_o + F_{col})/A_g$$

in decreasing order of accuracy, and $e_{ps} = e_{ns} = 0$. Other parameters or effects, such as steel relaxation, are essentially the same.

Unshored and shored composite beams containing both prestressed steel and non-prestressed tension steel The loss of prestress at any time including ultimate values is computed by Eq. (5-17) for both unshored and shored composite beams. Subscripts 1 and 2 are used to refer to the slab (or effect of the slab such as under slab dead load) and precast beam, respectively.

$$(\Delta f_s)_{i,t} = \overbrace{(nf_{ci})}^{(1)} + \overbrace{(nf_{ci})(k_r C_{s2})(1 - \Delta P_s/2P_o)}^{(2)}$$
$$+ \overbrace{(nf_{ci})[(k_r C_{t2})(1 - \Delta P_t/2P_o) - (k_r C_{s2})(1 - \Delta P_s/2P_o)]I_2/I_c}^{(3)}$$
$$+ \overbrace{k_r(\varepsilon_{sh})_t E_s/(1 + b_{11})}^{(4)} + \overbrace{(f_{sr})_t}^{(5)} - \overbrace{(mf_{cs1})}^{(6)}$$
$$- \overbrace{[(mf_{cs1})(k_r C_{t1})/(1 + b_{11})]I_2/I_c}^{(7)} - \overbrace{(f_{sds})}^{(8)} \tag{5-17}$$

It can be seen by inspection that Eq. (5-17) for composite beams reduces to Eq. (5-6) for noncomposite beams when $I_2/I_c = 1$, $C_{t2} = C_t$, and terms (6), (7), and (8)—effects of composite slab—are deleted. Note that term (2) and the second part of term (3) cancel.

Term (1) is the prestress loss due to elastic shortening. See term (1) of Eq. (5-4) for the calculation of f_{ci} and other comments.

Term (2) is the prestress loss due to concrete creep up to the time of slab casting. See term (2) of Eq. (5-6) and Eq. (5-7) or (5-9) for k_r. C_{s2} is the creep coefficient of the precast beam concrete at the time of slab casting. See term (2) of Eq. (5-4) for comments concerning the reduction factor, $(1 - \Delta P_s/2P_o)$.

Term (3) is the prestress loss due to concrete creep for any period following slab casting. C_{t2} is the creep coefficient of the precast beam concrete at any time after slab casting. The expressions in the bracket approximate the incremental creep effect under the variable prestress force that occurs after slab casting. See term (2) of Eq. (5-4) for comments on the reduction factors. The expression I_2/I_c modifies the initial stress (strain) and accounts for the effect of the composite section in restraining additional creep curvature (strain) after slab casting.

Term (4) is the prestress loss due to shrinkage of the precast beam concrete. See term (3) of Eqs. (5-4) and (5-6) for comments, and Eqs. (5-7) or (5-9) for k_r. Since this term refers to shrinkage of the precast beam only (assumed to be uniformly distributed), the small effect of b_{11} (especially for any period following slab casting) is assumed to refer to the precast beam section both before and after slab casting.

Term (5) is the prestress loss due to steel relaxation. See term (4) of Eq. (5-4) for comments and equations, and Eq. (5-6) for a comment pertaining to the effect of nonprestressed steel on relaxation.

Term (6) is the elastic prestress gain due to slab dead load, and m is the modular ratio at the time of slab casting. See Table 5-1 for values of m. $f_{cs1} = M_{S,Di}e/I$, where $M_{S,Di}$ refers to slab or slab plus diaphram dead load, and e, I refer to the precast beam section properties for unshored construction and the composite beam section properties for shored construction (transformed section properties or for most calculations, gross section properties).

Term (7) is the prestress gain due to creep under slab dead load. C_{t1} is the creep coefficient for the slab loading, where the age of the precast beam concrete at the time of slab casting is considered. For shored construction, k_r and b_{11} refer to the composite section, and I_2/I_c is deleted.

Term (8) is the prestress gain due to differential shrinkage and creep. $f_{sds} = mf_{cds}$, where $f_{cds} = Q_{DS}\,y_{cS}e/I$, and f_{cds} is the concrete stress at the steel c.g.s. See Chap. 2, Eqs. (2-18) to (2-21), etc., for Q_{DS}, and Fig. 2-1 for the definition of y_{cS}. I and the prestressed steel eccentricity, e, refer to the composite section (normally the gross section). Also see Chap. 2 for the effect of continuity on differential shrinkage, and the selection of appropriate differential shrinkage coefficients (Table 2-1).

For the case of composite beams with prestressed steel only (containing no principal nonprestressed steel), the loss of prestress is given by Eq. (5-17) with $k_r = 1$.

The compressive stress in the nonprestressed tension steel is given by Eq. (5-18).

$$(f_{ns})_{i,t} = \overbrace{(nf_{ci})}^{(1)} + \overbrace{(nf_{ci})C_{s2}/(1 + b_{22})}^{(2)} + \overbrace{[(nf_{ci})(C_{t2} - C_{s2})/(1 + b_{22})]I_2/I_c}^{(3)}$$
$$+ \overbrace{(\varepsilon_{sh})_t E_s/(1 + b_{22})}^{(4)} - \overbrace{(mf_{cs1})}^{(5)} - \overbrace{[(mf_{cs1})C_{t1}/(1 + b_{22})]I_2/I_c}^{(6)} - \overbrace{(f_{sds})}^{(7)} \quad (5\text{-}18)$$

where:

Term (1) is the compression due to elastic shortening. See term (1) of Eq. (5-10).

Term (2) is the compression due to concrete creep up to the time of slab casting. See term (2) of Eq. (5-11) and term (2) of Eq. (5-17).

Term (3) is the compression due to concrete creep for any period following slab casting. See term (3) of Eq. (5-17).

Term (4) is the compression due to concrete shrinkage. See term (3) of Eq. (5-10) and term (4) of Eq. (5-17).

Term (5) is the elastic tension due to slab dead load. This term is the same as term (6) of Eq. (5-17), except that e is replaced by e_{ns} in computing f_{cs1}.

Term (6) is the tension due to creep under slab dead load. This term is the same as term (7) of Eq. (5-17), except that e is replaced by e_{ns}, b_{22} replaces b_{11}, and k_r is deleted.

Term (7) is the tension due to differential shrinkage and creep. This term is the same as term (8) of Eq. (5-17), except that e is replaced by e_{ns}. Note that e_{ns} refers to the composite section for both unshored and shored construction in this term.

General equations at ultimate time for composite beams containing both prestressed steel and nonprestressed tension steel The loss of prestress in the prestressed steel, and compression in the nonprestressed tension steel, at ultimate time are given by Eqs. (5-19) and (5-20), respectively, for the case of composite beams containing both types of steel. These equations correspond to Eqs. (5-17) and (5-18), respectively, for any time, t.

$$
\begin{aligned}
(\Delta f_s)_u = {} & \overbrace{(nf_{ci})}^{(1)} + \overbrace{(nf_{ci})(\alpha_s k_r C_u)(1 - \Delta P_s/2P_o)}^{(2)} \\
& + \overbrace{(nf_{ci})(k_r C_u)[(1 - \Delta P_u/2P_o) - \alpha_s(1 - \Delta P_s/2P_o)]I_2/I_c}^{(3)} \\
& + \overbrace{k_r(\varepsilon_{sh})_u E_s/(1 + b_{11})_u}^{(4)} + \overbrace{(f_{sr})_u}^{(5)} - \overbrace{(mf_{cs1})}^{(6)} \\
& - \overbrace{[(mf_{cs1})(\beta_s k_r C_u)/(1 + b_{11})_u]I_2/I_c}^{(7)} - \overbrace{(f_{sds})_u}^{(8)}
\end{aligned}
\qquad (5\text{-}19)
$$

$$
\begin{aligned}
(f_{ns})_u = {} & \overbrace{(nf_{ci})}^{(1)} + \overbrace{(nf_{ci})(\alpha_s C_u)/(1 + b_{22})_u}^{(2)} \\
& + \overbrace{[(nf_{ci})(1 - \alpha_s)C_u/(1 + b_{22})]I_2/I_c}^{(3)} + \overbrace{(\varepsilon_{sh})_u E_s/(1 + b_{22})_u}^{(4)} \\
& - \overbrace{(mf_{cs1})}^{(5)} - \overbrace{[(mf_{cs1})(\beta_s C_u)/(1 + b_{22})_u]I_2/I_c}^{(6)} - \overbrace{(f_{sds})_u}^{(7)}
\end{aligned}
\qquad (5\text{-}20)
$$

Following an evaluation of the stresses, and hence forces, in the prestressed steel and nonprestressed tension steel, concrete stresses are computed using Eq. (5-12).

When prestressed steel only is used, the loss of prestress is given by Eq. (5-19), with $k_r = 1$.

The terms α_s and β_s are defined and tabulated in Table 5-3.

Design simplifications Approximate values of n, m, $\Delta P_s/P_o$, $\Delta P_u/P_o$, α_s, and β_s are given in Tables 5-1, 5-2, and 5-3. From Ref. 17 by Neville and Dilger, for typical values of ηC_u of say 1.0 to 2.5, values of b are found to vary from about

Table 5-1 Average modular ratios*

Modular ratios and time conditions		Normal-weight (w = 145 pcf)		Sand-lightweight (w = 120 pcf)		All-lightweight (w = 100 pcf)	
		M.C.	S.C.	M.C.	S.C.	M.C.	S.C.
n at release of prestress		7.3	7.3	9.8	9.8	12.9	12.9
m for indicated time between prestressing and sustained load application, including composite slab	3 weeks	6.1	6.2	8.1	8.3	10.7	10.9
	1 month	6.0	6.2	8.0	8.2	10.5	10.7
	2 months	5.9	6.1	7.9	8.2	10.2	10.6
	3 months	5.8	6.0	7.7	8.0	10.2	10.5

(For metric units, see metric conversion factors herein.)

* Based on f'_{ci} = 4000 to 4500 psi for both moist-cured (M.C.) and steam-cured (S.C.) concrete and Type I cement, up to 3-months f'_c = 6360 to 7150 psi (using Eq. (1-2) for moist-cured) and up to 3-months f'_c = 6050 to 6800 psi (using Eq. (1-4) for steam-cured); on E_c in Eq. (1-10) or (1-13); and on $E_s = 27 \times 10^6$ psi for 250 K strands or $E_s = 28 \times 10^6$ psi for 270 K strands.

Table 5-2 Typical loss of prestress ratios for different weight concrete*

Ratio	Normal-weight	Sand-lightweight	All-lightweight
$\Delta P_s/P_o$—for 3 weeks to 1 month between prestressing and sustained load application, including composite slab	0.10	0.12	0.14
$\Delta P_s/P_o$—for 2 to 3 months between prestressing and sustained load application, including composite slab	0.14	0.16	0.18
$\Delta P_u/P_o$	0.18	0.21	0.23

* Modify these ratios when necessary to take into account the effect of nonprestressed tension steel. Note that the ratios $\Delta P_s/P_o$ and $\Delta P_u/P_o$ are defined as the total loss (at the time of sustained load application and ultimate, respectively) minus the initial elastic loss divided by the prestress force after elastic loss. The different values for different weight concrete are due primarily to different initial strains (due to E_s) for normal stress levels.

Table 5-3 Average values of α_s and β_s*

Time between prestressing and sustained load application, including composite slab	α_s	β_s
3 weeks	0.38	0.85
1 month	0.44	0.83
2 months	0.54	0.78
3 months	0.60	0.75

* α_s refers to the part of the total creep that takes place before the sustained load application (given by $\alpha_s = C_{s2}/C_u = t^{0.60}/(10 + t^{0.60})$, as by Eq. (1-16)), and β_s (equal to the average creep correction factor for effect of loading age or $(\text{C.F.})_{LA}$ from Eqs. (1-20) and (1-21)) is the creep correction factor for the age of the beam concrete when the sustained load is applied. These factors apply for different weight concrete, and moist and steam curing.

0.72 to 0.14 for normal-weight concrete and 1.5 to 0.14 for lightweight concrete, as the steel percentage varies from 1 to 0.20 percent, respectively. The author believes a value of $b_{11} = b_{12} = b_{22} = 0.25$ for say one month to ultimate values will normally be satisfactory and on the conservative side for most design calculations for this somewhat secondary effect. When the steel percentage is relatively high and/or when lightweight concrete is used, this assumption can be overly conservative.

The longer prestress loss equations thus far presented could be greatly shortened by combining terms and substituting the approximate parameters referred to above, but are presented in the form of separate terms in order to show the separate effects or contributions to the behavior (such as those due to the prestress force, dead load, creep, shrinkage, etc., that occur both before and after sustained load application, including a composite slab).

For example, Eqs. (5-15), (5-16), (5-19), and (5-20) reduce to Eqs. (5-21*a,b,c*) to (5-26*a,b, c*) as indicated, when the following representative parameters are used: from Chap. 1 and Table 1-10, for 70 percent humidity, $C_u = 1.88$ and $(\varepsilon_{sh})_u = 546 \times 10^{-6}$ in/in (cm/cm) (size correction factor is neglected here, which is on the safe side except for small members—see footnote in Table 1-10 and Tables 1-8 and 1-9); using steam-cured concrete prestressed at age 1 day, and $(1 + b_{11})_u = (1 + b_{22})_u = 1.25$ for both noncomposite and composite sections (composite section value applies in certain terms for shored construction, otherwise the precast section applies); $E_s = 28 \times 10^6$ psi (1.97×10^6 kg/cm^2, 193 kN/mm^2), $f_{si} = 190\,000$ psi (13 360 kg/cm^2, 1310 N/mm^2), $(f_{sds})_u = 2\%\, f_{si} = 3800$ psi (267 kg/cm^2, 26.2 N/mm^2). Assuming a 2-month period between prestressing and the application of an additional sustained load including composite slab: from Table 5-3, $\alpha_s = 0.54$ and $\beta_s = 0.78$; and from Tables 5-1 and 5-2, for normal-weight concrete, $n = 7.3$, $m = 6.1$, $\Delta P_s/P_o = 0.14$, $\Delta P_u/P_o = 0.18$; for sand-lightweight concrete, $n = 9.8$, $m = 8.2$, $\Delta P_s/P_o = 0.16$, $\Delta P_u/P_o = 0.21$; and for all-lightweight concrete, $n = 12.9$, $m = 10.6$, $\Delta P_s/P_o = 0.18$, $\Delta P_u/P_o = 0.23$.

Ultimate loss of prestress for noncomposite beams, as by Eq. (5-15)

Normal-weight concrete

$$(\Delta f_s)_u = \overbrace{7.3 f_{ci}}^{(1)} + \overbrace{7.3(1.88)(1-0.09)k_r f_{ci}}^{(2)} + \overbrace{(546)(28)k_r/1.25}^{(3)}$$
$$+ \overbrace{0.075(190\,000\text{ psi})}^{(4)} - \overbrace{6.1 f_{cs}}^{(5)} - \overbrace{6.1(0.78 \times 1.88)k_r f_{cs}/1.25}^{(6)}$$
$$= \overbrace{7.3 f_{ci}}^{(1)} + \overbrace{12.49 k_r f_{ci}}^{(2)} + \overbrace{12\,230 k_r}^{(3)} + \overbrace{14\,250\text{ psi}}^{(4)}$$
$$- \overbrace{6.1 f_{cs}}^{(5)} - \overbrace{7.16 k_r f_{cs}}^{(6)} \qquad (5\text{-}21a)$$

(1 psi = 0.07031 kg/cm^2 = 0.006895 N/mm^2)

When $k_r = 1$ (no nonprestressed steel), Eq. (5-21*a*) reduces to Eq. (5-22*a*).

$$(\Delta f_s)_u = \overbrace{19.79 f_{ci}}^{(1+2)} + \overbrace{26\,480\text{ psi}}^{(3+4)} - \overbrace{13.26 f_{cs}}^{(5+6)} \qquad (5\text{-}22a)$$

Sand-lightweight concrete

$$(\Delta f_s)_u = 9.8 f_{ci} + 16.49 k_r f_{ci} + 12\,230 k_r + 14\,250\text{ psi}$$
$$- 8.2 f_{cs} - 9.62 k_r f_{cs} \qquad (5\text{-}21b)$$

When $k_r = 1$, $(\Delta f_s)_u = 26.29 f_{ci} + 26\,480\text{ psi} - 17.82 f_{cs}$ (5-22*b*)

All-lightweight concrete

$$(\Delta f_s)_u = 12.9 f_{ci} + 21.47 k_r f_{ci} + 12\,230 k_r + 14\,250\text{ psi}$$
$$- 10.6 f_{cs} - 12.43 k_r f_{cs} \qquad (5\text{-}21c)$$

When $k_r = 1$, $(\Delta f_s)_u = 34.37 f_{ci} + 26\,480\text{ psi} - 23.03 f_{cs}$ (5-22*c*)

See Eq. (5-13), also Eqs. (5-4) or (5-6), for term-by-term definitions of Eqs. (5-21*a,b,c*) and (5-22*a,b,c*).

Ultimate compression in nonprestressed tension steel for noncomposite beams, as by Eq. (5-16)

Normal-weight concrete

$$(f_{ns})_u = \overbrace{7.3 f_{ci}}^{(1)} + \overbrace{7.3(1.88) f_{ci}/1.25}^{(2)} + \overbrace{(546)(28)/1.25}^{(3)} - \overbrace{6.1 f_{cs}}^{(4)}$$

$$\overbrace{- 6.1(0.78 \times 1.88)f_{cs}/1.25}^{(5)}$$

$$= \overbrace{7.3f_{ci}}^{(1)} + \overbrace{10.98f_{ci}}^{(2)} + \overbrace{12\,230 \text{ psi}}^{(3)} - \overbrace{6.1f_{cs}}^{(4)} - \overbrace{7.16f_{cs}}^{(5)}$$

$$= \overbrace{18.28f_{ci}}^{(1+2)} + \overbrace{12\,230 \text{ psi}}^{(3)} - \overbrace{13.26f_{cs}}^{(4+5)} \qquad (5\text{-}23a)$$

Sand-lightweight concrete

$$(f_{ns})_u = 24.54f_{ci} + 12\,230 \text{ psi} - 17.82f_{cs} \qquad (5\text{-}23b)$$

All-lightweight concrete

$$(f_{ns})_u = 32.30f_{ci} + 12\,230 \text{ psi} - 23.03f_{cs} \qquad (5\text{-}23c)$$

See Eq. (5-14), also Eq. (5-10), for term-by-term definitions of Eqs. (5-23*a,b,c*).

Ultimate loss of prestress for unshored and shored composite beams, as by Eq. (5-19)

Normal-weight concrete

$$(\Delta f_s)_u = \overbrace{7.3f_{ci}}^{(1)} + \overbrace{7.3(0.54 \times 1.88)(1 - 0.070)k_r f_{ci}}^{(2)}$$

$$\overbrace{+ 7.3(1.88)[(1 - 0.090) - 0.54(1 - 0.070)]k_r f_{ci} I_2/I_c}^{(3)}$$

$$+ \overbrace{(546)(28)k_r/1.25}^{(4)} + \overbrace{0.075(190\,000)}^{(5)} - \overbrace{6.1f_{cs1}}^{(6)}$$

$$- \overbrace{6.1(0.78 \times 1.88)k_r f_{cs1} I_2/1.25I_c}^{(7)} - \overbrace{3800 \text{ psi}}^{(8)}$$

$$= \overbrace{7.3f_{ci}}^{(1)} + \overbrace{6.89k_r f_{ci}}^{(2)} + \overbrace{5.60k_r f_{ci} I_2/I_c}^{(3)} + \overbrace{12\,230k_r}^{(4)} + \overbrace{14\,250 \text{ psi}}^{(5)}$$

$$- \overbrace{6.1f_{cs1}}^{(6)} - \overbrace{7.16k_r f_{cs1} I_2/I_c}^{(7)} - \overbrace{3800 \text{ psi}}^{(8)} \qquad (5\text{-}24a)$$

When $k_r = 1$ (no nonprestressed steel), Eq. (5-24*a*) reduces to Eq. (5-25*a*).

$$(\Delta f_s)_u = \overbrace{(14.19 + 5.60I_2/I_c)f_{ci}}^{(1+2+3)} + \overbrace{22\,680\text{ psi}}^{(4+5+8)} - \overbrace{(6.1 + 7.16I_2/I_c)f_{cs1}}^{(6+7)} \qquad (5\text{-}25a)$$

Sand-lightweight concrete

$$(\Delta f_s)_u = 9.8f_{ci} + 9.15k_r f_{ci} + 7.34k_r f_{ci}I_2/I_c + 12\,230k_r + 14\,250\text{ psi}$$
$$- 8.2f_{cs1} - 9.62k_r f_{cs1}I_2/I_c - 3800\text{ psi} \qquad (5\text{-}24b)$$

When $k_r = 1$,

$$(\Delta f_s)_u = (18.95 + 7.34I_2/I_c)f_{ci} + 22\,680\text{ psi} - (8.2 + 9.62I_2/I_c)f_{cs1} \qquad (5\text{-}25b)$$

All-lightweight concrete

$$(\Delta f_s)_u = 12.9f_{ci} + 11.92k_r f_{ci} + 9.55k_r f_{ci}I_2/I_c + 12\,230k_r + 14\,250\text{ psi}$$
$$- 10.6f_{cs1} - 12.43k_r f_{cs1}I_2/I_c - 3800\text{ psi} \qquad (5\text{-}24c)$$

When $k_r = 1$,

$$(\Delta f_s)_u = (24.82 + 9.55I_2/I_c)f_{ci} + 22\,680\text{ psi} - (10.6 + 12.43I_2/I_c)f_{cs1} \qquad (5\text{-}25c)$$

See Eq. (5-19), also Eq. (5-17), for term-by-term definitions of Eqs. (5-24*a,b,c*) and (5-25*a, b, c*).

Ultimate compression in nonprestressed tension steel for unshored and shored composite beams, as by Eq. (5-20)

Normal-weight concrete

$$(f_{as})_u = \overbrace{7.3f_{ci}}^{(1)} + \overbrace{7.3(0.54 \times 1.88)f_{ci}/1.25}^{(2)} + \overbrace{7.3(1 - 0.54)(1.88)f_{ci}I_2/1.25I_c}^{(3)}$$
$$+ \overbrace{(546)(28)/1.25}^{(4)} - \overbrace{6.1f_{cs1}}^{(5)} - \overbrace{6.1(0.78 \times 1.88)f_{cs1}I_2/1.25I_c}^{(6)} - \overbrace{3800\text{ psi}}^{(7)}$$
$$= \overbrace{7.3f_{ci}}^{(1)} + \overbrace{5.93f_{ci}}^{(2)} + \overbrace{5.05f_{ci}I_2/I_c}^{(3)} + \overbrace{12\,230\text{ psi}}^{(4)}$$
$$- \overbrace{6.1f_{cs1}}^{(5)} - \overbrace{7.16f_{cs1}I_2/I_c}^{(6)} - \overbrace{3800\text{ psi}}^{(7)}$$
$$= \overbrace{(13.23 + 5.05I_2/I_c)f_{ci}}^{(1+2+3)} + \overbrace{8430\text{ psi}}^{(4+7)} - \overbrace{(6.1 + 7.16I_2/I_c)f_{cs1}}^{(5+6)} \qquad (5\text{-}26a)$$

Sand-lightweight concrete

$$(f_{ns})_u = (17.76 + 6.78 I_2/I_c) f_{ci} + 8430 \text{ psi} - (8.2 + 9.62 I_2/I_c) f_{cs1} \tag{5-26b}$$

All-lightweight concrete

$$(f_{ns})_u = (23.38 + 8.92 I_2/I_c) f_{ci} + 8430 \text{ psi} - (10.6 + 12.43 I_2/I_c) f_{cs1} \tag{5-26c}$$

See Eq. (5-20), also Eq. (5-18), for term-by-term definitions of Eqs. (5-26*a,b,c*).

Again, it can be seen by inspection that the loss of prestress by Eqs. (5-25*a,b,c*) and compression in nonprestressed steel by Eqs. (5-26*a,b,c*) for *composite beams* reduces, respectively, to the loss of prestress by Eqs. (5-22*a,b,c*) and compression in nonprestressed steel by Eqs. (5-23*a,b,c*) for *noncomposite beams* when $I_2/I_c = 1$, and the differential shrinkage and creep term (−3800 psi) is deleted in the composite beam equations.

When the average thickness of the part of the section containing the prestressing steel is less than about 4 in. or greater than about 10 to 12 in., the creep and shrinkage correction factors for size effect (see Table 1-8) should be considered, even for approximate calculations.

Since concrete stresses after losses are normally computed under live load in addition to all other loads, the author believes the *gain in prestress* due to live load should be taken advantage of in prestress loss calculations, as: $f_{sL} = nf_{cL}$, where n is the modular ratio referring to E_c say at 28 days, and $f_{cL} = M_L e/I$. However, it appears to be common practice to consider this effect as an additional margin of safety not included in calculations; hence this term has been omitted from the equations herein. Note the numerical values of the gain in prestress due to live load are 1.5 to 2.9 percent for the four subsequent examples, versus the total loss, excluding the live load effect, of 21.1 to 32.6 percent.

Simplified method, PCI Committee on Prestress Losses Similar simplified prestress loss equations have recently (May–June 1975) been recommended by the PCI Prestress Loss Committee[73] for normal and lightweight concrete; stress-relieved, low-relaxation, and high strength bar tendons; pre and post-tensioned members; and different initial tension in percent of f_{pu}, as follows. The method does not directly include the effect of nonprestressed tension steel or composite action.

N-SR-PRE-70	Total loss (ksi) $= 33.0 + 13.8 f_{ci} - 4.5 f_{cs}$
L-SR-PRE-70	$= 31.2 + 16.8 f_{ci} - 3.8 f_{cs}$
N-LR-PRE-75	$= 19.8 + 16.3 f_{ci} - 5.4 f_{cs}$
L-LR-PRE-75	$= 17.5 + 20.4 f_{ci} - 4.8 f_{cs}$
N-SR-POST-68.5	$= 29.3 + 5.1 f_{ci} - 3.0 f_{cs}$
L-SR-POST-68.5	$= 27.1 + 10.1 f_{ci} - 4.9 f_{cs}$
N-LR-POST-68.5	$= 12.5 + 7.0 f_{ci} - 4.1 f_{cs}$
L-LR-POST-68.5	$= 11.9 + 11.1 f_{ci} - 6.2 f_{cs}$

N-BAR-POST-70 $= 12.8 + 6.9f_{ci} - 4.0f_{cs}$

L-BAR-POST-70 $= 12.5 + 10.9f_{ci} - 6.0f_{cs}$

where f_{ci} = concrete stress at center of gravity of prestressing force immediately after transfer (same as herein), and f_{cs} = concrete compressive stress at center of gravity of prestressing force due to all permanent (dead) loads not used in computing f_{ci} (similar to herein). Also, N, L refer to normal-weight (about 145 pcf, 2320 kg/m^3, 22.8 kN/m^3) and lightweight (about 115 pcf, 1840 kg/m^3, 18.1 kN/m^3) concrete; SR = stress-relieved, LR = low-relaxation, BAR = high strength bar; PRE, POST refer to pre and post-tensioning; and the fourth term, as 70, indicates the initial tension in percent of f_{pu}. For pretensioned tendons it is the tension at which the tendons are anchored in the casting bed before concrete is placed, and for post-tensioned tendons it is the initial tension in the tendons at the critical location in the concrete member after losses due to friction and anchor set have been deducted.

The above equations are based on a volume/surface ratio, $v/s = 2.0$. For other values, the following adjustments are recommended by the PCI Committee:

v/s (in.)	= 1.0	2.0	3.0	4.0
Adjustment (percent)	+ 3.2	0	− 3.8	− 7.6

A step-by-step method is also outlined in the PCI Loss Committee[73] report.

Relaxation method The loss of prestress at any time including ultimate values is computed by Eq. (5-27), in which the creep and shrinkage terms are by Trost.[11,12]

$$(\Delta f_s)_{i,t} = \underbrace{(nf_{ci})}_{(1)} + \frac{\overbrace{nf_{ci}C_t}^{(2)} + \overbrace{(\varepsilon_{sh})_t E_s}^{(3)}}{1 + n(\rho_{pt})(1 + e^2/r_t^2)(1 + \eta C_t)} + \overbrace{(f_{sr})_t}^{(4)} \qquad (5\text{-}27)$$

where:

Term (1) is the prestress loss due to elastic shortening. See term (1) of Eq. (5-4).

Term (2) is the prestress loss due to concrete creep. The numerator represents the loss of prestress due to creep under constant stress, using the creep coefficient, C_t. The denominator is obtained in the derivation of Eq. (5-31) and includes the "relaxation coefficient – η" which takes into account the effect of creep under the decreasing prestress force. Typical values given by Neville and Dilger[17] range from 0.75 to 0.90, with an average value of $\eta = 0.88$. For an approximate evaluation of the somewhat secondary effect represented by the denominator, which has been expressed as $(1 + b_{11})$ in Eq. (5-5) using the gross-section properties, see the previous section on design simplifications.

Term (3) is the prestress loss due to concrete shrinkage. The numerator represents the loss of prestress due to a strain equal to the free shrinkage strain, $(\varepsilon_{sh})_t$. The denominator is obtained in the derivation of Eq. (5-31) and takes into account the reduction in the free shrinkage strain. See term (2) of Eq. (5-27) and term (3) of Eq. (5-4) for additional comments.

Term (4) is the prestress loss due to steel relaxation. See term (4) of Eq. (5-4) for a discussion of the difference between the constant-length relaxation (or intrinsic relaxation) of steel stress and the relaxation of steel stress under the decreasing strain condition in a prestressed concrete member. The author is of the opinion that the intrinsic relaxation should *not* be included in the numerator of Eq. (5-27) along with terms (2) and (3), as

$$\frac{\text{Constant-length relaxation stress}}{1 + n(\rho_{pt})(1 + e^2/r_t^2)(1 + \eta C_t)}$$

The derivation of this expression, which parallels the derivation of terms (2) and (3) of Eq. (5-27) on the basis of changes in time-dependent strain, does not appear to be correct when applied to relaxation. For example, if creep and shrinkage are zero the relaxation term should reduce, essentially, to the constant-length steel relaxation, which would not be the case in this expression.

The gross-section properties and P_o could also be used in Eq. (5-27) instead of the transformed-section properties and P_i to obtain approximately the same results.

Neglecting any variation in the concrete modulus of elasticity with time, and assuming that the concrete creep and shrinkage time functions are the same, terms (2) and (3) of Eq. (5-27) are derived by Eqs. (5-28) to (5-31). Assuming perfect bond between steel and concrete, the change in concrete and steel strain at the steel c.g.s. level due to creep and shrinkage is computed by Eq. (5-28), in which the relaxation coefficient, η, is introduced.

$$(\Delta\varepsilon_c)_t = (\Delta\varepsilon_s)_t = \frac{(\Delta f_s)_t}{E_s} = \frac{f_{ci}}{E_{ci}}C_t - \frac{(\Delta f_c)_t}{E_{ci}}(1 + \eta C_t) + (\varepsilon_{sh})_t \qquad (5\text{-}28)$$

in which the change in concrete stress, $(\Delta f_c)_t$, is given by Eq. (5-29).

$$(\Delta f_c)_t = \frac{(\Delta F_s)_t}{A_t} + \frac{(\Delta F_s)_t e^2}{I_t} = \frac{(\Delta F_s)_t}{A_t}(1 + e^2/r_t^2) = \frac{(\Delta f_s)_t A_{ps}}{A_t}(1 + e^2/r_t^2)$$

$$= (\Delta f_s)_t(\rho_{pt})(1 + e^2/r_t^2) \qquad (5\text{-}29)$$

Equation (5-30) is obtained by substituting Eq. (5-29) into Eq. (5-28).

$$\frac{(\Delta f_s)_t}{E_s} = \frac{f_{ci}}{E_{ci}}C_t - \frac{(\Delta f_s)_t}{E_{ci}}(\rho_{pt})(1 + e^2/r_t^2)(1 + \eta C_t) + (\varepsilon_{sh})_t \qquad (5\text{-}30)$$

Solving for $(\Delta f_s)_t$ in Eq. (5-30), one obtains Eq. (5-31), which represents terms (2) and (3) of Eq. (5-27).

$$(\Delta f_s)_t = \frac{n f_{ci} C_t + (\varepsilon_{sh})_t E_s}{1 + n(\rho_{pt})(1 + e^2/r_t^2)(1 + \eta C_t)} \qquad (5\text{-}31)$$

The relaxation method, given by Eq. (5-27) for noncomposite beams, has also been applied to composite beams by Dilger and Neville,[18] and by others.[17] In Ref. 19, Rao and Dilger found the method to be unsatisfactory under the decreasing compressive stresses found in composite beams.

AASHTO method The loss of prestress may be estimated by the following method[13,14] for prestressed members using 270 K, seven-wire, stress-relieved strand and normal-weight concrete. Both noncomposite and composite sections are included.

$$\begin{aligned} & \quad\;\; (1) \quad\;\; (2) \quad\;\; (3) \quad\;\; (4) \\ \Delta f_s &= SH + ES + CR_c + CR_s \\ &= \text{Total prestress loss, excluding friction} \end{aligned} \tag{5-32}$$

where:

Term (1) is the concrete shrinkage loss. Select the average ambient relative humidity for the geographic area and use a value of $SH = 5000$, $10\,000$, or $15\,000$ psi, derived as follows: for $H = 50$ percent, $SH = [(\varepsilon_{sh})_u$ of 600×10^{-6} in/in—versus an average value for steam-cured concrete in Table 1-10 of $702 \times 10^{-6}]$ (for average volume/surface ratio of 4, the size correction factor used is 0.77—versus 0.78 in Table 1-9) (E_s of 28×10^6) = (600)(0.77)(28) = 12 900 psi. For the other humidity ranges shown, the climatic shrinkage coefficients of 0.3 and 1.3 are used.

Humidity, H (%)	Calculated values	Approx. values recommended to reflect field conditions
100–75	$SH = 0.3 \times 12\,900 = 3\,900$ psi	$SH = 5\,000$ psi
75–25	$SH = 1.0 \times 12\,900 = 12\,900$ psi	$SH = 10\,000$ psi
25– 0	$SH = 1.3 \times 12\,900 = 16\,800$ psi	$SH = 15\,000$ psi

Term (2) is the elastic shortening loss, or $ES = nf_{cr} = (E_s/E_{ci})f_{cr} = (28/4)f_{cr} = 7f_{cr}$, where f_{cr} is the average concrete stress at the c.g.s. of the prestressing steel at the time of release. ES may be estimated using the following average values:

Sections with composite deck slab		Sections without composite deck slab	
f_{cr} (psi)	ES (psi)	f_{cr} (psi)	ES (psi)
1000 $\times 7 =$	7 000	600 $\times 7 =$	4000
1400	10 000	800	5500
1800	13 000	1000	7000

The level of prestress, f_{cr}, has been found to vary from area to area in the country. Values for sections without composite deck slabs, commonly known as box beams, vary from as low as 500 psi to 1500 psi. Average values of 600, 800, and 1000 psi were chosen as representative samples. Values for sections with composite deck slabs, generally I-sections, vary from 800 to 2500 psi. Again, average representative values of 1000, 1400, and 1800 psi were chosen. The average prestress level, f_{cr}, can be estimated, or determined more accurately.

Term (3) is the loss due to creep of concrete, or $CR_c = 16 f_{cd}$, where f_{cd} = average concrete compressive stress at the c.g.s. of the prestressing steel under full dead load. CR_c = [using average δ_u of 0.75×10^{-6} per psi for steam-cured concrete, in Eq. (1-19), $C_u = \delta_u E_{ci} = (0.75)(4) = 3.0$—versus an average value in Table A-3 of say 2.21 at 50 percent humidity] (for average volume/surface ratio of 4, the size correction factor used is 0.77—versus 0.80 in Table 1-9) $(n = E_s/E_{ci} = 7) f_{cd} = (3.0)(0.77)(7) f_{cd} = 16 f_{cd}$. CR_c may be estimated using the following average values. These values apply to sections that are both with and without composite deck slabs.

f_{cd} (psi)		CR_c (psi)
500	× 16 =	8 000
800		13 000
1200		19 000

Term (4) is the loss due to relaxation of prestressing steel. See Chap. 1, Eq. (1-35) for pretensioned members and Eq. (1-36) for post-tensioned members. Representative values are given in Table 1-12 in percent of initial tensioning stress—in this form by Eqs. (1-37) and (1-38), using $f_{si} = 190\,000$ psi.

See Eqs. (5-34) or (5-35) for the calculation of frictional losses by this method.

The AASHTO–PCI Committee[13] refers, for data regarding the properties and effects of lightweight aggregates and low-relaxation tendons, to other publications, such as Ref. 9. However, the author believes that Eq. (5-32) is equally applicable to 250 K strands and lightweight concrete, provided the n-value in terms (2) and (3) is modified accordingly, and the effect of low-relaxation steel is taken into account. The same nominal shrinkage and creep effects are deemed valid in Chap. 1 and by ACI Committee 209,[10] etc., for different weight concrete.

In the section of the Committee Report on Flexure,[13] it is suggested that in calculations of section properties, the transformed area of bonded reinforcement may be included in pretensioned members and in post-tensioned members after grouting; prior to bonding of tendons, areas of the open ducts shall be deducted.

PCI Handbook method For most designs, it is sufficiently accurate to estimate the total loss of prestress based on past performance of similar members. If loss calculations are required, the following method[2] may be used.

$$(\Delta f_s)_u = \overbrace{(n f_{cs})}^{(1)} + \overbrace{(n_c f_{cs})(1 - 0.34 n_c k_1)}^{(2)} + \overbrace{r_s(1 - 0.010 n_c k_1)}^{(3)} + \overbrace{(f_{sr})_u}^{(4)} \tag{5-33}$$

where:

Term (1) is the loss of prestress due to elastic shortening $= L_{es}$. $f_{cs} = P_0/A_c + P_0 e^2/I_g - M_{dl} e/I_g$, P_0 refers to the prestress force at transfer (after elastic loss), A_c is defined as the net area of concrete (presumably the gross area of pretensioned members could be used), and M_{dl} is the dead load moment of the prestressed member only. A rational analysis is indicated for the elastic loss of post-tensioned members.

Term (2) is the loss of prestress due to concrete creep $= L_c$. $n_c = nC_t$ (similar to Eq. (1-47) but without the elastic factor, 1)—values of n_c are determined from curves in the Handbook as a function of the volume/surface ratio, different weight concrete, moist and steam curing, and different loading ages; $k_l = (A_{ps}/A_c)(1 + e^2 A_c/I)$; and f_{cs} is the same as f_{cs} in term (1), except that M_{dl} shall include all dead load plus any portion of the live load which may be sustained. On sections which receive a cast-in-place topping, the moment of inertia of the composite section should be used in calculating stresses caused by loads applied after the topping concrete has cured. Also, P_0 may be further reduced, since part of the relaxation loss will have occurred.

The reduction factors, $(1 - 0.34n_c k_l)$ and $(1 - 0.010n_c k_l)$, in terms (2) and (3) were based on extensive computer calculations in which the progressive reduction in concrete stress was taken into account. Note the similarities and differences in these two terms compared to terms (2) and (3) of Eqs. (5-4) and (5-27), which were derived on a similar basis.

Term (3) is the loss of prestress due to concrete shrinkage $= L_s$. The parameter r_s is a shrinkage coefficient given in curve form in the Handbook as a function of the volume/surface ratio, different weight concrete, and concrete strength. Also see the comments in term (2) pertaining to this term. If transfer of the prestressing force occurs at an interval after the end of curing, a portion of the shrinkage has taken place. Advantage should be taken of the reduced amount of shrinkage, since the prestress losses due to shrinkage represent a sizeable portion of the total losses. This is especially desirable in post-tensioned members, since a considerable percentage of shrinkage occurs within the first month. The following values can be used for this purpose.

Time after end of curing, days:	1	3	5	7	10	20
Percentage of shrinkage left:	92	85	80	78	73	64

Term (4) is the prestress loss due to steel relaxation $= L_r = \alpha f_{py}$, where $f_{py} = 0.85 f_{pu}$ for stress-relieved steel, $f_{py} = 0.90 f_{pu}$ for stabilized steel, and f_{pu} is the ultimate strength as 250 ksi or 270 ksi. The parameter, α, is presented in tabular form in the Handbook as a function of $(L_c + L_s)/f_{py}$ and f_s/f_{py}, where $(L_c + L_s)$ is the combined loss due to creep and shrinkage and f_s is the stress remaining in the steel after the initial tension has been reduced by the relaxation in the strand from time of tensioning until time of release and by the instantaneous stress losses.

Other methods In 1975 the PCI Committee on Prestress Losses[73] presented a general method and a simplified method for computing losses. The general method involves a step-by-step procedure in which a minimum of four time intervals is recommended. The simplified procedure has already been summarized herein.

In a 1975 paper by Tadros, Ghali, and Dilger,[75] an analytical method suitable for hand calculations is presented for determining prestress loss, axial strain, curvature, and deflection. The method involves the use of a "recovery parameter" and a "relaxation parameter" which are based on step-by-step numerical calculations and tabulated in the form of design aids.

In 1975 Huang[77] also presented a direct method for estimating prestress losses in pretensioned members that avoids the need for step-by-step calculations. His procedure is similar to the simplified methods of the author and the PCI Committee on Prestress Losses, and is based on linking experimentally developed stress–strain–time relationships of steel and concrete materials.

Frictional Losses in Post-Tensioned Members

Frictional losses in post-tensioned members occur from angle changes in draped cables and from wobble of the ducts. Losses due to friction are computed using Eqs. (5-34) or (5-35) in the ACI Code[1] and Commentary[20] and the AASHTO[13,14] method, in which experimentally determined wobble and curvature friction coefficients, such as those in Table 5-4 and Table 5-5, are recommended. Such losses occur prior to anchoring but should be estimated for design and checked during stressing operations. Where rigid conduits are used, the wobble coefficient, K, can be considered as zero.

Referring to Fig. 5-2*a*, for a given differential angle of curvature, $d\alpha$, and the equilibrium of forces P, $(P - dP)$, and N,

$$dP = \mu N = \mu P\, d\alpha \qquad (a)$$

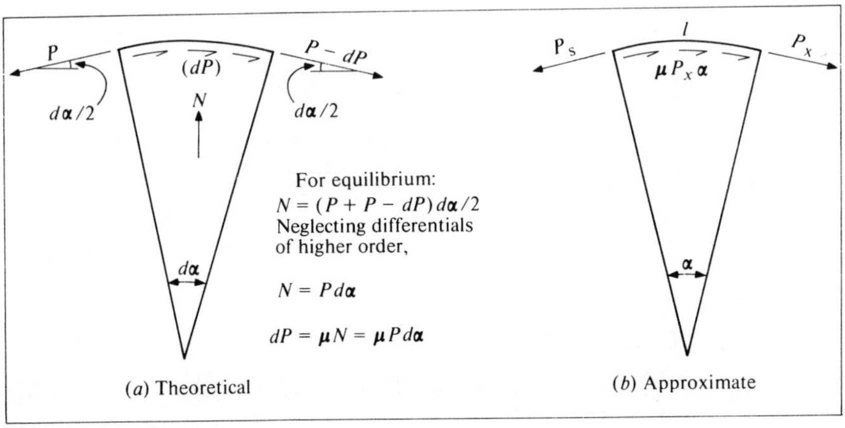

FIGURE 5-2
Diagrams of forces involved in the derivation of the frictional loss of prestress equations (5-34) and (5-35).

where μ is the curvature friction coefficient, and α is the angle change of the prestressing steel profile in radians from the jacking end to any point x. Cross-multiplying in Eq. (*a*) leads to Eq. (*b*).

$$dP/P = \mu\, d\alpha \tag{b}$$

Integrating both sides of Eq. (*b*),

$$\log_n P \Big]_{P_x}^{P_s} = \mu\alpha \tag{c}$$

or

$$\log_n P_s - \log_n P_x = \log_n (P_s/P_x) = \mu\alpha \tag{d}$$

where P_s is the steel force at the jacking end and P_x is the steel force at any point x. Raising the Napierian logarithm base, e, to the power of both sides of Eq. (*d*), yields Eq. (*e*).

Table 5-4 Friction coefficients, ACI Commentary[20]

			Wobble coeff. K, per ft	Curvature coeff. μ
Grouted tendons in metal sheathing		Wire tendons	0.0010–0.0015	0.15–0.25
		High str. bars	0.0001–0.0006	0.08–0.30
		7-wire strand	0.0005–0.0020	0.15–0.25
Unbonded tendons	Mastic-coated	Wire tendons	0.001–0.002	0.05–0.15
		7-wire strand	0.001–0.002	0.05–0.15
	Pre-greased	Wire tendons	0.0003–0.002	0.05–0.15
		7-wire strand	0.0003–0.002	0.05–0.15

Table 5-5 Friction coefficients, AASHTO[13,14]

Type of steel	Type of duct	K, per ft	μ
Wire or ungalvanized strand	Bright metal sheathing	0.0020	0.30
	Galvanized metal sheathing	0.0015	0.25
	Greased or asphalt-coated and wrapped	0.0020	0.30
	Galvanized rigid	0.0002	0.25
High str. bars	Bright metal sheathing	0.0003	0.20
	Galvanized metal sheathing	0.0002	0.15

$$P_s/P_x = e^{\mu\alpha} \tag{e}$$

Adding the experimentally determined wobble effect, Kl, and solving for P_x in Eq. (*e*), yields Eq. (5-34), in which Kl is dimensionless.

$$P_x = P_s/e^{(Kl + \mu\alpha)} \tag{5-34}$$

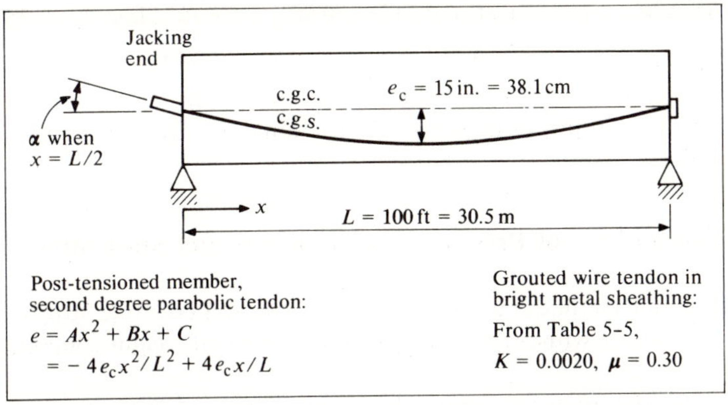

FIGURE 5-3
Beam details for the frictional loss of prestress example.

Referring to Fig. 5-2*b*, which represents a finite approximation of Fig. 5-2*a*, Eq. (*f*) is obtained in place of Eq. (*a*).

$$dP = \mu P_x \alpha \tag{f}$$

It follows from Fig. 5-2*b* that

$$P_s - P_x = \mu P_x \alpha \tag{g}$$

Adding the wobble effect and solving for P_x yields Eq. (5-35).

$$P_x = P_s/(1 + Kl + \mu\alpha) \tag{5-35}$$

Equation (5-35) may be used[1,14,20] when $(Kl + \mu\alpha)$ is not greater than 0.3. In Eqs. (5-34) and (5-35), l is the length of prestressing steel from the jacking end to any point x (such as at midspan).

It should be recognized that these procedures for computing frictional losses may provide only estimates of the actual situation in the field. When estimated frictional effects are determined to be significantly different than those assumed in design, the stressing force should be adjusted to meet design requirements at critical points. Tendons may be pulled from one or both ends of a member depending on the length of the member and the frictional losses incurred. A rule of thumb in the case of flat plates, for example, is that it is practical to pull from one end for tendon lengths up to about 150 ft (46 m).

Frictional loss example For the beam details given in Fig. 5-3, compute the frictional loss of prestress at midspan (l = 50 ft or 15.2 m) using Eqs. (5-34) and (5-35).

$$\alpha = (de/dx)_{x=0} = 4e_c/L = (4)(15)/1200 = 0.0500 \text{ radians}$$

$$(Kl + \mu\alpha) = (0.0020 \times 50 + 0.30 \times 0.0500) = 0.100 + 0.015 = 0.115$$

$$P_x/P_s = 1/e^{(Kl+\mu\alpha)} = 1/e^{0.115} = 1/1.12 = 0.89 \qquad \text{or } 11\% \text{ loss} \tag{5-34}$$

$$P_x/P_s = 1/(1 + Kl + \mu\alpha) = 1/(1 + 0.115) = 0.90 \qquad \text{or } 10\% \text{ loss} \qquad (5\text{-}35)$$

In simple and continuous beams, including those with reverse tendon curvatures, α is the total angle change from the jacking end to any point x, such as $\alpha = 2 \times 0.0500 = 0.100$ radians in this example for the frictional loss from one end to the other ($x = l = L$).

Numerical Examples of Loss of Prestress, and Concrete and Steel Stresses

Example 1 Using the four methods: ultimate midspan loss and concrete stresses, noncomposite double-T beam with prestressing steel only, normal-weight concrete, no topping.

See Table 5-6 for the material and section properties, parameters, and conditions of the problem. The loss of prestress results are tabulated term by term for comparison purposes in Table 5-9.

Table 5-6 Material and section properties, parameters, and conditions for Example 1

No topping

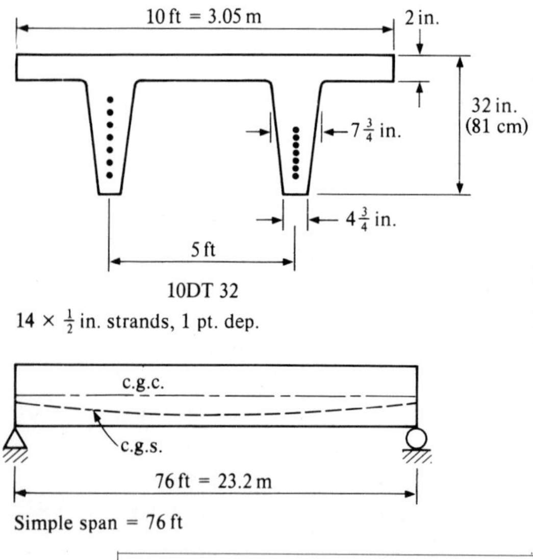

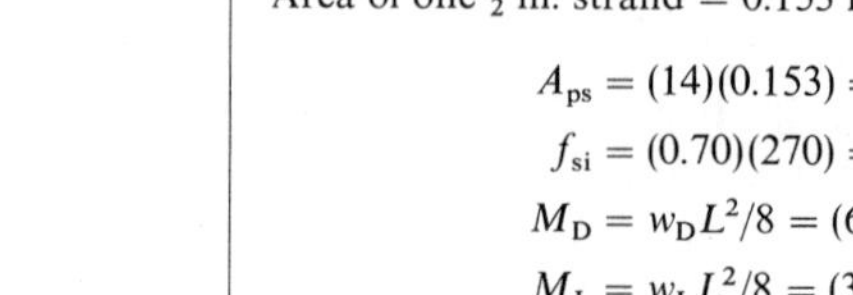

Simple span = 76 ft

Material properties

Steam-cured normal-weight concrete

$f_{pu} = 270$ ksi

$f'_{ci} = 4000$ psi $\qquad f'_c = 5000$ psi

Section prop. and loading, PCI Handbook

$A_g = 615\ \text{in}^2 \qquad y_b = 21.98$ in.

$I_g = 59\,720\ \text{in}^4$

$Z_t = 5960\ \text{in}^3 \qquad Z_b = 2717\ \text{in}^3$

$e_e = 11.12$ in. $\qquad e_c = y_b - 3.5$ in. $= 18.48$ in.

$w_D = 641$ #/ft $\qquad w_L = 30$ psf

Area of one ½ in. strand = 0.153 in^2 (Fig. 11.3.3, PCI Handbook)

$$A_{ps} = (14)(0.153) = 2.14\ \text{in}^2\ (13.8\ \text{cm}^2)$$

$$f_{si} = (0.70)(270) = 189\ \text{ksi}\ (13\,290\ \text{kg/cm}^2,\ 1303\ \text{N/mm}^2)$$

$$M_D = w_D L^2/8 = (641)(76)^2/8 = 462\,800\ \text{ft-\#}\ (64\,000\ \text{kg-m},\ 628\ \text{kN-m})$$

$$M_L = w_L L^2/8 = (300)(76)^2/8 = 216\,700\ \text{ft-\#}\ (29\,970\ \text{kg-m},\ 294\ \text{kN-m})$$

$$M_{D+L} = 679\,500\ \text{ft-\#}\ (93\,970\ \text{kg-m},\ 921\ \text{kN-m})$$

Assume 70 percent ambient relative humidity.

(For other metric units, see metric conversion factors herein.)

Approx. step-function method—Eq. (5-15), which is the ultimate loss form of Eq. (5-13), or Eq. (5-4) when applied to this case in which $f_{cs} = 0$ and $k_r = 1$; also compared with the results for typical conditions by the simplified Eq. (5-22a) Elastic loss—term (1), Eq. (5-15), as defined in Eq. (5-4):

Using the gross-section properties,

$$P_i = A_{ps} f_{si} = (2.14)(189) = 404 \text{ kips } (183\,300 \text{ kg}, 1797 \text{ kN})$$

$$\rho_{ps} = A_{ps}/A_g = 2.14/615 = 0.00348 \qquad n = 7.3 \qquad \text{(Table 5-1)}$$

$$\text{Trial } P_0 = P_i/(1 + 2n\rho_{ps}) = 404/(1 + 2 \times 7.3 \times 0.00348)$$
$$= 384 \text{ kips } (174\,200 \text{ kg}, 1708 \text{ kN})$$

$$f_{ci} = \frac{P_0}{A_g} + \frac{P_0 e_c^2}{I_g} - \frac{M_D e_c}{I_g} = \frac{384}{615} + \frac{(384)(18.48)^2}{59\,720} - \frac{(462.8)(12)(18.48)}{59\,720}$$
$$= 0.624 + 2.196 - 1.719 = 1.101 \text{ ksi } (77.4 \text{ kg/cm}^2, 7.59 \text{ N/mm}^2)$$

$$\text{Trial elastic loss} = (nf_{ci}) = (7.3)(1101) = 8040 \text{ psi}/1890 = 4.25\%$$

$$\text{Revised } P_0 = P_i - (nf_{ci})A_{ps} = 404 - (8.040)(2.14)$$
$$= 387 \text{ kips } (175\,500 \text{ kg}, 1721 \text{ kN})$$

$$\text{Revised } f_{ci} = 1.122 \text{ ksi } (78.9 \text{ kg/cm}^2, 7.74 \text{ N/mm}^2)$$

$$\text{Elastic loss} = (nf_{ci}) = (7.3)(1122) = 8190 \text{ psi}/1890 = 4.33\%$$

(Both trial and revised elastic loss = 4.3%)

Creep and shrinkage loss—terms (2) and (3), respectively, Eq. (5-15), as defined in Eq. (5-4):

For 70 percent humidity,

$$C_u = 1.88 \qquad (\varepsilon_{sh})_u = 546 \times 10^{-6} \text{ in/in (cm/cm)} \qquad \text{(Table 1-10)}$$

Approximate web thickness is about 6 in. (15.2 cm); hence the size correction factor for creep and shrinkage may be taken as unity (Table 1-8).

$$\Delta P_u/P_0 = 0.18 \qquad \text{(Table 5-2)}$$

$$E_s = 28 \times 10^6 \text{ psi } (1.97 \times 10^6 \text{ kg/cm}^2, 193 \text{ kN/mm}^2) \qquad \text{(Chap. 1)}$$

$$(1 + b_{11})_u = 1.25 \qquad \text{(Design simplifications)}$$

$$\text{Creep loss} = (nf_{ci})(k_r C_u)\left(1 - \frac{\Delta P_u}{2P_0}\right) = (8190)(1.88)(1 - 0.09)$$
$$= 14\,010 \text{ psi}/1890 = 7.41\%$$

$$\text{Shrinkage loss} = k_r(\varepsilon_{sh})_u E_s/(1 + b_{11})_u = (546)(28)/1.25$$
$$= 12\,230 \text{ psi}/1890 = 6.47\%$$

Relaxation loss—term (4), Eq. (5-15), as defined in Eq. (5-4):

$$\text{Relaxation loss} = (f_{sr})_u = 7.5\% \qquad f_{si} = (0.075)(189\,000)$$
$$= 14\,180 \text{ psi}/1890 = 7.50\%$$

$$\text{Total loss} = (\Delta f_s)_u = 8190 + 14\,010 + 12\,230 + 14\,180$$
$$= 48\,610 \text{ psi } (3418 \text{ kg/cm}^2, 335 \text{ N/mm}^2) \qquad (\text{Eq. (5-15)})$$
$$= 4.33 + 7.41 + 6.47 + 7.50 = 25.7\%$$

Since 70 percent humidity was assumed in this example and no size correction factors for creep and shrinkage were required, the result by the simplified Eq. (5-22*a*) is the same, or $(\Delta f_s)_u = 19.79 f_{ci} + 26\,480 - 13.26 f_{cs} = (19.79)(1122) + 26\,480 - 0 = 48\,680 \text{ psi}/1890 = 25.8\%$.

Using the simplified method of the PCI Loss Committee:[73]

$$\text{Total loss} = 33.0 + 13.8 f_{ci} - 4.5 f_{cs} = 33.0 + (13.8)(1.122) - 0$$
$$= 48.5 \text{ ksi}/1.89 = 25.7\%.$$

From Appendix C, Chap. 1, v/s = about 1.7 in. Hence, total loss = 25.7 + (0.3)(3.2) = 26.7%, which is about 1 percent higher than the above solutions.

Concrete stresses at transfer:

$$f_{top} = \frac{P_0}{A_g} - \frac{P_0 e_c}{Z_t} + \frac{M_D}{Z_t} = \frac{387\,000}{615} - \frac{387\,000 \times 18.48}{5960} + \frac{462\,800 \times 12}{5960}$$
$$= 629 - 1200 + 932 = 361 \text{ psi C. } (25.4 \text{ kg/cm}^2, 2.49 \text{ N/mm}^2)$$
$$f_{bot} = 629 + 1200(5960/2717) - 932(2.194) = 629 + 2633 - 2044$$
$$2.194$$
$$= 1218 \text{ psi C. } (85.6 \text{ kg/cm}^2, 8.40 \text{ N/mm}^2)$$

Concrete stresses after losses with live load:

$$f_{top} = \frac{P}{A_g} - \frac{P e_c}{Z_t} + \frac{M_D + M_L}{Z_t}$$

where

$$P = P_i(1 - 0.257) = (404)(0.743) = 300 \text{ kips } (136\,100 \text{ kg, } 1334 \text{ kN})$$
$$f_{top} = \frac{300\,000}{615} - \frac{300\,000 \times 18.48}{5960} + \frac{679\,500 \times 12}{5960} = 488 - 930 + 1368$$
$$= 926 \text{ psi C. } (65.1 \text{ kg/cm}^2, 6.38 \text{ N/mm}^2)$$
$$f_{bot} = 488 + 930(2.194) - 1368(2.194) = 488 + 2040 - 3002$$
$$= 474 \text{ psi T. } (33.3 \text{ kg/cm}^2, 3.27 \text{ N/mm}^2)$$

Relaxation method—Eq. (5-27) Elastic loss—term (1), Eq. (5-27):

$$\text{Elastic loss} = (n f_{ci}) = 8190 \text{ psi}/1890 = 4.33\% \qquad (\text{prev. calc.})$$

Creep and shrinkage loss—terms (2) and (3), respectively, Eq. (5-27):

The denominator is first computed using the gross-section properties.

Use

$$\rho_{ps} = 0.00348 \qquad \text{(prev. calc.)}$$
$$r_g^2 = I_g/A_g = 59\,720/615 = 97.11 \text{ in}^2$$
$$e^2/r_g^2 = 18.48^2/97.11 = 3.52$$
$$\eta = 0.88 \text{ (term 2, Eq. (5-27))}$$
$$n = 7.3 \qquad \text{(Table 5-1)}$$
$$1 + n(\rho_{ps})(1 + e^2/r_g^2)(1 + \eta C_u) = 1 + 7.3(0.00348)(1 + 3.52)$$
$$(1 + 0.88 \times 1.88) = 1 + 0.34 = 1.34$$
$$\text{Creep loss} = nf_{ci}C_u/1.34 = (8190)(1.88/1.34)$$
$$= 11\,490 \text{ psi}/1890 = 6.08\%$$
$$\text{Shrinkage loss} = (\varepsilon_{sh})_u E_s/1.34 = (546)(28)/1.34$$
$$= 11\,410 \text{ psi}/1890 = 6.04\%$$

Relaxation loss—term (4), Eq. (5-27):

$$(f_{sr})_u = 14\,180 \text{ psi}/1890 = 7.50\% \qquad \text{(prev. calc.)}$$
$$\text{Total loss} = (\Delta f_s)_u = 8190 + 11\,490 + 11\,410 + 14\,180$$
$$= 45\,270 \text{ psi } (3183 \text{ kg/cm}^2, 312 \text{ N/mm}^2) \qquad \text{(Eq. (5-27))}$$
$$= 4.33 + 6.08 + 6.04 + 7.50 = 24.0\%$$

The reduction factors in terms (2) and (3) for creep and shrinkage account for the small differences in the results by Eq. (5-15) of 25.7 percent loss, and by Eq. (5-27) of 24.0 percent loss. Otherwise the same parameters were used in these solutions.

Concrete stresses at transfer (see equations in previous calculations):

$$f_{top} = 361 \text{ psi C. } (25.4 \text{ kg/cm}^2, 2.49 \text{ N/mm}^2)$$
$$f_{bot} = 1218 \text{ psi C. } (85.6 \text{ kg/cm}^2, 8.40 \text{ N/mm}^2) \qquad \text{(same as prev. calc.)}$$

Concrete stresses after losses with live load (see equations in previous calculations):

$$P = P_i(1 - 0.240) = (404)(0.760) = 307 \text{ kips } (139\,300 \text{ kg, } 1366 \text{ kN})$$
$$f_{top} = 915 \text{ psi C. } (64.3 \text{ kg/cm}^2, 6.31 \text{ N/mm}^2)$$

versus 926 psi C. (65.1 kg/cm^2, 6.38 N/mm^2) in previous calculations.

$$f_{bot} = 414 \text{ psi T. } (29.1 \text{ kg/cm}^2, 2.85 \text{ N/mm}^2)$$

versus 474 psi T. (33.3 kg/cm^2, 3.27 N/mm^2) in previous calculations.

AASHTO method—Eq. (5-32) Shrinkage loss—term (1), Eq. (5-32):

$$SH = 10\,000 \text{ psi}/1890 = 5.29\%$$

Elastic loss—term (2), Eq. (5-32):

In prev. calc., $f_{cr} = f_{ci} = 1122$ psi (78.9 kg/cm^2, 7.74 N/mm^2)

$$\text{Select } ES = 7000 \text{ psi}/1890 = 3.70\%$$

Creep loss—term (3), Eq. (5-32):

In prev. calc., $f_{cd} = f_{ci} = 1122$ psi (78.9 kg/cm^2, 7.74 N/mm^2)

Select $CR_c = 18\,000$ psi/1890 = 9.52%

Relaxation loss—term (4), Eq. (5-32):

$$\begin{aligned} CR_s &= 20\,000 - 0.125(SH + ES + CR_c) \\ &= 20\,000 - 0.125(10\,000 + 7000 + 18\,000) \\ &= 15\,630 \text{ psi}/1890 = 8.27\% \qquad \text{(Eq. (1-35))} \\ \text{Total loss} &= \Delta f_s = 10\,000 + 7000 + 18\,000 + 15\,630 \\ &= 50\,630 \text{ psi } (3560 \text{ kg/cm}^2, 349 \text{ N/mm}^2) \\ &= 5.29 + 3.70 + 9.52 + 8.27 = 26.8\% \end{aligned}$$

Concrete stresses at transfer (see equations in previous calculations):

$$\begin{aligned} f_{top} &= 361 \text{ psi C. } (25.4 \text{ kg/cm}^2, 2.49 \text{ N/mm}^2) \\ f_{bot} &= 1218 \text{ psi C. } (85.6 \text{ kg/cm}^2, 8.40 \text{ N/mm}^2) \qquad \text{(same as prev. calc.)} \end{aligned}$$

Concrete stresses after losses with live load (see equations in previous calculations):

$$\begin{aligned} P &= P_i(1 - 0.268) = (404)(0.732) = 296 \text{ kips } (134\,300 \text{ kg}, 1317 \text{ kN}) \\ f_{top} &= 931 \text{ psi C. } (65.5 \text{ kg/cm}^2, 6.42 \text{ N/mm}^2) \end{aligned}$$

versus 926 psi C. and 915 psi C. in previous calculations.

$$f_{bot} = 507 \text{ psi T. } (35.6 \text{ kg/cm}^2, 3.50 \text{ N/mm}^2)$$

versus 474 psi T. and 414 psi T. in previous calculations.

PCI Handbook method—Eq. (5-33) Elastic loss—term (1), Eq. (5-33):

For $f_{ci} = 4000$ psi (281 kg/cm^2, 27.6 N/mm^2),

$$\begin{aligned} E_{ci} &= 3.6 \times 10^6 \text{ psi } (253\,000 \text{ kg/cm}^2, 24.8 \text{ kN/mm}^2) \\ & \qquad \text{(Fig. 11.3.2, PCI Handbook)} \\ E_s &= 27 \times 10^6 \text{ psi } (1.90 \times 10^6 \text{ kg/cm}^2, 186 \text{ kN/mm}^2) \\ n &= E_s/E_{ci} = 27/3.6 = 7.5 \\ \text{Elastic loss} &= (nf_{cs}) = (nf_{ci}) = (7.5)(1122, \text{ as prev. calc.}) \\ &= 8420 \text{ psi}/1890 = 4.46\% \end{aligned}$$

Creep and shrinkage loss—terms (2) and (3), respectively, Eq. (5-33):

Volume/surface ratio = 1.6 in. (4.1 cm) (Fig. 6.1.1, PCI Handbook)

$$n_c = 16.5 \qquad r_s = 14.8$$

(Figs. 5.3.1, 5.3.4, respectively, PCI Handbook)

$$\begin{aligned}
k_1 &= (A_{ps}/A_c)(1 + e^2 A_c/I) \\
&= (2.14/615)(1 + 18.48^2 \times 615/59\,720) = 0.0157 \\
\text{Creep loss} = L_c &= (n_c f_{cs})(1 - 0.34 n_c k_1) \\
&= (16.5 \times 1122)(1 - 0.34 \times 16.5 \times 0.0157) \\
&= 16\,880 \text{ psi}/1890 = 8.93\% \\
\text{Shrinkage loss} = L_s &= r_s(1 - 0.010 n_c k_1) \\
&= 14\,800(1 - 0.010 \times 16.5 \times 0.0157) \\
&= 14\,420 \text{ psi}/1890 = 7.63\%
\end{aligned}$$

Relaxation loss—term (4), Eq. (5-33):

$$f_{py} = 0.85 f_{pu} = (0.85)(270) = 230 \text{ ksi } (16\,200 \text{ kg/cm}^2,\ 1586 \text{ N/mm}^2)$$

$$(L_c + L_s)/f_{py} = (16\,880 + 14\,420)/230\,000 = 0.136$$

Assume

$$\begin{aligned}
f_s &= 189 - 15 = 174 \text{ ksi } (12\,250 \text{ kg/cm}^2,\ 1200 \text{ N/mm}^2) \\
f_s/f_{py} &= 174/230 = 0.76 \\
\alpha &= 0.042 \qquad \text{(Table 5.3.8, PCI Handbook)} \\
\text{Relaxation loss} = L_r &= \alpha f_{py} = (0.042)(230\,000) = 9660 \text{ psi}/1890 = 5.11\% \\
\text{Total loss} &= 8420 + 16\,880 + 14\,420 + 9660 \\
&= 49\,380 \text{ psi } (3472 \text{ kg/cm}^2,\ 340 \text{ N/mm}^2) \\
&= 4.46 + 8.93 + 7.63 + 5.11 = 26.1\%
\end{aligned}$$

Concrete stresses at transfer (see equations in previous calculations):

$$\begin{aligned}
\text{Revised } P_0 &= P_i - (n f_{cs}) A_{ps} = 404 - (8.420)(2.14) \\
&= 386 \text{ kips } (175\,000 \text{ kg},\ 1717 \text{ kN}) \\
f_{top} &= 359 \text{ psi C. } (25.2 \text{ kg/cm}^2,\ 2.48 \text{ N/mm}^2)
\end{aligned}$$

versus 361 psi C. (25.4 kg/cm^2, 2.49 N/mm^2) in previous calculations.

$$f_{bot} = 1210 \text{ psi C. } (85.1 \text{ kg/cm}^2,\ 8.34 \text{ N/mm}^2)$$

versus 1218 psi C. (85.6 kg/cm^2, 8.40 N/mm^2) in previous calculations.

Concrete stresses after losses with live load (see equations in previous calculations):

$$P = P_i(1 - 0.261) = (404)(0.739) = 299 \text{ kips } (135\,600 \text{ kg},\ 1330 \text{ kN})$$

$$f_{top} = 927 \text{ psi C. } (65.2 \text{ kg/cm}^2,\ 6.39 \text{ N/mm}^2)$$

versus 926, 915, and 931 psi C. in previous calculations.

$$f_{\text{bot}} = 482 \text{ psi T. } (33.9 \text{ kg/cm}^2, 3.32 \text{ N/mm}^2)$$

versus 474, 414, 507 psi T. in previous calculations.

As these comparisons show, the calculated concrete stresses by the different methods are in general agreement, with some differences being found due primarily to different assumptions for material parameters. Also, as can be seen in Table 5-9, the loss of prestress results as computed by the different methods are in relatively close agreement.

The gain in prestress due to live load in Example 1 is:

$$f_{sL} = nf_{cL} = nM_L e/I$$
$$= (7.2)(216\,700)(12)(18.48)/59\,720 = 2920 \text{ psi}/1890 = 1.5\%$$

The following allowable concrete stresses from the ACI Code[20] are noted:

At transfer—

$$f_{\text{comp}} = 0.60 f'_{ci} = 0.60(4000) = 2400 \text{ psi } (169 \text{ kg/cm}^2, 16.5 \text{ N/mm}^2)$$
$$f_{\text{tens}} = 3\sqrt{f'_{ci}} = 3\sqrt{4000} = 190 \text{ psi } (13.4 \text{ kg/cm}^2, 1.31 \text{ N/mm}^2)$$

After losses—

$$f_{\text{comp}} = 0.45 f'_c = 0.45(5000) = 2250 \text{ psi } (158 \text{ kg/cm}^2, 15.5 \text{ N/mm}^2)$$
$$f_{\text{tens}} = 6\sqrt{f'_c} = 6\sqrt{5000} = 424 \text{ psi } (29.8 \text{ kg/cm}^2, 2.92 \text{ N/mm}^2)$$

When deflections are checked—

$$f_{\text{tens}} = 12\sqrt{f'_c} = 12\sqrt{5000} = 849 \text{ psi } (59.7 \text{ kg/cm}^2, 5.85 \text{ N/mm}^2)$$

Modulus of rupture (Table 1-2)—

Recommended,

$$f_r = 7.84\sqrt{f'_c} = 7.84\sqrt{5000} = 555 \text{ psi } (39.0 \text{ kg/cm}^2, 3.83 \text{ N/mm}^2)$$

ACI Code,

$$f_r = 7.5\sqrt{f'_c} = 7.5\sqrt{5000} = 531 \text{ psi } (37.3 \text{ kg/cm}^2, 3.66 \text{ N/mm}^2)$$

Thus it is seen that all computed stresses by all methods are within the allowable values, with the exception of the bottom stress after losses under live load. In this case, the computed stress by three of the four methods was higher than the allowable stress of $6\sqrt{f'_c}$, but less than both of the modulus of rupture values.

Example 2 Same as Example 1 by the approximate step-function method, except using sand-lightweight concrete (120 pcf, 1922 kg/m^3, 18.9 kN/m^3).

All parameters and conditions in Table 5-6 apply to this problem, except $w_D = 530$ #/ft (789 kg/m, 7730 N/m) and $M_D = (530)(76)^2/8 = 382\,700$ ft-# (52 900 kg-m, 519 kN-m). The loss of prestress results are tabulated term by term for comparison purposes in Table 5-9.

Elastic loss: $n = 9.8$ (Table 5-1)

$$\text{Trial } P_0 = P_i(1 + 2n\rho_{ps}) = 404/(1 + 2 \times 9.8 \times 0.00348)$$
$$= 378 \text{ kips (171 000 kg, 1681 kN)}$$
$$f_{ci} = \frac{P_0}{A_g} + \frac{P_0 e_c^2}{I_g} - \frac{M_D e_c}{I_g} = \frac{378}{615} + \frac{(378)(18.48)^2}{59\,720} - \frac{(382.7)(12)(18.48)}{59\,720}$$
$$= 0.615 + 2.162 - 1.421 = 1.356 \text{ ksi (95.3 kg/cm}^2\text{, 9.35 N/mm}^2\text{)}$$

$$\text{Trial elastic loss} = (nf_{ci}) = (9.8)(1356) = 13\,290 \text{ psi}/1890 = 7.03\%$$
$$\text{Revised } P_0 = P_i - (nf_{ci})A_{ps} = 404 - (13.290)(2.14)$$
$$= 376 \text{ kips (170 600 kg, 1672 kN)}$$
$$\text{Revised } f_{ci} = 1.342 \text{ ksi (94.4 kg/cm}^2\text{, 9.25 N/mm}^2\text{)}$$
$$\text{Elastic loss} = (nf_{ci}) = (9.8)(1342) = 13\,150 \text{ psi}/1890 = 6.96\%$$

(Both trial and revised elastic loss = 7.0%)

Creep, shrinkage and relaxation loss: $\Delta P_u/P_0 = 0.21$ (Table 5-2)

$$\text{Creep loss} = (nf_{ci})(k_r C_u)\left(1 - \frac{\Delta P_u}{2P_0}\right)$$
$$= (13\,150)(1.88)(1 - 0.105)$$
$$= 22\,130 \text{ psi}/1890 = 11.71\%$$

Shrinkage loss and relaxation loss = same as Example 1

$$\text{Total loss} = (\Delta f_s)_u = 13\,150 + 22\,130 + 12\,230 + 14\,180$$
$$= 61\,690 \text{ psi (4337 kg/cm}^2\text{, 425 N/mm}^2\text{)} \qquad \text{(Eq. (5-15))}$$
$$= 6.96 + 11.71 + 6.47 + 7.50 = 32.6\%$$

Since 70 percent humidity was assumed in this example and no size correction factors for creep and shrinkage were required, the result by the simplified Eq. (5-22*b*) is the same, or

$$(\Delta f_s)_u = 26.29 f_{ci} + 26\,480 - 17.82 f_{cs}$$
$$= (26.29)(1342) + 26\,480 - 0 = 61\,760 \text{ psi}/1890 = 32.7\%$$

Using the simplified method of the PCI Loss Committee:[73]

$$\text{Total loss} = 31.2 + 16.8 f_{ci} - 3.8 f_{cs} = 31.2 + 16.8(1.342) - 0$$
$$= 53.7 \text{ ksi}/1.89 = 28.4\%$$

From Appendix C, Chap. 1, v/s = about 1.7 in. Hence, total loss = 28.4 + (0.3)(3.2) = 29.4%, which is about 3 percent lower than the above solutions.

Concrete stresses at transfer (see equations in Example 1):

$$f_{top} = 216 \text{ psi C. (15.2 kg/cm}^2\text{, 1.49 N/mm}^2\text{)}$$

versus 361 psi C. (25.4 kg/cm², 2.49 N/mm²) in Example 1.

$$f_{\text{bot}} = 1477 \text{ psi C. } (103.8 \text{ kg/cm}^2, 10.18 \text{ N/mm}^2)$$

versus 1218 psi C. (85.6 kg/cm^2, 8.40 N/mm^2) in Example 1.

Concrete stresses after losses with live load (see equations in Example 1):

$$P = P_i(1 - 0.326) = 404(0.674) = 272 \text{ kips } (123\,400 \text{ kg, } 1210 \text{ kN})$$

$$f_{\text{top}} = 806 \text{ psi C. } (56.7 \text{ kg/cm}^2, 5.56 \text{ N/mm}^2)$$

versus 926 psi C. (65.1 kg/cm^2, 6.38 N/mm^2) in Example 1.

$$f_{\text{bot}} = 356 \text{ psi T. } (25.0 \text{ kg/cm}^2, 2.45 \text{ N/mm}^2)$$

versus 474 psi T. (33.3 kg/cm^2, 3.27 N/mm^2) in Example 1.

The gain in prestress due to live load in Example 2 is:

$$f_{sL} = nf_{cL} = nM_L e/I$$
$$= (10.0)(216\,700)(12)(18.48)/59\,720 = 4050 \text{ psi}/1890 = 2.1\%$$

The bottom stress after losses with live load was less in Example 2 than in Example 1, because the reduction in dead-load stress for the sand-lightweight beam more than offset the reduction in stress resulting from the increase in prestress loss.

The allowable concrete stresses from the ACI Code[20] are the same for sand-lightweight concrete as for the normal-weight concrete case in Example 1. Thus it is seen that all computed stresses in Example 2 are within the allowable values.

Example 3 Ultimate midspan stresses in steel and concrete, noncomposite single-T beam, first containing only prestressing steel and second containing both prestressed and nonprestressed tension steel, sand-lightweight concrete (120 pcf, 1922 kg/m^3, 18.9 kN/m^3), with additional sustained loading applied some time after the transfer of prestress—approximate step-function method, Eqs. (5-15) and (5-16), which are the ultimate forms of Eqs. (5-13) and (5-14), respectively.

See Table 5-7 for the material and section properties, parameters, and conditions of the problem. The results are tabulated term by term in Table 5-10.

Beam containing only prestressing steel Delete the nonprestressed tension steel in Table 5-7 and use the same eccentricity for the prestressing steel only.

Elastic loss—term (1), Eq. (5-15), as defined in Eq. (5-4):

Using the gross-section properties,

$$P_i = A_{ps} f_{si} = (2.45)(189) = 463 \text{ kips } (210\,000 \text{ kg, } 2059 \text{ kN})$$

$$\rho_{ps} = A_{ps}/A_g = 2.45/782 = 0.00313 \qquad n = 9.8 \qquad \text{(Table 5-1)}$$

$$\text{Trial } P_0 = P_i/(1 + 2n\rho_{ps}) = 463/(1 + 2 \times 9.8 \times 0.00313)$$
$$= 436 \text{ kips } (197\,800 \text{ kg, } 1939 \text{ kN})$$

Table 5-7 Material and section properties, parameters, and conditions for Example 3

No topping

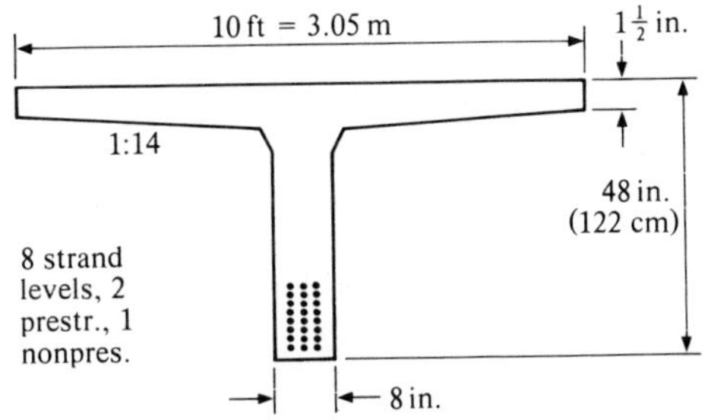

10LST 48

16 × ½ in. pres. strands, 1 pt. dep.
8 × ½ in. nonprestressed strands

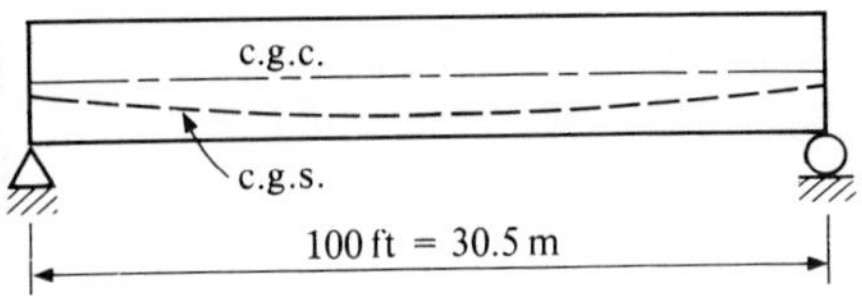

Simple span = 100 ft

Material properties

Steam-cured sand-lightweight concrete (120 pcf, 1922 kg/m³, 18.9 kN/m³)

$f_{pu} = 270$ ksi

$f'_{ci} = 4000$ psi $\quad f'_c = 5000$ psi

Section prop. and loading, PCI Handbook

$A_g = 782\ \text{in}^2 \qquad y_b = 35.19$ in.

$I_g = 168\,970\ \text{in}^4$

$Z_t = 13\,190\ \text{in}^3 \qquad Z_b = 4802\ \text{in}^3$

$e_c = y_b - 4.75 = 30.44$ in. for both prestressed and nonprestressed steel

$e_e = 9.86$ in. for prestressed steel

$w_D = 652$ #/ft

Superimposed $w_D = 10$ psf applied at age 1 month

$w_L = 30$ psf of which 10 psf is sustained and assumed to be applied at age 1 month

Area of one ½ in. strand = 0.153 in² (Fig. 11.3.3, PCI Handbook)

$$A_{ps} = (16)(0.153) = 2.45\ \text{in}^2 \qquad A_{ns} = (8)(0.153) = 1.22\ \text{in}^2$$

$$f_{si} = (0.70)(270) = 189\ \text{ksi}\ (13\,290\ \text{kg/cm}^2,\ 1303\ \text{N/mm}^2)$$

$$M_D = w_D L^2/8 = (652)(100)^2/8 = 815\,000\ \text{ft-\#}\ (112\,700\ \text{kg-m},\ 1105\ \text{kN-m})$$

Superimposed sustained mom.

$$M_S = (20 \times 10)(100)^2/8 = 250\,000\ \text{ft-\#}$$

Short-term live mom.

$$M_L = 250\,000\ \text{ft-\#}\ (34\,600\ \text{kg-m},\ 339\ \text{kN-m})$$

$$\text{Total moment} = M_D + M_S + M_L = 1\,315\,000\ \text{ft-\#}\ (181\,900\ \text{kg-m},\ 1783\ \text{kN-m})$$

Assume 70 percent ambient relative humidity

(For other metric units, see metric conversion factors herein.)

$$f_{ci} = \frac{P_0}{A_g} + \frac{P_0 e_c^2}{I_g} - \frac{M_D e_c}{I_g} = \frac{436}{782} + \frac{(436)(30.44)^2}{168\,970} - \frac{(815.0)(12)(30.44)}{168\,970}$$

$$= 0.558 + 2.798 - 1.762$$

$$= 1.187\ \text{ksi C.}\ (83.5\ \text{kg/cm}^2,\ 8.18\ \text{N/mm}^2)$$

Trial elastic loss = $(nf_{ci}) = (9.8)(1187) = 11\,630\ \text{psi}/1890 = 6.15\%$

Check: $\quad P_0 = P_i - (nf_{ci})A_{ps} = 463 - (11.630)(2.45)$

$$= 435\ \text{kips}\ (197\,300\ \text{kg},\ 1935\ \text{kN})$$

No revision is necessary.

Creep and shrinkage loss—terms (2) and (3), respectively, Eq. (5-15), as defined in Eq. (5-4):

For 70 percent humidity,

$$C_u = 1.88 \qquad (\varepsilon_{sh})_u = 546 \times 10^{-6} \text{ in/in (cm/cm)} \qquad \text{(Table 1-10)}$$

For 8 in. web, Creep $(\text{C.F.})_T = 0.96$ Shrinkage $(\text{C.F.})_T = 0.94$ (Table 1-8)

$$\text{Adjusted } C_u = (1.88)(0.96) = 1.80$$

$$\text{Adjusted } (\varepsilon_{sh})_u = (546 \times 10^{-6})(0.94) = 513 \times 10^{-6} \text{ in/in (cm/cm)}$$

$$\Delta P_u/P_0 = 0.21 \qquad \text{(Table 5-2)}$$

$$E_s = 28 \times 10^6 \text{ psi } (1.97 \times 10^6 \text{ kg/cm}^2, 193 \text{ kN/mm}^2) \qquad \text{(Chap. 1)}$$

$$(1 + b_{11})_u = 1.25 \text{ (design simplifications)} \qquad k_r = 1.00$$

$$\text{Creep loss} = (nf_{ci})(k_r C_u)\left(1 - \frac{\Delta P_u}{2P_0}\right) = (11\,630)(1.80)(1 - 0.105)$$

$$= 18\,740 \text{ psi}/1890 = 9.92\%$$

$$\text{Shrinkage loss} = k_r(\varepsilon_{sh})_u E_s/(1 + b_{11})_u = (513)(28)/1.25$$

$$= 11\,490 \text{ psi}/1890 = 6.08\%$$

Relaxation loss—term (4), Eq. (5-15), as defined in Eq. (5-4):

$$\text{Relaxation loss} = (f_{sr})_u = 7.5\% f_{si} = (0.075)(189\,000)$$

$$= 14\,180 \text{ psi}/1890 = 7.50\%$$

Elastic gain due to superimposed sustained load—term (5), Eq. (5-15), as defined in Eq. (5-13):

$$m = 8.2 \qquad \text{(Table 5-1, for load applied at age 1 month)}$$

$$f_{cs} = M_S e/I_g = (250\,000)(12)(30.44)/168\,970$$

$$= 540 \text{ psi } (38.0 \text{ kg/cm}^2, 3.72 \text{ N/mm}^2)$$

$$\text{Elastic gain} = (mf_{cs}) = (8.2)(540) = 4430 \text{ psi}/1890 = 2.34\%$$

Gain due to concrete creep under superimposed sustained load—term (6), Eq. (5-15), as defined in Eq. (5-13):

$$\beta_s = 0.83 \qquad \text{(Table 5-3, for load applied at age 1 month)}$$

$$\text{Gain due to creep} = (mf_{cs})(\beta_s k_r C_u)/(1 + b_{11})_u$$

$$= (4430)(0.83 \times 1.80)/1.25 = 5290 \text{ psi}/1890 = 2.80\%$$

$$\text{Total loss} = (\Delta f_s)_u = 11\,630 + 18\,740 + 11\,490 + 14\,180 - 4430 - 5290$$

$$= 46\,320 \text{ psi } (3257 \text{ kg/cm}^2, 319 \text{ N/mm}^2) \qquad \text{(Eq. (5-15))}$$

$$= 6.15 + 9.92 + 6.08 + 7.50 - 2.34 - 2.80 = 24.5\%$$

Using the simplified Eq. (5-22*b*), which is also based on 70 percent humidity but neglects any size correction factors for creep and shrinkage,

$$(\Delta f_s)_u = 26.29 f_{ci} + 26\,480 - 17.82 f_{cs}$$
$$= (26.29)(1187) + 26\,480 - (17.82)(540) = 48\,070 \text{ psi}/1890 = 25.4\%$$

The simplified result is on the safe side compared to that of Eq. (5-15) which is 24.5 percent.

Using the simplified method of the PCI Loss Committee:[73]

$$\text{Total loss} = 31.2 + 16.8 f_{ci} - 3.8 f_{cs} = 31.2 + 16.8(1.187) - 3.8(0.540)$$
$$= 49.0 \text{ ksi}/1.89 = 25.9\%$$

From Appendix C, Chap. 1, v/s = about 2.3 in. Hence, total loss = 25.9 − (0.3)(3.8) = 24.8%, which is in essential agreement with the above solutions.

Concrete stresses at transfer:

$$f_{top} = \frac{P_0}{A_g} - \frac{P_0 e_c}{Z_t} + \frac{M_D}{Z_t} = \frac{435\,000}{782} - \frac{435\,000 \times 30.44}{13\,190} + \frac{815\,000 \times 12}{13\,190}$$
$$= 556 - 1004 + 741 = 293 \text{ psi C. } (20.6 \text{ kg/cm}^2, 2.02 \text{ N/mm}^2)$$
$$f_{bot} = 556 + 1044(13\,190/4802) - 741(2.747)$$
$$2.747$$
$$= 556 + 2868 - 2036 = 1388 \text{ psi C. } (97.6 \text{ kg/cm}^2, 9.57 \text{ N/mm}^2)$$

Concrete stresses after losses with superimposed loads:

$$f_{top} = \frac{P}{A_g} - \frac{Pe_c}{Z_t} + \frac{M_D + M_S + M_L}{Z_t}$$

where $P = P_i(1 - 0.245) = (463)(0.755) = 350$ kips (158 800 kg, 1557 kN).

$$f_{top} = \frac{350\,000}{782} - \frac{350\,000 \times 30.44}{13\,190} + \frac{1\,315\,000 \times 12}{13\,190}$$
$$= 448 - 808 + 1196 = 841 \text{ psi C.}$$
$$f_{bot} = 448 + 808(2.747) - 1196(2.747) = 448 + 2220 - 3285$$
$$= 617 \text{ psi T. } (43.4 \text{ kg/cm}^2, 4.25 \text{ N/mm}^2)$$

The computed tensile stress of 617 psi is less than the ACI Code (318-71) allowable of $12\sqrt{f_c'} = 849$ psi (59.7 kg/cm², 5.85 N/mm²), but deflections would have to be checked, as shown later in this chapter. The other computed stresses are well within the ACI Code allowable values (shown in Example 1).

Beam containing both prestressing steel and nonprestressed tension steel

Prestressing steel Elastic loss—term (1), Eq. (5-15), as defined in Eq. (5-4):

Same as prev. prob.—P_0 = 435 kips (197 300 kg, 1935 kN)

Elastic loss = 11 630 psi/1890 = 6.15%

Creep and shrinkage loss—terms (2) and (3), respectively, Eq. (5-15), as defined in Eqs. (5-4) and (5-6):

In addition to prev. prob.—$k_r = 1/(1 + b_{12}) = 1/1.25 = 0.80$ (Eq. (5-7))

Also, $k_r = 1/[1 + (A_{ns}/2A_{ps})] = 1/(1 + 0.25) = 0.80$ (Eq. (5-9))

Use $k_r = 0.80$

From previous problem,

$$\text{Creep loss} = (0.80)(18\,740) = 14\,990 \text{ psi}/1890 = 7.93\%$$

$$\text{Shrinkage loss} = (0.80)(11\,490) = 9190 \text{ psi}/1890 = 4.86\%$$

Relxation loss—term (4), Eq. (5-15), as defined in Eq. (5-4):

Same as prev. prob.—relaxation loss = 14 180 psi/1890 = 7.50%

Elastic gain due to superimposed sustained load—term (5), Eq. (5-15), as defined in Eq. (5-13):

Same as prev. prob.—elastic gain = 4430 psi/1890 = 2.34%

Gain due to concrete creep under superimposed sustained load—term (6), Eq. (5-15), as defined in Eq. (5-13):

$$\text{From prev. prob.—gain due to creep} = (0.80)(5290) = 4230 \text{ psi}/1890 = 2.24\%$$

$$\begin{aligned}\text{Total loss} = (\Delta f_s)_u &= 11\,630 + 14\,990 + 9190 + 14\,180 - 4430 - 4230 \\ &= 41\,330 \text{ psi } (2906 \text{ kg/cm}^2, 285 \text{ N/mm}^2) \qquad \text{(Eq. (5-15))} \\ &= 6.15 + 7.93 + 4.86 + 7.50 - 2.34 - 2.24 = 21.9\%\end{aligned}$$

Using the simplified Eq. (5-21*b*), which is also based on 70 percent humidity but neglects any size correction factors for creep and shrinkage,

$$\begin{aligned}(\Delta f_s)_u &= 9.8 f_{ci} + 16.49 k_r f_{ci} + 12\,230 k_r + 14\,250 - 8.2 f_{cs} - 9.62 k_r f_{cs} \\ &= (9.8)(1187) + (16.49)(0.80)(1187) + 12\,230(0.80) \\ &\quad + 14\,250 - (8.2)(540) - (9.62)(0.80)(540) \\ &= 42\,740 \text{ psi}/1890 = 22.6\%\end{aligned}$$

The simplified result is on the safe side compared to that of Eq. (5-15) which is 21.9 percent.

Using the simplified method of the PCI Loss Committee:[73]

Total loss = 24.8% (same as without nonprestressed tension steel)

which is about 4 percent higher than the above solutions, since the effect of nonprestressed tension steel is not included.

Nonprestressed tension steel Compression due to elastic shortening—term (1), Eq. (5-16), as defined in Eq. (5-10):

Using the gross-section properties,

$$\text{Elastic compression} = (nf_{ci}) = 11\,630 \text{ psi } (818 \text{ kg/cm}^2, 80.2 \text{ N/mm}^2)$$

(Same as elastic loss in the prestressing steel, since the eccentricities are the same.)

Compression due to concrete creep—term (2), Eq. (5-16), as defined in Eq. (5-10):

$$\begin{aligned}\text{Creep compression} &= (nf_{ci})C_u/(1 + b_{22})_u = (11\,630)(1.80)/1.25 \\ &= 16\,750 \text{ psi } (1178 \text{ kg/cm}^2, 115.5 \text{ N/mm}^2)\end{aligned}$$

Compression due to concrete shrinkage—term (3), Eq. (5-16), as defined in Eq. (5-10):

$$\begin{aligned}\text{Shrinkage compression} &= (\varepsilon_{sh})_u E_s/(1 + b_{22})_u = (513)(28)/1.25 \\ &= 11\,490 \text{ psi } (808 \text{ kg/cm}^2, 79.2 \text{ N/mm}^2)\end{aligned}$$

Elastic tension due to superimposed sustained load—term (4), Eq. (5-16), as defined in Eq. (5-14):

$$\text{Elastic tension} = (mf_{cs}) = 4430 \text{ psi } (311 \text{ kg/cm}^2, 30.5 \text{ N/mm}^2)$$

(Same as elastic gain in the prestressing steel, since the eccentricities are the same.)

Tension due to concrete creep under superimposed sustained load—term (5), Eq. (5-16), as defined in Eq. (5-14):

$$\begin{aligned}\text{Tension due to creep} &= (mf_{cs})(\beta_s C_u)/(1 + b_{22})_u = (4430)(0.83 \times 1.80)/1.25 \\ &= 5290 \text{ psi } (372 \text{ kg/cm}^2, 36.5 \text{ N/mm}^2)\end{aligned}$$

$$\text{Total compression} = (f_{ns})_u = 11\,630 + 16\,750 + 11\,490 - 4430 - 5290 \quad \text{(Eq. (5-16))}$$

$$= 30\,150 \text{ psi } (2120 \text{ kg/cm}^2, 208 \text{ N/mm}^2)$$

Using the simplified Eq. (5-23*b*), which is also based on 70 percent humidity but neglects any size correction factors for creep and shrinkage,

$$\begin{aligned}(f_{ns})_u &= 24.54 f_{ci} + 12\,230 - 13.26 f_{cs} \\ &= (24.54)(1187) + 12\,230 - (13.26)(540) \\ &- 31\,740 \text{ psi } (2232 \text{ kg/cm}^2, 193 \text{ N/mm}^2)\end{aligned}$$

The simplified result is on the safe side compared to that of Eq. (5-16) which is 30 150 psi (2120 kg/cm², 208 N/mm²).

Concrete stresses at transfer—Eqs. (5-12), (5-3):

$$P_{ps} = P_0 = 435 \text{ kips } (197\,300 \text{ kg}, 1935 \text{ kN})$$

$$F_{ns} = (f_{ns})_i A_{ns} = (nf_{ci})A_{ns} = (11.630)(1.22) = 14.2 \text{ kips } (6441 \text{ kg}, 63.2 \text{ kN})$$

$$f_{top} = P_{ps}/A_g - P_{ps}e_{ps}/Z_t - F_{ns}A_g + F_{ns}e_{ns}/Z_t + M_D/Z_t \qquad \text{(Eq. (5-12))}$$

$$f_{top} = \frac{435\,000}{782} - \frac{435\,000 \times 30.44}{13\,190} - \frac{14\,200}{782} + \frac{14\,200 \times 30.44}{13\,190} + \frac{815\,000 \times 12}{13\,190}$$

$$= 556 - 1004 - 18 + 33 + 741 = 308 \text{ psi C. } (21.7 \text{ kg/cm}^2, 2.12 \text{ N/mm}^2)$$

$$f_{bot} = 556 + 1004(13\,190/4802) - 18 - 33(2.747) - 741(2.747)$$
$$2.747$$

$$= 556 + 2758 - 18 - 91 - 2036 = 1169 \text{ psi C. } (82.2 \text{ kg/cm}^2, 8.06 \text{ N/mm}^2)$$

Concrete stresses after losses with superimposed loads—Eqs. (5-12), (5-3):

$$P_{ps} = P = P_i(1 - 0.219) = 463(0.781) = 362 \text{ kips } (164\,200 \text{ kg}, 1610 \text{ kN})$$

$$F_{ns} = (f_{ns})_u A_{ns} = (30.150)(1.22) = 36.8 \text{ kips } (16\,690 \text{ kg}, 164 \text{ kN})$$

$$f_{top} = P_{ps}/A_g - P_{ps}e_{ps}/Z_t - F_{ns}/A_g + F_{ns}e_{ns}/Z_t + (M_D + M_S + M_L)/Z_t$$
(Eq. (5-12))

$$f_{top} = \frac{362\,000}{782} - \frac{362\,000 \times 30.44}{13\,190} - \frac{36\,800}{782} + \frac{36\,800 \times 30.44}{13\,190} + \frac{1\,315\,000 \times 12}{13\,190}$$

$$= 463 - 835 - 47 + 85 + 1196 = 862 \text{ psi C. } (60.6 \text{ kg/cm}^2, 5.94 \text{ N/mm}^2)$$

$$f_{pot} = 463 + 835(2.747) - 47 - 85(2.747) - 1196(2.747)$$

$$= 463 + 2294 - 47 - 233 - 3285 = 808 \text{ psi T. } (56.8 \text{ kg/cm}^2, 5.57 \text{ N/mm}^2)$$

The computed tensile stress of 808 psi (versus 617 psi or 43.4 kg/cm^2, 4.25 N/mm^2 in previous problem) is less than the ACI Code (318-71) allowable of $12\sqrt{f'_c} = 849$ psi (59.7 kg/cm^2, 5.85 N/mm^2), but deflections would have to be checked, as shown later in this chapter. Such usage of nonprestressed tension steel may be beneficial in connection with ultimate strength and/or camber control considerations. The other computed stresses are well within the ACI Code allowable values (shown in Example 1).

The gain in prestress due to live load in Example 3 is:

$$f_{sL} = nf_{cL} = nM_L e/I$$
$$= (10.0)(250\,000)(12)(30.44)/168\,970 = 5400 \text{ psi}/1890 = 2.9\%$$

The tension in the nonprestressed steel due to live load is also 5400 psi.

Example 4 Using the approximate step-function method and the AASHTO method—ultimate midspan loss of prestress and concrete stresses, composite AASHTO Type IV girder with prestressing steel only, normal-weight concrete.

See Table 5-8 for the material and section properties, parameters, and conditions of the problem. The loss of prestress results are tabulated term by term in Table 5-11.

Approx. step-function method—Eq. (5-19), which is the ultimate form of Eq. (5-17)
Elastic loss—term (1), Eq. (5-19), as defined in Eq. (5-17):

Using the gross-section properties,

$$P_i = A_{ps} f_{si} = (4.90)(189) = 926 \text{ kips } (420\,000 \text{ kg, } 4119 \text{ kN})$$

$$\rho_{ps} = A_{ps}/A_g = 4.90/789 = 0.00621 \qquad n = 7.3 \qquad \text{(Table 5-1)}$$

$$\text{Trial } P_0 = P_i/(1 + 2n\rho_{ps}) = 926/(1 + 2 \times 7.3 \times 0.00621)$$
$$= 849 \text{ kips } (385\,100 \text{ kg, } 3776 \text{ kN})$$

$$f_{ci} = \frac{P_0}{A_g} + \frac{P_0 e_c^2}{I_g} - \frac{M_D e_c}{I_g} = \frac{849}{789} + \frac{(849)(21.26)^2}{260\,740} - \frac{(657.6)(12)(21.26)}{260\,740}$$
$$= 1.076 + 1.472 - 0.643$$
$$= 1.905 \text{ ksi } (133.9 \text{ kg/cm}^2\text{, } 13.13 \text{ N/mm}^2)$$

$$\text{Trial elastic loss} = (nf_{ci}) = (7.3)(1905) = 13\,910 \text{ psi}/1890 = 7.36\%$$

$$\text{Revised } P_0 = P_i - (nf_{ci})A_{ps} = 926 - (13.91)(4.90)$$
$$= 858 \text{ kips } (389\,200 \text{ kg, } 3816 \text{ kN})$$

$$\text{Revised } f_{ci} = 1.931 \text{ ksi } (135.8 \text{ kg/cm}^2\text{, } 13.31 \text{ N/mm}^2)$$

$$\text{Elastic loss} = (nf_{ci}) = (7.3)(1931) = 14\,100 \text{ psi}/1890 = 7.46\%$$

Creep and shrinkage loss—terms (2), (3), and (4), Eq. (5-19), as defined in Eq. (5-17); also see terms (7) and (8):

For 40 percent humidity,

$$C_u = 2.35 \qquad (\varepsilon_{sh})_u = 780 \times 10^{-6} \text{ in/in (cm/cm)} \qquad \text{(Eqs. (1-16), (1-18))}$$

For 60 percent humidity,

$$\text{Creep (C.F.)}_H = 0.87 \qquad \text{Shrinkage (C.F.)}_H = 0.80 \qquad \text{(Table 1-7)}$$

For v/s ratio of 4.8 in. (12.2 cm) (Fig. C-1, Chap. 1),

$$\text{Creep (C.F.)}_T = 0.74 \qquad \text{Shrinkage (C.F.)}_T = 0.71 \qquad \text{(Table 1-9)}$$

$$\text{Adjusted } C_u = (2.35)(0.87)(0.74) = 1.51$$

$$\text{Adjusted } (\varepsilon_{sh})_u = (780 \times 10^{-6})(0.80)(0.71) = 443 \times 10^{-6} \text{ in/in (cm/cm)}$$

$$\Delta P_u/P_0 = 0.18 \qquad \text{(Table 5-2)}$$

$$E_s = 28 \times 10^6 \text{ psi } (1.97 \times 10^6 \text{ kg/cm}^2\text{, } 193 \text{ kN/mm}^2) \qquad \text{(Chap. 1)}$$

For deck slab cast 2 months after prestressing—

$$\Delta P_s/P_0 = 0.14 \qquad \text{(Table 5-2)} \qquad \alpha_s = 0.54 \qquad \text{(Table 5-3)}$$

$$(1 + b_{11})_u = 1.25 \qquad \text{(design simplifications)} \qquad k_r = 1.00$$

Table 5-8 Material and section properties, parameters, and conditions for Example 4

Unshored composite girder

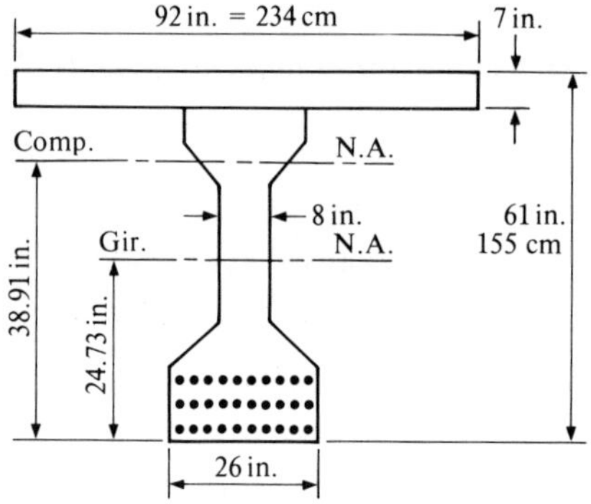

AASHTO IV

32 × $\frac{1}{2}$ in. strands, 2 pt. dep. at 3rd pts.

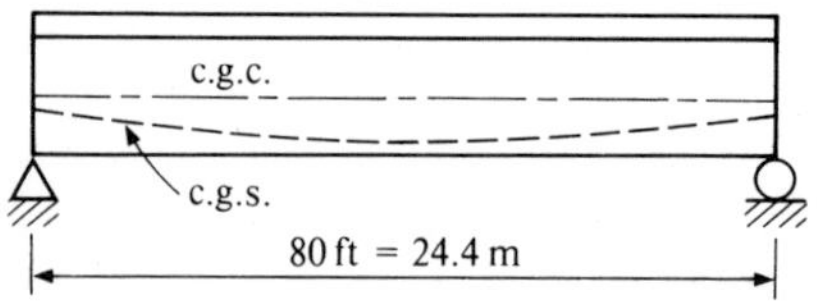

Simple span = 80 ft

Material properties

Steam-cured normal-weight concrete

$$f_{pu} = 270 \text{ ksi}$$

$$f'_{ci} = 4000 \text{ psi} \qquad f'_c = 5000 \text{ psi}$$

$$\text{Deck } f'_c = 4000 \text{ psi}$$

From Table 1-3,

$$E_{\text{Girder}}/E_{\text{Slab}} = E_2/E_1 = 3.89/3.64 = 1.07$$

Section prop. and loading, PCI Handbook

Girder:

$$\text{Spacing} = 7 \text{ ft } 8 \text{ in.} \qquad A_g = 789 \text{ in}^2$$

$$y_b = 24.73 \text{ in.} \qquad I_g = 260\,740 \text{ in}^4$$

$$Z_t = 8908 \text{ in}^3 \qquad Z_b = 10\,544 \text{ in}^3$$

$$e_e = 2.86 \text{ in.} \qquad e_c = 21.26 \text{ in.}$$

$$w_D = 822 \text{ \#/ft}$$

Section properties and loading, calculated—composite section

$$\text{Modified slab area} = 7 \times 92/1.07 = \underline{602 \text{ in}^2}$$

$$y_b = (789 \times 24.73 + 602 \times 57.50)/(789 + 602) = \underline{38.91 \text{ in.}}$$

$$I_g = 260\,740 + 789(38.91 - 24.73)^2 + 92 \times 7^3/(12 \times 1.07) + 602(57.50 - 38.91)^2 = \underline{629\,890 \text{ in}^4}$$

$$Z_t = 629\,890/(61.00 - 38.91) = \underline{28\,510 \text{ in}^3}$$

$$Z_b = 629\,890/38.91 = 16\,190 \text{ in}^3$$

$$\text{Slab D.L.} = w_S = (7.5 \times 92)(150/144) = \underline{719 \text{ \#/ft}} \quad \text{including } \tfrac{1}{2} \text{ in. w.s.}$$

HS 20-44 AASHTO loading, impact = 50/(80 + 125) = $\underline{0.25}$

Assume deck slab cast 2 months after prestressing

$$\text{Area of one } \tfrac{1}{2} \text{ in. strand} = \underline{0.153 \text{ in}^2} \quad \text{(Fig. 11.3.3, PCI Handbook)}$$

$$A_{ps} = (32)(0.153) = \underline{4.90 \text{ in}^2} \qquad f_{si} = (0.70)(270) = \underline{189 \text{ ksi}}$$

$$M_D = w_D L^2/8 = (822)(80)^2/8 = \underline{657\,600 \text{ ft-\#}}$$

$$M_{S,Di} = (719)(80)^2/8 + 50\,000 = \underline{625\,200 \text{ ft-\#}}$$

$$M_D + M_{S,Di} = \underline{1\,282\,800 \text{ ft-\#}}$$

$$\text{Interior girder } M_{L+I} = (1165\text{—AASHTO Table})(\tfrac{1}{2}\text{—single wheels})(1.25\text{—}L + I) \times (7.67/5.5\text{—AASHTO-S/5.5}) = \underline{1\,015\,400 \text{ ft-\#}}$$

Assume 60 percent ambient relative humidity

(For metric units, see metric conversion factors herein.)

$$\text{Creep loss up to slab casting} = (nf_{ci})(\alpha_s k_r C_u)\left(1 - \frac{\Delta P_s}{2P_0}\right)$$
$$= (14\,100)(0.54 \times 1.51)(1 - 0.070)$$
$$= 10\,690 \text{ psi}/1890 = 5.66\%$$

$$\text{Creep loss following slab casting} = (nf_{ci})(k_r C_u)\left[\left(1 - \frac{\Delta P_u}{2P_0}\right) - \alpha_s\left(1 - \frac{\Delta P_s}{2P_0}\right)\right]\frac{I_2}{I_c}$$
$$= (14\,000)(1.51)[(1 - 0.090) - 0.54(1 - 0.070)] \times \frac{260\,740}{629\,890}$$
$$= 21\,290[0.91 - 0.50](0.414)$$
$$= 3610 \text{ psi}/1890 = 1.91\%$$

$$\text{Shrinkage loss} = k_r(\varepsilon_{sh})_u E_s/(1 + b_{11})_u = (443)(28)/1.25$$
$$= 9920 \text{ psi}/1890 = 5.25\%$$

Relaxation loss—term (5), Eq. (5-19), as defined in Eq. (5-17):

$$\text{Relaxation loss} = (f_{sr})_u = 7.5 f_{si} = (0.075)(189\,000)$$
$$= 14\,180 \text{ psi}/1890 = 7.50\%$$

Elastic gain due to slab dead load—term (6), Eq. (5-19), as defined in Eq. (5-17):

$$m = 6.1 \qquad \text{(Table 5-1, for deck slab cast 2 months after prestressing)}$$
$$f_{cs1} = M_{S,Di}e/I = (625\,200)(12)(21.26)/260\,740$$
$$= 612 \text{ psi } (43.0 \text{ kg/cm}^2, 4.22 \text{ N/mm}^2)$$
$$\text{Elastic gain} = (mf_{cs1}) = (6.1)(612) = 3730 \text{ psi}/1890 = 1.97\%$$

Gain due to concrete creep under slab dead load—term (7), Eq. (5-19), as defined in Eq. (5-17):

$$\beta_s = 0.78 \qquad \text{(Table 5-3, for deck slab cast 2 months after prestressing)}$$

$$\text{Gain due to creep} = [(mf_{cs1})(\beta_s k_r C_u)/(1 + b_{11})_u]\frac{I_2}{I_c}$$
$$= [(3730)(0.78 \times 1.51)/1.25]\frac{260\,740}{629\,890}$$
$$= (3510)(0.414) = 1450 \text{ psi}/1890 = 0.77\%$$

Gain due to differential shrinkage and creep—term (8), Eq. (5-19), as defined in Eq. (5-17):

$$\text{Composite } I_g = 629\,890 \text{ in}^4 \ (26\,216\,000 \text{ cm}^4)$$
$$\text{Composite } e = 21.26 + (38.91 - 24.73) = 35.44 \text{ in. } (90.0 \text{ cm})$$
$$y_{cs} = 54 + 3.5 - 38.91$$
$$= 18.59 \text{ in. } (47.2) \qquad \text{(Fig. 2-1)}$$

$$D_u = 375 \times 10^{-6} \text{ in/in (cm/cm)} \qquad \text{(Table 2-1, for deck slab cast 2 months after prestressing and 70\% humidity)}$$

For 60 percent humidity,

$$D_u = (375 \times 10^{-6})(0.80/0.70) = 429 \times 10^{-6} \text{ in/in (cm/cm)} \qquad \text{(Table 1-7)}$$

$$E_1 = 3.64 \times 10^6 \text{ psi } (255\,900 \text{ kg/cm}^2, 25.1 \text{ kN/mm}^2) \qquad \text{(Table 1-3)}$$

$$Q_{DS} = (Q \text{ in Eq. (2-1)})/2 \qquad \text{(Eq. (2-21))}$$

$$= (DA_1E_1)/2 = (429)(7.5 \times 92)(3.64)/2$$

$$= 538\,800 \text{ lb } (244\,400 \text{ kg, } 2397 \text{ kN})$$

The wearing surface is included in this calculation (7.5 in., 19.1 cm) but is not included in the composite section properties as a matter of convenience for this rather secondary effect, in order to avoid two separate sets of calculations for the composite section properties.

$$f_{cds} = Q_{DS}\, y_{cs}\, e/I = (538\,800)(18.59)(35.44)/629\,890$$

$$= 564 \text{ psi } (39.7 \text{ kg/cm}^2, 3.89 \text{ N/mm}^2)$$

$$\text{Differential shrinkage and creep gain} = (f_{sds})_u = mf_{cds} = (6.1)(564)$$

$$= 3440 \text{ psi}/1890 = 1.82\%$$

$$\text{Total loss} = (\Delta f_s)_u$$

$$= 14\,100 + 10\,690 + 3610 + 9920 + 14\,180 - 3730 - 1450 - 3440$$

$$= 43\,880 \text{ psi } (3085 \text{ kg/cm}^2, 303 \text{ N/mm}^2) \qquad \text{(Eq. (5-19))}$$

$$= 7.46 + 5.66 + 1.91 + 5.25 + 7.50 - 1.97 - 0.77 - 1.82 = 23.2\%$$

Using the simplified Eq. (5-25*a*), which is based on 70 percent humidity and neglects any size correction factors for creep and shrinkage,

$$(\Delta f_s)_u = (14.19 + 5.60I_2/I_c)f_{ci} + 22\,680 - (6.1 + 7.16I_2/I_c)f_{cs1}$$

$$= (14.19 + 5.60 \times 0.414)(1931) + 22\,680 - (6.1 + 7.16 \times 0.414)(612)$$

$$= 31\,880 + 22\,680 - 5540 = 49\,020 \text{ psi}/1890 = 25.9\%$$

The simplified result is on the safe side compared to that of Eq. (5-19) which is 23.2 percent.

Using the simplified method of the PCI Loss Committee:[73]

$$\text{Total loss} = 33.0 + 13.8f_{ci} - 4.5f_{cs} = 33.0 + 13.8(1.931) - 4.5(0.612)$$

$$= 56.8 \text{ ksi}/1.89 = 30.1\%$$

From Appendix C, Chap. 1, v/s = about 4.8 in. (without composite slab). Hence, total loss = 30.1 − (say 8) = 22.1%, which is about 1 to 4 percent lower than the above solutions. The equation does not include the effect of composite action. Also, for these bridge girders the size effect appears to be quite large, as the 8 percent reduction in the computed loss of prestress.

Concrete stresses at transfer:

$$f_{top} = \frac{P_0}{A_g} - \frac{P_0 e_c}{Z_t} + \frac{M_D}{Z_t} = \frac{858\,000}{789} - \frac{858\,000 \times 21.26}{8908} + \frac{657\,600 \times 12}{8908}$$

$$= 1087 - 2048 + 886 = 75 \text{ psi T. } (5.3 \text{ kg/cm}^2, 0.52 \text{ N/mm}^2)$$

$$f_{bot} = 1087 + 2048(8908/10\,544) + 886(0.8448)$$

0.8448

$$= 1087 + 1730 - 748 = 2069 \text{ psi C. } (145.5 \text{ kg/cm}^2, 14.27 \text{ N/mm}^2)$$

Concrete stresses after losses with live load and differential shrinkage and creep effects:

At top of slab,

$$f_{top} = (M_{L+I})/\text{Composite } Z_t = (1\,015\,400)(12)/28\,510$$

$$= 427 \text{ psi C. } (30.0 \text{ kg/cm}^2, 2.94 \text{ N/mm}^2)$$

Converting to slab concrete,

$$f_{top} = 427/1.07 = 399 \text{ psi C. } (28.1 \text{ kg/cm}^2, 2.75 \text{ N/mm}^2)$$

Due to differential shrinkage and creep, Eqs. (2-2), (2-21),

$$f_{top} = -\frac{Q}{A_1} - \left(-\frac{Q}{A_c} - \frac{Q y_{cS} y_{cT}}{I_c}\right)\frac{E_1}{E_2}$$

$$= -\frac{538\,800}{7.5 \times 92} - \left(-\frac{538\,800}{789 + 602}\right.$$

1391

$$\left. - \frac{538\,800 \times 18.59 \times (61 - 38.91)}{629\,890}\right) \Big/ 1.07$$

$$= -781 + 738/1.07 = 91 \text{ psi T. } (6.4 \text{ kg/cm}^2, 0.63 \text{ N/mm}^2)$$

Combined $f_{top} = 399 - 91 = \underline{308 \text{ psi C. } (21.7 \text{ kg/cm}^2, 2.12 \text{ N/mm}^2)}$

At bottom of slab,

15.09

$$f_{1B} = (M_{L+I}) y_{cI}/I_c = (1\,015\,400)(12)(54 - 38.91)/629\,890$$

$$= 292 \text{ psi C. } (20.5 \text{ kg/cm}^2, 2.01 \text{ N/mm}^2)$$

Converting to slab concrete,

$$f_{1B} = 292/1.07 = 273 \text{ psi C. } (19.2 \text{ kg/cm}^2, 1.88 \text{ N/mm}^2)$$

Due to differential shrinkage and creep, Eqs. (2-2), (2-21),

$$f_{1B} = -\frac{Q}{A_1} - \left(-\frac{Q}{A_c} - \frac{Q y_{cS} y_{cI}}{I_c}\right)\frac{E_1}{E_2}$$

$$= -781 + 627/1.07 = 195 \text{ psi T. } (13.7 \text{ kg/cm}^2, 1.34 \text{ N/mm}^2)$$

Combined $f_{1B} = 273 - 195 = \underline{78 \text{ psi C. } (5.5 \text{ kg/cm}^2, 0.54 \text{ N/mm}^2)}$

At top of girder,

$$f_{2T} = \frac{P}{A_g} - \frac{Pe_c}{Z_t} + \frac{M_D + M_{S,Di}}{Z_t} + \frac{(M_{L+I})y_{cI}}{I_c}$$

where

$$P = P_i(1 - 0.232) = (926)(0.768) = 711 \text{ kips } (323\,000 \text{ kg, } 3163 \text{ kN})$$

$$f_{2T} = \frac{711\,000}{789} - -\frac{711\,000 \times 21.26}{8908} + \frac{1\,282\,800 \times 12}{8908} + \frac{1\,015\,400 \times 12 \times 15.09}{629\,890}$$

$$= 901 - 1697 + 1728 + 292 = 1224 \text{ psi C. } (86.1 \text{ kg/cm}^2, 8.44 \text{ N/mm}^2)$$

Due to differential shrinkage and creep, Eqs. (2-2), (2-21),

$$f_{2T} = Q/A_c + Qy_{cS}y_{cI}/I_c$$
$$= 538\,800/1391 + 538\,800 \times 18.59 \times 15.09/629\,890$$
$$= 387 + 240 = 627 \text{ psi C. } (44.1 \text{ kg/cm}^2, 4.32 \text{ N/mm}^2)$$

$$\text{Combined } f_{2T} = 1224 + 627 = \underline{1851 \text{ psi C. } (130.1 \text{ kg/cm}^2, 12.76 \text{ N/mm}^2)}$$

At bottom of girder,

$$f_{bot} = \frac{P}{A_g} + \frac{Pe_c}{Z_b} - \frac{M_D + M_{S,Di}}{Z_b} - \frac{M_{L+I}}{\text{Comp. } Z_b}$$

$$= 901 + \frac{711\,000 \times 21.26}{10\,544} - \frac{1\,282\,800 \times 12}{10\,544} - \frac{1\,015\,400 \times 12}{16\,190}$$

$$= 901 + 1434 - 1460 - 753 = 122 \text{ psi C. } (8.6 \text{ kg/cm}^2, 0.84 \text{ N/mm}^2)$$

Due to differential shrinkage and creep, Eqs. (2-2), (2-21),

$$f_{bot} = Q/A_c - Qy_{cS}y_{cB}/I_c = 387 - 538\,800 \times 18.59 \times 38.91/629\,890$$
$$= 387 - 619 = 232 \text{ psi T. } (16.3 \text{ kg/cm}^2, 1.60 \text{ N/mm}^2)$$

$$\text{Combined } f_{bot} = 232 - 122 = \underline{120 \text{ psi T. } (8.4 \text{ kg/cm}^2, 0.83 \text{ N/mm}^2)}$$

The computed stresses are all well within the AASHTO[14] allowable stresses.

AASHTO method—Eq. (5-32) Shrinkage loss—term (1), Eq. (5-32):

For 60 percent humidity, $SH = 10\,000 \text{ psi}/1890 = 5.29\%$

Elastic loss—term (2), Eq. (5-32):

$$f_{cr} = f_{ci} = 1931 \text{ psi } (135.8 \text{ kg/cm}^2, 13.31 \text{ N/mm}^2) \qquad \text{(previous problem)}$$

Use $\quad ES = 7f_{cr} = 7(1931) = 13\,500 \text{ psi}/1890 = 7.14\%$

Creep loss—term (3), Eq. (5-32):

Use $\quad f_{cd} = f_{ci} - M_S e/I = 1931 - 625\,200 \times 12 \times 21.26/260\,740$

$$= 1931 - 612 = 1319 \text{ psi } (92.7 \text{ kg/cm}^2, 9.09 \text{ N/mm}^2)$$

Use $\quad CR_c = 16f_{cd} = 16(1319) = 21\,100 \text{ psi}/1890 = 11.16\%$

Relaxation loss—term (4), Eq. (5-32):

$$CR_s = 20\,000 - 0.125(SH + ES + CR_c) \qquad \text{(Eq. (1-35))}$$
$$= 20\,000 - 0.125(10\,000 + 13\,500 + 21\,100)$$
$$= 20\,000 - 5570 = 14\,430 \text{ psi}/1890 = 7.63\%$$
$$\text{Total loss} = \Delta f_s = SH + ES + CR_c + CR_s \qquad \text{(Eq. (5-32))}$$
$$= 10\,000 + 13\,500 + 21\,100 + 14\,430$$
$$= 59\,030 \text{ psi } (4150 \text{ kg/cm}^2, 407 \text{ N/mm}^2)$$
$$= 5.29 + 7.14 + 11.16 + 7.63 = 31.2\%$$
$$P = P_i(1 - 0.312) = (926)(0.688) = 637 \text{ kips } (288\,900 \text{ kg}, 2833 \text{ kN})$$

From step-function solution:

$$f_{bot} = (901 + 1434)(637/711) - 1460 - 753$$
$$= 121 \text{ psi T. } (8.5 \text{ kg/cm}^2, 0.83 \text{ N/mm}^2)$$
$$\text{Combined } f_{bot} = 232 + 121 = \underline{353 \text{ psi T. } (24.8 \text{ kg/cm}^2, 2.43 \text{ N/mm}^2)}$$

The allowable tension at service load after losses by AASHTO[14] is:

(*a*) For members with bonded reinforcement

$$6\sqrt{f'_c} = 6\sqrt{5000} = 424 \text{ psi } (29.8 \text{ kg/cm}^2, 2.92 \text{ N/mm}^2)$$

(*b*) For severe corrosive exposure conditions, such as coastal areas

$$3\sqrt{f'_c} = 3\sqrt{5000} = 212 \text{ psi } (14.9 \text{ kg/cm}^2, 1.46 \text{ N/mm}^2)$$

The computed stress of 353 psi is less than (*a*) but greater than (*b*). The other computed stresses by this method are well within the AASHTO allowable stresses.

The gain in prestress due to live load in Example 4 is:

$$f_{sL} = nf_{cL} = nM_{L+I}e_{comp}/I_{comp}$$
$$= (7.2)(1\,015\,400)(12)(35.44)/629\,800$$
$$= 4940 \text{ psi}/1890 = 2.6\%$$

Comments on numerical examples and Tables 5-9, 5-10, and 5-11 The comments in this section refer to specific examples only, but the results and comparisons are thought to represent reasonably typical cases.

Example 1 Noncomposite double-T beam with prestressing steel only, normal-weight concrete, no topping.

The ultimate loss of prestress results by the four methods (approx. step-function—Eq. (5-15) with terms (5) and (6) deleted; relaxation—Eq. (5-27); AASHTO—Eq. (5-32); and PCI—Eq. (5-33)) are tabulated term by term for comparison purposes in Table 5-9. The total loss of prestress by the four methods varies from 24.0 percent by the relaxation method to 26.8 percent by the

AASHTO method, with the approx. step-function method and the PCI method in between. Also, for the normal-weight composite bridge girder in Example 4, Table 5-11, the total loss of prestress was 23.2 percent by the approx. step-function method and 31.2 percent by the AASHTO method. The AASHTO method was on the high side in both examples.

In Example 1 by the four methods, the following is shown in Table 5-9:

Elastic loss—3.7% by AASHTO to 4.5% by PCI with step-fn and relax. in between.
Creep loss—6.1% by relax. to 9.5% by AASHTO, with step-fn and PCI in between.
Shrink loss—5.3% by AASHTO to 7.6% by PCI, with step-fn and relax. in between.
Relax. loss—5.1% by PCI to 8.3% by AASHTO, with step-fn and relax. in between.

The computed midspan bottom fiber tensile stresses after losses with live load were 474, 414, 507, and 482 psi by the four methods, respectively; versus the allowable of $6\sqrt{f_c'} = 424$ psi (29.8 kg/cm^2, 2.92 N/mm^2) and $f_r = 7.5\sqrt{f_c'} =$ 531 psi (37.3 kg/cm^2, 3.66 N/mm^2) (ACI 318-71), or the author's recommended $f_r = 555$ psi (39.0 kg/cm^2, 3.83 N/mm^2) (Table 1-2). Although the Code allowable of 424 psi (29.8 kg/cm^2, 2.92 N/mm^2) is exceeded by three of the methods, this appears to be acceptable since the Code also has the upper limit allowable value computed tensile stress of $12\sqrt{f_c'} = 849$ psi (59.7 kg/cm^2, 5.85 N/mm^2). In this case the beam is uncracked. All other computed stresses are within the ACI

Table 5-9 Term-by-term loss of prestress for Examples 1 and 2

	Elastic loss		Creep loss		Shrinkage loss		Relaxation loss		Total loss	
Method	ksi	%	ksi	%	ksi	%	ksi	%	ksi	%
Example 1: ultimate midspan loss, noncomposite double-T, nor.-wt. conc.										
Approx. step-fn	8.19	4.33	14.01	7.41	12.23	6.47	14.18	7.50	48.61	25.7
Relaxation	8.19	4.33	11.49	6.08	11.41	6.04	14.18	7.50	45.27	24.0
AASHTO	7.00	3.70	18.00	9.52	10.00	5.29	15.63	8.27	50.63	26.8
PCI Handbook	8.42	4.46	16.88	8.93	14.42	7.63	9.66	5.11	49.38	26.1
* Example 2: ultimate midspan loss, noncomposite double-T, sand-lt.-wt.-conc.										
Approx. step-fn	13.15	6.96	22.13	11.71	12.23	6.47	14.18	7.50	61.69	32.6

(1 ksi = 70.31 kg/cm^2 = 6.895 N/mm^2.)

* Same as Example 1, except for different weight concrete.

Code allowable values. The differences in the loss of prestress and concrete stresses are due primarily to different assumptions for the material parameters in the four methods.

Example 2 Same as Example 1, except using sand-lightweight concrete (120 pcf, 1922 kg/m^3, 18.9 kN/m^3).

The ultimate loss of prestress results are tabulated term by term in Table 5-9. The total loss of prestress by the approx. step-function method was 25.7 percent using normal-weight concrete in Example 1 and 32.6 percent for the same problem using sand-lightweight concrete in Example 2. The difference is found in the elastic and creep terms because of different concrete moduli of elasticity and dead load.

The computed concrete stresses are all within the ACI Code allowable values, with the midspan bottom fiber tensile stress after losses with live load equalling 356 psi (25.0 kg/cm^2, 2.45 N/mm^2) versus 474 psi (33.3 kg/cm^2, 3.27 N/mm^2) by the same method for the normal-weight beam in Example 1. The reason for this is that the reduced dead load had a greater effect than the increased loss of prestress in Example 2. Since $f_r = 6.4\sqrt{f_c'} = 453$ psi (31.9 kg/cm^2, 3.12 N/mm^2) (ACI 318-71), or the author's recommended $f_r = 504$ psi (35.4 kg/cm^2, 3.48 N/mm^2) (Table 1-2), the beam in Example 2 is also uncracked.

Example 3 Noncomposite single-T beam first containing only prestressing steel and second containing both prestressed and nonprestressed steel, sand-lightweight concrete (120 pcf, 1922 kg/m^3, 18.9 kN/m^3), with additional sustained loading applied some time after the transfer of prestress.

The ultimate loss of prestress results for both cases (Eq. (5-15)), and the ultimate stress in the nonprestressed steel (Eq. (5-16)), are tabulated term by term in Table 5-10. The total loss of prestress (loss of stress in prestressing steel) is 24.5 percent in the beam with prestressing steel only and 21.9 percent in the same beam with both steels. The additional sustained loading assumed to be applied one month after the transfer of prestress had the effect of reducing the prestress loss by 5.1 percent in the beam with prestressing steel only and 4.6 percent in the same beam with both steels. In this case of the beam containing prestressing steel plus half as much nonprestressed tension steel, the loss of stress in the prestressing steel was 41.3 ksi (2906 kg/cm^2, 285 N/mm^2) and the compression in the nonprestressed steel was 30.2 ksi (2123 kg/cm^2, 208 N/mm^2),

The computed concrete stresses are all within the ACI Code allowable stresses, although the midspan bottom fiber tensile stresses after losses with all loading including live load are 617 psi (43.4 kg/cm^2, 4.25 N/mm^2) in the beam with prestressing steel only, and 808 psi (56.8 kg/cm^2, 5.57 N/mm^2) in the same beam with both steels; versus $f_r = 6.4\sqrt{f_c'} = 453$ psi (31.9 kg/cm^2, 3.12 N/mm^2) and the upper limit allowable $= 12\sqrt{f_c'} = 849$ psi (59.7 kg/cm^2, 5.85 N/mm^2) (ACI 318-71). Hence deflections must be checked in both cases using a cracked section analysis (Eqs. (5-60) to (5-73)).

Table 5-10 Term-by-term loss of prestress and nonprestressed steel stress for Example 3—ultimate midspan values, noncomposite single-T, sand-lightweight concrete, with additional sustained loading

Beam containing only prestressing steel

Loss of prestress	Elastic loss	Creep loss	Shrinkage loss	Relax. loss	Elastic gain, sus. load	Creep gain, sus. load	Total
ksi	11.63	18.74	11.49	14.18	−4.43	−5.29	46.32 ksi
%	6.15	9.92	6.08	7.50	−2.34	−2.80	24.5%

Beam containing both prestressing steel and nonprestressed steel

Loss of prestress	Elastic loss	Creep loss	Shrinkage loss	Relax. loss	Elastic gain, sus. load	Creep gain, sus. load	Total
ksi	11.63	14.99	9.19	14.18	−4.43	−4.23	41.33 ksi
%	6.15	7.93	4.86	7.50	−2.34	−2.24	21.9%

Nonprestressed steel compression	Elastic compression	Creep compression	Shrinkage compression	Elastic tension, sus. load	Creep tension, sus. load	Total
ksi	11.63	16.75	11.49	−4.43	−5.29	30.15

(1 ksi = 70.31 kg/cm^2 = 6.895 N/mm^2.)

Example 4 Unshored composite AASHTO Type IV girder with prestressing steel only, normal-weight concrete.

The ultimate loss of prestress results by the approx. step-function method (Eq. (5-19)) and by the AASHTO method (Eq. (5-32)) are tabulated term by term in Table 5-11. The total loss of prestress was 23.2 percent by the approx. step-function method and 31.2 percent by the AASHTO method, as noted in the discussion of Example 1. Also, the total loss by the PCI Loss Committee[73] simplified method was 22.1 percent.

The prestress loss percentages in Table 5-11 by the approx. step-function method show the elastic loss to be about 7.5 percent, the creep loss about 5.5 percent before slab casting and about 2 percent following slab casting, the total shrinkage loss about 5 percent, the relaxation loss about 7.5 percent, and the gain in prestress due to the elastic and creep effect of the slab dead load plus differential shrinkage and creep of about 4.5 percent; for a total loss of about 23 percent.

The computed concrete stresses are all within the AASHTO[14] allowable stresses, with the midspan bottom fiber tensile stresses after losses with live load plus impact equalling 120 psi (8.4 kg/cm^2, 0.83 N/mm^2) using prestress losses by the approx. step-function method and 353 psi (24.8 kg/cm^2, 2.43 N/mm^2) using

Table 5-11 Term-by-term loss of prestress for Example 4—ultimate midspan loss, composite AASHTO Type IV girder, normal-weight concrete

Method and units		Elastic loss	Creep loss before slab cast	Creep loss after slab cast	Shrinkage loss	Relaxation loss	Elastic gain due to slab	Creep gain due to slab	Gain due to diff. sh. and creep	Total loss
Approx. step-fn	ksi	14.10	10.69	3.61	9.92	14.18	−3.73	−1.45	−3.44	43.88 ksi
	%	7.46	5.66	1.91	5.25	7.50	−1.97	−0.77	−1.82	23.2%

Method and units		Shrinkage loss	Elastic loss	Creep loss	Relaxation loss	Total loss
AASHTO	ksi	10.00	13.50	21.10	14.43	59.03 ksi
	%	5.29	7.14	11.16	7.63	31.2%

(1 ksi = 70.31 kg/cm² = 6.895 N/mm².)

prestress losses by the AASHTO Method versus $6\sqrt{f_c'} = 424$ psi (29.8 kg/cm², 2.92 N/mm²).

The differential shrinkage and creep stresses in this example are roughly the same as those for the same girder and slab in Chap. 2, Table 2-4, but do vary somewhat because of different concrete moduli of elasticity, etc.

The gain in prestress due to live load was 1.5 to 2.9 percent for the four examples, versus the total loss, excluding the live load effect, of 21.1 to 32.6 percent. In Example 3 the stress in the nonprestressed tension steel under all dead and live loading was 30 150 − 5400 = 24 740 psi (1739 kg/cm², 171 N/mm²) compression (not tension under working load).

5-3 CAMBER AND DEFLECTION

Although various procedures have been outlined by Eqs. (5-4) to (5-35) for computing the loss of prestress and stress in nonprestressed steel, such documents as the ACI Code and Commentary,[20] the AASHTO Highway Bridge Specifications,[14] the PCI Handbook,[1] and the CEB Recommendations[21] do not prescribe methods for computing camber, except in the broadest of terms.

The formulation of the camber and deflection equations that follow is based on the step-function approach of the author and Ozell[2,3] for noncomposite and composite beams, and parallels in some respects the work of Corley, Sozen, and Siess,[22] and Pauw and Breen.[23] These papers were published in 1960–1961. The widely cited 1963 ACI Committee 435 paper[9] used this step-function relation (Eqs. (5-36) and (5-37) herein) and evaluated the effect of creep under variable stress both in an approximate single-step equation (Eq. (5-39) herein) and also by a summation or numerical integration technique involving the direct calculation of strains, curvatures, and deflections (as in Ref. 22).

More recent work on the approximate solutions includes that of the author, Shaikh, Meyers, Kripanarayanan, Chen, ACI Committee 209,[4-8,10,15] and Zundelevich, Lee, Chiu, and Hamada.[31,39,41,65,66] Similar approximate equations are outlined by Sinno and Furr[26,67] using a hyperbolic time-function, and by Mossiossian and Gamble[32] using the Trost[11,12] relaxation coefficient. More recent work on the numerical evaluation technique includes that of Sinno and Furr,[26,67] Mossiossian and Gamble,[32] and Rao and Dilger.[19] All of these references are discussed further in Sec. 5-5 (discussion of experimental results). Other contributions are listed at the end of the chapter. A more detailed historical review may be found in the 1972 University of Illinois Report.[32]

With regard to the determination of camber, the loss of prestress and change in stress in nonprestressed steel are effects that vary, not only with time, but also with location along the member. This is true for both straight and nonstraight steel profiles. While greater refinement would theoretically be possible by taking both of these effects into account, the usual practice of engineers appears to be to use the midspan steel forces in computing camber. A slightly closer approximation might be to use an average of the end and midspan values for straight tendons and one-point harping, but still the midspan value for two-point harping and draped tendons. In fact, usually only approximate values of loss of prestress, such as those given in Table 5-2, are used when losses are not calculated. Typical camber, deflection versus time curves are shown in Fig. 5-4 for both noncomposite and composite beams.

It may be noted that the camber and deflection equations could be greatly shortened by combining terms and substituting approximate parameters (such as those given in the previous section on design simplifications) for the secondary effects, but are presented in the form of separate terms in order to show the separate effects or contributions to the structural deformation, such as those due to the prestress force, dead load, creep, shrinkage, etc., and also that occur both before and after slab casting in the case of composite beams. The effects of nonprestressed tension steel and sustained loads other than a composite slab or topping applied some time after the transfer of prestress are also included.

Step-Function Relations and Approximate Step-Function Equation

For a noncomposite prestressed beam, the curvature expression corresponding to the initial plus time-dependent camber, deflection due to prestress and dead load was presented in the form of a step function by Eqs. (5-36) and (5-37).[2,3]

$$\phi = \frac{d^2y}{dx^2} = \frac{1}{E_{ci}I}[(1 + C_t)(M_x) - (1 + S_t)(P_0 e_x)] \tag{5-36}$$

where

$$S_t = -\sum_{n=0}^{t} \frac{k_n - k_{n-1}}{E_n/E_{ci}} + \sum_{n=0}^{t} (1 - k_n)(C_n - C_{n-1}) \tag{5-37}$$

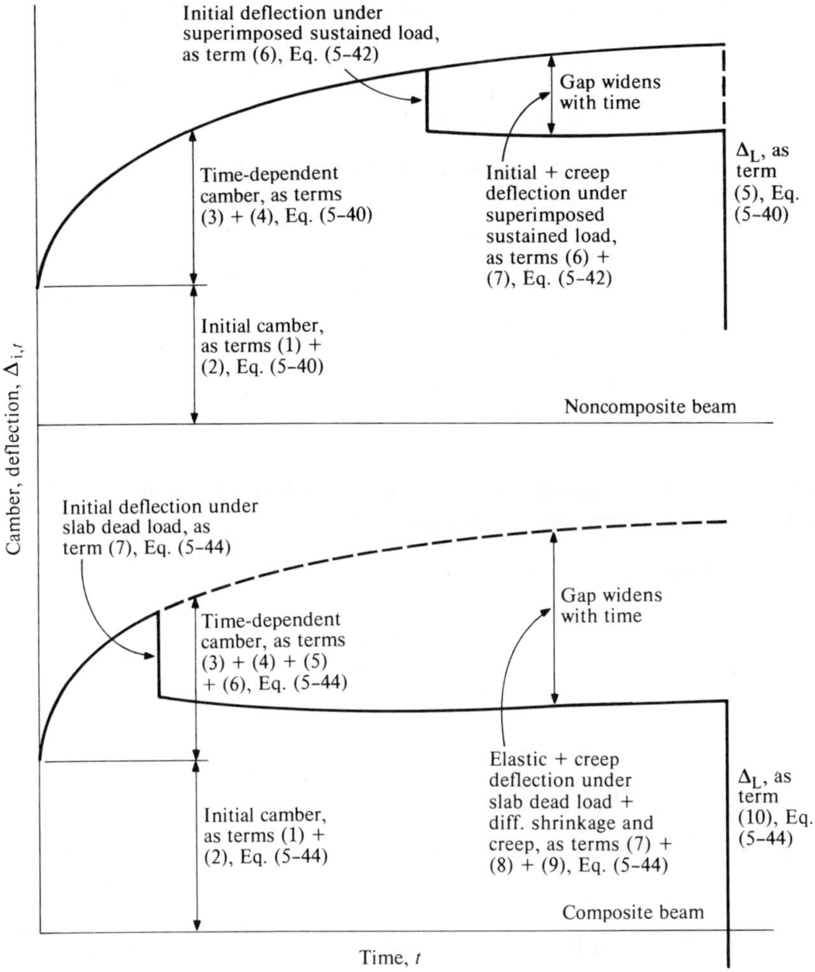

FIGURE 5-4
Typical camber, deflection versus time curves for noncomposite and composite beams.

In Eq. (5-36) the creep coefficient C_t modifies the constant sustained beam dead load curvature, $M_x/E_{ci}I$, and the summation coefficient S_t (a reduced creep coefficient to account for the loss of prestress and variable stress history) modifies the initial prestress moment curvature, $P_0e_x/E_{ci}I$.

The first term of Eq. (5-37) corresponds to the total "elastic" curvature (downward deflection) resulting from the loss of prestress, k_nP_0, for a given time $n = t$, in which $k_t = (P_0 - P_t)/P_0 = \Delta P_t/P_0$. If the variation in E_c with time is neglected so that $E_n = E_{ci}$, this term is simply equal to the prestress loss ratio, k_t. E_{ci}/E_n is less than unity, and a constant value for this term might be approximated as $0.9k_t$. The second term in Eq. (5-37) corresponds to the creep curvature (upward camber) resulting from the prestress force $(1 - k_n)P_0$ and the effective creep coefficient ΔC_t during a given increment of time. Assuming a parallel variation of k_t with C_t, this term can be approximated by $(1 - 0.5k_t)C_t$. Thus

$$S_t = -0.9k_t + (1 - 0.5k_t)C_t \tag{5-38}$$

Results by Eq. (5-38) agree with the experimental data of Refs. 2 and 3 within a few percent. As a practical matter, the factor 0.9 in the first term is dropped in subsequent equations herein and elsewhere, including the ACI Committee 435[9] equations. Thus the modifier for time-dependent camber due to prestress moment is given by the approximate step-function equation (5-39), which is used in the camber equations that follow.

$$S_t = -\frac{\Delta P_t}{P_0} + \left(1 - \frac{\Delta P_t}{2P_0}\right)C_t \tag{5-39}$$

Noncomposite Beams Containing Prestressed Steel Only or Both Prestressed Steel and Nonprestressed Tension Steel

Camber or deflection at any time including ultimate values is computed by Eq. (5-40).

$$\Delta_{i,t} = \underbrace{(\Delta_i)_{P_0}}_{(1)} - \underbrace{(\Delta_i)_D}_{(2)} + \underbrace{\left[-\frac{\Delta P_t}{P_0} + (k_r C_t)\left(1 - \frac{\Delta P_t}{2P_0}\right)\right](\Delta_i)_{P_0}}_{(3)} - \underbrace{(k_r C_t)(\Delta_i)_D}_{(4)} - \underbrace{\Delta_L}_{(5)} \tag{5-40}$$

where:

Term (1) is the initial camber due to the initial prestress moment after elastic loss, $P_0 e$. See Appendix K for common cases of prestress moment diagrams, with formulas for computing camber. For practical reasons, I_g is recommended instead of I_t for most purposes. For pretensioned members, P_0 can be estimated using $P_0 = P_i/(1 + 2n\rho_{ps})$, as determined in term (1) of Eq. (5-4). For post-tensioned members, P_0 is the prestress force at transfer. For continuous members, the effect of secondary moments due to prestressing, which normally results in a reduction in camber, should also be included.[25,27,56,58,59]

Term (2) is the initial dead-load deflection of the beam. For example, in its simplest form for ideal end conditions, $(\Delta_i)_D = KM_D L^2/E_{ci}I_g$. For practical reasons, I_g is recommended instead of I_t for most purposes. See the Notation for K and M formulas. For continuous beams uniformly loaded, the midspan deflection may be computed by Eq. (5-41).

$$\Delta_{\text{Midspan}} = \frac{5L^2}{48EI}[M_m + (M_a + M_b)/10] \tag{5-41}$$

where M_m is the midspan moment (normally positive) and M_a, M_b are the moments at the ends of the beam span (normally negative or zero).

Term (3) is the creep (time-dependent) camber of the beam due to the prestress moment. This expression includes the effects of creep and loss of prestress; that is, the creep effect under variable stress. ΔP_t refers to the total loss at any time minus the elastic loss. See Table 5-2 for approximate values of $\Delta P_t/P_0$ (in the form of $\Delta P_s/P_0$ and $\Delta P_u/P_0$) for this somewhat secondary effect at 3 weeks to 1 month, 2 to 3 months, and ultimate values. The effect of nonprestressed tension steel in reducing time-dependent camber (k_r) is given by Eqs. (5-57) to (5-59). When $A_{ns} = 0$, $k_r = 1$.

Term (4) is the dead-load creep deflection of the beam. Since creep under prestress and dead load takes place under their combined stress, the effect of nonprestressed tension steel, k_r, is included in this term as well.

Term (5) is the live-load deflection of the beam. See Fig. 5-7 and Eqs. (5-60) to (5-71) for the live-load deflection increment due to a live load into the cracking range (using I_e) or below the cracking load level (using I_g or I_t). $\Delta_L = KM_L L^2/E_c I$, or by Eq. (5-41).

The effect of compression steel in reducing time-dependent deflections will usually be negligible in prestressed beams, since the sustained load stress in any compression steel (top steel in positive bending moment regions) is usually quite small. When this is not the case, as in some composite beams, the effect can be estimated using the procedures for ordinary reinforced beams.

Effect of Sustained Loads Other Than a Composite Slab or Topping Applied Some Time After the Transfer of Prestress

When a sustained load other than a composite slab or topping (such as a girder supporting floor beams) is applied some time after the transfer of prestress, the camber or deflection is computed by Eq. (5-42).

$$\Delta_{i,t} = \overbrace{\Delta_{i,t} \text{ in Eq. (5-40)}}^{\text{Terms (1–5)}} - \overbrace{(\Delta_i)_S}^{(6)} - \overbrace{(k_r C_{ts})(\Delta_i)_S}^{(7)} \qquad (5\text{-}42)$$

where:

Term (6) is the initial deflection of the beam under the superimposed sustained load. For example, $(\Delta_i)_S = KM_S L^2/E_c I_g$ or by Eq. (5-41).

Term (7) is the creep deflection of the beam under the superimposed sustained load. C_{ts} is the creep coefficient, where the age of the beam concrete at the time of load application is taken into account. For ultimate values, term (7) becomes $(\beta_s k_r C_u)(\Delta_i)_S$. Since creep under prestress and all dead load takes place under their combined stress, the effect of nonprestressed tension steel, k_r, is included in this term as well. See Table 5-3 for representative values of β_s.

General Equation at Ultimate Time for Noncomposite Beams

$$\Delta_u = \underbrace{(\Delta_i)_{P_0}}_{(1)} - \underbrace{(\Delta_i)_D}_{(2)} + \underbrace{\left[-\frac{\Delta P_u}{P_0} + (k_r C_u)\left(1 - \frac{\Delta P_u}{2P_0}\right)\right](\Delta_i)_{P_0}}_{(3)}$$

$$- \underbrace{(k_r C_u)(\Delta_i)_D}_{(4)} - \underbrace{(\Delta_i)_S}_{(6)} - \underbrace{(\beta_s k_r C_u)(\Delta_i)_S}_{(7)} - \underbrace{\Delta_L}_{(5)} \qquad (5\text{-}43)$$

Equation (5-43) corresponds term by term to Eqs. (5-40) and (5-42).

Unshored and Shored Composite Beams Containing Both Prestressed Steel and Nonprestressed Tension Steel

Camber or deflection at any time including ultimate values is computed by Eq. (5-44) for unshored construction and Eq. (5-46) for shored construction. Subscripts 1 and 2 are used to refer to the slab (or effect of the slab such as under slab dead load) and precast beam, respectively.

Unshored construction

$$\Delta_{i,t} = \underbrace{(\Delta_i)_{P_0}}_{(1)} - \underbrace{(\Delta_i)_2}_{(2)} + \underbrace{\left[-\frac{\Delta P_s}{P_0} + (k_r C_{s2})\left(1 - \frac{\Delta P_s}{2P_0}\right)\right](\Delta_i)_{P_0}}_{(3)}$$

$$+ \underbrace{\left[-\frac{\Delta P_t - \Delta P_s}{P_0} + (k_r C_{t2})\left(1 - \frac{\Delta P_t}{2P_0}\right) - (k_r C_{s2})\left(1 - \frac{\Delta P_s}{2P_0}\right)\right](\Delta_i)_{P_0}\frac{I_2}{I_c}}_{(4)}$$

$$- \underbrace{(k_r C_{s2})(\Delta_i)_2}_{(5)} - \underbrace{k_r(C_{t2} - C_{s2})(\Delta_i)_2 I_2/I_c}_{(6)} - \underbrace{(\Delta_i)_1}_{(7)}$$

$$- \underbrace{(k_r C_{t1})(\Delta_i)_1 I_2/I_c}_{(8)} - \underbrace{\Delta_{DS}}_{(9)} - \underbrace{\Delta_L}_{(10)} \qquad (5\text{-}44)$$

It can be seen by inspection that Eq. (5-44) for composite beams reduces to Eq. (5-40) for noncomposite beams when $I_2/I_c = 1$, $C_{t2} = C_t$, and terms (7), (8), and (9)—effects of composite slab—are deleted.

Term (1) is the initial camber due to the initial prestress moment after elastic loss, $P_0 e$. See term (1) of Eq. (5-40) for comments.

Term (2) is the initial dead-load deflection of the precast beam. $(\Delta_i)_2 = KM_2L^2/E_{ci}I_g$ or by Eq. (5-41). See term (2) of Eq. (5-40) for comments.

Term (3) is the creep (time-dependent) camber of the beam, due to the prestress moment, up to the time of slab casting. See term (3) of Eq. (5-40) and terms (2) and (3) of Eq. (5-17) for comments.

Term (4) is the creep camber of the composite beam, due to the prestress moment, for any period following slab casting. See term (3) of Eq. (5-40) and terms (2) and (3) of Eq. (5-17) for comments.

Term (5) is the creep deflection of the precast beam up to the time of slab casting due to the precast beam dead load. See term (4) of Eq. (5-40) for comments.

Term (6) is the creep deflection of the composite beam for any period following slab casting due to the precast beam dead load. See term (3) of Eq. (5-17) for comments.

Term (7) is the initial deflection of the precast beam under slab dead load. $(\Delta_i)_1 = KM_1L^2/E_cI_g$ or by Eq. (5-41). See term (2) of Eq. (5-40) for comments. When diaphragms are used, add $(\Delta_i)_{1D}$ to $(\Delta_i)_1$. For two symmetrical diaphragms,

$$(\Delta_i)_{1D} = \frac{M_{1D}}{E_cI_g}\left(\frac{L^2}{8} - \frac{a^2}{6}\right) \tag{5-45}$$

where M_{1D} is the moment between diaphragms and a is $L/4$, $L/3$, etc., for diaphragms at the quarter points, third points, etc., respectively.

Term (8) is the creep deflection of the composite beam due to slab dead load. C_{t1} is the creep coefficient for the slab loading, where the age of the precast beam concrete at the time of slab casting is considered. See term (3) of Eq. (5-17) for comment concerning I_2/I_c.

Term (9) is the deflection due to differential shrinkage and creep. For simple spans, $\Delta_{DS} = \frac{4}{3}(Q_{DS}y_{cs}L^2)/8E_cI_c$, Eq. (2-22). See term (8) of Eq. (5-17) for comments and references, and Chap. 2, including continuous beams.

Term (10) is the live-load deflection of the composite beam, in which the gross-section flexural rigidity of the composite section, E_cI_c, is normally used. See term (5) of Eq. (5-40) for comments.

Shored construction

$$\Delta_{i,t} = \text{Eq. (5-44), with terms (7) and (8) modified as follows} \tag{5-46}$$

Term (7) is the initial deflection of the composite beam under slab dead load. $(\Delta_i)_1 = KM_1L^2/E_cI_c$, or by Eq. (5-41). See the Notation for K and M formulas.

Term (8) is the creep deflection of the composite beam under slab dead load $= (k_rC_{t1})(\Delta_i)_1$. The composite-section effect is already included in term (7).

It is assumed in Eqs. (5-44) and (5-46) that the precast beam supports the beam-plus-slab dead load in the case of unshored construction; and, in the case of shored construction, the precast beam supports its own dead load and the composite beam supports the slab dead load. The equations herein for the shored case assume more or less continuous shoring. For one-point shoring, for example, the initial slab dead load deflection should be calculated with the one-point shoring in place. It is suggested that the 28-day E_c for both slab and precast beam concretes, and the gross-section I (I_g) be used in computing the composite $I - I_c$.

For the case of composite beams with prestressed steel only (containing no principal nonprestressed steel), camber or deflection is computed by Eqs. (5-44) and (5-46), with $k_r = 1$.

A further refinement of Eq. (5-44) was demonstrated by Zundelevich[41] in which the increase in camber due to the initial and time-dependent *gain* in prestress under slab dead load is considered separately from the *loss* of prestress due to all other causes. This results in the additional terms for Eq. (5-44) as follows:

$$(\Delta_i)_{G_0} + \left[\frac{\Delta G_t}{G_0} + (k_r C_{t1})\left(1 + \frac{\Delta G_t}{2G_0}\right)\right](\Delta_i)_{G_0}\frac{I_2}{I_c} \tag{5-47}$$

where $(\Delta_i)_{G_0}$ is the initial camber due to the initial prestress gain under slab dead load, and $\Delta G_t/G_0$ is the time-dependent prestress gain following slab casting divided by the initial prestress gain due to slab dead load.

It is noted that the separate prestress loss (terms (1) to (5) in Eqs. (5-17) and (5-19)) and gain (terms (6) to (8) in Eqs. (5-17) and (5-19)) results are computed automatically in the prestress loss equations herein (see sample results in Table 5-11). However, this refinement will normally not be necessary, especially in ultimate camber calculations, since in many cases only representative or approximate values of the loss of prestress are needed for including the somewhat secondary effect in camber calculations. This is not to say that the effect of prestress loss on camber is a secondary effect; only that slight differences in loss percentages or separate loss and gain percentages represent a secondary effect on calculated camber results.

General Equations at Ultimate Time for Composite Beams

Unshored construction

$$\Delta_u = \overbrace{(\Delta_i)_{P_0}}^{(1)} - \overbrace{(\Delta_i)_2}^{(2)} + \overbrace{\left[-\frac{\Delta P_s}{P_0} + (\alpha_s k_r C_u)\left(1 - \frac{\Delta P_s}{2P_0}\right)\right](\Delta_i)_{P_0}}^{(3)}$$

$$+ \overbrace{\left\{-\frac{\Delta P_u - \Delta P_s}{P_0} + (k_r C_u)\left[\left(1 - \frac{\Delta P_u}{2P_0}\right) - \alpha_s\left(1 - \frac{\Delta P_s}{2P_0}\right)\right]\right\}(\Delta_i)_{P_0}\frac{I_2}{I_c}}^{(4)}$$

$$- \overbrace{(\alpha_s k_r C_u)(\Delta_i)_2}^{(5)} - \overbrace{(1 - \alpha_s)(k_r C_u)(\Delta_i)_2 I_2/I_c}^{(6)} - \overbrace{(\Delta_i)_1}^{(7)}$$

$$- \overbrace{(\beta_s k_r C_u)(\Delta_i)_1 I_2/I_c}^{(8)} - \overbrace{\Delta_{DS}}^{(9)} - \overbrace{\Delta_L}^{(10)} \tag{5-48}$$

Shored construction

Δ_u = Eq. (5-48), except that the composite moment of inertia is used in term (7) to compute $(\Delta_i)_1$, and the ratio I_2/I_c is deleted in term (8); see Eq. (5-46) for comments (5-49)

Equations (5-48) and (5-49) correspond term by term to Eqs. (5-44) and (5-46), respectively. The parameters, α_s and β_s, are defined and tabulated in Table 5-3.

Design Simplifications

The ultimate camber Eqs. (5-43), (5-48), and (5-49) reduce to Eqs. (5-50*a,b,c*) or (5-51*a,b,c*), (5-52*a,b,c*) or (5-53*a,b,c*), and (5-54*a,b,c*) or (5-55*a,b,c*), respectively, when the following representative parameters are used: from Chap. 1 and Table 1-10, for 70 percent humidity, $C_u = 1.88$ and $(\varepsilon_{sh})_u = 546 \times 10^{-6}$ in/in (cm/cm), (size correction factor is neglected here, which is on the safe side except for small members—see footnote in Table 1-10 and Tables 1-8 and 1-9). Assuming a 2-month period between prestressing and the application of an additional sustained load including composite slab, from Table 5-3, $\alpha_s = 0.54$ and $\beta_s = 0.78$; and from Table 5-2, for normal-weight concrete—$\Delta P_s/P_0 = 0.14$, $\Delta P_u/P_0 = 0.18$, for sand-lightweight concrete—$\Delta P_s/P_0 = 0.16$, $\Delta P_u/P_0 = 0.21$, and for all-lightweight concrete—$\Delta P_s/P_0 = 0.18$, $\Delta P_u/P_0 = 0.23$.

Ultimate camber for noncomposite beams, as by Eq. (5-43)

Normal-weight concrete

$$\Delta_u = \overbrace{(\Delta_i)_{P_0}}^{(1)} - \overbrace{(\Delta_i)_D}^{(2)} + \overbrace{(-0.18 + 1.71k_r)(\Delta_i)_{P_0}}^{(3)} - \overbrace{1.88k_r(\Delta_i)_D}^{(4)} - \overbrace{(\Delta_i)_S}^{(6)} - \overbrace{1.47k_r(\Delta_i)_S}^{(7)} - \overbrace{\Delta_L}^{(5)}$$

$$= \overbrace{(0.82 + 1.71k_r)(\Delta_i)_{P_0}}^{(1+3)} - \overbrace{(1 + 1.88k_r)(\Delta_i)_D}^{(2+4)} - \overbrace{(1 + 1.47k_r)(\Delta_i)_S}^{(6+7)} - \overbrace{\Delta_L}^{(5)} \qquad (5\text{-}50a)$$

When $k_r = 1$ (no nonprestressed steel), Eq. (5-50*a*) reduces to Eq. (5-51*a*).

$$\Delta_u = \overbrace{2.53(\Delta_i)_{P_0}}^{(1+3)} - \overbrace{2.88(\Delta_i)_D}^{(2+4)} - \overbrace{2.47(\Delta_i)_S}^{(6+7)} - \overbrace{\Delta_L}^{(5)} \qquad (5\text{-}51a)$$

Sand-lightweight concrete

$$\Delta_u = \text{same as Eq. (5-50}a\text{), except } (0.79 + 1.68k_r)(\Delta_i)_{P_0} \tag{5-50b}$$

When $k_r = 1$,

$$\Delta_u = \text{same as Eq. (5-51}a\text{), except } 2.47(\Delta_i)_{P_0} \tag{5-51b}$$

All-lightweight concrete

$$\Delta_u = \text{same as Eq. (5-50}a\text{), except } (0.77 + 1.66k_r)(\Delta_i)_{P_0} \tag{5-50c}$$

When $k_r = 1$,

$$\Delta_u = \text{same as Eq. (5-51}a\text{), except } 2.43(\Delta_i)_{P_0} \tag{5-51c}$$

See Eqs. (5-40) and (5-42) for term-by-term definitions of Eqs. (5-50*a,b,c*) and (5-51*a,b,c*). It can be seen in Eqs. (5-50*a,b,c*) and (5-51*a,b,c*) that the difference in the values used for the time-dependent prestress loss ratios (for different weight concrete) had a small effect on these camber equations. This tends to confirm the secondary nature of the effect of prestress loss (or differences thereof) on camber calculations, as previously discussed.

Ultimate camber for unshored composite beams, as by Eq. (5-48)

Normal-weight concrete

$$\begin{aligned}
\Delta_u &= \overbrace{(\Delta_i)_{P_0}}^{(1)} - \overbrace{(\Delta_i)_2}^{(2)} + \overbrace{(-0.14 + 0.94k_r)(\Delta_i)_{P_0}}^{(3)} \\
&\quad + \overbrace{(-0.04 + 0.77k_r)(\Delta_i)_{P_0} I_2/I_c}^{(4)} - \overbrace{1.02k_r(\Delta_i)_2}^{(5)} - \overbrace{0.86k_r(\Delta_i)_2 I_2/I_c}^{(6)} \\
&\quad - \overbrace{(\Delta_i)_1}^{(7)} - \overbrace{1.47k_r(\Delta_i)_1 I_2/I_c}^{(8)} - \overbrace{\Delta_{DS}}^{(9)} - \overbrace{\Delta_L}^{(10)} \\
&= \overbrace{(0.86 + 0.94k_r - 0.04 I_2/I_c + 0.77k_r I_2/I_c)(\Delta_i)_{P_0}}^{(1+3+4)} \\
&\quad - \overbrace{(1.00 + 1.02k_r + 0.86k_r I_2/I_c)(\Delta_i)_2}^{(2+5+6)} - \overbrace{(1.00 + 1.47k_r I_2/I_c)(\Delta_i)_1}^{(7+8)} \\
&\quad - \overbrace{\Delta_{DS}}^{(9)} - \overbrace{\Delta_L}^{(10)}
\end{aligned} \tag{5-52a}$$

When $k_r = 1$ (no nonprestressed steel), Eq. (5-52*a*) reduces to Eq. (5-53*a*).

$$\Delta_u = \overbrace{(1.80 + 0.73 I_2/I_c)(\Delta_i)_{P_0}}^{(1+3+4)} - \overbrace{(2.02 + 0.86 I_2/I_c)(\Delta_i)_2}^{(2+5+6)}$$

$$- \overbrace{(1.00 + 1.47 I_2/I_c)(\Delta_i)_1}^{(7+8)} - \overbrace{\Delta_{DS}}^{(9)} - \overbrace{\Delta_L}^{(10)} \qquad (5\text{-}53a)$$

See Eq. (5-44) for term-by-term definitions of Eqs. (5-52*a*) and (5-53*a*).

Sand-lightweight concrete

$$\Delta_u = \text{same as Eq. (5-52}a\text{), except}$$
$$(0.84 + 0.93k_r - 0.05 I_2/I_c + 0.75 k_r I_2/I_c)(\Delta_i)_{P_0} \qquad (5\text{-}52b)$$

When $k_r = 1$,

$$\Delta_u = \text{same as Eq. (5-53}a\text{), except } (1.77 + 0.70 I_2/I_c)(\Delta_i)_{P_0} \qquad (5\text{-}53b)$$

All-lightweight concrete

$$\Delta_u = \text{same as Eq. (5-52}a\text{), except}$$
$$(0.82 + 0.92k_r - 0.05 I_2/I_c + 0.74 k_r I_2/I_c)(\Delta_i)_{P_0} \qquad (5\text{-}52c)$$

When $k_r = 1$,

$$\Delta_u = \text{same as Eq. (5-53}a\text{), except } (1.74 + 0.69 I_2/I_c)(\Delta_i)_{P_0} \qquad (5\text{-}53c)$$

Again, it is seen by inspection that Eqs. (5-52*a,b,c*) and (5-53*a,b,c*) for composite beams reduce, respectively, to Eqs. (5-50*a,b,c*) and (5-51*a,b,c*) for noncomposite beams when $I_2/I_c = 1$, and the differential shrinkage and creep term is deleted in the composite beam equations.

Ultimate camber for shored composite beams, different weight concrete, as by Eq. (5-49)

$$\Delta_u = \text{same as Eqs. (5-52}a,b,c\text{), except as described in Eq. (5-49)} \qquad (5\text{-}54a,b,c)$$

When $k_r = 1$,

$$\Delta_u = \text{same as Eqs. (5-53}a,b,c\text{), except as described in Eq. (5-49)} \qquad (5\text{-}55a,b,c)$$

Grossly approximate method for composite beams Equation (5-56) is sometimes used[61] as a grossly approximate method for estimating camber of composite beams.

$$\Delta_u = \Delta_i + k_r C_u \Delta_i (I_2/I_c) - \Delta_L = [1 + \text{say } 2k_r(I_2/I_c)]\Delta_i - \Delta_L \qquad (5\text{-}56)$$

where $\Delta_i = (\Delta_i)_{P_0} - (\Delta_i)_2 - (\Delta_i)_1$, and $C_u = 1.88$ as above, or say 2. k_r is not included in Ref. 61.

Numerical Examples of Camber, Deflection

Example 1 Ultimate midspan camber, deflection; noncomposite double-T beam with prestressing steel only, normal-weight concrete, no topping—Eq. (5-43) with terms (6) and (7) deleted, which is the ultimate form of Eq. (5-40); also compared with results by the simplified Eq. (5-51*a*).

See Table 5-6 for the material and section properties, parameters, and conditions of the problem. The camber and deflection results are tabulated term by term in Table 5-12.

Initial camber due to prestress—term (1), Eq. (5-43), as defined in Eq. (5-40):

From Loss Example 1,

$$\text{Trial } P_0 = P_i/(1 + 2n\rho_{ps}) = 384 \text{ kips } (165\,000 \text{ kg, } 1619 \text{ kN})$$

$$\text{Revised } P_0 = 387 \text{ kips } (175\,500 \text{ kg, } 1721 \text{ kN})$$

$$E_{ci} = 3.64 \times 10^3 \text{ ksi } (256\,000 \text{ kg/cm}^2\text{, } 25.1 \text{ kN/mm}^2) \qquad \text{(Table 1-3)}$$

From case 3, Appendix K, for one-point harping,

$$(\Delta_i)_{P_0} = \frac{P_0(e_c - e_e)L^2}{12E_{ci}I_g} + \frac{P_0 e_e L^2}{8E_{ci}I_g}$$

$$= \frac{387(18.48 - 11.12)(76)^2(144)}{12(3.64 \times 10^3)(59\,720)} + \frac{387(11.12)(76)^2(144)}{8(3.64 \times 10^3)(59\,720)}$$

$$= 0.91 + 2.06 = 2.97 \text{ in. } (7.54 \text{ cm})$$

Initial dead-load deflection—term (2), Eq. (5-43), as defined in Eq. (5-40):

$$(\Delta_i)_D = \frac{KM_D L^2}{E_{ci}I_g} = \frac{(5/48)(462.8)(12)(76)^2(144)}{(3.64 \times 10^3)(59\,720)} = -2.21 \text{ in. } (5.61 \text{ cm})$$

Time-dependent camber due to prestress—term (3), Eq. (5-43), as defined in Eq. (5-40):

For 70 percent humidity, $C_u = 1.88$ (Table 1-10); approximate web thickness is about 6 in. (15.2 cm); hence the size correction factor for creep may be taken as unity (Table 1-8).

Approx. $\Delta P_u/P_0 = 0.18$ (Table 5-2) but use from Loss Example 1, calc.

$$\Delta P_u/P_0 = (48.61 - 8.19)(2.14)/387 = 0.22 \qquad k_r = 1 \text{ (no nonprestressed steel)}$$

$$\left[-\frac{\Delta P_u}{P_0} + (k_r C_u)\left(1 - \frac{\Delta P_u}{2P_0}\right)\right](\Delta_i)_{P_0} = [-0.22 + 1.88(1 - 0.11)]2.97$$

$$= 4.31 \text{ in. } (10.95 \text{ cm})$$

Dead-load creep deflection—term (4), Eq. (5-43), as defined in Eq. (5-40):

$$(k_r C_u)(\Delta_i)_D = (1.88)(2.21) = -4.15 \text{ in. } (10.54 \text{ cm})$$

Live-load deflection—term (5), Eq. (5-43), as defined in Eq. (5-40):

$$E_c = 3.89 \times 10^3 \text{ ksi } (273\,500 \text{ kg/cm}^2\text{, } 26.8 \text{ kN/mm}^2) \qquad \text{(Table 1-3)}$$

Beam is uncracked (Loss Example 1)

$$\Delta_L = \frac{KM_L L^2}{E_c I_g} = \frac{(5/48)(216.7)(12)(76)^2(144)}{(3.89 \times 10^3)(59\,720)}$$
$$= -0.97 \text{ in. } (2.46 \text{ cm})$$

$$\text{Initial camber} = 2.97 - 2.21 = 0.76 \text{ in. } (1.93 \text{ cm})$$
$$\text{Residual camber} = 2.97 - 2.21 + 4.31 - 4.15$$
$$= 0.92 \text{ in. } (2.34 \text{ cm})$$
$$\text{Residual camber + live-load deflection} = 0.92 - 0.97 = -0.05 \text{ in. } (0.13 \text{ cm})$$

ACI Code (318-71) checks:

$$\text{Time-dependent camber} + \Delta_L = 4.31 - 4.15 - 0.97 = -0.81 \text{ in. } (2.06 \text{ cm})$$
$$\Delta_L = -0.97 \text{ in. } (2.46 \text{ cm})$$

$L/180 = (76)(12)/180$	$= 5.07$ in. (12.88 cm) > 0.97 in. (2.46 cm)	OK
$L/360$	$= 2.53$ in. (6.43 cm) > 0.97 in. (2.46 cm)	OK
$L/480$	$= 1.90$ in. (4.83 cm) > 0.81 in. (2.06 cm)	OK
$L/240$	$= 3.80$ in. (9.65 cm) > 0.81 in. (2.06 cm)	OK

Hence the beam has no deflection limitations according to the ACI Code.

Using the simplified Eq. (5-51*a*),

$$\text{Residual camber} = 2.53(\Delta_i)_{P_0} - 2.88(\Delta_i)_D$$
$$= 2.53(2.97) - 2.88(2.21)$$
$$= 1.15 \text{ in. } (2.92 \text{ cm})$$
versus 0.92 in. (2.34 cm) by Eq. (5-43)

$$\text{Residual camber + live-load deflection} = 1.15 \text{ in.} - 0.97 \text{ in.} = 0.18 \text{ in. } (0.46 \text{ cm})$$
versus -0.05 in. (0.13 cm) by Eq. (5-43)

The difference of 0.23 in. (0.58 cm) in both comparisons is found in term (3) when using $\Delta P_u/P_0 = 0.18$ versus 0.22, respectively.

Example 2 Same as Example 1 except using sand-lightweight concrete (120 pcf, or 1920 kg/m^3, or 18.9 kN/m^3).

All parameters and conditions in Table 5-6 apply to this problem, except $w_D = 530$ lb/ft (789 kg/m, 7730 N/m) and $M_D = (530)(76)^2/8 = 382\,700$ ft-lb (52 900 kg-m, 518 900 N-m). The camber and deflection results are tabulated term by term in Table 5-12.

Initial camber due to prestress—term (1), Eq. (5-43), as defined in Eq. (5-40):

From Loss Example 2,

$$\text{Trial } P_0 = P_i/(1 + 2n\rho_{ps}) = 378 \text{ kips } (171\,500 \text{ kg}, 1680 \text{ N})$$
$$\text{Revised } P_0 = 376 \text{ kips } (170\,600 \text{ kg}, 1670 \text{ N})$$
$$E_{ci} = 2.63 \times 10^3 \text{ ksi } (185\,000 \text{ kg/cm}^2, 18.1 \text{ kN/mm}^2) \qquad \text{(Table 1-3)}$$

From Appendix K,

$$(\Delta_i)_{P_0} = \frac{P_0(e_c - e_e)L^2}{12E_{ci}I_g} + \frac{P_0 e_e L^2}{8E_{ci}I_g}$$

$$= \frac{376(18.48 - 11.12)(76)^2(144)}{12(2.63 \times 10^3)(59\,720)} + \frac{376(11.12)(76)^2(144)}{8(2.63 \times 10^3)(59\,720)}$$

$$= 1.22 + 2.77 = 3.99 \text{ in. } (10.13 \text{ cm})$$

Initial dead-load deflection—term (2), Eq. (5-43), as defined in Eq. (5-40):

$$(\Delta_i)_D = \frac{KM_D L^2}{E_{ci}I_g} = \frac{(5/48)(382.7)(12)(76)^2(144)}{(2.63 \times 10^3)(59\,720)} = -2.53 \text{ in. } (6.43 \text{ cm})$$

Time-dependent camber due to prestress—term (3), Eq. (5-43), as defined in Eq. (5-40):

$$C_u = 1.88 \qquad \text{(from Example 1)}$$

Approx. $\Delta P_u/P_0 = 0.21$ (Table 5-2) but use from Loss Example 2, calc.

$$\Delta P_u/P_0 = (61.69 - 13.15)(2.14)/376 = 0.27 \qquad k_r = 1$$

$$\left[-\frac{\Delta P_u}{P_0} + (k_r C_u)\left(1 - \frac{\Delta P_u}{2P_0}\right)\right](\Delta_i)_{P_0} = [-0.27 + 1.88(1 - 0.135)]3.99$$

$$= 5.43 \text{ in. } (13.79 \text{ cm})$$

Dead-load creep deflection—term (4), Eq. (5-43), as defined in Eq. (5-40):

$$(k_r C_u)(\Delta_i)_D = (1.88)(2.53) = -4.76 \text{ in. } (12.09 \text{ cm})$$

Live-load deflection—term (5), Eq. (5-43), as defined in Eq. (5-40):

$$E_c = 2.81 \times 10^3 \text{ ksi } (197\,600 \text{ kg/cm}^2, 19.4 \text{ kN/mm}^2) \qquad \text{(Table 1-3)}$$

Beam is uncracked (Loss Example 2)

$$\Delta_L = \frac{KM_L L^2}{E_c I_g} = \frac{(5/48)(216.7)(12)(76)^2(144)}{(2.81 \times 10^3)(59\,720)} = -1.34 \text{ in. } (3.40 \text{ cm})$$

Initial camber = 3.99 − 2.53 = 1.46 in. (3.71 cm)

Residual camber = 3.99 − 2.53 + 5.43 − 4.76
= 2.13 in. (5.41 cm)

Residual camber + live-load deflection = 2.13 − 1.34 = 0.79 in. (2.01 cm)

ACI Code (318-71) checks:

Time-dependent camber + Δ_L = 5.43 − 4.76 − 1.34 = −0.67 in. (1.70 cm)

$\Delta_L = -1.34$ in. (3.40 cm)

$L/180 = (76)(12)/180 = 5.07$ in. (12.88 cm) > 1.34 in. (3.40 cm) OK

$L/360$ = 2.53 in. (6.43 cm) > 1.34 in. (3.40 cm) OK

$$L/480 = 1.90 \text{ in. } (4.83 \text{ cm}) > 0.67 \text{ in. } (1.70 \text{ cm}) \qquad \text{OK}$$
$$L/240 = 3.80 \text{ in. } (9.65 \text{ cm}) > 0.67 \text{ in. } (1.70 \text{ cm}) \qquad \text{OK}$$

Hence the beam has no deflection limitations according to the ACI Code.

Using the simplified Eq. (5-51*b*),

$$\begin{aligned} \text{Residual camber} &= 2.47(\Delta_i)_{P_0} - 2.88(\Delta_i)_D \\ &= 2.47(3.99) - 2.88(2.53) \\ &= 2.57 \text{ in. } (6.53 \text{ cm}) \\ &\quad \text{versus } 2.13 \text{ in. } (5.41 \text{ cm}) \text{ by Eq. (5-43)} \\ \text{Residual camber + live-load deflection} &= 2.57 - 1.34 = 1.23 \text{ in.} \\ &\quad \text{versus } 0.79 \text{ in. } (2.01 \text{ cm}) \text{ by Eq. (5-43)} \end{aligned}$$

The difference of 0.44 in. (1.12 cm) in both comparisons is found in term (3) when using $\Delta P_u/P_0 = 0.21$ versus 0.27, respectively.

Example 3 Ultimate midspan camber, deflection; noncomposite single-T beam, first containing only prestressing steel and second containing both prestressed and nonprestressed steel, sand-lightweight concrete (120 pcf, or 1920 kg/m^3, or 18.9 kN/m^3), with additional sustained loading applied some time after the transfer of prestress—Eq. (5-43), which is the ultimate form of Eq. (5-40); also compared with results by the simplified Eqs. (5-50*b*) and (5-51*b*).

See Table 5-7 for the material and section properties, parameters, and conditions of the problem. The camber and deflection results are tabulated term by term for both cases in Table 5-13.

Beam containing only prestressing steel Initial camber due to prestress—term (1), Eq. (5-43), as defined in Eq. (5-40):

From Loss Example 3,

$$\begin{aligned} \text{Trial } P_0 &= P_i/(1 + 2n\rho_{ps}) = 436 \text{ kips } (197\,800 \text{ kg, } 1939 \text{ kN}) \\ \text{Revised } P_0 &= 435 \text{ kips } (197\,300 \text{ kg, } 1935 \text{ kN}) \\ E_{ci} &= 2.63 \times 10^3 \text{ ksi } (185\,000 \text{ kg/cm}^2, 18.1 \text{ kN/mm}^2) \qquad \text{(Table 1-3)} \end{aligned}$$

From Appendix K,

$$\begin{aligned} (\Delta_i)_{P_0} &= \frac{P_0(e_c - e_e)L^2}{12E_{ci}I_g} + \frac{P_0 e_e L^2}{8E_{ci}I_g} \\ &= \frac{435(30.44 - 9.86)(100)^2(144)}{12(2.63 \times 10^3)(168\,970)} + \frac{435(9.86)(100)^2(144)}{8(2.63 \times 10^3)(168\,970)} \\ &= 2.42 + 1.74 = 4.16 \text{ in. } (10.57 \text{ cm}) \end{aligned}$$

Initial dead-load deflection—term (2), Eq. (5-43), as defined in Eq. (5-40):

$$(\Delta_i)_D = \frac{KM_D L^2}{E_{ci}I_g} = \frac{(5/48)(815.0)(12)(100)^2(144)}{(2.63 \times 10^3)(168\,970)} = -3.30 \text{ in. } (8.38 \text{ cm})$$

Time-dependent camber due to prestress—term (3), Eq. (5-43), as defined in Eq. (5-40):

For 70 percent humidity, $C_u = 1.88$ (Table 1-10)

For 8 in. (20.3 cm) web,

$$\text{Creep } (\text{C.F.})_T = 0.96 \qquad \text{(Table 1-8)}$$
$$\text{Adjusted } C_u = (1.88)(0.96) = 1.80$$

Approx. $\Delta P_u/P_0 = 0.21$ (Table 5-2) but use from Loss Example 3, calc.

$$\Delta P_u/P_0 = (46.32 - 11.63)(2.45)/435 = 0.20 \qquad k_r = 1$$

$$\left[-\frac{\Delta P_u}{P_0} + (k_r C_u)\left(1 - \frac{\Delta P_u}{2P_0}\right)\right](\Delta_i)_{P_0} = [-0.20 + 1.80(1 - 0.10)]4.16$$
$$= 5.91 \text{ in. } (15.01 \text{ cm})$$

Dead-load creep deflection—term (4), Eq. (5-43), as defined in Eq. (5-40):

$$(k_r C_u)(\Delta_i)_D = (1.80)(3.30) = -5.94 \text{ in. } (15.09 \text{ cm})$$

Initial deflection under superimposed sustained load—term (6), Eq. (5-43), as defined in Eq. (5-42):

$$E_c = 2.81 \times 10^3 \text{ ksi } (197\,600 \text{ kg/cm}^2, 19.4 \text{ kN/mm}^2) \qquad \text{(Table 1-3)}$$
$$(\Delta_i)_S = \frac{K M_S L^2}{E_c I_g} = \frac{(5/48)(250.0)(12)(100)^2(144)}{(2.81 \times 10^3)(168\,970)} = -0.95 \text{ in. } (2.41 \text{ cm})$$

Creep deflection under superimposed sustained load—term (7), Eq. (5-43), as defined in Eq. (5-42):

$$\beta_s = 0.83 \text{ for load application at age 1 month (Table 5-3)}$$
$$(\beta_s k_r C_u)(\Delta_i)_S = (0.83 \times 1.80)(0.95) = -1.42 \text{ in. } (3.61 \text{ cm})$$

Live-load deflection—term (5), Eq. (5-43), as defined in Eq. (5-40):

Beam is cracked (Loss Example 3), hence use Eqs. (5-60), (5-61), and (5-63) or (5-71).

$$f_r = 504 \text{ psi } (35.4 \text{ kg/cm}^2, 3.48 \text{ N/mm}^2) \qquad \text{(Table 1-2)}$$
$$(M_L)_{cr} = \frac{P I_g}{A_g y_t} + Pe - M_{D+S} + \frac{f_r I_g}{y_t}$$
$$(y_t \text{ referring to the tension fiber, Eq. (5-63)})$$
$$= \frac{(350)(168\,970)}{(782)(35.19)12} + \frac{(350)(30.44)}{12} - 1065 + \frac{(0.504)(168\,970)}{(35.19)(12)}$$
$$= 204 \text{ ft-k } (28\,210 \text{ kg-m}, 277 \text{ kN-m})$$

and

$$(M_L)_{cr}/(M_L)_{max} = 204/250 = 0.816$$

Alternatively,

$$(M_L)_{cr}/(M_L)_{max} = (f_{pe} - f_D + f_r)/(f_{pe} - f_D + f_{to}) \qquad \text{(Eq. (5-71))}$$

From the stress calculations,

$$f_{pe} = 2668 \text{ psi } (188 \text{ kg/cm}^2, 18.4 \text{ N/mm}^2)$$
$$f_D = 2661 \text{ psi } (187 \text{ kg/cm}^2, 18.3 \text{ N/mm}^2)$$
$$f_{to} = 617 \text{ psi } (43 \text{ kg/cm}^2, 4.3 \text{ N/mm}^2)$$
$$(M_L)_{cr}/(M_L)_{max} = (2668 - 2661 + 504)/(2668 - 2661 + 617) = 0.819$$

which checks the previous value. Use $(M_{cr}/M_{max})^3 = 0.82^3 = 0.55$.

$$n = 28/2.81 = 9.96 \qquad d = 48 - 4.75 = 43.25 \text{ in. } (109.0 \text{ cm})$$
$$A_{ps} = 2.45 \text{ in}^2 \ (15.8 \text{ cm}^2)$$

Considering only rectangular flange in compression, 10 ft by 1.5 in. (3.05 m by 3.81 cm),

$$(120)(1.5)(kd - 0.75) = (9.96)(2.45)(43.25 - kd)$$
$$kd = 5.82 \text{ in. } (14.78 \text{ cm})$$
$$I_{cr} = (120)(1.5)^3/12 + (120)(1.5)(5.82 - 0.75)^2 + (9.96)(2.45)(43.25 - 5.82)^2$$
$$= 38\,860 \text{ in}^4 \ (1.617 \times 10^6 \text{ cm}^4) \qquad \text{Use}$$

Alternatively, considering all compression concrete, by trial $kd = 4.15$ in. (10.54 cm), which falls within the tapered part of the flange, and $I_{cr} = 40\,040$ in^4 (1.666×10^6 cm^4). This solution is only slightly higher than the much simpler solution above. Also

$$\text{Approx.}^{63} \ I_{cr} = nA_s d^2(1 - \sqrt{\rho_{ps}}) = 43\,100 \text{ in}^4 \ (1.794 \times 10^6 \text{ cm}^4)$$
$$I_e = (M_{cr}/M_{max})^3 I_g + [1 - (M_{cr}/M_{max})^3] I_{cr} \qquad \text{(Eq. (5-61))}$$
$$= (0.55)(168\,970) + (1 - 0.55)(38\,860)$$
$$= 110\,420 \text{ in}^4 \ (4.596 \times 10^6 \text{ cm}^4)$$
$$\Delta_L = \frac{KM_L L^2}{E_c I_e} = \frac{(5/48)(250.0)(12)(100)^2(144)}{(2.81 \times 10^3)(110\,420)}$$
$$= -1.45 \text{ in. } (3.68 \text{ cm}) \qquad \text{(Eq. (5-60))}$$
$$\text{Initial camber} = 4.16 - 3.30 = 0.86 \text{ in. } (2.18 \text{ cm})$$
$$\text{Residual camber} = 4.16 - 3.30 + 5.91 - 5.94 - 0.95 - 1.42$$
$$= -1.54 \text{ in. } (3.91 \text{ cm})$$

Residual camber + live-load deflection = $-1.54 - 1.45 = -2.99$ in. (7.59 cm)

which is $(L/1200)(2.99) = L/400$, and could be objectionable.

ACI Code (318-71) checks:

$$\text{Time-dependent camber} + \Delta_L = 5.91 - 5.94 - 0.95 - 1.42 - 1.45$$
$$= -3.85 \text{ in. } (9.78 \text{ cm})$$
$$\Delta_L = -1.45 \text{ in. } (3.68 \text{ cm})$$

$L/180 = (100)(12)/180$	$= 6.67$ in. (16.94 cm) $>$ 1.45 in. (3.68 cm)	OK
$L/360$	$= 3.33$ in. (8.46 cm) $>$ 1.45 in. (3.68 cm)	OK
$L/480$	$= 2.50$ in. (6.35 cm) $<$ 3.85 in. (9.78 cm)	NG
$L/240$	$= 5.00$ in. (12.70 cm) $>$ 3.85 in. (9.78 cm)	OK

Hence the beam should not be used for roofs or floors supporting or attached to nonstructural elements likely to be damaged by large deflections; otherwise it is satisfactory.

Using the simplified Eq. (5-51*b*),

$$\text{Residual camber} = 2.47(\Delta_i)_{P_0} - 2.88(\Delta_i)_D - 2.47(\Delta_i)_S$$
$$= 2.47(4.16) - 2.88(3.30) - 2.47(0.95)$$
$$= -1.57 \text{ in. } (3.99 \text{ cm})$$
$$\text{versus } -1.54 \text{ in. } (3.91 \text{ cm}) \text{ by Eq. (5-43)}$$
$$\text{Residual camber} + \text{live-load deflection} = -1.57 - 1.45 = -3.02 \text{ in. } (7.67 \text{ cm})$$
$$\text{versus } -2.99 \text{ in. } (7.59 \text{ cm}) \text{ by Eq. (5-43)}$$

The reason for this small differential is found in using $\Delta P_u/P_0 = 0.21$ versus 0.20, $C_u = 1.88$ versus 1.80, and $\beta_s = 0.78$ versus 0.83, respectively.

Beam containing both prestressing steel and nonprestressing tension steel

Initial camber due to prestress:

$$(\Delta_i)_{P_0} = 4.16 \text{ in. } (10.57 \text{ cm}) \qquad \text{(same)}$$

Initial dead-load deflection:

$$(\Delta_i)_D = -3.30 \text{ in. } (8.38 \text{ cm}) \qquad \text{(same)}$$

Time-dependent camber due to prestress:

Approx. $\Delta P_u/P_0 = 0.21$ (Table 5-2) but use from Loss Example 3, calc.

$$\Delta P_u/P_0 = (41.33 - 11.63)(2.45)/435 = 0.17$$
$$k_r = 1/(1 + A_{ns}/A_{ps}) = 1/(1 + 0.50)$$
$$= 0.67 \qquad \text{(Eq. (5-59))}$$

$$\left[-\frac{\Delta P_u}{P_0} + (k_r C_u)\left(1 - \frac{\Delta P_u}{2P_0}\right)\right](\Delta_i)_{P_0} = [-0.17 + (0.67 \times 1.80)(1 - 0.085)]4.16$$
$$= 3.88 \text{ in. } (9.86 \text{ cm})$$

Dead-load creep deflection:

$$(k_r C_u)(\Delta_i)_D = (0.67 \times 1.80)(3.30) = -3.98 \text{ in. } (10.11 \text{ cm})$$

Initial deflection under superimposed sustained load:

$$(\Delta_i)_S = -0.95 \text{ in. } (2.41 \text{ cm}) \qquad \text{(same)}$$

Creep deflection under superimposed sustained load:

$$(\beta_s k_r C_u)(\Delta_i)_S = (0.83 \times 0.67 \times 1.80)(0.95) = -0.95 \text{ in. } (2.41 \text{ cm})$$

Live-load deflection:

Beam is cracked (Loss Example 3), hence use Eqs. (5-60), (5-61), and (5-68) or (5-71).

$$f_r = 504 \text{ psi } (35.4 \text{ kg/cm}^2, 3.48 \text{ N/mm}^2) \qquad \text{(Table 1-2)}$$

$$(M_L)_{cr} = \frac{P_{ps} I_g}{A_g y_t} + P_{ps} e_{ps} - \frac{F_{ns} I_g}{A_g y} - F_{ns} e_{ns} - M_{D+S} + \frac{f_r I_g}{y_t} \qquad \text{(Eq. (5-68))}$$

$$= \frac{(362 - 36.8)(168\,970)}{(782)(35.19)(12)} + \frac{(362 - 36.8)(30.44)}{12} - 1065 + \frac{(0.504)(168\,970)}{(35.19)(12)}$$

$$= 128 \text{ ft-k } (17\,700 \text{ kg-m, } 174 \text{ kN-m})$$

and $\qquad (M_L)_{cr}/(M_L)_{max} = 128/250 = 0.512$

Alternatively,

$$(M_L)_{cr}/(M_L)_{max} = (f_{pe} - f_D + f_r)/(f_{pe} - f_D + f_{t0}) \qquad \text{(Eq. (5-71))}$$

From the stress calculations,

$$f_{pe} = 463 + 2294 - 47 - 233 = 2477 \text{ psi } (174 \text{ kg/cm}^2, 17.1 \text{ N/mm}^2)$$

$$f_D = 2661 \text{ psi } (187 \text{ kg/cm}^2, 18.3 \text{ N/mm}^2)$$

$$f_{t0} = 808 \text{ psi } (57 \text{ kg/cm}^2, 5.6 \text{ N/mm}^2)$$

$$(M_L)_{cr}/(M_L)_{max} = (2477 - 2661 + 504)/(2477 - 2661 + 808) = 0.513$$

which checks the previous value. Use $(M_{cr}/M_{max})^3 = 0.51^3 = 0.13$

Considering only rectangular flange in compression,

10 ft by 1.5 in. (3.05 m by 3.81 cm), $\qquad$ and $\qquad A_{ps} + A_{ns} = 3.67 \text{ in}^2$

Also

$$n = 28/2.81 = 9.96$$

$$d = 48 - 4.75 = 43.25 \text{ in. } (109.9 \text{ cm})$$

$$(120)(1.5)(kd - 0.75) = (9.96)(3.67)(43.25 - kd)$$

$$kd = 7.92 \text{ in. } (20.1 \text{ cm})$$

$$I_{cr} = (120)(1.5)^3/12 + (120)(1.5)(7.92 - 0.75)^2 + (9.96)(3.67)(43.25 - 7.92)^2$$
$$= 56\,160 \text{ in}^4 \; (2.337 \times 10^6 \text{ cm}^4) \qquad \text{Use}$$

Alternatively, considering all compression concrete, by trial $kd = 5.20$ in. (13.21 cm), which falls within the tapered part of the flange, and $I_{cr} = 58\,480$ in^4 (2.434×10^6 cm^4). This solution is only slightly higher than the much simpler solution above. Also

$$\text{Approx.}^{63}\; I_{cr} = nA_s d^2(1 - \sqrt{\rho_{ps}}) = 63\,700 \text{ in}^4 \; (2.651 \times 10^6 \text{ cm}^4)$$
$$I_e = (M_{cr}/M_{max})^3 I_g + [1 - (M_{cr}/M_{max})^3] I_{cr} \qquad \text{(Eq. (5-61))}$$
$$= (0.13)(168\,970) + (1 - 0.13)(56\,160)$$
$$= 70\,830 \text{ in}^4 \; (2.95 \times 10^6 \text{ cm}^4)$$

$$\Delta_L = \frac{KM_L L^2}{E_c I_e} = \frac{(5/48)(250.0)(12)(100)^2(144)}{(2.81 \times 10^3)(70\,830)} = -2.26 \text{ in. } (5.74 \text{ cm})$$

$$\text{Initial camber} = 4.16 - 3.30 = 0.86 \text{ in. } (2.18 \text{ cm})$$
$$\text{Residual camber} = 4.16 - 3.30 + 3.88 - 3.98 - 0.95 - 0.95$$
$$= -1.14 \text{ in. } (2.90 \text{ cm})$$

Residual camber + live-load deflection = $-1.14 - 2.26 = -3.40$ in. (8.64 cm)

which is $(L/1200)(3.40) = L/353$, and could be objectionable.

ACI Code (318-71) checks:

$$\text{Time-dependent camber} + \Delta_L = 3.88 - 3.98 - 0.95 - 0.95 - 2.26$$
$$= -4.26 \text{ in. } (10.82 \text{ cm})$$
$$\Delta_L = 2.26 \text{ in. } (5.74 \text{ cm})$$

$L/180 = (100)(12)/180$	$= 6.67$ in. (16.94 cm) $>$ 2.26 in. (5.74 cm)	OK
$L/360$	$= 3.33$ in. (8.46 cm) $>$ 2.26 in. (5.74 cm)	OK
$L/480$	$= 2.50$ in. (6.35 cm) $<$ 4.26 in. (10.82 cm)	NG
$L/240$	$= 5.00$ in. (12.70 cm) $>$ 4.26 in. (10.82 cm)	OK

Hence the beam should not be used for roofs or floors supporting or attached to nonstructural elements likely to be damaged by large deflections; otherwise it is satisfactory.

Using the simplified Eq. (5-50*b*),

$$\text{Residual camber} = (0.79 + 1.68k_r)(\Delta_i)_{P_0} - (1 + 1.88k_r)(\Delta_i)_D - (1 + 1.47k_r)(\Delta_i)_S$$

$$= (0.79 + 1.68 \times 0.67)(4.16) - (1 + 1.88 \times 0.67)(3.30) - (1 + 1.47 \times 0.67)(0.95)$$

$$= 7.99 - 7.46 - 1.88$$

$$= -1.35 \text{ in. } (3.43 \text{ cm})$$

versus -1.14 in. (2.90 cm) by Eq. (5-43)

$$\text{Residual camber + live-load deflection} = -1.35 - 2.26$$

$$= -3.61 \text{ in. } (9.17 \text{ cm})$$

versus -3.40 in. (8.64 cm) by Eq. (5-43)

The difference of 0.21 in. (0.53 cm) in both comparisons is found in using $\Delta P_u/P_0 = 0.21$ versus 0.17, $C_u = 1.88$ versus 1.80, and $\beta_s = 0.78$ versus 0.83, respectively.

Example 4 Ultimate midspan camber, deflection; unshored composite AASHTO Type IV girder with prestressing steel only, normal-weight concrete—Eq. (5-48), which is the ultimate form of Eq. (5-44); also compared with results by the simplified Eq. (5-53*a*).

See Table 5-8 for the material and section properties, parameters and conditions of the problem. The camber and deflection results are tabulated term by term in Table 5-14.

Initial camber due to prestress—term (1), Eq. (5-48), as defined in Eq. (5-44):

From Loss Example 4,

$$\text{Trial } P_0 = P_i/(1 + 2n\rho_{ps}) = 849 \text{ kips } (385\,100 \text{ kg, } 3776 \text{ kN})$$

$$\text{Revised } P_0 = 858 \text{ kips } (389\,200 \text{ kg, } 3816 \text{ kN})$$

$$E_{ci} = 3.64 \times 10^3 \text{ ksi } (256\,000 \text{ kg/cm}^2\text{, } 25.1 \text{ kN/mm}^2) \qquad \text{(Table 1-3)}$$

From case 9, Appendix K, for two-point harping,

$$(\Delta_i)_{P_0} = \frac{P_0(e_c - e_e)}{E_{ci}I_g}\left[\frac{L^2}{8} - \frac{a^2}{6}\right] + \frac{P_0 e_e L^2}{8E_{ci}I_g}$$

$$= \frac{858(21.26 - 2.86)}{(3.64 \times 10^3)(260\,740)}\left[\frac{80^2}{8} - \frac{26.67^2}{6}\right]144 + \frac{858(2.86)(80)^2(144)}{8(3.64 \times 10^3)(260\,740)}$$

$$= 1.63 + 0.30 = 1.93 \text{ in. } (4.90 \text{ cm})$$

Initial dead-load deflection of precast beam—term (2), Eq. (5-48), as defined in Eq. (5-44):

$$(\Delta_i)_2 = \frac{KM_2L^2}{E_{ci}I_g} = \frac{(5/48)(657.6)(12)(80)^2(144)}{(3.64 \times 10^3)(260\,740)} = -0.80 \text{ in. } (2.03 \text{ cm})$$

Time-dependent camber due to prestress up to the time of slab casting—term (3), Eq. (5-48), as defined in Eq. (5-44):

For 60 percent humidity, $C_u = 2.04$ (Table 1-10)

For v/s ratio of 4.8 in. (12.2 cm) (Fig. C-1, Chap. 1),

$$\text{Creep } (\text{C.F.})_T = 0.74 \qquad \text{(Table 1-9)}$$
$$\text{Adjusted } C_u = (2.04)(0.74) = 1.51$$

For deck slab cast 2 months after prestressing,

$$\Delta P_s/P_0 = 0.14 \qquad \text{(Table 5-2)}$$
$$\alpha_s = 0.54 \qquad \beta_s = 0.78 \qquad \text{(Table 5-3)}$$

Approx. $\Delta P_u/P_0 = 0.18$ (Table 5-2), but use from Loss Example 4, calc.

$$\Delta P_u/P_0 = (43.88 - 14.10)(4.90)/858 = 0.170 \qquad k_r = 1$$

$$\left[-\frac{\Delta P_s}{P_0} + (\alpha_s k_r C_u)\left(1 - \frac{\Delta P_s}{2P_0}\right)\right](\Delta_i)_{P_0} = [-0.14 + (0.54 \times 1.51)(1 - 0.07)]1.93$$
$$= 1.19 \text{ in. } (3.02 \text{ cm})$$

Time-dependent camber due to prestress following slab casting—term (4), Eq. (5-48), as defined in Eq. (5-44):

$$\left\{-\frac{\Delta P_u - \Delta P_s}{P_0} + (k_r C_u)\left[\left(1 - \frac{\Delta P_u}{2P_0}\right) - \alpha_s\left(1 - \frac{\Delta P_s}{2P_0}\right)\right]\right\}(\Delta_i)_{P_0}\frac{I_2}{I_c}$$
$$= \{-0.030 + (1.51)[(1 - 0.085) - 0.54(1 - 0.07)]\}(1.93)\frac{260\,740}{629\,890}$$
$$= 0.47 \text{ in. } (1.19 \text{ cm})$$

Dead-load creep deflection of the precast beam up to the time of slab casting—term (5), Eq. (5-48), as defined in Eq. (5-44):

$$(\alpha_s k_r)(\Delta_i)_2 = (0.54 \times 1.51)0.65 \text{ in. } (1.65 \text{ cm})$$

Creep deflection of the composite beam following slab casting due to the precast beam dead load—term (6), Eq. (5-48), as defined in Eq. (5-44):

$$(1 - \alpha_s)(k_r C_u)(\Delta_i)_2\frac{I_2}{I_c} = (1 - 0.54)(1.51)(0.80)\frac{260\,740}{629\,890} = -0.23 \text{ in. } (0.58 \text{ cm})$$

Initial deflection of the precast beam under slab plus diaphragm dead load—term (7), Eq. (5-48), as defined in Eq. (5-44):

$$E_c = 3.89 \times 10^3 \text{ ksi } (273\,500 \text{ kg/cm}^2,\ 26.8 \text{ kN/mm}^2) \qquad \text{(Table 1-3)}$$

$$(\Delta_i)_1 = \frac{KM_{S,Di}L^2}{E_c I_g} = \frac{(5/48)(625.2)(12)(80)^2(144)}{(3.89 \times 10^3)(260\,740)} = -0.74 \text{ in. } (1.88 \text{ cm})$$

Creep deflection of the composite beam under slab plus diaphragm dead load—term (8), Eq. (5-48), as defined in Eq. (5-44):

$$\beta_s = 0.78 \qquad \text{(See term (3))}$$

$$(\beta_s k_r C_u)(\Delta_i)_1 \frac{I_2}{I_c} = (0.78)(1.51)(0.74)\frac{260\,740}{629\,890} = -0.36 \text{ in. } (0.91 \text{ cm})$$

Differential shrinkage and creep deflection—term (9), Eq. (5-48), as defined in Eq. (5-44):

From Loss Example 4,

$$y_{cS} = 18.59 \text{ in. } (47.2 \text{ cm})$$

$$Q_{DS} = 538\,800 \text{ lb } (244\,000 \text{ kg, } 2397 \text{ kN})$$

$$\text{Composite } I_c = 629\,890 \text{ in}^4 \ (26.22 \times 10^6 \text{ cm}^4)$$

$$\Delta_{DS} = \frac{(4/3)Q_{DS}\, y_{cS}\, L^2}{8E_c I_c} = \frac{(4/3)(538.8)(18.59)(80)^2(144)}{8(3.89 \times 10^3)(629\,890)}$$

$$= -0.63 \text{ in. } (1.60 \text{ cm}) \qquad \text{(Eq. (2-22))}$$

Live-load deflection—term (10), Eq. (5-48), as defined in Eq. (5-44):

Beam is uncracked (Loss Example 4). Using Eq. (5-41),

$$\Delta_{L+I} = \frac{KM_{L+I}L^2}{E_c I_c} = \frac{(5/48)(1015.4)(12)(80)^2(144)}{(3.89 \times 10^3)(629\,890)}$$

$$= -0.50 \text{ in. } (1.27 \text{ cm})$$

$$\text{Initial camber} = 1.93 - 0.80 = 1.13 \text{ in. } (2.87 \text{ cm})$$

$$\text{Residual camber} = 1.93 - 0.80 + 1.19 + 0.47 - 0.65 - 0.23 - 0.74 - 0.36 - 0.63$$

$$= 0.18 \text{ in. } (0.46 \text{ cm})$$

$$\text{Residual camber + live-load and impact deflection} = 0.18 - 0.50$$

$$= -0.32 \text{ in. } (0.81 \text{ cm})$$

Using the simplified Eq. (5-53*a*),

$$I_2/I_c = 260\,740/62\,989 = 0.4139$$

$$\text{Residual camber} = (1.80 + 0.73 I_2/I_c)(\Delta_i)_{P_0} - (2.02 + 0.86 I_2/I_c)(\Delta_i)_2$$

$$- (1.00 + 1.47 I_2/I_c)(\Delta_i)_1 - \Delta_{DS}$$

$$= (1.80 + 0.73 \times 0.4139)(1.93) - (2.02 + 0.86 \times 0.4139)(0.80)$$

$$- (1.00 + 1.47 \times 0.4139)(0.74) - 0.63$$

$$= 4.05 - 1.90 - 1.19 - 0.63 = 0.33 \text{ in. } (0.84 \text{ cm})$$

versus 0.18 in. (0.46 cm) by Eq. (5-48)

$$\text{Residual camber + live-load and impact} = 0.33 - 0.50 = -0.17 \text{ in. } (0.43 \text{ cm})$$

versus -0.32 in. (0.81 cm) by Eq. (5-48)

The difference of 0.15 in. (0.38 cm) in both comparisons is found in using $C_u = 1.88$ versus 1.51, respectively.

Using the grossly approximate Eq. (5-56),

$$\Delta_i = (\Delta_i)_{P_0} - (\Delta_i)_2 - (\Delta_i)_1 = 1.93 - 0.80 - 0.74$$
$$= 0.39 \text{ in. } (0.99 \text{ cm})$$

$$\text{Residual camber} = [1 + 2(I_2/I_c)]\Delta_i = (1 + 2 \times 0.4139)0.39 = 0.71 \text{ in. } (1.80 \text{ cm})$$

Table 5-12 Term-by-term camber, deflection in inches (cm) for Examples 1 and 2

	Initial camber due to prestress	Initial defl. due to beam D.L.	Creep camber due to prestress	Creep defl. due to beam D.L.	Total
Example 1: ultimate midspan camber, defl., noncomposite, double-T, nor.-wt. concrete	2.97 (7.54)	−2.21 (5.61)	4.31 (10.95)	−4.15 (10.54)	0.92 (2.34)
Example 2: ultimate midspan camber, defl., noncomposite double-T, sand-lt.-wt. conc.	3.99 (10.13)	−2.53 (6.43)	5.43 (13.79)	−4.76 (12.09)	2.13 (5.41)

Example 1: live-load deflection = −0.97 in. (2.46 cm), Example 2: live-load deflection = −1.34 in. (3.40 cm).

Example 2 is the same as Example 1, except for different weight concrete.

Table 5-13 Term-by-term camber, deflection in inches (cm) for Example 3—ultimate midspan values, noncomposite single-T, sand-lightweight concrete, with additional sustained loading

	Initial camber due to prestr.	Initial defl. due to beam D.L.	Creep camber due to prestr.	Creep defl. due to beam D.L.	Initial defl. due to add. sus. load	Creep defl. due to add. sus. load	Total
Beam containing only prestressing steel	4.16 (10.57)	−3.30 (8.38)	5.91 (15.01)	−5.94 (15.09)	−0.95 (2.41)	−1.42 (3.61)	−1.54 (3.91)
Beam containing both prestressing steel and non-prestressed steel	4.16 (10.57)	−3.30 (8.38)	3.88 (9.86)	−3.98 (10.11)	−0.95 (2.41)	−0.95 (2.41)	−1.14 (2.90)

Beam containing only prestressing steel: live-load deflection = −1.45 in. (3.68 cm).
Beam containing both steels: live-load deflection = −2.26 in. (5.74 cm).
In both cases the beam was cracked, necessitating the use of I_e for Δ_L.

versus 0.18 in. (0.46 cm) by the general Eq. (5-48) and 0.33 by the simplified Eq. (5-53*a*).

Residual camber + live-load and impact = 0.71 − 0.50 = 0.21 in. (0.53 cm)

versus −0.32 in. (0.81 cm) by the general Eq. (5-48) and −0.17 in. (0.43 cm) by the simplified Eq. (5-53*a*).

It should be noted that the part of the time-dependent camber which occurs before the placement of nonstructural elements can be deducted from the total time-dependent camber, deflection plus live load deflection in the ACI Code check. This was not done in the examples herein, although one can tell by inspection that doing so would not alter any of the conclusions.

Effect of Partial Continuity Involving a Continuous Composite Deck Slab on Prestress Loss or Gain, and Camber or Deflection

It appears that a continuous composite deck slab has a relatively small effect on the loss or gain in prestress. Such a slab would affect primarily term (3)—creep loss after slab casting under prestress and precast beam dead load, term (7)—creep gain due to slab dead load, term (8)—differential shrinkage and creep, and for shored construction, term (6)—elastic gain due to slab dead load; all from Eqs. (5-17) or (5-19). As can be seen in Table 5-11 for the composite AASHTO Type IV girder, for example, prestress loss and gain results for the simple-span girder were 1.91 (term (3)) −0.77 (term (7)) −1.82 (term (8)) = −0.68 percent or 0.7 percent gain in prestress. The restraint provided by the continuous composite slab would serve to reduce this already very small value further. Judging from this typical example, it seems safe to assume that the effect of the continuous composite slab would normally have a negligible effect on the loss of prestress (also noted by Mossiossian and Gamble[32]), and that the effect could be either beneficial or nonbeneficial with regard to prestress loss (here slightly nonbeneficial since continuity would reduce the prestress gain).

Table 5-14 Term-by-term camber, deflection in inches (cm) for Example 4—ultimate midspan values, unshored composite girder, normal-weight concrete

Initial camber due to prestr.	Initial defl. due to beam D.L.	Creep camber up to slab cast	Creep camber after slab cast	Creep defl. up to slab cast	Creep defl. after slab cast	Elastic defl. due to slab D.L.	Creep defl. due to slab D.L.	Defl. due to diff. shrinkage and creep	Total
1.93 (4.90)	−0.80 (2.03)	1.19 (3.02)	0.47 (1.19)	−0.65 (1.65)	−0.23 (0.58)	−0.74 (1.88)	−0.36 (0.91)	−0.63 (1.60)	0.18 (0.46)

AASHTO Type IV girder, live-load plus impact deflection = −0.50 in. (1.27 cm).

A similar consideration of the effect of a continuous composite deck slab on camber or deflection shows that the effect may be worth calculating. Such a slab would affect primarily term (4)—creep camber due to prestress after slab casting, term (6)—creep deflection due to precast beam dead load after slab casting, term (8)—creep deflection due to slab dead load, term (9)—deflection due to differential shrinkage and creep, and for shored construction, term (7)—elastic deflection due to slab dead load; all from Eqs. (5-44), (5-46), (5-48), or (5-49). As can be seen in Table 5-14 for the composite AASHTO Type IV girder, for example, the camber and deflection results for a simple-span girder were 0.47 (term (4)) -0.23 (term (6)) -0.36 (term (8)) -0.63 (term (9)) $= -0.75$ in. (1.91 cm), or a downward deflection of 0.75 in. (1.91 cm). The restraint provided by the continuous composite deck slab would serve to reduce this value.

The effect of continuity in this problem is to restrain curvatures and deflections that occur after slab hardening. For the dead-load effects in the case of unshored construction, this can be accomplished by multiplying the simple-span values by the ratio of the deflection coefficients for uniform loading: or multiply terms (6), (8) by a factor (see Notation) of $(8/185)/(5/48) = 0.42$ for spans with one end continuous (ideal hinged–fixed beam) and $(1/32)/(5/48) = 0.30$ for spans with both ends continuous (ideal fixed–fixed beam). For parabolic prestress profiles with zero end eccentricity the same factors would apply (ideally) to term (4), and might be so used as an approximation for other profiles. The differential shrinkage and creep deflections of continuous composite members, with regard to term (9), are developed in Chap. 2, with the results summarized in Table 2-6. For example, the ratio of the deflection coefficients is $0.0528/0.125 = 0.42$ for a three-span continuous beam and $0.0481/0.125 = 0.38$ for a four-span continuous beam, as compared to a simple beam. Thus a reasonable approximation for the total effect of the continuous composite slab on camber or deflection in the case of unshored construction, or terms (4), (6), (8), (9), would appear to be to modify the simple-span camber or deflection by a factor of say 0.4 (above deflection $= 0.75 \times 0.4 = 0.30$ in. (0.76 cm)). A more detailed procedure for these calculations has been presented by Freyermuth.[27] In the case of shored construction, the slab dead load would be applied to the continuous composite member to begin with.

For live-load deflections the continuous composite member would be effective in resisting live-load moments. In connection with live-load deflections of cracked continuous members using the I-effective method, see Sec. 5-4 herein and the part entitled "I_e for continuous beams."

5-4 SPECIAL EFFECTS IN PARTIALLY PRESTRESSED MEMBERS

Many practical applications involving camber control and the control of cracking and deflections of partially prestressed members have been adopted following the work of Abeles[35–37,42–44] and others[6,33,45,46,47] since the 1940s. The term "partial" prestressing has come to refer to prestressed concrete members with tension or containing nonprestressed steel in addition to prestressed steel or both. Since any prestressed concrete beam can undergo tension if the load is high enough, other terms such as "limited" prestressing and "secondary or passive" reinforcement (because the nonprestressed steel is usually relatively inactive under working loads) have also been used.

Consideration is given in this section to the effect of nonprestressed tension steel in *reducing* time-dependent creep, shrinkage, loss of stress in prestressing steel, and camber; and on the deflection of cracked prestressed flexural members. These calculations have already been described in this chapter including the examples. For information on the design of partially prestressed members, the reader might refer to publications by Abeles,[44] Lin,[56] Salse[64] (presents a concise trial and error procedure for computing ultimate moments according to the ACI Code, 318-71, for both partially prestressed beams and columns), and others.

Effect of Nonprestressed Tension Steel in Reducing Time-Dependent Creep, Shrinkage, Loss of Stress in Prestressing Steel, and Camber

The relaxation method of Trost[11,12] for determining the loss of prestress due to concrete creep and shrinkage was derived by Eqs. (5-27) to (5-31). This procedure is then used in the calculation of shrinkage loss in the form of b_{11} (Eq. (5-5)), reduction of creep and shrinkage loss due to nonprestressed tension steel in the form of b_{12} (Eq. (5-8)), and stress in nonprestressed tension steel due to creep and shrinkage in the form of b_{22} (Eq. (5-11)). An approximate value of $b_{11} = b_{12} = b_{22} = 0.25$ was also suggested. The corresponding equations for the effect of compression steel in prestressed beams were mentioned by replacing e_{ps} with $-e_{ns}$ in the above equations. However, in most cases the sustained load stress is small in compression steel in prestressed beams, so that the effects of compression steel are not considered in detail in this chapter.

Examples of the effect of nonprestressed tension steel in reducing time-dependent camber, and of the behavior of cracked prestressed beams are shown in Table 5-15 and Figs. 5-5, 5-6 by Shaikh and the author.[6] Based on the following energy conditions, Eqs. (5-57) to (5-59) were developed and results compared with experimental data in Table 5-15:

(Work done by forces in prestressed and nonprestressed steel)
 − (Work done against the beam dead load)
 = (Change in internal strain energy)

Due to creep,
$$(k_r)_{cp} = \frac{1}{1 + \left(E_{ns}A_{ns}\int_0^L (\varepsilon'_{ci})^2\,dx\right)\Big/\left(E_{ps}A_{ps}\int_0^L (\varepsilon_{ci})^2\,dx\right)} \tag{5-57}$$

Due to shrinkage,
$$(k_r)_{sh} = \frac{1}{1 + E_{ns}A_{ns}/E_{ps}A_{ps}} \tag{5-58}$$

where E_{ns}, A_{ns}, and (ε'_{ci}) refer to the nonprestressed steel; E_{ps}, A_{ps}, and (ε_{ci}) refer to the prestressed steel; and k_r is the reduction factor for the effect of nonpre-

Table 5-15 Details of test beams and experimental (determined as the ratio of measured camber in each series from Fig. 5-5) versus calculated values of the reduction factor, k_r[6]

Series no	I			II			III			IV		
7d—f'_{ci}, psi	4120			4380			4830			4300		
28d—f'_c, psi	5400			5890			6570			5880		
Beam no.	B1	B2	B3	B1	B2	B3	B1	B2	B3	B1	B2	B3
Nonprestr. steel	#4	#4	#4	5/16	#4	#4			#5	3/8	#5	#4
Prestressing strand dia, in.	5/16	5/16	5/16	5/16	5/16	5/16	5/16	3/8	3/8	3/8	3/8	3/8
Meas. P_0, kips	29.8	29.0	30.1	20.2	20.0	19.7	30.5	29.8	29.8	25.2	25.8	24.4
A_{ps}, in^2	0.173	0.173	0.173	0.116	0.116	0.116	0.173	0.240	0.240	0.160	0.160	0.160
A_{ns}, in^2	0.200	0.400	0.600	0.058	0.200	0.400	0	0	0.310	0.080	0.310	0.600
$p = A_{ps}/bd$, %	0.45	0.45	0.45	0.30	0.30	0.30	0.45	0.60	0.60	0.40	0.40	0.40
$p' = A_{ns}/bd$, %	0.50	1.00	1.50	0.15	0.50	1.00	0	0	0.80	0.20	0.80	1.50
$A_{ns}/A_{ps} = p'/p$	1.15	2.30	3.46	0.50	1.73	3.46	0	0	1.30	0.50	1.94	3.76
$p\dfrac{f_{su}}{f'_c} + p'\dfrac{f_y}{f'_c}$	0.23	0.27	0.31	0.20	0.20	0.21	0.20	0.25	0.31	0.28	0.28	0.29
Meas. Q_u, kips	8.7	9.4	9.9	6.8	7.2	7.3	7.0	9.3	9.7	8.9	10.1	9.2
Experimental range of k_r in % for period *	100 to 100	70 to 77	60 to 65	100 to 100	62 to 66	32 to 40	100 to 100	70 to 80	43 to 50	100 to 100	78 to 85	39 to 45
Avg. exp. k_r in % *	100	74	63	100	65	37	100	77	48	100	83	42
Calc. value of k_r in %†	100	74	56	100	56	42	100	73	48	100	54	44

(For metric units, see metric conversion factors herein.)

All beams are 6 in. by 8 in., $d = 6.5$ in., span = 15 ft simply supported.

○ Prestressing steel
● Nonprestressed 33 ksi yield steel
Nonprestressed high strength steel
Nonprestressed 60 ksi yield steel

* The experimental values of k_r are simply the ratios of the time-dependent camber of the three beams in each series to the maximum time-dependent camber (of reference beam with smallest nonprestressed tension steel content) in the series.

† The calculated values of k_r were determined by the general procedure of Ref. 6, since Eqs. (5-57) to (5-59) apply only when the reference beam contains no nonprestressed steel, and Eq. (5-59) only when the eccentricities of the two steels are approximately the same; which is not the case for most of these beams.

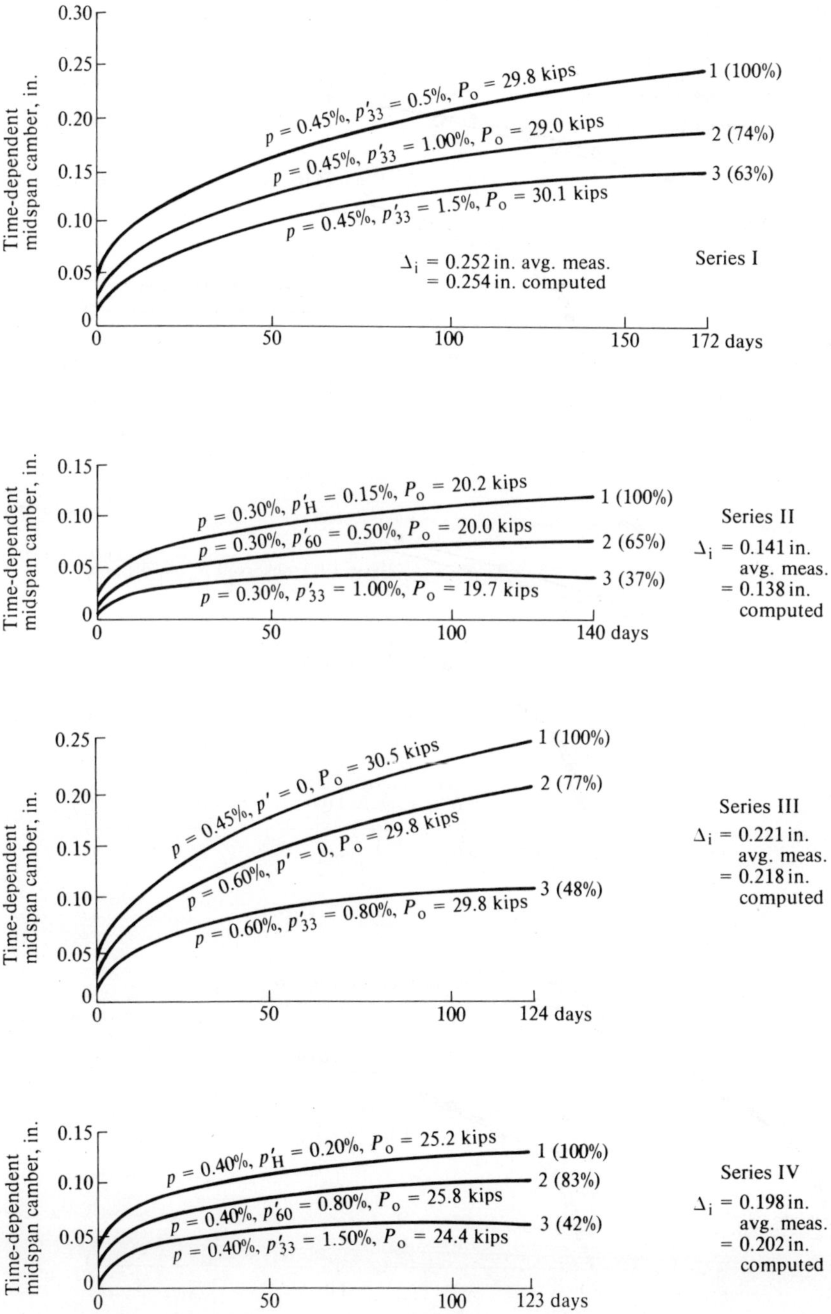

FIGURE 5-5
Experimental time-dependent (total minus initial) midspan camber versus time in days, from Ref. 6 (0.20 in. = 0.51 cm, 30 kips = 13 610 kg = 133 kN).

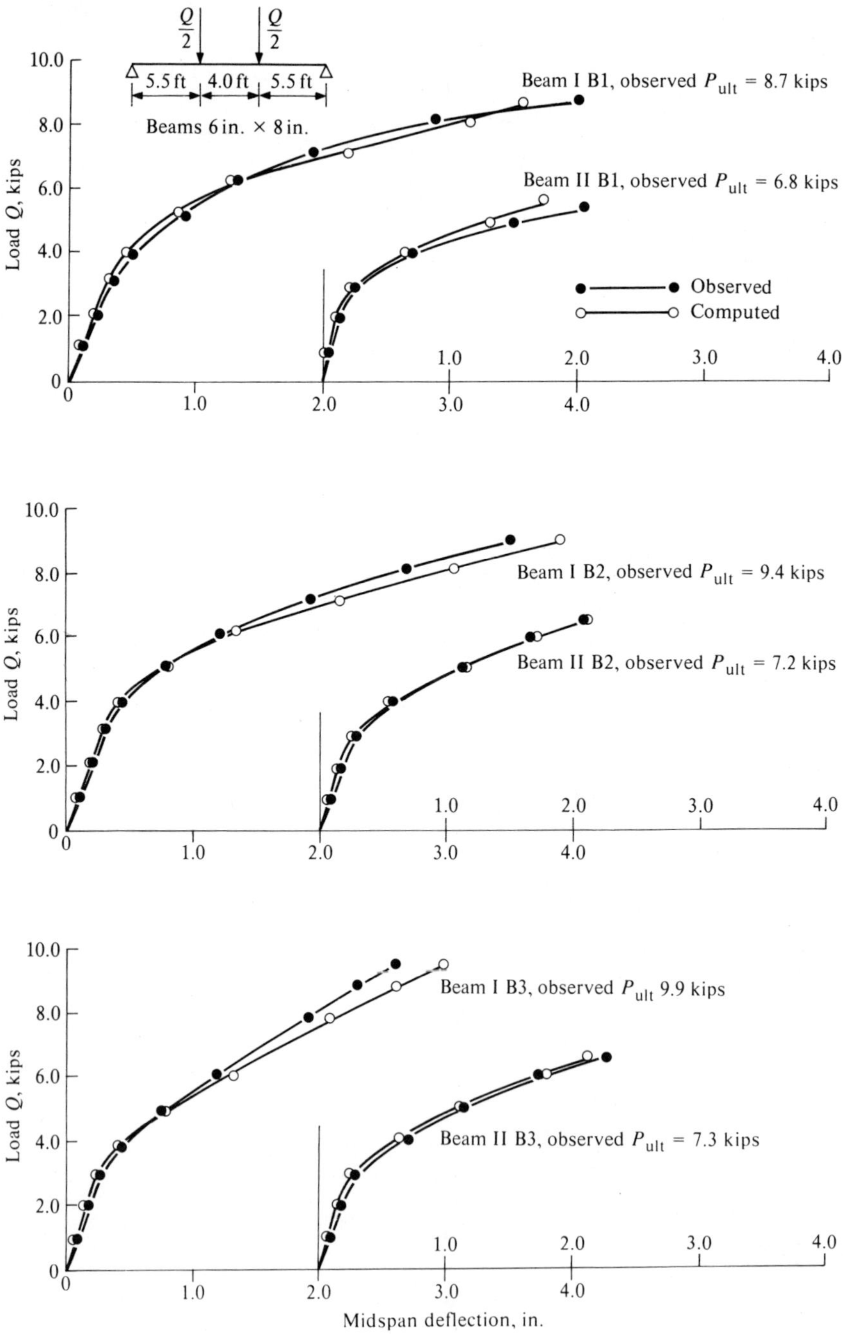

FIGURE 5-6
Experimental and computed (using I_e) load versus midspan deflection for the 12 beams (Table 5-15) of Ref. 6 (2.0 in. = 5.08 cm, 6 kips = 2720 kg = 26.7 kN).

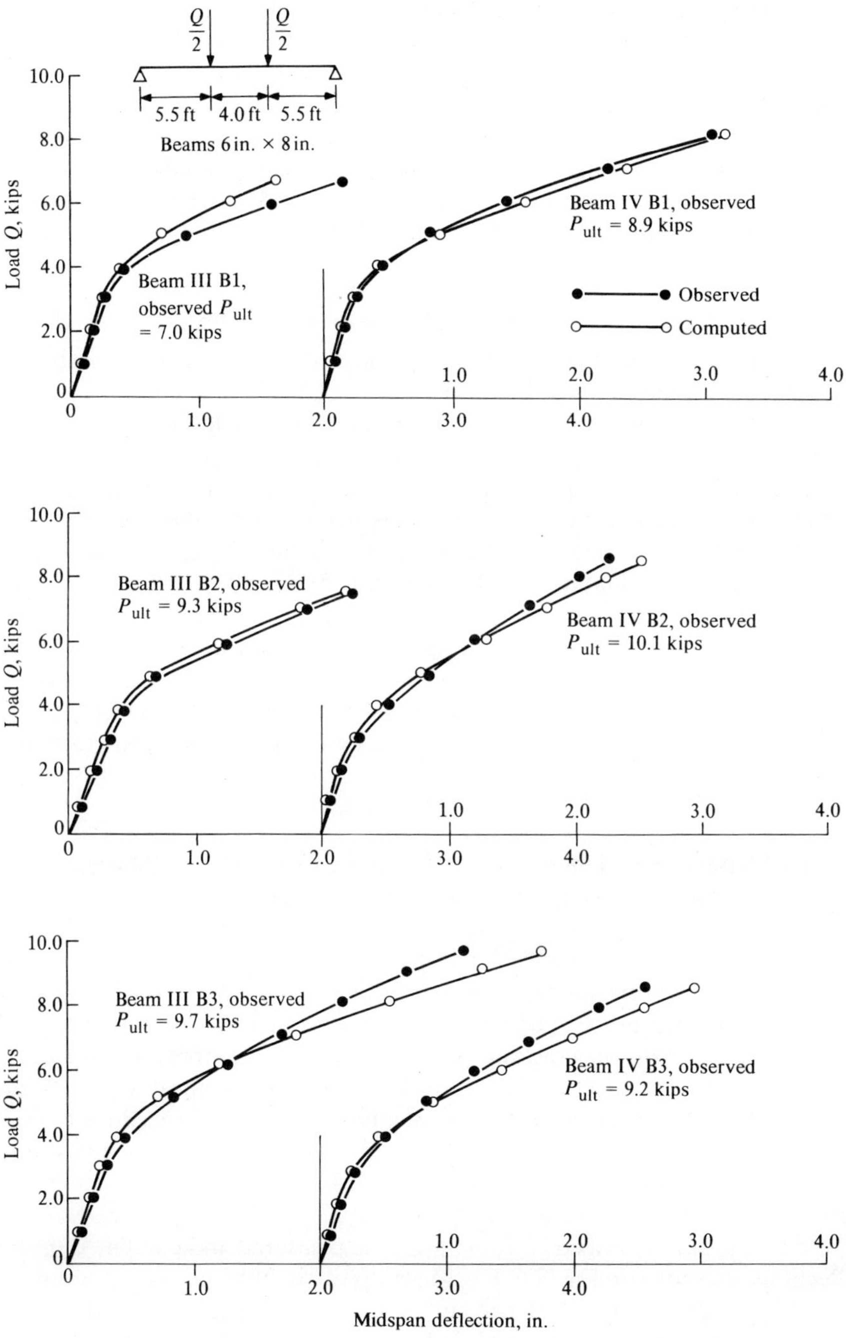

FIGURE 5-6 *(continued)*

stressed tension steel in *reducing* time-dependent creep, shrinkage, loss of stress in prestressing steel, and (approximately) camber. When the eccentricities of the two steels are equal ($e_{ns} = e_{ps}$) and $E_{ns} = E_{ps}$ (approximate design conditions in most cases), both Eqs. (5-57) and (5-58) reduce to Eq. (5-59). The reader might note the similarity in this approximate solution for the effect of nonprestressed tension steel in reducing time-dependent camber in prestressed beams and the approximate method given in Chap. 3 and used in the ACI Code (318-71) for the effect of compression steel in reducing time-dependent deflection of ordinary reinforced beams.

$$k_r = 1/(1 + A_{ns}/A_{ps}) \qquad A_{ns}/A_{ps} \leq 2 \text{ for camber} \tag{5-59}$$

The application of k_r in the calculation of loss of prestress and camber is accomplished by replacing all values of the creep coefficient and shrinkage strain with modified values—as $k_r C_t$ and $k_r(\varepsilon_{sh})_t$—in all terms, including those due to composite slab dead load, etc., since the creep strains occur under the combined loading, and simultaneously with the shrinkage strains. These modified expressions are included in the loss and camber equations herein. The reduction factor, k_r, can also be applied as a single reduction factor for time-dependent camber in grossly approximate calculations, as shown in Eq. (5-56) and by the comparisons with experimental data in Table 5-15. Although this procedure has a theoretical base for creep and shrinkage strains, the agreement with the data shown in Table 5-15, while quite good except for Series IV Beam B2, is only approximate; which appears to reflect the inexact nature of the physical problem and the fact that the experimental camber results include all time-dependent effects, not just creep and shrinkage alone.

Live-Load Deflection of Uncracked and Cracked Prestressed Beams, Either With or Without Nonprestressed Tension Steel

***I*-effective (I_e) method** The *I*-effective (I_e) method, initially developed by the author[24] for reinforced beams, was shown by Shaikh and the author[6] to apply equally well to bonded prestressed beams, either with or without nonprestressed tension steel, loaded into the cracking range. The effect of nonprestressed steel is included in I_{cr}, and the effect of the prestress force on the neutral axis location is not included in the moment of inertia calculations. This I_e procedure is used in the 1971 ACI Code for reinforced beams, slabs, and composite members, and in the 1971 PCI Handbook for satisfying the 1971 ACI Code provisions on deflections of partially prestressed members.

For the 12 beams described in Table 5-15, the measured and computed (using I_e) deflections are shown in Fig. 5-6. For loads up to about 80 percent of ultimate, with cracking loads at about 30 percent of ultimate, the maximum deviation between the measured and computed deflections in Fig. 5-6 is 19 percent. The measured modulus of rupture equal to about $11\sqrt{f'_c}$, and calculated value for E_c using the ACI Code equation (Eq. (1-10) herein) were used in obtaining the computed results. Additional verification of the procedure has been indicated by Nawy and Ptoyondy,[28,30] and Bennett and Veerasubramanian.[34] The latter authors found the I_e method to be "very accurate for the I section beam, but conservative for the monolithic and composite T sections" (computed deflections greater than measured deflections).

For computing the live-load deflection increment due to a live load into the cracking range (or below the cracking load level using I_g), as shown in Fig. 5-7, use Eqs. (5-60) or (5-41), and (5-61), with either of two methods described below for determining $(M_L)_{cr}/(M_L)_{max}$ in Eq. (5-61). $(M_L)_{cr}$ is the additional moment above the prestress and dead-load moment necessary to crack the beam, and $(M_L)_{max}$ is the maximum live-load moment. Computed results by the two methods are the same, as shown in Example 3 for the beam containing only prestressing steel, and also for the beam with both prestressed and nonprestressed steel.

$$\Delta_L = KM_L L^2/E_c I_e \tag{5-60}$$

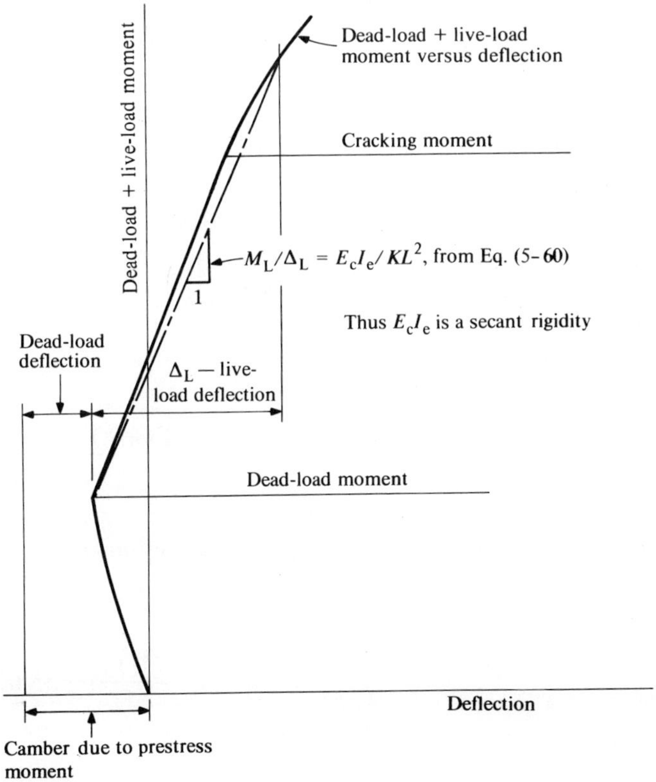

FIGURE 5-7
Illustration of moment–deflection terms and flexural rigidity for partially prestressed beams.

Equation (5-61) is used to compute I_e when $M_{max} \geq M_{cr}$, otherwise $I_e = I_g$. It is shown in Fig. 5-7 that $E_c I_e$ represents a secant rigidity expression.

$$I_e = (M_{cr}/M_{max})^3 I_g + [1 - (M_{cr}/M_{max})^3] I_{cr} \tag{5-61}$$

Method 1—direct determination of $(M_L)_{cr}$

Beams containing only prestressing steel

NONCOMPOSITE PRESTRESSED BEAMS

$$f_r = -\frac{P}{A_g} - \frac{Pey_t}{I_g} + \frac{M_{D+S}y_t}{I_g} + \frac{(M_L)_{cr}y_t}{I_g} \tag{5-62}$$

where y_t refers to the tension fiber. Solving for $(M_L)_{cr}$,

$$(M_L)_{cr} = \frac{PI_g}{A_g y_t} + Pe - M_{D+S} + \frac{f_r I_g}{y_t} \tag{5-63}$$

UNSHORED COMPOSITE PRESTRESSED BEAMS

$$f_r = -\frac{P}{(A_g)_2} - \frac{Pe_2(y_t)_2}{(I_g)_2} + \frac{(M_1 + M_2)(y_t)_2}{(I_g)_2} + \frac{(M_L)_{cr}(y_t)_c}{(I_g)_c} \tag{5-64}$$

where subscripts 2 and c refer to the precast and composite sections, respectively, and M_1 and M_2 refer to the slab and precast beam dead load moments, respectively. Solving for $(M_L)_{cr}$,

$$(M_L)_{cr} = \frac{P(I_g)_c}{(A_g)_2(y_t)_c} + Pe_2\frac{(y_t)_2(I_g)_c}{(y_t)_c(I_g)_2} - (M_1 + M_2)\frac{(y_t)_2(I_g)_c}{(y_t)_c(I_g)_2} + \frac{f_r(I_g)_c}{(y_t)_c} \tag{5-65}$$

SHORED COMPOSITE PRESTRESSED BEAMS

$$(M_L)_{cr} = [\text{same as Eq. (5-65), except delete } M_1 \text{ in the third term}] - M_1 \tag{5-66}$$

Beams containing both prestressing steel and nonprestressed tension steel

NONCOMPOSITE PRESTRESSED BEAMS

$$f_r = -\frac{P_{ps}}{A_g} - \frac{P_{ps}e_{ps}y_t}{I_g} + \frac{F_{ns}}{A_g} + \frac{F_{ns}e_{ns}y_t}{I_g} + \frac{M_{D+S}y_t}{I_g} + \frac{(M_L)_{cr}y_t}{I_g} \tag{5-67}$$

Solving for $(M_L)_{cr}$,

$$(M_L)_{cr} = \frac{P_{ps}I_g}{A_g y_t} + P_{ps}e_{ps} - \frac{F_{ns}I_g}{A_g y_t} - F_{ns}e_{ns} - M_{D+S} + \frac{f_r I_g}{y_t} \tag{5-68}$$

It is seen that Eq. (5-68) is the same as Eq. (5-63) for beams with prestressing steel only, except for the two F_{ns} terms.

UNSHORED COMPOSITE PRESTRESSED BEAMS It follows by inspection of Eqs. (5-63), (5-65), and (5-68), that Eq. (5-69) is the same as Eq. (5-65), except for the two F_{ns} terms.

$$(M_L)_{cr} = \frac{P_{ps}(I_g)_c}{(A_g)_2(y_t)_c} + P_{ps}e_{ps}\frac{(y_t)_2(I_g)_c}{(y_t)_c(I_g)_2} - \frac{F_{ns}(I_g)_c}{(A_g)_2(y_t)_c} - F_{ns}e_{ns}\frac{(y_t)_2(I_g)_c}{(y_t)_c(I_g)_2} - (M_1 + M_2)\frac{(y_t)_2(I_g)_c}{(y_t)_c(I_g)_2} + \frac{f_r(I_g)_c}{(y_t)_c} \tag{5-69}$$

SHORED COMPOSITE PRESTRESSED BEAMS

$$(M_L)_{cr} = [\text{same as Eq. (5-69), except delete } M_1 \text{ in the fifth term}] - M_1 \tag{5-70}$$

Method 2—stress ratio determination of $(M_L)_{cr}/(M_L)_{max}$ For all cases represented by Eqs. (5-62) to (5-70), the moment ratio in Eq. (5-61) for computing I_e can be determined by Eq. (5-71). From stress calculations for f_r, such as Eqs. (5-62), (5-64), and (5-67), and similar calculations for f_{t0}, one finds that Eq. (5-71) applies to unshored and shored composite beams as well as noncomposite beams. Since the equation is in the form of a ratio, the section modulus in both numerator and denominator cancel. Equation (5-71) was used in the development of the 1971 PCI Handbook chart to be used in the calculation of I_e.

For noncomposite or composite beams, with or without nonprestressed tension steel in addition to prestressing steel,

$$(M_L)_{cr}/(M_L)_{max} = (f_{pe} - f_D + f_r)/(f_{pe} - f_D + f_{t0}) \tag{5-71}$$

where f_r = modulus of rupture of concrete

f_{t0} = total computed tensile stress (above the cracking stress, f_r, but not to exceed the maximum allowable stress, which is $12\sqrt{f_c'}$ by the 1971 ACI Code). f_{t0} is a fictitious elastic stress

$f_{pe} = P/A_g + Pey_t/I_g$. In the case of composite beams, use the precast beam section properties. y_t refers to the tension fiber

$f_{pe} = P_{ps}/A_g + P_{ps}e_{ps}y_t/I_g - F_{ns}/A_g - F_{ns}e_{ns}y_t/I_g$ for beams with both prestressed steel and nonprestressed tension steel. In the case of composite beams, use the precast beam section properties

$f_D = M_{D+S}y_t/I_g$ for noncomposite prestressed beams

$f_D = (M_1 + M_2)(y_t)_2/(I_g)_2$ for unshored composite prestressed beams

$f_D = M_2(y_t)_2/(I_g)_2 + M_1(y_t)_c/(I_g)_c$ for shored composite prestressed beams.

These equations (from Ref. 72) assume that the precast beam supports the beam-plus-slab dead load in the case of unshored construction; and, in the case of shored construction, the precast beam supports its own dead load and the composite beam supports the slab dead load. When the uncracked transformed moment of inertia is used in deformation calculations of prestressed members, A_t and I_t replace A_g and I_g in these equations. In computing both uncracked and cracked transformed section properties, appropriate n-values should be used for nonprestressed tension steel and prestressed steel depending on the moduli of elasticity of the steel and concrete.

Sample calculations using Eqs. (5-61) and (5-71) Given: $f_r = 7.5\sqrt{f'_c}$, $6.4\sqrt{f'_c}$, and $5.6\sqrt{f'_c}$ for normal-weight, sand-lightweight, and all-lightweight concrete, respectively, and $f_{t0} = 12\sqrt{f'_c}$ (ACI Code 318-71). Assume conservatively—$f_{pe} - f_D =$ something less than $0.45f'_c =$ say 1500 psi (105 kg/cm^2, 10.3 N/mm^2), $f'_c = 5000$ psi (352 kg/cm^2, 34.5 N/mm^2), and $I_{cr}/I_g =$ a low value of say 0.20.

Calculation: substituting these numbers directly in Eqs. (5-71) and (5-61) yields $I_e/I_g = 0.71$, 0.66, and 0.62 for normal-weight, sand-lightweight, and all-lightweight concrete, respectively. Thus it is seen that a *maximum* reduction (for these conservative assumptions) in the effective I, or I_e, is about 29, 34, and 38 percent, respectively, for the live-load deflection increment (as shown in Fig. 5-7) for normal-weight, sand-lightweight, and all-lightweight concrete; based on the 1971 ACI Code maximum allowable stress for partially prestressed members.

I_e for continuous beams I_e in Eq. (5-61) represents an average I for simple spans. In this case, the moments in the equation refer to the maximum moment section, as at midspan for symmetrical simple spans. For continuous spans, the 1971 ACI Code recommends the use of an average of the positive and negative moment region values of I_e in the case of reinforced beams. The same may be used in the case of bonded prestressed beams. An improved value, as recommended by the author in Chap. 3 (based on ACI 435[38]), is obtained by using a weighted average of 0.70(positive moment I_e) + 0.30(negative moment I_e) for spans with both ends continuous, and 0.85(positive moment I_e) + 0.15(continuous end I_e) for spans with one end continuous. These apply for uniformly distributed loading, and the same section at both ends for spans continuous at both ends. In the case of unbonded beams loaded into the cracking range, this or other methods for cracked beam deflections should be used with extreme caution if at all.

Bilinear method Using I_g for the uncracked deflection increment and I_{cr} for the cracked deflection increment leads directly to Eqs. (5-72) and (5-73) for prestressed concrete beams.

When $(M_L)_{max} \le (M_L)_{cr}$: $$\Delta_L = KL^2\left[\frac{M_L}{E_c I_g}\right] \qquad (5\text{-}72)$$

When $(M_L)_{max} > (M_L)_{cr}$: $$\Delta_L = KL^2\left[\frac{(M_L)_{cr}}{E_c I_g} + \frac{M_L - (M_L)_{cr}}{E_c I_{cr}}\right] \qquad (5\text{-}73)$$

where $(M_L)_{cr}$ is the same as the previous equations. For the deflection coefficient, K, as defined herein, $M_L = (M_L)_{max}$.

Solving Eq. (5-73) for the cracked beams in Example 3:

Beam containing only prestressing steel

$$\Delta_L = \frac{(5/48)(100)^2(144)}{2.81 \times 10^3}\left[\frac{(204)(12)}{168\,970} + \frac{(250 - 204)(12)}{38\,860}\right] = 1.53 \text{ in. } (3.89 \text{ cm}) \qquad (5\text{-}73)$$

versus 1.45 in. (3.68 cm) in Example 3 using I_e for which $(M_L)_{max}/(M_L)_{cr} = 1/0.82 = 1.2$.

Beam containing both prestressing steel and nonprestressed tension steel

$$\Delta_L = \frac{(5/48)(100)^2(144)}{2.81 \times 10^3}\left[\frac{(128)(12)}{168\,970} + \frac{(250 - 128)(12)}{56\,160}\right] = 1.88 \text{ in. } (4.78 \text{ cm}) \qquad (5\text{-}73)$$

versus 2.26 in. (5.74 cm) in Example 3 using I_e for which $(M_L)_{max}/(M_L)_{cr} = 1/0.51 = 1.96$.

It might be noted that the I_e method is also essentially a "bilinear" method.

Comments

The procedures herein for taking into account the effect of nonprestressed tension steel include the stress or force in the nonprestressed steel when computing concrete stresses and prestress loss (such as Eqs. (5-6) to (5-9)), but not when computing camber (such as Eqs. (5-40) and (5-59)). In the latter case, this approximate method was developed on that basis, which is similar to the ACI Code (318-71) method for including the effect of compression steel in reducing time-dependent deflections of reinforced beams.

Chen[15] has suggested the following slight improvement in the use of I_e for cracked prestressed beams: instead of computing the more convenient total live-load deflection increment as shown in Fig. 5-7, compute only that part of the deflection increment above the "decompression" point, or point of zero bottom fiber stress, using the I_e formula. This is then the same as the reinforced concrete beam case for which the method was originally derived.

The relative effect of bonded steel on cracking is shown in Fig. 5-8 for three similar beams (6 in. by 8 in. by 15 ft, or 15.2 cm by 20.3 cm by 4.6 m) with approximately the same ultimate load capacity, but with varying amounts and strengths of nonprestressed tension steel, under a load of 8 kips (3630 kg, 35.6 kN) (about 85 percent of ultimate). Beams IV B 2 and B 3 contain four times and eight times as much nonprestressed steel as Beam IV B 1. The shallower crack pattern (and smaller deflection) are apparent for the beams with the greater bonded steel

Beam IV B1, Table 5-5
Figs. 5-5, 5-6
P_o = 25.2 kips
p = 0.40%
p'_H = 0.20%
$$p\frac{f_{su}}{f'_c} + p'_H\frac{f_y}{f'_c} = 0.28$$
Q_{ult} = 8.95 kips

Beam IV B2, Table 5-5
Figs. 5-5, 5-6
P_o = 25.8 kips
p = 0.40%
p'_{60} = 0.80%
$$p\frac{f_{su}}{f'_c} + p'_{60}\frac{f_y}{f'_c} = 0.28$$
Q_{ult} = 10.11 kips

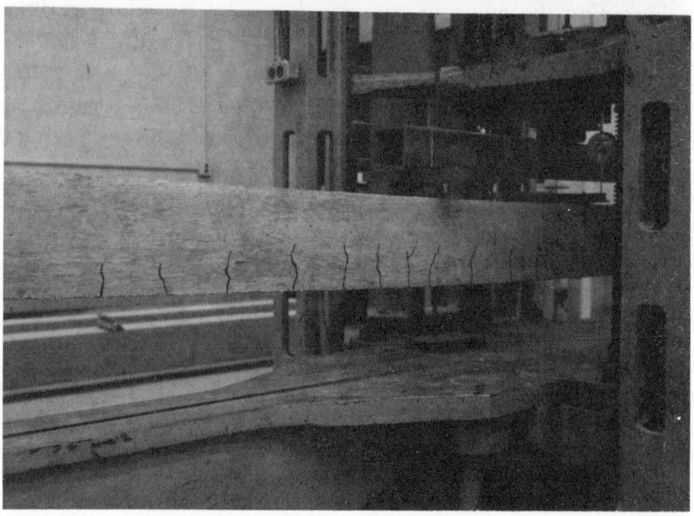

Beam IV B3, Table 5-5
Figs. 5-5, 5-6
P_o = 24.4 kips
p = 0.40%
p'_{33} = 1.50%
$$p\frac{f_{su}}{f'_c} + p'_{33}\frac{f_y}{f'_c} = 0.29$$
Q_{ult} = 9.23 kips

content, etc. Pictures of the same three beams at the 6 kip (2720 kg, 26.7 kN) load level are also shown by Shaikh and the author,[6] along with a quantitative description of the cracks.

The effect of nonprestressed tension steel is to reduce time-dependent camber, which in turn results in an increase in the net downward deflection under all loads.

Effect of Repeated Load Cycles Into the Cracking Range

Based on a study of 15 prestressed beams (nine noncomposite and six composite), and on data in the literature for reinforced beams, the same I_e procedure as given by Eq. (5-61) was shown by Kripanarayanan and the author[33] to apply as well for repeated load cycles into the cracking range. In this case, a modified I (I_{rep}, defined by Eq. (5-74)) was developed for the unloading and reloading back-to-the-previous-maximum-load-point parts of the load–deflection curve. A practical application of I_{rep} is discussed in Chap. 3 for reinforced beams (Eq. (3-29) and Example 4).

As described in Fig. 5-9, I_{rep} represents some value between I_e and I_g, as a function of the cracking, unloading (so-called repeat loading), and ultimate loads or moments.

$$I_{rep} = \psi I_e + (1 - \psi) I_g \qquad (5\text{-}74)$$

where $\psi = (Q_{ult} - Q_{rep})/(Q_{ult} - Q_{cr}) = (M_{ult} - M_{rep})/(M_{ult} - M_{cr})$

in which Q_{rep} or M_{rep} refers to the maximum load or moment in a given cycle (the repeat load or moment). It may be noted that no empirical constants except the third power in the I_e equation are required in Eq. (5-61) for I_e or in Eq. (5-74) for I_{rep}.

The rationale for defining ψ for short-time deflections is as follows. In Fig. 5-9 following unloading at point B, line BO (slope proportional to I_e) refers to the case of complete recovery, line BD parallel to line AO (slope proportional to I_g) refers to the case of ideal inelastic unloading, and line BE refers to the actual case somewhere between the other two for both unloading and reloading. For a severely cracked beam (Q_{rep} approaching Q_{ult}), there is no recovery due to the closure of cracks. For an uncracked beam ($Q_{rep} < Q_{cr}$), the recovery may be assumed to be total from Q_{rep} to 0. The value of ψ interpolates linearly between these two limits. For example:

(*a*) When $Q_{rep} \le Q_{cr}$, $\psi = 1$, and $I_{rep} = I_e = I_g$. This is a condition of total recovery.

(*b*) When $Q_{rep} > Q_{cr}$, ψ varies between 1 and 0, and I_{rep} is between I_e and

FIGURE 5-8
Three similar beams with approximately the same ultimate load capacity, but with varying amounts of nonprestressed tension steel, under a load of 8 kips (3630 kg, 35.6 kN) or about 85% of ultimate.[6]

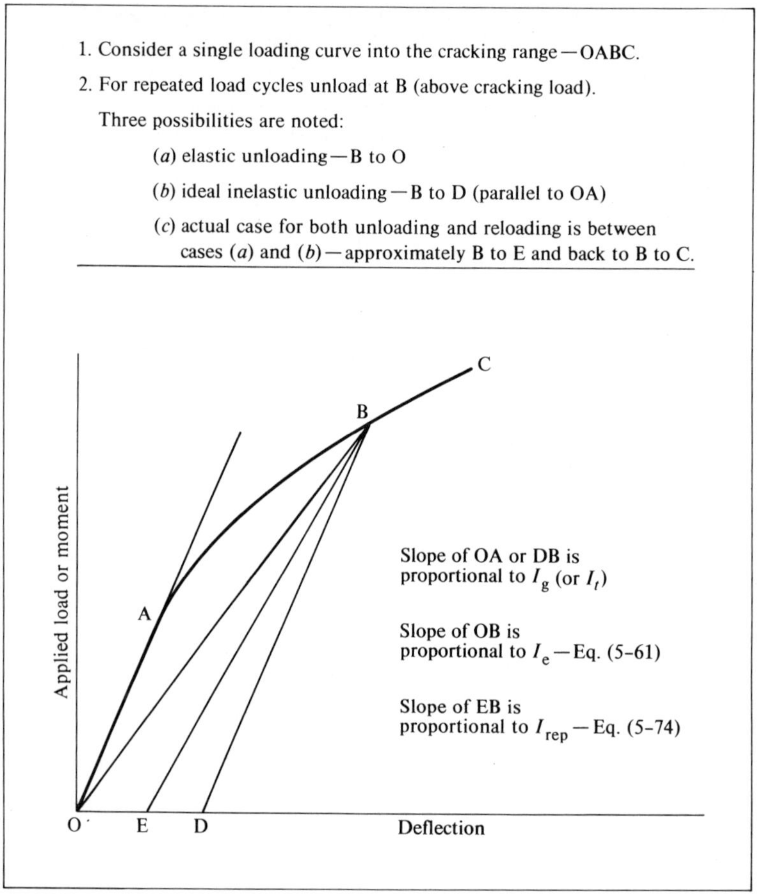

FIGURE 5-9
Description of load or moment versus deflection curve and moments of inertia in repeated load cycles.

I_g. This is a condition of some inelastic recovery due to the cracks being closed.

(c) When Q_{rep} approaches Q_{ult}, $\psi = 0$, and $I_{rep} = I_g$. This is a condition of no inelastic recovery due to the cracks being closed. This may also be considered as a condition of maximum residual deflection under short-time loading.

In this approximate procedure the hysteresis loop is neglected.

Shown in Fig. 5-10 are three load–deflection cycles to increasing load levels above the cracking load for one of the prestressed beams in Ref. 33. The following observations are noted regarding this generally typical case:

1. The predicted results using I_e and I_{rep} agree quite well with the measured load–deflection curves, except for the final loading to failure.

2. In Cycle 1 the approximate straight line up to cracking, followed by the

curved part of the diagram, followed by the approximate straight line unloading, can be seen in Fig. 5-10—as described in Fig. 5-9.

3. In Cycles 2 and 3 the approximate straight line reloading to the previous maximum-load-point, followed by the curved part of the diagram can be seen in Fig. 5-10—as described in Fig. 5-9.

4. An envelope of the curved parts of each cycle approximates the single loading curve in the cracking range, which tends to justify the use of I_e for computing deflections, even when a structural member will be subjected to repeated loading in the cracking range.

5. The increasing residual deflections in each cycle were attributed in part to the effects of cracking, but primarily to nonrecoverable creep effects due to the time involved in the loading cycles.

With reference to comparative repeated load tests of bonded and unbonded partially prestressed girders, Magura and Hognestad[45] found that pretensioned bridge girders with a concrete tension of about 700 psi (49.2 kg/cm^2, 4.83 N/mm^2) showed no significant detrimental effects from flexural cracks under repeated loading, whereas similarly stressed and cracked post-tensioned bridge girders showed serviceability distress and reduced flexural capacity from load repetitions.

The effect of repeated load cycles is discussed further in Chap. 3.

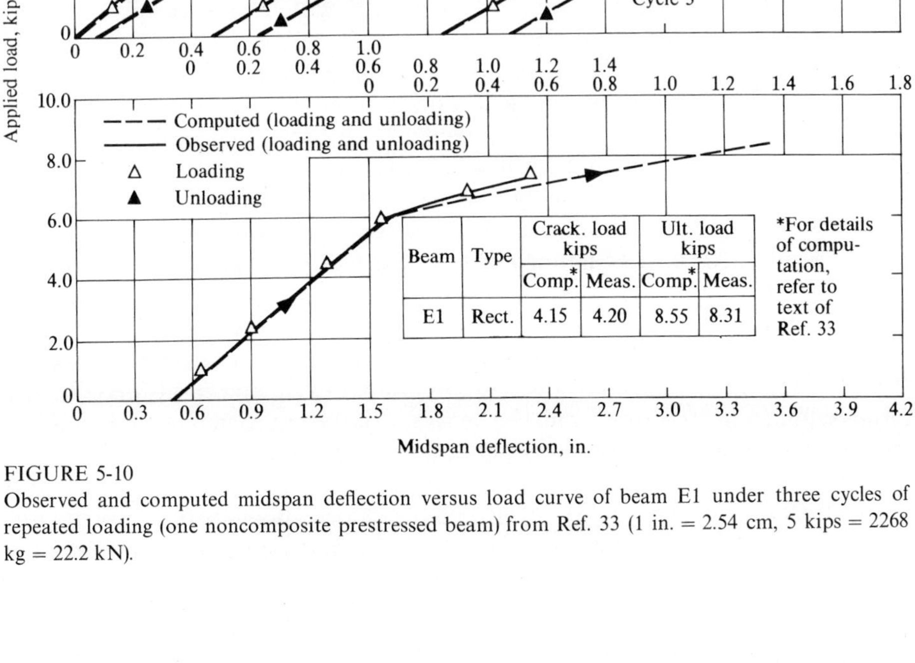

Beam	Type	Crack. load kips		Ult. load kips	
		Comp.*	Meas.	Comp.*	Meas.
E1	Rect.	4.15	4.20	8.55	8.31

FIGURE 5-10
Observed and computed midspan deflection versus load curve of beam E1 under three cycles of repeated loading (one noncomposite prestressed beam) from Ref. 33 (1 in. = 2.54 cm, 5 kips = 2268 kg = 22.2 kN).

5-5 DISCUSSION OF EXPERIMENTAL RESULTS

Comments in this section will be limited to fully prestressed members since reference to experimental results of partially prestressed members was included in Sec. 5-4. Published analytical and experimental studies of prestress concrete deformation exceed a hundred in number and cannot be reviewed here. Hence only a few papers relating to the calculation or prediction of losses, camber, and deflection will be mentioned (also see Refs. 47 to 55, and 68 to 70, etc.).

In 1960–1961 relative satisfaction with the calculation of camber using the approximate approach versus measured camber was indicated by the author and Ozell[2,3] for 10 post-tensioned laboratory beams (two in field environment), five of which were composite; and by Pauw and Breen[23] for two composite post-tensioned girders of a pedestrian bridge. Agreement with measured results did, of course, depend on the use of appropriate material parameters, such as by tests.

Also in 1961 Corley, Sozen, and Siess[22] presented a numerical integration procedure for determining curvatures and deflections using strain–time functions and both the rate-of-creep and superposition methods of creep analysis. Using experimental material parameters in the calculations, generally satisfactory concurrence was obtained with measured camber and deflection for two unloaded and two loaded laboratory beams, and one full-size bridge girder.

The 1963 ACI Committee 435 paper[9] used the step-function relation of the author and Ozell[2,3] (Eq. (5-39) herein), for which two solutions were presented; the approximate solution and the summation of numerical integration solution. As explained in the Committee paper, both of these solutions are based on the rate-of-creep method in which the creep coefficient is applied to initial or elastic deformation. However, both of these solutions also involve the superposition method as separate calculations are made for prestress loss, and camber or deflection caused by prestress and transverse loads, with the separate effects being combined to obtain the net result.

The Committee paper further states that, although the superposition method may predict the time-dependent deflections caused by the variable effects of prestress better than the rate-of-creep method, as shown by Corley, Sozen, and Siess,[22] its use is laborious and worthwhile only if the basic creep, shrinkage, and relaxation data are known reliably and accurately. Also, when the data on the time-dependent properties of the materials are not reliable, deflections should be calculated for critical combinations of the expected upper and lower bounds of the creep, shrinkage, and relaxation versus time curves.

With regard to comparisons for several groups of experimental beams, the Committee concluded that the numerical integration method involving a step-by-step summation technique for any desired number of time intervals may predict results within 5 to 10 percent, with the accuracy of the approximate method being somewhat less than this. The approximate method was deemed sufficiently accurate for most design purposes.

In 1968 Sinno and Furr[26,67] measured the deformational behavior of 19 full-sized prestressed concrete bridge beams and compared the results with computed

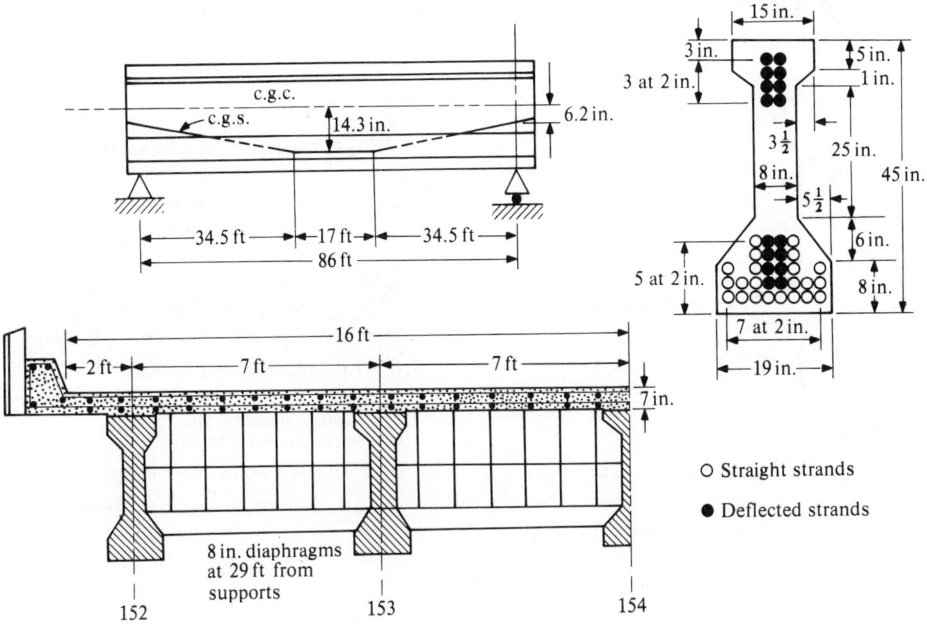

FIGURE 5-11
Sand-lightweight concrete composite prestressed bridge girders with normal-weight concrete deck slab[7,8] (for metric units, see metric conversion factors herein).

values using a one-step approximate method with a hyperbolic time-function, and also by a step-by-step computerized method. Known initial and time-dependent material properties were also used. They found predicted camber by both methods to be in excellent agreement with measured values for normal-weight concrete beams and shorter (40 ft, or 12.2 m) lightweight concrete beams, and a consistent slight overestimation of predicted values for longer (56 ft, or 17.1 m) lightweight concrete beams. The study found that all beams attained ultimate camber within 150 to 200 days. This corroborated the observations of the author and Ozell[2,3] that "total camber appears to reach its ultimate value relatively early (100 to 200 days) compared to the long-time shrinkage and creep strains that occur." This apparently is due to the offsetting effects of camber growth resulting directly from creep on the one hand, and reduction in camber resulting from the loss of prestress due to all time-dependent effects on the other.

Two papers by the author, Meyers, and Kripanarayanan[7,8] in 1970–1971 involved five noncomposite and four composite laboratory beams and five composite bridge girders of an Iowa bridge, all constructed of sand-lightweight concrete (124 pcf, or 1990 kg/m^3, or 19.5 kN/m^3). The bridge girders and resulting measured and computed camber are shown in Figs. 5-11 and 5-12 using the approximate step-function method. The agreement was considered to be reasonably good, but not precise, and probably represents the nature of the correlation

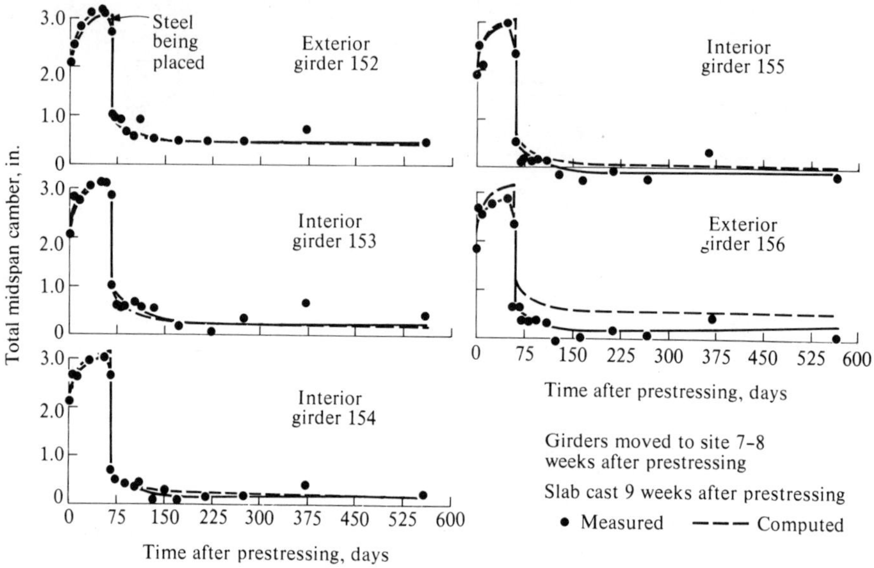

FIGURE 5-12
Measured and computed camber (using Eq. (5-44) herein with experimental material parameters) for the sand-lightweight bridge girders[7,8] (1 in. = 2.54 cm).

that might be expected at best for this type of behavior; that is, when experimentally determined material parameters are used in the calculations.

Other results from the literature were analyzed by the approximate step-function method bringing the total to 27 laboratory beams and 10 actual structural members. The comparisons between measured and computed loss of prestress and camber are shown in Figs. 5-13 and 5-14 using both experimental and general or average (as in Chap. 1) material parameters. The authors concluded that results by the approximate step function method will normally agree with actual results within ± 15 percent when using experimentally determined material parameters. The use of general or average material parameters predicted results that agreed with measured results in the range of ± 30 percent. With some knowledge of the time-dependent behavior of concrete using local aggregates and under local conditions, it was concluded that one should normally be able to predict loss of prestress and camber within about ± 20 percent using the procedures herein without testing for material properties in advance. It may also be noted in Figs. 5-13 and 5-14 that most of the results were considerably better than these limits, with a few points outside the limit-bands.

The following was further observed in Ref. 7:

> The loss of prestress for the sand-lightweight concrete bridge girders was of the order of 27 to 29% at 560 days and 29 to 31% ultimately. It was determined that loss percentages for bridges under similar conditions using normal weight concrete would be of the order of 25%, and those using all-lightweight concrete of the order of 35%. These losses were somewhat larger than

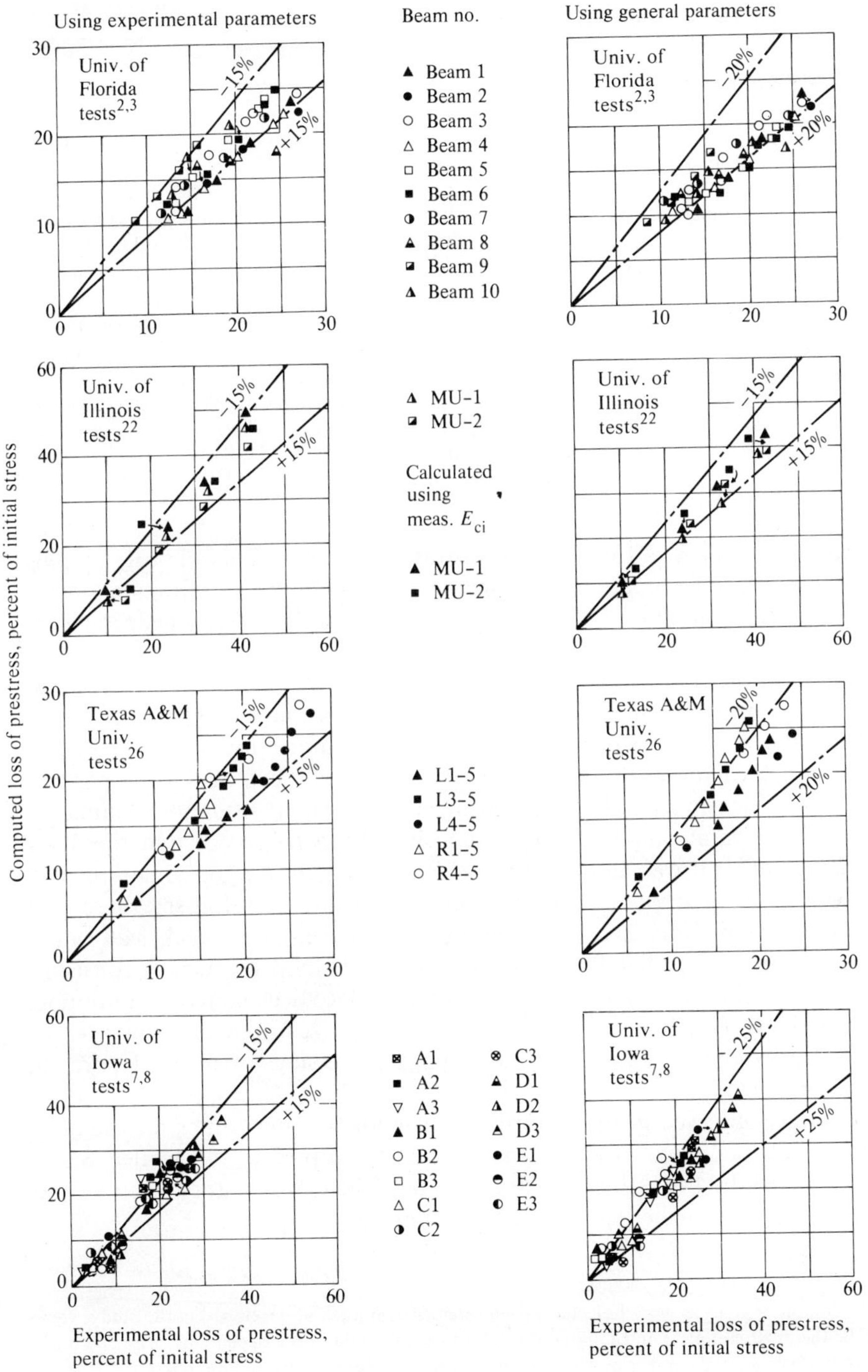

FIGURE 5-13
Comparison of experimental and computed (using Eqs. (5-6) and (5-17) herein) midspan loss of prestress at various ages.[8]

might be expected because a relatively large release strength of $f'_{ci} = 4500$ psi (316 kg/cm^2, 31.0 N/mm^2) and release stress of $0.60f'_{ci} = 2700$ psi (190 kg/cm^2, 18.6 N/mm^2) was used. Higher losses for the lighter concretes are due primarily to the lower concrete modulus of elasticity and not necessarily to greater creep and shrinkage.

The ultimate loss of prestress at midspan would be about 2% less (as 28% loss versus 30% loss), and the ultimate camber about 0.10 in. (0.25 cm) less, for a deck slab cast 3 weeks after prestressing, compared to the actual 9-week slab. This serves to point out the relatively small but still beneficial effect of casting a composite slab as early as possible, as noted also by Corley, Sozen, and Siess.[22] It was also noted that there are offsetting effects for different slab casting schedules. An earlier slab tends to reduce total creep deformation (which causes upward camber) by forming an earlier composite section, but it also reduces differential shrinkage and creep deformation (which causes downward deflection).

All five of the bridge girders in Fig. 5-11 show an increase in measured camber in Fig. 5-12 of about 0.4 in. (1.0 cm) between 300 and 370 days (which starts in April). This appears to be due to higher temperatures and is consistent with the observations of Delarue.[49]

In his discussion of Ref. 8, Libby[71] demonstrated the convenience of programming the approximate step-function relations so that engineers can investigate the effect of several variables on loss of prestress and long-term deflections at nominal cost.

The study of Mossiossian and Gamble[32] in 1972 used computer analysis methods involving the summation or numerical integration technique, which were based on the rate-of-creep and superposition methods for predicting time-dependent deformations. In addition, revised rate-of-creep and superposition methods were developed. They also outlined approximate equations similar to those herein for noncomposite and composite members, except that the Trost relaxation coefficient was used instead of the step-function approximation. Predicted deformations were compared with measured values for a full-size three-span prestressed I-beam Illinois highway bridge with a composite deck slab which provided continuity for live loads, and with a $\frac{1}{8}$-scale model structure consisting of a single row of prestressed I-beams and cast-in-place deck concrete continuous over three spans. The total loss of prestress of the full-size bridge of normal-weight concrete was about 21 percent of the initial prestressing force just before the release of prestress.

The study observed in the historical review for treating composite members that Branson's approach has been the one most accepted in the United States, and that the method is simple but does not take the entire history of stress into account. The study further concluded the following:

The computer analysis, based on the revised rate-of-creep method developed in the study, seems to be the most complete and reliable procedure for predicting time-dependent stresses, strains, loss of prestress, and camber of noncomposite and composite prestressed members.

The approximate procedure developed in the study, based on the superposition method, can be applied for predicting expected higher and lower bounds of time-dependent stresses, strains, loss of prestress, and camber of noncomposite and composite prestressed members, using only single-step slide rule calculations.

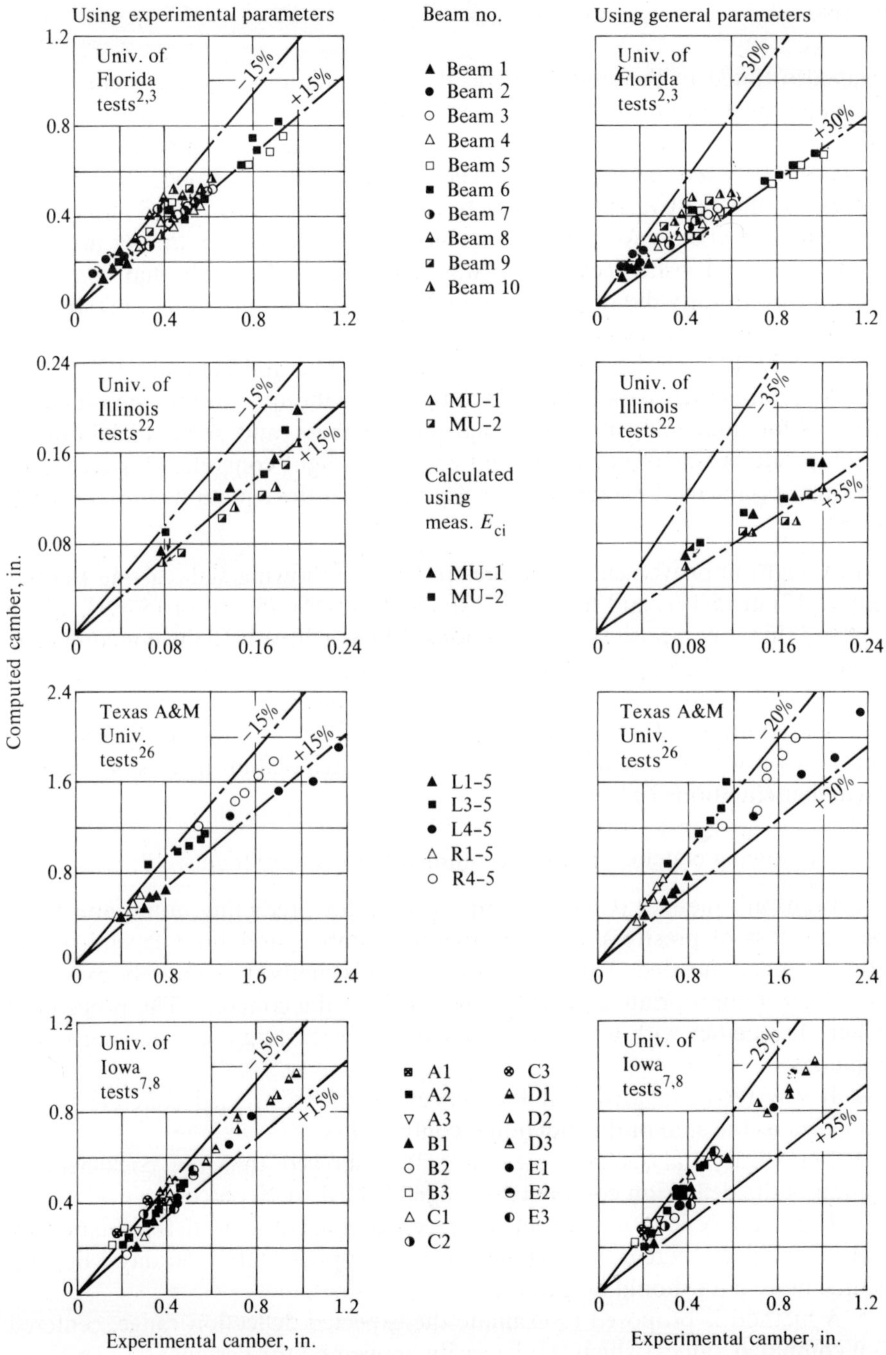

FIGURE 5-14
Comparison of experimental and computed (using Eqs. (5-40) and (5-44) herein) midspan camber at various ages[8] (1 in. = 2.54 cm).

It was also noted in the study that results using the approximate method with hyperbolic function of Sinno and Furr[26,67] and the approximate method in the study using the relaxation coefficient were approximately the same, and that the correlation between measured and calculated values was good. The loss of prestress of the lightweight concrete beam used from Refs. 26, 67 was about 24 percent at 300 days. With regard to the effect of partial continuity involving a continuous composite deck slab, it was indicated in the study that composite action and continuity did not have a significant effect on variation of prestressing force after casting of deck concrete.

The 1972–1974 studies by Zundelevich, Lee, Chiu, and Hamada[31,39,41,65,66] involved a comprehensive analytical investigation of the approximate step-function equations for noncomposite and composite members, and some 18 laboratory noncomposite beams of normal-weight and lightweight concrete, plus creep and shrinkage specimens. Two refinements were made in the equations for composite beams:

1. A slight improvement in the prestress term following slab casting (term 3 of Eq. (5-17) or (5-19); and term 4 of Eq. (5-44), (5-46), (5-48), or (5-49)), which now results in the composite beam equations reducing directly to the noncomposite beam equations when $I_2/I_c = 1$, $C_{t2} = C_t$, and the terms for the effects of the slab dead load and differential shrinkage and creep are deleted.
2. The addition of Eq. (5-47) herein, which appears to be a decidedly secondary effect, as acknowledged by the authors, and is not included in the loss of prestress and camber equations herein.

The following conclusions and recommendations were made:

1. Branson's method (Chaps. 1 and 5 herein) for predicting initial and time-dependent loss of prestress, axial shortening, camber, and time-dependent deflection is recommended. The data show that this method accurately estimates these effects if appropriate coefficients are used for the concrete. The procedures in Chap. 1, together with test data, were used in evaluating creep and shrinkage parameters.
2. It was recommended that the deflections of prestressed concrete beams be modeled as the sum of the individual components.
3. Statistical analysis of the limited data showed that the coefficient of variation of the deflection components ranged from 1 to 10 percent.
4. The effect of loading age on the creep coefficient is well represented by Eq. (1-20), Chap. 1 herein, for loading prior to approximately 56 days. Beyond this time more data should be gathered.
5. A method is proposed to evaluate the expected deflection range, centered about computed values, which can be easily applied by the designer.

In 1973 Rao and Dilger[19] reported an analytical and experimental investigation of the time-dependent behavior of composite prestressed composite beams in which a new rational method called the "varying stiffness approach" was presented. This is a step-by-step computerized procedure in which normal strain and curvature are the unknowns in each interval of time. Forces are expressed in terms of these deformations using the stiffness and compatibility conditions. Thus the problem is reduced to finding the two unknown deformations from the two equations of equilibrium. The forces are then determined by using these two deformations and the force–displacement relations of each interval. The experimental phase of the study included eight laboratory and field composite prestressed beams.

The following conclusions and recommendations were made:

1. The time-dependent parameters based on the CEB recommendations, and on Branson's standard curves (Chap. 1 procedure herein) gave very close results under no stress or constant stress. Under varying stress also, the total deformations according to these two approaches were approximately the same.
2. The approximate Trost relaxation approach does not give accurate results for this kind of problem involving creep under decreasing stress levels.
3. Dilger's concept of modified superposition principle in stress removal is verified, but its use showed only slight improvement, as only about 25% of the stress is removed. As such, McHenry's superposition principle is recommended for the analysis of composite prestressed beams; whereas Dilger's modification is recommended for problems in which complete or nearly complete recovery of stress takes place.
4. The role of shear or web reinforcement and deck reinforcement depends on the presence and magnitude of superimposed dead load. In the absence of superimposed dead load, the presence of web reinforcement with a minimum of deck reinforcement is recommended. But in the presence of superimposed dead load, deck reinforcement with a minimum of web reinforcement is recommended for better long-time performance.
5. The "varying stiffness approach" is far simpler than the conventional method and leads to the same results as the method given in the book by Neville and Dilger.[17] Results by this method were found to be in fairly good agreement with the experimental results. This method of analysis can also be used in hand calculations, and is better suited for use with a computer than the conventional approaches.

The measured deflections of double-Ts and hollow-cored slabs with high tension allowed in the concrete were reported in 1974 by Wright and Burns.[74] Results were compared with a numerical analysis using tensile stresses of $6\sqrt{f'_c}$, $12\sqrt{f'_c}$, and $16\sqrt{f'_c}$. The double-Ts were found to be more efficient in controlling post-cracking deflections, as might be expected. The hollow-cored slabs were found to deflect very rapidly after cracking and tended to develop deflection problems when $12\sqrt{f'_c}$ was used. Overall, the double-Ts met the ACI Code deflection requirements for the allowable tensile stress of $12\sqrt{f'_c}$ under full live load. A composite topping was shown to add to the post-cracking stiffness of the double-Ts. The following specimens were used:

Double-Ts:	8DT12—	span 23 (refers to 8 ft or 2.44 m flange width, 12 in. or 30.5 cm depth, 23 ft or 7.0 m span)
	8DT12 + 2—	span 36 (+2 refers to 2 in., or 5.1 cm topping)
	8DT24—	span 32
	8DT24 + 2—	span 60
Hollow-cored slabs:	4HC8—	span 24 (refers to 4 ft or 1.22 m width, 8 in. or 20.3 m depth, 24 ft or 7.3 m span)
	4HC12—	span 32
	4HC12—	span 44

In a 1975 paper by Tadros, Ghali, and Dilger,[75] an analytical method suitable for hand calculations is presented for determining time-dependent prestress loss, axial strain, curvature, and deflection of prestressed members, with computed results comparing favorably with experimental results from seven test beams. The method involves the use of a "recovery parameter" and a "relaxation parameter" which are based on step-by-step numerical calculations and tabulated in the form of design aids. These step-by-step computations employed a time-variation function for creep, shrinkage, relaxation, and the modulus of elasticity of concrete, which were presented by the authors in another paper.[76] Two numerical examples are included.

In the discussion of prestress loss and camber data herein, it should be remembered that only camber data can be obtained *directly*. In the case of prestress loss data, experimental results are obtained only *indirectly*, since steel relaxation is not a strain phenomenon and must be determined separately, even though steel relaxation does influence the beam strains. In addition, most experimental results used in studying loss of stress in the prestressing steel has been based on concrete strain measurements on the beam faces.

REFERENCES

1. Prestressed Concrete Institute, "PCI Design Handbook," 1-1 to 12-8, Prestressed Concrete Institute, Chicago, Illinois, 1971.
2. D. E. Branson, "Time-Dependent Effects in Non-Composite and Composite Prestressed Concrete Beams," Ph.D. dissertation, University of Florida, June 1960.
3. D. E. Branson and A. M. Ozell, Camber in Prestressed Concrete Beams, *ACI Journal*, Proceedings **57**, 12, 1549–1574, June 1961.
4. D. E. Branson, Time-Dependent Effects in Composite Concrete Beams, *ACI Journal*, Proceedings **61**, 2, 213–230, Feb. 1964.
5. D. E. Branson, Design Procedures for Computing Deflections, *ACI Journal*, Proceedings **65**, 9, 730–742, Sept. 1968.
6. A. F. Shaikh and D. E. Branson, Non-Tensional Steel in Prestressed Concrete Beams, *PCI Journal*, **15**, 1, 14–36, Feb. 1970.
7. D. E. Branson, B. L. Meyers, and K. M. Kripanarayanan, Time-Dependent Deformation of Noncomposite and Composite Prestressed Concrete Structures (paper presented by D.E.B. at the 40th Annual Meeting, Highway Research Board, Washington, D.C., Jan. 1970; based on Iowa State Highway Commission and University of Iowa Research Reports 69-1, Feb. 1969, and 70-6, Aug. 1970), *National Academy of Sciences—National Academy of Engineering, Highway Research Record*, No. 324: Symposium on Concrete Deformation, 15–43, 1970.

8. D. E. Branson and K. M. Kripanarayanan, Loss of Prestress, Camber and Deflection of Non-Composite and Composite Prestressed Concrete Structures (four papers presented by D.E.B. on various aspects at the 6th Congress, Fédération Internationale de la Précontrainte, Prague, June 1970; and at Design Seminars sponsored by the Prestressed Concrete Institute, Chicago, Jan. 1971 and June 1971, and by the California Division of Highways and Ceramic Lightweight Aggregate Association, Sacramento, Mar. 1971), *PCI Journal,* **16,** 5, 22–52, Sept.–Oct. 1971.
9. Subcommittee 5, ACI Committee 435, Subcommittee Chairman, A. C. Scordelis; Committee 435 Chairman, D. E. Branson; and M. A. Sozen, Deflections of Prestressed Concrete Members, *ACI Journal,* Proceedings **60,** 12, 1697–1728, Dec. 1963. Republished in "ACI Manual of Concrete Practice," Part 2, 1967, 1972.
10. Subcommittee 2, ACI Committee 209, Subcommittee Chairman, D. E. Branson, Committee 209 Chairman, J. R. Keeton, Prediction of Creep, Shrinkage and Temperature Effects in Concrete Structures (paper presented by D.E.B. at the Annual ACI Convention, New York, Apr. 1970), in "Designing for Effects of Creep, Shrinkage and Temperature in Concrete Structures," 51–93, ACI Publication SP 27-3 (symposium volume), 1971.
11. H. Trost, Effect of the Principle of Superposition on Creep and Relaxation Problems in Concrete and Prestressed Concrete (in German), *Beton- und Stahlbetonbau (Berlin-Wilmersdorf),* **62,** 10, 230–238, Oct. 1967; and **62,** 11, 261–269, Nov. 1967.
12. H. Trost, Calculation of Creep and Relaxation Effects in Prestressed Concrete Structures, *International Association for Bridge and Structural Engineering, Symposium on Design of Concrete Structures for Creep, Shrinkage and Temperature Changes, Madrid,* 191–198, 1970.
13. AASHTO–PCI Joint Committee, Chairmen, W. Henneberger and W. B. Bennett, Jr., "Interim Specification, Prestressed Concrete, Section 6," 1–38, Prestressed Concrete Institute, Chicago, Illinois, Nov. 1970.
14. AASHTO, "Standard Specifications for Highway Bridges," 11th edn, 1–469, 341 National Press Building, Washington, D.C., 20004, 1973.
15. C. I. Chen, "Integrated Procedures for Predicting the Initial and Time-Dependent Deformation of 'Reinforced' Concrete Structures," Ph.D. thesis, Civil Engineering Department, University of Iowa, Dec. 1973.
16. R. H. Evans and E. W. Bennett, "Pre-Stressed Concrete," 1–294, John Wiley & Sons, New York, 1958.
17. A. M. Neville and W. Dilger, "Creep of Concrete: Plain, Reinforced, and Prestressed," 1–622, North-Holland Publishing Company, Amsterdam, 1970.
18. W. Dilger and A. M. Neville, "Effects of Creep and Shrinkage in Composite Members," *Second Australasian Conference on the Mechanics of Structures and Materials.*
19. V. J. Rao and W. Dilger, "Time-Dependent Behavior of Composite Prestressed Concrete Beams," University of Calgary, Calgary, Alberta, Canada, 1973.
20. American Concrete Institute, "Building Code Requirements for Reinforced Concrete," 1–78, and Commentary, 1–96, American Concrete Institute, Detroit, Michigan, 1971.
21. Comité Européen du Béton—Fédération Internationale de la Précontrainte, "International Recommendations for the Design and Construction of Concrete Structures," 1–80, Cement and Concrete Association, London, June 1970.
22. W. G. Corley, M. A. Sozen, and C. P. Siess, Time-Dependent Deflections of Prestressed Concrete Beams, *Highway Research Board, Bulletin* 307, 1–25, 1961.
23. Adrian Pauw and John E. Breen, Field Testing of Two Prestressed Concrete Girders, *Highway Research Board, Bulletin* 307, 42–63, 1961.
24. D. E. Branson, Instantaneous and Time-Dependent Deflections of Simple and Continuous Reinforced Concrete Beams, *Alabama Highway Department, Bureau of Public Roads, HPR Report* 7, Part 1, 1–78, Aug. 1963 (1965).
25. C. L. Freyermuth and R. A. Shoolbred, Posttensioned Prestressed Concrete, *Portland Cement Association Report,* 1–24, 1967.
26. R. Sinno, "The Time-Dependent Deflections of Prestressed Concrete Bridge Beams," Ph.D. dissertation, Civil Engineering Department, Texas A & M University, Jan. 1968.
27. C. L. Freyermuth, Design of Continuous Highway Bridges with Precast Prestressed Concrete Girders, *PCI Journal,* **14,** 2, 14–36, Apr. 1969.

28. E. G. Nawy and J. G. Potyondy, Moment Rotation, Cracking and Deflection of Spirally Bound, Pretensioned Prestressed Concrete Beams, *Rutgers University, Engineering Research Bulletin* 51, 1–97, 1970.
29. D. E. Branson, Camber and Deflection of Beams, *PCI International Design Seminars, Chicago,* 1–38, Jan. 1971 and June 1971.
30. E. G. Nawy and J. G. Potyondy, authors' closure and discussion by D. E. Branson of Flexural Cracking Behavior of Pretensioned Prestressed Concrete I- and T-beams, *ACI Journal,* May 1971; discussion and closure in *ACI Journal,* Proceedings **68,** 11, 873–877, Nov. 1971.
31. S. Zundelevich, D. L. N. Lee, and A. N. L. Chiu, Camber and Deflection Behavior of Prestressed Concrete Beams, *University of Hawaii Technical Report* CE71-R1, 1–98, Feb. 1972.
32. V. Mossiossian and W. L. Gamble, Time-Dependent Behavior of Noncomposite and Composite Prestressed Concrete Structures Under Field and Laboratory Conditions, *University of Illinois Structural Research Series* No. 385, 1–517, May 1972.
33. K. M. Kripanarayanan and D. E. Branson, Short-Time Deflections of Beams Under Single and Repeated Load Cycles, *ACI Journal,* Proceedings **69,** 2, 110–117, Feb. 1972.
34. E. W. Bennett and N. Veerasubramanian, Behavior of Nonrectangular Beams With Limited Prestress After Flexural Cracking, *ACI Journal,* Proceedings **69,** 9, 533–542, Sept. 1972. Discussion by D. E. Branson and C. I. Chen in *ACI Journal,* Mar. 1973.
35. P. W. Abeles, Partial Prestressing and Possibilities for Its Practical Application, *PCI Journal,* **4,** 1, 35–51, June 1959.
36. P. W. Abeles, Studies of Crack Widths and Deformation Under Sustained and Fatigue Loading, *PCI Journal,* **10,** 6, 43–52, Dec. 1965.
37. P. W. Abeles, E. I. Brown, II, and J. O. Woods, Jr., Preliminary Report on Static and Sustained Loading Tests, *PCI Journal,* **13,** 4, 12–32, Aug. 1968.
38. Subcommittee 7, ACI Committee 435, Subcommittee Chairmen, R. S. Fling and A. F. Shaikh, Committee 435 Chairmen, B. L. Meyers and G. M. Sabnis, Deflections of Continuous Beams, *ACI Journal,* Proceedings **70,** 12, 781–787, Dec. 1973.
39. S. Zundelevich, H. S. Hamada, and A. N. L. Chiu, Time-Dependent Behavior of Hawaiian Aggregate Concrete Subject to Repeated Loadings, *Highway Research Board, Record* 428, 32–50, 1973.
40. D. E. Branson, The Deformation of Noncomposite and Composite Prestressed Concrete Members, *ACI Committee 435 Symposium Paper, ACI Convention, San Francisco,* Apr. 1974.
41. S. Zundelevich, Variability of Deflections of Simply Supported Prestressed Concrete Beams, *ACI Committee 435 Symposium Paper, ACI Convention, San Francisco,* Apr. 1974.
42. P. W. Abeles, Static and Fatigue Tests on Partially Prestressed Concrete Construction, *ACI Journal,* Proceedings **50,** 7, 361–376, Dec. 1954.
43. P. W. Abeles, Partial Prestressing in England, *PCI Journal,* **8,** 1, 51–72, Feb. 1963.
44. P. W. Abeles, Design of Partially Prestressed Concrete Beams, *ACI Journal,* Proceedings **64,** 10, 669–677, Oct. 1967.
45. D. Magura and E. Hognestad, Tests of Partially Prestressed Concrete Girders, *Journal, ASCE Structural Division,* Proceedings **92,** ST1, 327–343, Feb. 1966.
46. N. H. Burns, Moment Curvature Relationships for Partially Prestressed Concrete Beams, *PCI Journal,* **9,** 1, 52–63, 1964.
47. S. G. Hutton and R. E. Loov, Flexural Behavior of Prestressed, Partially Prestressed and Reinforced Concrete Beams, *ACI Journal,* Proceedings **63,** 12, 1401–1408, Dec. 1966.
48. P. Zia and J. F. Stevenson, Creep of Concrete Under Nonuniform Stress Distribution and Its Effect on Camber of Prestressed Concrete Beams, *North Carolina State Highway Commission and Bureau of Public Roads, Project* ERD-110-R, 1–110, June 1964.
49. J. Delarue, "Fluage et Béton Précontraint," *RILEM Colloquium,* Munich, Nov. 1958.
50. J. Warawaruk, M. A. Sozen, and C. P. Siess, Strength and Behavior, in Flexure of Prestressed Concrete Beams, *University of Illinois, Engineering Experiment Station Bulletin* 464, 1–105, Aug. 1962.
51. G. L. Rogers, Validity of Certain Assumptions in the Mechanics of Prestressed Concrete, *ACI Journal,* Proceedings **49,** 7, 317–330, Dec. 1953.
52. M. W. Ruiz, "Effect of Repeated Loads on the Rotation Capacity of Reinforced Concrete Beams," dissertation, Cornell University, Sept. 1968.

53. J. R. Keeton, Study of Creep in Concrete, Phases 1–5, *U.S. Naval C.E. Lab., Port Hueneme, California, Technical Reports* R333-I, II, III, 1965.
54. W. L. Gamble, Field Investigation of a Continuous Composite Prestressed I-Beam Highway Bridge Located in Jefferson County, Illinois, *University of Illinois, Civil Engineering Studies, Structural Research Series* No. 360, June 1970.
55. R. A. Breckenridge, P. J. Valent, and S. L. Bugg, Effects of Long-Time Loads on Prestressed Concrete Beams, *U.S. Naval C.E. Lab., Port Hueneme, California, Technical Report* R-518, 1–43, Mar. 1967.
56. T. Y. Lin, "Design of Prestressed Concrete Structures," 2nd edn, 1–614, John Wiley & Sons, 1963.
57. James R. Libby, "Modern Prestressed Concrete," 1–516, Van Nostrand Reinhold Co., New York, 1971.
58. Post-Tensioning Committee, Western Concrete Reinforcing Steel Institute, "Post-Tensioned Box Girder Bridges, Design and Construction," 1–106, Prestressed Concrete Institute, 1969.
59. J. J. Shideler and E. V. Lyon, "Post-Tensioned Prestressed Concrete: An Annotated Bibliography," 1–145, Prestressed Concrete Institute, Jan. 1972.
60. C. K. Wang and C. G. Salmon, "Reinforced Concrete Design," 2nd edn, 1–934, Index Educational Publishers, New York, 1973.
61. H. K. Preston, "Practical Prestressed Concrete," 1–340, McGraw-Hill Book Co., New York, 1960.
62. H. K. Preston and N. J. Sollenberger, "Modern Prestressed Concrete," 1–337, McGraw-Hill Book Co., New York, 1967.
63. PCI Committee on Allowable Stresses, Chairman G. G. Goble, Allowable Tensile Stresses for Prestressed Concrete, *PCI Journal*, **15**, 1, 37–42, Feb. 1970.
64. E. A. B. Salse, "Prestressed Concrete Design Requirements for Flexure," Chap. 23, 23-1 to 23-21, notes on ACI 318-71, Building Code Requirements with Design Applications, Portland Cement Association, 1972.
65. H. S. Hamada, S. Zundelevich, and A. N. L. Chiu, Time-Dependent Behavior of Concrete Made with Hawaiian Aggregates, *Highway Research Board, Record* 400, 37–54, 1972.
66. S. Zundelevich, C. T. Hamada, and A. N. L. Chiu, Camber and Deflection Behavior of Prestressed Concrete Beams, Phase II, *University of Hawaii Technical Report* CE73-R1, 1–123, June 1973.
67. R. Sinno and H. L. Furr, Hyperbolic Functions for Prestress Loss and Camber, *Journal, ASCE Structural Division*, Proceedings **96**, ST4, 803–821, Apr. 1970.
68. A. M. Mattock, Precast-Prestressed Concrete Bridges, 5. Creep and Shrinkage Studies, *PCA Research and Development Laboratories, Skokie, Illinois, Bulletin* 46, May 1961.
69. U. Finsterwalder, Results of Creep and Shrinkage Measurements on Prestressed Concrete Structures, *Beton und Stahlbetonbau Journal* (in German), **50**, 44–50, 1955; **53**, 136–144, 1958.
70. F. Leonhardt, "Prestressed Concrete, Design and Construction," Wilhelm Ernst und Sohn, Berlin, 1964.
71. J. R. Libby, discussion of Ref. 8, *PCI Journal*, **17**, 1, 109–110, Jan.–Feb. 1972.
72. D. E. Branson, The Deformation of Noncomposite and Composite Prestressed Concrete Members (paper presented at the Annual ACI Convention, San Francisco, Apr. 1974), *ACI Publication SP 43-4, Deflections of Concrete Structures*, 83–127, 1974.
73. PCI Committee on Prestress Losses, Chairman, H. K. Preston, Recommendations for Estimating Prestress Losses, *PCI Journal*, **20**, 4, 43–75, July–Aug. 1975.
74. H. E. Wright, Jr., and N. H. Burns, Deflection of Double Tees and Hollow-Cored Slabs with High Tension Allowed in Concrete, *PCI Journal*, **19**, 4, 100–118, July–Aug. 1974.
75. M. K. Tadros, A. Ghali, and W. H. Dilger, Time-Dependent Prestress Loss and Deflection in Prestressed Concrete Members, *PCI Journal*, **20**, 3, 86–98, May–June 1975.
76. A. Ghali, M. K. Tadros, and W. H. Dilger, Accurate Evaluation of Time-Dependent Deflection of Prestressed Concrete Frames, *ACI Publication SP 43, Deflections of Concrete Structures*, 357–375, 1974.
77. Ti Huang, Direct Method for Estimating Prestress Losses, *National Research Council, Concrete and Steel Bridges, Transportation Research Record* 547, 37–46, 1975.

APPENDIX
K

COMMON CASES OF PRESTRESS MOMENT DIAGRAMS WITH FORMULAS FOR COMPUTING CAMBER

	Prestressed beam / P_oe moment diagram	Midspan camber due to P_oe moments
1.	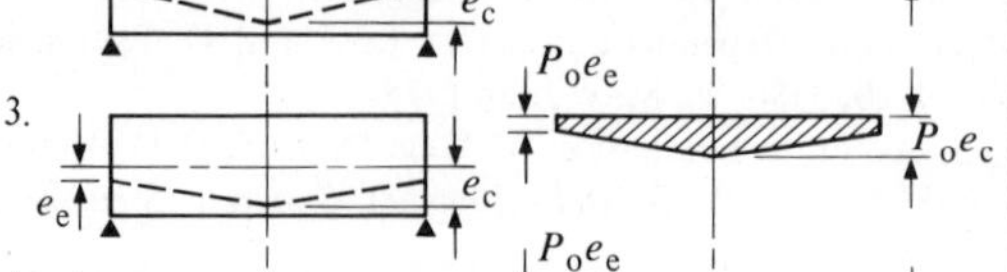	$(\Delta_i)_{P_o} = P_oeL^2/8E_{ci}I_g$
2.	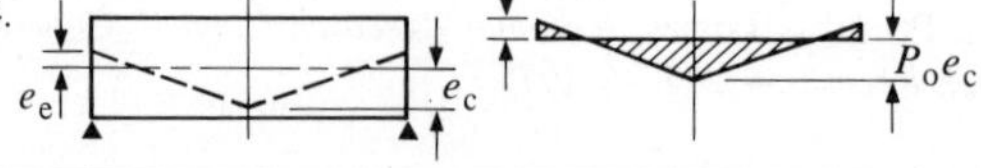	$(\Delta_i)_{P_o} = P_oe_cL^2/12E_{ci}I_g$
3.	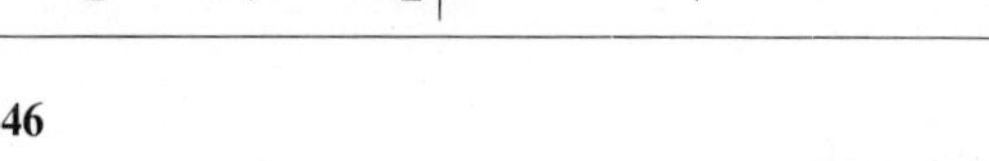	$(\Delta_i)_{P_o} = \dfrac{P_o(e_c - e_e)L^2}{12E_{ci}I_g} + \dfrac{P_oe_eL^2}{8E_{ci}I_g}$
4.		$(\Delta_i)_{P_o} = \dfrac{P_o(e_c + e_e)L^2}{12E_{ci}I_g} - \dfrac{P_oe_eL^2}{8E_{ci}I_g}$

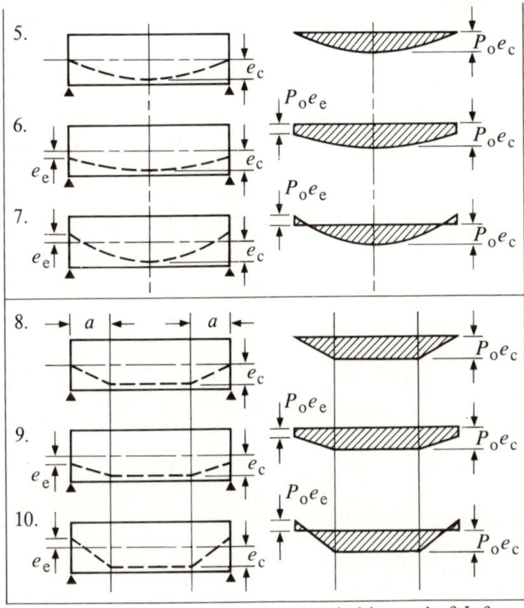

5. $$(\Delta_i)_{P_o} = 5P_oe_cL^2/48E_{ci}I_g$$

6. $$(\Delta_i)_{P_o} = \frac{5P_o(e_c - e_e)L^2}{48E_{ci}I_g} + \frac{P_oe_eL^2}{8E_{ci}I_g}$$

7. $$(\Delta_i)_{P_o} = \frac{5P_o(e_c + e_e)L^2}{48E_{ci}I_g} - \frac{P_oe_eL^2}{8E_{ci}I_g}$$

8. $$(\Delta_i)_{P_o} = \frac{P_oe_c}{E_{ci}I_g}\left(\frac{L^2}{8} - \frac{a^2}{6}\right)$$

9. $$(\Delta_i)_{P_o} = \frac{P_o(e_c - e_e)}{E_{ci}I_g}\left(\frac{L^2}{8} - \frac{a^2}{6}\right) + \frac{P_oe_eL^2}{8E_{ci}I_g}$$

10. $$(\Delta_i)_{P_o} = \frac{P_o(e_c + e_e)}{E_{ci}I_g}\left(\frac{L^2}{8} - \frac{a^2}{6}\right) - \frac{P_oe_eL^2}{8E_{ci}I_g}$$

For practical reasons I_g is recommended instead of I_t for most purposes.

FIGURE K-1
Common cases of prestress moment diagrams with formulas for computing camber.

CHAPTER

SIX

CONTROL OF DEFLECTIONS

6-1 INTRODUCTION

The Deflection Problem

Prior to the 1950s allowable stresses were lower than those permitted today, and reinforced concrete structures were, thus, often uncracked or only slightly cracked. Also, walls and partitions tended to be more massive and helped to carry structural loads, even though they were not designed to do so. This situation usually resulted in relatively large structural members and relatively small deflections. However, with the present-day use of higher strength materials, the strength or load-factor method of design, and the resulting shallower sections and more slender members, the prediction and control of deflections has become increasingly important.

The following are three principal types of deflection problems commonly found in practice.

1. Where excessive deflections result in either esthetic or functional problems. Examples include objectionable visual sagging, ponding of water on roofs which can result in accelerated deterioration, and the prevention of proper operation of machinery, sliding doors, etc.

2. Where structural members are attached to other structural elements or nonstructural elements that may be damaged by large deflections. Examples include excessive rotation at joints causing unsightly cracks which may also permit the entrance of water, and the damage of partitions, cladding, ceilings, and other finishes.

3. Where dynamic effects due to excessive deflections or insufficient stiffness cause discomfort. Examples include the perceptible ("spongy") vertical motion of building and bridge floor systems, and the perceptible lateral motion of high-rise buildings.

Very little information had been developed on type 3 in the literature and in the various codes of practice, although all three types of problems are included in this chapter.

In an extensive study by Mayer and Reusch[1] involving a survey conducted in Europe from 1960 to 1965, the building damage caused by the deflection of reinforced concrete building components was summarized as follows:

Partition wall damage	106 times
Excessive slab deflection	60
Cracks in reinforced concrete components	39
Horizontal cracks in the vicinity of slab supports	29
Damage to plaster and finishes	13
Isolated deflection damage	12
	259 times

The total number of possible types of damage was greater than the reported 181 cases of damage, since often more than one possibility could be observed in a single instance. Damage was most frequently observed in two-way reinforced plates and ribbed slabs. It was also established that the problems were attributable to design errors 137 times (45 percent), size deficiencies 76 times (25 percent), errors in workmanship 56 times (19 percent), and calculation errors 33 times (11 percent). Frequently more than one cause contributed to the same damage. It was also noted that in countries where low strength concrete and steel are still in use, deflection damage was practically unknown.

Partition wall damage constituted the most important group. This included not only cracks in the walls, but also large gaps in the vicinity of baseboards, and horizontal cracks at the joint between wall and ceiling. In addition, complaints about jammed doors must be included in this group. The second most important damage group stems from excessive slab deflections, and includes the formation of troughs in roofs and complaints about slanting furniture. A further damage group consisted of plaster and finishing damage due to excessive slab deflections. As a separate damage group may be listed all cases of damage due to excessive rotation of the reinforced concrete edge beams. This includes horizontal cracks above or below ceiling supports, as well as excessive lateral bending deflections of steel or concrete columns. Included in a further damage group were the cracks in reinforced concrete components themselves.

A 1971 study by Plewes[2] of reinforced and prestressed building and bridge deflection problems in North America is summarized in Table 6-1. Various types of difficulty are noted in addition to those above.

The problem of controlling deflections arises, not only from one's inability to accurately predict structural deformation in more or less ideal laboratory situations, but also from the many uncertainties associated with construction.

Table 6-1 Examples of reinforced and prestressed building and bridge deflection problems in North America, from a study by Plewes[2]

Type of structure	Type of construction	Amount of deflection	Result of deflection	Contributors' remarks
Office	Flat plate 20 ft × 25 ft bays Thickness 8 in.	2 in. at c (L/135)	Walls out of plumb Doors out of square Furniture rolled Tenants moved	Poor curing Some of deflection during construction 8 in. said to be too thin
Bridge	Continuous box girder, 7 ft deep, 104-134-134-104 spans	3 in. (L/535)	Railings and curbs out of line	Camber inadvertently omitted
Bridge	Post-tensioned CIP slab, 2 span contin., 110-110, 51 in. deep		Spoiled appearance	Friction losses less than assumed in design
Multi-storey drying house 150°F	6 in. simple slab 20 ft span	$1\frac{3}{4}$ in. (L/137)	Posts of drying racks buckled	Slab should have been at least 9 in. thick. Heat arrested normal concrete strength gain
Penthouses (several)	Various simple slabs		Curling at corners	Warping due to differential temperature and humidity
Pulp silo roof	30 ft dia.—12 in. thick at centre, 8 in. at edge	8 in. (L/45)	Cracked tile walls	Differential temperature—30°F o/s, 120°F i/s
School	Rigid frames 42 ft span	Double calc. defl.		Inadequate theory (1951)
Class room	Waffle slabs	$2\frac{1}{2}$ to 3 in.	Cracked partitions	Slabs said to be too thin (within ACI 1951 Code limits)

26 bldgs. dormitories	Simple 8 in. slabs, 19 ft span on brick walls	Up to 2 in. ($L/114$)	Lifting at corners and distortion of walls	1957 design changed for subsequent buildings
Unknown	14 in. waffle slab 60 ft span	5–6 in. ($L/120$)	Partitions and door frames racked	"Design by aggressive and popular 'modern' architect engineer"
Office	5 in. pan joists 25 in. o.-c. 25 ft spans Masonry-bearing walls	$\frac{3}{4}$ in. ($L/400$)	Cracks in block partitions at doors Crushing of blocks at corners Cracks in o/s masonry walls	Too early placement of partitions Lack of continuity at ends of spans Excessive water
School	6 in. pan joists, 36 in. o.-c., 32 ft span Masonry-bearing walls	$1\frac{1}{2}$ in. ($L/256$)	Cracks below partitions	
Office	8 in. flat plate 16 in. spandrels	$\frac{1}{2}$ in.	Cracks in partitions Cracks around columns	Partitions built too early
Apartment building	One-way slab 26 in. span, 6 in. thick	Unknown	Visible "scalloped" effect of edge of slab	
Office building	Flat plate, *h* unknown 25 ft × 25 ft bays	4 in. ($L/75$)	Visible in shimming of counters	"Cheap office building"
Office building	Flat plate 26 ft × 26 ft bays *h* unknown	2 in. ($L/156$)	Dishing of floor noticeable to staff Filing cabinets and desks needed shimming	Apparent creep deformation with positive rather than negative bending within foot or two of some columns
Apartment	Concrete shear walls One-way continuous floor slabs 7 in. thick Spans up to 23 ft No exterior spandrels	Unknown	Excessive visible deflections Difficulty fitting precast panels	Deflection of balconies cantilevered from floor slabs required remedial measures

Table 6-1—(*continued*)

Type of structure	Type of construction	Amount of deflection	Result of deflection	Contributors' remarks
Subway entrance canopy	Cantilever slab Tapered—avg. $h = 5\frac{1}{2}$ in., Span = 11 ft	3 in. ($L/44$)	Spoiled appearance	Upward deflection of tail span contributed to defl. Passed load test easily
Canopy	Prestressed concrete 14 ft span, 7 in. thick	4 in. down ($L/42$)	Spoiled appearance	Deflected down when expected to go up; differential shrinkage probable cause; deflection of tail spans also involved
Office	Flat plate 25 ft × 25 ft	2 in. ($L/150$)	Filing cabinets etc. tipped	
Hospital	Waffle slab	Unknown	Cleaning water ran to centre of slab	

(From paper by G. W. Plewes,[2] with permission.)

These various factors were discussed at the end of Chap. 3 under the heading "Practical considerations."

An excellent discussion of deflection problems and treatment in various countries may be found in a 1974 paper by Clarke, Neville, and Houghton-Evans[3] (see Appendix L). Also presented in the paper are numerous drawings and descriptions of different types of damage and other problems caused by excessive deformation.

Approaches for Controlling Deflections

1. The use of compression steel and minimum tension steel percentages in reinforced members and nonprestressed tension steel in prestressed members (to control camber).

The method of using relatively small tension steel percentages in the design of reinforced members (which in turn results in relatively deep beams) to minimize deflections was discussed in Chap. 3 (preceding Eqs. (3-15), (3-16)) in connection with the 1963 ACI Code. The tension steel ratio $0.18f_c'/f_y$ was shown to be less than half the balanced ratio (ultimate strength design), and was judged to be sufficiently low to minimize deflection problems in most cases. This approach remains a useful one, in the author's opinion, and is noted in the strength design of the deflection examples 1, 2, and 3 of Chap. 3.

Alternatively, structural members will normally be of sufficient size so that deflections will be within acceptable limits when the tension steel reinforcement used in the positive moment zone does not exceed the following percentages of that in the balanced condition: for members not supporting or not attached to nonstructural elements likely to be damaged by large deflections—35 percent for rectangular and 40 percent for T or box beams, of the balanced ratio; and for members supporting or attached to nonstructural elements likely to be damaged by large deflections—25 percent for rectangular and 30 percent for T or box beams, of the balanced ratio.

The use of compression steel to control time-dependent deflections was discussed in Chap. 3, Eqs. (3-66) to (3-75), with the procedure recommended by the author using Eqs. (3-70) to (3-75). As described in Chap. 4, these procedures can also be applied to the case of two-way construction. In the case of prestressed members, the effect of nonprestressed tension steel on prestress loss and stress calculations is given in Chap. 5 by Eqs. (5-7) or (5-9), and the effect on camber calculations is given by Eqs. (5-57) to (5-59).

2. The use of maximum span–depth ratios and minimum depths.
3. The use of calculated and allowable deflections.

Various methods of controlling deflections by approaches 2 and 3 are presented in the remainder of this chapter.

4. By appropriate construction practices.

In addition to the use of concrete with maximum strength and stiffness properties and minimum creep and shrinkage properties—to the extent that is feasible—deflections can be minimized in other ways by appropriate construction practices. One example of this is to delay form removal (to minimize creep deformation) as long as possible, and then to install partitions as late as possible, since this would tend to minimize the creep and shrinkage deflection that could cause damage to the partitions.

Another procedure for controlling deflections of reinforced concrete members is, of course, precambering. This is commonly done to compensate for elastic deflections due to dead load. The same practice can also be used for part of the long-term effects as well. In addition, camber and deflection of prestressed members can be controlled to a considerable extent at the design stage as well as during construction.

6-2 MAXIMUM SPAN–EFFECTIVE DEPTH OR SPAN–THICKNESS RATIOS AND MINIMUM EFFECTIVE DEPTHS OR THICKNESSES

Controlling deflections by the indirect means of maximum span–effective depth or span–thickness ratios and minimum effective depths or thicknesses for typical cases is essentially a rule-of-thumb approach based largely on experience. Even when analytical methods are used to derive such limiting values, the calculations are based on selected allowable deflections, as shown later in this chapter on analytical procedures for determining span–depth ratios. Also, when various modifying factors or multipliers are applied to basic span–depth ratios, as in the 1971 ACI Code[4] and 1972 British Standard Code,[5] the *basic ratios* are still based primarily on experience, as stated in the 1972 British Handbook on the Unified Code for Structural Concrete.[6]

This method of controlling deflections is simpler than by the use of calculated and allowable deflections. However, unless a relatively large number of variables is included in the procedure, which would diminish its simplicity, such an approach usually must be quite conservative and/or with limited applicability. This is increasingly true as deflections become more critical.

A number of past and present methods for controlling deflections by means of maximum span–effective depth or span–thickness ratios and minimum effective depths or thicknesses are presented in this section.

Specified Values of Maximum Span–Effective Depth or Span–Thickness Ratios and Minimum Effective Depths or Thicknesses for Reinforced Concrete Structures

As used below, $L =$ span, $L' =$ distance between points of zero moment, $d =$ effective depth, and $h =$ overall thickness.

1959 CEB study[7]

Austria:	Slabs with one-way steel	Max. $L/d = 35$
	Two-way slabs without restraint ($L =$ shorter span)	50
	Two-way slabs with restraint ($L =$ shorter span)	60
England:	Beams on two supports	Max. $L/h = 20$
	Continuous beams	25
	Cantilever beams	10
	Slabs with one-way steel—	
	simply supported	30
	continuous	35
	Two-way reinforced slabs—	
	simply supported	35
	continuous	40
	cantilever	12
Germany:	Same as in Austria	
Italy:	Slabs	Max. $L/d = 30$
Poland:	Main beams	Max. $L/h = 15$
	Secondary beams	20
	Simply supported one-way slabs	35
	Restrained one-way slabs	40
	Simply supported two-way slabs	45
	Restrained two-way slabs	50
Sweden:	Beams	Max. $L'/h = 24$
	Flat slabs with caps ($L =$ longer span)	$L/d = 32$
	Roofs ($L =$ longer span)	$L/d = 40$
	Massive slabs supporting nonbearing walls, where M is the maximum span moment and σ is the allowable concrete tensile stress in bending	Min. $d = \sqrt{7M/\sigma}$
Turkey:	Simply supported two-way slabs	Max. $L/d = 50$
	Restrained or continuous two-way slabs	60
Yugoslavia:	Rectangular or T-beams	Max. $L/d = 20$
	One-way slabs	$L'/d = 30$
	Two-way reinforced slabs	$L'/d = 35$

1963 ACI Code (318-63)[8] The 1963 ACI Code stated that the minimum thicknesses in Table 6-2 shall apply to flexural members of normal-weight concrete, except when calculations of deflections prove that lesser thicknesses may be used without adverse effects.

Table 6-2 Minimum thickness of flexural members unless deflections are computed*

	Minimum thickness, h			
Member	Simply supported	One end continuous	Both ends continuous	Cantilever
One-way slabs	$L/25$	$L/30$	$L/35$	$L/12$
Beams	$L/20$	$L/23$	$L/26$	$L/10$

* 1963 ACI Code (318-63)[8]

It can be seen that the values for different European countries in the 1959 CEB study[7] and those in the 1963 ACI Code, Table 6-2, were comparable. However, in subsequent years such values have been found to be inadequate to cover many cases in design, so that either more conservative values or more complex procedures or more restricted usage is required. For example, in both the 1971 ACI Code and 1972 British Standard Code the span–depth or span–thickness ratios and minimum depths or thicknesses are more conservative than those above, and they also include several modifying factors or multipliers for different conditions. In addition, the 1971 ACI Code minimum thickness table applies only to members *not* supporting or attached to partitions or other construction likely to be damaged by large deflections.

1967 study by Mayer and Reusch[1] A somewhat more complex approach for controlling deflections by means of span–effective depths or slenderness ratios is described in a study by Mayer and Reusch,[1] and proposed for the German Reinforced Concrete Code, as follows.

Slenderness ratios are specified for a given set of conditions, such as loading, constant moment of inertia along the span, creep and shrinkage behavior, and structural configuration. For other structural configurations, an equivalent member having an equivalent span length is introduced, such that one of the following two groups of conditions describing equality of deformation and stress in the actual and the substitute members has to be satisfied at any one time:

(*a*) Deformation conditions
- Same absolute deflection
- Same relative deflection (deflection/span ratio)
- Same maximum slope of the deflection curve
- Same maximum end rotation

(*b*) Stress conditions
- Same loading
- Same stress at the center section
- Same curvature at the center section (M/EI), or
- Same maximum stress

For the equivalent span length, the span of a simply supported beam with constant moment of inertia should be chosen in such a manner that under uniformly distributed loading the same ratio of center deflection to span length and the same curvature at the center section exist as in the beam under consideration.

An extensive discussion of the arguments which enter into the choice of the most suitable measures of comparison is given by Mehmel.[9] This approach is similar to the analytical procedures for determining span–depth ratios near the end of this section.

1967 recommendation for ACI building code provision on deflections by ACI committee 435[10] The following brief but rather comprehensive provision on span–thickness limitations of reinforced concrete beams and slabs was recommended by ACI Committee 435, Deflections: the maximum span–thickness ratios, L/h, stipulated in Table 6-3 shall normally apply, except when the calculation of deflections shows that greater values may be used without adverse effects.

The author believes that Table 6-3 is still useful as a simple guide for the determination of preliminary sizes, for example. Thicknesses by this method tend to be conservative and to represent an upper bound of practical thicknesses for most loading conditions. Table 6-3 has been expanded and modified[45] in Appendix M.

Table 6-3 Maximum span–thickness ratios, L/h, of beams and slabs unless deflections are computed—ACI 435[10,11]*

Simply supported roof and floor construction *not intended* (first L/h value in each case) or *intended* (second L/h value in each case) to support or to be attached to partitions or other construction likely to be damaged by large deformations

† Roof slab	† Floor slab	Roof beam or ribbed roof slab	Floor beam or ribbed floor slab
24, 14	18, 12	18, 12	14, 10

* This table does not apply to prestressed concrete or composite members.

The L/h values in the table apply to normal-weight concrete. Use 0.8 of the tabulated values for lightweight concrete.

The L/h values in the table apply to simply supported members. Multiply the tabulated values by 0.4 for cantilever members, 1.3 for members with one end continuous, and 1.6 for members with both ends continuous.

For square two-way slabs supported on all four sides by walls, or by beams at least 3 times as deep as the slab, multiply the tabulated values by 1.4. Span–thickness ratios of rectangular two-way slabs should be between those of one-way slabs and two-way square slabs. Similarly multiply the tabulated values by 1.3 for flat plates (added by author).

† Refers to one-way slabs; and other slabs when modified as described above.

1970 American Institute of Steel Construction[12] For steel–concrete composite beams, the AISC Commentary suggests a limiting span–depth ratio of $800/F_y$ for fully stressed beams in order to prevent many deflection problems. The corresponding span–depth ratios are 22 for $F_y = 36$ ksi, and 16 for $F_y = 50$ ksi.

1971 ACI Code (318-71)[4]

Solid one-way slabs and beams or ribbed one-way slabs The 1971 ACI Code states that the minimum thickness in Table 6-4 (also see the expanded Table 6-5, which includes the Code modifying factors for different weight concrete and steel yield strengths) shall apply for one-way construction unless the computation of deflection indicates that lesser thickness may be used without adverse effects. However, these tables apply only to members *not* supporting or attached to partitions or other construction likely to be damaged by large deflections.

It can be seen in Table 6-5, and in the sample correction factors in the footnote of Table 6-4, for example, that larger minimum thicknesses correspond to lower weight concrete and higher yield strength steel. These provisions were illustrated in the deflection examples 1, 2, and 3 of Chap. 3 and the examples of this chapter.

Nonprestressed two-way construction

Minimum thickness requirements, except when deflections are checked The minimum thicknesses of slabs or other two-way construction for floors designed in accordance with the provisions of the ACI Code, and having a ratio of long to short span not exceeding two, shall be governed by Eqs. (6-1), (6-2), and (6-3)—ACI Code equations (9-6), (9-7), and (9-8)—and the other provisions below.

$$h = \frac{L_n(800 + 0.005f_y)}{36\,000 + 5000\beta[\alpha_m - 0.5(1 - \beta_s)(1 + 1/\beta)]} \tag{6-1}$$

but not less than (required lower limit)

$$h = \frac{L_n(800 + 0.005f_y)}{36\,000 + 5000\beta(1 + \beta_s)} \tag{6-2}$$

The thickness need not be more than (permissible upper limit)

$$h = \frac{L_n(800 + 0.005f_y)}{36\,000} \tag{6-3}$$

where L_n = length of clear span in long direction of two-way construction, measured face-to-face of columns in slabs without beams and face-to-face of beams or other supports in other cases

α = ratio of flexural stiffness of beam section to the flexural stiffness of a width of slab bounded laterally by the centerline of the adjacent panel, if any, on each side of the beam

α_m = average value of α for all beams on the edges of a panel

β = ratio of clear spans in long to short direction of two-way construction

β_s = ratio of length of continuous edges to total perimeter of a slab panel.

However, the thickness shall be not less than the following values:

For slabs without beams or drop panels—5 in.
For slabs without beams, but with drop panels—4 in.
For slabs having beams on all four edges with a value of α_m at least equal to 2.0—3.5 in.

Drop panels For slabs without beams but with drop panels extending in each direction, from the centerline of support, a distance equal to at least one-sixth the span length measured from center to center of supports in that direction, and a projection below the slab of at least $h/4$, the thickness required by Eqs. (6-1), (6-2), or (6-3) may be reduced by 10 percent.

Table 6-4 Minimum thickness of beams or one-way slabs unless deflections are computed—ACI Code (318-71)[4]

Member	Minimum thickness, h			
	Simply supported	One end continuous	Both ends continuous	Cantilever
	Members not supporting or attached to partitions or other construction likely to be damaged by large deflections			
Solid one-way slabs	$L/20$	$L/24$	$L/28$	$L/10$
Beams or ribbed one-way slabs	$L/16$	$L/18.5$	$L/21$	$L/8$

The values given in this table shall be used directly for nonprestressed reinforced concrete members made with normal-weight concrete (w = 145 pcf) and Grade 60 reinforcement. For other conditions, the values shall be modified as follows:

(*a*) For structural lightweight concrete having unit weights in the range 90–120 pcf, the values in the table shall be multiplied by $1.65 - 0.005w$ but not less than 1.09, where w is the unit weight in pcf.

(*b*) For nonprestressed reinforcement having yield strengths other than 60 000 psi, the values in the table shall be multiplied by $0.4 + f_y/100\,000$.

f_y =	40	50	60	75	w =	90	100	110	120
Sample corr. fac. =	0.80	0.90	1.00	1.15		1.20	1.15	1.10	1.09

(From ACI Code,[4] with permission.)

Table 6-5 1971 ACI Code minimum thicknesses as expanded from Table 6-4 to include the modifying factors for different weight concrete and steel yield strengths

		Simply supported		One end continuous		Both ends continuous		Cantilever	
Conc. unit wt. pcf	Steel yield str. ksi	Solid one-way slabs	Beams or ribbed one-way slabs	Solid one-way slabs	Beams or ribbed one-way slabs	Solid one-way slabs	Beams or ribbed one-way slabs	Solid one-way slabs	Beams or ribbed one-way slabs
90	40	*L*/20.8	*L*/16.7	*L*/25.0	*L*/19.3	*L*/29.2	*L*/21.9	*L*/10.4	*L*/8.3
	50	*L*/18.5	*L*/14.8	*L*/22.2	*L*/17.1	*L*/25.9	*L*/19.4	*L*/9.3	*L*/7.4
	60	*L*/16.7	*L*/13.3	*L*/20.0	*L*/15.4	*L*/23.3	*L*/17.5	*L*/8.3	*L*/6.7
	75	*L*/14.5	*L*/11.6	*L*/17.4	*L*/13.4	*L*/20.3	*L*/15.2	*L*/7.3	*L*/5.8
100	40	*L*/21.7	*L*/17.4	*L*/26.1	*L*/20.1	*L*/30.4	*L*/22.8	*L*/10.9	*L*/8.7
	50	*L*/19.3	*L*/15.5	*L*/23.2	*L*/17.9	*L*/27.1	*L*/20.3	*L*/9.7	*L*/7.7
	60	*L*/17.4	*L*/13.9	*L*/20.9	*L*/16.1	*L*/24.4	*L*/18.3	*L*/8.7	*L*/7.0
	75	*L*/15.1	*L*/12.1	*L*/18.2	*L*/14.0	*L*/21.2	*L*/15.9	*L*/7.6	*L*/6.1
110	40	*L*/22.7	*L*/18.2	*L*/27.3	*L*/21.0	*L*/31.8	*L*/23.9	*L*/11.4	*L*/9.1
	50	*L*/20.2	*L*/16.2	*L*/24.3	*L*/18.7	*L*/28.3	*L*/21.2	*L*/10.1	*L*/8.1
	60	*L*/18.2	*L*/14.6	*L*/21.8	*L*/16.8	*L*/25.5	*L*/19.1	*L*/9.1	*L*/7.3
	75	*L*/15.8	*L*/12.7	*L*/19.0	*L*/14.6	*L*/22.1	*L*/16.6	*L*/7.9	*L*/6.3
120	40	*L*/22.9	*L*/18.4	*L*/27.5	*L*/21.2	*L*/32.1	*L*/24.1	*L*/11.5	*L*/9.2
	50	*L*/20.4	*L*/16.3	*L*/24.5	*L*/18.9	*L*/28.5	*L*/21.4	*L*/10.2	*L*/8.2
	60	*L*/18.4	*L*/14.7	*L*/22.0	*L*/17.0	*L*/25.7	*L*/19.3	*L*/9.2	*L*/7.3
	75	*L*/16.0	*L*/12.8	*L*/19.2	*L*/14.8	*L*/22.3	*L*/16.8	*L*/8.0	*L*/6.4
>120	40	*L*/25.0	*L*/20.0	*L*/30.0	*L*/23.1	*L*/35.0	*L*/26.3	*L*/12.5	*L*/10.0
	50	*L*/22.2	*L*/17.8	*L*/26.7	*L*/20.6	*L*/31.1	*L*/23.3	*L*/11.1	*L*/8.9
Table 6-4 →	60	*L*/20.0	*L*/16.0	*L*/24.0	*L*/18.5	*L*/28.0	*L*/21.0	*L*/10.0	*L*/8.0
	75	*L*/17.4	*L*/13.9	*L*/20.9	*L*/16.1	*L*/24.4	*L*/18.3	*L*/8.7	*L*/7.0

Edge beams At discontinuous edges, an edge beam shall be provided having a stiffness such that the value of α is at least 0.80, or the minimum thickness required by Eq. (6-1), (6-2), or (6-3), or the crop panel section shall be increased by at least 10 percent in the panel having a discontinuous edge.

Comments The ACI Code thickness requirements apply to *floor* slabs only. Equations (6-1) to (6-3) and the other thickness requirements provide a means for considering the effect of interior or edge panels, panel shape, span ratios, size of supporting columns, the presence of drop panels, the use of higher strength steels, and the stiffening effect of beams on the deflection of slabs in terms of minimum thicknesses. These thicknesses are consistent with those found in practice to provide satisfactory control of deflections for flat slabs, flat plates, and two-way slabs supported on stiff beams. The equations provide for a transition from slabs on stiff beams to slabs without beams, and also include a term to adjust the thickness as a function of steel yield strength.

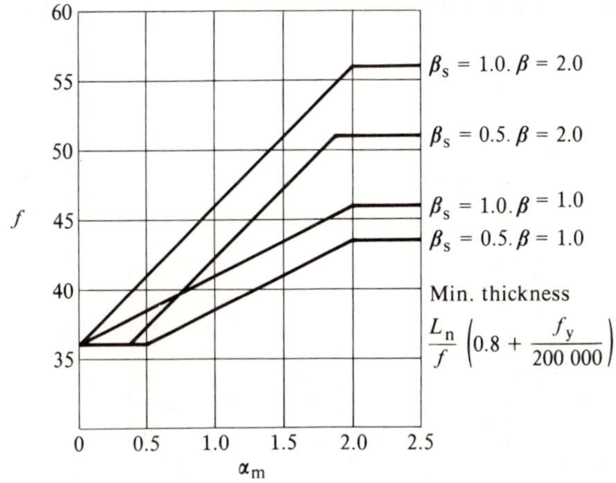

FIGURE 6-1
1971 ACI Code minimum thicknesses for two-way construction by Eqs. (6-1), (6-2), and (6-3). (*From chapter by Denkins, Portland Cement Association,*[13] *with permission.*)

The effect of yield strength in these equations is different than in the case of beams and one-way slabs (footnote to Table 6-4) because the degree of cracking has been observed to be less in two-way construction than in beams and one-way slabs, with a consequent smaller effect of steel stress or strain on the stiffness of the structural member. Lightweight concrete is not mentioned in this provision.

For edge panels, either an edge beam must be provided or the minimum thickness must be increased by 10 percent. On the other hand, the required minimum thickness may be reduced by 10 percent where drop panels with minimum specified dimensions are used.

Figure 6-1 and Table 6-6 are design aids[13] to assist in determining minimum thicknesses by Eqs. (6-1) to (6-3). As indicated in Table 6-6, for example, Eq. (6-3) will control for all flat plates and flat slabs without beams, since $\alpha = \alpha_m = 0$, in which case from Fig. 6-1,

$$\left.\begin{array}{llll} \text{Min. } h = L_n/36.0 & \text{for} & f_y = 40 \text{ ksi} \\ \qquad\qquad 34.3 & & \qquad 50 \\ \qquad\qquad 32.7 & & \qquad 60 \\ \qquad\qquad 30.6 & & \qquad 75 \end{array}\right\} \tag{6-4}$$

Also, as shown in Table 6-6, it will never be necessary to check more than two equations and in most cases not more than one equation need be checked. These minimum thicknesses for two-way slab systems are also tabulated in Table O-1 from the CRSI Handbook.[44]

Although the ACI Code minimum thickness two-way provisions apply only to floor slabs, they may frequently be used for *roof* slabs as well. While roofs do not normally support partitions, roof deflections could affect partitions

Table 6-6 Flow diagram for 1971 ACI Code minimum thicknesses for two-way construction by Eqs. (6-1), (6-2), and (6-3)

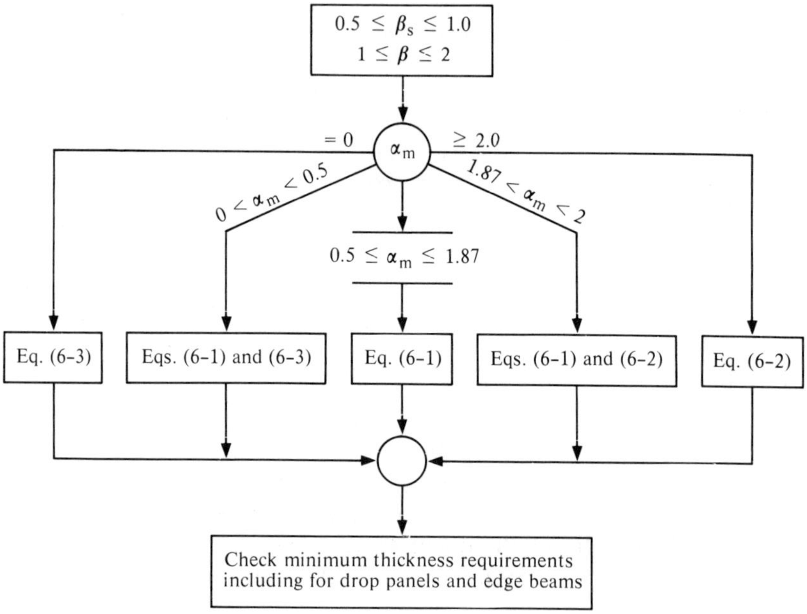

(From chapter by Denkins, Portland Cement Association,[13] with permission.)

built tightly underneath. Also, ponding can represent an additional problem for roofs. Another point is that roofs with comparable loads in multistory buildings are generally built with the same thickness as the floors below for economy of reinforcement.

The designer has the option of using thinner slabs than those resulting from the minimum thickness requirements, provided the computed deflections satisfy the ACI Code limitations.

For slab systems without beams, it is advisable to initially check the shear strength of the slab in the vicinity of the columns, as this effect frequently controls the slab thickness requirements.

Composite members

Shored construction If composite members are supported during construction in such a manner that, after removal of temporary supports, the dead load is resisted by the full composite section, the composite member may be considered equivalent to a cast-in-place member for the purposes of deflection calculation. For nonprestressed members, the portion of the member in compression shall determine whether the values given in Table 6-4 for normal weight or lightweight concrete shall apply. If deflection is calculated, account should be taken of the curvatures resulting from differential shrinkage of the precast and cast-in-place components, and of the axial creep effects in a prestressed concrete member.

Unshored construction If the thickness of a nonprestressed precast member meets the requirements of Table 6-4, deflection need not be computed. If the thickness of a nonprestressed composite member meets the requirements of Table 6-4, deflection occurring after the member becomes composite need not be calculated, but the long-time deflection of the precast member should be investigated for the magnitude and duration of load prior to the beginning of effective composite action.

It should be noted, for example, that the effects of load level and compression steel are not included in any of the ACI Code minimum thickness procedures.

1972 British Standard Code of Practice[5,6] (Extracts from CP 110 Part 1[5] are reproduced by permission of BSI, 2 Park Street, London W1A 2BS.)

Based primarily on the work of Beeby and Miles,[14] and Beeby,[15,16] the British Code specifies a set of basic span–effective depth ratios in Table 6-7, together with various modifying factors given in Tables 6-8 to 6-10. The approach used in deriving the modifying factors was to accept the basic ratios used since the 1957 Code, with only minor adjustments, as having proved satisfactory for normal circumstances by 15 years' practical experience, and then to assess the relative effect of the following variables by analysis: tension steel percentage (0.25 to 3 percent), compression steel percentage (0 percent to the tension steel percentage), ratio of permanent load to total load (0.6 to 1.0), free shrinkage strain (up to 1200×10^{-6}), creep coefficient (up to 4.5), concrete cube strength (22.5, 30, 45 N/mm^2 = 3260, 4350, 6350 psi), and steel strength (f_y = 200 to 500 N/mm^2 = 29 000 to 72 500 psi).

To cover the practical range of variables adequately, a computer was used, and the deflections under service load in excess of 2000 beams were calculated. The results from these calculations were used to isolate the significant variables and to produce relationships for modifying the basic span–effective depth ratios to allow for their influence. The two significant variables were found to be the percentage of tension steel and compression steel. Other variables, while having some influence, were considered to be not sufficiently significant to be taken into account in such an approximate method.

Deflection of normal-weight reinforced concrete beams Deflections may be calculated and compared with the serviceability requirements of the Code, but in all normal cases the deflection of a beam will not be excessive if the ratio of its span to effective depth is not greater than the appropriate value determined below.

Table 6-7 Basic span–effective depth ratios for rectangular beams of normal-weight concrete

Support conditions	Ratio
Cantilever	7
Simply supported	20
Continuous	26

(From British Standards Institution,[5] with permission.)

Table 6-8 Special span–effective depth ratios for rectangular beams

Span		Cantilever	Simply supp.	Continuous
m	ft			
10	32.8	Value to be justified by calculation	20	26
12	39.4		18	23
14	45.9		16	21
16	52.5		14	18
18	59.1		12	16
20	65.6		10	13

(From British Standards Institution,[5] with permission.)

Table 6-9 Modification factor for tension reinforcement

Service stress, f_s		$100A_s/bd$							
N/mm^2	ksi	0.25	0.50	0.75	1.00	1.50	2.00	2.50	≥ 3.00
145 ($f_y = 250$)	21.0 ($f_y = 36.3$)	2.00	1.98	1.62	1.44	1.24	1.13	1.06	1.01
150	21.8	2.00	1.91	1.58	1.41	1.22	1.11	1.04	0.99
200	29.0	2.00	1.46	1.26	1.15	1.02	0.94	0.89	0.85
238 ($f_y = 410$)	34.5 ($f_y = 59.5$)	1.60	1.23	1.09	1.00	0.90	0.84	0.80	0.77
246 ($f_y = 425$)	35.7 ($f_y = 61.6$)	1.55	1.20	1.06	0.98	0.88	0.83	0.79	0.76
250	36.3	1.52	1.18	1.05	0.97	0.87	0.82	0.78	0.75
267 ($f_y = 460$)	38.7 ($f_y = 66.7$)	1.41	1.11	0.99	0.92	0.84	0.78	0.75	0.72
290 ($f_y = 500$)	42.1 ($f_y = 72.5$)	1.27	1.03	0.92	0.86	0.79	0.74	0.71	0.68
300	43.5	1.22	0.99	0.90	0.84	0.77	0.72	0.69	0.67

The service stress may be estimated from the equation

$$f_s = 0.58 \frac{f_y A_{s\,req}}{A_{s\,prov}} \times \frac{1}{\beta_b}$$

where f_s is the estimated service stress in the tension steel
f_y is the characteristic strength of the steel
$A_{s\,req}$ is the area of tension steel required at midspan to resist the moment due to ultimate loads (at support for cantilever)
$A_{s\,prov}$ is the area of tension steel provided at midspan (at support for cantilever)
β_b is the ratio of the resistance moment provided at midspan to the midspan moment taken from the maximum moments diagram before redistribution.

If the percentage of redistribution is not known but the design ultimate moment at midspan is obviously the same or greater than the elastic ultimate moment, the stress f_s in Table 6-9 may be taken as $0.58f_y$.

(From British Standards Institution,[5] with permission.)

Table 6-10 Modification factor for compression reinforcement

$100A'_s/bd$	Factor	$100A'_s/bd$	Factor
0.25	1.07	1.5	1.33
0.50	1.14	2.0	1.40
0.75	1.20	≥ 3.0	1.50
1.0	1.25		

Intermediate values may be interpolated.

The area of compression steel A'_s used in Table 6-10 may include all bars in the compression zone, even those not effectively tied with links.

(From British Standards Institution,[5] with permission.)

Span–effective depth ratio for rectangular beams The basic span–effective depth ratios for rectangular beams are given in Table 6-7. These are based on limiting the deflection to span/250, and this should normally avoid damage to finishes and partitions for beams of up to 10 m (33 ft) span.

Table 6-7 should only be used for spans not exceeding 10 m (33 ft) if the engineer is satisfied that a deflection of span/250 is acceptable. When it is necessary further to restrict the deflection, to avoid damage to finishes or partitions, Table 6-8 should be used for spans exceeding 10 m (33 ft).

If the engineer considers that limits other than those for which the tables have been produced are more appropriate, the tables of basic ratios can be modified to suit the chosen limits. This may be done by multiplying the basic ratio by the ratio of the required deflection to the deflection for which the table was derived. Thus, if the total deflection is to be limited to span/a' instead of span/250, the figures in Table 6-7 should be multiplied by $250/a'$. Similarly, if the deflection occurring after construction of the partitions is to be limited to some absolute value b', rather than to 20 mm, the figures in Table 6-8 can be adjusted by multiplying them by $b'/20$.

The category "continuous" in Tables 6-7 and 6-8 may be taken to apply to any situation where at least one end of the beam is continuous.

Deflection is influenced by the amount of tension reinforcement and its stress and therefore the span–effective depth ratios should be modified according to the area of reinforcement provided and its service stress at the centre of the span (or at the support in the case of a cantilever). Values of span–effective depth ratio obtained from Table 6-7 or 6-8 should therefore be multiplied by the appropriate factor obtained from Table 6-9.

Compression reinforcement also influences deflection and the value of the span–effective depth ratio obtained from Table 6-7 or 6-8, modified by the factor obtained from Table 6-9, may be multiplied by a further factor obtained from Table 6-10.

Permissible span–effective depth ratios obtained from Tables 6-7 to 6-10 take account of normal creep and shrinkage deflection. If it is expected that creep or shrinkage of the concrete may be particularly high (e.g. if the free shrinkage strain is expected to be greater than 750×10^{-6} or the creep coefficient greater than 4) or if other abnormally adverse conditions are expected, the permissible span–effective depth ratio should be suitably reduced. A reduction of more than 15 percent is unlikely to be required.

Span–effective depth ratio for flanged beams For a flanged beam the span–effective depth ratio may be determined as in the case of rectangular beams, but when the web width is less than 0.3 times the effective flange width, the final ratio obtained should be multiplied by 0.8. For values of web width to effective flange width greater than 0.3, this factor may be increased linearly from 0.8 to 1.0 as the ratio of web width to effective flange width increases to unity. Reinforcement percentages should be based on the effective flange width b of the beam multiplied by d.

Deflection of normal-weight reinforced concrete solid slabs Deflections may be calculated and compared with the serviceability requirements of the Code, but in all normal cases it will be sufficient to restrict the span–effective depth ratio. The appropriate ratios for a solid slab may be obtained from Tables 6-7 and 6-8, modified by Tables 6-9 and 6-10. Only the reinforcement at the centre of the span in the width of slab under consideration should be considered to influence deflection. The ratio for a two-way spanning slab should be based on the shorter span and its amount of reinforcement in that direction.

Note that, with the exception of cantilever slabs, it is always the midspan steel percentage which is used in assessing the factor from Table 6-9. It is worth bearing in mind that the ratios of span to effective depth have been arranged to ensure that the deflections of flexural members do not exceed the specified limits relative to their supports. If the support for a slab consists of beams which will themselves deflect under load, the total deflection of the whole system might not be satisfactory even though each member considered individually was satisfactory relative to its own supports. It might be unwise, therefore, to design both the beams and slabs in such a system to be on the limit of the ratios of span to effective depth.

Deflection of normal-weight reinforced concrete panels For slabs with drops of total width in both directions at least equal to one-third of the respective spans the provisions for solid slabs can be applied directly. Otherwise the span–effective depth ratios obtained for rectangular beams should be multiplied by 0.9.

In calculating the modification factor for tension reinforcement from Table 6-9, the average steel percentage across the whole width of the panel at midspan should be used. Where the spans in the two directions differ significantly, the ratio of span to effective depth need be checked only for the shorter span. Where drops are used, the ratio of span to effective depth should still be assessed on the basis of the section at midspan and the total span.

Deflection of lightweight reinforced concrete beams Deflection of lightweight aggregate concrete beams may be calculated using the appropriate value of E_c as described in the Code. Alternatively, span–effective depth ratios may be obtained as in the case of normal-weight concrete and multiplied by a factor of 0.85.

Deflection of lightweight reinforced concrete slabs Deflection of lightweight aggregate concrete slabs may be calculated using the appropriate value of E_c as described in the Code. Alternatively, the same recommendations as for normal-weight concrete slabs may be used for any slab subject to a characteristic imposed load of 4 kN/m² (21 psf) or less. For slabs supporting a higher characteristic imposed load the span–effective depth ratios obtained for normal-weight concrete slabs should be multiplied by a factor of 0.85.

1973 AASHTO Highway Bridge Specifications[17]

Superstructure depth recommendations As a means of controlling long-time deflections due to creep and shrinkage, it is recommended that superstructure depths for various bridge types be not less than the value obtained from the equations in Table 6-11. Depths less than these may be used when specific consideration is given to limiting long-time deflections by use of compression reinforcement or some other method.

Table 6-11 Recommended minimum thickness* for constant depth members†

Slabs with main reinforcement parallel and transverse to traffic S = effective span‡ in feet	$t_{min} = 0.33 + \frac{S}{30}$ in feet but not less than 0.542 feet
T-beams S = actual span in feet	$t_{min} = 0.5 + \frac{S}{18}$ in feet
Box girders S = actual span in feet	$t_{min} = 0.5 + \frac{S}{20}$ in feet

(From AASHTO,[17] with permission.)

* Recommended values for continuous spans; simple spans should have about 10 percent greater thickness.

† When variable depth members are used, table values may be adjusted to account for change in relative stiffness of positive and negative moment sections.

‡ Slabs monolithic with beams or walls (without haunches), S = clear span. Slabs supported on steel stringers, S = distance between edges of flanges plus ½ of the stringer flange width. Slabs supported on timber stringers, S = clear span plus ½ thickness of stringer.

1974 paper by Clarke, Neville, and Houghton-Evans[3] A summary of methods in various countries for controlling deflections is briefly summarized in Appendix L, from Ref. 3.

Specified Values of Span–Depth Ratios and Depths for Prestressed Concrete Structures

In the case of prestressed members, the word depth is normally used to refer to the overall thickness.

1963 book, T. Y. Lin[18] (from John Wiley & Sons, Inc.,[18] with permission) In estimating the depth of a prestressed section, an approximate rule is to use 70 percent of the corresponding depth for conventional reinforced concrete construction. Some other empirical rules are also available. For example, the thickness of prestressed slabs may vary from $L/35$ for heavy loads to $L/55$ for light loads. The depth of beams of the usual proportions can be approximated by Eq. (6-5).

$$h = k\sqrt{M} \tag{6-5}$$

where h = depth of beam in inches, M = maximum bending moment in ft-k, and k = a coefficient varying from 1.5 to 2.0. Such empirical rules apply only under average conditions and should be used merely as a preliminary guide.

Higher ratios than for reinforced members are possible because deflections can be much better controlled in prestressed design. On the other hand, when these ratios get too high, camber and deflection become quite sensitive to variations in loadings, in material properties, in magnitude and location of prestress, and in environmental conditions. Furthermore, the effects of vibration become more pronounced.

Approximate limits of span–depth ratios for buildings For cantilever solid slabs, a span–depth ratio of 18 for floors and 20 for roofs has been found to be satisfactory. But cantilevers are sensitive to deflections and vibrations, and greater care should be taken. For example, a camber in the anchor span would usually produce a dip in the cantilever. Span–depth ratios for simple and continuous spans are given in Table 6-12.

Generally speaking, when span–depth ratios are some 10 percent below the values tabulated in Table 6-12, problems of camber, deflection, and vibration should not occur unless the loadings are extremely heavy and vibratory in nature. Occasionally, these ratios can be exceeded by 10 percent or more, if careful study would justify and ensure proper behavior.

The above values are intended for both hard-rock concrete and lightweight concrete, but should be reduced by about 5 percent for lightweight concrete having E_c less than 3×10^6 psi. For long spans (say, in excess of about 70 ft) and for heavy loads (say, live loads over 100 psf), the above values should be reduced by 5 to 10 percent. For in-place concrete in composite action with the precast

Table 6-12 Approximate limits for span–depth ratios, from Lin

	Simple spans		Continuous spans	
	Roof	Floor	Roof	Floor
One-way solid slabs	48	44	52	48
Two-way solid slabs (supported on columns only)	44	40	48	44
Two-way waffle slabs (3 ft waffles)	36	32	40	36
Two-way waffle slabs (12 ft waffles)	32	28	36	32
One-way slabs with small cores	46	42	50	46
One-way slabs with large cores	44	40	48	44
Double Ts and single Ts (side by side)	40	36	44	40
Single Ts (spaced 20 ft centers)	32	28	36	32

(From John Wiley & Sons, Inc.,[18] with permission.)

elements, the total depth may be considered in computing the above span–depth ratios.

Approximate limits of span–depth ratios for bridges Little experience has been obtained for *railway bridges* of prestressed concrete to justify any limitation on their span–depth ratios; the usual ratios have been in the range of 10 to 14 for box sections up to 100 ft or more. For simple-span *highway bridges* of the I-beam type, up to about 200 ft, a span–depth of 20 is considered conservative, 22 to 24 is normal, while 26 to 28 would be the critical limit. Box sections can have ratios about 5 to 10 percent higher than I-beams, while T-sections spaced far apart should have ratios about 5 to 10 percent lower than I-beams.

1971 PCI Handbook[19] Typical span to depth ratios of flexural precast prestressed concrete members are:

Hollow-core floor slabs	30 to 40
Hollow-core roof slabs	40 to 50
Stemmed floor slabs	25 to 35
Stemmed roof slabs	35 to 40
Beams	10 to 20

The required depth of a beam or slab is influenced by the ratio of live load to total load. Where this ratio is high, deeper sections may be needed.

There are no set span–depth limits for post-tensioned members, but the typical values given in Table 6-13 are provided as a guide to the preliminary sizing of members. These ratios may be increased if calculations verify that deflection, camber, and vibration frequency and amplitude are not objectionable.

1972 British Standard Code of Practice[5] No span–depth provisions are specified for Class 1, No tensile stresses, and Class 2, Tensile stresses, but no visible cracking. For Class 3, Tensile stresses with surface cracks within specified limits,

Table 6-13 Typical span–depth ratios*

	Simple spans		Continuous spans	
	Roof	Floor	Roof	Floor
One-way solid slabs	45	40	50	45
Two-way solid slabs (supported on columns only)			45	40
Two-way waffle slabs (36 in. pans)	35	30	40	35
Beams	30	26	35	30
One-way joists	38	35	42	38

(From PCI Handbook,[19] with permission.)

* For post-tensioned members.

where the permanent load is greater than 25 percent of the design load the basic span–effective depth ratios given in Tables 6-7 and 6-8 only should be complied with unless more rigorous calculations based on the moment–curvature relationship are made.

Numerical Examples of Maximum Span–Effective Depth or Span–Thickness Ratios and Minimum Effective Depths or Thicknesses

Example 1 Simple span beam, $L = 20$ ft (6.1 m), normal-weight concrete, $f_y = 60$ ksi, $h = 14$ in. (35.6 cm), $d = 12$ in. (30.5 cm). (Same as Example 1, Chap. 3, Table 3-2.)

1963 ACI Code

$$h_{\min} = L/20 \qquad \text{(Table 6-2)}$$

$$h_{\min} = 240/20 = \underline{12.0 \text{ in. } (30.5 \text{ cm})} < 14 \text{ in. } (35.6 \text{ cm})$$

Deflections need not be checked.

1967 ACI Committee 435 Report Assume *floor* construction *not intended* to support or to be attached to partitions or other construction likely to be damaged by large deformations.

$$\text{Max. } L/h = 14 \qquad \text{(Table 6-3)}$$

$$h_{\min} = 240/14 = \underline{17.1 \text{ in. } (43.4 \text{ cm})} > 14 \text{ in. } (35.6 \text{ cm})$$

Deflections must be checked.

Assume *roof* construction *not intended,* etc.

$$\text{Max. } L/h = 18 \qquad \text{(Table 6-3)}$$
$$h_{\min} = 240/18 = \underline{13.3 \text{ in. } (33.8 \text{ cm})} < 14 \text{ in. } (35.6 \text{ cm})$$

Deflections need not be checked.

Assume *floor* construction *intended,* etc.

$$\text{Max. } L/h = 10 \qquad \text{(Table 6-3)}$$
$$h_{\min} = 240/10 = \underline{24.0 \text{ in. } (61.0 \text{ cm})} > 14 \text{ in. } (35.6 \text{ cm})$$

Deflections must be checked.

Assume *roof* construction *intended,* etc.

$$\text{Max. } L/h = 12 \qquad \text{(Table 6-3)}$$
$$h_{\min} = 240/12 = \underline{20.0 \text{ in. } (50.8 \text{ cm})} > 14 \text{ in. } (35.6 \text{ cm})$$

Deflections must be checked.

1971 ACI Code Assume the member *is not* supporting or attached to partitions or other construction likely to be damaged by large deflections.

$$h_{\min} = L/16 \qquad \text{(Table 6-4)}$$
$$= 240/16 = \underline{15.0 \text{ in. } (38.1 \text{ cm})} > 14 \text{ in. } (35.6 \text{ cm})$$

Deflections must be checked.

The minimum thickness provisions of the 1971 ACI Code do not include the case of a member that *is* supporting or attached to partitions or other construction likely to be damaged by large deflections.

British Code From Table 3-2,

$$f_s = \frac{M}{A_s jd} = \frac{(65)(12)}{(2.37)(10.30)} = 32.0 \text{ ksi}$$

$$100A_s/bd = 1.98 \qquad \text{Basic } L/d = 20 \qquad \text{(Table 6-7)}$$
$$\text{Tension steel factor} = 0.89 \qquad \text{(Table 6-9)} \qquad L/d = (20)(0.89) = 17.8$$
$$d = L/17.8 = 240/17.8 = \underline{13.5 \text{ in. } (34.3 \text{ cm})} > 12 \text{ in. } (30.5 \text{ cm})$$

Alternatively using $f_y = 60$ ksi in Table 6-9:

$$\text{Tension steel factor} = 0.84 \qquad \text{and} \qquad L/d = 16.8$$
$$d = \underline{14.3 \text{ in. } (36.3 \text{ cm})} > 12 \text{ in. } (30.5 \text{ cm})$$

Hence in both cases deflections may be excessive and should be checked.

The British Code does not distinguish between roof and floor members, nor does it refer to the attachment of nonstructural elements, but it does include the effect of service stress in the steel (or steel yield strength) and percentage of tension and compression steel.

Example 2 Three-span continuous one-way slab, $L = 20$ ft (6.1 m), sand-lightweight concrete (120 pcf), $f_y = 40$ ksi, $h = 8$ in. (20.3 cm), $d = 6.75$ in. (17.1 cm). (Same as Example 2, Chap. 3, Table 3-4.)

1963 ACI Code Did not apply to lightweight concrete.

1967 ACI Committee 435 Report From Table 6-3:

$$\text{Lightweight concrete factor} = 0.8$$
$$\text{End span factor} = 1.3$$
$$\text{Center span factor} = 1.6$$

Assume *floor* construction *not intended* to support or to be attached to partitions or other construction likely to be damaged by large deformations.

$$\text{Max. } L/h = 18 \qquad \text{(Table 6-3)}$$
$$\text{End span } L/h = (18)(0.8)(1.3) = 18.7$$
$$\text{End span } h_{\min} = 240/18.7 = \underline{12.8 \text{ in. (32.5 cm)}} > 8 \text{ in. (20.3 cm)}$$
$$\text{Center span } L/h = (18)(0.8)(1.6) = 23.0$$
$$\text{Center span } h_{\min} = 240/23.0 = \underline{10.4 \text{ in. (26.4 cm)}} > 8 \text{ in. (20.3 cm)}$$

Hence deflections must be checked in both spans.

Assume *roof* construction *not intended,* etc.

$$\text{Max. } L/h = 24 \qquad \text{(Table 6-3)}$$
$$\text{End span } L/h = (24)(0.8)(1.3) = 25.0$$
$$\text{End span } h_{\min} = 240/25.0 = \underline{9.6 \text{ in. (24.4 cm)}} > 8 \text{ in. (20.3 cm)}$$
$$\text{Center span } L/h = (24)(0.8)(1.6) = 30.7$$
$$\text{Center span } h_{\min} = 240/30.7 = \underline{7.8 \text{ in. (19.8 cm)}} < 8 \text{ in. (20.3 cm)}$$

Hence deflections must be checked in the end span but need not be checked in the center span.

For *roof* and *floor* construction *intended,* etc., it is obvious by inspection that deflections must be checked in both spans (which is usually the case).

1971 ACI Code Assume the member is *not* (only case included in 1971 Code) supporting or attached to partitions or other construction likely to be damaged by large deflections. From Table 6-5 for $w = 120$ pcf, $f_y = 40$ ksi:

$$\text{End span } h_{\min} = L/27.5 \qquad \text{Center span } h_{\min} = L/32.1$$
$$\text{End span } h_{\min} = 240/27.5 = \underline{8.7 \text{ in. } (22.1 \text{ cm})} > 8 \text{ in. } (20.3 \text{ cm})$$
$$\text{Center span } h_{\min} = 240/32.1 = \underline{7.5 \text{ in. } (19.1 \text{ cm})} < 8 \text{ in. } (20.3 \text{ cm})$$

Hence deflections must be checked in the end span but need not be checked in the center span.

British Code

$$\text{End span } f_s = 23.5 \text{ ksi} \qquad \text{Center span } f_s = 21.7 \text{ ksi}$$

(calculated from Tables 3-4, 3-5). From Table 3-4:

$$\text{End span pos. } 100A_s/bd = 0.765 \qquad \text{Center span pos. } 100A_s/bd = 0.494$$

No pos. Compression steel. Note that this method refers only to the positive region steel for continuous members. From Table 6-7:

$$\text{Basic } L/d = 26$$

From Table 6-9:

$$\text{End span tension steel factor} = 1.50$$
$$\text{Center span tension steel factor} = 1.91$$
$$\text{Lightweight conc. factor} = 0.85$$
$$\text{End span } L/d = (26)(0.85)(1.50) = 33.2$$
$$\text{End span } d = 240/33.2 = \underline{7.2 \text{ in. } (18.3 \text{ cm})} > 6.75 \text{ in. } (17.1 \text{ cm})$$
$$\text{Center span } L/d = (26)(0.85)(1.91) = 42.2$$
$$\text{Center span } d = 240/42.2 = \underline{5.7 \text{ in. } (14.5 \text{ cm})} < 6.75 \text{ in. } (17.1 \text{ cm})$$

Hence deflections may be excessive and should be checked for the end span but need not be checked in the center span (same as 1971 ACI Code). The British Code method is based on an allowable deflection of $L/250$, and does not refer to the question of roof or floor usage or the attachment of nonstructural elements.

Alternatively using $f_y = 40$ ksi in Table 6-9:

$$\text{End span tension steel factor} = 1.62$$
$$\text{Center span tension steel factor} = 1.98 \qquad \text{(Table 6-9)}$$
$$\text{End span } L/d = 35.8$$
$$\text{End span } d = \underline{6.7 \text{ in. } (17.0 \text{ cm})} < 6.75 \text{ in. } (17.1 \text{ cm})$$
$$\text{Center span } L/d = 43.8$$
$$\text{Center span } d = \underline{5.5 \text{ in. } (14.0 \text{ cm})} < 6.75 \text{ in. } (17.1 \text{ cm})$$

Hence in this case, deflections will not be excessive in either span.

Example 3 Continuous T-beam, $L_n = 29$ ft (8.8 m), normal-weight concrete, $f_y = 40$ ksi, $h = 25$ in. (63.5 cm), pos. $d = 22.5$ in. (57.2 cm). (Same as Example 3, Chap. 3, Table 3-7.)

1963 ACI Code Assume all spans continuous at both ends.

$$h_{min} = L/26 \qquad \text{(Table 6-2)}$$
$$= 348/26 = \underline{13.4 \text{ in. } (34.0 \text{ cm})} < 25 \text{ in. } (63.5 \text{ cm})$$

Deflections need not be checked.

1967 ACI Committee 435 Report Assume all spans continuous at both ends. From Table 6-3:

$$\text{Continuity factor} = 1.6$$

Assume construction *not intended* to support or to be attached to partitions or other construction likely to be damaged by large deformations.

$$\text{Floor max. } L/h = 14 \qquad \text{Roof max. } L/h = 18 \qquad \text{(Table 6-3)}$$
$$\text{Floor max. } L/h = (14)(1.6) = 22.4$$
$$\text{Floor } h_{min} = 348/22.4 = \underline{15.5 \text{ in. } (39.4 \text{ cm})} < 25 \text{ in. } (63.5 \text{ cm})$$
$$\text{Roof max. } L/h = (18)(1.6) = 28.8$$
$$\text{Roof } h_{min} = 348/28.8 = \underline{12.1 \text{ in. } (30.7 \text{ cm})} < 25 \text{ in. } (63.5 \text{ cm})$$

Hence deflections need not be checked for either roof or floor usage.

Assume construction *intended* to support or to be attached, etc.

$$\text{Floor max. } L/h = 10 \qquad \text{Roof max. } L/h = 12 \qquad \text{(Table 6-3)}$$
$$\text{Floor max. } L/h = (10)(1.6) = 16.0$$
$$\text{Floor } h_{min} = 348/16.0 = \underline{21.8 \text{ in. } (55.4 \text{ cm})} < 25 \text{ in. } (63.5 \text{ cm})$$
$$\text{Roof max. } L/h = (12)(1.6) = 19.2$$
$$\text{Roof } h_{min} = 348/19.2 = \underline{18.1 \text{ in. } (46.0 \text{ cm})} < 25 \text{ in. } (63.5 \text{ cm})$$

Hence deflections need not be checked for either roof or floor usage.

1971 ACI Code Assume all spans continuous at both ends. Assume the member is *not* (only case included in the 1971 Code) supporting or attached to partitions or other construction likely to be damaged by large deflections. From Table 6-5 for $w > 120$ pcf, $f_y = 40$ ksi:

$$h_{min} = L/26.3$$
$$= 348/26.3 = \underline{13.2 \text{ in. } (33.5 \text{ cm})} < 25 \text{ in. } (63.5 \text{ cm})$$

Deflections need not be checked.

British Code

$$\text{End span } f_s = 23.6 \text{ ksi} \qquad \text{Interior span } f_s = 20.6 \text{ ksi}$$

(calculated from Tables 3-7, 3-8, 3-9). From Table 3-7:

$$\text{Pos. } 100A_s/bd = 0.25 \text{ in both spans} \qquad \text{No pos. compression steel}$$

Note that this method refers only to the positive region steel for continuous members. From Table 6-7:

$$\text{Basic } L/d = 26$$

From Table 6-9:

$$\text{Tension steel factor} = 2.00$$

For flanged beam, $b'/b = 15/90 = 0.17 < 0.3$

Hence Flange factor $= 0.8$

$$L/d = (26)(2.00)(0.8) = 41.6$$
$$d = 348/41.6 = \underline{8.4 \text{ in. } (21.3 \text{ cm})} < 22.5 \text{ in. } (57.2 \text{ cm})$$

Hence deflections will not be excessive.

Alternatively using $f_y = 40$ ksi in Table 6-9: the tension steel factor $= 2.00$ for both spans. Hence the same results as above are found.

It may be noted that the span–depth provisions for controlling deflections were satisfied for this continuous T-beam example by all of the methods above. Also, in most cases the minimum depths or thicknesses were considerably lower than the actual designed depth for the T-beam. In Chap. 3 all computed deflections were found to be satisfactory for all conditions checked as well.

Example 4 Interior girder of simple-span T-beam or deck girder bridge, $L = 60$ ft (18.3 m), normal-weight concrete, $f_y = 40$ ksi, $h = 36.25$ in. (92.1 cm). (Same as Example 4, Chap. 3, Table 3-11.)

AASHTO Highway Bridge Specifications

$$S = 60 \text{ ft } (18.3 \text{ m})$$
$$t_{min} = 0.5 + S/18 \qquad \text{(Table 6-11)}$$
$$= 0.5 + 60/18 = 0.5 + 3.33 = 3.83 \text{ ft}$$
$$= \underline{46.0 \text{ in. } (116.8 \text{ cm})} > 36.25 \text{ in. } (92.1 \text{ cm})$$

Hence consideration must be given to limiting long-time deflections by use of compression reinforcement or some other method according to AASHTO. The girder was designed by the load-factor method, and deflections were calculated by several methods in Chap. 3.

Example 5 Unshored reinforced concrete composite beam, simple span, $L = 30$ ft (9.1 m), all concrete normal-weight, $f_y = 40$ ksi, precast $h = 14$ in. (35.6 cm), composite $h = 17.5$ in. (44.5 cm). (Same as composite reinforced concrete example in Chap. 3, Table 3-29.)

1971 ACI Code Assume the member *is not* (only case included in 1971 Code) supporting or attached to partitions or other construction likely to be damaged by large deflections.

$$h_{min} = L/20.0 \qquad \text{(Table 6-5)}$$
$$= 360/20.0 = \underline{18 \text{ in. } (45.7 \text{ cm})} > \text{precast } h = 14 \text{ in. } (35.6 \text{ cm})$$

Hence precast beam deflections must be checked.

$$h_{min} = \underline{18 \text{ in. } (45.7 \text{ cm})} > \text{composite } h = 17.5 \text{ in. } (44.5 \text{ cm})$$

Hence composite beam deflections must be checked.

Example 6 Solid slab without (flat plate) and with (two-way slab) edge beams, square panels, $L = 18$ ft $= 216$ in. (6.7 m), normal-weight concrete, $f_y = 60$ ksi, say $h = 8$ in. (20.3 cm), $d = 6.75$ in. (17.1 cm).

1967 ACI Committee 435 Report Assume *floor* slab *not intended* to support or to be attached to partitions or other construction likely to be damaged by large deformations.

$$\text{Basic } L/h = 18 \qquad \text{(Table 6-3)}$$

Without edge beams (flat plate)

$$\text{Factor} = 1.3 \qquad \text{(Table 6-3)}$$
$$\text{Max. } L/h = (18)(1.3) = 23.4$$
$$h_{min} = 216/23.4 = \underline{9.2 \text{ in. } (23.4 \text{ cm})} > 8 \text{ in. } (20.3 \text{ cm})$$

With edge beams (two-way slab)

$$\text{Factor} = 1.4 \qquad \text{(Table 6-3)}$$
$$\text{Max. } L/h = (18)(1.4) = 25.2$$
$$h_{min} = 216/25.2 = \underline{8.6 \text{ in. } (21.8 \text{ cm})} > 8 \text{ in. } (20.3 \text{ cm})$$

Hence deflections must be checked for this slab both without and with edge beams. This simple method tends to be conservative and to represent an upper bound for slab thicknesses for most loading conditions.

1971 ACI Code Method applies to floor slabs only, although it might be used for roof slabs as well in many cases, $f_y = 60$ ksi.

Without edge beams (flat plate)

$$h_{min} = 5 \text{ in.} \qquad \alpha_m = 0$$

for the case of no edge beams. Hence, from Fig. 6-1 (or according to Table 6-6, Eq. (6-3)),

$$L_n/h = (L_n/36)(0.8 + 0.3) = L_n/32.7$$
$$h_{min} = 216/32.7 = 6.61 \text{ in.} > 5 \text{ in.}$$

However, without edge beams at the discontinuous edges the thickness must be increased by 10 percent; therefore,

$$h_{min} = \underline{7.3 \text{ in. (18.5 cm)}} < 8 \text{ in. (20.3 cm)}$$

Also, from Table D-1,

$$h_{min} = L_n/29.75 = 216/29.75 = 7.3 \text{ in. (18.5 cm)}$$

With edge beams (two-way slab)

For a corner panel, $\beta_s = 1/2$
For a square panel, $\beta = 1.0$
Assume $\alpha_m \geq 2.0$
From Fig. 6-1, $f = 43.5$
and

$$h_{min} = (L_n/43.5)(0.8 + 0.3) = L_n/39.5 = 216/39.5$$
$$= \underline{5.5 \text{ in. (14.0 cm)}} < 8 \text{ in. (20.3 cm)}$$

Hence deflections need not be checked for this slab both without and with edge beams. Using Table D-1 (even though $\alpha = 0.8$ in the table versus say 2.0 here),

$$h_{min} = L_n/35.00 = 216/35.00 = 6.2 \text{ in. (15.7 cm)}$$

See Chap. 4 for additional examples of slabs.

British Code This method does not distinguish between one-way and two-way construction, nor between slabs without and with edge beams.

$$f_y = 60 \text{ ksi} \qquad \text{Basic } L/d = 26 \qquad \text{(Table 6-7)}$$

Assume $100A_s/bd = 1.0$ and no compression steel. From Table 6-7, the tension steel factor = 1.00.

$$d = L/26 = 216/26 = \underline{8.3 \text{ in. (21.1 cm)}} > 6.75 \text{ in. (17.1 cm)}$$

Hence deflections may be excessive and should be checked.

Example 7 Noncomposite double-T beam with prestressing steel only, $L = 76$ ft $= 912$ in. (23.2 m), normal-weight concrete, $h = 32$ in. (81.3 cm). (Same as Example 1, Chap. 5, Table 5-6.)

T. Y. Lin The following approximate limits are given.

$$\text{Roof } L/h = 40 \qquad \text{Floor } L/h = 36 \qquad \text{(Table 6-12)}$$

Reduce by 5 to 10 percent for spans in excess of about 70 ft (use 10 percent).

Roof: $L/h = (40)(0.9) = 36.0$

$h_{min} = 912/36.0 = \underline{25.3 \text{ in. (64.3 cm)}} < 32$ in. (81.3 cm)

Floor: $L/h = (36)(0.9) = 32.4$

$h_{min} = 912/32.4 = \underline{28.1 \text{ in. (71.4 cm)}} < 32$ in. (81.3 cm)

Hence the design thickness would be satisfactory for both roof and floor construction.

PCI Handbook The following are typical span–depth ratios of precast prestressed concrete members.

Stemmed roof slabs: $L/h = 35$ to 40

Typical $h = 912/35$ to $912/40$

$= \underline{26.1 \text{ in. (66.3 cm) to } 22.8 \text{ in. (57.9 cm)}} < 32$ in. (81.3 cm)

Hence the design thickness would be satisfactory for roof construction.

Stemmed floor slabs: $L/h = 25$ to 35

Typical $h = 912/25$ to $912/35$

$= \underline{36.5 \text{ in. (92.7 cm) to } 26.1 \text{ in. (66.3 cm)}}$ versus 32 in. (81.3 cm)

Hence the design thickness is bracketed by these typical thicknesses in the case of floor construction.

British Code Span–depth provisions apply only to cracked prestressed members.

Example 8 Same as Example 7 except using sand-lightweight concrete.

T. Y. Lin For approximate limits, reduce the normal-weight concrete span–depth ratios in Example 7 by 5 percent for lightweight concrete with E_c less than 3×10^6 psi.

Roof: $L/h = (36.0)(0.95) = 34.2$

$h_{min} = 912/34.2 = \underline{26.7 \text{ in. (70.4 cm)}} < 32$ in. (81.3 cm)

Floor: $L/h = (32.4)(0.95) = 30.8$

$h_{min} = 912/30.8 = \underline{29.6 \text{ in. (75.2 cm)}} < 32$ in. (81.3 cm)

Hence the design thickness would be satisfactory for both roof and floor construction using lightweight as well as normal-weight concrete.

PCI Handbook Does not refer specifically to span–depth ratios for lightweight concrete members. See Example 7 for normal-weight concrete members.

British Code Does not apply.

Example 9 Noncomposite single-T beam, first containing only prestressing steel and second containing both prestressed and nonprestressed tension steel, $L =$ 100 ft = 1200 in. (30.5 m), sand-lightweight concrete, with additional sustained loading applied some time after the transfer of prestress, $h = 48$ in. (121.9 cm). (Same as Example 3, Chap. 5, Table 5-7.)

The span–depth methods by T. Y. Lin and the PCI Handbook do not refer specifically to partially prestressed members. The British Code refers to span–depth ratios only for cracked members.

T. Y. Lin The following approximate limits are given. Reduce the span–depth ratios in Table 6-12 by 5 to 10 percent (use 10 percent) for spans in excess of about 70 ft, and 5 percent for lightweight concrete with E_c less than 3×10^6 psi.

Single-Ts side by side

Roof: $L/h = 40$ Floor: $L/h = 36$ (Table 6-12)

Roof: $L/h = (40)(0.90)(0.95) = 34.2$

$h_{\min} = 1200/34.2 = \underline{35.1 \text{ in. } (89.2 \text{ cm})} < 48 \text{ in. } (121.9 \text{ cm})$

Floor: $L/h = (36)(0.90)(0.95) = 30.8$

$h_{\min} = 1200/30.8 = \underline{39.0 \text{ in. } (99.1 \text{ cm})} < 48 \text{ in. } (121.9 \text{ cm})$

Hence the design thickness required was considerably in excess of the approximate minimum thicknesses for both roof and floor construction.

Single-Ts spaced 20 ft centers

Roof: $L/h = 32$ Floor: $L/h = 28$ (Table 6-12)

Roof: $L/h = (32)(0.90)(0.95) = 27.4$

$h_{\min} = 1200/27.4 = \underline{43.8 \text{ in. } (111.3 \text{ cm})} < 48 \text{ in. } (121.9 \text{ cm})$

Floor: $L/h = (28)(0.90)(0.95) = 23.9$

$h_{\min} = 1200/23.9 = \underline{50.2 \text{ in. } (127.5 \text{ cm})} > 48 \text{ in. } (121.9 \text{ cm})$

Hence the design thickness required was greater than the approximate minimum thickness for roof construction and slightly smaller than the approximate minimum thickness for floor construction.

PCI Handbook The following are typical span–depth ratios of precast prestressed concrete members of normal-weight concrete only.

Stemmed roof slabs: $L/h = 35$ to 40

$$\text{Typical } h = 1200/35 \text{ to } 1200/40$$
$$= \underline{34.3 \text{ in. } (87.1 \text{ cm}) \text{ to } 30.0 \text{ in. } (76.2 \text{ cm})} < 48 \text{ in. } (121.9 \text{ cm})$$

Stemmed floor slabs: $L/h = 25$ to 35

$$\text{Typical } h = 1200/25 \text{ to } 1200/35$$
$$= \underline{48.0 \text{ in. } (121.9 \text{ cm}) \text{ to } 34.3 \text{ in. } (87.1 \text{ cm})} \leq 48 \text{ in. } (121.9 \text{ cm})$$

Hence the design thickness required was considerably greater than the typical thickness range for stemmed roof slabs, and equal to the upper end of the typical thickness range for stemmed floor slabs. However, these results should be viewed in the context that the single-T in the example was constructed of sand-lightweight concrete.

British Code

$$L/d = 20 \qquad \text{(Table 6-7)}$$
$$\text{Actual } d = 48 - 4.75 = 43.25 \text{ in. } (109.0 \text{ cm})$$
$$d = L/20 = 1200/20 = \underline{60 \text{ in. } (152.4 \text{ cm})} > 43.25 \text{ in. } (109.0 \text{ cm})$$

Hence the design effective depth is considerably less than that suggested for the avoidance of deflection problems. Note that this is the same basic span–effective depth ratio as for reinforced beams, which is also applied to certain cracked prestressed beams.

Example 10 Composite AASHTO Type IV girder with prestressing steel only, $L = 80$ ft $= 960$ in. (24.4 m), normal-weight concrete, precast $h = 54$ in. (137.2 cm), composite $h = 61$ in. (154.9 cm). (Same as Example 4, Chap. 5, Table 5-8.)

T. Y. Lin

Precast section (I-beam)

$$\text{Conservative } L/h = 20 \qquad h = 960/20 = \underline{48 \text{ in. } (121.9 \text{ cm})} < 54 \text{ in. } (137.2 \text{ cm})$$
$$\text{Normal } L/h = 22 \text{ to } 24$$
$$h = 960/22 \text{ to } 960/24$$
$$= \underline{43.6 \text{ in. } (110.7 \text{ cm}) \text{ to } (40.0 \text{ in. } (101.6 \text{ cm})}$$
$$< 54 \text{ in. } (137.2 \text{ cm})$$
$$\text{Critical } L/h = 26 \text{ to } 28$$
$$h = 960/26 \text{ to } 960/28$$
$$= \underline{36.9 \text{ in. } (93.7 \text{ cm}) \text{ to } 34.3 \text{ in. } (87.1 \text{ cm})} < 54 \text{ in. } (137.2 \text{ cm})$$

The designed section is deeper than all of these cases.

Composite section (T-beam) Reduce above L/h values by 5 to 10 percent (use 10 percent) for T-beams.

$$\text{Conservative } L/h = (20)(0.90) = 18$$

$$h = 960/18 = \underline{53.3 \text{ in. } (135.4 \text{ cm})} < 61 \text{ in. } (154.9 \text{ cm})$$

$$\text{Normal } L/h = (22)(0.90) \text{ to } (24)(0.90) = 19.8 \text{ to } 21.6$$

$$h = 960/19.8 \text{ to } 960/21.6$$

$$= \underline{48.5 \text{ in. } (123.2 \text{ cm}) \text{ to } 44.4 \text{ in. } (112.8 \text{ cm})}$$

$$< 61 \text{ in. } (154.9 \text{ cm})$$

$$\text{Critical } L/h = (26)(0.90) \text{ to } (28)\ (0.90) = 23.4 \text{ to } 25.2$$

$$h = 960/23.4 \text{ to } 960/25.2$$

$$= \underline{41.0 \text{ in. } (104.1 \text{ cm}) \text{ to } 38.1 \text{ in. } (96.8 \text{ cm})}$$

$$< 61 \text{ in. } (154.9 \text{ cm})$$

The designed section is deeper than all of these cases.

Observations pertaining to Examples 1 to 10, and the different methods for evaluating maximum span–depth or span–thickness ratios and minimum depths or thicknesses 1. The indirect method of controlling deflections by means of maximum span–depth or span–thickness ratios or minimum depths or thicknesses is primarily a *rule-of-thumb* approach. Even when basic factors together with various modifying factors are used to determine limiting values, the procedure is based largely on experience. In addition, analytical procedures for determining limiting span–depth ratios or depths, or for controlling deflections by means of calculated versus limiting deflections, are based on nominal allowable deflections, which in turn are *also* based on experience.

Beams and solid or ribbed one-way slabs, by the methods of the 1963 ACI Code, 1967 ACI Committee 435, 1971 ACI Code and British Code 2. All of the four methods provide for simple, continuous, and cantilever spans.

3. The 1963 ACI Code method applied to normal-weight concrete only, and did not distinguish between members either attached or not attached to non-structural elements. Minimum thicknesses by this method tend to be on the low or liberal side as compared with the other three methods.

4. The 1967 ACI Committee 435 method is the only one of the four methods that specifically distinguishes between *roof* and *floor* members either *intended* or *not intended* to support or to be attached to partitions or other construction likely to be damaged by large deformations. This method includes the effect of lightweight concrete but not different steel strengths. It also does not include the effect of different load levels or compression steel. Minimum thicknesses by this method tend to be on the high or conservative side as compared to the other three methods. In virtually all practical designs, deflections will have to be checked by this method when the member is *intended* to support or to be attached to partitions or other construction likely to be damaged by large deformations.

5. The 1971 ACI Code method applies only to the case of members *not supporting* or attached to partitions or other construction likely to be damaged by large deflections. It is the only one of the four methods that specifically refers to composite members. This method takes into account the effects of lightweight concrete (the only one to include the effect of different weight lightweight concrete) and different steel yield strengths, but not different load levels nor the effect of compression steel.

6. The British Code does not refer to the question of members either attached or not attached to nonstructural elements. It is expressed in terms of basic span–depth ratios based on experience for typical members, with modifying factors for the steel service stress (or yield strength), percentages of tension and compression steel, and the use of normal-weight or lightweight concrete. The span–effective depths by this method refer to a nominal allowable deflection of $L/250$, with the provision for a direct modification based on other allowable deflections.

7. Examples 1, 2, and 3 referred to a simple-span, normal-weight rectangular beam; a continuous sand-lightweight one-way slab; and a continuous normal weight T-beam. These were designed by the strength method. Only the T-beam required no deflection calculations at all based on the span–depth provisions of all four methods. Although their bases and conditions are different, the limitations of the 1971 ACI Code and British Code were comparable for Examples 1 and 2, but somewhat different for the T-beam in Example 3.

8. In Example 4, the simple-span, normal-weight bridge girder was designed by the load-factor method. For this example, the designed thickness was somewhat less than the minimum thickness by the AASHTO method—unless consideration is given to the use of compression steel or some other method for controlling deflections.

9. In Example 5, the simple-span, normal-weight unshored reinforced concrete composite beam was designed by the strength method. For this example, the designed thicknesses of both the precast and composite sections were less than the minimum thicknesses by the 1971 ACI Code, in which case deflections must be checked. It was found in Chap. 3 that some, but not all, of the 1971 ACI Code deflection limitations were satisfied.

Two-way construction, either without or with edge beams, by the methods of the 1967 ACI Committee 435 Report, 1971 ACI Code, and British Code 10. The 1967 ACI Committee 435 method represents a simple method that distinguishes between *roof* and *floor* members whether *intended* or *not intended* to support or to be attached to partitions or other construction likely to be damaged by large deformations. The method includes the effect of lightweight concrete but not different load levels or compression steel. Thicknesses by this method tend to be on the high or conservative side compared with the 1971 ACI Code method, and to represent an upper bound of practical thicknesses for most loading conditions.

11. The 1971 ACI Code method applies to floor, but not roof, construction, and to normal-weight, but not lightweight, concrete. However, as previously discussed, it may be used for roof slabs as well in many cases. This method does

not include the effect of different load levels nor the effect of compression steel. The method provides for a consideration of the effect of interior or edge panels, panel shape, span ratios, size of supporting columns, the presence of drop panels, the use of higher strength steels, and the stiffening effect of beams. The equations provide for a transition from slabs on stiff beams to slabs without beams. The 1971 ACI Code method provides for minimum thicknesses that are lower or more liberal than the other two methods. It is also a more detailed procedure.

12. The British Code essentially does not distinguish between one-way and two-way construction, nor between slabs without and with edge beams. The same procedure for determining limiting span–depth ratios is applicable to solid slabs as to beams. Thus, typical depths by this method tend to be on the high or conservative side for two-way construction.

13. Example 6 referred to a solid slab both without (flat plate) and with (two-way slab) edge beams using normal-weight concrete. For this example, deflections need to be checked by the 1967 ACI Committee 435 and British Code methods for both cases of without and with edge beams, but need not be checked by the 1971 ACI Code method.

Prestressed members, by the methods of T. Y. Lin, PCI Handbook, British Code

14. The span–depth methods by T. Y. Lin and the PCI Handbook do not refer specifically to partially prestressed members. Also, the PCI method does not refer specifically to lightweight concrete members. The British Code refers to span–depth ratios only for cracked prestressed members, in which case the same ratios are used as for nonprestressed members.

15. Example 7 refers to a noncomposite double-T beam with prestressing steel only, and using normal-weight concrete. Using both the Lin method and the PCI Handbook method, the minimum thicknesses were on the low or liberal side compared to the designed thicknesses for both roof and floor construction, except for part of the L/h range for floor construction by the PCI method.

16. Example 8 is the same as Example 7, except using sand-lightweight concrete. The minimum thicknesses using the Lin method were again less than the designed thicknesses in this lightweight concrete example, as in the case of Example 7 for the same beam with normal-weight concrete, for both roof and floor construction.

17. Example 9 refers to a noncomposite single-T beam, first containing only prestressing steel and second containing both prestressed and nonprestressed tension steel, and using sand-lightweight concrete. Using the span–thickness ratios of Lin, the minimum thicknesses were on the low or liberal side for both roof and floor construction, as compared to the designed thickness, except for floor construction for single-Ts spaced 20 ft apart.

18. For the partially prestressed single-T beam of Example 9, the minimum depth by the British Code was somewhat on the high, or conservative, side compared with the designed depth, as might be expected, since this provision refers to the same span–depth ratios and minimum depths for partially prestressed (or cracked) members as for nonprestressed members.

19. Example 10 refers to a composite AASHTO Type IV girder with prestressing steel only, and using normal-weight concrete. Using the span–thickness ratios of Lin, the typical thicknesses were all (including so-called conservative, normal, and critical ratios) on the low or liberal side compared with the designed thicknesses for both the precast and composite sections.

Analytical Procedure for Determining Span–Depth Ratios Using I_e

Short-time deflections Using average beam proportions, the L/h curves in Fig. 6-2 were derived by the author[11] (see Appendix N) for singly reinforced rectangular beams and one-way slabs, and for singly reinforced T-beams, for the following conditions: $\Delta = L/360$, normal-weight concrete, uniformly distributed loading, and for different idealized boundary conditions, steel percentages, and concrete strengths.

For other values of Δ/L; for example, for $\Delta = L/180$:

$$(L/h)_{\Delta = L/180} = (L/h)_{\Delta = L/360}(360/180) \tag{6-6}$$

For sand-lightweight concrete, including all strengths ($w = 120$ pcf), and using Eqs. (1-7) and (1-13):

$$(L/h)_{\text{sand-lightweight}} = (L/h)_{\text{normal-weight}}(0.80) \tag{6-7}$$

For all-lightweight concrete, including all strengths ($w = 100$ pcf), and using Eqs. (1-7) and (1-13):

$$(L/h)_{\text{all-lightweight}} = (L/h)_{\text{normal-weight}}(0.69) \tag{6-8}$$

Dead-load creep plus shrinkage plus live-load deflections As derived by the author[11] and modified by Salmon,[20]

$$(L/h)_{\text{D cp+sh+L}} = (L/h)_{\text{short-time D+L}}\left[\frac{C_L + 1}{C_L + 1 + (k_r T - 1)(I_{e_{D+L}}/I_{e_D})}\right] \tag{6-9}$$

where $C_L = M_L/M_D$ and $k_r T$ is defined in Chap. 3 and the Notation. When $(I_e)_{D+L} = (I_e)_D$, Eq. (6-9) reduces to Eq. (6-10).

$$(L/h)_{\text{D cp+sh+L}} = (L/h)_{\text{short-time D+L}}\left[\frac{C_L + 1}{C_L + k_r T}\right] \tag{6-10}$$

These procedures are derived in Appendix N. Although the results in Fig. 6-2 were developed for singly reinforced rectangular and T-beams, they may be used for doubly reinforced beams as well, in which the parameter $k_r T$ takes into account the time-dependent effect of the compression steel. The author believes the curves in Fig. 6-2 can be especially useful in estimating L/h for comparative cases. It might be noted that Eqs. (6-6) to (6-10) can be used for the conditions given in Table 6-15 from the ACI Code (318-71).

A similar procedure based on I_g, I_e, and I_{cr} was derived by Fling for the ACI Committee 435 Report[21] and plotted in terms of L/h, M/bd^2, L/Δ, and a parameter $C_w L$, where C_w = ratio of the total load to weight of concrete beam (stem) alone. Two curves were presented, but only for $E_c = 3.5 \times 10^6$ psi and F = ratio of long-time to short-time deflection = 3. Several examples by this method are included in Ref. 21.

Comments on Fig. 6-2 1. The steel percentage $p = (100)A_s/bd$ in all cases except for the T-beam negative moment sections, where $p = (100)A_s/b'd$. The same values for p were assumed for both positive and negative moment sections in all con-

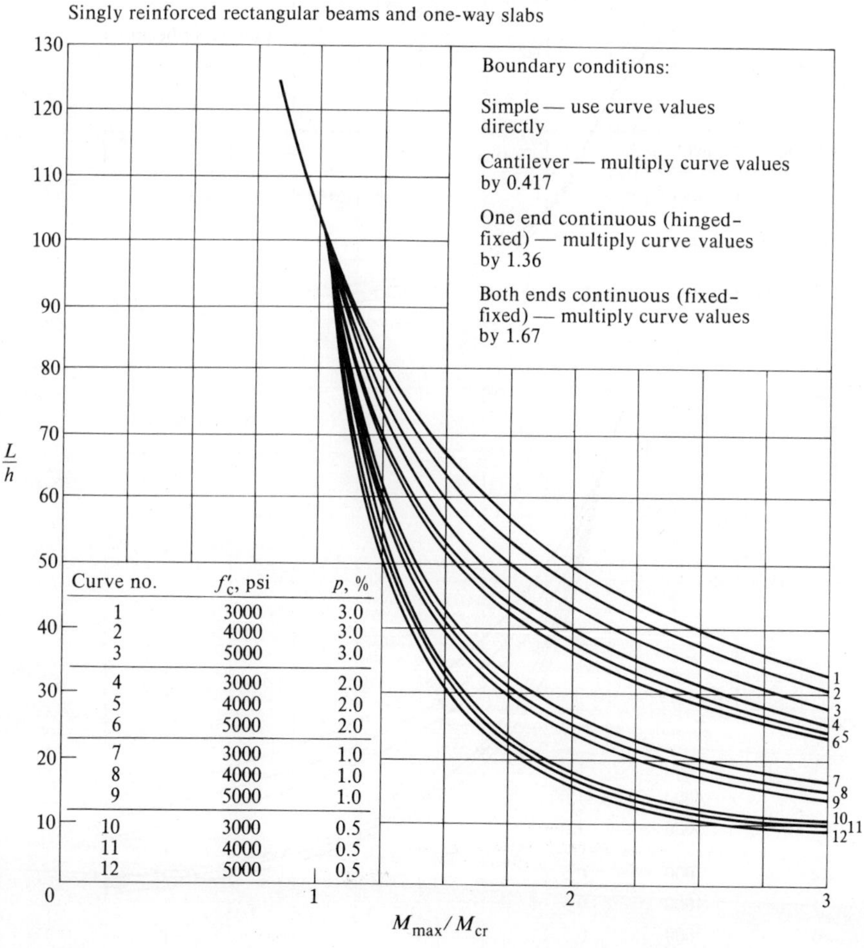

FIGURE 6-2
L/h versus M_{max}/M_{cr} curves for the conditions: $\Delta = L/360$, normal-weight concrete, uniformly distributed short-time loading, and for different boundary conditions, steel percentages, concrete strengths.[11]

tinuous beam solutions. However, in using these curves, it is suggested that the positive region steel percentage be used in all cases.

2. M_{max}, M_{cr}, and y_t all refer to the numerical maximum moment region (positive for simple beams, negative for the other cases), except that an average of the I_e (which include M_{cr}/M_{max} parameters) was used for the continuous beams.

3. The L/h values are greater for the lower-strength concrete because the n value is greater. The compensator in this is the M_{cr}—abscissa parameter (M_{cr} is lower for lower-strength concrete).

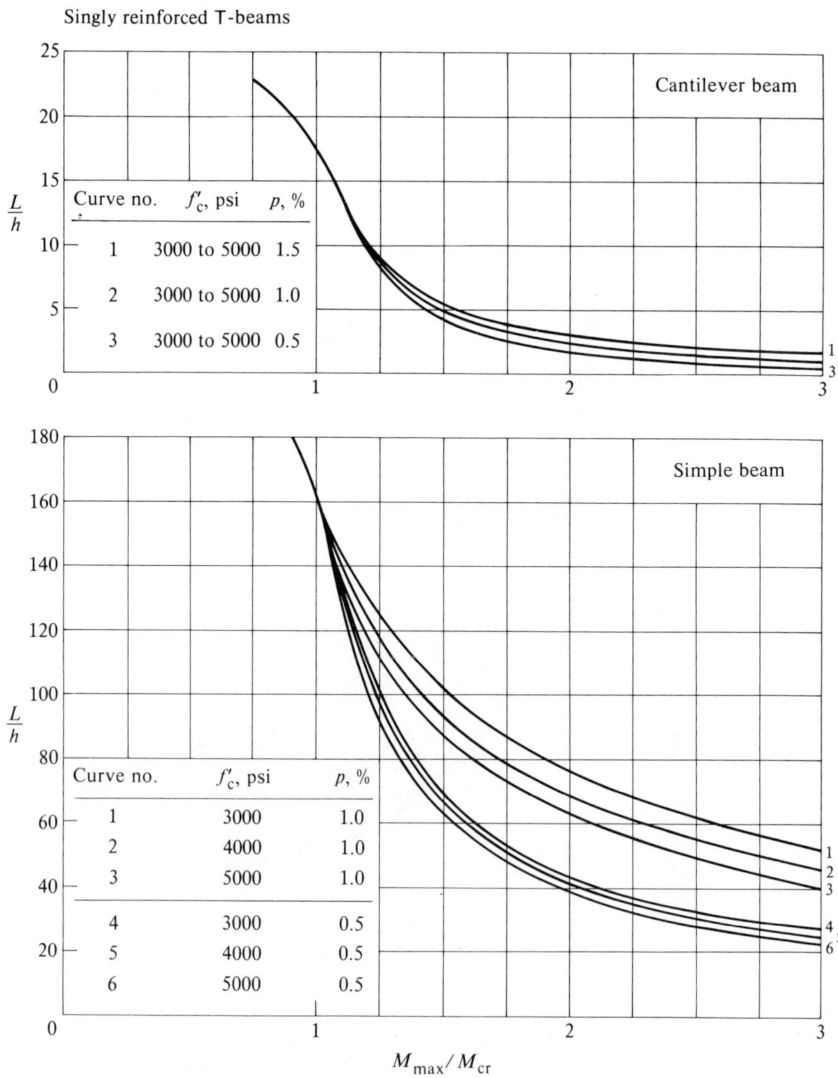

Figure 6-2 (*continued*)

4. The L/h values appear to be abnormally high for the case of the simple T-beam—higher than for the continuous beams. The reason for this is that M_{cr} in the abscissa refers to the positive moment section for the simple beam (where $y_t = \frac{4}{5}h$ in the solution) and to the negative moment section (where $y_t = \frac{1}{5}h$ in the solution) for the other T-beam cases.

Example 11 Simple span beam, normal-weight concrete, $L/h = 240/14 = 17.1$, $f'_c = 4000$ psi, $p = 1.98$ percent, $M_D = 15.0$ ft-k, $M_L = 50.0$ ft-k. (Same as Example 1, Chap. 3, Table 3-2.)

From Example 1, Chap. 3,

$$M_{cr} = 13.50 \text{ ft-k} \qquad (I_e)_{D+L}/(I_e)_D = 1355/2168 = 0.625$$

$$C_L = M_L/M_D = 50.0/15.0 = 3.33$$

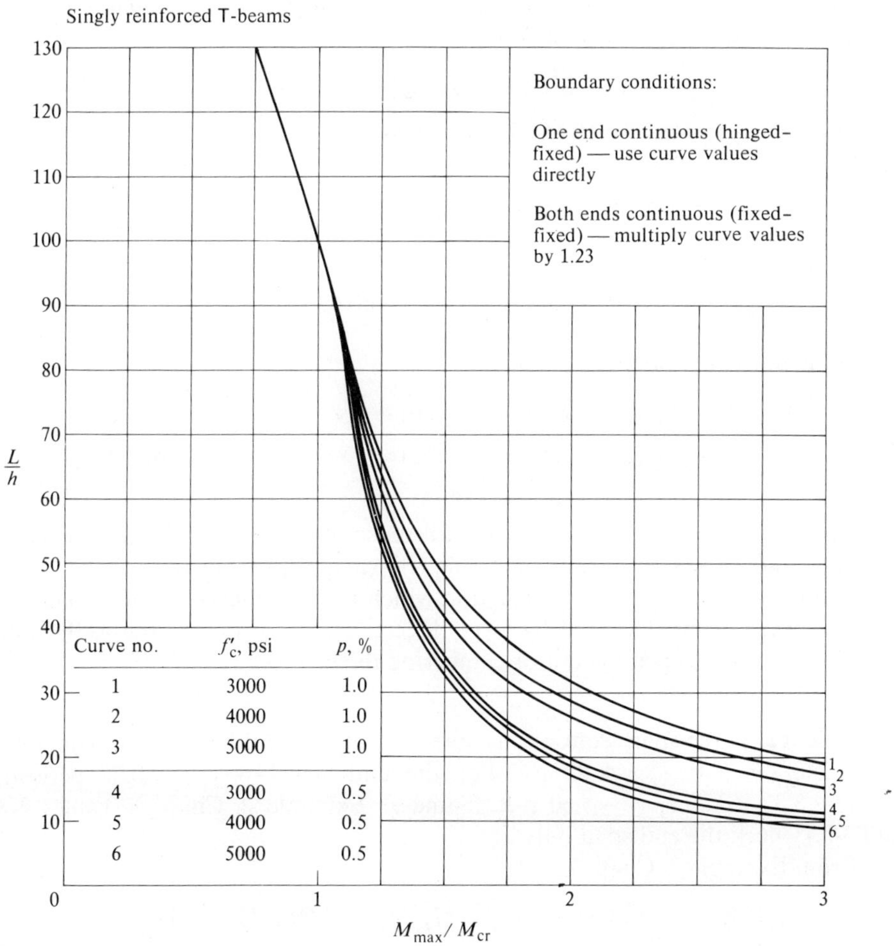

Figure 6.2 (*continued*)

For initial dead-load plus live-load deflection

$$M_{max}/M_{cr} = (15.0 + 50.0)/13.50 = 65.0/13.50 = 4.81$$

From Fig. 6-2 for $f_c' = 4000$ psi, $p = 1.98$ percent, and $M_{max}/M_{cr} = 4.81$,

$$\text{Max.}\,(L/h)_{\Delta=L/360} = \underline{12 \text{ (off chart)}}$$

Using Eq. (6-6), $\text{Max.}\,(L/h)_{\Delta=L/180} = (12)(360/180) = \underline{24}$

For live-load deflection and allowable $\Delta = L/360$ and $L/180$ (Table 6-15)

$\text{Max.}\,(L/h)_{\Delta=L/360} = (12)(M_{D+L}/M_L) = (12)(65.0/50.0) = \underline{15.6} < 17.1$ Not OK

$\text{Max.}\,(L/h)_{\Delta=L/180}$ $(24)(65.0/50.0) = \underline{31.2} > 17.1$ OK

For dead-load creep plus shrinkage plus live-load deflection and allowable $\Delta = L/480$ and $L/240$ (Table 6-15) From above,

$$\text{Max.}\,(L/h)_{\Delta=L/360} = 12$$

Using Eq. (6-9) and $k_r T = 2.0$ (ACI Code, 318-71),

$$(L/h)_{D\,cp+sh+L} = (L/h)_{\text{short-time } D+L}\left[\frac{C_L + 1}{C_L + 1 + (k_r T - 1)(I_{e_{D+L}}/I_{e_D})}\right] \tag{6-9}$$

$$= 12\,\frac{3.33 + 1}{3.33 + 1 + (2.0 - 1)(0.625)} = 10.48$$

Using Eq. (6-6) for $L/480$ and $L/240$,

$\text{Max.}\,(L/h)_{\Delta=L/480} = (10.48)(360/480) = \underline{7.9} < 17.1$ Not OK

$\text{Max.}\,(L/h)_{\Delta=L/240} = (10.48)(360/240) = \underline{15.7} < 17.1$ Not OK

Comment These results are consistent with those of Chaps. 3 and 6 for the same example, and shows by means of calculated span–depth ratios that the beam should not be used for roof construction which is not attached to nonstructural elements likely to be damaged by large deflections, according to the ACI Code (318-71), without deepening and/or cambering the beam.

Example 12 Three-span continuous one-way slab, sand-lightweight concrete, $L/h = 240/8 = 30.0$, $f_c' = 4000$ psi. For the end span—pos. $p = 0.77$ percent, $M_D = -3.32$ ft-k, $M_L = -5.60$ ft-k. (Same as Example 2, Chap. 3, Tables 3-4 and 3-5.) Check the end span only.

From Example 2, Chap. 3,

$$M_{cr} = 4.80 \text{ ft-k} \qquad C_L = M_L/M_D = 5.60/3.32 = 1.69$$

$$\text{Avg.}\,(I_e)_{D+L}/(I_e)_D = 292/512 = 0.570 \qquad \text{(Table 3-6)}$$

For initial dead-load plus live-load deflection

$$M_{max}/M_{cr} = (3.32 + 5.60)/4.80 = 8.92/4.80 = 1.86$$

From Fig. 6-2 for $f'_c = 4000$ psi, $p = 0.77$ percent and $M_{max}/M_{cr} = 1.86$,

Max. $(L/h)_{\Delta = L/360}$

$$= 23(1.36 \text{ for end span})(0.80 \text{ by Eq. (6-7) for lightweight concrete}) = \underline{25.0}$$

Using Eq. (6-6),

$$\text{Max.}\,(L/h)_{\Delta = L/180} = (25.0)(360/180) = \underline{50.0}$$

For live-load deflection and allowable $\Delta = L/360$ and $L/180$ (Table 6-15)

Max. $(L/h)_{\Delta = L/360} = (25.0)(M_{D+L}/M_L) = (25.0)(8.92/5.60) = \underline{39.8} > 30.0$ OK

Max. $(L/h)_{\Delta = L/180} = (50.0)(8.92/5.60) = \underline{79.6} > 30.0$ OK

For dead-load creep plus shrinkage plus live-load deflection and allowable $\Delta = L/480$ and $L/240$ (Table 6-15) From above,

$$\text{Max.}\,(L/h)_{\Delta = L/360} = 25.0$$

Using Eq. (6-9) and $k_r T = 1.68$ (Chap. 3, by ACI Code),

$$(L/h)_{D\,cp+sh+L} = (L/h)_{\text{short-time } D+L}\left[\frac{C_L + 1}{C_L + 1 + (k_r T - 1)(I_{e_{D+L}}/I_{e_D})}\right] \tag{6-9}$$

$$= (25.0)\,\frac{1.69 + 1}{1.69 + 1 + (1.68 - 1)(0.570)} = 21.8$$

Using Eq. (6-6) for $L/480$ and $L/240$,

Max. $(L/h)_{\Delta = L/480} = (21.8)(360/480) = \underline{16.4} < 30.0$ Not OK (only limitation)

Max. $(L/h)_{\Delta = L/240} = (21.8)(360/240) = \underline{32.7} > 30.0$ OK

Comment These results are consistent with those of Chapters 3 and 6 for the same example (using the ACI Code method and $k_r T = 1.68$).

Example 13 Continuous T-beam, normal-weight concrete, $L_n/h = (29)(12)/25 = 13.9$, $f'_c = 3000$ psi. For the exterior span—pos $p = 0.25$ percent, $M_D = -126.2$ ft-k, $M_L = -168.2$ ft-k. (Same as Example 3, Chap. 3, Table 3-7.) Check the exterior span only.

From Example 3, Chap. 3,

$$\text{Neg. } M_{cr} = -186.5 \text{ ft-k} \qquad C_L = M_L/M_D = 168.2/126.2 = 1.33$$

$$\text{Avg. } (I_e)_{D+L}/(I_e)_D = 22\,400/34\,500 = 0.649 \qquad \text{(Table 3-10)}$$

For initial dead-load plus live-load deflection

$$M_{max}/M_{cr} = (126.2 + 168.2)/186.5 = 294.4/186.5 = 1.58$$

From Fig. 6-2 for $f'_c = 3000$ psi, $p = 0.25$ percent, and $M_{max}/M_{cr} = 1.58$,

$$\text{Max. } (L/h)_{\Delta = L/360} = 20(1.36 \text{ for exterior span}) = \underline{27.2}$$

Using Eq. (6-6),

$$\text{Max. } (L/h)_{\Delta = L/180} = (27.2)(360/180) = \underline{54.4}$$

For live-load deflection and allowable $\Delta = L/360$ and $L/180$ (Table 6-15)

$$\text{Max. } (L/h)_{\Delta = L/360} = (27.2)(M_{D+L}/M_L) = (27.2)(294.4/168.2) = \underline{47.6} > 13.9 \quad \text{OK}$$
$$\text{Max. } (L/h)_{\Delta = L/180} = (54.4)(294.4/168.2) = \underline{95.2} > 13.9 \quad \text{OK}$$

For dead-load creep plus shrinkage plus live-load deflection and allowable $\Delta = L/480$ and $L/240$ (Table 6-15) From above,

$$\text{Max. } (L/h)_{\Delta = L/360} = 27.2$$

Using Eq. (6-9) and $k_r T = 1.61$ (Chap. 3, by ACI Code),

$$(L/h)_{D\,cp+sh+L} = (L/h)_{\text{short-time } D+L}\left[\frac{C_L + 1}{C_L + 1 + (k_r T - 1)(I_{e_{D+L}}/I_{e_D})}\right]$$

$$= (27.2)\frac{1.33 + 1}{1.33 + 1 + (1.61 - 1)(0.649)} = 23.3 \qquad (6\text{-}9)$$

Using Eq. (6-6) for $L/480$ and $L/240$,

$$\text{Max. } (L/h)_{\Delta = L/480} = (23.2)(360/480) = \underline{17.4} > 13.9 \quad \text{OK}$$
$$\text{Max. } (L/h)_{\Delta = L/240} = (23.2)(360/240) = \underline{34.8} > 13.9 \quad \text{OK}$$

Comment These results are consistent with those of Chaps. 3 and 6 for the same example (using the ACI Code method and $k_r T = 1.61$), which shows that the beam has no deflection limitations according to the ACI Code (318-71).

Observations pertaining to the analytical procedure for determining span–depth ratios using I_e, and Examples 11 to 13 1. Although the results for all four allowable deflection cases in Examples 11 to 13 were consistent with the allowable deflection results in Chap. 3 (also shown in this chapter), the author believes this method should be used with caution. The primary reason for this is that the charts used in such an analytical procedure necessarily refer to average or typical beam proportions and conditions, which may be significantly different from the actual properties and conditions in specific cases.

2. The calculation time involved in such an approach for determining span–depth ratios can be of the same order of magnitude as the direct calculation of deflections. Thus, it may be preferable to use relatively simple span–depth or

span–thickness procedures for preliminary design and deflection control of beams, followed by some method of directly checking deflections.

3. The use of analytical procedures for evaluating span–depth or span–thickness ratios, etc., including the use of Fig. 6-2, can be beneficial in comparative evaluations of different members, and different allowable deflection cases, with regard to deflection control.

4. Regardless of the method to be used, it should be clear that one can calculate the span–depth ratio based on an allowable deflection–span ratio, or the deflection directly for a given beam, loading, and other pertinent conditions, with presumably equal facility. The latter determination provides a direct evaluation of the deflection properties of the structure, while the former may be more useful in the proportioning of members.

6-3 ALLOWABLE DEFLECTIONS

The proper control of deformations in structures involves a consideration of various displacements, deflections, rotations, and both amplitude and frequency of vibrations, etc., compared with usable limits based on structural integrity, and serviceability, esthetic, and physiological requirements. However, the primary approach used by most engineers refers to placing limits on computed deflections.[1–45]

As discussed by Plewes and Garden[22] in their interesting paper, an earlier allowable deflection of $L/480$ used by European engineers was increased by American engineers in the nineteenth century to $L/360$. This value has been widely used to this date, and still is regarded as adequate in many cases for steel, timber, and concrete beams with regard to plaster and masonry cracking, visibility of sags, and springiness of floors. However, it offers no guarantee that plaster will not crack, for example, and part of its success has been due to the fact that buildings seldom receive their full design loads.

For relatively critical situations allowable deflections of $L/480$, $L/720$, and $L/1000$ have also been used. On the other hand, deflections of the order of $L/180$ are often allowed for roofs having no finishing materials on their underside, although damage to roofing and ponding can also present problems. In addition, as spans increase the absolute deflection tends to become critical rather than the deflection–span ratio. It is thus apparent that the subject of allowable deflections relates primarily to experience with different types of structures, nonstructural elements, short-term and long-term loading, and other conditions.

For example, the ACI Committee 435 paper[21] describes various deflection limits of $L/180$ to $L/600$; and absolute limits of 0.15 in. (3.8 mm) offset per story for lateral building movement, 0.30 in. (7.7 mm) for masonry and plaster walls, and 1 in. (25.4 mm) for metal partitions (see Table 6-14). The ACI Code[4] (Table 6-15) deflection limits vary from $L/180$ to $L/480$, which also applies to prestressed members. The Czechoslovakian Code[23] (Tables 6-16 to 6-18) deflection limits vary from $L/200$ to $L/750$ (up to $L/1200$ for prestressed members), with various rotations and absolute slope and deflection limits provided. The British Code[5] uses $L/250$ to $L/350$ for its allowable deflections, plus an absolute deflection limit of 20 mm (0.79 in.), and $L/300$ for the maximum camber of prestressed members. In most codes, the computed deflection may exceed the allowable deflection, provided the member is cambered by the difference.

These and other methods for controlling deflections by means of allowable values are presented in this section, along with a number of examples and observations. In addition, a brief summary of various deflection limitations in different countries is given in Appendix L.[3]

When considering computed versus allowable deflections, one is concerned with the possibility of consistently miscalculating deflections. On the one hand, deflections may tend to be overestimated[14] primarily because the full design loads are seldom applied to the structure, plus the fact that the effects of walls and other elements in stiffening the structure are usually not taken into account. On the other hand, deflections may be underestimated[24] because of the effect of construction loads, and high creep and/or shrinkage behavior that was not anticipated.

The lateral movement of buildings should also be considered in relation to finishes such as cladding,[25] although lateral deformation may frequently be more critical from the standpoint of vibration.

Specified Values of Allowable Deflections for Reinforced Concrete Members

1963 ACI Code (318-63)[8] The following allowable deflections apply to reinforced concrete building members when deflections are computed by the 1963 ACI Code method using I_g or I_{cr}, as described in Chap. 3.

Maximum limits for immediate deflection due to live load:

1. For roofs which do not support plastered ceilings —$L/180$
2. For roofs which support plastered ceilings or for floors which do not support partitions —$L/360$

   For a floor or roof construction intended to support or to be attached to partitions or other construction likely to be damaged by large deflections of the floor, the *allowable limit* for the *sum of the immediate deflection due to live load and the additional deflection due to shrinkage and creep under all sustained loads* —$L/360$

1966 Bureau of Public Roads[34] The following limitations apply to the instantaneous deflections of simple and continuous reinforced concrete bridge spans caused by live load plus impact. The short-time dead-load deflection is normally used in the preparation of dead-load camber diagrams.

Simple spans Solid slabs (main reinforcement parallel to traffic), T-girders, box-girders, spans 15.0 ft $\leq L \leq$ 70.0 ft.

$$\Delta_L \leq \frac{w_L}{w_T(4.555 - 0.0364L)^2} \tag{6-11}$$

but shall be not more than the following deflections:

Urban areas ($L >$ 42.4 ft):

$$\Delta_L \leq \frac{w_L}{9.0w_T} \tag{6-12}$$

Rural areas ($L >$ 70.0 ft):

$$\Delta_L \leq \frac{w_L}{4.0w_T} \tag{6-13}$$

Continuous spans Solid slabs (main reinforcement perpendicular to traffic).

Span $L \leq$ 15.0 ft:

$$\Delta_L \leq \frac{w_L}{w_T} \frac{(15.60 - 6.69L_1/L)^3}{(327.6 - 12.24L)^2} \tag{6-14}$$

Spans $L >$ 15.0 ft: the calculated deflection, Δ_L, shall be obtained from the equation for deflection of solid slabs with main reinforcement parallel to traffic.

Solid slabs (main reinforcement parallel to traffic), T-girders, box-girders.

$$\Delta_L \leq \frac{w_L}{w_T} \frac{(15.60 - 6.69L_1/L)^3}{(163.44 - 1.307L)^2} \tag{6-15}$$

but shall be not more than the following deflections:

Urban areas:

$$\Delta_L \leq \frac{w_L}{w_T}(0.688 - 0.295L_1/L)^3 \tag{6-16}$$

Rural areas:

$$\Delta_L \leq \frac{w_L}{w_T}(0.902 - 0.387L_1/L)^3 \tag{6-17}$$

The following equations shall be used in computing the maximum deflection in the span caused by live load plus impact Simple spans:

$$\Delta_L = \frac{M_L L^2}{10E_c I_t} \tag{6-18}$$

Continuous spans (end span):

$$\Delta_L = \frac{M_{CL} L^2(0.0126 - 0.0057L_1/L)}{E_c I_t(0.125 - 0.045L_1/L)} \tag{6-19}$$

and when $L_1/L = 1.0$,

$$\Delta_L \simeq \frac{M_{CL} L^2}{12E_c I_t} \tag{6-20}$$

Notation (units are foot and pound)

Δ_L = Maximum deflection due to live load plus impact
w_L = Equivalent uniform live load plus impact, per unit length of span based on the length of the structure (span L for simple spans; sum of end and interior spans for continuous structures)
w_T = Total dead load plus equivalent uniform live load plus impact, per unit length of span
L = Length of simple span, or length of end span of continuous structure
L_1 = Length of interior span next to end span of continuous structure
M_L = Maximum moment in simple span, at working load level due to w_L
M_{CL} = Maximum positive moment in end span of continuous structure due to w_L, or $M_{CD} \simeq w_L L^2(0.125 - 0.045 L_1/L)$
E_c = Modulus of elasticity of concrete
I_t = Moment of inertia of the transformed cracked section in span L $= A_s n d^2 j(1 - k)$

1968 ACI Committee 435 Paper[21] On the basis of the results of a committee questionnaire sent to practising engineers, it appeared that the design for deflection of reinforced concrete members had not kept pace with the design for strength. Many engineers apparently had vague ideas on how to compute deflections and on the significance of the computed results. Hence the committee formulated the allowable deflections in Table 6-14 as a guide.

For the purpose of controlling deflections and the tendency of structural members to vibrate, the following four categories were included in the table: (1) sensory acceptability, (2) serviceability of the structure, (3) effect on nonstructural elements, and (4) effect on structural elements.

Sensory acceptability tends to be a matter of personal opinion. Therefore, limits will vary considerably with the social culture and use of the structure. Deflection limits for serviceability of the structure are more easily defined since they can be related to the intended use of the structure. Movements affecting the nonstructural elements of construction, such as walls and ceilings, must be limited to prevent cracking or other damage. In some cases excessive deflection can prevent doors, folding partitions, and other movable elements from operating properly. Similarly, movements may affect the structural element itself. In this case, the movement must be limited to prevent structural behavior from being different than that assumed in the design. If this is not possible then the deflection must be considered as part of the design for strength.

Structural deflections are usually thought of as linear movements or a vertical translation of the member at midspan or in cantilever beams at the unsupported end. Generally, these movements do cause the most serious problems. However, angular distortions or rotations without linear movements can also cause similar problems. An example of such a rotational problem is a wall crossing the end of a beam subject to large end rotations, where wall and beam are at right

Table 6-14 Deflection limitations for various examples[21]

Reasons for limiting deflections	Examples	Deflection limitations[a]	Portion of total deflection on which the deflection limitation is based
1. Sensory acceptability			
1.1 Visual	Droopy cantilevers and sag in long-span beams	By personal preference[b]	Total deflection[c]
1.2 Tactile	Vibrations of floors that can be felt	$L/360$[d]	Full live load
	Lateral building vibrations	No recommendation[e]	Gust portion of wind
1.3 Auditory	Vibrations producing audible noise	Not permitted	
2. Serviceability of structure			
2.1 Surfaces which should drain water	Roofs, outdoor decks	$L/240$[f]	Total deflection
2.2 Floors which should remain plane	Gymnasia and bowling alleys	$L/360$ + camber[g,h] or $L/600$[h]	Total deflection Incremental deflections after floor is installed[j]
2.3 Members supporting sensitive equipment	Printing presses and certain building mechanical equipment	Manufacturer's recommendations	Incremental deflections after equipment is leveled[j]
3. Effect on nonstructural elements			
3.1 Walls	Masonry and plaster	$L/600$ or 0.30 in. (7.6 mm) max.[k,h] or $\phi = 0.00167$ rad.[l] See Ref. 35	Incremental deflections after walls are constructed[j]
	Metal movable partitions and other temporary partitions	$L/240$ or 1 in. (25.4 mm) max.[h]	Incremental deflections after walls are constructed[j]
	Lateral building movement	0.15 in. (3.8 mm) offset[m] per story $0.002 \times$ (height) See Fig. 6-2	5 min. sustained wind load
	Vertical thermal movements	$L/300$ or 0.60 (15.2 mm) max.[n] See Ref. 36, Fig. 6-4	Full temperature differential movement

Table 6-14—(*continued*)

Reasons for limiting deflections	Examples	Deflection limitations[a]	Portion of total deflection on which the deflection limitation is based
3.2 Ceilings	Plaster Unit ceilings such as acoustic tile	$L/360^{h,p}$ $L/180^{h,p}$	Incremental deflections after ceiling is built[j]
3.3 Adjacent building elements supported by other members	Windows, walls and folding partitions on unyielding supports below the deflecting member	Absolute deflection limited by tolerances built into the element in question	Incremental deflection after building element in question is constructed[j]
4. Effect on structural elements			
4.1 Deflections causing instability of primary structure	Arches and shells Long columns	Effect of deflections on the stresses and stability of the structure should be taken into account in the structural design of the element	
4.2 Deflections causing different force system or change in stresses in some other element	Beam bearing rotation on masonry wall	Effect of deflections on the stresses and stability of the structure should be taken into account in the structural design of the element	
4.3 Deflections causing dynamic effects	Resonant vibrations which increase static deflections and stresses such as those produced by wind, dancing, moving loads and machinery	Dynamic deflections should be added to static deflections and the total should be less than the limitations imposed for other reasons	

(From ACI Journal,[21] with permission.)

[a] Deflection limitations are given for members supported at both ends and for cantilevers, except as noted. It is assumed that the supports do not move.

[b] For excessive deflections, correct total deflection by camber. Overcorrection by camber is desirable.

[c] Total deflection is the sum of all individual computed deflections due to all loadings plus those due to time dependent effects.

[d] Noticeable floor vibrations depend upon the frequency as well as the amplitude. In addition, the damping efficiency is important since vibrations which die out within a few cycles will be felt only as a single movement, if at all.[37]

[e] Limitations may be based on the lateral acceleration (or deceleration) of the building. Secondary vibrations caused by the wind sway might also be a problem and require limitation in some manner.[38] More research in this area is necessary.

[f] Surfaces should be sloped or total deflections corrected by camber to prevent ponding of water.

[g] The deflection of floors which should remain plane may be partially compensated by camber so that any possible incremental deflection as well as the camber itself does not produce a floor which deviates from plane by more than $L/360$ either upward or downward.

[h] The span L should be considered either parallel to or perpendicular to the direction of stress, whichever is shorter (see Fig. 6-5).

[j] Incremental deflection is the change in elastic deflection over a period of time produced by the addition of new loads or subtraction of existing loads during that time, plus the change in deflection produced by creep and shrinkage during that time.

[k] This limitation assumes that the deflection computation does not give credit to composite action between the member and the wall, or what is commonly called "arching action." If composite action is computed,[39] the deflection limitations should be considerably less than that tabulated. The tabulated limits might still allow some visible damage to weak fragile walls and walls which cannot act compositely with the structural member (e.g., walls with large openings in the span).

[l] Rotation of any member supporting a wall, at the point of support of the wall. This is equivalent to a limitation of 1/600th of the height of the wall.

[m] This limitation applies to the lateral deflection between adjacent floors caused by the story wind shear but does not include the deflection caused by axial lengthening and shortening of the columns. It also applies to the vertical offset deflection between two shear walls in the same plane (see Fig. 6-3).

[n] L is the distance between the exterior column and the first interior column (see Fig. 6-4 and Ref. 40).

[p] Deflection of cantilevers may be twice as high as the tabulated ratios.

angles. In this case, rotation could crack the wall. In some cases, such as plaster ceilings, the linear movement *per se* does not cause damage. Here beam curvature is the proper measure of potential damage. Linear movement, rather than beam curvature, is usually the proper index of damage to walls and partitions because they usually cannot follow the beam curvature due to their excessive rigidity.

Linear movements may be limited to an absolute value or to a fraction of the span depending upon the reasons for the limitation. In other cases, the permissible movement may be a function of the frequency of vibration or the rate of damping. The deflection limitations in Table 6-14 do not apply to movements caused by earthquakes, since most building codes permit some damage when severe earthquakes occur.

1970 American Institute of Steel Construction[12] Beams and girders, including composite steel–concrete members, supporting plastered ceilings shall be so proportioned that the maximum live-load deflection does not exceed $L/360$. The dead-load deflection for typical floor beams in office buildings is limited to $1\frac{1}{2}$ in. as a suggestion.

1971 ACI Code (318-71)[4] When the minimum thicknesses given in the Code are exceeded or do not apply, the allowable deflections in Table 6-15 are specified for one-way beams and slabs, two-way nonprestressed construction, composite beams, and prestressed beams. This table is the result of an effort by ACI

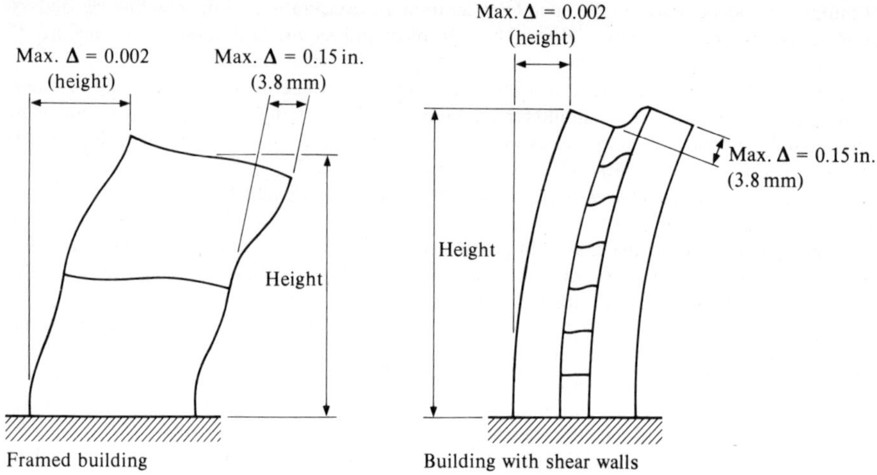

FIGURE 6-3
Lateral deflection limitations for tall buildings (see Table 6-14, Sec. 3.1.3). (*From ACI Journal*,[21] *with permission.*)

Committee 318 to simplify the more extensive set of limitations that would be required to cover various types of structures and conditions (such as the ACI Committee 435, Table 6-15).

For the purpose of calculating deflections, the I_e method is used for all reinforced cases. For prestressed members, any bilinear moment–deflection relationship may be used, including the I_e method.

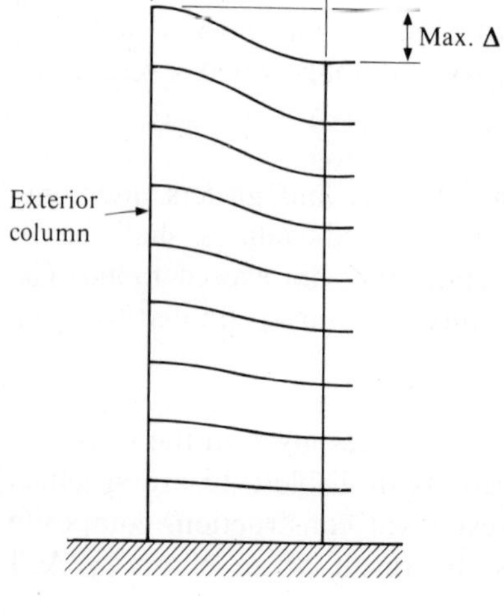

FIGURE 6-4
Vertical thermal movement limitations for tall buildings with exposed columns (see Table 6-14, Sec. 3.1.4). (*From ACI Journal*,[21] *with permission.*)

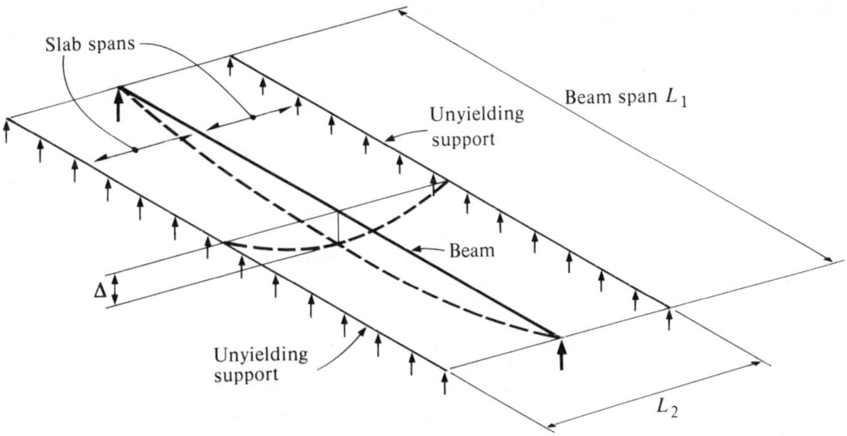

FIGURE 6-5
Beam deflection limitation. The deflection–span limitation for the beam on dimension L_2 when the bowling alley, wall, ceiling or other nonstructural element is oriented in the direction of the slab span (see Table 6-14, Secs. 2.2, 3.1.1, 3.1.2, 3.2, and Note *h*). (*From ACI Journal,*[21] *with permission.*)

1971 Czechoslovakian Code[23] Limiting values of slopes and deflections in terms of span-ratios and absolute values are given in Tables 6-16 to 6-18 for typical cases.

1972 British Standard Code of Practice[5,6] (with permission) 1. The final deflection (including the effects of temperature, creep and shrinkage), measured below the as-cast level of the supports of floors, roofs and all other horizontal members, should not, in general, exceed $L/250$.

2. Partitions and finishes will be affected only by that part of the deflection (including the effects of temperature, creep, and shrinkage) that takes place after the construction of the partitions or the application of the finishes. Information is lacking, but a limit of span/350, or 20 mm (0.79 in.), whichever is the lesser, is suggested.

3. If finishes are to be applied to prestressed concrete members, the total upward deflection should not exceed $L/300$ unless uniformity of camber between adjacent units can be ensured.

(*a*) Span/250 as the total deflection below the horizontal represents the optical sag beyond which awareness of occupants starts. It follows the recommendations of the Structural Engineers' Committee on the Testing of Structures[41] and is consistent with Mayer's[42] findings from a survey in Germany of structures in which deflection had caused damage or led to complaints.

(*b*) Span/350 as the limiting deflection produced by the imposed load, long-term effects of creep and shrinkage, and temperature to avoid damage to partitions, with an absolute limit of 20 mm (0.79 in.) for the larger spans, is obviously a function of the types of partition used. The values suggested are consistent with the ISE report[41] and the 1963 ACI Code;[8] it is worth noting that in the 1971

Table 6-15 Maximum allowable computed deflections[4]

<table>
<tr><th>Type of member</th><th>Deflection to be considered</th><th>Deflection limitation</th></tr>
<tr><td>Flat roofs not supporting or attached to nonstructural elements likely to be damaged by large deflections</td><td>Immediate deflection due to the live load</td><td>* $L/180$</td></tr>
<tr><td>Floors not supporting or attached to nonstructural elements likely to be damaged by large deflections</td><td>Immediate deflection due to the live load</td><td>$L/360$</td></tr>
<tr><td>Roof or floor construction supporting or attached to nonstructural elements likely to be damaged by large deflections</td><td rowspan="2">That part of the total deflection which occurs after attachment of the nonstructural elements, the sum of the long-time deflection due to all sustained loads and the immediate deflection due to any additional live load†</td><td>‡ $L/480$</td></tr>
<tr><td>Roof or floor construction supporting or attached to nonstructural elements not likely to be damaged by large deflections</td><td>§ $L/240$</td></tr>
</table>

(From ACI Code,[4] with permission.)

* This limit is not intended to safeguard against ponding. Ponding should be checked by suitable calculations of deflection, including the added deflections due to ponded water, and considering long-time effects of all sustained loads, camber, construction tolerances, and reliability of provisions for drainage.

† The long-time deflections shall be determined in accordance with Secs. 9.5.2.3 or 9.5.4.2 of the Code, but may be reduced by the amount of deflection which occurs before attachment of the nonstructural elements. This amount shall be determined on the basis of accepted engineering data relating to the time–deflection characteristics of members similar to those being considered.

‡ This limit may be exceeded if adequate measures are taken to prevent damage to supported or attached elements.

§ But not greater than the tolerance provided for the nonstructural elements. This limit may be exceeded if camber is provided so that the total deflection minus the camber does not exceed the limitation.

ACI Code[4] the limit has been reduced to $L/480$. Sometimes partition fixings will allow deflections greater than those indicated in (*b*) and the criterion in (*a*) will apply. The choice of partition and its method of fixing may well condition the criterion adopted.

Using the British Code method, deflections are calculated separately from initial, creep, and shrinkage curvatures.

1973 AASHTO Highway Bridge Specifications[17] (with permission) The following provisions apply to steel bridges including composite steel–concrete bridges.

Members having simple or continuous spans shall be designed so that the deflection due to live load plus impact shall not exceed 1/800 of the span, except on bridges in urban areas used in part by pedestrians whereon the ratio preferably shall be 1/1000.

The deflection of cantilever arms due to live load plus impact shall be limited to 1/300 of the cantilever arm except for the case including pedestrian use, where the ratio preferably shall be 1/375.

When spans have cross-bracing or diaphrams sufficient in depth or strength to insure lateral distribution of loads, the deflection may be computed for the standard H or HS loading, considering all beams or stringers as acting together and having equal deflection.

The moment of inertia of the gross cross-sectional area shall be used for computing the deflections of beams and girders. When the beam or girder is a part of a composite member, the live load may be considered as acting upon the composite section.

1974 paper by Clarke, Neville, and Houghton-Evans[3] A summary of the methods in various countries for controlling deflections is briefly summarized in Appendix L, from Ref. 3.

Table 6-16 Limit deformation of beams supporting crane-runways and of industrial floors[23]

Structural elements			Limit deformation*		Deformation considered for the check
			Slope of deflection curve at the point of load	Deflection at midspan	
Beams of gantries for	Hand cranes		0.0030	$L/500$	Deformation after placing and leveling of rails
	Electrically powered with max. load	Less than 50 t	0.0025	$L/600$	
		Over 50 t	0.0020	$L/750$	
	Suspended cranes†		0.0040	$L/400$	
Beams of industrial floors	With railway traffic	Regular	0.0025	$L/600$	Deformation after leveling of floors‡
		Narrow-gage	0.0040	$L/400$	
	Without railway traffic	Girders	0.0040	$L/400$	
		Other beams	0.0060	$L/250$	

* The checks of the slope and the deflection are considered as equivalent. The more convenient procedure is to be chosen.

† The deformations of rails are checked independently according to respective codes.

‡ For example: long-term part of deformation due to dead load and prestressing, and short-term deformation due to live load.

Table 6-17 Limit deformations governed by esthetical and physiological requirements[23]

Structural elements			Limit deformations[a]: Slope of deflection curve at the point of load[b]	Limit deformations[a]: Deflection at midspan	Deformation considered for the check
Ceilings and walls with compact (for example covered by plaster) surfaces[c]			0.0090	$L/350$	Deformation after finishing the surface[d]
Ceilings and walls mountable surfaces (for example suspended ceilings)			0.0150	$L/200$	
Floors made out of prestressed elements			—	− 1.5 cm[e]	Short-term part of deflection due to prestressing and due to dead load
Axially loaded elements with lining			—	Depends on the kind of surface	After finishing the surface
Floors in apartment and public buildings and elements of outside walls	$L_v \leq 6$ m[f]		0.0150[g]	$L_v/200$[h]	After finishing the ceiling or the floor[i]
	$L_v \geq 11$ m[f]		0.0100[g]	$L_v/300$[h]	
Floors under rooms crowded by people	Flat roofs and terraces	Reinforced concrete	0.0150	$L_v/200$	Short-term deformation due to movable live load
		Prestressed concrete	0.0090	$L_v/350$	
	Other	Reinforced concrete	0.0090	$L_v/350$	
		Prestressed concrete	0.0055	$L_v/600$	
Floors of gymnasia, dancing rooms, etc.[j]	To avoid vibration	Reinforced concrete	0.0045	$L_v/700$	Short-term deformation due to live, movable load
		Prestressed concrete	0.0025	$L_v/1200$	
	To ensure plane surface of floor		0.0055	$L_v/600$	After finishing the floors
Floors under all utility spaces[k]			0.0210	$L_v/150$	

[a] If both limit values (slope and deflection) are given, they are considered as equivalent.

[b] If not specified otherwise, the slope is considered above the support point.

[c] Depending on the finishing of surface more accurate criteria can be established (for example the limit curvature).

[d] For example: long-term part of deformation due to dead load and prestressing, total deformation due to the weight of partition walls (if the ceiling was made before building of these), total deformation due to the live load.

[e] Maximum negative deflection (upright) due to the prestressing and deformation due to the dead load.

[f] For the visible part of span, L_v, between 6 m and 11 m, limit values shall be found using the linear interpolation.

[g] Maximum visible slope.

[h] L_v is visible part of span.

[i] Depending on the ceiling the camber shall be also included or possibly the short-term part of the deflection due to prestressing shall be also included. In case of the outside walls it usually concerns the deformation due to temperature gradients.

[j] If needed, check shall be made for both criteria.

[k] Only if the limit value given for the other cases in the table is not decisive.

Table 6-18 Limit deformations governed by structural requirements[23]

Structural elements			Limit deformations[a]: Slope of deflection curve at the point of load[b]	Limit deformations[a]: Deflection at midspan[c]	Deformation considered for the check
Floors under partition walls	Compact partition walls	Parallel to the plane of bending	0.005[d] max. 2.5/L	$L/600$ max. 0.8 cm	Deformation after placing of partition walls
		Perpendicular to the plane of bending	0.0017[e]	—	
	Flexible partition walls, vertically movable partition walls		0.013[f] max. 8/L	$L/250$ max. 2.5 cm	
Horizontal elements deformed due to temperature differences in columns			—	$L/300$[g] max. 1.5 cm	Deformation due to temp. diff.
Columns of the frames (total and relative horizontal displacements of girders)			—	$H/500$, max. 0.4 cm in a storey[h]	Due to wind effects
Structural elements adjacent to window and door frames, movable walls, etc.			As specified for the built-in element		After placing the element

[a] If both limit values are specified, it is considered that the checks of the slope and deflection are equivalent.
[b] If not specified otherwise, the slope above the support is considered.
[c] If not specified otherwise, the deflection at midspan is considered.
[d] Span L shall be given in cantilevers.
[e] Slope of the bed of the partition wall at direction perpendicular to its plane.
[f] For example: long-term part of deformation due to the weight of element, due to the weight of the wall and prestressing plus total loading due to live load, etc.
[g] Relative difference in extension or shortening of vertical elements; the span L refers to the horizontal elements.
[h] H is the total height of the building.

Long span construction In long span construction and in the case of large incremental deflections due to long-time sustained loads, special consideration should be given to providing sliding joints, flexible connections, or other measures to prevent damage. In buildings for example, absolute deflection limits for dead plus live load deflections might be taken as 2 in. (5.1 cm) for floors and 3 in. (7.6 cm) for roofs.

Lateral deflection limitations for high-rise concrete buildings Lateral deflection limitations for tall buildings were given by the ACI Committee 435[21] in Fig. 6-3 and Table 6-14. In a more recent paper by Fintel,[43] a survey of the deflections computed by advanced methods of about 20 high-rise reinforced concrete buildings containing shear walls shows a drift ratio of between 1/1000 and 1/2500 (described as extremely rigid) depending on the slenderness ratio of the building and on the amount and layout of shear walls. The survey also shows that the real drift ratio of the building is in the order of 1/1000 if a more advanced computation method is applied. The Uniform Building Code, the National Building Code of Canada, and ACI Committee 435 have recommended a drift limit of 1/500.

Allowable Deflections for Prestressed Concrete Members

Camber and deflection are usually not a problem in prestressed members, except with regard to differential camber of adjacent units, or when such deformation may damage attached nonstructural elements. However, prestressed members in some cases are sensitive to unsatisfactory vibrations.

The same allowable deflections for reinforced members apply to prestressed members in the 1971 ACI Code. For both uncracked members, and when the computed tensile stress exceeds say $6\sqrt{f_c'}$ but is not greater than $12\sqrt{f_c'}$, computed deflections must be shown to be within the allowable values in Table 6-15.

The same deflection limits for reinforced members also apply to prestressed members in the British Code. In addition, the British Code provides that when finishes are to be applied to prestressed members, the total upward deflection should not exceed $L/300$ unless uniformity of camber between adjacent units can be ensured.

Numerical Examples of Allowable Deflections

Because of the fact that very specialized conditions and cases are given in Table 6-14 by ACI Committee 435 and in Tables 6-16 to 6-18 by the Czechoslovakian Code, these provisions are not included in these examples. For the purpose of comparison and study, the following examples of computed deflections are compared with allowable deflections by the 1971 ACI Code and British Code for building examples, and the Bureau of Public Roads and AASHTO for bridge examples.

Example 1 Simple span beam, $L = 20$ ft (6.1 m), normal-weight concrete, $f_y = 60$ ksi, $h = 14$ in. (35.6 cm), $d = 12$ in. (30.5 cm). (Same as Example 1, Chap. 3, Table 3-2.)

ACI Code Computed deflections from Chap. 3 compared with four cases in Table 6-15.

Δ_L = 0.81 in. (20.6 mm) by I_e method of the author and ACI Code (Table 3-3)

$< L/180$ = 1.33 in. (33.8 mm) OK

$< L/360$ = 0.67 in. (17.0 mm) OK

$\Delta_t + \Delta_L$ = 1.17 in. (29.7 mm) by author's method and 1.09 in. (27.7 mm) by ACI Code method (Table 3-22)

$> L/480$ = 0.50 in. (12.7 mm) Not OK unless cambered by the difference

$> L/240$ = 1.00 in. (25.4 mm) Not OK unless cambered by the difference

British Code Computed deflections from Chap. 3 using the author's separate initial, creep, and shrinkage method, compared with Cases (1) $\Delta_{\text{Total}} \leq L/250$, and (2) $\Delta_t + \Delta_L \leq L/350$ or 20 mm (0.79 in.).

$$\Delta_{\text{Total}} = \underline{1.31 \text{ in } (33.3 \text{ mm}) \qquad (\text{Table 3-22})}$$

$> L/250 = 0.96$ in. (24.4 mm) Not OK unless cambered by the difference

$$\Delta_t + \Delta_L = \underline{1.17 \text{ in. } (29.7 \text{ mm}) \qquad (\text{Table 3-22})}$$

$> L/350 = 0.69$ in. (17.5 mm) or 0.79 in. (20 mm) Not OK for partitions

Example 2 Three-span continuous one-way slab, $L = 20$ ft (6.1 m), sand-lightweight concrete (120 pcf), $f_y = 40$ ksi, $h = 8$ in. (20.3 cm), $d = 6.75$ in. (17.1 cm). (Same as Example 2, Chap. 3, Table 3-4.)

ACI Code Computed deflections from Chap. 3 compared with four cases in Table 6-15.

End span

$\Delta_L =$ 0.55 in. (14.0 mm) by author's method and 0.46 in. (11.7 mm) by ACI Code method (Table 3-23)

$< L/180 = 1.33$ in. (33.8 mm) OK

$< L/360 = 0.67$ in. (17.0 mm) OK

$\Delta_t + \Delta_L =$ 1.14 in. (29.0 mm) by author's method and 0.91 in. (23.1 mm) by ACI Code method (Table 3-23)

$> L/480 = 0.50$ in. (12.7 mm) Not OK unless cambered by the difference

$> L/240 = 1.00$ in. (25.4 mm) Not OK by author's method and OK by ACI Code method

Center span

$\Delta_L =$ 0.20 in. (5.1 mm) by author's method and 0.23 in. (5.8 mm) by ACI Code method (Table 3-23)

$< L/180 = 1.33$ in. (33.8 mm) OK

$< L/360 = 0.67$ in. (17.0 mm) OK

$\Delta_t + \Delta_L =$ 0.41 in. (10.4 mm) by author's method and 0.37 in. (9.4 mm) by ACI Code method (Table 3-23)

$< L/480 = 0.50$ in. (12.7 mm) OK

$< L/240 = 1.00$ in. (25.4 mm) OK

British Code Computed deflections from Chap. 3 using the author's separate initial, creep, and shrinkage method, compard with (1) $\Delta_{\text{Total}} \leq L/250$, and (2) $\Delta_t + \Delta_L \leq L/350$ or 20 mm (0.79 in.).

End span

$$\Delta_{\text{Total}} = \underline{1.27 \text{ in. } (32.3 \text{ mm}) \quad (\text{Table 3-23})}$$
$$> L/250 = 0.96 \text{ in. } (24.4 \text{ mm}) \quad \text{Not OK unless cambered by the difference}$$
$$\Delta_t + \Delta_L = \underline{1.14 \text{ in. } (29.0 \text{ mm}) \quad (\text{Table 3-23})}$$
$$> L/350 = 0.69 \text{ in. } (17.5 \text{ mm}) \text{ or } 0.79 \text{ in. } (20 \text{ mm}) \quad \text{Not OK for partitions}$$

Center span

$$\Delta_{\text{Total}} = \underline{0.42 \text{ in. } (10.7 \text{ mm}) \quad (\text{Table 3-23})}$$
$$< L/250 = 0.96 \text{ in. } (24.4 \text{ mm}) \quad \text{OK}$$
$$\Delta_t + \Delta_L = \underline{0.41 \text{ in. } (10.4 \text{ mm}) \quad (\text{Table 3-23})}$$
$$< L/350 = 0.69 \text{ in. } (17.5 \text{ mm}) \text{ or } 0.79 \text{ in. } (20 \text{ mm}) \quad \text{OK for partitions}$$

Example 3 Continuous T-beam, $L_n = 29$ ft (8.8 m), normal-weight concrete, $f_y = 40$ ksi, $h = 25$ in. (63.5 cm), pos. $d = 22.5$ in. (57.2 cm). (Same as Example 3, Chap. 3, Table 3-7.)

ACI Code Computed deflections from Chap. 3 compared with four cases in Table 6-15.

Exterior span

$$\Delta_L = \underline{0.26 \text{ in. } (6.6 \text{ mm}) \text{ by author's method and ACI Code method}}$$
$$\underline{(\text{Table 3-23})}$$
$$< L_n/180 = 1.93 \text{ in. } (49.0 \text{ mm}) \quad \text{OK}$$
$$< L_n/360 = 0.97 \text{ in. } (24.6 \text{ mm}) \quad \text{OK}$$
$$\Delta_t + \Delta_L = \underline{0.59 \text{ in. } (15.0 \text{ mm}) \text{ by author's method and ACI Code method}}$$
$$\underline{(\text{Table 3-23})}$$
$$< L_n/480 = 0.73 \text{ in. } (18.5 \text{ mm}) \quad \text{OK}$$
$$< L_n/240 = 1.45 \text{ in. } (36.8 \text{ mm}) \quad \text{OK}$$

Interior spans

$$\Delta_L = \underline{0.25 \text{ in. } (6.4 \text{ mm}) \text{ by author's method and ACI Code method}}$$
$$\underline{(\text{Table 3-23})}$$
$$< L_n/180 = 1.93 \text{ in. } (49.0 \text{ mm}) \quad \text{OK}$$
$$< L_n/360 = 0.97 \text{ in. } (24.6 \text{ mm}) \quad \text{OK}$$
$$\Delta_t + \Delta_L = \underline{0.53 \text{ in. } (13.5 \text{ mm}) \text{ by author's method and ACI Code method}}$$
$$\underline{(\text{Table 3-23})}$$
$$< L_n/480 = 0.73 \text{ in. } (18.5 \text{ mm}) \quad \text{OK}$$
$$< L_n/240 = 1.45 \text{ in. } (36.8 \text{ mm}) \quad \text{OK}$$

British Code Computed deflections from Chap. 3 using the author's separate initial, creep, and shrinkage method, compared with (1) $\Delta_{\text{Total}} \leq L/250$, and (2) $\Delta_t + \Delta_L \leq L/350$ or 20 mm (0.79 in.).

Exterior span

$$\Delta_{\text{Total}} = \underline{0.69 \text{ in. } (17.5 \text{ mm})} \qquad \underline{(\text{Table 3-23})}$$
$$< L_n/250 = 1.39 \text{ in. } (35.3 \text{ mm}) \qquad \text{OK}$$
$$\Delta_t + \Delta_L = \underline{0.59 \text{ in. } (15.0 \text{ mm})} \qquad \underline{(\text{Table 3-23})}$$
$$< L_n/350 = 0.99 \text{ in. } (25.1 \text{ mm}) \qquad \text{OK for partitions}$$

Interior spans

$$\Delta_{\text{Total}} = \underline{0.60 \text{ in. } (15.2 \text{ mm})} \qquad \underline{(\text{Table 3-23})}$$
$$< L_n/250 = 1.39 \text{ in. } (35.3 \text{ mm}) \qquad \text{OK}$$
$$\Delta_t + \Delta_L = \underline{0.53 \text{ in. } (13.5 \text{ mm})} \qquad \underline{(\text{Table 3-23})}$$
$$< L_n/350 = 0.99 \text{ in. } (25.1 \text{ mm}) \qquad \text{OK for partitions}$$

Example 4 Interior girder of simple-span T-beam or deck girder bridge, $L =$ 60 ft (18.3 m), normal-weight concrete, $f_y = 40$ ksi, $h = 36.25$ in. (92.1 cm). (Same as Example 4, Chap. 3, Table 3-11.)

Bureau of Public Roads Computed live load plus impact deflection from Chap. 3 using the BPR method and the author's I_e method (also used by AASHTO) compared with Eqs. (6-11) and (6-12).

$$\Delta_{L+I} = \underline{1.04 \text{ in. } (26.4 \text{ mm}) \text{ by BPR method and } 1.11 \text{ in. } (28.2 \text{ mm}) \text{ by author's and}}$$
$$\underline{\text{AASHTO method}} \qquad (\text{Table 3-24})$$
$$< \frac{w_{L+I}}{w_T(4.555 - 0.0364L)^2} = \frac{476.3}{(882.7)(4.555 - 0.0364 \times 60)^2} \qquad (\text{Eq. (6-11)})$$
$$= 0.0960 \text{ ft} = 1.15 \text{ in. } (29.2 \text{ mm}) \qquad \text{OK by general criterion.}$$

For urban areas (which includes pedestrian travel),

$$L > 42.4 \text{ ft} \qquad (\text{Eq. (6-12)})$$

$$\Delta_{L+I} > w_{L+I}/9w_T = (476.3/(9)(882.7) = 0.072 \text{ ft} = 0.86 \text{ in. } (21.8 \text{ mm}) \qquad \text{Not OK}$$

No deflection limitation is specified for the time-dependent or total deflection.

Example 5 Unshored reinforced concrete composite beam, simple span, $L = 30$ ft (9.1 m), all-concrete normal-weight, $f_y = 40$ ksi, precast $h = 14$ in. (35.6 cm), composite $h = 17.5$ in. (44.5 cm). (Same as composite reinforced concrete example in Chap. 3, Table 3-29.)

ACI Code Computed deflections from Chap. 3, Table 3-31 (including allowable deflections in Note *) compared with four cases in Table 6-15.

$$\Delta_L = \underline{0.66 \text{ in. } (1.68 \text{ mm})}$$
$$< L/180 = 2.00 \text{ in. } (50.8 \text{ mm}) \qquad \text{OK}$$
$$< L/360 = 1.00 \text{ in. } (25.4 \text{ mm}) \qquad \text{OK}$$
$$\Delta_t + \Delta_L = \underline{1.34 \text{ in. } (34.0 \text{ mm}) \text{ computed in Note *, Table 3-31}}$$
$$> L/480 = 0.75 \text{ in. } (19.1 \text{ mm}) \qquad \text{Not OK unless cambered by the difference}$$
$$< L/240 = 1.50 \text{ in. } (38.1 \text{ mm}) \qquad \text{OK}$$

British Code Computed deflections from Chap. 3, Table 3-31 compared with (1) $\Delta_{\text{Total}} \leq L/250$, and (2) $\Delta_t + \Delta_L \leq L/350$ or 20 mm (0.79 in.).

$$\Delta_{\text{Total}} = \underline{2.67 \text{ in. } (67.8 \text{ mm})}$$
$$> L/250 = 1.44 \text{ in. } (36.6 \text{ mm}) \qquad \text{Not OK unless cambered by the difference}$$
$$\Delta_t + \Delta_L = \underline{1.34 \text{ in. } (34.0 \text{ mm}) \text{ computed in Note *, Table 3-31}}$$
$$> L/350 = 1.03 \text{ in. } (26.2 \text{ mm}) \text{ or } 0.79 \text{ in. } (20 \text{ mm}) \qquad \text{Not OK for partitions}$$

Example 6 Noncomposite double-T beam with prestressing steel only, $L = 76$ ft = 912 in. (23.2 m), normal-weight concrete, $h = 32$ in. (81.3 m). (Same as Example 1, Chap. 5, Table 5-6.)

ACI Code Computed camber (+) and deflection (−) from Chap. 5 compared with four cases in Table 6-15.

$$\Delta_L = \underline{-0.97 \text{ in. } (24.6 \text{ mm})} \qquad \text{Time-dependent camber} + \Delta_L = \underline{-0.81 \text{ in. } (20.6 \text{ mm})}$$
$$L/180 = (76)(12)/180 = 5.07 \text{ in.} > 0.97 \text{ in.} \qquad \text{OK}$$
$$L/360 = 2.53 \text{ in.} > 0.97 \text{ in.} \qquad \text{OK}$$
$$L/480 = 1.90 \text{ in.} > 0.81 \text{ in.} \qquad \text{OK}$$
$$L/240 = 3.80 \text{ in.} > 0.81 \text{ in.} \qquad \text{OK}$$

Hence the beam has no camber or deflection limitations.

British Code Computed camber (+) and deflection (−) from Chap. 5 compared with three cases of the British Code.

$$\Delta_{\text{Total}} = \underline{-0.05 \text{ in. } (1.3 \text{ mm})} < L/250 = 3.65 \text{ in. } (92.7 \text{ mm}) \qquad \text{OK}$$
$$\Delta_t + \Delta_L = \underline{-0.81 \text{ in. } (20.6 \text{ mm})} < L/350 = 2.61 \text{ in. } (66.3 \text{ mm}) \qquad \text{OK}$$
$$\text{Total upward defl.} = \underline{0.92 \text{ in. } (23.4 \text{ mm})} < L/300 = 3.04 \text{ in. } (77.2 \text{ mm}) \qquad \text{OK}$$

Hence the beam has no camber or deflection limitations.

Example 7 Same as Example 6 except using sand-lightweight concrete.

ACI Code Computed camber (+) and deflection (−) from Chap. 5 compared with four cases in Table 6-15.

$$\Delta_L = \underline{-1.34\text{ in. }(34.0\text{ mm})} \qquad \text{Time-dependent camber} + \Delta_L = \underline{-0.67\text{ in. }(17.0\text{ mm})}$$

$L/180 = (76)(12)/180$	$= 5.07$ in. > 1.34 in.	OK
$L/360$	$= 2.53$ in. > 1.34 in.	OK
$L/480$	$= 1.90$ in. > 0.67 in.	OK
$L/240$	$= 3.80$ in. > 0.67 in.	OK

Hence the beam has no camber or deflection limitations.

British Code Computed camber (+) and deflection (−) from Chapter 5 compared with three cases of the British Code.

$$\Delta_{\text{Total}} = \underline{0.79\text{ in. }(2.01\text{ mm})} < L/250 = 3.65\text{ in. }(92.7\text{ mm}) \qquad \text{OK}$$

$$\Delta_t + \Delta_L = \underline{-0.67\text{ in. }(17.0\text{ mm})} < L/350 = 2.61\text{ in. }(66.3\text{ mm}) \qquad \text{OK}$$

$$\text{Total upward defl.} = \underline{2.13\text{ in. }(54.1\text{ mm})} < L/300 = 3.04\text{ in. }(77.2\text{ mm}) \qquad \text{OK}$$

Hence the beam has no camber or deflection limitations.

Example 8 Noncomposite single-T beam, first containing only prestressing steel and second containing both prestressed and nonprestressed tension steel, $L =$ 100 ft = 1200 in. (30.5 m), sand-lightweight concrete, with additional sustained loading applied some time after the transfer of prestress, $h = 48$ in. (121.9 cm). (Same as Example 3, Chap. 5, Table 5-7.)

ACI Code Computed camber (+) and deflection (−) from Chap. 5 compared with four cases in Table 6-15.

Beam containing only prestressing steel

$$\Delta_L = \underline{-1.45\text{ in. }(36.8\text{ mm})} \qquad \text{Time-dependent camber} + \Delta_L$$

$$= \underline{-3.85\text{ in. }(97.8\text{ mm})}$$

$L/180 = (100)(12)/180$	$= 6.67$ in. > 1.45 in.	OK
$L/360$	$= 3.33$ in. > 1.45 in.	OK
$L/480$	$= 2.50$ in. < 3.85 in.	Not OK
$L/240$	$= 5.00$ in. > 3.85 in.	OK

Beam containing both prestressing steel and nonprestressed tension steel

$$\Delta_L = \underline{-2.13\text{ in. }(54.1\text{ mm})} \qquad \text{Time-dependent camber} + \Delta_L$$

$$= \underline{-4.06\text{ in. }(103.1\text{ mm})}$$

$$L/180 = (100)(12)/180 = 6.67\text{ in.} > 2.13\text{ in.} \qquad \text{OK}$$

$$L/360 = 3.33 \text{ in.} > 2.13 \text{ in.} \qquad \text{OK}$$
$$L/480 = 2.50 \text{ in.} < 4.06 \text{ in.} \qquad \text{Not OK}$$
$$L/240 = 5.00 \text{ in.} > 4.06 \text{ in.} \qquad \text{OK}$$

Hence, for both cases, the beam should not be used for roofs or floors supporting or attached to nonstructural elements *likely* to be damaged by large deflections; otherwise it is satisfactory.

British Code Computed camber (+) and deflection (−) from Chap. 5 compared with three cases of the British Code.

Beam containing only prestressing steel

$$\Delta_{\text{Total}} = \underline{-2.99 \text{ in.} (75.9 \text{ mm})} < L/250$$
$$= 4.80 \text{ in.} (121.9 \text{ mm}) \qquad \text{OK}$$
$$\Delta_t + \Delta_L = \underline{-3.85 \text{ in.} (97.8 \text{ mm})} > L/350$$
$$= 3.43 \text{ in.} (87.1 \text{ mm}) \qquad \text{Not OK for partitions}$$
$$\text{Total upward defl.} = \underline{-1.54 \text{ in.} (39.1 \text{ mm})}$$

or a downward deflection due to the superimposed sustained load. Hence $L/300$ is not applicable.

Beam containing both prestressing steel and nonprestressed tension steel

$$\Delta_{\text{Total}} = \underline{-3.20 \text{ in.} (81.3 \text{ mm})} < L/250$$
$$= 4.80 \text{ in.} (121.9 \text{ mm}) \qquad \text{OK}$$
$$\Delta_t + \Delta_L = \underline{-4.06 \text{ in.} (103.1 \text{ mm})} > L/350$$
$$= 3.43 \text{ in.} (87.1 \text{ mm}) \qquad \text{Not OK for partitions}$$
$$\text{Total upward defl.} = \underline{-1.07 \text{ in.} (27.2 \text{ mm})}$$

or a downward deflection due to the superimposed sustained load. Hence $L/300$ is not applicable.

Hence, for both cases, the beam should not be used with partitions; otherwise it is satisfactory. This is similar to the ACI Code conclusion.

Example 9 Composite AASHTO Type IV girder with prestressing steel only, $L = 80$ ft $= 960$ in. (24.4 m), normal-weight concrete, precast $h = 54$ in. (137.2 cm), composite $h = 61$ in. (154.9 cm). (Same as Example 4, Chap. 5, Table 5-8.)

AASHTO Computed live load plus impact deflection $\leq L/800$, $L/1000$:

$$\Delta_{L+I} = \underline{0.50 \text{ in.} (12.7 \text{ mm})} < L/800 = (80)(12)/800 = 1.20 \text{ in.} (30.5 \text{ mm}) \qquad \text{OK}$$
$$< L/1000 = 0.96 \text{ in.} (24.4 \text{ mm}) \qquad \text{Also OK for bridges in urban areas used in part by pedestrians}$$

Examples of slab without beams (flat plate) and slab with beams (two-way slab) See Chap. 4.

Observations pertaining to the allowable deflections of Examples 1 to 9

Reinforced members, by the ACI Code and British Code methods for buildings and the BPR method for bridges 1. Example 1 referred to a simple-span, normal-weight, rectangular beam designed by the strength method with a steel percentage of 1.98 percent. Based on the allowable deflections of the ACI Code, the member could be used only when not supporting or attached to nonstructural elements likely to be damaged by large deflections. Based on the allowable deflections of the British Code, the member could be used only after cambering by a specified amount, and could not be used with partitions.

2. Example 2 referred to a three-span continuous sand-lightweight one-way slab designed by the strength method with a steel percentage of 0.49 to 0.93 percent. Based on the allowable deflections of the ACI Code, for both the end and center spans, there were no deflection limitations (with one marginal exception when using the author's method of calculation). Based on the allowable deflections of the British Code, there were no deflection limitations for the center span, but the end span could be used only after cambering by a specified amount, and could not be used with partitions.

3. Example 3 referred to a continuous normal-weight T-beam designed by the strength method with a steel percentage of +0.25 percent. There were no deflection limitations for either exterior or interior spans by both the ACI and British Codes.

4. Example 4 referred to an interior girder of a simple-span normal-weight T-beam bridge designed by the load-factor method with a steel percentage of 0.94 percent. Based on the BPR method for live load plus impact deflection, the bridge met the general deflection criterion, but did not meet the criterion for urban areas which includes pedestrian travel. No deflection limitation is specified for the time-dependent or total deflection for bridges.

5. Example 5 referred to a simple-span, normal-weight unshored reinforced concrete composite beam designed by the strength method. Based on the allowable deflections of the ACI Code, the beam could be used for all cases except for roof or floor construction supporting or attached to nonstructural elements *likely* to be damaged by large deflections. Based on the allowable deflections of the British Code, the member could be used only after cambering by a specified amount, and could not be used with partitions.

6. In comparing the allowable deflections by the ACI Code ($L/180$ to $L/480$) and the British Code ($L/250$ to $L/350$), the ACI Code is somewhat more specific about the effect of nonstructural elements, and provides a broader range of allowable deflections. In all of the reinforced examples except the center span of the one-way continuous slab and the T-beam of Example 3, the members could not be used without cambering, and could not be used with partitions, according to the British Code. The ACI Code is somewhat less restrictive than the British Code, except for the condition allowing a deflection of only $L/480$.

Prestressed members, by the ACI Code and British Code methods for buildings and the AASHTO method for bridges 7. Examples 6 and 7 refer to a noncomposite double-T beam with, prestressing steel only using first normal-weight concrete and then sand-lightweight concrete. There were no deflection limitations for either case by both the ACI and British Codes. This is typically the case for fully prestressed members.

8. Example 8 referred to a partially prestressed noncomposite single-T beam, first containing only prestressed steel and second containing both prestressed and nonprestressed tension steel, and using sand-lightweight concrete. Based on the ACI Code both with and without nonprestressed tension steel, the member could be used for all cases except when supporting or attached to nonstructural elements *likely* to be damaged by large deflections. Based on the British Code both with and without nonprestressed tension steel, the member could be used generally, except when attached to partitions. Thus, the two codes reached a similar conclusion regarding this partially prestressed member.

9. Example 9 referred to a composite AASHTO Type IV girder with prestressing steel only, and using normal-weight concrete. Based on the allowable live load plus impact deflection by AASHTO, the design was satisfactory for general use, and also for bridges in urban areas used in part by pedestrians.

6-4 CONCLUDING REMARKS

Because of the high degree of uncertainty involved in the computation of deflections and in the selection of appropriate limiting values, it is considered to be not only desirable but essential to use a relatively simple calculation procedure in order to guard against the tendency of placing undue reliance on computed results. In this connection, it should be borne in mind that concrete deformation is probabilistic and our knowledge is inadequate even to provide mean value functions and variance.

Regardless of the method to be used, it should be clear that one can calculate the span–depth, span–thickness ratio or depth, thickness based on an allowable deflection–span ratio—or the deflection directly—for a given beam, loading, and other pertinent conditions, with presumably equal facility. The latter determination provides a direct evaluation of the deflection properties of a structure, while the former may be more useful in the initial proportioning of members with regard to deflection control.

The indirect method of controlling deflections by means of maximum span–depth, span–thickness ratios or minimum depths, thicknesses is primarily a *rule-of-thumb* approach in its simplest form. Even when basic factors together with various modifying factors for different conditions are used, or when detailed analytical calculations are made, this procedure is still based largely on experience. However, the method of controlling deflections by means of calculated versus allowable deflections is based on nominal limiting deflections, which in turn are *also* based primarily on experience.

The author prefers to use simple span–depth, span–thickness ratios or minimum depths, thicknesses in the preliminary proportioning of members with regard to deflection control, but then to check deflections directly by means of a relatively simple calculation procedure and allowable deflections that have become reasonably well understood. This procedure seems to be preferable from both a physical and an intuitive point of view in controlling deflections.

Primarily because of the uncertainties involved in connection with loading histories (including the effect of construction loads), variable creep and shrinkage behavior for different structures, etc., procedures for computing deflections should be simple and should be thought of as determining representative deflections (which may be taken as conservative, average, or whatever), rather than as predicting deflections in every case. Of course, there must be a reasonable correlation between the two for a broad class of structures. Limiting deflections are then prescribed to be consistent with the computed values.

REFERENCES

1. H. Mayer and H. Reusch, Building Damage Caused by Deflection of Reinforced Concrete Building Components, *Deutscher Ausschuss für Stahlbeton,* **193,** 1967; trans. J. H. Rainer, *National Research Council of Canada, Technical Translation* 1412, 1–115, Ottawa, 1970.
2. W. G. Plewes, Deflections—Is There a Problem? *ACI Canadian Capital Chapter Symposium, Montreal,* 5–14, Oct. 1971.
3. C. V. Clarke, A. M. Neville, and W. Houghton-Evans, Deflections—Problems and Treatment in Various Countries, *ACI Publications SP-43, Deflections of Concrete Structures,* 129–178, 1974.
4. American Concrete Institute, "Building Code Requirements for Reinforced Concrete," 1–78, and Commentary, 1–96, American Concrete Institute, Detroit, Michigan, 1971.
5. British Standards Institution, "Code of Practice for The Structural Use of Concrete, Part 1: Design, Materials, Workmanship," CP 110, 1–155, British Standards Institution, London, 1972.
6. Cement and Concrete Association, "Handbook on the Unified Code for Structural Concrete," 1–153, Cement and Concrete Association, London, 1972.
7. Comité Européen du Béton (CEB), Chairman, I. S. Soretz, Deformations, *Comité Européen du Béton, Bulletin d'Information* 16, 1–27, Mar. 1959.
8. American Concrete Institute, "Building Code Requirements for Reinforced Concrete," 1–144, and Commentary, 1–91, American Concrete Institute, Detroit, Michigan, 1963.
9. A. Mehmel, Uber eine sinnvolle Beschrankung der Durchbiegung von Stahlbetonbauteilen (On a Meaningful Deflection Limitation of Reinforced Concrete Building Components), *Der Bauingenieur,* **8,** 293–300, 1961.
10. Subcommittee 4, ACI Committee 435, Subcommittee and Committee Chairman, D. E. Branson, Recommendation for ACI Building Code Provision on Deflections, *ACI Committee 435 Report, submitted to ACI Committee 318,* 1–5, Aug. 1967.
11. D. E. Branson, Design Procedures for Computing Deflections, *ACI Journal,* Proceedings **65,** 9, 730–742, Sept. 1968.
12. American Institute of Steel Construction, "Manual of Steel Construction," 7th edn., 1–1 to 7–12, American Institute of Steel Construction, New York, 1970.
13. R. F. Denkins, "Serviceability Requirements—Control of Deflections," Chap. 6, 6–1 to 6–16, notes on ACI 318-71, Building Code Requirements with Design Applications, Portland Cement Association, 1972.
14. A. W. Beeby and J. R. Miles, Proposals for the Control of Deflection in the New Unified Code for Structural Concrete, *Concrete,* **2,** 12, 511–513, Dec. 1968; **3,** 3, 101–110, Mar. 1969.

15. A. W. Beeby, Modified Proposals for Controlling Deflections by Means of Ratios of Span to Effective Depth, *Cement and Concrete Association, Technical Report* SBN 7210 0816 X, 1–19, London, Apr. 1971.
16. A. W. Beeby, "A Note on Studies of the Calculation and Limitation of Deflection Carried Out at the Cement and Concrete Association," Cement and Concrete Association, London, individual communication, 1–18, Sept. 1972.
17. AASHTO, "Standard Specifications for Highway Bridges," 11th edn, 1–469, 341 National Press Building, Washington, D.C., 20004, 1973.
18. T. Y. Lin, "Design of Prestressed Concrete Structures," 2nd edn, 1–614, John Wiley & Sons, 1963.
19. Prestressed Concrete Institute, "PCI Design Handbook," 1-1 to 12-8, Prestressed Concrete Institute, Chicago, Illinois, 1971.
20. C. K. Wang and C. G. Salmon, "Reinforced Concrete Design," 2nd edn, 1–934, Intext Press, Inc., New York, 1973.
21. Subcommittee 1, ACI Committee 435, Subcommittee Chairman, R. S. Fling, ACI Committee Chairman, D. E. Branson, Allowable Deflections, *ACI Journal,* Proceedings **65,** 6, 433–444, June 1968.
22. W. G. Plewes and G. K. Garden, Deflections of Horizontal Structural Members, *Canadian Building Digest,* 54-1 to 54-4, Division of Building Research, National Research Council, Ottawa, June 1964.
23. Czechoslovak Code, Design of Concrete Structures, *Czechoslovak Code,* CSN 731201, 1971.
24. J. L. Heiman, Long-Term Deformations in the Tower Building, Australia Square, Sydney, *ACI Journal,* Proceedings **70,** 4, 277–284, Apr. 1973.
25. W. G. Plewes, Cladding Problems Due to Frame Movements, *Canadian Building Digest,* 125-1 to 125-4, Division of Building Research, National Research Council, Ottawa, May 1970.
26. R. H. Wood, Studies in Composite Construction: Part 1—The Composite Action of Brick Panel Walls Supported on Reinforced Concrete Beams, *National Building Studies Research, Paper* 13, 1–25, H.M.S.O., London, 1952.
27. F. A. Blakey, Deflection as a Design Criterion in Concrete Buildings, *Institution of Engineers of Australia—Civil Engineering Transactions,* **CE5,** 2, 55–60, Sept. 1963.
28. M. D. Vanderbilt, M. A. Sozen, and C. P. Siess, Deflections of Reinforced Concrete Floor Slabs, *University of Illinois, Civil Engineering Studies, Structural Research Series* No. 263, 1–287, Apr. 1963.
29. D. A. Sawyer, Ponding of Rainwater on Flexible Roof Systems, *Journal of the Structural Division, ASCE,* **93,** ST1, 127–147, Feb. 1967.
30. A. E. Salama and M. L. Moody, Analysis of Beams and Plates for Ponding Loads, *Journal of the Structural Division, ASCE,* **93,** ST1, 109–126, Feb. 1967.
31. D. A. Sawyer, Roof–Structural, Roof–Drainage Interactions, *Journal of the Structural Division, ASCE,* **94,** ST1, 175–198, Jan. 1968.
32. J. Chinn, H. Mansouri, and S. F. Adams, Ponding of Liquids on Flat Roofs, *Journal of the Structural Division, ASCE,* **95,** ST1, 797–807, May 1969.
33. B. L. Meyers and G. M. Sabnis, Code Criteria for Deflection Limitation in Reinforced Concrete Structures, *ACI Publication SP-43, Deflections of Concrete Structures,* 1–14, 1974.
34. Bureau of Public Roads, "Strength and Serviceability Criteria, Reinforced Concrete Bridge Members—Ultimate Design," 1–81, U.S. Department of Commerce, Bureau of Public Roads, Aug. 1966.
35. Task Committee on Reinforced Masonry Design and Practice, Committee on Masonry and Reinforced Concrete, Reinforced Masonry Design and Practice, *Proceedings, ASCE,* **87,** ST8, 87–138, Dec. 1961.
36. P. Weidlinger, Temperature Stresses in Tall Reinforced Concrete Buildings, *Civil Engineering—ASCE,* **34,** 8, 58–61, Aug. 1964.
37. Kenneth H. Lenzen, Vibration of Steel Joist—Concrete Slab Floor Systems, *Studies in Engineering Mechanics, University of Kansas, Report* 16, Aug. 1962. (Has a discussion and extensive bibliography on the effect of vibrations on human occupants.)

38. Fu-Kuei Chang, Wind and Movement in Tall Buildings, *Civil Engineering—ASCE*, **37,** 8, 70–72, Aug. 1967.
39. S. Rosenhaupt and Y. Sokal, Masonry Walls on Continuous Beams, *Proceedings, ASCE*, **91,** ST1, 155–171, Feb. 1965.
40. Fazlur R. Kahn and Mark Fintel, Effect of Column Exposure in Tall Structures—Analysis for Length Changes of Exposed Columns, *ACI Journal*, Proceedings **63,** 8, 843–864, Aug. 1966.
41. Committee on Testing, "The Testing of Structures," Institution of Structural Engineers, 1964.
42. H. Mayer, Bauschaden als Folge der Durchbiegung von Stahlbetonbauteilen, *Materialprufungsamt für das Bauwesen der Technischen Hochschule, Report* 68, München, 1966.
43. Mark Fintel, Deflections of High-Rise Concrete Buildings, *ACI Journal*, Proceedings **72,** 7, 324–328, July 1975.
44. P. F. Rice (ed.), "CRSI Handbook," 2nd edn, 1-1 to 15-51, Concrete Reinforcing Steel Institute, Chicago, 1975.
45. Building Code Subcommittee, ACI Committee 435, Subcommittee Chairman, D. E. Branson, Committee 435 Chairman, G. M. Sabnis, Revised ACI Code and Commentary Provision on Deflections, Section 9.5, *ACI Committee 435 Report, submitted to ACI Committee 318*, 1–26, Dec. 1975.

APPENDIX

L

BRIEF SUMMARY OF METHODS IN VARIOUS COUNTRIES FOR CONTROLLING DEFLECTIONS, FROM REF. 3

(From ACI SP 43,[3] with permission.)

See the paper by Clarke, Neville, and Houghton-Evans[3] for the detailed Code references.

Australia CA 2—1963

Reinforced concrete members should have sufficient stiffness so as not to impair the strength or produce cracks in finishes or partitions. Deflection must be considered when the net ratio of reinforcement exceeds $0.18f'_c/f_y$. Under normal circumstances span–depth ratios are as follows:

Beams simply supported	20
Beams continuous	25
Beams cantilevered	10
Slabs one-way simply supported	30
Slabs one-way continuous	35
Slabs two-way simply supported	35
Slabs two-way continuous	40

$E_c = 1 \times 10^6 + 500f'_c$ (psi). Gross transformed section considered in calculating instantaneous deflection. Average I of positive and negative moment regions considered for continuous beams. Multiplier for long-term deflection: for $A'_s \leq A_s/2$, 3; for A'_s from $A_s/2$ to A_s, 2.

CA 35—1963 Where finishes are to be applied the upward deflection is not to exceed $L/300$.

DOC 1589 draft—1970 Minimum depths:

	Cantilever	Simply supported	Cont. one end	Cont. both ends
Solid slabs	$L/10$	$L/18$	$L/23$	$L/29$
Beams and ribbed slabs	$L/8$	$L/14$	$L/18$	$L/22$

$E_c = w^{1.5}33\sqrt{f'_c}$ Long-term multiplier $= 3 - 1.2A_s/A_s$

Austria ONORM B 4001—1965

Calculated deflection under total load for floor slabs, floor beams and ribbed floors must not be greater than $L/300$ or $L/500$ for supports.

ONORM B4200 Part 8 1971 Span–depth ratios:

	$L \leq 4$ m (13.25 ft)	$L \geq 8$ m (26.5 ft)
Simply supported beams and ribbed slabs	15	10
Simply supported slabs	30	20
Continuous beams and ribbed slabs	25	20
Continuous slabs	40	30

Interpolate linearly between 4 and 8 m.

Short-term deflections are calculated by the CEB equation (Eqs. (3-23), (3-24), Chap. 3 herein), with long-term $\Delta = 3\Delta_1 + 1.5\Delta_2$, as defined in Fig. 3-3 herein.

Belgium NBN 15—1963

General statement recommends that deflection should not cause difficulties from slope, damage to finishes, percolation of water, cracking of partitions including a reminder about creep. In the absence of laboratory data, use:

$E_c = 250\,000$ kg/cm² (3.56×10^6 psi) for 100–200 kg/cm² (1420–2850 psi)
$= 450\,000$ kg/cm² (6.4×10^6 psi) for 350 kg/cm² (5000 psi)

Multiplier for creep is proportional to concrete strength as:

$$100\text{–}200\ \text{kg/cm}^2\ (1420\text{–}2850\ \text{psi}) = 4$$
$$200\text{–}350\ \text{kg/cm}^2\ (2850\text{–}5000\ \text{psi}) = 3$$
$$350\ \text{kg/cm}^2\ (5000\ \text{psi})\ \text{and above} = 2$$

The limit of deflection for partition supporting members is as: instantaneous deflection $\leq L/2500$ – nature of partition should be considered—using a multiplier of 4, or the long-term deflection $\leq L/625$.

Brazil ABNT—NB1—1960

Limits deflection under accidental load to $\leq L/1000$ instantaneously or $L/500$ in long-term. Nothing else is given.

Canada CSA A23.3—1970 (as U.S.A.)

General statement—deflection should not affect strength/serviceability of the structure—the code gives a list of things which should be considered.

Span–depth ratios—same as 1963 ACI Code (Table 6-2 herein). These values are for f_y up to 40 000 psi (276 N/mm²); for $f_y > 40\,000$, use table values times $0.5 + 0.0125 f_y$. E_c, section properties, and long-term multipliers are the same as the 1963 ACI Code.

Czechoslovakia C SN 732001—1970

The code gives no span–depth ratios, and recommends the calculation of deflection in every case. See Tables 6-16 to 6-18 herein for a comprehensive list of deflection limits. For calculation, the gross transformed section is taken into account.

Denmark

No mention of deflection.

France BA68—1968

Probably the most comprehensive code.

$E_c = 2100\sqrt{\sigma'}$, bars, where σ' = concrete strength in bars (1 bar = 14.5 psi)
E_c for long-term deflection $= E_v = 7000\sqrt{\sigma'}$, bars
For 4500 psi concrete, $E_c = 5.35 \times 10^6$ psi, and $E_v = 1.78 \times 10^6$ psi
I = gross section, neglecting steel $\quad n = 15 \quad$ Curvature $= (\varepsilon b' + \varepsilon a)/h$

For each type of construction the code suggests that deflection be calculated or kept within the following span–depth ratios.

For hollow tile floors

$h_t/L \geq (M_t = \text{max. midspan mom.})/(15M_0 = 15 \text{ max. midspan mom. if S.S.})$

provided $\rho = A_s/b_0h \leq 36/\sigma_y$

where $b_0 = $ width of ribs $h_t = $ height of ribs

Only valid when $h_t/L \leq 1/22.5$

For beams supporting slabs

$h_t/L \geq M_t/10M_0$ provided $A_s/b_0h \leq 43/\sigma_y$

Minimum value of $h_t/L \geq 1/16$

$I = I_t/(1 + \lambda\mu)$ $I_t = $ gross transformed I

$\lambda = \lambda_i = \sigma_b/72(2 + 3b_0/b)\rho$ for ST loads

$\lambda = \lambda_v = \sigma_b/180(2 + 3b_0/b)\rho$ for LT loads

$= 1 - [5\sigma_b/(4\rho\sigma_a + 3\sigma_b)]$

$\sigma_b = $ tensile stress in concrete in bars

$\sigma_a = $ tensile stress in steel in bars

$\rho = A_s/b_0h$ $b_0 = $ width of compression face $n = 15$

Curvature $1/r_0$ and $1/r_\infty$ calculated using I_i and I_v, corresponding to λ_i and λ_v. Deflection $= ML^2/(10EI)$, $M = $ deflection moment.

Deflection limits: the increase in deflection due to permanent load should not exceed $L/500$ if $L \leq 5$ m (16.4 ft), or 0.5 cm + $L/1000$ if $L > 5$ m. For cantilevers $\Delta = ML^2/(4EI)$ + rotation at root of cantilever. Deflection limit $= L/250$ if $L \leq$ 2.5 m (8.2 ft) or 0.5 cm + $L/500$ if $L > 2.5$ m.

Long-term deflection $= f_{g\infty} - f_{j0} + f_{q0} - f_{g0}$, where f_{g0} and $f_{g\infty} = $ deflection due to permanent load $f_{j0} = $ deflection due to permanent loads applied at erection of partitions, and $f_{q0} = $ deflection due to all loads and imposed loads supported by the beam.

For slabs $h_0/L_x \geq M_t/(20M_x)$, where $M_t \geq 0.75M_x$, $A_s/bh \leq 20/\sigma_y$. If not, calculate.

Germany DIN 1045—1968

There is little about deflection in this standard except that, when deflection is calculated, creep, shrinkage, and cracking should be considered.

The span–depth approach is based on a study by Mayer and Reusch,[1] as discussed in the text. The span–depth ratio limit of 35 is imposed for slabs carrying partitions. Span–depth expression is $L_i/h \leq 150/L_i$ (all in meters), $L_i = \alpha L = $ span of simply supported member of constant I and the same curvature. $\alpha = 1$ for slabs simply supported, 2.4 for cantilevers, 0.8 for one end continuous and 0.6 for both ends continuous.

For f'_c = 100 kg/cm^2	(1424 psi),	E_c = 220 000 kg/cm^2	(3.13×10^6 psi)
200	2842	250 000	3.56
300	4270	300 000	4.27
400	5700	340 000	4.84
500	7100	370 000	5.26
600	8500	390 000	5.55

Hungary MSZ 18022 Sheet 1-61—1965

$E_c = 550\,000K/(K + 200)$, where K = concrete strength in kg/cm^2, Gross I used.

Deflection limit for floors is $L/500$ under normal load and $L/200$ under full load.

India IS 456—1964

Similar to British CP 114. General statement to the effect that flexural members should possess adequate stiffness. Span–depth ratios:

Cantilevers	12	Slabs simply supported one way	30
Beams simply supported	20	Slabs simply supported two way	35
Beams continuous	25	Slabs continuous one way	35
Beams cantilever	10	Slabs continuous two way	40

International Draft ISO/TC 98/SC4-14E—1973

Consists of a very comprehensive list of deflection limits for different cases similar to the Czech Code. No span–depth ratios are given as it is not intended for use as a structural standard—not specifically concrete.

Differential settlement limit $L/560$. Floors supporting partitions $L/500$ for partitions easily cracked; $L/350$ for more resilient partitions, maximum of 10 mm (0.4 in.). To avoid damage to roof covering and glazing $L/125$. Visual limit $L/250$ or 30 mm (1.2 in.).

Netherlands GBV—1962

$$\text{Depth of slabs with smooth bars } h \geq \sigma_a/30\,000 \times \frac{P}{P+q} \times L_{\min} \geq L_{\min}/35$$

$$\text{In slabs with deformed bars } h \geq \sigma_a/35\,000 \times \frac{P}{P+q} \times L_{\min} \geq L_{\min}/35$$

$$\text{Beams with smooth bars } h \geq \frac{\sigma_a + 15\sigma_b}{42\,000} \times \frac{P}{P+q} \times L_{\min}$$

$$\text{Beams with deformed bars } h \geq \frac{\sigma_a + 15\sigma_b}{50\,000} \times \frac{P}{P+q} \times L_{\min}$$

σ_a = max. tensile stress in steel (in kg/cm^2)

σ_b = calculated max. compressive stress in the concrete (in kg/cm^2)

P = live load q = dead load

L_{min} is effective span if $P/(P + q) < \frac{1}{3}$ then take as $\frac{1}{2}$. Simply supported $L_{min} = 0.85L$, continuous $= 0.7L$, cantilever $= 2L$. No deflection limits given.

New Zealand NZSS 1900 Chap. 9.3—1964

The ratio of the effective span of a beam to its effective depth shall not exceed the less of the two following ratios: (1) 20 × the permissible max. tensile stress in the steel ÷ the actual stress in the steel; (2) 20 × the permissible stress in the concrete ÷ the actual compressive stress in the concrete. For continuous beams the effective span is the distance between points of contraflexure. No other recommendations.

Norway 427A, Part 2—1962

$E_c = 400K + 120\,000$ Kp/cm^2 (1 Kp = 9.81 N) for short-term loads, where K is the concrete strength in Kp/cm^2.

For long-term loads $E_c = E_c/3$, E of steel $= 2\,100\,000$ kp/cm^2 (29.9×10^6 psi). Deflection limits $L/200$ for walls, $L/300$ for floor slabs and beams. For calculation assume concrete carries no tension and the above elastic moduli.

Span–depth ratios:

Slabs one-way simply supported	30	Slab must not be <6 cm (2.36 in.) for roofs or <8 cm (3.15 in.) for all other slabs
Slabs one-way one end continuous	35	
Slabs one-way both ends continuous	40	
Two-way slab simply supported	35	Two-way slabs $L \leq 1.5L_y$
Two-way slab one end fixed	40	Two-way slabs $L \leq 1.5L_y$
Two-way slab both ends fixed	45	Two-way slabs $L \leq 1.5L_y$
Beams simply supported	20	
Beams continuous one end	23	
Beams continuous both ends	26	

Poland PN 56/B—03260—1956

One-way slabs, i.e. $L_x : L_y \geq 2$, min. thickness for roofs = 6 cm (2.36 in.). Slabs supporting ceilings 7 cm (2.76 in.), industrial slabs 8 cm (3.15 in.). Slabs over passageways 12 cm (4.72 in.).

Thickness should be at least: one-way simply supported $L/35$, one-way continuous $L/40$, two-way simply supported $L/45$, two-way continuous $L/50$.

PN 66/B—03320—1966 Slabs supporting plastered ceilings: deflection limit given as $L/300$ at right angles to the plane of bending or $L/400$ in plane of bending.

Beams supporting plastered ceilings: $L/400$ at right angles to plane of bending or $L/400$ in plane of bending. Crane girder beams: $L/500$ at right angles to plane of bending or $L/750$ in plane of bending.

When uncracked consider gross section. When cracked use transformed cracked section. For long-term deflection multiply instantaneous values by $1 + C_t$. Absolute limit of deflection $L/200$ or $L/300$ for large spans. Complex calculation method proposed, similar to Russian.

Portugal Directo No. 47223—1967

Similar to British CP 110. Calculation method uses a bi-linear moment curvature expression. For I neglect steel. Span–depth ratio $= L/h = k_1 \times k_2$:

	Slabs k_1	Beams k_1	Steel type	Slabs k_2	Beams k_2
Slabs simply supported one-way	1.0	1.0	A24	35	25
Slabs continuous one-way	1.2	1.2	A40	30	20
Slabs built in one-way	2.0	2.0	A50	25	17
Simply supported two-way slabs	1.7		A60	20	15
Cantilever	0.3	0.3			

Romania STAS—8000—1967

Similar to U.S.S.R.

Deflection limits:

Runway beams for hand operated travelling cranes	$L/500$
Runway beams for electric operated travelling cranes	$L/750$
Roofs $L \leq 7$ m (23 ft)	$L/200$
$L > 7$ m	$L/300$
Floors $L \leq 5$ m (16.4 ft)	$L/200$
$5 \text{ m} \leq L \leq 7$ m	$L/300$
$L > 7$ m	$L/400$

Also takes duration of loading into account. One-half the above ratios for cantilevers. For plastered ceilings deflection under temporary load should not exceed $L/360$.

South Africa—1966

Deflection shall not be such as to endanger stability or impair proper function of a building. Span–depth ratios:

Simply supported one-way slabs	$L/30$	Cantilever slabs	$L/12$
Continuous one-way slabs	$L/35$	Simply supported beams	$L/20$
Simply supported two-way slabs	$L/35$	Continuous beams	$L/25$
Continuous two-way slabs	$L/40$	Cantilever beams	$L/10$

These depths may be altered if deflection nowhere exceeds $L/360$.

Spain

Similar to Netherlands. $E_c = 21\,000;\ 14\,000;\ 8500\sqrt{\sigma_{bk}}$;—instantaneously; LT load in humid; dry climate, respectively, where σ_{bk} = concrete crushing strength in kg/cm^2.

LT $\Delta = K \times$ ST Δ, $K = 2$ if $A_s' = 0$, 1.2 if $A_s' = A_s/2$, and 0.8 if $A_s' = A_s$.

$$\text{Deflection limit } \Delta/L \leq h/500 \geq \frac{10\sigma_{bk} + \sigma_{aK}}{60\,000} \times \frac{G}{G+Q} \times L_i$$

Sweden—1967

The height of the member is made proportional to the applied moment and a property of the material, ignoring span.

Deflection $\leq L/400$. Deflection after erection of wall must be $< L/500$.

For $f_c' =$ 150 kp/cm^2	(2140 psi)	$E_c =$ 70 000 kp/cm^2	(1×10^6 psi)
200–250 kp/cm^2	(2848–3560 psi)	85 000 kp/cm^2	(1.21×10^6 psi)
300–350 kp/cm^2	(4270–5000 psi)	105 000 kp/cm^2	(1.5×10^6 psi)
400 kp/cm^2	(5700 psi)	120 000 kp/cm^2	(1.7×10^6 psi)

$$\text{Min. } d = \sqrt{\frac{7 \times \text{max. design mom.}}{\text{modulus of rupture}}}$$

Modulus of rupture varies between 20 and 40 kp/cm^2 (285–570 psi).

Switzerland Sia 162—1968

$E_c = 19\,000\sqrt{\sigma_{Bw}}$ in kg/cm^2, $n = 10$, $E_s = 2\,100\,000$ kg/cm^2 (29.9×10^6 psi). For creep take $E_c/3$. Multiplier for initial deflection is 2.5 to 3.

United Kingdom (British Code) CP 110—1972

Calculation method proposed asks to calculate deflection in cracked and uncracked states (such as shown in Fig. 3-3 and Eqs. (3-23), (3-24) of Chap. 3 herein) and to choose whichever value gives the largest deflection. $E_c = E_c \times 1/(1 + C_t)$ for long-term effects.

Value for concrete in tension to be taken as 1 N/mm^2 (145 psi) in short-term, reducing to 0.55 N/mm^2 (79.6 psi) in the long-term, at the tensile steel level, which varies linearly to the neutral axis. Shrinkage curvature is also allowed for. Method is integration of curvature along the span.

See text, Chap. 6 herein, for allowable deflections and span–depth ratios.

U.S.A. ACI 318-71—1971

See text, Chaps. 3–5 herein for calculation methods, and Chap. 6 herein for allowable deflections and span–depth ratios.

U.S.S.R. Chull II-B. 1–62—1962

Deflection limits:

Crane beams hand operated		$L/500$
Crane beams electric operated		$L/600$
Flat slab and roofs	$L \leq 7$ m (23 ft)	$L/200$
	$L > 7$ m	$L/300$
Ribbed slabs and stairways	$L \leq 5$ m (16.4 ft)	$L/200$
	$5 \text{ m} \leq L \leq 7$ m	$L/300$
	$L > 7$ m	$L/400$

APPENDIX

M

MINIMUM THICKNESS OF BEAMS OR ONE-WAY SLABS USED IN ROOF AND FLOOR CONSTRUCTION, AS DESCRIBED BY ACI COMMITTEE 435

Table M-1 was expanded and modified from Table 6-3.

Table M-1 Minimum thickness of beams or one-way slabs used in roof and floor construction, as described by ACI Committee 435[45]*†

	Members not supporting or attached to nonstructural elements likely to be damaged by large deflections				Members supporting or attached to nonstructural elements likely to be damaged by large deflections			
Member	Simply supported	One end cont.	Both ends cont.	Cantilever	Simply supported	One end cont.	Both ends cont.	Cantilever
Roof slab‡	$l/22$	$l/28$	$l/35$	$l/9$	$l/14$	$l/18$	$l/22$	$l/5.5$
Floor slab	$l/18$	$l/23$	$l/28$	$l/7$	$l/12$	$l/15$	$l/19$	$l/5$
Roof beam‡ or ribbed roof slab‡	$l/18$	$l/23$	$l/28$	$l/7$	$l/12$	$l/15$	$l/19$	$l/5$
Floor beam or ribbed floor slab	$l/14$	$l/18$	$l/21$	$l/5.5$	$l/10$	$l/13$	$l/16$	$l/4$

* The span length l is in inches.

† The values given in this table shall be used directly for nonprestressed reinforced concrete members made with normal-weight concrete ($w = 145$ pcf) and Grade 60 reinforcement. For other conditions, the values shall be modified as follows.

For structural lightweight concrete having unit weights in the range 90–120 pcf, the values in the table shall be multiplied by $1.65 - 0.005w$ but not less than 1.09, where w is the unit weight in pcf.

For nonprestressed reinforcement having yield strengths other than 60 000 psi, the values in the table shall be multiplied by $0.4 + f_y/100\,000$.

‡ Refers to roofs subjected to normal snow or construction live loads only, and with minimal ponding problems.

Table M-1—*(continued)*

f_y (ksi) =	40	50	60	75	w =	90	100	110	120
Sample corr. fac. =	0.80	0.90	1.00	1.15	=	1.20	1.15	1.10	1.09

APPENDIX

N

DERIVATION OF EQS. (6-6) TO (6-10) FOR DETERMINING SPAN–DEPTH RATIOS USING I_e[10,20]

General Formulas

$$\Delta_i = KM_{max}L^2/E_cI_e \tag{N-1}$$

where M_{max} = maximum moment due to all dead and live load. Positive and negative moments are considered separately for beams continuous at one or both ends in computing I_e. Also, $f_r = 7.5\sqrt{f'_c}$ for normal-weight concrete.

Multiplying Eq. (N-1) by f_rI_g/y_tM_{cr}, which is equal to unity,

$$\Delta_i = K\frac{M_{max}L^2}{E_cI_e}\frac{f_rI_g}{y_tM_{cr}} \tag{N-2}$$

Solving Eq. (N-2) for L/y_t, and letting $C_w = E_c/f_r$,

$$\frac{L}{y_t} = \frac{\Delta_i}{L}\frac{C_w}{K}\frac{M_{cr}}{M_{max}}\frac{I_e}{I_g} \tag{N-3}$$

When $M_{cr}/M_{max} \geq 1$, $I_e = I_g$. When $M_{cr}/M_{max} < 1$, I_e = Eq. (3-19). Let

$$m = I_e/I_g = (M_{cr}/M_{max})^3 + [1 - (M_{cr}/M_{max})^3](I_{cr}/I_g) \tag{N-4}$$

Approximate values of m and y_t are derived below for the positive and negative moment regions of rectangular and T-beams. For Fig. 6-2, an average of the positive and negative moment region values for I_e is used for beams continuous at one or both ends. On this basis, Eq. (N-3) is plotted in the form of L/h versus M_{max}/M_{cr} for $\Delta_i/L = 1/360$, normal-weight concrete, uniformly distributed short-time loading, and for different boundary conditions, steel percentages, and concrete strengths. The modifying factor for cantilevers $= K_{ss}/K_{cant} = (5/48)/(1/4) = 0.417$, is the theoretical ratio for these statically determinate cases. The modifying factors for continuous beams in Fig. 6-2 are based on approximate ratios of $1/(0.4 + 0.6 \times 72/128) = 1.36$ for one end continuous and $1/(0.4 + 0.6 \times 8/24) = 1.67$ for both ends continuous (should be used only when M_{max} = midspan moment in Eq. (N-1)). However, the author has found that the ratio of the K values for ideal end conditions for statically indeterminate beams in Eq. (N-1) is not reasonable for most practical cases, and the values of 1.36 and 1.67 (also $1.67/1.36 = 1.23$) are selected for Fig. 6-2 on the basis of judgement. They are similar to the factors of 1.3 and 1.6 in Note ‡, Table 6-3, of ACI Committee 435.

In this procedure M_{max}, M_{cr}, and y_t refer to the numerical maximum moment and corresponding sections (which are negative moment sections for all cases except the simple beam), but an average m value is used for the positive and negative moment regions in the case of continuous beams.

For other values of Δ_i/L, Eq. (6-6) follows from Eq. (N-3), since L/y_t, and thus L/h, is directly proportional to Δ_i/L.

For other than normal-weight concrete, Eqs. (6-7) and (6-8) follow from Eq. (N-3), since L/y_t, and thus L/h, is directly proportional to $C_w = E_c/f_r$ (M_{cr} is computed separately for each problem). Using Eqs. (1-7) and (1-13) for concrete strengths from 2500 to 8000 psi, $C_w = 6014$ to 7875 when $w = 145$ pcf, 4780 to 6236 when $w = 120$ pcf, and 4148 to 5415 when $w = 100$ pcf. However, $(C_w)_{120}/(C_w)_{145}$ = approx. constant of 0.795, and $(C_w)_{100}/(C_w)_{145}$ = approx. constant of 0.690. Hence the modifying values of 0.80 and 0.69 are used in Eqs. (6-7) and (6-8), respectively.

Formulas for Dead-Load Creep Plus Shrinkage Plus Live Load

For dead-load deflection, $M_{max} = M_D$, $m = m_D$, and Eq. (N-3) becomes

$$\frac{(\Delta_i)_D}{L} = \frac{K}{C_w}\frac{M_D}{M_{cr}}\frac{L}{y_t}\frac{1}{m_D} \tag{N-5}$$

For dead-load plus live-load deflection,

$$\frac{(\Delta_i)_{D+L}}{L} = \frac{K}{C_w}\frac{M_{D+L}}{M_{cr}}\frac{L}{y_t}\frac{1}{m_{D+L}} \tag{N-6}$$

For live-load deflection,

$$\frac{(\Delta_i)_L}{L} = \frac{(\Delta_i)_{D+L}}{L} - \frac{(\Delta_i)_D}{L} \tag{N-7}$$

From Eqs. (N-5) to (N-7), and letting $C_L = M_L/M_D$,

$$\frac{(\Delta_i)_L}{L} = \frac{K}{C_w}\frac{L}{y_t}\frac{M_{D+L}}{M_{cr}}\left[\frac{1}{m_{D+L}} - \frac{1}{m_D(1 + C_L)}\right] \tag{N-8}$$

Using $\Delta_{cp+sh} = k_r T(\Delta_i)_D$,

$$\frac{\Delta_{cp+sh}}{L} + \frac{(\Delta_i)_L}{L} = \frac{K}{C_w}\frac{L}{y_t}\frac{M_{D+L}}{M_{cr}}\left[\frac{1}{m_{D+L}} + \frac{(k_r T - 1)}{m_D(1 + C_L)}\right]$$

$$= \frac{K}{C_w}\frac{L}{y_t}\frac{M_{D+L}}{M_{cr}}\left[\frac{1 + C_L + (k_r T - 1)(m_{D+L}/m_D)}{1 + C_L}\right]\frac{1}{m_{D+L}} \qquad \text{(N-9)}$$

Solving for L/y_t,

$$(L/y_t)_{D\,cp+sh+L} = \frac{\Delta}{L}\frac{C_w}{K}\frac{M_{cr}}{M_{D+L}}\left[\frac{C_L + 1}{C_L + 1 + (k_r T - 1)(m_{D+L}/m_D)}\right] m_{D+L} \qquad \text{(N-10)}$$

where

$$\Delta/L = \frac{\Delta_{cp+sh}}{L} + \frac{(\Delta_i)_L}{L}$$

From Eq. (N-3),

$$(L/y_t)_{\text{short-time}\,D+L} = \frac{\Delta}{L}\frac{C_w}{K}\frac{M_{cr}}{M_{D+L}} m_{D+L} \qquad \text{(N-11)}$$

Hence, for a given Δ/L in Eqs. (N-10) and (N-11), the combined effect of dead-load creep plus shrinkage plus live load can be computed in terms of the initial dead plus live-load value by Eq. (N-12). From Eqs. (N-10) and (N-11),

$$(L/y_t)_{D\,cp+sh+L} = (L/y_t)_{\text{short-time}\,D+L}\left[\frac{C_L + 1}{C_L + 1 + (k_r T - 1)(m_{D+L}/m_D)}\right] \qquad \text{(N-12)}$$

Since y_t is proportional to h, Eqs. (6-9) and (6-10) follow from Eq. (N-12).

Approximate Values of m and y_t Used in Fig. 6-2

Singly reinforced rectangular beams

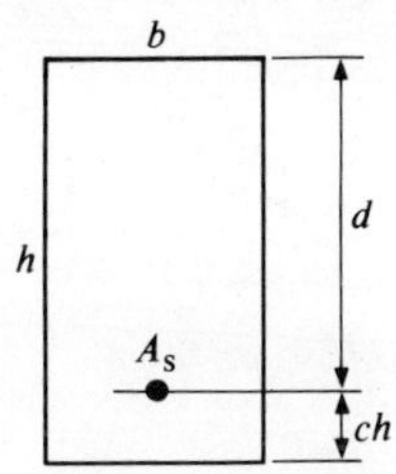

Singly reinforced rectangular beam section

$$n = E_s/E_c \qquad \rho = A_s/bd \qquad \text{Use } c = 0.1$$

$$k = \sqrt{(\rho n)^2 + 2\rho n} - \rho n$$

$$I_g = bh^3/12$$

$$I_{cr} = b(kd)^3/3 + nA_s(d - kd)^2$$

$$= 0.729bh^3[k^3/3 + n\rho(1 - k)^2]$$

$$I_{cr}/I_g = 8.75\,[k^3/3 + n\rho(1 - k)^2]$$

$$m = \text{Eq. (N-4) and } I_{cr}/I_g$$

$y_t = h/2$. See Eqs. (N-3) and (N-4) for computing L/h.

Singly reinforced T-beams

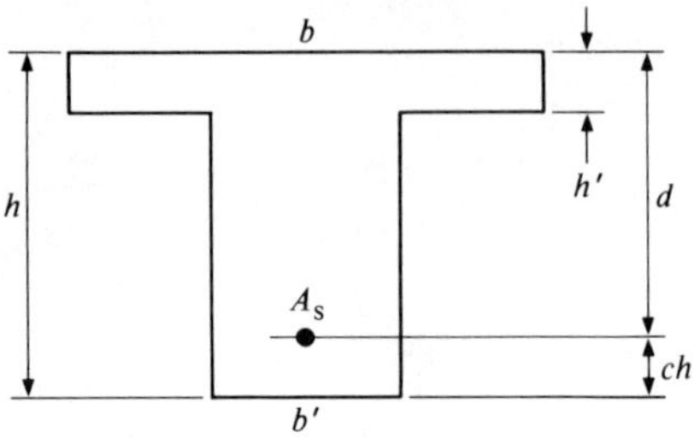

Singly reinforced
T-beam section

$n = E_s/E_c \qquad \rho = A_s/bd$ for pos. mom.

$\rho = A_s/b'd$ for neg. mom. $\qquad$ Use $c = 0.1$

Assume $\bar{y}_{bot} = \frac{4}{5}h$, $b' = \frac{1}{4}b$, $h' = \frac{1}{5}h$; hence the centroid of the cracked and gross sections are assumed to be at the bottom of the flange. The method is not used for $\rho > 0.01$.

$$I_g = \frac{b'(\frac{4}{5}h)^3}{3} + \frac{b(h')^3}{3} = 0.181b'h^3$$

Positive moment region:

$$I_{cr} = b(h')^3/3 + nA_s(d - h')^2 = b'h^3(0.0107 + 1.764n\rho)$$

$$I_{cr}/I_g = 5.51(0.0107 + 1.764n\rho) \qquad m = \text{Eq. (N-4) and } I_{cr}/I_g$$

Negative moment region: see corresponding equation for rectangular beam, except $\rho = A_s/b'd$.

Get $\qquad I_{cr} = 0.729b'h^3[k^3/3 + n\rho(1 - k)^2]$

and $\qquad I_{cr}/I_g = 4.015[k^3/3 + n\rho(1 - k)^2] \qquad m = $ Eq. (N-4) and I_{cr}/I_g

Span–thickness ratios:

Cantilever beams	$y_t = \frac{1}{5}h$
Simple beams	$y_t = \frac{4}{5}h$
Beams with one end continuous	$y_t = \frac{1}{5}h$
Beams with both ends continuous	$y_t = \frac{1}{5}h$

(for the last two:) Because M_{max} and M_{cr} refer to neg. mom. in Eq. (N-3), and y_t is introduced with M_{cr}

See Eqs. (N-3) and (N-4) for computing L/h.

APPENDIX

O

ACI CODE MINIMUM THICKNESSES FOR TWO-WAY SLAB SYSTEMS

Table O-1 ACI Code[4] minimum thicknesses for two-way slab systems* (expressed as fraction of longer clear span, L_n)

Two-way construction	Minimum $h =$ Eq. (6-1)	Minimum $h =$ Eq. (6-2)	—but need not exceed Eq. (6-3)
Solid flat plate			
Square interior panel (Min. $h = 5$ in.)	$L_n/32.73$†	$L_n/41.8$	$L_n/73$†
Square edge panel (Min. $h = 5$ in.)	$L_n/28.7$	$L_n/36.98$	$L_n/29.75$†
Square corner panel (Min. $h = 5$ in.)	$L_n/27.7$	$L_n/35.95$	$L_n/29.75$†
Solid flat plate with edge beams (stiffness such that $\alpha = 0.8$)			
Square edge panel	$L_n/35.23$†	$L_n/40.68$	$L_n/32.73$
Square corner panel	$L_n/35.00$†	$L_n/39.55$	$L_n/32.73$
Solid flat slab with drop panels (length $> L_n/3$; depth $\geq 1.25h$)			
Square interior panel (Min. $h = 4$ in.)	$L_n/36.37$†	$L_n/46.56$	$L_n/36.37$†
Square edge panel (Min. $h = 4$ in.)	$L_n/31.59$	$L_n/40.68$	$L_n/32.73$†
Square corner panel (Min. $h = 4$ in.)	$L_n/30.45$	$L_n/39.55$	$L_n/32.73$†
Solid flat slab with drop panels and edge beam $\alpha = 0.8$			
Square edge panel	$L_n/39.14$†	$L_n/45.20$	$L_n/36.37$
Square corner panel	$L_n/37.88$†	$L_n/43.94$	$L_n/36.37$
Solid flat plate			
Rectangular interior panel ($\beta = L_{n1}/L_{n2} = 2.0$; Min. $h = 5$ in.)	$L_n/32.73$†	$L_n/50.91$	$L_n/32.73$†
Rectangular edge panel ($\beta = L_{n1}/L_{n2} = 2.0$; Min. $h = 5$ in.)	$L_n/28.20$	$L_n/44.21$	$L_n/29.75$†
Rectangular corner panel ($\beta = L_{n1}/L_{n2} = 2.0$; Min. $h = 5$ in.)	$L_n/26.65$	$L_n/42.15$	$L_n/29.75$†

Table O-1—(*continued*)

Solid flat plate with edge beams (*stiffness such that* $\alpha = 0.8$)			
Rectangular edge panel } $\beta = L_{n1}/L_{n2} = 2.0$	$L_n/38.30^\dagger$	$L_n/48.64$	$L_n/32.73$
Rectangular corner panel }	$L_n/36.59^\dagger$	$L_n/46.36$	$L_n/32.73$
Solid flat slab with drop panels (*length* $> L_n/3$; *depth* $\geq 1.25h$)			
Rectangular interior panel } $\beta = L_{n1}/L_{n2} = 2.0$	$L_n/36.37^\dagger$	$L_n/56.57$	$L_n/36.37^\dagger$
Rectangular edge panel } Min. $h = 4$ in.	$L_n/31.02$	$L_n/48.64$	$L_n/32.73^\dagger$
Rectangular corner panel }	$L_n/29.32$	$L_n/46.36$	$L_n/32.73^\dagger$
Solid flat slab with drop panels and edge beam $\alpha = 0.8$			
Rectangular edge panel } $\beta = L_{n1}/L_{n2} = 2.0$	$L_n/42.55^\dagger$	$L_n/54.04$	$L_n/36.37$
Rectangular corner panel }	$L_n/40.66^\dagger$	$L_n/51.52$	$L_n/36.37$

(From CRSI Handbook,[44] P. F. Rice, Editor, with permission.)

* Minimum thickness unless deflection is computed (ACI Code 318-71), from Eqs. (6-1) to (6-3) herein.

† Controlling limit.

NAME INDEX

SUBJECT INDEX